Structural Optimization

Dynamic and seismic applications

T0179322

Structural Optimization
Dynamic and seismic applications

Franklin Y. Cheng and
Kevin Z. Truman

CRC Press
Taylor & Francis Group
Boca Raton London New York

CRC Press is an imprint of the
Taylor & Francis Group, an **informa** business

A SPON PRESS BOOK

CRC Press
Taylor & Francis Group
6000 Broken Sound Parkway NW, Suite 300
Boca Raton, FL 33487-2742

First issued in paperback 2019

© 2010 Franklin Y. Cheng and Kevin Z. Truman
CRC Press is an imprint of Taylor & Francis Group, an Informa business

No claim to original U.S. Government works

ISBN-13: 978-0-415-42370-0 (hbk)
ISBN-13: 978-0-367-86513-9 (pbk)

Typeset in Sabon by
Integra Software Services Pvt. Ltd, Pondicherry, India

British Library Cataloguing-in-Publication Data
A catalogue record for this book is available
from the British Library

Library of Congress Cataloging in Publication Data
Cheng, Franklin Y.
Structural optimization : dynamic and seismic applications /
Franklin Y. Cheng and Kevin Z. Truman.
p. cm.
Includes bibliographical references and index.
1. Live loads.   2. Structural design.   I. Truman,
Kevin Z.   II. Title.
TA654.3.C48 2010
624.1′7713--dc22
2009045605

Contents

Preface

The major challenge in today's structural engineering is to design better structures with an integrated approach and consideration of lower cost and required performance against damaging effects of dynamic forces and earthquakes. Structural optimization is based on rigorous mathematical formulation and computation algorithms for sizing structural elements and synthesizing systems. The book is a comprehensive presentation of optimization strategies currently in vogue and provides sufficient examples on how to use these theories for designing elements and systems. A specific emphasis is given to dynamic loads, in particular, seismic forces. This book is a useful reference for researchers and practising engineers working in the structural engineering field. It is also a key resource for senior undergraduates and all postgraduate students to find an organized collection of information for structural design. The advent of fast computations and large data-storage capacity has now made it possible to use structural optimization, not only in a research mode, but also in a professional mode. The algorithms and extensive numerical examples are provided to illustrate how optimization can be used and how to make logical design choices when designing for dynamic or seismic loads. Many professionals check very few alternative designs, whereas structural optimization will provide a suite of acceptable designs on its path to the mathematical optimal design. The design professional can then choose one that is better than his/her original design (possibly not the mathematically optimal design), yet still a practical design. This book illustrates how to use optimization procedures to perform parametric studies that will enhance the professional's understanding of the effects of dynamic loads and how to influence structural responses to these loads.

This book has 12 chapters and 11 appendices of which the junior author wrote four chapters: 6, 8, 9 and 10. Key features of this book are:

1. Complete mathematical formulations and numerical procedures for the topics presented;
2. New technologies;
3. Design guidelines and examples based on official building codes;

4. Detailed figures and illustrations;
5. Extensive numerical examples and references.

This book has been prepared with the following emphasis:

1. The book functions as a self-study unit. Essential information on structural formulations, optimization techniques, numerical algorithms, and so forth, is given in detail.
2. Step-by-step numerical examples are provided. These serve to illustrate mathematical formulations and to interpret physical representations, enabling the reader to understand the formulas vis-a-vis their applications.
3. Each chapter discusses a specific topic, and the topic areas are comprehensive.

Acknowledgements

I would like to sincerely thank my former mentors, Profs. C.K. Wang and T.C. Huang for their inspiration and influence on my lifetime of academic career. I have also been fortunate to have received advice, encouragement, and help from my long-time friends, Drs. S.C. Liu, K.P. Chong, Alfred H.S. Ang, Henry Yang, V.B. Venkayya and N. Khot, who have kept me active during the course of my research.

The book consolidates results from a number of research projects carried out at the Missouri University of Science and Technology, MST (formerly known as the University of Missouri-Rolla, UMR), that were awarded by several funding agencies including the National Science Foundation (NSF), Intelligent Systems Center at MST, and the Curators' Professorship, particularly to Chancellor, Dr J.F. Carney III and the Department Chair, Dr William Schonberg. The sustained support and encouragement of these agencies and the individuals are gratefully acknowledged. My deep gratitude also goes to my collaborator Dr Kevin Z. Truman who has taken time from his busy administrative work to prepare four chapters: 6, 8, 9 and 10. This book contains a number of research reports co-authored with my former Ph.D. students, M.E. Botkin, D.S. Juang, D. Srifuengfung, Li Dan and C.C. Chang. Their intelligence and hard work significantly helped the project's success.

I heartfully thank my wife, Betrice (Pi Yu), who has been my pillar and comforter for 45 years, my children and their spouses, George and Annie, Deborah and Craig, as well as my grandchildren, Hau Yih, Yu Bi and Yuan Bi, for their love which I treasure tremendously.

Franklin Y. Cheng

I would like to acknowledge the support of many individuals that have provided their intelligence, mentorship and moral support in the 30 years of development of this work. Many hours of devotion to research in the use of structural optimization towards the solution of a variety of problems from a cadre of graduate students made my contributions to this book possible. The students are David Petruska, Alan Hoback, Chi-Chi Jan, Leif Johnson, Tony Hurd, Mark Hendel and Gus Terlaje. A special thanks to Gus as he helped produce and compile many of the components for this book. Several key colleagues have invested numerous hours in mentoring me and leading me to a successful professional career. In particular, my Ph.D. advisor and co-author, Dr Franklin Y. Cheng, has shown tremendous patience and provided an immeasurable amount of energy and opportunities in helping me reach my professional goals. Dr Phillip L. Gould, teacher, mentor, supervisor and close personal friend, has been a pillar of strength and optimism throughout my career. Last, my family has been so extraordinarily supportive throughout my academic and research career. My wife, Katina, for more than 30 years, has supported and encouraged me to stay the course. My children, Zane, Kameryn and Stephanie, have given of their time in order for me to have the time to pursue my research, to be active in my profession, to be a teacher of many and to be an academic administrator. I thank all of these individuals; they have all contributed to my success and the possibility of contributing to this book.

Kevin Z. Truman

Authors

Franklin Y. Cheng, PE, distinguished member (formerly honorary member) of ASCE, joined the University of Missouri-Rolla (now Missouri University of Science and Technology, MST) as assistant professor in 1966. In 1987, the Board of Curators of the University appointed him Curators' Professor, the highest professorial position in the system comprising four campuses, and Curators' Professor Emeritus in 2000. He is a former senior investigator, Intelligent Systems Center, University of Missouri-Rolla. Dr Cheng received four honorary professorships abroad and chaired seven of his 24 National Science  Foundation (NSF) delegations to various countries for research and development cooperation. He has also been the director of international earthquake engineering symposia and numerous state-of-the-art short courses. His work has warranted grants from several funding agencies including more than 30 from the NSF. He has served as either chairman or member of 37 professional societies and committees, 12 of which are ASCE groups. He was the first chair of the Technical Administrative Committee on Analysis and Computation, initiated the Emerging Computing Technology Committee and Structural Control Committee. He also initiated and chaired the Stability Under Seismic Loading Task Group of the Structural Stability Research Council (SSRC).

Dr Cheng has served as a consultant for Martin Marietta Energy Systems Inc., Los Alamos National Laboratory, and Martin & Huang International, among others. The author, co-author, or editor of 26 books and over 250 publications, Dr Cheng's authorship includes three textbooks, *Smart Structures: Innovative Systems for Seismic Response Control*, *Matrix Analysis of Structural Dynamics: Applications and Earthquake Engineering*, and *Dynamic Structural Analysis*. Dr Cheng is the recipient of numerous honours, including the MSM-UMR Alumni Merit, ASCE State-of-the-Art twice, the Faculty Excellence, and the Haliburtan Excellence awards. In 2007, he was elected as 565th honorary member of ASCE with a history since 1852. After receiving a BS degree (1960) from the National Cheng-Kung University, Taiwan, and a MS degree (1962) from the University of Illinois at Urbana-Champaign, he gained industrial experience with C.F. Murphy and Sargent & Lundy in Chicago, Illinois. Dr Cheng received a Ph.D. degree (1966) in civil engineering from the University of Wisconsin-Madison.

Kevin Z. Truman was raised in north-western Illinois. He attended Aledo High School which led him to a nearby college, Monmouth College, and the awarding of Bachelor of Arts degrees in Physics and Mathematics in 1979. He then attended Washington University in St Louis and earned Bachelor of Science and Master of Science degrees in Civil Engineering in 1979 and 1981, respectively. While teaching at Washington University, he commuted to University of
Missouri-Rolla (now Missouri University of Science and Technology) to work with Dr Franklin Y. Cheng in the area of three-dimensional structural optimization subject to static and dynamic loads. Upon completing his Ph.D. in 1985, Dr Truman continued his academic career at Washington University in St Louis. He spent the next 23 years there as a professor, chairman and Assistant Dean. While at Washington University in St Louis, he won numerous awards for teaching including the Missouri Governor's Award, had an extensive research career with the U.S. Army Corps of Engineers as well as several other government agencies, was awarded the Albert P. and Blanche Y. Greensfelder Professorship in Civil Engineering and was a co-recipient of the 1998 ASCE State-of-the-Art-Award. In 2008, Dr Truman moved to the University of Missouri-Kansas City to become Dean of the School of Computing and Engineering.

Dr Truman has had over 120 publications, numerous funded research projects totalling more than $6 million, lectured throughout the world on engineering education, structural engineering, seismic-resistant design of structures and massive concrete structural analysis, design and construction. His desire to couple mathematics with structural engineering and his love for teaching led to a research thrust of structural optimization that has been applied to design, topology problems, damage detection and structural identification throughout his 28-year academic career.

Notations

A_a = coefficient of effective ground acceleration
A_i = cross-sectional area of member i
A_v = coefficient of effective ground velocity
A' = required sectional area of the plate at beam-column connections
a = ground acceleration normalized by g
a_{max} = maximum expected normalized ground acceleration in the lifetime of the building
a^* = design normalized ground acceleration
b = width of wide-flange section
b_j = limitation imposed on the jth behaviour constraint
C_{con} = total connection costs
C_{cl} = total construction costs of a building
C_D = total non-structural damage repair costs in the lifetime of a structure
C_d = deflection amplification factor
C_{EX} = total extra charge of structural members
$(C_{nc})_i$ = construction cost of damage items on storey i
C_p = total painting cost; or wind pressure coefficient
$(C_{pc})_i$ = cost of the steel plates at connections of beam i
C_{pl} = unit price of the plate at connections
C_{pt} = unit price of painting
$(C_r)_i$ = non-structural damage repair cost of storey i
C_{TB} = total basic charge of structural members
C_w = unit price of welding
$(C_{wc})_i$ = welding cost of each beam i
$[M_s]_i$ = mass matrix member i in global coordinates
d = depth of wide-flange section
E = modulus of elasticity
I_i = moment of inertia of member i
J = set of design variables that is not constrained by the lower bound
J_o = set of design variables constrained by the lower bound
$[K]$ = stiffness matrix of a structure

$[K]_i$ = stiffness matrix of member i

$[K_g]_i$ = geometric matrix of member i

M_p' = plastic moment capacity induced by the plates at beam-column connections

n_s = Total number of floors

P_i' = derivative of $P - \Delta$ forces acting on member i

p = forcing frequency based on Rayleigh quotient

$\{Q_j\}$ = vector of virtual load in jth direction and zero values for others of a structure

$\{\bar{Q}_j\}_{ei}$ = nodal forces of member i in local coordinates due to the virtual load vector

$\{\bar{Q}_j\}_{gi}$ = nodal forces of member i in global coordinates due to the virtual load vector

$\{\bar{Q}_{gi}\}_{ei}$ = geometric forces vector of member i in local coordinates due to the virtual displacement resulting from the virtual load vector $\{Q_j\}$

$\{q_j\}$ = vector of virtual displacements resulting from the virtual load vector $\{Q_j\}$

R = response modification factor

$\{R\}$ = load vector

$\{r\}$ = displacement vector

$\{\bar{r}\}_{ei}$ = static displacement vector corresponding to the local coordinate of member i

$\{\bar{r}\}_{gi}$ = static displacement vector in global coordinates corresponding to the degrees of freedom of member i

$\{r(x,t)\}$ = dynamic displacement vector

$\{r(x,t)\}_{ei}$ = dynamic displacement vector corresponding to the local coordinates of member i

r_x = radius of gyration of wide-flange sections

$\{dr\}_i$ = change in displacement vector, $\{r\}$, due to the change in the design variable x_i

Δr_j = total influence on the displacement of jth degree of freedom due to the change of design variables x

dr_{ji} = change of displacement in jth degree of freedom due to the change in design variable x_i

S_1, S_2, S_3 = soil profile types

S_a = spectral acceleration

S_d = spectral displacement

S_v = spectral velocity

s = step size for the contribution based on constraint gradients

U = strain energy of a system

u_i = strain energy of member i

$u_j(x)$ = static constraint function for the jth direction

W = total gravity load applied on a building

x_i = ith design variable
x_i^o = lower bound of ith design variable
dx_i = change of design variable x_i
$y_j(x)$ = jth behaviour constraint
Z_x = plastic modulus of wide-flange sections
α_a = amplification factor of spectral acceleration of Newmark's spectra
α_d = amplification factor of spectral displacement of Newmark's spectra
α_i = ith relative design variable
α_v = amplification factor of spectral velocity of Newmark's spectra
$\delta\alpha_i$ = departure of new relative design variable from present design member i
β = friction of critical damping for the coupled structure–foundation system; or damping ratio of a structure
Λ = scaling factor
Ω, Ω_1 = constant of proportionalities of the constraint-gradient method for displacement constraint and stress constraint, respectively
$\Delta\sigma_j$ = total influence on the stress of member j due to change of design variable x_i
$d\sigma_{ji}$ = influence on the stress of member j due to change of design variable x_i
σ_y = yielding stress of beam
σ_y' = yielding stress of the plate at beam-column connection
$\{\phi_j\}$ = eigenvector of jth mode
ℓ_i = length of member i
ℓ_{bw} = welded length of the bottom plate at beam-column connections
ℓ_{tw}' = welded length of the top plate at beam-column connections
ρ_i = mass density of member i
η_i = ratio of the cross-sectional area, A_i, to the moment of inertia I_i, for ith member
λ_j = jth Lagrange multiplier
ε_y = yielding strain of the plate at beam-column connections
ξ = damage ratio
χ = ratio of the construction cost of damage items to the construction of a structure
ω_j = undamped natural frequency of jth mode
ΔF_{ijk} = change in internal force i due to a change in design parameter j under the action of loading condition k
F_a = allowable axial stress
F_b = allowable bending stress
ΔM_{ijk} = change in internal moment i due to a change in design parameter j under the action of loading condition k
ΔX_{ijk} = change in deformation i due to a change in design parameter j under the action of loading condition k

A = area

A_I = influence area

$[A]_m^T$ = transpose of a member static matrix

$[C]$ = damping matrix

C_D = influence coefficient which transforms dead load intensity
D into the desired load effect (displacement of internal force)

C_L = influence coefficient which transforms live load intensity L into
the desired load effect (displacement of internal force)

C_E = influence coefficient which transforms earthquake load intensity
E into the desired load effect (displacement of internal force)

C_I = initial construction cost

C_p = percentage constant of $\bar{C}_S$

C_T = total structural cost

C_S = influence coefficient of structural response

D = dead load intensity

E = earthquake load intensity which may be the shear force at the
structural base in Uniform Building Code

E_m = elastic modulus

EUDL = equivalent uniform distributed load

f_1 = yielding failure in column failure

f_2 = bending instability in column failure

f_3 = lateral-torsional buckling in column failure

f_4 = buckling about weak axis in column failure

F_{Eux} = lateral force applied to level x due to a unit seismic load
intensity

F_{Ex} = lateral force applied to level x

$\{F\}_m$ = internal forces in the member

F_y = yield strength of material

g_j = jth constant

I = moment of inertia

$[K]^{-1}$ = inversion of stiffness matrix

$[K]_G$ = geometric stiffness matrix

KL = effective length

$[K]_s$ = linear elastic stiffness matrix

L = live load intensity

L_f = expected failure cost

$[\bar{m}]$ = mass matrix

$\bar{M}$ = mean of applied moment

σ_M^2 = variance of applied moment

V_M = coefficient of variation of applied moment

M_{cr} = critical moment

$\bar{M}_{cr}$ = mean critical moment

$\sigma_{M_{cr}}^2$ = variance of critical moment

$V_{M_{cr}}$ = coefficient of variation of critical moment

M_y = yield moment
$\bar{M}_y$ = mean yield moment
$\sigma^2_{M_y}$ = variance of yield moment
V_{M_y} = coefficient of variation of yield moment
$O(\ )$ = objective function
$P(\ )$ = penalty function
P_{cr} = axial load capacity of columns
$\bar{P}_{cr}$ = mean critical load
$\sigma^2_{P_{cr}}$ = variance of critical load
$V_{P_{cr}}$ = coefficient of variation of critical load
P_E = Euler buckling load
$\bar{P}_E$ = mean Euler buckling load
$\sigma^2_{P_E}$ = variance of Euler buckling load
V_{P_E} = coefficient of variation of Euler buckling load
P_f = probability of failure
P_{fT} = system probability of failure
$PN(\ )$ = standard cumulative normal function
P_r = reliability
P_y = yield load
$\bar{P}_y$ = mean yield load
$\sigma^2_{P_y}$ = variance of yield load
V_{P_y} = coefficient of variation of yield load
Q = function in terms of R and S
$\bar{Q}, \sigma_Q$ = mean and standard deviation of Q
q' = load intensity
r = random parameters
R = structural resistance
$\bar{R}$ = mean structural resistance
σ_R = standard deviation of structural resistance
V_R = coefficient of variation of structural resistance
r^k_p = penalty value
S = structural response
S_c = elastic section modulus
S_D = dead load effect
S_E = earthquake load effect
S_L = live load effect
$[S]_m$ = member stiffness matrix
$\bar{s}_{max}$ = mean of peak response
T = structural fundamental period
T_i = recurrence equation
u = structural displacement
$\dot{u}$ = structural velocity
$\ddot{u}$ = structural acceleration
u_D = displacement due to dead load intensity

u_E	= displacement due to earthquake load intensity
u_L	= displacement due to live load intensity
$\{u\}_m$	= corresponding external displacements in a member
V_D	= coefficient of variation of dead load intensity
V_E	= coefficient of variation of earthquake load intensity
V_L	= coefficient of variation of live load intensity
W	= weight
x^k	= design variables
x^*	= optimal solution
Y_n	= generalized coordinate
$y_{\omega n}$	= spectral displacement
α	= relaxation factor
β	= safety factor
ρ	= correlation coefficient
Λ	= scaling factor
λ_j	= jth Lagrangian multiplier
ζ_n	= critical damping ratio coefficient for nth mode
Γ_n	= participation factor for nth mode
ω_n	= structural natural frequency for nth mode
$\{\Phi\}_n$	= nth mode shape matrix
A_i	= cross-sectional area for the ith structural element
A	= matrix of coefficients for quadratic formulations
A_s	= area of steel in a reinforced concrete element
A_t	= transformed area for a reinforced concrete element
C	= vector of coefficients for quadratic formulations
C_s	= seismic response coefficient
C_{vx}	= IBC vertical force distribution factor
D	= constant for cubic interpolation
E_s	= Young's modulus for steel
E_c	= Young's modulus for concrete
F_x	= xth level horizontal code-based force
F_t	= top-storey horizontal code-based force
L	= length of a structural element
$L(\delta,\lambda)$	= Lagrangian function
M_x	= moment about the cross-sectional x-axis
M_y	= moment about the cross-sectional y-axis
$[M]$	= mass matrix for a structural system
$\{P\}$	= structural load vector or component
R_w	= structural-framing-related ductility factor
$[R]$	= mapping matrix for linking design variables
S_{DS}	= site location and soil condition factor
S_p	= stiffness multiplier for pile groups
S_x	= section modulus about the elemental x-axis
S_y	= section modulus about the elemental y-axis
S_{ij}	= jth secondary design variable for the ith element

$\{S_a\}_j$ = spectral acceleration for the jth natural frequency (or period)

T_i = optimality criteria for element i

V = base shear force

V_j = jth modal base shear

V_{xi} = x-direction second-order shear at the ith node for a column

V_{yi} = y-direction second-order shear at the ith node for a column

b_f = flange width for a steel W or HP section

f_s = working stress for steel

f_c = working stress for concrete

g = gravitational constant

t_i = ith elemental thickness

u = structural responses for structural systems, displacements or stress

$\{v\}_j$ = virtual displacement vector

W_i = weightless constraint scaling factor

w_x = xth level portion of the seismic dead weight W

δ = design variable vector or design variable vector component, often a cross-sectional property, thickness, etc.

δ_{xe} = IBC drift determined for the xth level by an elastic analysis

$\delta(t)$ = deflection as a function of thickness in topology problems

λ = Lagrange multiplier

$\{\phi\}_j$ = eigenvector or mode shape associated with the jth natural frequency

φ = orientation of a battered pile relative to the x–y plane or the z-axis

$\{v\}_i$ = virtual displacement vector for ith element

ω_j = jth natural frequency

Γ = 0 or 1 in the quasi-Newton methods

$\{\Lambda\}$ = vector of linked design variables

Π = IBC structural importance factor

θ_i = rotational displacement at the ith node for beam elements

Abbreviations

AL	auxiliary lock
ATC	Applied Technology Council
BOCA	Building Officials Code Administration
CPGA	corps pile group analysis
CQC	complete quadratic combination
FEM	finite element method
GOC	generalized optimality criteria
IBC	International Building Code
NEHRP	National Earthquake Hazard Reduction Program
NIM	number of independent measurements
SHM	structural health monitoring
SRSS	square root of the sum of the squares
UBC	Uniform Building Code
USACE	United States Army Corps of Engineers

1 Introduction

1.1 Background

It has long been recognized that there is a need for the development of more efficient design methods with computer programs in civil engineering. The methods should be more rational and reliable than those that are commonly used and based on conventional trial-and-error processes. The rapid progress in computer technology has, however, led to the development of some sophisticated computer programs that can be used to analyze and design complex structures. When these programs are used for analysis and design, the relative stiffnesses of the constituent members must first be assumed from which the design variables are then modified based on the designer's intuition, experience and skills. If these preliminary stiffnesses are not correctly assumed, in spite of the program's sophistication, repeated analyses can only yield improved design results. Apparently, this does not utilize the total design technology and expectation. This is particularly true in the case of civil engineering building structures. Optimum design of civil engineering structures needs the support from several technological disciplines in order to fulfil its expectation such as building code provisions and cost functions. In the case of a seismic structural design, additional disciplinary parameters are necessary such as the random nature of earthquake excitations and the cost function related to the structural damage. Optimization application in other engineering disciplines is quite successful including airplane structures [1] and automobiles [2]. With this expectation, the book has several emphases: (a) Presentation of modern optimization techniques which can be commonly used in various civil engineering constructions. (b) For the convenience of practitioners, the matrix displacement method is reviewed adequately for 2-D and 3-D building systems using modern digital computers. (c) Various formulations of seismic excitations including elastic and inelastic responses with deterministic and nondeterministic design parameters. (d) Several objective functions are formulated including structural weight, combination of initial and failure cost, as well as life-cycle cost. (e) Extensive numerical results are provided so that the reader can thoroughly follow the equations derived and comprehend optimization results.

1.2   Scope of the book

This book has 12 chapters which are briefly introduced as follows:

Chapter 2: This chapter introduces the fundamentals of linear programming which encompasses the mathematical derivation of equations. In order for the reader to grasp the derivation concept, a simple algebraic problem is used to illustrate the derivation procedures. The mathematical equations are further interwoven with numerical algorithms for convenience of computer programming. As illustrated numerical examples, the Simplex method passes from vertex to vertex on the boundary of the feasible polyhedron, repeatedly increasing the objective function until either an optimal solution is found, or it is established that no solution exists. In general engineering practice of hundreds of variables, the method is highly efficient. Problems of thousands (or even more) of variables can be routinely solved using the Simplex method on modern computers. The method undoubtedly requires significant computing time; however, efficient and sophisticated implementations are available in the form of computer software packages.

Chapter 3: This chapter is focused on using linear programming technique presented in Chapter 2 to synthesize constituent members of an elastic structure for minimum weight of a system. The structural system includes intermediate trusses, continuous beams, and frameworks for which the displacement method is briefly reviewed and employed in formulating various constraint equations and objective functions. Since the displacement and stress equations of an intermediate structure are highly nonlinear, the concept of Taylor series is adapted in the matrix formulation to linearize constraint and objective functions.

Chapter 4: Although many practical problems in optimization can be formulated as linear programs, the real world is full of problems that are not so easy. If either the objective function or any one of the inequalities is nonlinear, then the optimization problem is said to be a *nonlinear program*. Although several varieties of nonlinearity offer special techniques of solution, this chapter introduces the nonlinear optimization with emphasis on the selected algorithms which will be used in later chapters.

Chapter 5: This chapter is solely developed for rigid frames which are commonly constructed in civil engineering. Therefore emphases are placed on matrix formulations of stiffness, mass and second-order matrices, eigen-solutions, calculations of various gradients, step-by-step procedures for feasible direction optimization, and numerical examples and design results. This chapter is intended to establish solid foundation for dynamic analysis and mathematical optimization.

Chapter 6: Gradient-based optimization techniques have been a mainstay for unconstrained mathematical optimization procedures of which the concepts coupled with their mathematical rigour are important and useful in certain constrained optimization algorithms. This chapter provides the necessary theory as well as examples to illustrate the effectiveness of the different gradient-based, unconstrained optimization techniques.

Chapter 7: The modern optimization algorithms of various optimality-criteria methods are extensively used in several chapters of this book. This chapter starts with the following items: formulation of a structural model suitable for finite element analysis and energy distribution, with consideration of coupling ground motions and the $P - \Delta$ effect for 2-D frames; derivation of a primary recursion formula based on energy distribution; derivation of a secondary recursion formula based on constraint gradients; development of minimum cost design; derivation and selection of seismic response spectra; observation of the influence of interacting earthquake motions and building-code provision on structural stiffness requirements; and comparison of the optimum solutions based on a minimum weight and minimum cost design.

Chapter 8: The generalized optimality-criteria method is based on the foundation presented in Chapter 7. This method, however, offers the different techniques for recursion, scaling and Lagrange multiplier determination which are presented extensively in this chapter.

Chapter 9: This chapter focuses on the use of generalized optimality criteria as a means of optimizing both two-dimensional and three-dimensional structural systems subject to static and dynamic, essentially seismic, loads. The seismic loads in this chapter are reflected in code-based static equivalent (lateral) loads, code-based modal loads, response spectra-based loads and time-history accelerograms.

Chapter 10: This chapter covers topological design, pile foundations, damage detection and structural identification. In this chapter an amalgamation of examples where the optimality-criteria techniques presented in Chapter 8 have been modified to solve several non-traditional structural problems are presented. The unique quality of using optimality-criteria techniques is that they can be modified to accommodate the types of, and relationship between, the design variables, the analysis methods, the constraints, and the objective for the optimization.

Chapter 11: In the structural optimization field as presented in the previous chapters, most of the optimization techniques are generally developed on the premise that the design variables, resistances, responses and loadings are deterministic. In recognition of the random nature of seismic structural problems and the advances that have been made in reliability analyses of structures, this chapter is focused on optimum design of nondeterministic structures with consideration of safety levels of a system with respect to its various failure modes, uncertainties in the dead and live loads as well as seismic forces, and random parameters in responses and resistances. The cost function includes initial construction costs of structural and non-structural elements and expected costs of failure at various safety levels.

Chapter 12: This chapter develops a constrained multi-objective optimization method in the form of a robust, practical, problem-independent algorithm to investigate the effect of multi-objective optimization on structural design. A multi-objective optimum procedure comprises constructing an analysis model, deciding objective functions and constraints, and selecting mathematical programming techniques. The solution technique presented is based on a Pareto genetic algorithm. Genetic algorithm search procedures display the intelligent characteristics of utilizing and/or learning from information generated from previous stages which are useful for parallel computation.

Much work has been done in recent years on optimum design of civil engineering structures. The reader may find useful literature of selected monographs listed in the References of this chapter [3–10].

References

1. Morris, A.J., ed., *Foundation of Structural Optimization: A Unified Approach*, John Wiley and Sons, New York, 1982.
2. Bennett, J.A. and Botkin, M.E., eds, *The Optimum Shape: Automated Structural Design*, General Motors Research Laboratories, International Symposium Proceedings, Plenum Press, New York, 1986.
3. Arora, J.S., ed., *Guide to Structural Optimization*, ASCE, Reston, Virginia, 1997.
4. Burn, S.A., ed., *Recent Advances in Optimal Structural Design*, ASCE, Reston, Virginia, 2002.
5. Cheng, F.Y., ed., *Recent Developments in Structural Optimization*, ASCE, Reston, Virginia, 1986.
6. Cheng, F.Y. and Zizhi, F., eds, *Computational Mechanics in Structural Engineering – Recent Development and Future Trends*, Elsevier Applied Science, London, 1992.
7. Cheng, F.Y. and Gu, Y., eds, *Computational Mechanics in Structural Engineering – Recent Developments*, Elsevier, London, 1999.
8. Cheng, F.Y., Introduction of Optimum Structural Design and its recent developments, *Frontier Technologies for Infrastructures Engineering*, Chen, S.S. and Ang, A.H.-S., eds, CRC Press, Taylor & Francis Group, Boca Raton, FL, 2009.
9. Frangopol, D.M., ed., *Optimal Performance of Civil Engineering Infrastructure System*, ASCE, Reston, Virginia, 1998.
10. Lev, O.E., ed., *Structural Optimization: Recent Developments and Applications*, ASCE, Reston, Virginia, 1981.

2 Fundamentals of linear programming

2.1 Introduction

Linear programming was developed as a discipline in the 1940s, motivated initially by the need to solve complex planning problems in wartime operations. Its development accelerated rapidly in the post-war period as many industries found valuable uses for linear programming. The founders of the subject are generally regarded as George B. Dantzig, who devised the Simplex method in 1947, and John von Neumann, who established the theory of duality that same year [1]. Problems from many areas are amenable to attack by linear programming: project developments, product mixing, engineering scheduling, engineering design, airline crew scheduling, shipping or telecommunication networks, oil refining and blending, and stock and bond portfolio selection.

This chapter introduces the fundamentals of linear programming which encompasses the mathematical derivation of equations. In order for the reader to grasp the derivation concept, a simple algebraic problem is used to illustrate the derivation procedures. The mathematical equations are further interwoven with numerical algorithms for convenience of computer programming. The numerical algorithm is then demonstrated with a typical problem of financial planning. Also derived is primal and dual correspondence of which primal and dual relationship merit is clearly shown with an investment case. The chapter ends with concluding remarks on future development. In the subsequent two chapters linear programming is applied to structure design.

2.2 Mathematical derivation

The linear programming problem can be generally formulated as: given a set of m linear equations or inequalities in n variables, find the non-negative values of the variables which will satisfy the constraints and minimize or maximize some linear function of the variables.

Minimize or maximize $\qquad z = c_1 x_1 + c_2 x_2 + c_n x_n$
Subject to $\qquad a_{i1}x_1 + a_{i2}x_2 + \cdots a_{in}x_n\{\geq, =, \leq\}d_i, \quad i=1, m$
$$x_j \geq 0, \quad j=1, n$$

$$(2.1)$$

Any set of x_j which satisfies the constraints is called a solution, while any solution which satisfies the non-negativity restrictions is called a *feasible solution*. A feasible solution with no more than m positive x's and the remaining variables equal to zero is called a *basic feasible solution*. Finally, any feasible solution which optimizes the objective function is called an *optimal feasible solution*.

Note it is not necessary to have an optimal feasible solution for a given set of linear equations; three possible cases may exist: *unbound, indeterminate,* and *no feasible solution*. These three cases are given in Eqs. (2.2) through (2.4) in which the number of variables is two; we may thus solve the minimization problems by a semigraphical procedure as shown in Fig. 2.1 [2, 3].

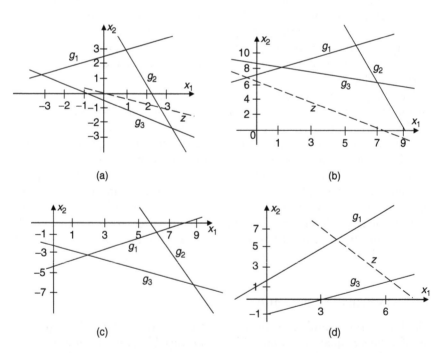

Figure 2.1 Characteristics for non-existence of linear programming solutions: (a) empty set, (b) indeterminate, (c) no feasible solution, (d) unbound.

Case 1 – *Empty set*

$$\left.\begin{array}{ll}\text{Minimize} & z = 2x_1 + 7x_2 \\ & x_1 - 2x_2 \geq -5 \dots\dots\dots g_1(x) \\ \text{Subject to} & 5x_1 + 2x_2 \leq 11 \dots\dots\dots g_2(x) \\ & x_1 + 2x_2 \geq -1 \dots\dots\dots g_3(x)\end{array}\right\} \qquad (2.2)$$

The results may be observed from Fig. 2.1a: the feasible region is in the first quadrant; however, the solution may also exist in the second quadrant with negative x_1. Thus, the case is called *empty set*.

Case 2 – *Indeterminate*

$$\left.\begin{array}{ll}\text{Minimize} & z = x_1 + 2x_2 \\ & x_1 - 2x_2 \geq -14 \dots\dots\dots g_1(x) \\ \text{Subject to} & 5x_1 + 2x_2 \leq 50 \dots\dots\dots g_2(x) \\ & x_1 + 2x_2 \geq 18 \dots\dots\dots g_3(x)\end{array}\right\} \qquad (2.3)$$

Observing Fig. 2.1b notice that z is parallel to $g_3(x)$; the solution is thus *indeterminate*.

Case 3 – *No feasible solution*

$$\left.\begin{array}{ll}\text{Minimize} & z = 2x_1 + 7x_2 \\ & x_1 - 2x_2 \geq 8 \dots\dots\dots g_1(x) \\ \text{Subject to} & 5x_1 + 2x_2 \leq 28 \dots\dots\dots g_2(x) \\ & x_1 + 2x_2 \geq -4 \dots\dots\dots g_3(x)\end{array}\right\} \qquad (2.4)$$

The solution of Eq. (2.4) is drawn in Fig. 2.1c in the fourth quadrant and it violates the criterion of non-negative design variable. The solution is therefore called *non-feasible*.

Case 4 – *Unbound*

The scenario may be observed from Fig. 2.1d in which $-x_1 + x_2 \geq 2$, $3x_1 - x_2 \geq -3$ for g_1 and g_3 respectively; and g_2 does not exist. Since g_1 and g_3 are diverging in the positive region; then the objective function cannot have a finite value. The case is called *unbound*.

In conclusion, any general linear programming algorithm must be formulated to indicate the existence of each of these conditions. To develop a computational method of solution for the linear programming problem, it is desirable to convert the inequalities in the constraints into equalities [4, 5].

Maximization

$$\sum_{j=1}^{n} a_{pj}x_j \leq d_p \qquad (2.5)$$

For this inequality, the non-negative variable x_{n+p} is expressed as

$$x_{n+p} = d_p - \sum a_{pj} x_j \geq 0 \tag{2.6}$$

or

$$\sum_{j=1}^{n} a_{pj} x_j + x_{n+p} = d_p \tag{2.6a}$$

x_{n+p} is called a *slack variable.*

Minimization

$$\sum_{j=1}^{n} a_{qj} x_j \geq d_q \tag{2.7}$$

$$x_{n+q} = \sum_{j=1}^{n} a_{qj} x_j - d_q \geq 0 \tag{2.8}$$

or

$$\sum_{j=1}^{n} a_{qj} x_j - x_{n+q} = d_q \tag{2.8a}$$

x_{n+q} is called a *surplus variable.*

The optimization problem may be generally stated as follows

$$\text{Maximize } z = [c]_{1 \times p} \{x\}_{p \times 1} = \sum_{i=1}^{p} c_i x_i \tag{2.9}$$

in which z = value of the objective function, $[c]$ = cost coefficients, and $\{x\}$ = problem variables.

$$\text{Subject to } [A]_{m \times p} \{x\}_{p \times 1} = \sum_{i=1}^{p} \{a_i\}_{m \times 1} x_i = \{d\}_{m \times 1} \tag{2.10}$$

where $[A]$ = coefficients associated with the problem variables in the constraint set (includes the basis matrix $[B]_{m \times m}$), and $\{d\}$ = coefficients on the right-hand side of the constraints.

The initial basic feasible solution may be expressed as

$$[B]_{m \times m} \{x_b\}_{m \times 1} = \{d\}_{m \times 1} \tag{2.11}$$

in which $\{x_b\}$ = variables in basis. The initial value of the objective function is

$$z = [c_b]_{1 \times m} \{x_b\}_{m \times 1} = \sum_{i=1}^{m} c_{Bi \times Bi} \tag{2.12}$$

where $[c_b]$ = cost coefficients in the basis. The solution procedure is to improve the initial solution by interchanging variables in and out of the basis. This involves four fundamental mathematical operations: (a) to find the variable removed, (b) to determine the variable entered, (c) to calculate other coefficients, and (d) to find a new solution. The operations of (a) through (d) are detailed hereafter. Since the basis $[B]_{m \times m}$ must consist of m linearly independent columns $\{b_i\}_{m \times 1}$ from $[A]_{m \times p}$, there exists a set of coefficients $\{y_j\}_{m \times 1}$ such that the column $\{b_i\}_{m \times 1}$ of $[B]_{m \times m}$ as

$$\{a_j\}_{m \times 1} = \sum_{i=1}^{m} \{b_i\} y_{ij} = [B]_{m \times m} \{y_j\}_{m \times 1} \tag{2.13}$$

from which

$$\{y_j\}_{m \times 1} = [B]_{m \times m} \{a_j\}_{m \times 1} \tag{2.14}$$

Let the non-basic variable x_j be inserted into the basis to replace the basic variable x_{Br}, we rewrite as

$$\{a_j\}_{m \times 1} = \sum_{i=1}^{m} \{b_i\}_{m \times 1} y_{ij} + \{b_r\}_{m \times 1} y_{rj} \tag{2.15}$$

from which

$$\{b_r\}_{m \times 1} = \frac{1}{y_{rj}} \{a_j\}_{m \times 1} - \sum_{\substack{i=1 \\ i \neq r}}^{m} \frac{y_{ij}}{y_{rj}} \{b_i\}_{m \times 1}, \quad y_{rj} \neq 0 \tag{2.16}$$

(a) To find the variable removed

From Eq. (2.11),

$$\sum_{\substack{i=1 \\ i \neq r}} \{b_i\}_{m \times 1} x_{bi} + \{b_r\}_{m \times 1} x_{Br} = \{d\}_{m \times 1} \tag{2.17}$$

Substituting Eq. (2.16) into Eq. (2.17) yields

$$\sum_{\substack{i=1 \\ i \neq r}}^{m} \{b_i\}_{m \times 1} x_{Bi} + x_{Br} \frac{1}{y_{rj}} \{a_j\}_{m \times 1} - x_{Br} \sum_{\substack{i=1 \\ i \neq r}}^{m} \frac{y_{ij}}{y_{rj}} \{b_i\}_{m \times 1} = \{d\}_{m \times 1} \tag{2.18}$$

or

$$\sum_{\substack{i=1 \\ i\neq r}}^{m}\left(x_{Bi}-x_{Br}\frac{y_{ij}}{y_{rj}}\right)\{b_i\}_{m\times 1}+\frac{x_{Br}}{y_{rj}}\{a_j\}_{m\times 1}=\{d\}_{m\times 1} \tag{2.18a}$$

from which we may observe that the new values of the basis valuables are

$$\boxed{\begin{aligned} x'_{Bi} &= x_{Bi}-x_{Br}\frac{y_{ij}}{y_{rj}} \\ x'_{Br} &= \frac{x_{Br}}{y_{rj}} \end{aligned}} \tag{2.19}$$

Note: the equations shown in the box above as well as hereafter in this section are those corresponding to the equations in the numerical algorithm of Section 2.3. In order to achieve a basic feasible solution, we have

$$\left.\begin{aligned} x_{Bi}-x_{B}\frac{y_{ij}}{y_{rj}}\geq 0,\quad i\neq r,\quad y_{rj}\neq 0 \\ \frac{x_{Br}}{y_{rj}}\geq 0 \end{aligned}\right\} \tag{2.20}$$

Since the initial solution was defined to be feasible, the second condition in the above requires

$$y_{rj}>0 \tag{2.21}$$

and the first condition can be rewritten

$$\frac{x_{Bi}}{y_{ij}}-\frac{x_{Br}}{y_{rj}}\geq 0 \tag{2.22}$$

Thus, the new solution will only be a basic feasible solution if the column r removed from the basis is chosen to satisfy the following criterion:

$$\boxed{\frac{x_{Br}}{y_{rj}}=\min_{i}\left\{\frac{x_{Bi}}{y_{ij}},\quad y_{ij}>0\right\}} \tag{2.23}$$

(b) To determine the variable entered (i.e., the objective function must be improved)

From Eq. (2.12)

$$z=\sum_{i=1}^{m}c_{Bi}x_{Bi}=\sum_{\substack{i=1 \\ i\neq r}}^{m}c_{Bi}x_{Bi}+c_{Br}x_{Br} \tag{2.24}$$

Substituting Eq. (2.19) into the above and noting that $c_{Br} = c_j$

$$z' = \sum_{\substack{i=1 \\ i \neq r}}^{m} c_{Bi} \left(x_{Bi} - x_{Br} \frac{y_{ij}}{y_{rj}} \right) + c_j \frac{x_{Br}}{y_{rj}} \tag{2.25}$$

Since the term for $i = r$ is zero, we have

$$z' = \sum_{i=1}^{m} c_{Bi} \left(x_{Bi} - x_{Br} \frac{y_{ij}}{y_{rj}} \right) + c_j \frac{x_{Br}}{y_{rj}} \tag{2.26}$$

or

$$z' = z + \frac{x_{Br}}{y_{rj}} \left(c_j - \sum_{i=1}^{m} c_{Bi} y_{ij} \right) \tag{2.26a}$$

It is convenient to define a new quantity as

$$z_j = \sum_{i=1}^{m} c_{Bi} y_{ij} \tag{2.27}$$

Thus, Eq. (2.26a) becomes

$$z' = z + \frac{x_{Br}}{y_{rj}} (c_j - z_j) \tag{2.28}$$

The new value of the objective function is dependent upon x_{Br}/y_{rj} and $c_j - z_j$. The most logical choice for choosing the variable to enter the basis would be to select the variable x_j according to

$$\frac{x_{Br}}{y_{rj}}(c_j - z_j) = \max_i \left\{ \frac{x_{Br}}{y_{ri}}(c_i - z_i) \right\}, \quad c_i - z_i > 0 \tag{2.29}$$

or in general

$$\boxed{\begin{aligned} c_j - z_j &= \max_i \{c_i - z_i\}, \quad c_i - z_i > 0 \\ z_j - c_j &= \min_i \{z_i - c_i\}, \quad z_i - c_i < 0 \end{aligned}} \tag{2.30}$$

Let j be denoted by k, then

$$
\begin{aligned}
c_k - z_k &= \max_j \left\{ c_j - z_j \right\}, \quad c_j - z_j > 0 \\
z_k - c_k &= \max_j \left\{ z_j - c_j \right\}, \quad z_j - c_j < 0
\end{aligned}
\tag{2.31}
$$

$$
\frac{x_{Br}}{y_{rj}} = \min_i \left\{ \frac{x_{Bi}}{y_{ik}}, \quad y_{ik} > 0 \right\}
\tag{2.32}
$$

$$
\begin{aligned}
x'_{Bi} &= x_{Bi} - x_{Br} \frac{y_{ij}}{y_{rj}}, \quad i \neq r, j = k \\
x'_{Br} &= \frac{x_{Br}}{y_{rj}}, \quad i = r, j = k
\end{aligned}
\tag{2.33}
$$

(c) To calculate other coefficients

Substituting Eq. (2.16) into Eq. (2.15)

$$
\{a_j\}_{m \times 1} = \sum_{\substack{i=1 \\ i \neq r}}^{m} \left(y_{ij} - y_{rj} \frac{y_{ij}}{y_{rk}} \right) \{b_i\}_{m \times 1} + \frac{y_{rj}}{y_{rk}} \{a_k\}_{m \times 1}
\tag{2.34}
$$

which gives

$$
\begin{aligned}
y'_{ij} &= y_{ij} - y_{ik} \frac{y_{rj}}{y_{rk}}, \quad i \neq r \\
y'_{rj} &= \frac{y_{rj}}{y_{rk}}
\end{aligned}
\tag{2.35}
$$

(d) To find a new solution

The new $z_j - c_j$ is

$$
z'_j - c_j = \sum_{i=1}^{m} c'_{Bi} y'_{ij} - c_j
\tag{2.36}
$$

where c_j is not affected due to the change of the basis. Substituting Eq. (2.35) into Eq. (2.36) and noting that $c'_{Bi} = c_{Bi}$ for $i \neq r$ and $c'_{Br} = c_k$,

$$
z'_j - c_j = \left(\sum_{i=1}^{m} c_{Bi} \left(y_{ij} - y_{ik} \frac{y_{rj}}{y_{rk}} \right) + \frac{y_{rj}}{y_{rk}} c_k - c_j \right)
\tag{2.37}
$$

Since the term for $i = r$ is zero, then

$$z'_j - c_j = \left(\sum_{i=1}^{m} c_{Bi} y_{ij} - c_j \right) - \frac{y_{rj}}{y_{rk}} \left(\sum_{i=1}^{m} c_{Bi} y_{ik} - c_k \right) \tag{2.38}$$

or

$$\boxed{z'_j - c_j = (z_j - c_j) - \frac{y_{rj}}{y_{rk}} (z_k - c_k)} \tag{2.39}$$

The new solution is obtained from Eq. (2.26a) as

$$\boxed{z' = z - \frac{x_{Br}}{y_{rj}} (z_k - c_k)} \tag{2.40}$$

2.3 Detailed illustration

The iterative process developed in the foregoing theory is illustrated in detail by means of a numerical example as follows:

$$\left. \begin{array}{ll} \text{Maximize} & z = 2x_1 - 4x_2 \\ & 3x_1 + 5x_2 \geq 15 \\ \text{Subject to} & 4x_1 + 9x_2 \leq 36 \\ & x_1, x_2 \geq 0 \end{array} \right\} \tag{a}$$

First, we convert the constraints of the above equations into equalities by adding slack or surplus variables as

$$\left. \begin{array}{l} 3x_1 + 5x_2 - x_3 = 15 \\ 4x_1 + 9x_2 + x_4 = 36 \\ x_1, x_2, x_3, x_4 \geq 0 \end{array} \right\} \tag{b}$$

Let us now define the $[A]$ matrix and its constituent vectors, $\{a_j\}, j = 1, 2, 3, 4$ as

$$[A] = \begin{bmatrix} 3 & 5 & -1 & 0 \\ 4 & 9 & 0 & 1 \end{bmatrix} \begin{Bmatrix} x_1 \\ x_2 \\ x_3 \\ x_4 \end{Bmatrix}, \quad \{b\} = \begin{Bmatrix} 15 \\ 36 \end{Bmatrix} \tag{c}$$

$$\{a_1\} = \begin{Bmatrix} 3 \\ 4 \end{Bmatrix}, \quad \{a_2\} = \begin{Bmatrix} 5 \\ 9 \end{Bmatrix}, \quad \{a_3\} = \begin{Bmatrix} -1 \\ 0 \end{Bmatrix}, \quad \{a_4\} = \begin{Bmatrix} 0 \\ 1 \end{Bmatrix} \tag{d}$$

Take $\{a_1\}$ and $\{a_4\}$ as our initial basis. This means that

$$x_1\{a_1\}+x_4\{a_4\}=\{b\} \tag{e}$$

$$x_1\begin{Bmatrix}3\\4\end{Bmatrix}+x_4\begin{Bmatrix}0\\1\end{Bmatrix}=\begin{Bmatrix}15\\36\end{Bmatrix} \tag{f}$$

Therefore, $x_1=5$, $x_4=16$, $x_2=x_3=0$ and our initial basic feasible solution is $\{x_B\}=[5\ 16]^T$. Now we shall express the vectors not in the basis, i.e., $\{a_2\}$ $\{a_3\}$ in terms of $\{a_1\}$ and $\{a_4\}$,

$$\{a_2\}=y_{12}\{a_1\}+y_{42}\{a_4\} \tag{g}$$

$$\{a_3\}=y_{13}\{a_1\}+y_{43}\{a_4\} \tag{h}$$

$$\begin{Bmatrix}5\\9\end{Bmatrix}=y_{12}\begin{Bmatrix}3\\4\end{Bmatrix}+y_{42}\begin{Bmatrix}0\\1\end{Bmatrix} \tag{i}$$

This yields

$$\{y_2\}=[y_{12}\ \ y_{42}]^T=[5/3\ \ 7/3]^T \tag{j}$$

Similarly,

$$\begin{Bmatrix}-1\\0\end{Bmatrix}=y_{13}\begin{Bmatrix}3\\4\end{Bmatrix}+y_{43}\begin{Bmatrix}0\\1\end{Bmatrix} \tag{k}$$

which yields

$$\{y_3\}=[y_{13}\ \ y_{43}]^T=[-1/3\ \ 4/3]^T \tag{l}$$

Let us now compute z_j for $\{a_2\}$ and $\{a_3\}$

$$z_j=[c_B]\{y_j\}$$

$$=\sum_i^m c_{Bi}y_{ij}, [c_B]=[2\ \ 0],\quad c_1=2,\quad c_2=-4,\quad c_3=0,\quad c_4=0 \tag{m}$$

$$z_2=[2\ \ 0]\begin{bmatrix}5/3\\7/3\end{bmatrix}=\frac{10}{3} \tag{n}$$

$$z_3=[2\ \ 0]\begin{bmatrix}-1/3\\4/3\end{bmatrix}=-\frac{2}{3} \tag{o}$$

$$z_2-c_2=\frac{10}{3}+4=\frac{22}{3} \tag{p}$$

$$z_3-c_3=-\frac{2}{3}-0=-\frac{2}{3} \tag{q}$$

Since only $z_3 - c_3 < 0$, $\{a_3\}$ will be chosen to enter the basis. Let us now determine which vector must be removed. The rule is

$$\min_i \left\{ \frac{x_{Bi}}{y_{i3}}, \quad y_{i3} > 0 \right\} \tag{r}$$

We have only two choices x_1/y_{13} or x_4/y_{43}. However, since $y_{13} = -1/3 < 0$, we can only consider x_4/y_{43}. Hence, $\{a_4\}$ leaves the basis. The original value of the objective function was

$$z = [c_B]\{x_B\} = [2 \quad 0] \begin{Bmatrix} 5 \\ 16 \end{Bmatrix} = 10 \tag{s}$$

According to Eq. (2.40)

$$z' = z + \frac{x_4}{y_{43}}(c_3 - z_3) = 10 + \frac{16}{4/3}\left(\frac{2}{3}\right) = 18 \tag{t}$$

and as we see $z' > z$. Let us now repeat our previous work and express the vectors not in the basis, $\{a_2\}$ and $\{a_4\}$, but in terms of $\{a_1\}$ and $\{a_3\}$ our basic vectors.

$$\{a_2\} = y_{12}\{a_1\} + y_{32}\{a_3\} \tag{u}$$

$$\{a_4\} = y_{11}\{a_1\} + y_{34}\{a_3\} \tag{v}$$

$$\begin{Bmatrix} 5 \\ 9 \end{Bmatrix} = y_{12} \begin{Bmatrix} 3 \\ 4 \end{Bmatrix} + y_{32} \begin{Bmatrix} -1 \\ 0 \end{Bmatrix} \tag{w}$$

This yields

$$\{y_2\} = [y_{12} \quad y_{32}] = [9/4 \quad 7/4] \tag{x}$$

Similarly,

$$\begin{Bmatrix} 0 \\ 1 \end{Bmatrix} = y_{14} \begin{Bmatrix} 3 \\ 4 \end{Bmatrix} + y_{34} \begin{Bmatrix} -1 \\ 0 \end{Bmatrix} \tag{y}$$

$$\{y_4\} = [y_{14} \quad y_{34}] = [1/4 \quad 3/4]^T \tag{z}$$

Next, we compute z_j for $\{a_2\}$ and $\{a_3\}$, noting that $[c'_B] = [2 \quad 0]$, $c_1 = 2$, $c_2 = -4$, $c_3 = 0$, $c_4 = 0$

$$z_2 = [c'_B]\{y_2\} = [2 \quad 0]\left\{ \begin{array}{c} 9/4 \\ 7/4 \end{array} \right\} = \frac{9}{2} \tag{aa}$$

$$z_4 = [c'_B]\{y_4\} = [2 \quad 0]\left\{ \begin{array}{c} 1/4 \\ 3/4 \end{array} \right\} = \frac{1}{2} \tag{bb}$$

$$z_2 - c_2 = \frac{9}{2} + 4 = \frac{17}{2} > 0 \tag{cc}$$

$$z_4 - c_4 = \frac{1}{2} - 0 = \frac{1}{2} > 0 \tag{dd}$$

Since all $z_j - c_j \geq 0$ for vectors not in the basis, we have the optimal solution to the linear programming problem, which is obtained as follows

$$x_1\{a_1\} + x_3\{a_3\} = \{b\} \tag{ee}$$

$$x_1 \left\{ \begin{array}{c} 3 \\ 4 \end{array} \right\} + x_3 \left\{ \begin{array}{c} -1 \\ 0 \end{array} \right\} = \left\{ \begin{array}{c} 15 \\ 36 \end{array} \right\} \tag{ff}$$

$$x_1 = 9, \quad x_3 = 12, \quad z = 18 \tag{gg}$$

2.4   Numerical algorithm

From the sample illustration in the previous section, a numerical algorithm is presented using *Simplex tableau* which is based on the mathematical derivation in Section 2.2 but arranged in a different order for convenience of computer programming. The general tableau form may be expressed in Eq. (2.41). For illustration let us use the optimization problem in Eq. (2.42) to be expressed in a tableau as shown in Eq. (2.43).

	c_1		c_n	$c_{n+1}=0$	$c_{n+p}=0$	
Variables in Basis	a_1		a_n	$a_{n+1}=b_1$	$a_p=b_m$	x_B
x_{n+1}	$y_{11}=a_{11}$	----	$y_{1n}=a_{1n}$	$y_{1,n+1}=1$ ----	$y_{1p}=0$	$x_{B1}=d_1$
x_{n+2}	$y_{21}=a_{21}$	----	$y_{2n}=a_{2n}$	$y_{2,n+1}=0$ ----	$y_{2p}=0$	$x_{B2}=d_2$
$\vdots$	$\vdots$		$\vdots$	$\vdots$	$\vdots$	$\vdots$
x_p	$y_{m1}=a_{m1}$	----	$y_{mn}=a_{mn}$	$y_{m,n+1}=0$ ----	$y_{mp}=1$	$x_{Bm}=d_m$
$z_j - c_j$	$0-c_1$	----	$0--c_n$	$0-0$	$0-0$	$z=0$

$$\tag{2.41}$$

$$\left. \begin{array}{ll} \text{Maximize} & z = 5x_1 + x_2 \\ \text{Subject to} & -x_1 + x_2 \leq 2 \\ & 3x_1 - x_2 \leq 6 \\ & x_1 + x_2 \leq 6 \\ & x_1, x_2 \geq 0 \end{array} \right\} \tag{2.42}$$

c	5	1	0	0	0	
Variables in Basis	a_1	a_2	a_3	a_4	a_5	x_B
x_3	-1	1	1	0	0	2
x_4	3	-1	0	1	0	6
x_5	1	1	0	0	1	6
$z_j - c_j$	-5	-1	0	0	0	0

$$(2.43)$$

Numerical procedures are illustrated in the following iterative cycles

(1) Select the variable to enter the basis (See Eq. (2.31)).

$$\boxed{z_k - c_k = \min_j \left\{ z_j - c_j \right\}, \quad z_j - c_j < 0}$$

$$j = 1, \quad z_1 - c_1 = -5$$
$$j = 2, \quad z_2 - c_2 = -1$$

which indicate variable x_1 should enter the basis, thus $k = 1$. Note equation numbers assigned to the equations in boxes of this section correspond to those derived in Section 2.2.

(2) Select the variable to leave the basis (See Eq. (2.32))

$$\boxed{\frac{x_{Br}}{y_{rk}} = \min_i \left\{ \frac{x_{Bi}}{y_{ik}}, \quad y_{ik} > 0 \right\}}$$

$$i = 1, \quad \frac{x_{B1}}{y_{11}} = \frac{2}{-1} = -2 \quad \text{(we don't consider this, because } y_{11} < 0)$$

$$i = 2, \quad \frac{x_{B2}}{y_{21}} = \frac{6}{3} = 2 \quad \text{(this is the minimum, therefore } r = 2,$$
$$x_{B2} \text{ should leave the basis)}$$

$$i = 3, \quad \frac{x_{B3}}{y_{31}} = \frac{6}{1} = 6$$

(3) The new value of the basis variables can be obtained by (See Eq. (2.33))

$$\boxed{\begin{aligned} x'_{Bi} &= x_{Bi} - x_{Br} \frac{y_{ij}}{y_{rj}}, \quad i \neq r \\ x'_{Br} &= \frac{x_{Br}}{y_{rj}}, \quad i = r \end{aligned}}$$

For this case $r = 2, j$ (actually is k) $= 1$

$$i = 1, \quad x'_{B1} = x_{B1} - \frac{x_{B2}y_{11}}{y_{21}} = 2 - 6\left(\frac{-1}{3}\right) = 2 + 2 = 4$$

$$i = r = 2, \quad x'_{B2} = \frac{x_{B2}}{y_{21}} = \frac{6}{3} = 2$$

$$i = 3, \quad x'_{B3} = x_{B3} - x_{B2}\left(\frac{y_{31}}{y_{21}}\right) = 6 - 6\left(\frac{1}{3}\right) = 6 - 2 = 4$$

(4) The new values of $\{y\}$ (See Eq. (2.35))

$$\boxed{\begin{aligned} y'_{ij} &= y_{ij} - y_{ik}\frac{y_{rj}}{y_{rk}}, \quad i \neq r \\ y'_{rj} &= \frac{y_{rj}}{y_{rk}} \end{aligned}}$$

$r = 2, k = 1$

$$y'_{21} = \frac{y_{21}}{y_{22}} = 1, \quad y'_{22} = \frac{y_{22}}{y_{21}} = \frac{-1}{3}, \quad y'_{23} = 0, \quad y'_{24} = \frac{1}{3}, \quad y'_{25} = 0$$

$$y'_{ij} = y_{ij} - y_{i1}\left(\frac{y_{2j}}{y_{21}}\right) \quad i = 1, 3, 4; \ j = 1, 2, 3, 4, 5$$

$$y'_{11} = y_{11} - y_{11}\left(\frac{y_{21}}{y_{21}}\right) = (-1) - (-1)\left(\frac{3}{3}\right) = 0$$

$$y'_{12} = 1 - (-1)\left(\frac{-1}{3}\right) = \frac{2}{3}$$

$$y'_{31} = 1 - (1)\left(\frac{3}{3}\right) = 0$$

$$y'_{32} = 1 - (1)\left(\frac{-1}{3}\right) = \frac{4}{3}$$

$$\vdots$$

(5) The new values of $z'_j - c_j$ (See Eq. (2.39))

$$\boxed{z'_j - c_j = (z_j - c_j) - \left(\frac{y_{rj}}{y_{rk}}\right)(z_k - c_k)}$$

$r = 2, k = 1$

$$z_1' - c_1 = (z_1 - c_1) - \left(\frac{y_{21}}{y_{21}}\right)(z_1 - c_1) = (-5) - \left(\frac{3}{3}\right)(-5) = 0$$

$$z_2' - c_2 = (z_2 - c_2) - \left(\frac{y_{22}}{y_{21}}\right)(z_1 - c) = (-1) - \left(\frac{-1}{3}\right)(-5) = \frac{-8}{3}$$

$\vdots$

(6) The new values of z' (See Eq. (2.40))

$$z' = z - \left(\frac{x_{Br}}{y_{rk}}\right)(z_k - c_k)$$

$r = 2, k = 1; z' = z - (x_{B2}/y_{21})(z_1 - c_1) = 0 - (6/3)(-5) = 10$
The complete revised tableau becomes

c	5	1	0	0	0	
Variables in Basis	a_1'	a_2'	a_3'	a_4'	a_5'	x_B'
x_3	0	2/3	1	1/3	0	4
x_1	1	−1/3	0	1/3	0	2
x_5	0	4/3	0	−1/3	1	4
$z_j' - c_j$	0	−8/3	0	5/3	0	10

which shows that x_4 is removed with x_1 entered. At the end of the cycle, the following criteria should be checked:

a) Are all $(z_j - c_j) \geq 0$? No, $z_2 - c_2 = -8/3$, therefore this solution is not optimal.

b) For each $z_j - c_j < 0$, are all $y_{ij} \leq 0$? No, for $z_2 - c_2 = -8/3$, $y_{12} = 2/3$, $y_{32} = 4/3$.

Therefore, a finite solution may exist and a new basis change can be performed.

Second Cycle – repeat the process as given previously

(1) Check $(z_k - c_k) = \min\{z_k - c_k\}$, $z_j - c_j < 0$. From Eq. (2.41), $k = 2$.
(2) Check $x_{Br}/y_{rk} = \min_i\{x_{Bi}/y_{ik}, y_{ik} > 0\}$ for $k = 2$,

$$x_{B1}/y_{12} = 4/(2/3) = 6, \quad x_{B2}/y_{22} = 2/(-1/3) = -6,$$

$$x_{B3}/y_{32} = 4/(4/3) = 12/4 = 3.$$

Therefore, $r = 3$.

(3) Check $x'_{Bi} = x_{Bi} - x_{Br}(y_{ij}/y_{rj})$, and $x'_{Br} = x_{Br}/y_{rj}$, $i \neq r$

For $r = 3$, $j = k = 2$

$$x'_{B1} = 4 - 4(3/2)/(4/3) = 2, \quad x'_{B2} = 2 - 4(-1/3)/(4/3) = 3,$$
$$x'_{B3} = 4/(4/3) = 3$$

(4) Check $y'_{ij} = y_{ij} - y_{ik}(y_{rj}/y_{rk})$, and $y'_{rj} = y_{rj}/y_{rk}$, $i \neq r$

For $r = 3$, $k = 2$, $i = 1, 2$; $j = 1, 2, 3, 4, 5$

$$y_{rk} = 4/3$$
$$y'_{11} = 0 - 2/3(0/(4/3)) = 0$$
$$y'_{13} = 1 - 2/3(0/(4/3)) = 1$$

$$\vdots$$

(5) Check $z'_j - c_j = (z_j - c_j) - (y_{rj}/y_{rk})(z_k - c_k)$

For $r = 3$, $k = 2$

$$z'_1 - c_1 = 0 - (0/(4/3))(-8/3) = 0$$
$$z'_2 - c_2 = -8/3 - (-8/3)(4/3)/(4/3) = 0$$

$$\vdots$$

(6) Check $z' = z - (x_{Br}/y_{rk})(z_k - c_k)$

$$z' = 10 - (4/(4/3))(-8/3) = 18$$

The complete revised tableau is given in Eq. (2.44) as:

c	5	1	0	0	0	
Variables in Basis	a''_1	a''_2	a''_3	a''_4	a''_5	x''_B
x_3	0	0	1	1/2	−1/2	2
x_1	1	0	0	1/4	1/4	3
x_2	0	1	0	−1/4	3/4	3
$z''_j - c_j$	0	0	0	1	2	18

(2.44)

Examination of the above reveals that $z_j - c_j \geq 0$, the solution is optimal; $x_1 = 3$, $x_2 = 3$, and $z = 5x_1 + x_2 = 18$.

2.5 Sample case for financial planning

A developer has the alternatives of building two-, three- and four-bedroom houses. He wishes to establish the number of each that will maximize his profit, subject to the following conditions:

1 The total budget for the project cannot exceed $9,000,000.
2 The total number of units cannot exceed 500.
3 The maximum percentage of each, based on a market survey, is

2 bedrooms   20%
3 bedrooms   60%
4 bedrooms   40%

4 Building costs including land, architecture and engineering fees, land-scaping, etc. are

2-bedroom unit $20,000
3-bedroom unit $25,000
4-bedroom unit $30,000

5 Net profit after interest, etc. is

2 bedrooms $2,000
3 bedrooms $3,000
4 bedrooms $4,000

The objective is to find the best combination of the three kinds of houses to obtain the maximum profit.

Solution:

Let x_1 be the number of 2-bedroom units, x_2 be the number of 3-bedroom units, and x_3 be the number of 4-bedroom units. Hence, the constraints are established as:

$$\left.\begin{array}{c} 20,000x_1 + 25,000x_2 + 30,000x_3 \leq 9,000,000 \\ x_1 + x_2 + x_3 \leq 500 \\ x_1 \leq 0.2(x_1 + x_2 + x_3) \\ x_2 \leq 0.6(x_1 + x_2 + x_3) \\ x_3 \leq 0.4(x_1 + x_2 + x_3) \\ x_1 \geq 0, x_2 \geq 0, x_3 \geq 0 \end{array}\right\} \quad \text{(a)}$$

or

$$\left.\begin{array}{c} 4x_1 + 5x_2 + 6x_3 \leq 1,800 \\ x_1 + x_2 + x_3 \leq 500 \\ 4x_1 - x_2 - x_3 \leq 0 \\ -3x_1 + 2x_2 - 3x_3 \leq 0 \\ -2x_1 - 2x_2 + 3x_3 \leq 0 \end{array}\right\} \quad \text{(b)}$$

Maximize $z = 2,000x_1 + 3,000x_2 + 4,000x_3$ \qquad (c)

The Simplex tableau form is

c	2,000	3,000	4,000	0	0	0	0	0	
Variables in Basis	a_1	a_2	a_3	a_4	a_5	a_6	a_7	a_8	
x_4	4	5	6	1	0	0	0	0	1,800
x_5	1	1	1	0	1	0	0	0	500
x_6	4	−1	−1	0	0	1	0	0	0
x_7	−3	2	−3	0	0	0	1	0	0
x_8	−2	−2	3	0	0	0	0	1	0
z_j-c_j	−2,000	−3,000	−4,000	0	0	0	0	0	0

(d)

First cycle

c	2,000	3,000	4,000	0	0	0	0	0	
Variables in Basis	a'_1	a'_2	a'_3	a'_4	a'_5	a'_6	a'_7	a'_8	x'_B
x_2	8	9	0	1	0	0	0	−2	1,800
x_5	5/3	5/3	0	0	1	0	0	−1/3	500
x_6	10/3	−5/3	0	0	0	1	0	1/3	0
x_7	−5	0	0	0	0	0	1	1	0
x_3	−2/3	−2/3	1	0	0	0	0	1/3	0
z'_j-c_j	−14,000/3	−17,000/3	4,000/3	0	0	0	0	0	0

(e)

Second cycle

c	2,000	3,000	4,000	0	0	0	0	0	
Variables in Basis	a'_1	a'_2	a'_3	a'_4	a'_5	a'_6	a'_7	a'_8	x'_B
x_2	8/9	1	0	1/9	0	0	0	−2/9	200
x_5	5/27	0	0	−5/27	1	0	0	1/27	500/3
x_6	130/27	0	0	5/27	0	1	0	−1/27	1,000/3
x_7	−5	0	0	0	0	0	1	1	0
x_3	−2/27	0	1	2/27	0	0	0	5/27	400/3
z'_j-c_j	10,000/27	0	0	17,000/27	0	0	0	2,000/27	340,000/3

(f)

The optimal solution from Eq. (f) is $x_1 = 0$, $x_2 = 200$, $x_3 = 400/3$ which produce the highest profit of $z = 340,000/3$.

2.6 Primal and dual correspondence in linear programming

The standard form for a linear programming problem is generally written as follows [2, 4]

$$\left.\begin{array}{ll} \text{Maximize} & z = [c]_{1 \times n} \{x\}_{n \times 1} \\ \text{Subject to} & [A]_{k \times n} \{x\}_{n \times 1} \leq \{b\}_{k \times 1} \\ & \{x\} \geq 0 \end{array}\right\} \tag{2.45}$$

This is called the *primal formulation* of the problem. Notice that $k \leq n$ and the maximization can also be treated, if desired, as a minimization by multiplying by -1.0.

If a new linear function is defined as:

$$Z = [b]_{1 \times k}^{T} \{w\}_{k \times 1} \tag{2.46}$$

such that

$$[A]_{n \times k}^{T} \{w\}_{k \times 1} \geq \{c\}_{n \times 1}^{T} \tag{2.47}$$

and that

$$\{w\} \geq 0 \tag{2.48}$$

It can be proved that the new function Z is related to the primal formulation by a *dual formulation*. From the primal formulation and Eq. (2.47) we have

$$z = [c]\{x\} \tag{2.49}$$

$$[[A]^{T}\{w\}]^{T} \geq [c] \tag{2.50}$$

Using

$$[[A]^{T}\{w\}]^{T} = [w]^{T}[A] \tag{2.51}$$

Eq. (2.47) becomes

$$[w]^{T}[A] \geq [c] \tag{2.52}$$

From Eqs. (2.49) and (2.52) we have

$$[w]^{T}[A]\{x\} \geq z \tag{2.53}$$

Based on Eq. (2.45)

$$[A]\{x\} \leq \{b\}$$

premultiplying the above with $[w]^T$ yields

$$[w]^T[A]\{x\} \leq [w]^T\{b\} \tag{2.54}$$

From Eq. (2.46)

$$[w]^T\{b\} \equiv Z \tag{2.55}$$

Comparing Eqs. (2.54) with (2.55) and (2.53) yields

$$Z \geq [w]^T[A]\{x\} \geq z \tag{2.56}$$

which indicates

$$Z \geq z \tag{2.57}$$

By examining the relationship between Z and z, it can be concluded that Z can only be equal to z when Z is at its minimum and z is at its maximum. Since the purpose of the primal formulation is to search for the maximum solution for $z(x)$, then the minimum solution for $Z(w)$ will also serve to satisfy the solution of the primal formulation. As a result, the problem of minimizing $Z(w)$ is the dual form of the problem of maximizing $z(x)$ and they are both the same problem.

2.7   Physical interpretation of the primal and dual relationship

A farmer has a 150-acre farm. From last year's crops, he has an abundant supply of potatoes, corn, wheat and hay. He wants to apportion his 150 acres among the four crops to yield a maximum return on the market. He has an outside job which prevents him from working more than 50 h a week on his farm during the 10 wk required for planting, cultivating and harvesting. Data on the amounts of seed required, time needed to tend the crop and expected dollar return is tabulated for each crop as follows [2].

Crops	Acres per pound of seed	Total hours of labour per pound of seed	Dollar return per pound of seed
Potatoes	0.1	0.6	4
Corn	0.2	0.5	6
Wheat	0.3	0.4	6
Hay	0.4	0.3	5

With this information we can write the optimization problem in the standard form of Eq. (2.43).

Let x_1, x_2, x_3, x_4 be the number of pounds of seed planted of potatoes, corn, wheat and hay respectively. Then our linear program is

$$
\begin{aligned}
\text{Maximize} \quad & z = 4x_1 + 6x_2 + 6x_3 + 5x_4 \\
\text{Subject to} \quad & 0.1x_1 + 0.2x_2 + 0.3x_3 + 0.4x_4 \le 150 \\
& 0.6x_1 + 0.5x_2 + 0.4x_3 + 0.3x_4 \le 500 \\
& x_1 \ge 0, x_2 \ge 0, x_3 \ge 0, x_4 \ge 0
\end{aligned}
\tag{2.58}
$$

These algebraic statements ask that the return on seed investment be maximized subject to the requirements that only 150 acres are available to plant and only 500 h are available to plant, tend and harvest the crops.

If the farmer knew how to solve this linear programming problem, he would have the optimal allocation of crops and the amount of return the solution will produce. Unfortunately, he does not know how to find the optimal solution so he turns to a banker for help. The banker offers to rent the 150 acres from the farmer and furthermore offers to hire him for 500 h during the planting, cultivating and harvest seasons. The rent, w_1, and the hourly wage, w_2, dollars per hour are to be determined by a mutually agreeable scheme. The values of w_1 and w_2 were decided upon as follows: the outlay of cash that the banker must produce is to be minimized, subject to the condition that for each pound of seed planted, the farmer will receive at least as much in land rental and hourly wage as he would have obtained if he himself had sold the produce from the seed. The banker now has his own linear programming to formulate and solve:

$$
\begin{aligned}
\text{Minimize} \quad & Z = 150w_1 + 500w_2 \\
\text{Subject to} \quad & 0.1w_1 + 0.6w_2 \ge 4 \\
& 0.2w_1 + 0.5w_2 \ge 6 \\
& 0.3w_1 + 0.4w_2 \ge 6 \\
& 0.4w_1 + 0.3w_2 \ge 5 \\
& w_1 \ge 0, w_2 \ge 0
\end{aligned}
\tag{2.59}
$$

We now have two distinct linear programming problems and they represent two points of view on one economic situation. The set of linear inequalities and linear objective function of Eq. (2.59) is not in the standard form as Eq. (2.58) is standard. However, the two problems are related by a *duality theory* and Eq. (2.59) does represent a second standard form. To be more concise, we will rewrite Eq. (2.58) in matrix notation as:

$$
\begin{aligned}
\text{Maximize} \quad & z = [c]\{x\} \\
\text{Subject to} \quad & [A]\{x\} \le \{b\} \\
& \{x\} \ge 0
\end{aligned}
\tag{2.60}
$$

The second form of the same problem is written

$$
\begin{aligned}
\text{Minimize} \quad & Z = [b]\{w\} \\
\text{Subject to} \quad & [A]^T\{w\} \geq \{c\} \\
& w \geq 0
\end{aligned}
\left.\rule{0pt}{6em}\right\} \tag{2.61}
$$

which is the fundamental theorem in linear programming that at *the point x* that maximizes z and w* that minimizes Z*, the two functions are actually equal in value. This result is intuitively evident as we consider the two objectives of the farmer and the banker. Let us now consider how the farmer or the banker might solve the problem. We will take a careful look at the banker's version because it contains only two variables, thereby admitting a graphical representation as shown in Fig. 2.2. The four lines that represent the constraints on the four crops are drawn in solid lines; the level contours of the objective function $[A]^T\{w\} \geq \{c\}$, $w \geq 0$, hold as the (dual) feasible region. Of all of the feasible points, that formed by the intersection of the corn and potato constraints gives the minimum value of $Z(w)$, i.e., the solution point. The solution of the banker's problem is $w_1 = \$160/7$ per acre rent and $w_2 = \$20/7$ per hour wages which results in the total outlay of $Z = 150w_1 + 500w_2 = \$34,000/7$.

We now can ask several questions about the farmer's problem. First, we know that his maximum return will be \$34,000/7 if he plants the optimal crop allocation. We ask if it is possible to obtain the optimal planting schedule knowing only the rental rate and wage rate. Our answer comes directly from Eq. (2.58) where we observe that the wheat and hay constraints are irrelevant. We refer to those constraints which determine the solution point as *active constraints;* the remaining constraints are said to be

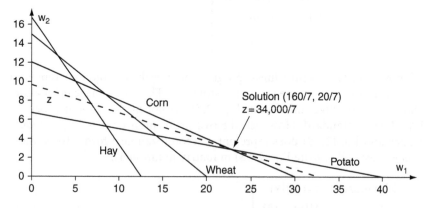

Figure 2.2 Graphical solution of farmer and banker investment.

inactive. Since the potato and corn constraints are active, their corresponding seed amounts x_1 and x_2 must be determined. The inactive constraints must have multipliers of zero in the inequality, then we have

$$\left.\begin{array}{l} 0.1x_1 + 0.2x_2 = 150 \\ 0.6x_1 + 0.5x_2 = 500 \end{array}\right\} \qquad (2.62)$$

The farmer should plant $x_1 = 2,500/7$ lbs of potatoes and $x_2 = 4,000/7$ lbs of corn. His maximum return on his seed of course will be $z = 4x_1 + 6x_2 = \$34,000/7$.

This observation which has allowed us to obtain all of the *primal* information from the *dual* solution is a result of the *complementary slackness principle* which tells us that inactive constraints in the primal/dual imply that the corresponding variables are zero in the primal/dual solution. Thus, we are free to choose either the primal or dual program upon which to direct our attack.

2.8 Concluding remarks

As illustrated in the previous sections, the Simplex method passes from vertex to vertex on the boundary of the feasible polyhedron, repeatedly increasing the objective function until either an optimal solution is found, or it is established that no solution exists. In general engineering practice with hundreds of variables, the method is highly efficient. Problems of thousands (or even more) of variables can be routinely solved using the Simplex method on modern computers. The method undoubtedly requires significant computing time; however, efficient and sophisticated implementations are available in the form of computer software packages.

In 1984, Narendra Karmarkar introduced an *interior-point method* for linear programming which does not require passing from vertex to vertex, but passes only through the interior of the feasible region [6]. Though the property is easy to state, the analysis of the interior-point method is a subtle subject which is much less easily understood than the behaviour of the Simplex method. The method claimed by the author is comparable with the Simplex method in most, though not all, applications.

References

1. Dantzig, G.B., *Linear Programming and Extension*, Princeton University Press, Princeton, New Jersey, 1963.
2. Cheng, F.Y., Venkayya, V.B., Khachaturian, N., *et al.*, *Computer Methods of Optimal Structural Design*, Vol. I and II, Fifth Annual One-Week Short Course Notes, University of Missouri-Rolla, Rolla, MO, 1976.
3. Wang, C.K., *Computer Methods in Advanced Structural Analysis*, Intext Education Publishers, New York, 1973.

4. Cheng, F.Y., *Linear Programming – Prime and Dual Firms*, Honorary Professor Lecture Series, presented at and published by Harbin Architectural and Civil Engineering Institute, China, 1981.
5. Hadley, G., *Linear Programming*, Addison-Wesley Publishing Co., Reading, MA, 1962.
6. Karmarkar, N., A new polynomial-time algorithm for linear programming, *Combinatorica*, Vol. 4, No. 4, pp. 373–396, 1984.

3 Linear programming optimization of elastic structural systems

3.1 Introduction

This chapter is focused on using linear programming techniques presented in Chapter 2 to synthesize constituent members of an elastic structure for the minimum weight of the system. The structural system includes indeterminate trusses, continuous beams and frameworks for which the displacement method is briefly reviewed and employed in formulating various constraint equations and objective functions. Since the displacement and stress equations of an indeterminate structure are highly nonlinear, the concept of the Taylor series technique is adapted in the matrix formulation to linearize constraints and objective functions. In Section 3.2 the difference between rigid frames and elastic frames is defined and then the matrix formulation based on the displacement method is reviewed for convenience to the reader. In Sections 3.3 through 3.5 are developed mathematical formulations and numerical solutions for various structures subjected to static loads. The optimization is extended in Section 3.7 for dynamic responses. Concluding remarks are in Section 3.8.

3.2 Characteristics of elastic structures

3.2.1 Elastic frames versus rigid frames

The fundamental difference between elastic frames and rigid frames is due to the consideration of the constituent member's axial deformation: axial deformation is included in elastic frames' members but neglected in rigid frames. Therefore, the number of degrees of freedom (d.o.f.) is quite different for those two systems. Note that the d.o.f. represent *independent movements of each of the structure's nodes*. For a plane framework, there are three d.o.f. at each node of an elastic frame; one rotation and two linear displacements; however, in a rigid frame there is one rotational d.o.f. at each node, but linear displacements have to be associated with *independent sidesway* because the neglected axial deformation can't allow linear d.o.f.

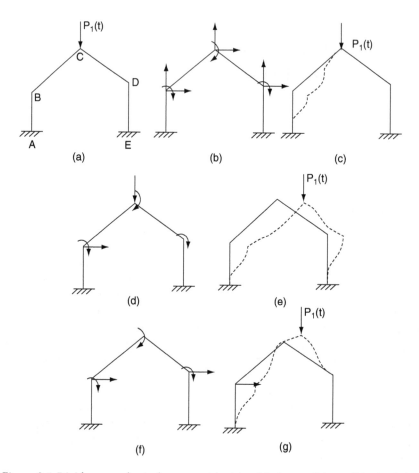

Figure 3.1 Rigid versus elastic frame models: (a) gable frame, (b) possible d.o.f. for elastic frame model, (c) deformation due to sidesway at B, (d) possible d.o.f. for rigid frame model, (e) deformation due to sidesway at B, (f) possible d.o.f. for rigid frame model, (g) deformation due to sidesway at B.

automatically assigned at the node. The difference between these two systems' d.o.f. is illustrated in Fig. 3.1a of a gable frame whose elastic system shown in (b) has nine d.o.f. – three rotational and six linear at the nodes B, C and D when the horizontal d.o.f. at B moves which does not cause other d.o.f. to move as sketched in (c). However, when the same gable is modulated as a rigid frame then it has only five d.o.f. – three rotational and two linear; the linear displacements are corresponding independent sideways as shown in (d) or (f). For the case (d), when the horizontal displacement at B moves as shown in (e), the horizontal d.o.f. at D can move but node C can displace horizontally only as rigid-body motion. For the case (f), the horizontal motion at B will deform the frame shown in (g) where node C can move but not D. In the comparison of the two systems, the rigid frame model has fewer d.o.f. and requires much less computing time and computer storage, and furthermore can realistically represent building structures of which the axial deformation of columns and girders is negligible. But the displacement matrix formulation and the computer input coding are not easy. Optimization of rigid frames is presented in various chapters such as Chapters 5 and 11. On the contrary, elastic systems are convenient in both the displacement matrix formulation and computer input coding which are commonly modelled for complex structures with finite element analysis. The next section reviews the detailed matrix formulation as well as linear optimization techniques for elastic structures which are also designed with other optimization methods in later chapters including Chapters 7 and 10.

3.2.2 Matrix formulation for displacement method

(A) Elastic frames

The gable frame shown in Fig. 3.2a is used to illustrate the matrix formulation of the displacement method. The main tasks are summarized as: step (1) assign the number of d.o.f. as 0,1,2,...,14,0 in which 0 d.o.f. corresponds to the boundary where no movement is possible; step (2) assign numbers for the constituent members, for instance, the number for member CD is given as n; step (3) select the origin of the coordinates x, y of the frame where point A is chosen for convenience, thus other points are measured in reference to A, for instance, the location at C is x_c, y_c and the member length $L_{BC} = \left[(x_c - x_B)^2 + (y_c - y_B)^2\right]^{1/2}$; step (4) formulate the system stiffness matrix and perform structural analysis which are briefly explained in the following six items of (I) through (VI).

(I) Member's stiffness coefficients: a typical member is shown in (b) of which the two ends are identified with i and j where the three independent forces are the axial force in tension, F_a, and positive bending moments, M_i and M_j. Shears are depending on moments through the equilibrium condition of $(M_i + M_j)/L_n$. The member's positive deformations corresponding to the independent forces are e_a, e_i and e_j as sketched in (c). Note that e_i and e_j are measured from the chord to the tangent in a clockwise direction. The chord is a line connecting two displaced positions, i' and j', as shown in (e) (without sidesway), or (f) (with sidesway). Using Hooke's law, one may find the force–deformation relationship as the stiffness matrix of the member in Eq. (3.1).

$$\left\{\begin{array}{c} F_a \\ M_i \\ M_j \end{array}\right\} = \frac{E_n}{L_n} \left[\begin{array}{ccc} A_n & 0 & 0 \\ & 4I_n & 2I_n \\ \text{sym} & & 4I_n \end{array}\right] \left\{\begin{array}{c} e_a \\ e_i \\ e_j \end{array}\right\} \tag{3.1}$$

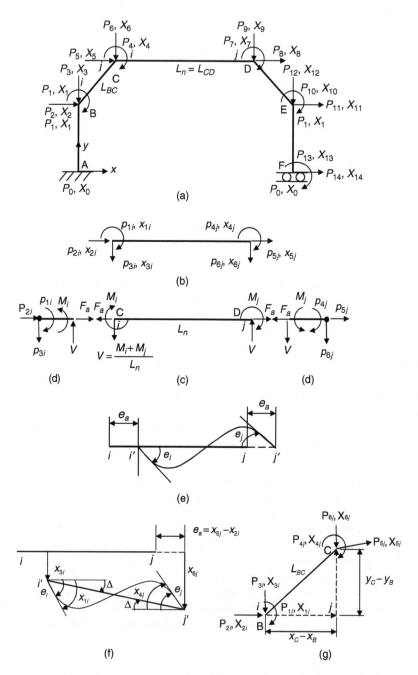

Figure 3.2 Elastic frame: (a) given structure, (b) local d.o.f. at member ends, (c) internal forces at member ends, (d) free-body diagrams at member ends, (e) internal deformations, (f) relationship between internal deformations and the local d.o.f. at member ends, (g) inclined member.

or

$$\{F\}_n = [s]_n \{d\}_n \tag{3.1a}$$

in which $[s]_n$ is composed of stiffness coefficients on member n.

(II) Member's kinetic matrix: the internal deformation of a member related to the member's nodal displacements may be established using the compatibility condition as shown in (d). The compatibility equation is formulated based on the general case of (f) as

$$e_a = x_{5j} - x_{2i}; \quad e_i = x_{1i} - \frac{(x_{6j} - x_{3i})}{L_n}; \quad e_j = x_{4j} - \frac{(x_{6j} - x_{3i})}{L_n} \tag{3.2}$$

In matrix form

$$\begin{Bmatrix} e_a \\ e_i \\ e_j \end{Bmatrix} = \begin{bmatrix} 0 & -1 & 0 & 0 & 1 & 0 \\ 1 & 0 & 1/L_n & 0 & 0 & -1/L_n \\ 0 & 0 & 1/L_n & 1 & 0 & -1/L_n \end{bmatrix} \begin{Bmatrix} x_{1i} \\ x_{2i} \\ x_{3i} \\ x_{4j} \\ x_{5j} \\ x_{6j} \end{Bmatrix} \tag{3.3}$$

or

$$\{d\}_n = [b]_n \{x\}_n \tag{3.3a}$$

in which $[b]_n$ is the kinematic matrix of member n. $\{d\}_n$ and $\{x\}_n$ represent respectively the internal deformations and nodal displacements at the member's ends i, j. The relationship between the local d.o.f. $\{x\}_n$ of a member (say CD) and the global d.o.f. of the structure $\{X\}$ is

$$\begin{bmatrix} x_{1i} & x_{2i} & x_{3i} & x_{4j} & x_{5j} & x_{6j} \end{bmatrix} = \begin{bmatrix} X_4 & X_5 & X_6 & X_7 & X_8 & X_9 \end{bmatrix} \tag{3.3b}$$

For an inclined member such as BC shown in (g), one must transform the nodal displacements in the internal deformations' directions as

$$\begin{Bmatrix} e_a \\ e_i \\ e_j \end{Bmatrix} = \begin{bmatrix} 0 & -\cos & \sin & 0 & \cos & -\sin \\ 1 & \sin/L_{BC} & \cos/L_{BC} & 0 & -\sin/L_{BC} & -\cos/L_{BC} \\ 0 & \sin/L_{BC} & \cos/L_{BC} & 1 & -\sin/L_{BC} & -\cos/L_{BC} \end{bmatrix}$$

$$\begin{bmatrix} x_{1i} & x_{2i} & x_{3i} & x_{4j} & x_{5j} & x_{6j} \end{bmatrix}^T \tag{3.4}$$

where

$$\sin = \frac{(y_c - y_b)}{L_{BC}}; \quad \cos = \frac{(x_c - x_b)}{L_{BC}} \tag{3.4a}$$

for a horizontal member; then $\sin = 0$ and $\cos = 1$, consequently Eq. (3.4) become Eq. (3.3).

(III) Member's equilibrium matrix: the equilibrium condition between the external forces at the node and internal forces may be established using the free-body diagrams shown in (d) as

$$
\left.
\begin{array}{lll}
p_{1i} = M_i, & p_{2i} = -F_a, & p_{3i} = \dfrac{(M_i + M_j)}{L_n} \\[2ex]
p_{4j} = M_j, & p_{5j} = F_a, & p_{6j} = -\dfrac{(M_i + M_j)}{L_n}
\end{array}
\right\}
\tag{3.5}
$$

in matrix form

$$
\begin{Bmatrix} p_{1i} \\ p_{2i} \\ p_{3i} \\ p_{4j} \\ p_{5j} \\ p_{6j} \end{Bmatrix}
=
\begin{bmatrix}
0 & 1 & 0 \\
-1 & 0 & 0 \\
0 & 1/L_n & 1/L_n \\
0 & 0 & 1 \\
1 & 0 & 0 \\
0 & -1/L_n & -1/L_n
\end{bmatrix}
\begin{Bmatrix} F_a \\ M_i \\ M_j \end{Bmatrix}
\tag{3.6}
$$

or

$$
\{P\}_n = [a]_n \{F\}_n
\tag{3.6a}
$$

where $[a]_n$ is the equilibrium matrix of a typical member n which is the transpose of the kinematic matrix of Eq. (3.3) as

$$
[a]_n = [b]_n^T
\tag{3.7}
$$

For an inclined member shown in (g), the equilibrium matrix is also the transpose of the kinematic matrix in Eq. (3.4).

(IV) Member's stiffness matrix: the relationship between the external nodal forces, $\{p\}_n$, and the displacements, $\{x\}_n$, can be obtained by substituting Eq. (3.3a) into Eq. (3.1a) and the result is substitut into Eq. (3.6a) with Eq. (3.7) as

$$
\{p\}_n = [b]_n^T [s]_n [b]_n \{X\}_n
\tag{3.8}
$$

or

$$
\{p\}_n = [K_E]_n \{x\}_n
\tag{3.9}
$$

where

$$
[K_E]_n = [b]_n^T [s]_n [b]_n
\tag{3.10}
$$

(V) Structural stiffness of a structural system: the structural stiffness matrix $[K]$ is assembled from the member's stiffness matrix $[K_E]$

$$[K] = \sum_{n=1}^{N} [K_E]_n \qquad (3.11)$$

Note that the relationship of the local d.o.f. of each member and the global d.o.f. of a structure is already established as shown in Eq. (3.3b); Eq. (3.11) is merely mapping the two d.o.f. systems by the computer program in an automatic process through which the global d.o.f. $\{X\}$ is simultaneously completed. Similarly, the local matrix $\{p\}$ is also automatically generated using the input of externally applied loads at the nodes including various load conditions. The relationship between external forces and nodal displacements of a structure is finally achieved as

$$\{P\} = [K]\{X\} \qquad (3.12)$$

Note that for a large structure with a huge number of global d.o.f. and many members, steps (1) and (2) are not needed because the computer program can automatically generate coordinates at each node. This technique is commonly used in finite element analysis programs. Note also that zero global d.o.f. are assigned at a boundary without movements so that the local d.o.f. assigned at a boundary without movements does not need any space in the computer program.

(VI) Response analysis: we first find the global displacements from Eq. (3.12) as

$$\{X\} = [K]^{-1}\{P\} \qquad (3.13)$$

then local displacements and internal forces of an individual member, n, are found from Eqs. (3.3a) and (3.1a) for which the relationship between $\{d\}_n$ and $\{x\}_n$ is already available during the mapping process, thus

$$\{F\}_n = [s]_n [b]_n \{x\}_n \qquad (3.14)$$

Eqs. (3.12) and (3.13) are simple illustrations of the static loading case. For dynamic loads, we should consider other parameters such as forces due to mass motions and characteristics of the applied forces, which will be discussed in detail later when the case is presented.

(B) *Continuous beams*

The main difference between continuous beams and elastic frames is that the axial deformation is not considered in beams because a beam with

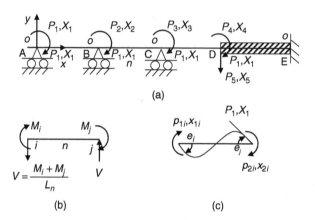

Figure 3.3 Continuous beam: (a) given structure with global d.o.f., (b) internal forces at member ends, (c) local d.o.f. and internal deformations at member ends.

roller support can move freely in the axial direction. A typical continuous beam is sketched in Fig. 3.3a with five d.o.f. – four rotational and one sidesway. The sidesway d.o.f. is assigned herewith because member CE is not uniform and has two different moments of inertia. Consequently, a structural node is added at D. The internal forces of a member n are sketched in (b) where M_i and M_j are independent moments, $V = (M_i + M_j)/L_n$ is shear. The internal deformations e_i and e_j and the nodal displacements x_{1i} and x_{2i} are shown in (c) for which $x_{1i} = X_2$ and $x_{2i} = X_3$. Note that e_i and e_j are the slopes of the deformed member and no d.o.f. of linear displacement is for this member. However, the formulation for members CD and DE should involve global linear displacement, X_5, as for elastic frames without axial d.o.f. The displacement matrix formulation of continuous beams can be modified from that for elastic frames by omitting the terms associated with axial deformation. Thus, Eqs. (3.1) and (3.3) should change to Eqs. (3.15) and (3.16), respectively as

$$\begin{Bmatrix} M_i \\ M_j \end{Bmatrix} = \frac{E_n I_n}{L_n} \begin{bmatrix} 4 & 2 \\ \text{sym} & 4 \end{bmatrix} \begin{Bmatrix} e_i \\ e_j \end{Bmatrix} \tag{3.15}$$

$$\begin{Bmatrix} e_i \\ e_j \end{Bmatrix} = \begin{bmatrix} 1 & 0 \\ 0 & 1 \end{bmatrix} \begin{Bmatrix} x_{1i} \\ x_{2j} \end{Bmatrix} \tag{3.16}$$

The structural stiffness matrix and response analysis are similar to Eqs. (3.12) through (3.14).

(C) Indeterminate trusses

The formulations of indeterminate trusses are also similar to those for elastic frames except that no bending moment and rotation are involved. A typical indeterminate truss is given in Fig. 3.4a for which the global d.o.f. is

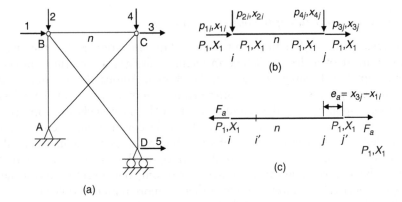

Figure 3.4 Intermediate truss: (a) given structure with global d.o.f., (b) local d.o.f. at member ends, (c) displacement of local d.o.f. induced internal deformations and forces.

assigned as five linear displacements, zero d.o.f. are for hinge support A and vertical displacement at roller support D. Following the procedures used for elastic frames, a typical member, n, (member BC) has four global d.o.f. as $X_1, .., X_4$ which correspond to the local d.o.f. x_{1i}, x_{2i}, x_{3j} and x_{4j}. The internal deformation $e_a = x_{3j} - x_{1i}$. Thus, Eqs. (3.1) and (3.3) should be modified as Eqs. (3.17) and (3.18), respectively, according to the displacement diagrams sketched in (b) and (c)

$$F_a = \frac{A_n E_n}{L_n} \tag{3.17}$$

$$e_a = \frac{x_{3j} - x_{1i}}{L_n} \tag{3.18}$$

If the member has slope, AC, Eq. (3.18) should be changed from Eq. (3.4) as

$$e_a = \begin{bmatrix} -c_s & s_n & c_s & -s_n \end{bmatrix} \begin{bmatrix} x_{1i} & x_{2i} & x_{3j} & x_{4j} \end{bmatrix}^T \tag{3.19}$$

and the compatibility matrix becomes

$$[b]_n = \begin{bmatrix} -c_s & s_n & c_s & -s_n \end{bmatrix} \tag{3.19a}$$

The structural stiffness matrix formulation and response analysis are identical to those for continuous beams using the procedures in Eqs. (3.12) through (3.14). In structural optimization, a significant amount of computation effort is for structural analysis. Section 3.2 is thus presented for preparing optimization procedures in the later sections of this chapter.

3.3   Objective functions

3.3.1   Characteristics of the objective function

The objective (cost) function is a function of design variables. It is the merit measurement of a design. When an engineer designs a structure, numerous acceptable (feasible) designs can be chosen. Among them, some need less monetary investment, some offer greater safety factors, some have less dynamic response than others, and so on. To make the best choice, appropriate criteria to compare various designs should be constructed. These criteria are called the *objective functions* of an optimum design problem. Their values can be minimized or maximized by an optimal set of design variables within the feasible design space. When there is only one objective, it is scalar (single objective) optimization; when there is more than one objective, it is multi-objective optimization. Selecting objective functions in an optimum design problem is highly important. This selection decides the optimum (search) direction and optimum results of a design. For example, minimizing structural weight usually decreases structural stiffness while minimizing top-storey displacements increases structural stiffness. If the constraints are the same, the structural properties and cost of the two optimal results are different although they are in the same feasible region. Objective functions are constructed according to various optimization purposes. Several objective functions are briefly introduced here which will be presented in detail in later chapters where some of the functions are used.

A) *Structural Cost*: this function consists of material, fabrication and installation costs, etc. Minimizing structural cost makes a design more competitive, and is generally the aim of an investor. For buildings, reinforced concrete structures should use structural cost in order to consider framework cost and different prices of concrete and steel.

B) *Structural Weight*: to better utilize materials and reduce costs of a structure, structural weight can be chosen as the objective function. For automobiles and airplanes, minimum weight is a major concern – reduction in weight means saving fuel. For buildings, structural weight is a good measurement of structural cost in comparison with other costs of connections, member shapes, etc. which are generally a small percentage of the total.

C) *Strain Energy*: in a given structure, when the internal work (strain energy) done by the stresses and the strains has a minimum value, the structure can have an optimal result.

D) *Potential Energy*: minimizing the structural potential energy can reduce the effects of external forces and increase the safety level of a structure.

E) *Displacements*: to minimize changes in the shape of a structure under the action of different loading conditions, displacements at selected points or regions of the structure are taken as the objectives.

F) *Seismic Input Energy*: when a structure is subjected to earthquake exci-
    tations, reducing the dynamic response and damage caused by seismic
    loads is the main design consideration. Decreasing dynamic response
    can be done by optimizing such factors as *earthquake input energy,
    acceleration of mass, fundamental period*, etc.

3.3.2 *Minimum weight of indeterminate structures*

Indeterminate structures have highly nonlinear constraint equations such as
for displacements and stresses. The minimum weight objective function of
the system is also nonlinear. This chapter aims to optimize these structures
with the linear optimization (L.P.) algorithms presented in Chapter 2. Therefore, we have to linearize the nonlinear equations as shown in Eq. (2.1)
using the Simplex method.

  Because the stresses of a truss, beam and framework are different, that
is, axial force, bending moment, and combined axial force with bending
moment are associated with the constituent members of trusses, beams and
frameworks, respectively, the cost constraint equations are not the same for
all these three constructed systems.

  The objective function is minimum weight which is very important in the
field of engineering practice. For instance, automobiles and airplanes should
be designed to have their structural mass as low as possible and yet keep a
high performance standard in order to conserve fuel. In civil engineering,
minimum weight is a good criterion for steel structures for which other
costs of connection, welding, painting, etc. are small percentages of the cost
of structural members as presented later in Chapter 7.

  Based on the aforementioned characteristics of trusses, beams and
frameworks, the constraints, objective functions and linearized equations
are presented separately for the individual systems-objective function in
Section 3.3.2, constraints in Sections 3.4 and 3.5. The structural weight
is the summation of individual members' weights, which is a function of
the members' length, cross-sectional area, and the material weight. The
length and material are not changed, therefore the only variable is the cross-
sectional area of each element regardless of the cross-sectional shapes such
as tubes, wide flanges, rectangular, etc.

  Let N = total number of elements; R_j = constant of element j incorporating material unit weight, γ, and length of the element, L_j ($R_j = \gamma L_j$); and
z_j^o = initial assumed parameter for element j (such as area: $A_j^o = z_j^o$); one may
write the total weight of an initial assumed truss as follows:

$$W^o = \sum_{j=1}^{N} R_j z_j^o \tag{3.20}$$

The optimum process takes a finite number of revised design parameters for achieving an optimal weight by assuming Δz_j for the change in

total weight ΔW. If W is the new total weight of the revised system, one may have

$$W = W^o + \Delta W = \sum_{j=1}^{N} R_j \left(z_j^o + \Delta z_j \right) \tag{3.21}$$

where Δz_j is a variable and may be positive or negative depending on the initial assumptions. In order to ensure that the solution vector of the linear programming problem will be non-negative, we may rewrite Eq. (3.21) as

$$W = \sum_{j=1}^{N} R_j z_j^o \left(1 + \frac{\Delta z_j}{z_j^o} \right) \tag{3.22}$$

Let $u_j = 1 + \Delta z_j / z_j^o$ in which Δz_j will be chosen to force $u_j \geq 0$. Thus, Eq. (3.22) becomes

$$W = \sum_{j=1}^{N} R_j z_j^o u_j \tag{3.23}$$

For a truss design problem, the objective function may take the following form

$$W = \sum_{j=1}^{N} \gamma L_j A_j^o u_j = [C]_{N\times1}^T \{u\}_{N\times1} \tag{3.24}$$

Since the unit weight γ is constant for all members, it need not be considered in the objective function. Note that the constituent members of a truss are subjected to axial force-tension or compression, for which the cross-section is the only design variable. Thus, Eq. (3.24) works well for truss design.

3.3.3 Minimum weight of beams and frameworks

As presented in Eqs. (3.1) and (3.15), members of frames are subjected to the interaction of the axial force and bending moments; continuous beams' internal forces are mainly bending moments. For resistance bending moments, the most economical cross-section of a member is a wide flange as shown in Fig. 3.5 for which cross-sectional area A, moment of inertia I, and section modulus S may be expressed as

$$A = \left(b_f, t_f \right) 2 + d t_w \tag{3.25}$$

$$I = \frac{t_w d_w^3}{12} + 2 b_f t_f \left(\frac{d_w + t_f}{2} \right)^2 \tag{3.26}$$

Figure 3.5 Wide flange section.

$$S = \left(\frac{I}{\left(\frac{d}{2} + t_f \right)} \right) \tag{3.27}$$

In the above equations, four unknown parameters, b_f, t_f, d and t_w associated with a cross-section should be optimized in order to achieve an optimal section to maximize resistance to axial stress and bending stress as

$$f_a = \frac{F_a}{A}; \quad f_b = \frac{M}{S} \tag{3.28a}$$

In steel structural design, various wide flanges are available in the AISC manual, the four parameters are given for A, I and S. Therefore, for a given beam depth, S is also available.

For a minimum weight objective function, the only design variable is the member's cross-sectional area A, as shown in Eq. (3.24). Therefore, S must be expressed in terms of A. Considering practical steel construction, variations of member depth are limited to a small range. For instance, columns between floors and beams at each floor can be practically the same for connection convenience and cost. It is therefore possible to unify the relationship between A and S by a modification as

$$q = \frac{S}{Ad} \tag{3.28b}$$

For practical construction, d varies around 10–21 in. and the members' weight varies around 21–100 lbs ft^{-1} for which the change of q is very small, in the range of 0.32 to 0.37. Therefore, for simplicity $q = 0.34$ can be used for practical structural members, as

$$S = 0.34Ad \tag{3.29}$$

in which d can be predetermined for practical construction, thus S and A have a linear relationship in the design. For built-up sections, we need to consider the four parameters in the design variables of A and I, for which the optimization procedures are presented in Chapter 7.

The objective function of continuous beams and frameworks may be modified from Eq. (3.24) as

$$W = \gamma \sum_{j=1}^{N} L_j \frac{S_j^o}{q_j d_j} u_j \tag{3.30}$$

3.4   Constraints of stresses, displacements and move limits

Since frameworks are composed of beam-columns which are subjected to the interaction of axial forces and bending moment, for a beam-column, the following interaction relationship may be used to serve as the stress-limiting criterion at any section.

$$\frac{F_{new}}{(F_{new})_{all}} + \frac{M_{new}}{(M_{new})_{all}} \leq 1 \tag{3.31}$$

in which, F_{new} = axial force on the new section; M_{new} = bending moment in the new section; $(F_{new})_{all}$ = allowable axial force on the new section = $F_a(A + \Delta A)$; $(M_{new})_{all}$ = allowable bending moment on the new section = $F_b(S + \Delta S)$; F_a = allowable axial stress in tension or compression; and F_b = allowable bending stress. Let F^o = acting force in the originally assumed design parameter (z^o), F_{new} = acting force in the newly selected parameter (z), and $(F_{new})_{all}$ = acting force in the newly selected parameter; one may express the constraints as follows:

$$\begin{aligned} F_{new} &\leq (F_{new})_{all}, & F^o &\geq 0 \\ F_{new} &\geq - (F_{new})_{all}, & F^o &< 0 \end{aligned} \tag{3.32}$$

If allowable stress is signified by F_a, then

$$(F_{new})_{all} = F_a (z^o + \Delta z) = F_a z^o u \tag{3.33}$$

where $u = 1 + \Delta z / z^o$. Let $z^o = A^o$, then $u = 1 + \Delta A / A^o$ and

$$(F_{new})_{all} = F_a A^o u \tag{3.34}$$

Similarly, let M_o = acting moment on the originally assumed design parameter (z^o), M_{new} = acting moment on the newly selected parameter (z) and $(M_{new})_{all}$ = allowable moment on the newly selected parameter; one may express the constraints for bending moment

$$M_{new} \leq (M_{new})_{all} = F_b (S^o + \Delta S) = F_b S^o u \tag{3.35}$$

in which $S^o = q d A^o$, $\Delta S = q d (\Delta A)$, $u = 1 + \Delta S / S^o$. Thus,

$$M_{new} \leq F_b \frac{q d}{A^o} u \tag{3.36}$$

The equation of an indeterminate structure's internal forces at each member is complex with nonlinear relationships in terms of the design variables of the system. In order to use the linear programming technique, we must linearize the nonlinear relationship based on the Taylor series shown below:

$$F(x) = F(x^o) + \frac{\partial F(x)}{\partial x_1}(x_1 - x_1^o) + \frac{\partial^2 F(x)}{\partial x_1^2}(x_1 - x_1^o) + \dots \tag{3.37}$$

Neglecting the higher-order terms, we can employ the above series to express

$$\Delta F_i = F_i - F_i^o = \sum_{j=1}^{N} \frac{\partial F_i}{\partial z_j} \Delta z_j \tag{3.38}$$

Similarly, the change of moment i affected by changes of other members can be expressed as

$$\Delta M_i = M_i - M_i^o = \sum_{j=1}^{N} \frac{\partial M_i}{\partial z_j} \Delta z_j \tag{3.39}$$

Substituting Eqs. (3.32), (3.34), (3.36), (3.38) and (3.39) into Eq. (3.31) yields

$$\frac{F_{ik}^o + \sum_{j=1}^{N} \frac{\partial F_{ik}}{\partial A_j} \Delta A_j}{F_a \left(A_i^o + \Delta A_i\right)} + \frac{M_{ik}^o + \sum_{j=1}^{N} \frac{\partial M_{ik}}{\partial S_j} \Delta S_j}{F_b \left(S_i^o + \Delta S_i\right)} \leq 1 \tag{3.40}$$

The stress constraints shown above have the design parameters A and S which should be transformed to a new variable u for the linear programming requirement of non-negative solution vector. The formulation can be done in two steps: first, by taking the incremental change in the design variable as 100 per cent or $\partial A_j = A_j^o$; $\partial S_j = S_j^o$; and two notations in Eq. (3.40) are modified as $\partial F_{ik} = \Delta F_{ijk}$ and $\partial M_{ik} = \Delta M_{ijk}$. Consequently, Eq. (3.40) may be written as

$$\frac{F_{ik}^o + \sum_{j=1}^{N} \Delta F_{ijk} \frac{\Delta A_j}{A_j^o}}{A_i^o + \Delta A_i} + \frac{M_{ik}^o + \sum_{j=1}^{N} \Delta M_{ijk} \frac{\Delta A_j}{A_j^o}}{qd\frac{F_b}{F_a}\left(A_i^o + \Delta A_i\right)} \quad \begin{array}{l} \leq F_a, \quad F_{ik}^o \geq 0 \\ \geq F_a, \quad F_{ik}^o < 0 \end{array} \tag{3.41}$$

Note that the sign of the inequality must change if F_a is negative for compression.

The second step is to add $\sum\limits_{j=1}^{N}\Delta F_{ijk}$ and $\sum\limits_{j=1}^{N}\Delta M_{ijk}$ to both sides of Eq. (3.41) as

$$F_{ik}^{o}+\sum_{j=1}^{N}\Delta F_{ijk}\frac{\Delta A_{j}}{A_{j}^{o}}+\sum_{j=1}^{N}\Delta F_{ijk}+\frac{M_{ik}^{o}}{qd\frac{F_{b}}{F_{a}}}+\sum_{j=1}^{N}\frac{\Delta M_{ijk}\frac{\Delta A_{j}}{A_{j}^{o}}}{qd\frac{F_{b}}{F_{a}}}+\sum_{j=1}^{N}\frac{\Delta M_{ijk}}{qd\frac{F_{b}}{F_{a}}}$$

$$\leq F_{a}\left(A_{i}+\Delta A_{i}\right)+\sum_{j=1}^{N}\Delta F_{ijk}+\sum_{j=1}^{N}\frac{\Delta M_{ijk}}{qd\frac{F_{b}}{F_{a}}}$$

$$\geq \text{same} \qquad\qquad (3.42)$$

ΔF_{ijk} and ΔM_{ijk} signify the change in force i due to a change in design parameter j of 100 per cent under the loading condition k. Therefore, $\sum\limits_{j=1}^{N}\Delta F_{ijk}$ and $\sum\limits_{j=1}^{N}\Delta M_{ijk}$ on the right-hand side of Eq. (3.42) are zero; because when all design variables are changed by an equal amount of percentage, there is no change in the relative stiffness of the constituent members and the force distribution does not change. Using $u_{j}=1+\Delta A_{j}/A_{j}^{o}$ in Eq. (3.42) which then becomes

$$F_{ik}^{o}+\frac{M_{ik}^{o}}{qd\frac{F_{b}}{F_{a}}}+\sum_{j=1}^{N}\left(\Delta F_{ijk}+\frac{\Delta M_{ijk}}{qd\frac{F_{b}}{F_{a}}}\right)u_{j}$$

$$\leq F_{a}A_{i}^{o}u_{i}+\sum_{j=1}^{N}\left(\Delta F_{ijk}+\frac{\Delta M_{ijk}}{qd\frac{F_{b}}{F_{a}}}\right), \qquad \begin{array}{l}F_{ik}^{o}\geq 0\\ F_{ik}^{o}<0\end{array} \qquad (3.43)$$

$$\geq \text{same}$$

In order to express this in a linear programming standard form, let us group the known and unknown terms to the left and right, respectively, of Eq. (3.43). We finally have the stress-constraint equations for the following three structural systems:

(I) Elastic frameworks

$$\left[\left[\left[\sum_{j=1}^{N}\left(\Delta F_{ijk}+\frac{\Delta M_{ijk}}{qd\frac{F_{b}}{F_{a}}}\right)u_{j}-F_{a}A_{i}^{o}u_{i}\right]\begin{array}{l}\leq -F_{ik}^{o}-\dfrac{M_{ik}^{o}}{qd\frac{F_{b}}{F_{a}}}\quad F_{ik}^{o}\geq 0\\[3mm] \geq -F_{ik}^{o}-\dfrac{M_{ik}^{o}}{qd\frac{F_{b}}{F_{a}}}\quad F_{ik}^{o}<0\end{array}\right]i=1,\text{nf}\right]k=1,\text{nc}\right]$$

$$(3.44)$$

where nf = No. of axial forces, nc = No. of loading conditions.

(II) Continuous beams or rigid frames (without consideration of axial deformation)

$$\left[\left[\left[\sum_{j=1}^{N} \Delta M_{ijk} u_j - F_b s_i^o u_i \begin{array}{cc} \leq -M_{ik}^o & M_{ik}^o \geq 0 \\ \geq -M_{ik}^o & M_{ik}^o < 0 \end{array}\right] \quad i=1, \text{nm}\right] \quad k=1, \text{nc}\right]$$

(3.45)

where nm = No. of moments.

(III) Trusses

$$\left[\left[\left[\sum_{j=1}^{N} \Delta F_{ijk} u_j - F_a A_i^o u_i \begin{array}{cc} \leq -F_{ik}^o & F_{ik}^o \geq 0 \\ \geq -F_{ik}^o & F_{ik}^o < 0 \end{array}\right] \quad i=1, \text{nf}\right] \quad k=1, \text{nc}\right]$$

(3.46)

Note again in Eqs. (3.44) through (3.46), ΔF_{ijk} and ΔM_{ijk} signify the force i or moment i due to a change in design parameter j of 100 per cent under the action of loading condition, k.

3.4.1 Displacement constraints

Let X_{new} = new displacements, X_{all} = allowable displacements, and X^o = initial displacements. The displacement constraints may be written as

$$X_{\text{new}} \begin{array}{cc} \leq X_{\text{all}} & X^o \geq 0 \\ \geq X_{\text{all}} & X^o < 0 \end{array}$$

(3.47)

Any new displacement X_i in a modified system (due to the change of member i) may be approximately obtained as

$$X_i^o + \sum_{j=1}^{N} \frac{\partial X_i}{\partial z_j} \begin{array}{cc} \leq (X_i)_{\text{all}} & X_i^o \geq 0 \\ \geq (X_i)_{\text{all}} & X_i^o < 0 \end{array}$$

(3.48)

If the above constraint is restricted to all displacements np and all loading conditions nc, Eq. (3.48) becomes

$$\left[\left[\left[X_{ik}^o + \sum_{j=1}^{N} \frac{\partial X_{ik}}{\partial z_j} \Delta z_j \begin{array}{cc} \leq (X_{ik})_{\text{all}} & X_{ik}^o \geq 0 \\ \geq (X_{ik})_{\text{all}} & X_{ik}^o < 0 \end{array}\right] \quad i=1, \text{np}\right] \quad k=1, \text{nc}\right]$$

(3.49)

where the design parameter Δz may be replaced by u by adding $\sum_{j=1}^{N} \Delta X_{ijk}$ to both sides of Eq. (3.49). As the stress constraint formulation, ∂z_j is the incremental change in a design variable as 100 per cent. Thus, the relation-ship between $\{P\}$ and $\{X\}$ in Eq. (3.12) should be modified to include the change influence. Using Eq. (3.8), we may rewrite Eq. (3.12) as

$$\{P\} = \sum_{n=1}^{N} [B]_n^T [s + \Delta s]_n [B]_n \{X + \Delta X\}$$

$$= \left[\sum_{n=1}^{N} [B]_n^T [s] [B]_n + \sum_{n=1}^{N} [B]_n^T [\Delta s] [B]_N \right] \{X + \Delta X\} \qquad (3.50)$$

$$= [[K] + [\Delta K]] \{X + \Delta X\} = [K]\{X\} + [\Delta K]\{X\} + [K]\{\Delta X\}$$

$$+ [\Delta K]\{\Delta X\}$$

Considering $\{P\} = [K]\{X\}$ and neglecting the higher-order term $[\Delta K]\{\Delta X\}$ for the Taylor series in Eq. (3.37), we then have the incremental displacement

$$\sum_{j=1}^{N} \Delta X_j = -[K]^{-1} \left(\sum_{j=1}^{N} \Delta K_j \right) \{X\} = -\{X\} \qquad (3.51)$$

The displacement constraint may finally be written as

$$\left[\left[\left[\sum_{j=1}^{N} \Delta X_{ijk} u_j \quad \begin{matrix} \leq (X_{ik})_{\text{all}} - 2X_{ik}^o, & X_{ik}^o \geq 0 \\ \geq (X_{ik})_{\text{all}} - 2X_{ik}^o, & X_{ik}^o < 0 \end{matrix} \right] \quad i = 1, \text{np} \right] \quad k = 1, \text{nc} \right] \tag{3.52}$$

The above equation can be applied to the three structural systems as

(I) Elastic Structures: the system has both rotation and linear displace-ments at each node. In general, the rotation is much less significant than the structural lateral displacements which are usually considered in practice.

(II) Continuous Beams: the allowable displacement is the nodal rotation; unless the beam is modified with linear displacement between adjacent supports, then both allowable rotation and linear displacement should be involved.

(III) Trusses: the system has only linear displacements which are allowable at the nodes.

3.4.2 Side constraints

The side constraints are used to serve two purposes: for the move limits of u as well as lower and upper bounds of design variable size. Since we use the design variable change by 100 per cent in the formulation, in order to eliminate errors in the linear optimization process, we must use the move limits on u as

$$u_i \geq (\text{lb})_i$$
$$u_i \leq (\text{ub})_i \tag{3.53}$$

where $(\text{lb}) = $ lower bound and $(\text{ub}) = $ upper bound. A practical number of 0.80 and 1.20 may be effectively used for lower and upper bounds, respectively. In practical design, the constituent members of a system must have reasonable lower and upper bounds of members' sizes. For instance, a double crossing bracing member can have a little amount of internal force for which the cross-sectional area should be very small. However, considering the slenderness ratio required by the code specification, a lower bound must be assigned to the cross-sectional area. Eq. (3.53) is also used to specify the individual member's size.

3.5 Primal and dual forms for constraints

In order to use the Simplex method presented in Chapter 2, all the constraints in Section 3.4 should be combined to a standard primal as

$$\begin{bmatrix} \Delta F \\ \Delta X \\ I \\ I \end{bmatrix}_{nr \times n} \{u\}_{n \times 1} + [I']_{nr \times nr} \{u'\}_{nr \times 1} = \begin{Bmatrix} F' \\ X' \\ ub \\ lb \end{Bmatrix}_{nr \times 1} \tag{3.54}$$

where, nr = number of rows = $(\text{nf or nm})\,(\text{nc}) + (\text{np})\,(\text{nc}) + n + n$; $\Delta F = $ the matrix of coefficients in the stress constraints = $(\text{nf} \times \text{nc ornm} \times \text{nc}) \times n$; $\Delta X = $ the matrix of coefficients in the displacement constraints = $(\text{np} \times \text{nc}) \times n$; $I = $ identity matrix = $n \times n$; $I' = $ the diagonal matrix consisting of positive or negative unit coefficients corresponding to slack and surplus variables = $\text{nr} \times \text{nr}$; $u = $ solution vector of multipliers in design parameters = $n \times 1$; $u' = $ solution vector of slack and surplus variables = $\text{nr} \times 1$; $F' = $ right-hand side of stress constraints = $-F_{ik}^o$ (Eqs. (3.44) and (3.46)) or $-M_{ik}^o$ (Eq. (3.45)); $X' = $ right-hand side of deformation constraints = $(X_{jk})_{\text{all}} - 2X_{jk}^o$, $j = 1$, np; $k = 1$, nc; ub = upper bounds on the allowable percentage changes in design variables = $n \times n$; lb = lower bounds on the allowable percentage changes in design variables = $n \times n$.

The dual problem can be written:

$$\text{Maximize } Z = \begin{bmatrix} F' \\ X' \\ ub \\ lb \end{bmatrix}^T_{1 \times nr} \{w\}_{nr \times 1} \tag{3.55}$$

$$\text{Subject to } \begin{bmatrix} \Delta F \\ \Delta X \\ I \\ I \end{bmatrix}^T_{nr \times n} \{w\}_{nr \times 1} \le (C)_{n \times 1} \tag{3.56}$$

in which $\{w\}_{nr \times 1}$ = the dual variables; $(C)_{n \times 1}$ = coefficients in Eq. (3.24). At the optimum solution, the maximum value of Z should equal the minimum value of W and the solutions to the primal problem are found in the dual tableau.

3.6   Numerical illustrations of optimization procedures

3.6.1   Example for indeterminate trusses

The truss shown in Fig. 3.6a is subjected to two loading conditions nc = 1 and nc = 2. Let $l = 10$ ft and $E = 2,500 \, \text{k ft}^{-1} \text{in.}^{-1}$. Formulate stress-constraint and displacement equations and then use longhand to illustrate the optimization procedure for stress constraints only with $F_a = 20$ and 15 ksi for tension and compression, respectively. No side constraint is considered.

Solution:

The nodal d.o.f. of P_1, X_1 and P_2, X_2 as well as member numbers are sketched in Fig. 3.6b. Since this is an indeterminate structure, we must assume the cross-sectional area for each member in order to analyze the system. Let $A_1 = A_2 = A_3 = 1 \, \text{in.}^2$, then the initial structural weight is measured

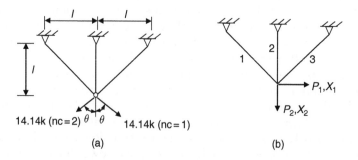

Figure 3.6 Intermediate truss: (a) given truss and loading conditions, (b) nodal displacement and member number.

by volume as $V = 10(1)l + 2(1.414)l(1) = 38.284l$. The matrices shown in Eq. (3.12) are used in the analysis for various optimization cycles.

First cycle

$$[P] = \begin{bmatrix} 10 & -10 \\ 10 & 10 \end{bmatrix} \tag{a}$$

$$[K] = [B]^T [S][B] = \begin{bmatrix} 0.707 & 0 \\ \text{sym} & 1.707 \end{bmatrix} \frac{E}{l} \tag{b}$$

in which

$$[B] = \sum [B_n] = \begin{bmatrix} 0.707 & 0.707 \\ 0 & 1 \\ -0.707 & 0.707 \end{bmatrix},$$

$$[S] = \sum [s] = \frac{A_i E}{L_i} = \begin{bmatrix} 0.707 & 0 & 0 \\ & 1 & 0 \\ \text{sym} & & 0.707 \end{bmatrix} \frac{E}{l} \tag{c, d}$$

The nodal displacements and internal forces formulated in Eqs. (3.13) and (3.14) are then obtained as shown below in Eqs. (e) and (f), respectively.

$$[X] = [K]^{-1}[P] = \frac{l}{E} \begin{bmatrix} 14.140 & -14.140 \\ 5.854 & 5.854 \end{bmatrix} \tag{e}$$

$$[F] = \begin{bmatrix} 10.000 & -4.139 \\ 5.854 & 5.854 \\ -4.139 & 10.000 \end{bmatrix} \tag{f}$$

Using $F_a = 20\,\text{ksi}$ for tension and $15\,\text{ksi}$ for compression in Eq. (f) yields the fully stressed cross-sectional area as

$$\frac{[F]}{F_a} = \begin{bmatrix} 0.500 & 0.276 \\ 0.290 & 0.290 \\ 0.276 & 0.500 \end{bmatrix} \tag{g}$$

which reveals that the largest required area is 0.5 which is used for all the members in the next analysis so that none of the members will be overstressed. Thus, the new areas are modified from the initial assumed areas as

$$\begin{bmatrix} A_1 \\ A_2 \\ A_3 \end{bmatrix} = \begin{bmatrix} 1 \\ 1 \\ 1 \end{bmatrix} 0.5 = \begin{bmatrix} 0.5 \\ 0.5 \\ 0.5 \end{bmatrix} \tag{h}$$

Repeating the procedure in Eqs. (b) through (d) yields

$$[X]^* = [K]^{*-1}[P] = \frac{l}{E} \begin{bmatrix} 28.28 & -28.28 \\ 11.716 & 11.716 \end{bmatrix} \tag{i}$$

for which

$$[S]^* = \begin{bmatrix} 0.35355 & & \\ & 0.5 & \\ & & 0.35355 \end{bmatrix} \frac{E}{l}, \quad [K]^* \begin{bmatrix} 0.35355 & 0 \\ 0 & 0.85355 \end{bmatrix} \frac{E}{l}$$

(j, k)

The new internal forces $[F]^*$ should be the same as those in Eq. (f) because the same change is for all the members and, therefore, there is no redistribution in the internal forces.

In order to find ΔF_{ijk} in Eq. (3.46), we first find ΔX, then ΔF due to an individual member's change with 100 per cent of stiffness of S^*. For the first member change of $\Delta S_1 = 0.35355E/l$, we have

$$[\Delta X]_1 = -[K]^{*-1}[\Delta K]_1^*[X]^* = \frac{l}{E}\begin{bmatrix} -20 & 8.28 \\ -8.28 & 3.43 \end{bmatrix}$$

(l)

in which

$$[\Delta K]_1^* = \begin{bmatrix} 0.17677 & 0.17677 \\ -0.17677 & -0.17677 \end{bmatrix}\frac{l}{E}$$

(m)

Similarly, for the 2nd and 3rd members,

$$[\Delta X]_2 = \frac{l}{E}\begin{bmatrix} 0 & 0 \\ -6.86 & -6.86 \end{bmatrix}, \quad [\Delta X]_3 = \frac{l}{E}\begin{bmatrix} -8.28 & 20 \\ 3.43 & -8.28 \end{bmatrix}$$

(n, o)

Finally, the internal forces due to these changes are

$$[\Delta F]_1 = [S]^*[B][\Delta X]_1 + [\Delta S_1][B][X]^* = \begin{bmatrix} 2.9274 & -1.2162 \\ -4.1436 & 1.7112 \\ 2.9274 & -1.2162 \end{bmatrix}$$

(p)

$$[\Delta F]_2 = \begin{bmatrix} -1.7112 & -1.7112 \\ 2.4324 & 2.4324 \\ -1.7112 & -2.42 \end{bmatrix}, \quad [\Delta F]_3 = \begin{bmatrix} -1.2161 & 2.9274 \\ 1.7112 & -4.1436 \\ -1.2162 & 2.9274 \end{bmatrix}$$

(q, r)

The stress constraints are formulated for the two loading conditions in accordance with Eq. (3.46) as

for nc = 1

$$\begin{bmatrix} (2.9274 - (20)0.5)u_1 & -1.7112u_2 & -1.2162u_3 \le -10 \\ -4.1436u_1 & +(2.4324 - (20)0.5)u_2 & +1.7112u_3 \le -5.854 \\ 2.9274u_1 & -1.7112u_2 & +(-1.2162 - (-15)0.5)u_3 \le 4.139 \end{bmatrix}$$

(s)

for $nc = 2$

$$\begin{bmatrix} (-1.2162 - (-15)0.5)\,u_1 & -1.7112u_2 & +2.9274u_3 \geq 4.139 \\ 1.7112u_1 & +(2.4324 - (20)0.5)\,u_2 & -4.1436u_3 \leq -5.854 \\ -1.2162u_1 & -2.42u_2 & +(2.9274 - (20)0.5)\,u_3 \leq -10 \end{bmatrix}$$

(t)

The displacement constraints shown in Eq. (3.49) may be similarly written for the loading conditions:

for $nc = 1$

$$\begin{bmatrix} \dfrac{l}{E}(-20)\,u_1 & & -8.28\dfrac{l}{E}u_3 \leq 1.5 - \dfrac{2l}{E}\,(28.28) \\[2ex] \dfrac{l}{E}(-8.28)\,u_1 & +\dfrac{l}{E}(-6.86)\,u_2 & +\dfrac{l}{E}(3.43)\,u_3 \leq 1.5 - \dfrac{2l}{E}\,(11.716) \end{bmatrix}$$

(u)

for $nc = 2$

$$\begin{bmatrix} \dfrac{l}{E}(8.28)\,u_1 & & +20\dfrac{l}{E}u_3 \leq 1.0 - \dfrac{2l}{E}\,(-28.28) \\[2ex] \dfrac{l}{E}(3.43)\,u_1 & -\dfrac{l}{E}(6.86)\,u_2 & +\dfrac{l}{E}(8.28)\,u_3 \leq 1.5 - \dfrac{2l}{E}\,(11.716) \end{bmatrix}$$

(v)

where $(X_{ik})_{all} = 1.0$ and 1.5 for X_1 and X_2, respectively; and $E = 2{,}500$ kip ft^{-1} in.$^{-1}$. No move limit constraint is imposed for this example.

Considering symmetry in the structural configuration and the loadings, we may reduce the number of design variables by assuming $u_1 = u_3$. Thus, the first three equations for the stress constraints in Eq. (s) become

$$\left.\begin{aligned} 8.2888u_1 + 1.7112u_2 &\geq 10 && (1) \\ 2.3424u_1 + 7.5676u_2 &\geq 5.854 && (2) \\ 8.7162u_1 - 1.7112u_2 &\geq 4.139 && (3) \end{aligned}\right\}$$

(w)

Instead of the Simplex method, we can solve Eq. (w) by using the graph shown in Fig. 3.7 as

$$u_1 = 1.122, \quad u_2 = 0.414$$

(x)

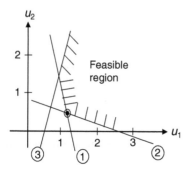

Figure 3.7 Solution at first optimization cycle.

The objective function as minimum structural volume is

$$V = 2\left(10\sqrt{2}\right)(0.5)\,u_1 + 10\,(0.5)\,u_2 = 2\left(10\sqrt{2}\right)(0.5)\,(1.122)$$
$$+ 10\,(0.5)\,(0.414) = 17.926\,\text{in.}^3 \tag{y}$$

in which

$$A_1 = A_3 = 0.5\,(1.122) = 0.561\,\text{in.}^2, \quad A_2 = 0.5\,(0.414) = 0.207\,\text{in.}^2 \tag{z}$$

Second cycle

Using the new areas of Eq. (z) in Eqs. (e), (f) and (g) yields

$$[X] = \frac{l}{E}\begin{bmatrix} 25.253 & -25.253 \\ 16.586 & 16.586 \end{bmatrix} \tag{aa}$$

$$[F] = \begin{bmatrix} 11.720 & -2.426 \\ 3.430 & 3.430 \\ -2.426 & 11.720 \end{bmatrix}; \quad \frac{[F]}{F_a} = \begin{bmatrix} 0.5860 & 0.1617 \\ 0.1715 & 0.1715 \\ 0.1617 & 0.5860 \end{bmatrix} \tag{bb, cc}$$

In order to push one of the members to be fully stressed, we can use the ratio of the largest area as $0.5860/0.561 = 1.0445$, Eq. (cc) is modified as

$$A_1 = A_3 = 0.586\,\text{in.}^2, \quad A_2 = 0.2162\,\text{in.}^2 \tag{dd}$$

which are used in Eqs. (q) through (w). The results of the second cycle are

$$u_1 = 0.689, \quad u_2 = 2.970 \tag{ee}$$

The new areas and structural volume are

$$\begin{bmatrix} A_1 \\ A_2 \\ A_3 \end{bmatrix} = \begin{bmatrix} (0.586)\,0.689 \\ (0.216)\,2.970 \\ (0.586)\,0.689 \end{bmatrix} = \begin{bmatrix} 0.4042 \\ 0.6422 \\ 0.4042 \end{bmatrix} \tag{ff}$$

$$V = 2\,(14.14)\,0.4042 + (10)\,0.6422 = 17.853 \text{ in.}^3 \tag{gg}$$

The design results for the next few cycles are sketched in Fig. 3.8 which shows fluctuating behaviour because no move limit is imposed as a side constraint for linearization of the nonlinear problem. If $B_u = 2$ is used, the fluctuation can be greatly reduced.

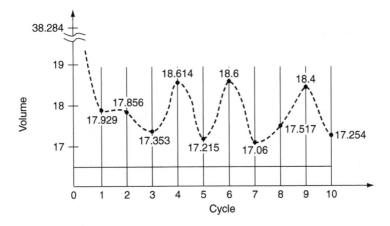

Figure 3.8 Solution of ten cycles without side constraints of move limits.

3.6.2 Example for continuous beams

The indeterminate beam shown in Fig. 3.9a is used to illustrate how to formulate constraint equations. The optimization procedures are similar to the example in Section 3.6.1. Assume the member depth for segment AB and BC are $d = 16$ in. and 12 in., respectively; $E = 29,000\,\text{ksi}, q = 0.35$, and allowable bending stress, $F_b = \pm 20\,\text{ksi}$.

Solution:

The nodal d.o.f. and the internal moments for each segment are sketched in Figs. 3.9b and c, respectively. For Eq. (3.12), the individual matrices are expressed in Eq. (a) through (f) as

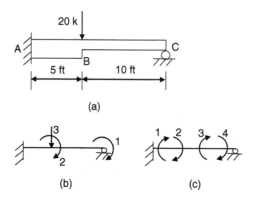

Figure 3.9 Intermediate beam: (a) given beam and loading, (b) nodal d.o.f., (c) internal moments.

$$\{P\} = \begin{bmatrix} 0 & 0 & 20 \end{bmatrix}^T \tag{a}$$

$$[B]^T = \begin{bmatrix} 0 & 0 & 0 & 1 \\ 0 & 1 & 1 & 0 \\ -1/5 & -1/5 & 1/10 & 1/10 \end{bmatrix}, \quad [S] = \begin{bmatrix} [S]_a & 0 \\ 0 & [S]_b \end{bmatrix} \tag{b, c}$$

$$[S]_a = \begin{bmatrix} 4EI_a/L_a & 2EI_a/L_a \\ \text{sym} & 4EI_a/L_a \end{bmatrix}, \quad [S]_b = \begin{bmatrix} 4EI_b/L_b & 2EI_b/L_b \\ \text{sym} & 4EI_b/L_b \end{bmatrix} \tag{d, e}$$

where I_a and I_b are moments of inertia of segments AB and BC, respectively.

$$[K] = [B]^T [S] [B] = \begin{bmatrix} 5,800S_b & 2,900S_b & 870S_b \\ & (15,467S_a + 5,800S_b) & (-4,640S_a + +870S_b) \\ \text{sym} & & (1,856S_a + 174S_b) \end{bmatrix} \tag{f}$$

in which the selection modulus of members a and b are $S_a = I_a/(d_a/2), S_b = I_b/(d_b/2)$. Select $q = 0.35$ for $S = qdA$; then the cross-sectional areas are

$$A_a = \frac{S_a}{0.35\,(16)} = 0.17857S_a; \quad A_b = \frac{S_b}{0.35\,(12)} = 0.2381S_b \tag{g}$$

The objective function is

$$V = 5\,(12)\,(0.17857S_a) + 10\,(12)\,(0.2381S_b)$$

$$= 10.7143S_a + 28.5714S_b \,(\text{in.}^3) \tag{h}$$

First cycle

Assume $S_a = 10 \text{ in.}^3$, $S_b = 10 \text{ in.}^3$, then we have

$$[K] = 10^3 \begin{bmatrix} 58 & 29 & 8.7 \\ & 212.67 & -37.7 \\ \text{sym} & & 20.3 \end{bmatrix}; \quad \{X\} = 10^{-4} \begin{Bmatrix} -5.0582 \\ 4.2134 \\ 19.8564 \end{Bmatrix} \quad \text{(i, j)}$$

$$\{F\} = \begin{Bmatrix} M_1 \\ M_2 \\ M_3 \\ M_4 \end{Bmatrix} = \begin{Bmatrix} -59.55 \\ -26.965 \\ 26.965 \\ 0 \end{Bmatrix} 12; \quad \begin{Bmatrix} S_a \\ S_b \end{Bmatrix} = \begin{Bmatrix} 2.9775 \\ 1.3483 \end{Bmatrix} 12 \quad \text{(k, l)}$$

Eq. (l) is obtained from Eq. (k) by dividing the allowable stress $F_b = \pm 20 \text{ ksi}$. Since previously $S = 10$, we change the members in order to have at least one segment fully stressed as

$$\begin{Bmatrix} S_a \\ S_b \end{Bmatrix} = \begin{Bmatrix} 35.73 \\ 35.73 \end{Bmatrix} \text{in.}^3 \quad \text{(m)}$$

Using the new design variables, we repeat the procedures illustrated in the previous example for $\{\Delta X\}$ and $\{\Delta F\}$ as

$$[S]^* = \begin{bmatrix} 46{,}053 & 23{,}026.5 & 0 & 0 \\ & 46{,}053 & 0 & 0 \\ & & 17{,}269.5 & 8{,}634.75 \\ \text{sym} & & & 17{,}269.5 \end{bmatrix} 12 \quad \text{(n)}$$

$$\{\Delta X\}_1 = 10^{-4} \begin{Bmatrix} 7.61 \\ -3.62 \\ -38.62 \end{Bmatrix} \frac{1}{12}; \quad \{\Delta X\}_2 = 10^{-4} \begin{Bmatrix} 9.48 \\ -10.53 \\ -28.07 \end{Bmatrix} \frac{1}{12} \quad \text{(o, p)}$$

$$\{\Delta F\}_1 = \begin{Bmatrix} -14.534 \\ 9.716 \\ -9.69 \\ 0 \end{Bmatrix} 12; \quad \{\Delta F\}_2 = \begin{Bmatrix} 14.534 \\ -9.716 \\ 9.69 \\ 0 \end{Bmatrix} \quad \text{(q, r)}$$

The stress constraints are formulated as

$$\left. \begin{array}{l} -14.533u_1 + 14.534u_2 - (-20)\,2.9775u_1 \geq 59.55 \\ 9.716u_1 - 9.716u_2 - (-20)\,2.9775u_1 \geq 26.965 \\ -9.69u_1 + 9.69u_2 - (20)\,2.9775u_2 \leq -26.965 \end{array} \right\} \quad \text{(s)}$$

or

$$45.017u_1 + 14.534u_2 \geq 59.55$$
$$69.266u_1 - 9.716u_2 \geq 26.965$$
$$9.69u_1 + 49.86u_2 \geq 26.965$$

(t)

and the displacement constraints are

$$\frac{1}{12}\left(10^{-4}(7.61)u_1 + 10^{-4}(9.48)u_2\right) \geq -(X_{all}) - 2(-1.708)\,10^{-3}\left(\frac{1}{12}\right)$$

$$\frac{1}{12}\left(10^{-4}(-3.62)u_1 + 10^{-4}(-10.53)u_2\right) \leq (X_{all}) - 2(1.4151)\,10^{-5}\left(\frac{1}{12}\right)$$

$$\frac{1}{12}\left(10^{-4}(-38.62)u_1 + 10^{-4}(-28.07)u_2\right) \leq (X_{all}) - 2(6.669)\,10^{-3}\left(\frac{1}{12}\right)$$

(u)

Considering the stress constraint only, one can solve Eq. (t) as

$$u_1 = 1.2251, \quad u_2 = 0.3$$

(v)

The section moduli at this design cycle are

$$S_a = 1.2251\,(35.73) = 43.77 \text{ in.}^3; \quad A_a = 0.1786\,(43.77) = 7.82 \text{ in.}^2$$

(w)

$$S_b = 0.3\,(35.73) = 10.719 \text{ in.}^3; \quad A_b = 0.2381\,(10.719) = 2.553 \text{ in.}^2 \quad \text{(x)}$$

The optimal volume shown in Eq. (h) becomes

$$V = 10.7143\,(43.77) + 28.5714\,(10.719) = 775.216 \text{ in.}^3$$

(y)

The following design procedures are similar to those shown above.

3.6.3   Example for elastic frames

The sample structure is given in Fig. 3.10a subjected to loading condition as shown. Let $d = 12$ in., $E = 29,000$ ksi, $F_b = \pm 20$ ksi, $F_a = 20$ ksi (tension) and $F_a = 15$ ksi (compression) for both members. Formulate constraint equations and find the structural volume in the first optimization cycle with consideration of the stress constraint only.

Solution:

The nodal d.o.f. and internal forces are sketched in Figs. 3.10b and c, respectively. The structural stiffness matrix is formulated as

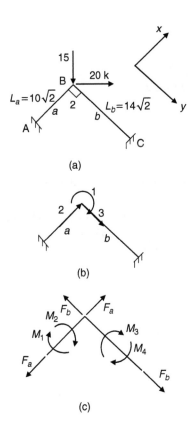

(a)

(b)

(c)

Figure 3.10 Intermediate elastic frame: (a) given structure with loading, (b) nodal displacements, (c) internal forces.

$$[K] = \begin{bmatrix} 4\,(S_1)_a + 4\,(S_1)_b & 6\dfrac{(S_1)_b}{L_b} & -6\dfrac{(S_1)_a}{L_b} \\ & S_2^a + 12\dfrac{(S_1)_b}{L_b^2} & 0 \\ \text{sym} & & 12\dfrac{(S_1)_a}{L_a^2} + (S_2)_b \end{bmatrix} \quad (a)$$

where

$$(S_1)_a = \frac{S_a d_a}{2L_a} = \frac{12S_a}{(2)\,(12)\,10\sqrt{2}} = \frac{S_a}{20\sqrt{2}}$$

$$(S_2)_a = \frac{S_a}{q d_a L_a} = \frac{S_a}{0.35\,(12)\left(10\sqrt{2}\right)(12)} = \frac{S_a}{504\sqrt{2}}$$

$$(S_1)_b = \frac{12 S_b}{2\,(12)\,14\sqrt{2}} = \frac{S_b}{28\sqrt{2}}$$

$$(S_2)_b = \frac{S_b}{0.35\,(12)\left(14\sqrt{2}\right)(12)} = \frac{S_b}{705.6\sqrt{2}}$$

Assume

$$S_a = 14.142 \text{ in.}^3, \quad S_b = 19.8 \text{ in.}^3$$

for the initial design; the loading and structural stiffness matrices are

$$\{P\} = \begin{Bmatrix} P_1 \\ P_2 \\ P_3 \end{Bmatrix} = \begin{Bmatrix} 0 \\ 3.536 \\ 24.749 \end{Bmatrix}; \quad [K] = \begin{bmatrix} 116,002 & -365.2 & -512.65 \\ & 578.48 & 0 \\ \text{sym} & & 581.47 \end{bmatrix}$$

$$(b,\,c)$$

The nodal displacements are obtained as

$$\begin{Bmatrix} X_1 \\ X_2 \\ X_3 \end{Bmatrix} = [K]^{-1}\,\{P\} = 10^{-4} \begin{bmatrix} 2.086 \\ 62.3 \\ 427.46 \end{bmatrix} \begin{matrix} \text{rad} \\ \text{in.} \\ \text{in.} \end{matrix} \qquad (d)$$

The internal forces of individual members are:

For member *a*

$$\begin{Bmatrix} F_a \\ M_1 \\ M_2 \end{Bmatrix} = E \begin{Bmatrix} (S_2)_a\,X_2 \\ 2\,(S_1)_a\,X_1 - 6\frac{(S_1)_a}{L_a}X_3 \\ 4\,(S_1)_a\,X_1 - 6\frac{(S_1)_a}{L_a}X_3 \end{Bmatrix} = \begin{Bmatrix} 3.592\,\text{k} \\ -16.594\,\text{k-in.} \\ -9.815\,\text{k-in.} \end{Bmatrix}; \qquad (e)$$

For member *b*

$$\begin{Bmatrix} F_b \\ M_3 \\ M_4 \end{Bmatrix} = E \begin{Bmatrix} -(S_2)_b\,X_3 \\ 4\,(S_1)_b\,X_1 - 6\frac{(S_1)_b}{L_b}X_2 \\ 2\,(S_1)_b\,X_1 - 6\frac{(S_1)_b}{L_b}X_2 \end{Bmatrix} = \begin{Bmatrix} -24.35\,\text{k} \\ 9.816\,\text{k-in.} \\ 3.76\,\text{k-in.} \end{Bmatrix} \qquad (f)$$

Using the given allowable stresses for Eqs. (e) and (f) yields

$$\begin{Bmatrix} A_a \\ S_1 \\ S_2 \end{Bmatrix} = \begin{Bmatrix} 3.592/20 \\ |16.594/20| \\ |9.815/20| \end{Bmatrix} = \begin{Bmatrix} 0.18 \text{ in.}^2 \\ 0.83 \text{ in.}^3 \\ 0.49 \text{ in.}^3 \end{Bmatrix};$$

$$\begin{Bmatrix} A_b \\ S_3 \\ S_4 \end{Bmatrix} = \begin{Bmatrix} |24.35/15| \\ 9.816/20 \\ 3.76/20 \end{Bmatrix} = \begin{Bmatrix} 1.623\,\text{in.}^2 \\ 0.49\,\text{in.}^3 \\ 0.19\,\text{in.}^3 \end{Bmatrix} \qquad \text{(g, h)}$$

In order to unify the design variables between A and S, we can use $S_a = qdA_a = 0.35(12)(0.18) = 0.8$ in.^{3} for member a and $S_b = qdA_b = 0.35\,(12)\,(1.623) = 6.82$ in.^{3} for member b. Convert A_a and A_b to S_a and S_b respectively; then observe which will be fully stressed. It is found that S_b converted from A_b should be used in order to achieve at least one fully stressed member and the others are not overstressed. The section modulus of member a is then $S_1 = 14.14/19.8\,(6.82) = 4.87$. The factor to modify all the members is $6.82/19.8 = 0.3447$. Thus, follow the procedures in the previous examples to modify the stiffness and to find incremental displacements $[\Delta X]$ and forces $[\Delta F]$ as follows:

First cycle

$$\{X\}^* = 10^{-4} \begin{Bmatrix} 6.056 \\ 180.871 \\ 1241.0 \end{Bmatrix} \begin{matrix} \text{rad} \\ \text{in.} \\ \text{in.} \end{matrix} \qquad \text{(i)}$$

$$\{\Delta X\}_1 = -[K]^{*-1}[\Delta K]_1 \{X\}^* = 10^{-4} \begin{Bmatrix} 1.854 \\ -179.3 \\ 548.21 \end{Bmatrix} \qquad \text{(j)}$$

$$\{\Delta X\}_2 = -[K]^{*-1}[\Delta K]_2 \{X\}^* = 10^{-4} \begin{Bmatrix} -7.91 \\ -1.571 \\ -1789.21 \end{Bmatrix} \qquad \text{(k)}$$

where $[\Delta K]_1$ is due to $S_a = 14.14$, $S_b = 0$ and $[\Delta K]_2$ is based on $S_b = 19.8$, $S_a = 0$. The incremental internal forces due to member a change are

$$\begin{Bmatrix} \Delta F_a \\ \Delta M_1 \\ \Delta M_2 \end{Bmatrix}_1 = \begin{Bmatrix} 0.0344 \\ -24.42 \\ -15.79 \end{Bmatrix}; \quad \begin{Bmatrix} \Delta F_b \\ \Delta M_3 \\ \Delta M_4 \end{Bmatrix}_1 = \begin{Bmatrix} -10.86 \\ 5.972 \\ 4.118 \end{Bmatrix} \qquad \text{(l, m)}$$

and the incremental forces due to member b change are

$$\begin{Bmatrix} \Delta F_a \\ \Delta M_1 \\ \Delta M_2 \end{Bmatrix}_2 = \begin{Bmatrix} -0.0344 \\ 24.42 \\ 15.79 \end{Bmatrix}; \quad \begin{Bmatrix} \Delta F_b \\ \Delta M_3 \\ \Delta M_4 \end{Bmatrix}_2 = \begin{Bmatrix} 10.86 \\ -5.972 \\ -4.118 \end{Bmatrix} \qquad \text{(n, o)}$$

Using Eqs. (j) through (o), we can formulate constraint equations in Eqs. (p) through (r).

Displacement constraints

$$1.854u_1 - 7.91u_2 \leq X_{\text{all}} - 2\,(6.056)$$
$$\left.\begin{array}{l} -179.3u_1 - 1.571u_2 \leq X_{\text{all}} - 2\,(180.871) \\ 548.21u_1 - 1789.21u_2 \leq X_{\text{all}} - 2\,(1241) \end{array}\right\} \tag{p}$$

Stress constraints

For member *a*

$$\left[-24.42 + \dfrac{0.0344}{\dfrac{20}{0.35\,(12)\,(20)}}\right]u_1 + \left[24.42 + \dfrac{0.0344}{\dfrac{20}{0.35\,(12)\,(20)}}\right]$$
$$u_2 - (20)\,4.87u_1 \geq 16.594 - \dfrac{3.592}{\dfrac{20}{0.35\,(12)\,(20)}} \tag{q}$$

For member *b*

$$\left[5.972 + \dfrac{-10.86}{\dfrac{20}{0.35\,(12)\,(20)}}\right]u_1 + \left[-5.972 + \dfrac{10.86}{\dfrac{15}{0.35\,(12)\,(20)}}\right]$$
$$u_2 - (20)\,6.82u_2 \leq -9.816 - \dfrac{-24.35}{\dfrac{15}{0.35\,(12)\,(20)}} \tag{r}$$

The above equations are rewritten as

$$72.83u_1 + 24.565u_2 \geq 31.68 \tag{s}$$
$$-66.788u_1 + 203.2u_2 \geq 146.176 \tag{t}$$

Considering stress constraints only and then solving Eqs. (s) and (t), we have

$$u_1 = 0.1732, \quad u_2 = 0.7763 \tag{u}$$

Thus, the section moduli are

$$S_a = (4.87)\,0.1732 = 0.843\,\text{in.}^3; \quad S_b = (6.82)\,0.7763 = 5.294\,\text{in.}^3 \quad\quad \text{(v)}$$

The optimal objective at this design cycle is

$$\begin{aligned} V &= \frac{(12)\,10\sqrt{2}}{12}\left(\frac{4.87}{0.35}\right)u_1 + \frac{(12)\,14\sqrt{2}}{12}\left(\frac{6.82}{0.35}\right)u_2 \\ &= 196.78\,(0.1732) + 385.8\,(0.7763) = 333.57\,\text{in.}^3 \end{aligned} \quad\quad \text{(w)}$$

3.7 Optimization for dynamic response

3.7.1 Modelling for dynamic analysis and design

The dynamic response of structures is becoming increasingly important in the design of multi-storey building for earthquake motions, spacecraft for minimum frequency requirements, and large aircraft for critical weight problems. The mathematical formulation for dynamic analysis has three well-known models as *lumped mass*, *consistent mass* and *distributed mass* [1]. Here, the emphasis is on the lumped mass model. The model is illustrated in Fig. 3.11a which has two members AB and BC with fixed support at A and roller support at C. Member AB is non-uniform in cross-section and member BC is prismatic. The lumped mass model assumes that one-half member mass is lumped at each end. A member can be divided into several segments (elements) for the purpose of serving: (a) for a system having non-uniform members which are then divided into a number of elements of which each is treated as prismatic with approximate moment of inertia, I, and area, A_a in stiffness formulation; (b) for achieving accuracy of frequencies and vibration modes. It is worthwhile to point out: (a) even though dividing a system's members into more elements can yield more accurate solutions, theoretically the solution based on an infinite number of elements can simulate the distributed mass model; but one must consider that the increasing number of elements will increase d.o.f. and computational efforts; (b) whether dividing into more elements can affect the solution of the first few modes which are practically needed for building design.

Consider the rigid frame shown in Fig. 3.12a which is divided into four elements. The number of d.o.f. of the system is seven as four d.o.f. in rotations signified by nr and three linear d.o.f. represented by ns. The motion of a mass can be in both rotational and transverse directions. However, the rotational inertia force is much smaller than that associated with the transverse and is usually therefore neglected in a lumped model [1]. A typical element is shown in Fig. 3.11c, where the local d.o.f. are slopes θ_i, θ_j and the

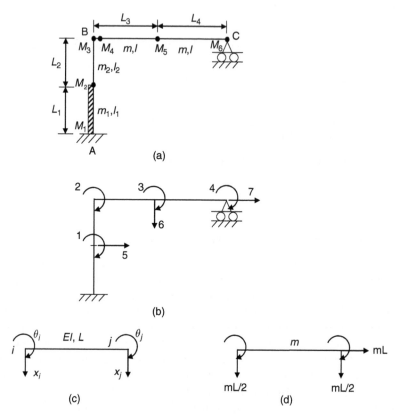

Figure 3.11 Dynamic structure: (a) lumped mass model, (b) d.o.f. for model, (c) typical element n with local d.o.f., (d) typical member with lumped masses.

displacements X_i, X_j are all in the positive direction. The force–deformation relationships of the elements are

$$
\left.
\begin{aligned}
M_i &= \frac{4EI}{L}\theta_i + \frac{2EI}{L}\theta_j + \frac{6EI}{L^2}x_i - \frac{6EI}{L^2}x_j \\
M_j &= \frac{2EI}{L}\theta_i + \frac{4EI}{L}\theta_j + \frac{6EI}{L^2}x_i - \frac{6EI}{L^2}x_j \\
V_i &= \frac{6EI}{L^2}\theta_i + \frac{6EI}{L^2}\theta_j + \frac{12EI}{L^3}x_i - \frac{12EI}{L^3}x_j \\
V_j &= -\frac{6EI}{L^2}\theta_i - \frac{6EI}{L^2}\theta_j - \frac{12EI}{L^3}x_i - \frac{12EI}{L^3}x_j
\end{aligned}
\right\}
\tag{3.57}
$$

which represents the element's stiffness. Expressing the coefficients in matrix form yields the stiffness matrix of the element as

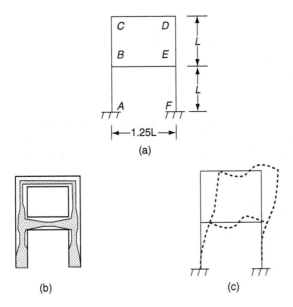

Figure 3.12 Two-storey one-bay rigid frame: (a) given structure, (b) optimal area distribution, (c) first mode shape.

$$[s]_n = \frac{EI}{L^3} \begin{pmatrix} 4L^2 & 2L^2 & 6L & -6L \\ & 4L^2 & 6L & -6L \\ \text{sym} & & 12 & 12 \\ & & & 12 \end{pmatrix} \qquad (3.58a)$$

As discussed in Eq. (3.10),

$$[K_E]_n = [b]^T_n [S]_n [b]_n \qquad (3.58b)$$

in which the compatibility matrix $[b]_n$ should be established based on the relationship between the global d.o.f. and local d.o.f. shown in Fig. 3.11b and c, respectively [1].

The mass of the element is shown in Fig. 3.11d where the masses are lumped at X_i and X_j each with half of the element mass, mL/2, without considering rotatory inertia. If the element has a rigid-body motion, then the total mass of the element, mL, can induce an inertia force along the motion. The element's mass matrix is shown in Eq. (3.59) as

$$[M_E]_n = \begin{bmatrix} 0 & 0 & 0 & 0 & 0 \\ & 0 & 0 & 0 & 0 \\ & & mL/2 & 0 & 0 \\ & \text{sym} & & mL/2 & 0 \\ & & & & mL \end{bmatrix} \qquad (3.59)$$

which is a diagonal matrix in which the first four rows (or columns) correspond to the d.o.f. in Fig. 3.11c and the last is associated with rigid-body motion. Employing Eq. (3.58), we can formulate the structural stiffness matrix of the system as

$$[K] = \sum [K_E] = \begin{bmatrix} K_{11} & K_{12} \\ \overline{sym} & \overline{K_{22}} \end{bmatrix} \begin{matrix} nr \\ ns \end{matrix} \qquad (3.60)$$

Similarly, we can also formulate the system's mass matrix as

$$[M] = \sum [M_E] = \begin{bmatrix} 0 & & & 0 & \\ & M_2 & & & 0 \\ & & M_5 & & \\ \text{sym} & & & M_3 + M_4 & \\ & & & +M_5 + M_6 & \end{bmatrix} \begin{matrix} nr \\ \\ ns \end{matrix} \qquad (3.61)$$

where the lumped masses are $M_2 = (m_1 L_1 + m_2 L_2)/2$, $M_3 = m_2 L_2/2$, $M_4 = mL/2$, $M_5 = m(L_3 + L_4)/2$, and $M_6 = mL_4/2$. The lumped mass associated with X_7 is due to a rigid motion of the system; consequently, M_3, M_4, M_5, and M_6 induce an inertia force in that direction. $[M]$ is a diagonal matrix in which the sub-matrix $[0]_{nr \times (nr+ns)}$ is always zero because no rotatory inertia is considered. Therefore, the system's mass matrix has only the mass coefficients in the sub-matrix $[M_d]_{ns \times ns}$.

The dynamic equilibrium matrix for free-vibration of the structure may be expressed as

$$[K]\{x\} + [M]\{\ddot{x}\} = 0 \qquad (3.62)$$

where $\{x\}$ represents response-associated linear displacements and $\{\ddot{x}\}$ signifies accelerations. Since the inertia force of an individual mass may be expressed as $m\ddot{x} = \omega^2 mX \sin \omega t$, Eq. (3.62) can be modified as

$$[K]\{X\} - \omega^2 [M]\{X\} \qquad (3.63)$$

where ω = natural frequency, and $\{X\}$ = mode shapes of free vibration [1]. Employing Eqs. (3.60) and (3.61) in Eq. (3.63) yields

$$[K_{11}]_{nr \times nr} \{X\}_{nr} + [K_{12}]_{nr \times ns} \{X\}_{ns} = \omega^2 [0] \{X\}_{nr} \tag{3.64}$$

and

$$[K_{21}]_{ns \times nr} \{X\}_{nr} + [K_{22}]_{ns \times ns} \{X\}_{nr} = \omega^2 [M_d]_{ns \times ns} \{X\}_{ns} \tag{3.65}$$

Eq. (3.64) can be used to express the relationship between rotational modes $\{X\}_{nr}$ and the transverse modes as

$$\{X\}_{nr} = -[K_{11}]_{nr \times nr}^{-1} [K_{12}]_{nr \times ns} \{X\}_{ns} \tag{3.66}$$

Substituting Eq. (3.66) into Eq. (3.64), we have

$$-[K_{21}]_{ns \times nr} [K_{11}]_{nr \times nr}^{-1} [K_{12}]_{nr \times nr} \{X\}_{ns} + [K_{22}]_{ns \times ns} \{X\}_{ns} = \omega^2 [M_d]_{ns \times ns} \{X\}_{ns} \tag{3.67}$$

which is condensed as

$$\left([K_{22}]_{ns \times ns} - [K_{21}]_{ns \times nr} [K_{11}]_{nr \times nr}^{-1} [K_{12}]_{nr \times nr}\right) \{X\}_{ns} = \omega^2 [M_d]_{ns \times ns} \{X\}_{ns} \tag{3.68}$$

or

$$[K_c]_{ns \times ns} \{X\}_{ns} = \omega^2 [M_d]_{ns \times ns} \{X\}_{ns} \tag{3.69}$$

Eq. (3.69) has much fewer dimensions than that for Eq. (3.63) and is therefore preferred to be used for eigen-solutions, ω and $\{X\}$. After $\{X\}_{ns}$ is known, it is then substituted into Eq. (3.66) for $\{X\}_{nr}$.

3.7.2 Linearization of objective function and constraints

The objective is to minimize structural weight and to satisfy the required fundamental frequency and strength of the system. The merit function may be

$$W = \sum_{j=1}^{EL} \gamma L_j A_j \tag{3.70}$$

where EL = no. of elements. The constraints are the minimum frequency and minimum element size. The former is to control dynamic response such as resonance induced due to machinery vibration and the latter is due to the required member size prior to static-load analysis or construction necessity [2, 3, 4]. The aforementioned constraints may be expressed as

$$\omega \geq \omega^l \tag{3.71}$$
$$A_j \geq A_j^l \tag{3.72}$$

where the superscript l signifies the lowest possible solution. The cross-sectional area A_j can relate to the section modulus, S_j as presented in Eq. (3.29). Since frequency is highly nonlinear in terms of cross-section, the Taylor series is used here to achieve the linear objective function and constraints as

$$W = W^o + \Delta W = \sum_{j=1}^{EL} (A_j^o + \Delta A_j) \gamma L_j \tag{3.73}$$

The incremental area can be conveniently changed by imposing a percentage number of the solution variable as

$$-\delta A_j^o \leq \Delta A_j \leq \delta A_j^o \tag{3.74}$$

where δ is suggested to be 0.02–0.20. Thus, the constraint on member size of Eq. (3.72) becomes

$$A_j^o + \Delta A_j \geq A_j^l \tag{3.75}$$

Similarly, the frequency constraint can be expressed in linear form as

$$\omega = \omega^o + \Delta\omega \approx \omega^o + \sum_{j=1}^{EL} \frac{\partial \omega}{\partial A_j} \Delta A_j \geq \omega^l \tag{3.76}$$

where $\partial \omega / \partial A_j$ is the rate of frequency change with respect to area j. Since the required minimum frequency is constant, that means the previous design step satisfied the requirement, we have $\Delta\omega = 0$. Numerically this can be accomplished by scaling the design variables based on the ratio of frequency at the design step and the required frequency. Thus, the second term in Eq. (3.76) is changed as

$$\frac{\partial \omega}{\partial A_j} \Delta A_j = \frac{\partial \omega}{\delta A_j^o} \Delta A_j \geq 0 \tag{3.77}$$

Note that standard linear programming theory requires a positive solution variable that is to be achieved by following Eq. (3.22) using

$$u_j = \frac{(1 + \Delta A_j)}{A_j^o} \tag{3.78}$$

Consequently, the objective function and constraints may be rewritten as follows:

$$\text{Minimize } W = \sum_{j=1}^{EL} A_j^o L_i \gamma u_j \tag{3.79}$$

Subject to $\sum_{j=1}^{\text{EL}} \Delta\omega_j u_j \geq \sum_{j=1}^{\text{EL}} \Delta\omega_j$ (3.80)

$A_j^o u_j \geq A_j^l \quad j=1,\ \text{EL}$ (3.81)

$1 - \delta \leq u_j \leq 1 + \delta \quad j=1,\ \text{EL}$ (3.82)

$u_j \geq 0 \quad j=1,\ \text{EL}$ (3.83)

In order to eliminate the redundancy of Eqs. (3.82) and (3.83), we can introduce the design variable as

$$z_j = u_j + \delta - 1$$ (3.84)

then Eqs. (3.79) through (3.83) become

Minimize $W = \sum_{j=1}^{\text{EL}} A_j^o \gamma L_j z_j + \text{constant}$ (3.85)

Subject to $\sum_{j=1}^{\text{EL}} \Delta\omega_j z_j \geq \delta \sum_{j=1}^{\text{EL}} \Delta\omega_j$ (3.86)

$A_j^o z_j \geq A_j^l - A_j^o (1 - \delta)$ (3.87)

$0 \leq z_j \leq 2\delta$ (3.88)

Converting the inequalities in Eqs. (3.86) through (3.88), we can then use the Simplex algorithm in prime form that can also be changed to dual form as discussed in Section 3.5.

3.7.3 Sensitivity analysis of frequency

Eq. (3.86) requires sensitivity analysis, $\Delta\omega_j$, of the frequency ω, due to a Δ percentage change in area, A_j. Let us use the perturbation technique that perturbating ΔA_j^o as one of the design parameters in Eq. (3.63) we have [3]

$$\left[[K] + [\Delta K_E]_j\right]\left[\{X\} + \{\Delta X\}_j\right] = (\omega + \Delta\omega_j)^2 \left[[M] + [\Delta M_E]_j\right]\left[\{X\} + \{\Delta X\}_j\right]$$ (3.89)

in which $[K_E]_j$ and $[M_E]_j$ are associated with element j change; other elements remain unchanged. Expanding the above yields

$$[K]\{X\} + [K]\{\Delta X\}_j + [\Delta K_E]_j\{X\} + [\Delta K_E]_j\{\Delta X\}_j = \omega^2 [M]\{X\}$$

$$+ \omega^2 [M]\{\Delta X\}_j + \omega^2 [\Delta M_E]_j\{X\} + \omega^2 [\Delta M_E]_j\{\Delta X\}_j + 2\omega\Delta\omega_j [M]\{X\}$$

$$+ 2\omega\Delta\omega_j [M]\{\Delta X\}_j + 2\omega\Delta\omega_j [\Delta M_E]_j \{X\} + 2\omega\Delta\omega_j [\Delta M_E]_j \{\Delta X\}_j$$

$$+ \Delta\omega_j^2 [M]\{X\} + \Delta\omega_j^2 [M]\{\Delta X\}_j + \Delta\omega_j^2 [\Delta M_E]_j \{X\}$$

$$+ \Delta\omega_j^2 [\Delta M_E]_j \{\Delta X\}_j \tag{3.90}$$

in which the higher-order terms are negligible. We may rewrite Eq. (3.90) as

$$[K]\{X\} + [K]\{\Delta X\}_j + [\Delta K_E]_j \{X\} = \omega^2 [M]\{X\} + \omega^2 [M]\{\Delta X\}_j$$
$$+ \omega^2 [\Delta M_E]_j \{X\} + 2\omega\Delta\omega_j [M]\{X\} \tag{3.91}$$

Subtract Eq. (3.63) from Eq. (3.91) and then rearrange the corresponding terms

$$\left[[K] - \omega^2 [M] \right]\{\Delta X\} = 2\omega\Delta\omega_j [M]\{X\} - \left([\Delta K_E]_j - \omega^2 [\Delta M_E]_j \right)\{X\} \tag{3.92}$$

Employing $[K] - \omega^2 [M] = 0$ in the above yields

$$\Delta\omega_j [M]\{X\} = \frac{1}{2\omega} \left[[\Delta K_E]_j - \omega^2 [\Delta M_E]_j \right]\{X\} \tag{3.93}$$

Let us use the following normalized model [1] in Eq. (3.93)

$$\{X\}^T [M]\{X\} = 1 \tag{3.94}$$

We finally obtain the frequency-sensitive equation as

$$\Delta\omega_j = \frac{1}{2\omega} \{X\}^T \left[[\Delta K_E]_j - \omega^2 [\Delta M_E]_j \right]\{X\} \tag{3.95}$$

in which the determination of frequency-sensitivity coefficients can be effectively done element-by-element in Eqs. (3.58) and (3.59). Find $\{X\}_{ns}$ and $\{X\}_{nr}$ from Eq. (3.69) and Eq. (3.66), respectively; then the complete normalized eigenvector can be obtained as

$$\{X\} = \left(\{X\}_{nr} \{X\}_{ns} \right)^T \tag{3.96}$$

which is used in Eq. (3.95).

3.7.4   Sample observations of optimization results

The two-storey and one-bay rigid frame shown in Fig. 3.12a is designed for the required fundamental frequency $\omega \geq 24\,\mathrm{rad\,s^{-1}}$. The constituent members are prismatic tubes which are divided into elements of which each has a length of $0.1L, L = 12\,\mathrm{ft}$. The optimum solution is $\omega = 24\,\mathrm{rad\,s^{-1}}$, the cross-sectional areas are shown in Fig. 3.12b, and the fundamental mode shape is sketched in Fig. 3.12c. The optimization results reveal: (A) a larger cross-section is needed for first-floor columns at the fixed support; (B) the girder's cross-section is less than that of the columns; (C) the first-floor girder is larger and has more variations in cross-section than the top floor; and (D) girder CD is basically uniform in cross-section. This simple frame indicates a few important design parameters in practice: (A) structural design for earthquakes should have strong columns and weak girders; (B) horizontal transverse inertia forces contribute significantly more than the vertical inertia forces on dynamic response; (C) the lumped mass model is very practical for tall building design in which each girder can be treated as one element; (D) the structural response to earthquake excitations is dominated by the first few modes, mostly the fundamental mode; and (E) control of natural frequency is of paramount importance in harmony motion induced by machinery vibration.

3.8   Concluding remarks

This chapter shows how to apply the linear programming algorithm in both prime and dual forms to nonlinear optimization of structures for both static and dynamic forces. As presented in Chapter 2, linear programming optimization can effectively solve problems with thousands of design variables. It can conveniently be used for detailed design of repeatedly constructed structures such as typical bridges, gable frames and precast members for which some sample cases will be presented in Chapter 4.

References

1. Cheng, F.Y., *Matrix Analysis of Structural Dynamics – Applications and Earthquake Engineering*, Marcel Dekker, Inc., New York; CRC Press/Taylor Francis Group, Boca Raton, FL, 2001.
2. McCart, B.R., Haug, E.J. and Streeter, T.D., Optimal design of structures with constraints on natural frequency, *Structural Dynamics at the AIAA Structural Dynamics and Aeroelasticity Specialists Conference*, American Institute of Aeronautics and Astronautics, 1920, New York, April 1969.
3. Romstad, K.M. and Wang, C.K., Optimum design of dynamic response, *Symposium Proceedings on Structural Dynamics*, Vol. 1, A.2.1-a.2.18, Loughborough, 1970.
4. Zarghamee, M.S., Optimum frequency of structures, *AIAA Journal*, Vol. 6, No. 4, 1968.

4 Introduction to nonlinear programming

4.1 Introduction

The general nonlinear optimization problem may be expressed as

$$\left. \begin{array}{l} \text{Maximize } F(x_i) \quad i = 1, 2, \ldots, n \\ \text{Subject to } g_j(x) \leq b \quad j = 1, 2, \ldots, m \end{array} \right\} \tag{4.1}$$

Although many practical problems in optimization can be formulated as linear programs, the real world is full of problems that are not so easy. If either the objective function or any one of the inequalities is nonlinear, then the optimization problem is said to be a *nonlinear program*. Several varieties of nonlinearity offer special techniques of solution. We will mention a few of these but we will concentrate upon methods that have been reasonably successful in solving general problems.

Before we delve into techniques, it is instructive to consider some of the characteristics of nonlinearity. First, we review the geometric nature of the *linear programming* problem in order to develop an intuitive feeling about the consequences of nonlinearity. The boundary of the constraint set is made up of hyperplanes; the feasible region is a *convex set*. That is, from any two points of the set x_a and x_b, a third point

$$x = \mu x_a + (1 - \mu) x_b \quad 0 < \mu < 1 \tag{4.2}$$

must also be a member of the set. Furthermore, the level contours of the objective function are also hyperplanes and the maximum or minimum must be on one of the many finite corners of the feasible set of points. A sketch of such a problem is shown in Fig. 4.1a.

First let us examine the consequences of making one of the constraints nonlinear while leaving the rest of the problem unchanged. Fig. 4.1b shows one constraint curved outward so that the feasible set of points remains convex. Again, the solution is at a boundary point but no longer must it lie at a corner. If we consider the same constraint bent inwards as in Fig. 4.1c, we see that the feasible region is no longer convex. The solution just happens to be at a corner point, but if we look at the neighbouring corner which

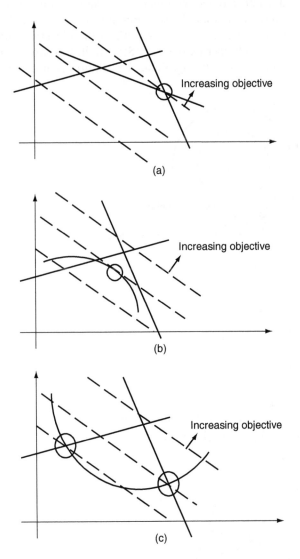

Figure 4.1 Linear versus nonlinear: (a) linear program, (b) nonlinear program –
convex, (c) nonlinear program – not convex.

is also circled in the figure, we see that in its immediate neighbourhood,
it has all of the characteristics of a solution. That is, no *feasible* point in
its neighbourhood will increase the objective function. We call this case a
local maximum to distinguish it from the true maximum, termed *global*.
Since many of our algorithms cannot determine whether a local solution is
actually global, we must accept the results and use some external criterion
for making such a decision.

We have maintained a linear objective function in the preceding examples. One of its characteristics is that the solution appears at a boundary
point; that is, some constraint is always active. When the objective function is nonlinear, this may no longer be the case. Several illustrations of the
possibilities are shown in Fig. 4.2. In Fig. 4.2a, the solution is at a corner

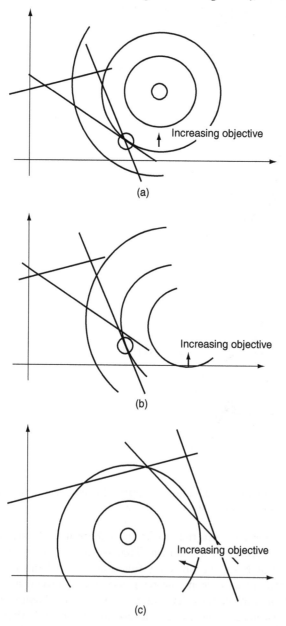

Figure 4.2 Nonlinear problem with nonlinear objective function: (a) solution at
corner, (b) solution at boundary, (c) solution at interior.

point; in Fig. 4.2b, the solution is on the boundary but not at a corner. In these two examples, at least one constraint is active. In Fig. 4.2c, the solution point is an interior point of the feasible region. None of the constraints are active and we can say that the solution is unconstrained. That is, the objective function has a local maximum within the feasible region, a concept that is meaningless when the function is linear. Most of the peculiarities we have illustrated complicate matters considerably when they occur together.

Several special cases that have their own well developed theory may be used when applicable. One of these is *quadratic programming* which treats a purely quadratic objective function subject to linear constraints. This type of problem can occur in linearly constrained regression analysis and as an approximation to more complicated nonlinear problems. Quadratic programs can be solved by modified linear programming techniques in reference [1]. A second class of problems whose objective function and constraints are made up of sums of terms in the form

$$Cx_1^1 x_2^2 \ldots x_n^n \qquad\qquad (4.3)$$

where C and x_i are positive and i are real numbers, can also be solved by the special techniques of *geometric programming*. Geometric programming is a theory based upon the inequality between the arithmetic and geometric means and is applicable to some engineering design problems. A complete discussion of geometric programming is given in reference [2] by the originators of the idea. For the focus of this book, we will concentrate on two types of approach to the most general nonlinear programming problems. The first of these is probably the most powerful and converts all problems into the unconstrained type for which very effective algorithms are available. The second method concentrates on the constraint set. Both are based upon the traditional mathematical approach of replacing a difficult problem by a sequence of more tractable problems.

4.2 Kuhn–Tucker theorem

Before developing techniques to use for locating a solution, it is worthwhile studying some of the characteristics of an optimal point. A theory that suggests numerical tests for optimality will also give information about solution sensitivity and may lead to algorithmic methods for obtaining the optimum. Such a theory was presented by Kuhn and Tucker [3] and the conditions for optimality are now known as the *Kuhn–Tucker conditions*. Their basic result is given in the following theorem. Assume $F(x)$ is to be maximized subject to the constraints $g_j(x) \le b_j$, $j = 1, 2, \ldots, m$, $x \ge 0$ and $F(x)$ and $g_j(x)$ are differentiable and satisfy certain regularity conditions (to prevent singularities on the boundary). Then x^* can be an optimal solution to the nonlinear program only if there exist m numbers $\lambda_1^*, \lambda_2^*, \ldots, \lambda_m^*$, such that the conditions listed below are satisfied [4].

i) If $x_i^* > 0$, $\dfrac{\partial F(x)}{\partial x_i} - \displaystyle\sum_{j=1}^{m} \lambda_j \dfrac{\partial g_j(x)}{\partial x_i} = 0$ at $x_i = x_i^*$, $i = 1, 2, \ldots, n$

$$(4.4a)$$

ii) If $x_i^* = 0$, $\dfrac{\partial F(x)}{\partial x_i} - \displaystyle\sum_{j=1}^{n} \lambda_j \dfrac{\partial g_j(x)}{\partial x_i} \leq 0$ at $x_i = x_i^*$, $i = 1, 2, \ldots, n$

$$(4.4b)$$

iii) If $\lambda_j^* > 0$, $g_j(x^*) - b_j = 0$ $j = 1, 2, \ldots, m$ (4.4c)

iv) If $\lambda_j^* = 0$, $g_j(x^*) - b_j \leq 0$ $j = 1, 2, \ldots, m$ (4.4d)

v) $x_i^* \geq 0$ (4.4e)

vi) $\lambda_j^* \geq 0$ (4.4f)

This theorem and its proof are included in one form or another in almost every test on optimization. One of the approaches to this theory is as a generalization of *Lagrange multipliers*. We consider the Lagrange function

$$L(x, \lambda) = F(x) - \sum_{j=1}^{m} \lambda_j (g_j(x) - b_j) \qquad (4.5)$$

Note that $g_j(x)$ is assumed to be less than or equal to b_j, therefore the negative sign is employed as $-\lambda_j(g_j(x) - b_j)$. If $g_j(x)$ is selected to be $\geq b_j$, then the positive sign should be used.

The Kuhn–Tucker conditions indicate that $L(x, \lambda)$ has a saddle point at the optimum. That is, $L(x^*, \lambda^*)$ is a maximum with respect to x and a minimum with respect to λ.

$$L(x, \lambda^*) \leq L(x^*, \lambda^*) \leq L(x^*, \lambda) \qquad (4.6)$$

The conditions also set forth a *complementary slackness principle* similar to that of linear programming. We must also note that the Kuhn–Tucker conditions are necessary but not sufficient. If any of the tests fail, the point is not optimal. However, only if additional conditions are imposed can we claim that they are sufficient to guarantee an optimum. For example, if $F(x)$ is concave and the constraint set is convex, the conditions are also sufficient for x^* to be the solution.

4.3   Characteristics of nonlinear optimization

We mentioned above that it is possible for an optimization problem to have its solution at a point which is interior to the constraint set when the objective function is nonlinear. At least in the neighbourhood of this solution, we may act as if there were no constraints existing. This certainly simplifies the problem although finding a maximum of a function of many variables is far from being a trivial effort. Of the many varieties of algorithms which are available to solve the unconstrained problems, we can distinguish three general types: direct search which makes use of function values only; gradient methods which use first derivatives; and second derivative or 'Newton-like' methods. Actually, the lines which separate these techniques are blurred by the fact that derivatives can be approximated by differences.

4.3.1 Description of unconstrained minimization

The unconstrained minimization problem is a very important case. It is the origin of the basic theory underlying nonlinear programming methods [5]. Many constrained problems are solved by solving a sequence of unconstrained problems. The following gives the outline of the unconstrained minimization problem:

$$\left. \begin{array}{l} \text{Minimize} \quad F(x) \\ \text{Subject to} \quad \overline{x}_i \geq x_i \geq \underline{x}_i \end{array} \right\} \tag{4.7}$$

Because only side constraints are considered we call it a *quasi-constrained problem*. Applying the Kuhn–Tucker (KC) condition yields

$$\frac{\partial F(x)}{\partial x_i} = 0 \quad \text{if } \underline{x}_i \leq x_i \leq \overline{x}_i \tag{4.8}$$

$$\frac{\partial F(x)}{\partial x_i} > 0 \quad \text{if } x_i = \underline{x}_i \tag{4.9}$$

$$\frac{\partial F(x)}{\partial x_i} < 0 \quad \text{if } x_i = \overline{x}_i \tag{4.10}$$

The problem in Eqs. (4.8) through (4.9) is called a *convex programming problem* when the functions $F(x)$ and $g_i(x)$ are convex. Then the feasible region is a convex set. The local solution is also global.

Two special cases of convex problems:

Case 1 – Quadratic *Programming Problem*

Minimizing a positive semi-definite quadratic form subject to linear constraints

$$\text{Minimize} \quad F(x) = \frac{1}{2} \{x\}^T [A] \{x\} - \{b\}^T \{x\} \tag{4.11}$$

$$\left. \begin{array}{l} \text{Subject to} \quad \sum_{i=1}^{n} c_{ij} x_i \geq c_j \\ \overline{x}_i \geq x_i \geq \underline{x}_i \quad i = 1, 2, \ldots, n \end{array} \right\} \tag{4.12}$$

This problem mainly serves as a reference for establishing the convergence properties of the various algorithms available. In the unconstrained case the quadratic problem is especially important and it leads to many fundamental concepts such as conjugate directions and quadratic termination.

Case 2 – Separable *Programming Problem*

In this problem duality results can be readily implemented into efficient algorithms

$$\text{Minimize} \quad F(x) = \sum_{i=1}^{n} F_i(x_i) \tag{4.13}$$

$$\left. \begin{array}{l} \text{Subject to} \quad g_j(x) = \sum_{i=1}^{n} g_{ji}(x_i) \geq 0 \quad j=1,2,\ldots,m \\ \bar{x}_i \geq x_i \geq \underline{x}_i \quad i=1,2,\ldots,n \end{array} \right\} \tag{4.14}$$

Each function $\{F_i(x_i), g_{ji}(x_i)\}$ depends only on the single variable x_i. Separable functions have several computationally important properties. In particular the *Hessian matrix* of such a function is diagonal. Unconstrained optimization will be discussed in more detail later in this book.

4.3.2 *Primal versus dual problem statement and dual solution algorithm*

Define the primal points satisfying the side constraints as

$$X = \{X = \underline{x}_i \leq x_i \leq \bar{x}_i \quad i=1,2,\ldots,n\} \tag{4.15}$$

Define a set of dual points satisfying the non-negativity conditions as

$$\Lambda = \{\lambda = \lambda_j \geq 0 \quad j=1,2,\ldots,m\} \tag{4.16}$$

The primal problem is expressed as

$$\text{Minimize } F(x) \quad \text{for} \quad x \in X \tag{4.17}$$

$$\text{Subject to } g_j(x) \geq 0 \quad j=1,2,\ldots,m \tag{4.18}$$

When $F(x)$ is strictly convex and $g_j(x)$ are concave, there exists a unique dual problem. The solution of the dual problem can be obtained through a two-phase procedure as follows:

$$\begin{array}{cc} \max & \min & L(x,\lambda) \\ \lambda \in \Lambda & x \in X \end{array} \tag{4.19}$$

where

$$L(x,\lambda) = F(x) - \sum_{j=1}^{m} \lambda_j g_j(x) \tag{4.20}$$

The dual problem can be written

$$\left.\begin{array}{ll} \text{Maximize} & l(\lambda) \\ \text{Subject to} & \lambda_j \geq 0 \quad j = 1, 2, \ldots, m \end{array}\right\} \tag{4.21}$$

and the dual function is

$$l(\lambda) = \min_{x \in X} L(x,\lambda) = F[x(\lambda)] - \sum_{j=1}^{m} \lambda_j g_j[x(\lambda)] \tag{4.22}$$

In the primal problem there are n variables, m general constraints and $2n$ side constraints, while in the dual problem there are only m variables and m non-negativity constraints. The relationships may be summarized as

	Primal problem	Dual problem
Variables	n	m
General constraints	m	m
Side constraints	$2n$	

The dual problem is therefore quasi-unconstrained and can be solved by using the algorithm of the *steepest descent method*. This requires the gradient of the dual function to be known for which $\nabla l(\lambda)$ can easily be obtained from primal constraints of Eq. (4.22) as

$$\begin{aligned}
\frac{\partial l(\lambda)}{\partial \lambda_j} &= \sum_{i=1}^{n} \frac{\partial F[x(\lambda)]}{\partial x_i} \frac{\partial x_i}{\partial \lambda_j} - \sum_{j=1}^{m} \lambda_j \sum_{i=1}^{n} \frac{\partial g_j[x(\lambda)]}{\partial x_i} \frac{\partial x_i}{\partial \lambda_j} - g_j[x(\lambda)] \\
&= \sum_{i=1}^{n} \frac{\partial F[x(\lambda)]}{\partial x_i} \frac{\partial x_i}{\partial \lambda_j} - \sum_{i=1}^{n} \sum_{j=1}^{m} \lambda_j \frac{\partial g_j[x(\lambda)]}{\partial x_i} \frac{\partial x_i}{\partial \lambda_j} - g_j[x(\lambda)] \\
&= \sum_{i=1}^{n} \sum_{j=1}^{m} \lambda_j \frac{\partial g_j[x(\lambda)]}{\partial x_i} \frac{\partial x_i}{\partial \lambda_j} - \sum_{i=1}^{n} \sum_{j=1}^{m} \lambda_j \frac{\partial g_j[x(\lambda)]}{\partial x_i} \frac{\partial x_i}{\partial \lambda_j} - g_j[x(\lambda)] \\
&= -g_j[x(\lambda)] \tag{4.23}
\end{aligned}$$

where $x(\lambda)$ denotes the primal point which minimizes $L(x,\lambda)$ over X for a given λ. Once the primal constraints are evaluated the first derivatives for the dual problem are available without additional computation.

$$\frac{\partial^2 l}{\partial \lambda_j \partial \lambda_k} = -\frac{\partial g_j [x(\lambda)]}{\partial \lambda_k} \tag{4.24}$$

We could use a *Newton-type method* to solve the dual problem. This needs the second derivatives of the dual function to be evaluated. Let N represent the matrix of the primal constraints, that is,

$$N = [\nabla g_1, \ldots, \nabla g_m] \tag{4.25}$$

Then the Hessian of the dual function is given by

$$H = \nabla^2 l = -N^T \tilde{G}^{-1} N \tag{4.26}$$

where $\tilde{G}$ is the Hessian matrix of the Lagrangian function, restricted to the free primal variables. A primal variable is said to be free if it has not taken on its lower- or upper-bound value. Introducing the set of indices as

$$\tilde{I} = \{i : \underline{x}_i \le x_i \le \overline{x}_i\} \tag{4.27}$$

then the matrix $\tilde{G}$ may be expressed as

$$\tilde{G}_{ik} = \begin{cases} \dfrac{\partial^2 L}{\partial x_i \partial x_k} & \text{for } ik \in \tilde{I} \\ 0 & \text{otherwise} \end{cases} \tag{4.28}$$

To compute $l(\lambda)$ it is necessary to find the x that minimizes the Lagrangian in Eq. (4.22). For the separate problem the dual function is

$$l(\lambda) = \sum_{i=1}^{n} \left\{ \min_{\underline{x}_i \le x_i \le \overline{x}_i} \left[F_i(x_i) - \sum_{j=1}^{m} \lambda_j g_{ji}(x_i) \right] \right\} \tag{4.29}$$

The Hessian matrix is

$$\frac{\partial^2 l}{\partial \lambda_j \partial \lambda_k} = -\sum_{i \in \tilde{I}} \frac{n_{ij} n_{ik}}{G_{ii}} \tag{4.30}$$

n_{ij} are the elements of N

$$n_{ij} = \frac{\partial g_{ji}}{\partial x_i} \tag{4.30a}$$

G_{ii} are the diagonal elements of G as

$$G_{ii} = \frac{\partial^2 L}{\partial x_i^2} \quad G_{ij} = 0 \quad i \neq j \tag{4.30b}$$

Example 4.3.1 – Solve the following problem by using dual formulation

Minimize $\quad F(x) = x_1^2 + 2x_1 + x_2^2$

Subject to $\quad g(x) = x_1 + x_2 - 3 \geq 0; \quad 1 \leq x_1 \leq 2; \quad 1 \leq x_2 \leq 4$

Solution:

For the given problem, we can write

$$L(x, \lambda) = F(x) - \sum_{j=1}^{m} \lambda_j - g_j(x) = \left(x_1^2 + 2x_1 + x_2^2\right) - \lambda(x_1 + x_2 - 3) \tag{a}$$

and express it as the separable problem

$$F_1(x_1) = x_1^2 + 2x_1; \quad F_2(x_2) = x_2^2 \tag{b}$$

for which the dual function is

$$l(\lambda) = \min\left\{x_1^2 + 2x_1 - \lambda x_1\right\} + \min\left\{x_2^2 - \lambda(x_2)\right\} \tag{c}$$

In order to find the limits, we calculate the following

$$\frac{\partial l}{\partial x_1} = 2x_1 + 2 - \lambda = 0; \quad \frac{\partial l}{\partial x_2} = 2x_2 - \lambda = 0 \tag{d, e}$$

from which

$$x_1 = 1, \quad \lambda = 4; \quad x_1 = 2, \quad \lambda = 6 \tag{f, g}$$

$$x_2 = 1, \quad \lambda = 2; \quad x_2 = 4, \quad \lambda = 8 \tag{h, i}$$

Therefore

$$0 \leq \lambda < 2, \quad x_1 = x_2 = 1; \quad 2 \leq \lambda < 4, \quad x_1 = 1 \text{ and } x_2 = \frac{\lambda}{2} \tag{j, k}$$

$$4 \leq \lambda < 6, \quad x_1 = \frac{\lambda}{2} - 1 \text{ and } x_2 = \frac{\lambda}{2}; \quad 6 \leq \lambda \leq 8 \, x_1 = 2 \text{ and } x_2 = \frac{\lambda}{2}$$

$$\tag{l, m}$$

$$8 < \lambda; \quad x_1 = 2 \text{ and } x_2 = 4 \tag{n}$$

Plugging the above into $L(x, \lambda)$ yields $l(\lambda)$; for instance, when $x_1 = x_2 = 1$ of Eq. (j), we have

$$l(\lambda) = \left(1^2 + 2(1) + 1^2\right) - \lambda(1 + 1 + 3) = 4 + \lambda. \text{ Thus}$$

$$l(\lambda) = \begin{cases} 4 + \lambda & 0 \le \lambda < 2 \\ 3 + 2\lambda - \dfrac{\lambda^2}{4} & 2 \le \lambda < 4 \\ -\dfrac{\lambda^2}{2} + 4\lambda - 1 & 4 \le \lambda < 6 \\ -\dfrac{\lambda^2}{2} + \lambda + 8 & 6 \le \lambda \le 8 \\ 24 - 3\lambda & \lambda > 8 \end{cases} \tag{o}$$

$$\frac{\partial l(\lambda)}{\partial \lambda} = \begin{cases} 1 & 0 \le \lambda < 2 \\ 2 - \dfrac{\lambda}{2} & 2 \le \lambda < 4 \\ -\lambda + 4 & 4 \le \lambda < 6 \\ -\lambda + 1 & 6 \le \lambda \le 8 \\ -3 & \lambda > 8 \end{cases} \tag{p}$$

For $\partial l/\partial \lambda = 0$ and $\lambda \ge 0$, we have $\lambda = 4$ and

$$l(\lambda) = -\frac{\lambda^2}{2} + 4\lambda - 1 = -\frac{16}{2} + 4(4) - 1 = 7 \tag{q}$$

Thus, the optimal results are

$$x_1 = \frac{\lambda}{2} - 1 = 1; \quad x_2 = \frac{\lambda}{2} = 2; \quad F(x) = 1^2 + 2(1) + 2^2 = 7; \quad F(x) = l(\lambda) \tag{r}$$

4.4 Description of the descent approach

Nonlinear optimization methods can be generally classified into unconstrained optimization methods and constrained optimization methods. Structural design problems are commonly formulated for constrained optimization which, however, can be solved by employing unconstrained optimization algorithms in the constrained optimization formulation. Within this Section 4.4, only the algorithms are selected and discussed in order to prepare the reader for using them in later chapters such as Chapters 5 and 11. Various topics in unconstrained optimization are also covered in Chapter 6.

Because structural design involves numerous variables, most algorithms for solving minimization problems are iterative. Therefore, an initial estimation of the solution is needed which is then followed by iterations. Nearly all unconstrained minimization methods are iterative such as the descent methods described hereafter for a typical cycle between kth and $(k+1)$th.

$$F(x_i^{(k+1)}) < F(x_i^{(k)}) \quad i = 1, 2, \ldots, n \tag{4.31}$$

Minimize $F(x)$ along successive search direction $s^{(k)}$, so that

$$x_i^{(k+1)} = x_i^{(k)} + \alpha^{(k)} s_i^{(k)} \tag{4.32a}$$

or

$$\Delta x_i^{(k)} = \alpha^{(k)} s_i^{(k)} \tag{4.32b}$$

$s^{(k)}$ is a *descent direction*. For a sufficiently small quantity $\alpha > 0$, the following equation should hold

$$F(x_i^{(k)} + \alpha s_i^{(k)}) < F(x_i^{(k)}) \tag{4.33}$$

Let $F(x)$ be differentiable, then using a linear Taylor's expansion to approximate the left side of Eq. (4.33), we have

$$\{s^{(k)}\}^T \{\nabla F(x^{(k)})\} < 0 \tag{4.34}$$

where $\nabla F(x_i^{(k)})$ is the gradient of $F(x_i)$ at $x_i^{(k)}$. Eq. (4.34) is called the *descent condition* for which the descent direction is also called the *downhill direction*, i.e. trying to reach the bottom of a hill from a high point. Consequently, the method is called the *descent method*. Since $\{F(x^{(k)})\}$ is a known quantity, the search direction $\{s^{(k)}\}$ must be calculated to satisfy Eq. (4.34), by using iteration procedures. Each iteration involves two parts: (A) calculate the downhill direction $s^{(k)}$ at $x^{(k)}$; (B) evaluate a step-length $\alpha^{(k)}$, then $x^{(k+1)} = x^{(k)} + \alpha^{(k)} s^{(k)}$. Most often the step-length $\alpha^{(k)}$ is estimated so as to minimize the objective function along the search direction $s^{(k)}$. The two parts in an iteration are discussed in detail in Sections 4.4.1 and 4.4.2.

4.4.1 Evaluation of step size $\alpha^{(k)}$

For an optimization problem with several variables, the direction-finding problem must be solved first. Then a step size must be determined by searching for the minimum of the cost function along the search direction. The process is often called the *one-dimensional search* or *line search* problem. The question is how the one-dimensional line search works for multi-dimensional problems.

Let us consider that a search direction $s_i^{(k)}$ has been found. Then Eq. (4.32b) indicates that scalar $\alpha^{(k)}$ is the only unknown which, for convenience, is replaced by α. The cost function $F(x)$ can be expressed as $F(x_i^{(k+1)}) = F(x_i^k + \alpha s_i^{(k)})$. Since $s^{(k)}$ is known, the right side becomes a function of the scalar parameter α only. Thus, the *one-dimensional minimization problem* is established to find $\alpha_k = \alpha$ such that $F(\alpha)$ is minimized as follows:

$$F(x_i^{(k+1)}) = F(x_i^{(k)} + \alpha s_i^{(k)}) = \phi(\alpha) \tag{4.35a}$$

where $\phi(\alpha)$ is the new function with α as the only independent variable. Note that at $\alpha = 0, F(0) = F(x_i^{(k)})$ which is the current value of the cost function. If $x_i^{(k)}$ is not a minimum point, then it is possible to find a descent direction $s_i^{(k)}$ at the point and reduce the cost function further. Recall that a small move along $s_i^{(k)}$ reduces the cost function. Therefore, using Eqs. (4.31) and (4.35a), the descent condition for the cost function can be expressed as the inequality:

$$\phi(\alpha) < \phi(0) \tag{4.35b}$$

Let $\phi(\alpha)$ denote the objective function along the line $s_i^{(k)}$

$$\phi(\alpha) = F(x_i^{(k)} + \alpha s_i^{(k)}) \tag{4.36a}$$

α should satisfy

$$F(x_i^{(k+1)}) = \min \phi(\alpha) \tag{4.36b}$$

Thus, in each iteration

$$\phi'(\alpha) = 0 \tag{4.37}$$

which should be solved as

$$\phi'(\alpha) = \sum_{i=1}^{n} \frac{\partial F}{\partial x_i} \{(x^{(k)} + \alpha s^{(k)})\} \frac{\partial (x_i^{(k)} + \alpha s_i^{(k)})}{\partial \alpha} = 0 \tag{4.38a}$$

$$\{\nabla F(x^{(k)} + \alpha s^{(k)})\}^T \{s^{(k)}\} = 0 \tag{4.38b}$$

and

$$x^{(k+1)} = x^{(k)} + \alpha s^{(k)} \tag{4.38c}$$

Therefore

$$\{\nabla F^{(k+1)}\}^T \{s^{(k)}\} = 0 \tag{4.39}$$

which indicates that the gradient $\nabla F^{(k+1)}$ at $x^{(k+1)}$ is orthogonal to the search direction $s^{(k)}$. Thus, $s^{(k)}$ is a target for the contour of F on which $F(x) = F(x^{(k+1)})$ as shown in Fig. 4.3. One may observe that

$$\left.\begin{array}{ll} \phi'(\alpha) < 0 & \text{if } \alpha < \alpha^{(k)} \\ \phi'(\alpha) > 0 & \text{if } \alpha > \alpha^{(k)} \end{array}\right\} \tag{4.40}$$

which is very useful to trap the optimal step-length $\alpha^{(k)}$. The process of finding $\alpha^{(k)}$ by estimating α as a minimum of $\phi(\alpha)$ is called *line search* as

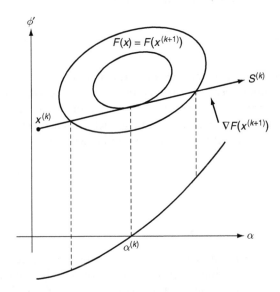

Figure 4.3 Relationships among $\alpha^{(k)}, \nabla F\left(x^{(k+1)}\right)$ and $x^{(k)}$.

mentioned earlier. Most line searches are carried out by using iteration procedures which are terminated when the convergence criteria are met. Since this is a simple function of α, we can also use necessary and sufficient conditions to solve for the optimum step length. Therefore, the second derivative of $\phi(\alpha)$ may be expressed as

$$\phi''(\alpha) = s^{(k)T} H[x^{(k)} + \alpha s^{(k)}] s^{(k)} \tag{4.41}$$

where H denotes the Hessian of the objective function. More discussions on the search are included in Chapter 6.

4.4.2   *Calculation of downhill direction*

In Section 4.4.1 it is assumed that a search direction in the design space was known before tackling the problem of step-size determination. Now the question is how to determine the search direction s. Several methods are available for determining a descent direction, among them the *steepest descent method* or the *gradient method* is the simplest, and probably the best known numerical method for unconstrained optimization. Presented herein is the steepest descent method which is a *first-order method* since only the gradient (that is why the method is also called the gradient method) of the cost function is calculated and used to evaluate the search direction. The *second-order methods* using the Hessian of the function can also be used in determining the search direction and are presented in Chapter 6.

The gradient of the cost function may be expressed as

$$\{\nabla F(x^{(k)})\} = \left[\frac{\partial F(x^{(k)})}{\partial x_i}\right]^T \tag{4.42}$$

which reveals a very important property that the gradient at a point x points in the direction of maximum increase of the cost function. Any small move in the negative gradient direction will result in the maximum local rate of decrease of the cost function. The negative gradient vector then represents a *direction of steepest descent* for the cost function and is written as

$$\{s^{(k)}\} = -\{F(x^{(k)})\} \tag{4.43}$$

which yields

$$\{s^{(k)}\}^T\{\nabla F(x^{(k)})\} < 0 \quad \text{if} \quad \{\nabla F(x^{(k)})\} \neq 0 \tag{4.44}$$

So, from the point $x^{(k)}$, we search along the direction of negative gradient to a minimum point $x^{(k+1)}$ on this line. It converges fast for nearly circular contours. If the contours are very elongated, the convergence will be slow as shown in Fig. 4.4. Note that Eqs. (4.43) and (4.44) can be expressed in vector form as

$$\{s\}^T\{\nabla F\} = -||\nabla F||^2 < 0 \tag{4.45}$$

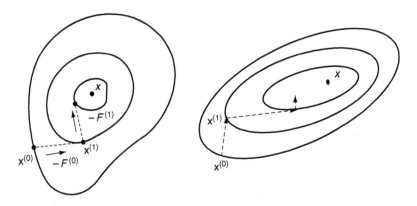

Figure 4.4 Convergence behaviour of steepest descent method.

4.5   Optimization strategies for steepest descent method

Based on Section 4.4.1, the optimization strategies are presented here in detail which can then be directly applied in Chapter 5. The strategies are divided into three distinct parts: Sections 4.5.1 through 4.5.3 [6].

4.5.1 Taking a step

When the constraint equations are evaluated and it is found that none are violated or bounded, a step can be taken in the direction of the negative gradient of the cost function as $\{d^{(k)}\} = \{\nabla F(x^{(k)})\}$.

The new set of design variables is shown in Eq. (4.38c) as

$$x^{(k+1)} = x^{(k)} + \alpha s^{(k)}$$

in which $\{s\} = -\{\nabla F\}$ (also $\{d\}$), and α can be calculated for a given percentage of weight or cost reduction. Based on Eqs. (4.35a) and (4.36a), let $F(x^{(k)} + \alpha s)$ be approximated by a linear function

$$F(\alpha) \cong a + b\alpha \tag{4.46}$$

when α is zero, the function represents the initial weight of the structure, which is known. This will be one 'boundary condition' in the solution for a trial α. In equation form

$$\overline{F} \equiv F(0) = F(x^{(k)}) \tag{4.47}$$

As shown in Eqs. (4.36a) through (4.39), the change in weight (herein 'weight' represents the cost function) with respect to α can be expressed in general form as

$$\overline{F} = \left. \frac{dF}{d\alpha} \right|_{\alpha=0} = \{s\}^T \{\nabla F\} \tag{4.48}$$

The new weight can now be approximated by a first-order Taylor expansion

$$F(\alpha) \cong \overline{F}'(\alpha) + \overline{F} \tag{4.49}$$

This new weight should be a specified percentage smaller than the initial weight; so select some α_t such that

$$\overline{F} + \overline{F}'\alpha_t = F_o - \mu F_o \tag{4.50}$$

where μ is the fractional reduction in the weight. Solving for α_t

$$\alpha_t = -\frac{\mu F_o}{\{s\}^T \{\nabla F\}} \tag{4.51}$$

$\{s\}$ in Eq. (4.51) will be, as noted above, equal to the negative weight gradient, for an unbounded move. If the design, $x^{(k)}$, is bounded then $\{s\}$ will be a new feasible direction vector which will be discussed in Section 4.5.3.

If {s} is nearly perpendicular to −{∇F}, the direction will need to be magnified in order to produce the desired decrease in weight. If {s} is nearly parallel to −{∇F}, then α_t will obviously be smaller. The new set of design variables can now be compared with the side constraints, the limits on the design variables, to see if any are violated; α_t is merely reduced in the {s} direction so that this variable is bounded. Obviously in this case the decrease in the weight will be less than the desired percentage. At this time a new analysis is made and the procedure outlined above is again carried out.

4.5.2   Correcting the step

Should it be found, after reanalyzing the new design, that one or more response quantities have been violated, α_t will need to be reduced so that the most violated quantity is just equal to its limit. The desired reduction can be handled in one of two ways. The first approach is to assume the response to be a linear function of α, as shown in Fig. 4.5. At α equal to zero, the value of the response quantity is smaller than the present limit, causing the value of the constraint to be negative because of the violated constraint. At α equal to α_t, the value of the violated constraint will be a positive number. A general rule, then, can be observed which is valuable in the logic of computer programming to distinguish between feasible and violated designs. The feasible design results in all constraint calculations being negative, whereas in a violated design one or more of these computations result in positive numbers.

It is desired to find the value α_l that makes g_j equal to zero. In order to make the constraint easier to 'hit', an allowable tolerance is assumed to be ε, and thus a bounded constraint is defined by

$$-\varepsilon \le g_j \le 0 \tag{4.52}$$

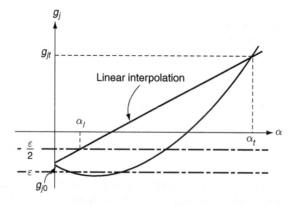

Figure 4.5 Violated constraint value plotted as a function of step length, α.

Given the two known values of the most violated constraint

$$g_{jo} \equiv g_j(x^{(k)}) < 0 \tag{4.53}$$

$$g_{jt} \equiv g_j(x^{(k)} + \alpha_t s) > 0 \tag{4.54}$$

as noted above, approximate the constraint by a linear function of α

$$g_j \cong a + b\alpha \tag{4.55}$$

from which

$$g_{jo} = a \tag{4.56}$$

and

$$g_{jt} = a + b\alpha_t \tag{4.57}$$

Solving for b

$$b = \frac{g_{jt} - g_{jo}}{\alpha_t} \tag{4.58}$$

and putting back into Eq. (4.57), we have

$$g_{jo} + \frac{g_{jt} - g_{jo}}{\alpha_t}\alpha_I = -\frac{\varepsilon}{2} \tag{4.59}$$

if the desired 'target' is the centre of the tolerance zone. Solving for α_I

$$\alpha_I = \frac{-g_{jo} - \dfrac{\varepsilon}{2}}{g_{jt} - g_{jo}}\alpha_t \tag{4.60}$$

The denominator in Eq. (4.60) will always be a positive number. If g_{jo} is less than $\varepsilon/2$ (absolute value-wise), which is the case when the previous design is bounded, then the numerator is also positive resulting in a negative value of α_I. Since a negative α_I would produce an increase in weight, a quadratic interpolation scheme must be used when the most violated constraint in the present design was bounded in the previous design. It is derived as follows:

Approximate $g_j(\alpha)$ by a quadratic

$$g_j \simeq a + b\alpha + c\alpha^2 \tag{4.61}$$

Since the boundary conditions are known from Eqs. (4.53) and (4.54), the slope is also known; the constants can be found. Thus

$$g_j(0) \equiv g_{jo} = a \tag{4.62}$$

$$\frac{dg_j}{d\alpha} = \{\nabla g_j(0)\}^T \{s\} \equiv g'_{jo} = b \tag{4.63}$$

$$g_j(\alpha_t) \equiv g_{jt} = a + b\alpha_t + c\alpha_t^2 \tag{4.64}$$

from which

$$c = \frac{g_{jt} - g_{jo} - g'_{jo}\alpha_t}{\alpha_t^2} \tag{4.65}$$

α_I can be obtained by setting g equal to $-\varepsilon/2$, which is the interpolating target.

$$g_j \cong a + b\alpha_I + c\alpha_I^2 = -\frac{\varepsilon}{2} \tag{4.66}$$

from which α_I can be obtained in terms of a, b and c

$$\alpha_I = \frac{-b + \left[b^2 - 4\left(a + \frac{\varepsilon}{2}\right)c\right]^{1/2}}{2c} \tag{4.67}$$

4.5.3   Finding a direction

After a successful interpolation has been made, all response quantities are less than the present limits and at least one will be bounded. The structure must be redesigned so that the weight (cost function) is decreased and the bounded constraint is made smaller. The problem is to find a new feasible direction $\{s\}$ to be used in Eq. (4.38c) which satisfies the condition as shown in Fig. 4.6. In order to satisfy the conditions imposed on $\{s\}^{(k+1)}$ which must satisfy the geometric conditions shown in Fig. 4.6, that is, angle ϕ_1, between the gradient of $F(\{\nabla F\}$ or $\{s\}^{(k)})$ and $\{s\}^{(k+1)}$ will be tangent to the constraint resulting in an immediate violation for any finite value of α. These restrictions can be expressed mathematically as

$$\cos\phi_1 = \frac{\{s\}^{(k+1)T}\{\nabla F\}}{|s||\nabla g|} < 0 \tag{4.68}$$

and

$$\cos\phi_2 = \frac{\{s\}^{(k+1)T}\{\nabla g\}}{|s||\nabla g|} < 0 \tag{4.69}$$

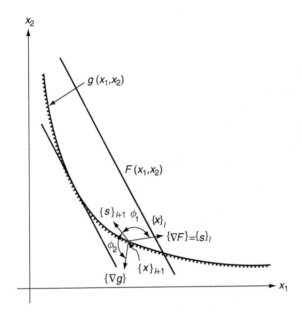

Figure 4.6 Condition for finding a new feasible direction.

If $\{s\}^{(k+1)}$, $\{\nabla F\}$ and $\{\nabla g\}$ are normalized ($\{s\}^T\{s\} = 1, \{\nabla g\}^T\{\nabla g\} = 1$) then $|s|$, $|\nabla g|$ and $|\nabla F|$ will be unity reducing Eqs. (4.68) and (4.69) to

$$\{s\}^{(k+1)T}\{\nabla g\} < 0 \qquad\qquad (4.70)$$

and

$$\{s\}^{(k+1)T}\{\nabla F\} < 0 \qquad\qquad (4.71)$$

Any random choice of $\{s\}^{(k+1)}$ that satisfies Eqs. (4.70) and (4.71) is sufficient to improve $\{x\}^{k+1}$. There are, however, methods to find $\{s\}^{k+1}$ which will lead to a minimum in the least number of steps. One of them is presented in the following paragraph.

Let Eqs. (4.70) and (4.71) be modified as expressed in Eqs. (4.72) and (4.73), respectively,

$$\{s\}^T\{\nabla g\}_j + \{\theta\}_j\beta \le 0 \quad j \text{ contained in } J \qquad\qquad (4.72)$$

$$\{s\}^T\{\nabla F\} + \beta \le 0 \quad \text{where } \beta \to \beta_{\max} \qquad\qquad (4.73)$$

where j is the number of bounded constraints and J is the total number of constraints. Also, $\{s\}$ must be bounded so that it is not forced to zero as β is allowed to increase. This is accomplished through a normalization process

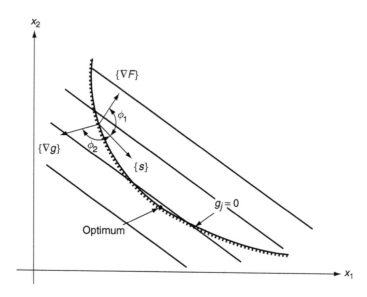

Figure 4.7 Obtuse angles which must exist between gradient vectors and the feasible
direction vector.

which forces the components of $\{s\}$ to lie between $+1$ and -1. All the elements of $\{\theta\}_j$ are unity for general curved constraints. It can be seen from
Fig. 4.7 why Eqs. (4.72) and (4.73) must hold. $\{s\}$, the feasible direction
vector, must point away from the constraint surface, $g_j = 0$, so that for some
finite α, although perhaps small, g_j is not violated. Geometrically the angle
ϕ_2 must be larger than $90°$, since $\phi_2 = 90°$ would allow $\{s\}$ to lie along the
tangent to $g_j = 0$ which would result in a violated constraint for any positive
value of α. From Eq. (4.72)

$$\{s\}^T\{\nabla g\}_j < 0$$

Eq. (4.72) ensures this to be true if $\{\theta\}_j$ and β are positive quantities. The
components of $\{\theta\}_j$ are chosen to be either 0 or 1; zero for linear constraints
in which case ϕ_2 may go to $90°$, or 1 for nonlinear constraints. The $\{\theta\}_j$
vector tends to 'push' $\{s\}$ away from the constraint as much as necessary
so that a relatively long step can be taken. For very curved constraints, $\{\theta\}_j$
may need to be increased beyond 1, but in this work it was not found to be
necessary. Similarly ϕ_1 in Fig. 4.7 is required to be greater than $90°$ so that
the weight is decreased during the step α. According to Eqs. (4.43) as well
as (4.73)

$$\{s\}^T\{\nabla F\} < 0$$

Again, Eq. (4.73) ensures the satisfaction of Eq. (4.34) as long β is positive. β is chosen to be as large as possible while still satisfying Eqs. (4.72)

and (4.73). In this light Eqs. (4.72) and (4.73) describe the classical linear programming problem where β is maximized subject to the constraints shown. The result will be the very best feasible direction vector, $\{s\}$, for the combination of constraint and weight gradients in the given stage of the design. Other methods have been proposed for finding $\{s\}$, but it is thought by many that the linear-programming approach is the most efficient and straightforward. By observation of Eqs. (4.72) and (4.73) it is obvious that for the condition shown in Fig. 4.7 as the optimum weight is approached, $\{\nabla F\}$ and $\{\nabla g\}$ will become collinear, i.e. lie in a straight line. When this is the case, β will approach signifying the optimum solution. Of course, in the finite arithmetic of the digital computer, β will never exactly equal zero, so that in each step of the computer program it is necessary to compare β with a present limit ultimately resulting in a termination of the program.

If two or more constraints are active when the direction-finding problem is encountered, one of the two situations depicted in Figs. 4.8 and 4.9 will result. The figures can be thought of, for generality's sake, as a two-dimensional subspace defined by $\{\nabla g\}_1$ and $\{\nabla g\}_2$. It can be seen from Fig. 4.8 that there may exist some vector, $\{s\}$, which satisfies Eqs. (4.71) and (4.72), since it is possible to produce a direction vector which makes an obtuse angle with $\{\nabla F\}$ and $\{\nabla g\}_i$. Hence, even a local minimum is not reached. This is not the case, however, in Fig. 4.9, because any proposed $\{s\}$ that makes an obtuse angle with $\{\nabla F\}$ will result in an acute angle with $\{\nabla g\}_2$. And, since no feasible direction vector can be found, at least a local minimum has been reached. If convex constraints are involved, then the global minimum has also been found.

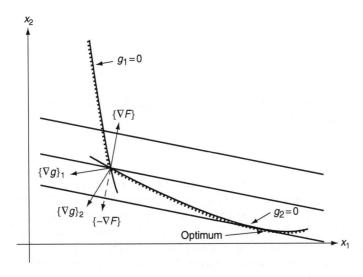

Figure 4.8 Non-optimum constraint intersection: new feasible direction vector exists.

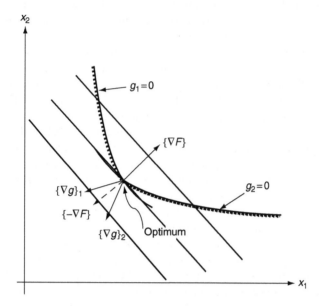

Figure 4.9 Optimum constraint intersection: no feasible direction exists and Kuhn–Tucker conditions for minimum are satisfied.

What is obvious geometrically is also confirmed theoretically by the well-known Kuhn–Tucker conditions for a relative minimum, which are

$$\frac{\delta W}{\partial \delta_i} + \sum_j \lambda_j \frac{\partial g_j}{\partial \delta_i} = 0 \qquad \begin{matrix} i = 1, \dots, n \\ j = 1, \dots, \text{no. of active constraints} \end{matrix} \tag{4.74}$$

and

$$\lambda_j \geq 0 \tag{4.75}$$

where λ_j is a set of scalar multipliers. For the specific, two-active-constraint condition, Eq. (4.74) reduces to

$$-\{\nabla W\} = \lambda_1 \{\nabla g\}_1 + \lambda_2 \{\nabla g\}_2 \tag{4.76}$$

Only for the case shown in Fig. 4.9 will both λ_1 and λ_2 be positive since in this case $\{\nabla F\}$ can be expressed as a non-negative linear combination of $\{\nabla g\}_1$ and $\{\nabla g\}_2$. The convenient aspect of the feasible-direction approach is that, due to this theoretical link, as soon as the direction-finding portion of the design method has resulted in a *no feasible direction* solution, then the Kuhn–Tucker conditions are also satisfied and the minimum weight design has been obtained.

Example 4.5.1 – Let Fig. 4.6 be represented by

$$F(x) = 2x_1 + x_2, \text{ and } g(x) = x_1 x_2 - 1 = 0 \qquad \text{(a, b)}$$

Find the minimization problem by using (A) the Lagrange-multiplier method and (B) the descent method.

(A) The Lagrange function is

$$L = F(x) + \lambda g(x) \qquad \text{(c)}$$

For the three unknowns problem, we can find the solution by using the following condition

$$\frac{\partial L}{\partial \lambda} = x_1 x_2 - 1 = 0 \qquad \text{(d)}$$

$$\frac{\partial L}{\partial x_1} = 2 + \lambda x_2 = 0 \qquad \text{(e)}$$

$$\frac{\partial L}{\partial x_2} = 1 + \lambda x_1 = 0 \qquad \text{(f)}$$

Solving Eqs. (d), (e) and (f)

$$x_1 = 0.707, \quad x_2 = 1.414, \quad \lambda = 1.414 \qquad \text{(g)}$$

Substituting x_1 and x_2 into the given objective function gives the optimal solution as

$$F_{min} = 2.828 \qquad \text{(h)}$$

(B) The descent method

Since the method is based on an iteration procedure, let us assume, at the kth iteration, $\alpha = 0.293$, $\{x\}^{(k)} = \left\{ \begin{matrix} 2 \\ 1 \end{matrix} \right\}$, then

$$\{x\}^{(k+1)} = \left\{ \begin{matrix} 2 \\ 1 \end{matrix} \right\} + 0.293 \left\{ \begin{matrix} -2 \\ -1 \end{matrix} \right\} = \left\{ \begin{matrix} 1.414 \\ 0.707 \end{matrix} \right\} \qquad \text{(i)}$$

$$\{s\}^{(k)} = \{-\nabla F\} = - \left\{ \begin{matrix} \partial F / \partial x_1 \\ \partial F / \partial x_2 \end{matrix} \right\} = \left\{ \begin{matrix} -2 \\ -1 \end{matrix} \right\} \qquad \text{(j)}$$

Eq. (j) is the negative of the gradient to the objective function. Now that the constraint has been reached, a new feasible direction $\{s\}^{(k+1)}$ must be found. $\{s\}^{(k+1)}$ can be calculated using information obtained from the gradient of F and from the constraint, g, at the point, $\{x\}^{(k+1)}$, which must be chosen so that the new point, $\{x\}^{(k+2)}$, will result in a smaller value of F and satisfaction of the given constraint. Any point that does not satisfy the constraint requirement is termed a point which violates the constraint. Let

$$\{s\}^{(k+1)} = \left\{ \begin{array}{c} 1 \\ -1 \end{array} \right\} \text{ and } \alpha = -0.707 \qquad (k)$$

then

$$\{x\}^{(k+2)} = \left\{ \begin{array}{c} 1.414 \\ 0.707 \end{array} \right\} - 0.707 \left\{ \begin{array}{c} 1 \\ -1 \end{array} \right\} = \left\{ \begin{array}{c} 0.707 \\ 1.414 \end{array} \right\} \qquad (l)$$

which is the minimum solution as obtained by Lagrange multipliers. A simple test to ensure that the minimum has been reached is to calculate $\{\nabla g\}$ at $\{x\}_{\min}$ and compare with $\{\nabla F\}$.

$$\nabla g(@\{x\}_{\min}) = \left\{ \begin{array}{c} x_2 \\ x_1 \end{array} \right\} = \left\{ \begin{array}{c} 1.414 \\ 0.707 \end{array} \right\} = 0.707 \left\{ \begin{array}{c} 2 \\ 1 \end{array} \right\} \qquad (m)$$

and

$$\nabla F(@\{x\}_{\min}) = \left\{ \begin{array}{c} 2 \\ 1 \end{array} \right\} \qquad (n)$$

Since both gradients are in the same direction, Eqs. (4.70) and (4.71) cannot be satisfied and $\{\nabla g\}$ and $\{\nabla F\}$ are collinear. Thus, a minimum result is ensured.

4.6 Gradient-projection method

4.6.1 Description of the method

The *gradient-projection method* is similar to the feasible-direction method based on the first-order information for calculations. However, the main difference is that the feasible-direction method requires the solution of linear-programming methods to search the descent direction, but the gradient-projection method doesn't need that approach. The basic procedures of the gradient-projection method are revealed in Fig. 4.10 and are further explained as follows:

(1) Choose an initial (feasible) point $\{A_0\}$ as a start point.
(2) If the point is inside the feasible region, the steepest descent direction for the cost function is used until a constraint boundary is encountered.

(3) When the point is on the boundary, a direction that is tangent to the constraint surface is used to change the design. This direction is computed by projecting the steepest descent direction for the cost function on the tangent plane.

(4) Since the direction is tangent to the constraint surface, the new point will be infeasible. A series of correction steps needs to be executed to reach the feasible plane.

The next two Sections 4.6.2 and 4.6.3 include the derivation of the *project matrix* [P] for single and multiple linear constraints, and discussion of general procedures for nonlinear constraints.

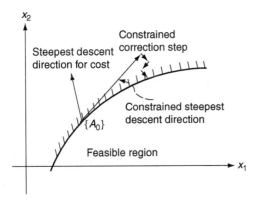

Figure 4.10 Description of the projection method.

4.6.2 *Projection matrix for single and multiple linear equality constraints*

Consider a simple case of minimizing $F(x)$ subject to a single linear equality constraint as

$$\{B\}^T \{x\} = d \tag{4.77}$$

given a feasible point $x^{(t)}$ at which the gradient is not equal to zero. We want to seek a search direction that lies on the constraint and is a descent direction; such a direction could be obtained by projecting the negative of $\{\nabla F(x^{(t)})\}$ perpendicularly on to the constraint surface. Recall that the statement $\{B\}^T\{s\} = 0$ implies that $\{s\}$ is a feasible direction (parallel to the constraint surface) and $\{B\}$ is perpendicular to the plane. If we scale the constraint so that $\|\{B\}\| = 1$, then the vector $\{B\}$ will be normal to the constraint surface; all vectors perpendicular to the constraint surface must be parallel to $\{B\}$. The component of $\{s\}$ that is parallel to the constraint surface $\{s''\}$ must satisfy $\{B\}^T\{s''\} = 0$. Since any vector can be written as a vector sum as

$$\{s\} = \{s'\} + \{s''\} \tag{4.78}$$

where $\{s'\} = r\{B\}$ and r is to be determined, $\{s''\}$ should satisfy $\{B\}^T\{s''\} = 0$. Multiplying $\{B\}^T$ on both sides of Eq. (4.78) yields

$$\{B\}^T\{s\} = \{B\}^T r\{B\} + \{B\}^T\{s''\} = r\left(\{B\}^T\{B\}\right) + 0 \tag{4.79}$$

from which

$$r = \frac{\{B\}^T\{s\}}{\{B\}^T\{B\}} = \left(\{B\}^T\{B\}\right)^{-1}\{B\}^T\{s\} \tag{4.80}$$

Substituting r in Eq. (4.80) into (4.78), we have

$$\{s\} = r\{B\} + \{s''\} = \{B\}\left(\{B\}^T\{B\}\right)^{-1}\{B\}^T\{s\} + \{s\} \tag{4.81}$$

solving for $\{s\}$

$$\{s\} = \{s\} - \{B\}\left(\{B\}^T\{B\}\right)^{-1}\{B\}^T\{s\} = \left([I] - \{B\}\left(\{B\}^T\{B\}\right)^{-1}\{B\}^T\right)\{s\} \tag{4.82}$$

where

$$[P] = \left([I] - \{B\}\left(\{B\}^T\{B\}\right)^{-1}\{B\}^T\right) \tag{4.83}$$

which is called the *projection matrix* that serves to project the vector $\{s\}$ into the plane of constraint along the direction of the normal $\{B\}$ to the constraint surface.

Eq. (4.83) can be extended to multiple constraints with m linear equations as

$$B_m^T x_m = d_m \quad m = 1, 2, \ldots, M \tag{4.84}$$

or

$$[A]\{x\} = [D] \tag{4.84a}$$

Following Eq. (4.78), we have

$$\{s\} = \{s'\} + \{s''\} = r[A] + \{s''\} \tag{4.85}$$

where $[A]$ is the matrix whose rows consist of the vector of B_m^T. Multiplying Eq. (4.85) by $[A]$ then solving for r, we have

$$r = ([A][A]^T)^{-1}[A]\{s\} \tag{4.86}$$

The projection matrix can be similarly found as

$$[P] = \left([I] - [A]^T ([A][A]^T)^{-1}[A]\right)\{s\} \tag{4.87}$$

Note that $[A][A]^T$ exists only if the vectors of B_m are independent; $[P]$ is symmetric and positive definite; and $\{s\} = -[P]\{\nabla F\}$ is a descent direction.

4.6.3 General procedures for nonlinear constraints

Observe that B_m is used to define a projection parallel to the constraint surface for a linear case, even though it can also define a direction orthogonal to the constraint surface. However, for a nonlinear case, both directions are needed; the former serves to search in a direction tangent to the constraint surface, and the latter which is the projection parallel to the constraint normals can help to define a direction for returning to the constraint surface. The general situation may be observed from Fig. 4.11 of a single-constraint case. $\nabla h\left(x^{(t)}\right)$ is the gradient vector at $x^{(t)}$, which may or may not always point towards the surface, labelled (I) and (II), respectively, in the figure. This problem, however, can be solved by finding the gradient of h to define a direction $\nabla h\left(F^{(t)}\right)$ that is likely to return to the surface (labelled III) with the algorithm as follows: let $I^{(t)}$ be the set of constraints active at $x^{(t)}$, and $F^{(t)}$ be the infeasible point in the direction $S^{(t)}$ from $x^{(t)}$, the general expression can be

$$s = [A]^T\left(F^{(t)}\right)r \tag{4.88}$$

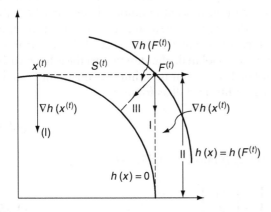

Figure 4.11 Optimization steps of a nonlinear constraint case.

where the column of $[A]^T (F^{(t)})$ comprises the gradients vector $\nabla h_m (F^{(t)})$, $m = 1, 2, \ldots, M$. Now, apply Newton's method to determine r by using the usual Taylor-series approximation

$$0 = h\left((F^{(t)}) + [A]^T r\right) = h\left(F^{(t)}\right) + [A]\left(F^{(t)}\right)[A]^T\left(F^{(t)}\right)r \qquad (4.89)$$

then use the chain rule

$$\nabla r h = \frac{dh}{dx}\frac{dx}{dr} \qquad (4.90)$$

Solving for r, we obtain

$$r = -\left([A][A]^T\right)^{-1}h \qquad (4.91)$$

Substituting Eq. (4.91) into (4.88), we have

$$F^{(t+1)} - F^{(t)} = s^{(t)} = -[A]^T\left([A][A]^T\right)^{-1}h \qquad (4.92a)$$

$$F^{(t+1)} = F^{(t)} - [A]^T\left([A][A]^T\right)^{-1}h \qquad (4.92b)$$

where $F^{(t+1)}$ and $F^{(t)}$ correspond to the current point and infeasible point, respectively. Apparently, the method requires a sequence of trial steps which are recommended as

$$t_j = \left(\frac{1}{2}\right)^j; \quad j = 0, 1, 2, 3, \ldots \qquad (4.93)$$

i.e. the sequence of trial step sizes is given as $j = 0$, $t_0 = (1/2)^0 = 1$; $j = 1$, $t_1 = (1/2)^1 = 1/2$; $j = 2$, $t_2 = (1/2)^2 = 1/4$; etc. Thus, we start with the trial step size as $t_0 = 1$. If a certain descent condition is not satisfied, the trial step is taken as half of the previous step size. During the constraint-correction steps, it must be ensured that $F(x^{(k+1)}) < F(x^{(k)})$. If one or more of the behaviour constraints are violated, the design variables have to be corrected until the most violated constraint is within the constraint bounds. We introduce a quadratic curve-fitting skill as follows:

$$g_k = ax^2 + bx + c \qquad (4.94)$$

Substituting $(x_1, g_1)\,(x_2, g_2)\,(x_3, g_3)$ into Eq. (4.94), we have

$$\begin{bmatrix} x_1^2 & x_1 & 1 \\ x_2^2 & x_2 & 1 \\ x_3^2 & x_3 & 1 \end{bmatrix} \begin{Bmatrix} a \\ b \\ c \end{Bmatrix} = \begin{Bmatrix} g_1 \\ g_2 \\ g_3 \end{Bmatrix} \qquad (4.95)$$

and then solve for a, b and c. Let

$$g_k = ax^2 + bx + c = \frac{\varepsilon}{2} \tag{4.96}$$

from which we solve for x, that is the correct value of the new step.

The gradient-projection algorithm can be summarized as:

Step 0 – Evaluate tall inequality constraints at $x^{(t)}$ to identify the active set

$$I^{(t)} = \left\{ j : g\left(x^{(t)}\right) \le \varepsilon_2; \quad j = 1, 2, \ldots, J \right\} \tag{4.97}$$

Step I – Calculate the projection matrix $[P]$, and the descent direction $\{s\} = -[P]\{\nabla F\}$

Step II – If $\|s^{(t)}\| > \varepsilon_1$, go to step III, otherwise calculate the multipliers

$$u = \left([A][A]^T\right)^{-1} [A]\{\nabla F\} \tag{4.98}$$

and find

$$u_m = \min \left\{ u_l : l \in I^{(t)} \right\} \tag{4.99}$$

If $|u_m| < \varepsilon_1$, then terminate. Otherwise, delete inequality constraint m from $I^{(t)}$ and go to step I.

Step III – Start with an initial step size α_0, determine $F(\alpha_0)$ and correct it, then back to Step I.

Note that the significant computation effort of this algorithm is the update of $\left([A][A]^T\right)^{-1}$ and $[P]$ for each time the active constraint set is modified. However, there is a way to avoid this problem which may be found in reference [7].

4.6.4 Numerical illustrations of structural design

Example 4.6.1 – Optimize the truss shown in Fig. 4.12 for which the design parameters are given:

$$\left. \begin{array}{l} E = 1 \times 10^4 \, \text{ksi} \\ \rho = 0.1 \, \text{lbs in.}^{-2} \\ A_{\min} = 0.1 \, \text{in.}^2 \\ \text{Stress constraint:} \, -25 \le \sigma \le 25 \, \text{ksi} \end{array} \right\} \tag{a}$$

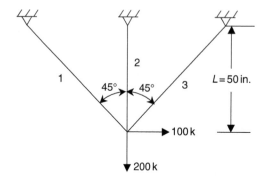

Figure 4.12 Truss for Examples 4.6.1 and 4.8.4.

Solution:

Following Eqs. (3.8) through (3.12), we have

$$[A_s] = \begin{bmatrix} \cos 45^\circ & 0 & -\cos 45^\circ \\ \sin 45^\circ & 1 & \sin 45^\circ \end{bmatrix}; \quad [S_s] = \begin{bmatrix} \dfrac{A_1 E}{L_1} & 0 & 0 \\ & \dfrac{A_2 E}{L_2} & 0 \\ \text{sym} & & \dfrac{A_3 E}{L_3} \end{bmatrix}$$

(b, c)

$$[K] = [A_s][S_s][A_s]^T = \begin{bmatrix} \dfrac{E}{2}\left(\dfrac{A_1}{L_1} + \dfrac{A_3}{L_3}\right) & \dfrac{E}{2}\left(\dfrac{A_1}{L_1} - \dfrac{A_3}{L_3}\right) \\ \dfrac{E}{2}\left(\dfrac{A_1}{L_1} - \dfrac{A_3}{L_3}\right) & \dfrac{E}{2}\left(\dfrac{A_1}{L_1} + \dfrac{A_3}{L_3}\right) + \dfrac{A_2 E}{L_2} \end{bmatrix};$$

$$\{P_s\} = \left\{ \begin{array}{c} 100 \\ 200 \end{array} \right\} k$$

(d, e)

where $L_1 = 50\sqrt{2}$ in., $L_2 = 50$ in., $L_3 = 50\sqrt{2}$ in. The displacement $\{X\}$, internal forces $\{F\}$ and the associated stress $\{\sigma\}$ are calculated from the following equations:

$$\{X\} = [K]^{-1}\{P_s\}; \quad \{F_s\} = [S_s][A_s]^T\{X\}; \quad \{\sigma\} = \left\{ \dfrac{F_s}{A_i} \right\}, \quad i = 1,2,3$$

(f, g, h)

Note that the original notations of static matrix and load matrix are added with subscript s as $[A_s]$, $[S_s]$ and $\{P_s\}$, respectively, for the purpose of avoiding the confusion of the notations of constraint equations $[A]$, feasible direction $\{s\}$, and projection matrix $[P]$ used in the method as shown in Eq. (4.87). The objective function is

$$F = \rho_1 A_1 l_1 + \rho_2 A_2 l_2 + \rho_3 A_3 l_3 = 5\sqrt{2}A_1 + 5A_2 + 5\sqrt{2}A_3 \tag{i}$$

Based on the initial given data, the member stresses are

$$\{\sigma\} = \begin{Bmatrix} 6.464 \\ 5.858 \\ -0.606 \end{Bmatrix} \text{ksi} \tag{j}$$

which reveal that all the constraints are not active, the descent direction is

$$\{s\} = -\{\nabla F\} = \begin{Bmatrix} -5\sqrt{2} \\ -5 \\ -5\sqrt{2} \end{Bmatrix} \tag{k}$$

and

$$\alpha_T = \frac{-\mu F_0}{\{S\}^T \{\nabla F\}} = \frac{-\mu(382.8)}{\begin{Bmatrix} 7.07 \\ 5 \\ 7.07 \end{Bmatrix}^T \begin{Bmatrix} 7.07 \\ 5 \\ 7.07 \end{Bmatrix}} = \frac{382.8}{125}\mu \tag{l}$$

where

$$\mu = \left| \frac{\sigma_{\text{actual}} - \sigma_{\text{allow}}}{\sigma_{\text{allow}}} \right| = \left| \frac{6.464 - 25}{25} \right| = 0.741424 \tag{m}$$

Let

$$\alpha_T = \frac{382.8}{125}\mu = 2.270537 \tag{n}$$

then

$$\begin{Bmatrix} A_1 \\ A_2 \\ A_3 \end{Bmatrix} = \begin{Bmatrix} 20 \\ 20 \\ 20 \end{Bmatrix} + 2.270537 \begin{Bmatrix} -7.07 \\ -5 \\ -7.07 \end{Bmatrix} = \begin{Bmatrix} 3.94488 \\ 8.64732 \\ 3.94488 \end{Bmatrix} \tag{o}$$

$$\begin{Bmatrix} \sigma_1 \\ \sigma_2 \\ \sigma_3 \end{Bmatrix} = \begin{Bmatrix} 26.6684 \\ 17.4875 \\ -9.18095 \end{Bmatrix} \leftarrow \text{violates the constraint and } (G_{TK}) = 1.6684$$

$$\tag{p}$$

Let

$$\alpha_T' = \frac{\alpha_T}{2} = 1.1352685 \tag{q}$$

then

$$\begin{Bmatrix} A_1 \\ A_2 \\ A_3 \end{Bmatrix} = \begin{Bmatrix} 11.9724 \\ 14.3237 \\ 11.9724 \end{Bmatrix} \tag{r}$$

Let $\alpha'' = \frac{3}{4}\alpha_T = 1.702903$, then

$$\begin{Bmatrix} A_1 \\ A_2 \\ A_3 \end{Bmatrix} = \begin{Bmatrix} 20 \\ 20 \\ 20 \end{Bmatrix} + 1.702903 \begin{Bmatrix} -7.07 \\ -5 \\ -7.07 \end{Bmatrix} = \begin{Bmatrix} 7.95877 \\ 11.4855 \\ 7.95877 \end{Bmatrix} \tag{s}$$

Repeat the process of finding $[K]$, $[K]^{-1}$ and $\{X\}$, then

$$\begin{Bmatrix} \sigma_1 \\ \sigma_2 \\ \sigma_3 \end{Bmatrix} = \begin{Bmatrix} 14.7281 \\ 11.6868 \\ -3.64118 \end{Bmatrix}, \quad G_{TK} = -10.2719 \tag{t}$$

Use the quadratic curve fitting shown in Fig. 4.13 where

$$G(\alpha) = a\alpha^2 + b\alpha + c \tag{u}$$

$$\left. \begin{array}{lll} \alpha_1 = 1.13526, & G_{TK1} = -14.7059 \\ \alpha_2 = 1.702903, & G_{TK2} = -10.2719 \\ \alpha_3 = 2.270537, & G_{TK3} = 1.6084 \end{array} \right\} \tag{v}$$

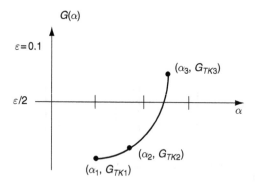

Figure 4.13 Quadratic curve fitting in Example 4.6.1.

from which

$$\begin{Bmatrix} (1.13526)^2 & 1.13526 & 1 \\ (1.702903)^2 & 1.702903 & 1 \\ (2.270537)^2 & 2.270537 & 1 \end{Bmatrix} \begin{Bmatrix} a \\ b \\ c \end{Bmatrix} = \begin{Bmatrix} -14.7059 \\ -10.2719 \\ 1.6084 \end{Bmatrix} \qquad \text{(w)}$$

Solving for a, b and c yields

$$G(\alpha) = 2.5089\alpha^2 - 4.8852\alpha - 18.5356 \qquad \text{(x)}$$

Let

$$G(\alpha) = \frac{\varepsilon}{2} = 0.05 \qquad \text{(y)}$$

then solve for

$$\alpha = 2.270537 \qquad \text{(z)}$$

Revise $\{A\}$ as

$$\begin{Bmatrix} A_1 \\ A_2 \\ A_3 \end{Bmatrix} = \begin{Bmatrix} 20 \\ 20 \\ 20 \end{Bmatrix} + 2.270537 \begin{Bmatrix} -7.07 \\ -5 \\ -7.07 \end{Bmatrix} = \begin{Bmatrix} 4.26603 \\ 8.8744 \\ 4.26603 \end{Bmatrix} \qquad \text{(aa)}$$

We find new stresses as

$$\begin{Bmatrix} \sigma_1 \\ \sigma_2 \\ \sigma_3 \end{Bmatrix} = \begin{Bmatrix} 24.98565 \\ 16.81953 \\ -8.1655 \end{Bmatrix} \qquad \text{(bb)}$$

which indicates that member 1 is the most stressed and

$$\sigma_1 - \sigma_{\text{allow}} = 24.98565 - 25 = 0.014 \leq \frac{\varepsilon}{2} = 0.05 \qquad \text{(cc)}$$

We can then say that the design point is on the constraint surface and start the iterations.

First cycle – Step I

From the start point

$$\begin{Bmatrix} A_1 \\ A_2 \\ A_3 \end{Bmatrix} = \begin{Bmatrix} 4.26603 \\ 8.8744 \\ 4.26603 \end{Bmatrix}, \text{ and } F = 104.702 \text{ lbs} \qquad \text{(dd)}$$

Since constraint g_1 is active, we find

$$\{\nabla g_1\} = -[S_s'][A_s]^T[K]^{-1}\frac{\partial[K]}{\partial A_i}\{X\}$$

$$= -100 \left\{ \begin{array}{cc} 1 & 1 \\ 0 & 2 \\ -1 & 1 \end{array} \right\} \left\{ \begin{array}{cc} 0.00166 & 0 \\ 0 & 4.205 \times 10^{-4} \end{array} \right\}$$

$$50\sqrt{2} \left\{ \begin{array}{cc} 1 & 1 \\ 1 & 1 \end{array} \right\} \left\{ \begin{array}{c} 0.16575 \\ 0.00841 \end{array} \right\}$$

$$= \left\{ \begin{array}{c} -3.67125 \\ -1.48577 \\ 2.18549 \end{array} \right\}$$

(ee)

Note that ∇g_2 is not needed for it is not active. Since the weight function is

$$F = 5\sqrt{2}A_1 + 5A_2 + 5\sqrt{2}A_3$$

we have

$$\{\nabla F\} = \left[\begin{array}{ccc} \dfrac{\partial F}{\partial A_1} & \dfrac{\partial F}{\partial A_2} & \dfrac{\partial F}{\partial A_3} \end{array} \right]^T = \left\{ \begin{array}{c} 7.07 \\ 5 \\ 7.07 \end{array} \right\}$$

(ff)

$$[A] = \{\nabla g_1\}^T = [-3.5680 \quad -1.1453 \quad 2.21147]$$

$$[A][A]^T = [20.4619]$$

(gg, hh)

$$[P] = [I] - [A]^T\{[A][A]^T\}[A]$$

$$= \left[\begin{array}{ccc} 1 & 0 & 0 \\ 0 & 1 & 0 \\ 0 & 0 & 1 \end{array} \right] - \left[\begin{array}{c} -3.5680 \\ -1.1453 \\ 2.2147 \end{array} \right] \frac{1}{20.461967}$$

$$[-3.568 \quad -1.145 \quad 2.2147]$$

(ii)

$$= \left[\begin{array}{ccc} 0.3413 & -0.26657 & 0.39212 \\ & 0.89212 & 0.15869 \\ \text{sym} & & 0.766573 \end{array} \right]$$

The descent direction is

$$\{s\} = -[P]\{\nabla F\} = - \left[\begin{array}{ccc} 0.3413 & -0.26657 & 0.39212 \\ & 0.89212 & 0.15869 \\ \text{sym} & & 0.766573 \end{array} \right]$$

$$\left\{ \begin{array}{c} 7.07 \\ 5 \\ 7.07 \end{array} \right\} = \left\{ \begin{array}{c} -3.85325 \\ -3.69774 \\ -8.98663 \end{array} \right\}$$

(jj)

Step II – Since $||S|| = 10.4537 >> 0$, continue to Step III
Step III – Follow Eq. (4.93)

$$\alpha_0 = \frac{1}{4} = 0.25 \tag{kk}$$

then

$$F^{(0)} = \left\{ \begin{array}{c} 4.26603 \\ 8.8744 \\ 4.26603 \end{array} \right\} + \left(\frac{1}{4}\right) \left\{ \begin{array}{c} -3.85325 \\ -3.69774 \\ -8.98663 \end{array} \right\} = \left\{ \begin{array}{c} 3.3027 \\ 7.95 \\ 2.0194 \end{array} \right\} \tag{ll}$$

and

$$[K] = \begin{bmatrix} 376.3 & 0 \\ \text{sym} & 1960.302 \end{bmatrix}, \quad [K]^{-1} = \begin{bmatrix} 0.00269 & 0 \\ \text{sym} & 5.143 \times 10^{-4} \end{bmatrix} \tag{mm, nn}$$

$$\{X\} = [K]^{-1}\{P_s\} = \left\{ \begin{array}{c} 0.2439 \\ 0.0905 \end{array} \right\} \tag{oo}$$

Thus

$$\left\{ \begin{array}{c} \sigma_1 \\ \sigma_2 \\ \sigma_3 \end{array} \right\} = \left\{ \begin{array}{c} 33.438 \\ 18.091 \\ -15.347 \end{array} \right\} \text{ violate and } h_1 = 8.438 \tag{pp}$$

$$\{\nabla g_1\} = \left\{ \begin{array}{c} -6.983 \\ -1.845 \\ 5.138 \end{array} \right\}; \quad [A] = \{\nabla g_1\}^T \tag{qq, rr}$$

Correct Step

$$F^{(1)} = F^{(0)} - [A]^T \left([A][A]^T\right)^{-1} h_1$$

$$= \left\{ \begin{array}{c} 3.3027 \\ 7.95 \\ 2.0194 \end{array} \right\} - \left\{ \begin{array}{c} -6.983 \\ -1.845 \\ 5.138 \end{array} \right\}$$

$$\left([-6.983 \quad -1.845 \quad 5.138] \left\{ \begin{array}{c} -6.983 \\ -1.845 \\ 5.138 \end{array} \right\} \right)^{-1} 8.438 \tag{ss}$$

$$= \left\{ \begin{array}{c} 4.053 \\ 8.148 \\ 1.468 \end{array} \right\}, \text{ and } F = 79.779 \text{ lbs.}$$

Repeat the calculations of $[K], [K]^{-1}, \{X\}$ then

$$\begin{Bmatrix} \sigma_1 \\ \sigma_2 \\ \sigma_3 \end{Bmatrix} = \begin{Bmatrix} 29.826 \\ 15.837 \\ -13.989 \end{Bmatrix} \text{ still violate and } h_1 = 4.826;$$

$$\{\nabla g_1\} = \begin{Bmatrix} -5.711 \\ -1.159 \\ 4.551 \end{Bmatrix} \tag{tt, uu}$$

$$[A] = \{\nabla g_1\}^T = [-5.711 \quad -1.159 \quad 4.551] \tag{vv}$$

$$F^{(1)} = F^{(0)} - [A]^T \left([A][A]^T \right)^{-1} h_1$$

$$= \begin{Bmatrix} 3.3027 \\ 7.95 \\ 2.0194 \end{Bmatrix} - \begin{Bmatrix} -5.711 \\ -1.159 \\ 4.551 \end{Bmatrix}$$

$$\left([-5.711 \quad -1.159 \quad 4.551] \begin{Bmatrix} -5.711 \\ -1.159 \\ 4.551 \end{Bmatrix} \right)^{-1} 4.826 \tag{ww}$$

$$= \begin{Bmatrix} 4.5568 \\ 8.251 \\ 1.067 \end{Bmatrix}, \text{ and } F = 81.02 \text{ lbs.}$$

Following the correcting procedure and after several correcting iterations, we have

$$F^{(2)} = \begin{Bmatrix} 5.515 \\ 8.356 \\ 0.213 \end{Bmatrix}, \text{ and } F = 82.637 \text{ lbs.} \tag{xx}$$

Repeating the process, we have the optimum results shown in Fig. 4.14.

4.7 Linearization method

Another approach to the optimization of a nonlinear problem is to replace it with a sequence of linear programming problems. The numerical method for this approach is to compute design change by using Taylor's expansion

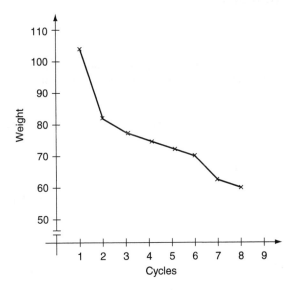

Figure 4.14 Optimum solution of eight cycles.

for the cost and constraint functions. The concept has been illustrated in Chapter 3 and is further applied hereafter. Let $\{x^{(k)}\}$ be the design estimate at the kth iteration and $\{\Delta x^{(k)}\}$ be the change in design. Taylor's expansion of the cost and constraint functions about the point $x^{(k)}$ may be expressed as:

$$\text{Minimize } F\left(x_i^{(k)} + \Delta x_i^{(k)}\right) \cong F\left(x_i^{(k)}\right) + \nabla F^T\left(x_i^{(k)}\right) \Delta x_i^{(k)} \quad i = 1, 2, \ldots, n$$

$$(4.100)$$

$$\text{Subject to } g_j\left(x_i^{(k)} + \Delta x_i^{(k)}\right) \cong g_j\left(x_i^{(k)}\right) + \nabla g_j^T\left(x_i^{(k)}\right) \Delta x_i^{(k)} \leq 0$$

$$j = 1, \ldots, m \qquad (4.101)$$

where $\nabla F\left(x_i^{(k)}\right)$ and $\nabla g_j\left(x_i^{(k)}\right)$ are gradients of the objective or cost function and jth inequality constraint, respectively. For equality constraints, the gradient can be similarly expressed as Eq. (4.101). Let Eq. (4.101) be expressed in detail as:

$$\left.\begin{array}{l} g_1(x) = g_1(x^0) + \dfrac{\partial g_1(x^0)}{\partial x_1}(x_1 - x_1^0) + \ldots + \dfrac{\partial g_1(x^0)}{\partial x_n}(x - x_n) \\[2mm] \quad . \\ \quad . \\ \quad . \\ g_m(x) = g_m(x^0) + \dfrac{\partial g_m(x^0)}{\partial x_1}(x_1 - x_1^0) + \ldots + \dfrac{\partial g_m(x^0)}{\partial x_n}(x - x_n) \end{array}\right\} \quad (4.102)$$

which is expressed in matrix form as

$$[A]\{x - x^0\} \le -\{g^0\} \tag{4.103}$$

where $\{x^0\}$ = column matrix of expansion point values; $\{g^0\}$ = column matrix of the nonlinear constraints evaluated at the expansion point, and

$$[A] = \left[\frac{\partial g_j^0}{\partial x_i}\right] \quad i = 1, \ldots, n \text{ and } j = 1, \ldots, m \tag{4.104}$$

Let Eq. (4.103) be rearranged as

$$[A]\{x\} \le \{B\} \tag{4.105}$$

where

$$\{B\} = [A]\{x^0\} - \{g^0\} \tag{4.106}$$

Similarly, Eq. (4.100) can be expressed in detail as

$$\left.\begin{aligned}
F(x) &= F(x^0) + \frac{\partial F(x^0)}{\partial x_1}(x_1 - x_1^0) + \frac{\partial F(x^0)}{\partial x_2}(x_2 - x_2^0) \\
&\quad + \cdots + \frac{\partial F(x^0)}{\partial x_n}(x_n - x_n^0) \\
&= F(x^0) + \frac{\partial F(x^0)}{\partial x_1}x_1 + \frac{\partial F(x^0)}{\partial x_2}x_2 + \cdots + \frac{\partial F(x^0)}{\partial x_n}x_n \\
&\quad - \frac{\partial F(x^0)}{\partial x_1}x_1^0 + \frac{\partial F(x^0)}{\partial x_2}x_2^0 + \cdots + \frac{\partial F(x^0)}{\partial x_n}x_n^0
\end{aligned}\right\} \tag{4.107}$$

which in matrix form is

$$F(x) = [C]\{x\} \tag{4.108}$$

Note that the second group of the right-hand side of Eq. (4.107) is constant which does not affect optimization of the cost function. Thus, Eqs. (4.105) and (4.107) are in linear optimization form and can be solved by using the standard Simplex method with iteration procedures as shown in Chapters 2 and 3.

4.7.1   *Design of a typical continuous steel beam*

This is to show that the method is very useful for designing a structure of repeat production for which efficient computer programs can be developed

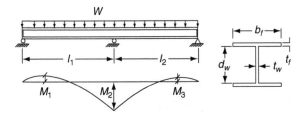

Figure 4.15 Typical two-span steel beam.

for routine design. A simple example of a steel highway bridge for two-lane traffic is illustrated in Fig. 4.15 for which the load, w; span length, l_1 and l_2; allowable bending stress, F_b; and yielding stress, F_y are given. The aim is to determine the cross-sectional dimension. For practical construction, the upper limit of the flange width, b_f, flange thickness, t_f, as well as the upper limit of web thickness, t_w, and web depth, d_w, are required. Let the design variables be

$$\{x_1\ x_2\ x_3\ x_4\} = \{t_f\ b_f\ d_w\ t_w\} \tag{4.109}$$

Then the cost function as weight of the beam may be expressed as:

$$F(x) = \gamma \left[(2b_f t_f + d_w t_w)(l_1 + l_2) \right] \tag{4.110}$$

where γ is the steel weight per unit volume. From Eq. (4.107)

$$\left. \begin{aligned} \frac{\partial F(x^0)}{\partial x_1} &= \frac{\partial F(x^0)}{\partial t_f} = \gamma 2 b_f^0 (l_1 + l_2) \\ \frac{\partial F(x^0)}{\partial x_2} &= \frac{\partial F(x^0)}{\partial b_f} = \gamma 2 t_f^0 (l_1 + l_2) \\ &\vdots \end{aligned} \right\} \tag{4.111}$$

The constraints for allowable bending are expressed as

$$\left. \begin{aligned} g_1(x) &= \frac{M_1 c}{I_x} - F_b \le 0 \\ g_2(x) &= \frac{M_2 c}{I_x} - F_b \le 0 \\ &\vdots \end{aligned} \right\} \tag{4.112}$$

Thus, a typical equation in Eq. (4.102) is

$$g_j(x) = g_j(x^0) + \frac{\partial g_j(x^0)}{\partial x_1}(x_1 - x_1^0) + \cdots + \frac{\partial g_j(x^0)}{\partial x_4}(x_4 - x_4^0) \tag{4.113}$$

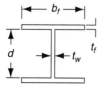

Figure 4.16 Optimum design of a built-up steel beam section.

which involves cross-sectional properties b_f, t_f, d_w, t_w, of the derivatives of $\partial g_i/\partial x_i$. Combining all the components in Eqs. (4.111) and (4.112) yields a linear optimization formulation as Eq. (4.105).

Example 4.7.1 – Consider a wide-flange section as shown in Fig. 4.16 subjected to $M = 1,500$ in.-k, $V = 15$ k. Let $I =$ moment of inertia of the cross-section, $A_w = t_w(d + 2t_f)$ as web area, then $f_b = Mc/I$ where $C = (t_f + d)/2$ and $f_v = V/A_w$. Assume $f_b \leq 0.6F_y$, $f_v \leq 0.4F_y$, $t_w \geq 0.1195$ in., $t_w \leq 0.376$ in., $b_f \geq 4.9$ in., $b_f \leq 8.1$ in., $t_f \geq 0.24$ in., $t_f \leq 0.6$ in., $d \geq 9.9$ in. and $d \leq 40.05$ in. Optimize the built-up steel beam section for minimum weight.

Solution:

The behaviour constraints are

$$g(1) = 0.6F_y - f_b, \quad g(2) = 0.4F_y - f_v \qquad \text{(a, b)}$$

and the side constraints are

$$g(3) = t_w - 0.1196, \quad g(4) = 0.376 - t_w \qquad \text{(c, d)}$$
$$g(5) = b_f - 4.9, \quad g(6) = 8.1 - b_f \qquad \text{(e, f)}$$
$$g(7) = t_f - 0.24, \quad g(8) = 0.6 - t_f \qquad \text{(g, h)}$$
$$g(9) = d - 9.9, \quad g(10) = 40.05 - d \qquad \text{(i, j)}$$

The minimum weight of the section may be expressed in terms of cross-sectional properties as

$$F = 2t_f b_f + dt_w \qquad \text{(k)}$$

Eqs. (a) through (j) can be expressed by a Taylor's series as shown by Eqs. (4.103) and (4.106), and Eq. (k) is also similarly expressed in Eqs. (4.107) and (4.108); then they become a linear programming problem as

$$\text{Minimize} \quad F(x) = [C]\{x\} \qquad \text{(l)}$$
$$\text{Subject to} \quad [A]\{x\} \leq \{B\} \qquad \text{(m)}$$

using the Simplex procedures presented in Chapter 2 yields the final results as

$$F = 8.229 \, \text{in.}^2 \tag{n}$$

$$b_f = 5.32 \, \text{in.}, \quad t_f = 0.36 \, in. \tag{o, p}$$

$$t_w = 0.18 \, in., \quad d = 24.25 \, in. \tag{q, r}$$

This method is practical for some structural members or systems which are repeatedly constructed such as precast members, gable frames, two- or three-span bridges. Special optimization computer programs can be written for the particular structural design with official code specifications and cost functions. The Simplex optimization subroutine is available in almost all computing centres and can easily handle a number of variables. Thus, computer programs can be used for the company's various projects and can accelerate production time with an increase in profits.

4.8   Interior- and exterior-penalty-function algorithms and other transformation methods

Both interior and exterior function methods are commonly classified as *transformation methods*. The so-called transformation is used to describe any method that solves the constrained optimization by transforming it to an unconstrained optimization problem. In comparison to optimization problems formulated previously, they are commonly classified as a *primal method* which implies that primal formulation works directly on the original problem by searching through the feasible region for an optimal solution. Since the transformation method is to solve a constrained optimization problem by using a non-constrained optimization technique, the basic idea is to construct a *composite function* using an objective function and a constraint function with certain parameters (called *penalty parameters*). Therefore, the transformation method can also be called the *penalty-function method* such as an *interior-penalty-function method* and *exterior-penalty-function method*, among others. The penalty function is a procedure for approximating a constrained optimization problem. The approximation is accomplished by adding to the objective function a term that prescribes a high penalty value for violation of constraints. The penalty parameter, r, determines the severity of the penalty and, consequently, the degree to which the unconstrained formulation approximates the original constrained problem. The interior-penalty-function method is extensively used in Chapter 11. Various algorithms for penalty-function methods are introduced in Sections 4.8.1 through 4.8.5.

4.8.1   *Interior-penalty-function method*

We first formulate a composite function, $\psi(x, r)$, to be minimized as

$$\text{Minimize } \psi(x,r) = F(x) + r\phi(x), \quad r > 0 \tag{4.114}$$

where $\phi(x)$ is the penalty function expressed in terms of constrained equations, $g_j(x) \geq 0$, $j = 1, \ldots, m$; $\phi(x)$ is established with the concept that when $\phi(x) > 0$, the design variables are always in the feasible region, when $\phi(x)$ approaches to infinity, then the design variables, x, approach to the boundary of the feasible region as $g_j(x) = 0$. In other words, assume the jth constraint is active at x^*, when x approaches to x^* as $g_j(x) \to 0$, then $\phi(x)$ approaches to ∞; since r approaches to 0, $r\phi(x)$ will be concurrent so that Eq. (4.114) becomes the minimization of $F(x)$. The convergence can briefly be expressed as:

$$\lim_{k \to \infty} \left(x^{(k)}, r^{(k)} \right) = \lim_{k \to \infty} F\left(x^{(k)} \right) = F(x^*) \tag{4.115}$$

which is illustrated graphically in Fig. 4.17 where $r^{(2)} < r^{(1)}$ and $F(x^*) < F\left(x^{(2)} \right) < F\left(x^{(1)} \right)$.

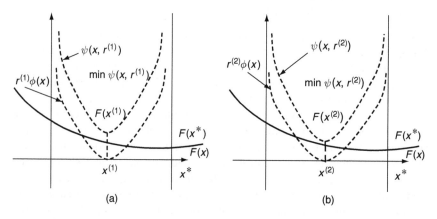

(a)   (b)

Figure 4.17 Convergence of interior-penalty-function method: (a) at $x^{(1)}r^{(1)}$, (b) at $x^{(2)}r^{(2)}$.

Two interior penalty functions commonly used are

Logarithmic function

$$\phi(x) = -\sum_{j=1}^{m} \ln \left[g_j(x) \right] \tag{4.116}$$

Inverse function

$$\phi(x) = \sum_{j=1}^{m} \frac{1}{g_j(x)} \tag{4.117}$$

which are employed in the following numerical examples.

Example 4.8.1 – Apply the logarithmic function for the interior-penalty-function method for the problem data given in Eqs. (a) and (b) as

Minimize $F(x) = x_1 + x_2$ (a)

Subject to $2x_1^2 + 2x_2^2 - 1 \geq 0; \quad x_1 + x_2 + 1 \geq 0$ (b)

Solution:

Let the penalty function be expressed by the logarithmic function as

$$\phi(x) = -\sum_{j=1}^{m} \ln\left[g_j(x)\right] = -\left[\ln\left(2x_1^2 + 2x_2^2 - 1\right) + \ln\left(x_1 + x_2 + 1\right)\right] \tag{c}$$

The original problem is then transformed to the following composite function as the unconstrained minimization problem:

$$\psi(x, r) = F(x) + r\phi(x) = x_1 + x_2 - r$$
$$\left[\ln\left(2x_1^2 + 2x_2^2 - 1\right) + \ln\left(x_1 + x_2 + 1\right)\right] \tag{d}$$

For simplicity of demonstration, let us differentiate the above with respect to x_1 and x_2 as

$$\frac{\partial \psi}{\partial x_1} = 1 - \frac{r}{2x_1^2 + 2x_2^2 - 1} 4x_1 - \frac{r}{x_1 + x_2 + 1} = 0 \tag{e}$$

$$\frac{\partial \psi}{\partial x_2} = 1 - \frac{r}{2x_1^2 + 2x_2^2 - 1} 4x_2 - \frac{r}{x_1 + x_2 + 1} = 0 \tag{f}$$

Solving (e) and (f) gives

$$x_1 = x_2 \tag{g}$$

Substituting (g) in (e), we have

$$1 - \frac{4rx_1}{4x_1^2 - 1} - \frac{r}{2x_1 + 1} = 0 \tag{h}$$

from which

$$\frac{r}{2x_1 - 1} = 1 \tag{i}$$

then

$$x_1 = \frac{r+1}{2} \tag{j}$$

Let r approach to 0, then the optimal point is at

$$x_1^* = \frac{1}{2}, \quad x_2^* = \frac{1}{2} \tag{k}$$

and the optimal solution is

$$F(x^*) = \frac{1}{2} + \frac{1}{2} = 1 \tag{l}$$

Example 4.8.2 – Apply the inverse function for the interior-penalty-function method for the following problem given in Eqs. (a) and (b) as

$$\text{Minimize} \quad F(x) = (x_1 - 1)^2 + (x_2 - 1)^2 \tag{a}$$

$$\text{Subject to} \quad \begin{cases} x_1 + x_2 - 2 \geq 0 \\ 1 - x_1 \geq 0 \\ 1 - x_2 \geq 0 \end{cases} \tag{b}$$

Solution:

The interior penalty function is now expressed by the inverse function as

$$\phi(x) = \sum_{j=1}^{m} \frac{1}{g_j(x)} = \frac{1}{x_1 + x_2 - 2} + \frac{1}{1 - x_1} + \frac{1}{1 - x_2} \tag{c}$$

which is transformed as

$$\psi(x,r) = F(x) + r\phi(x) = (x_1 - 1)^2 + (x_2 - 1)^2$$

$$+ r\left[\frac{1}{x_1 + x_2 - 2} + \frac{1}{1 - x_1} + \frac{1}{1 - x_2}\right] \tag{d}$$

For simplicity of demonstration, let us find the close-form solution as follows

$$\frac{\partial \psi}{\partial x_1} = 2(x_1 - 1) + \frac{r}{(x_1 + x_2 - 2)^2} + \frac{-r}{(1 - x_1)^2} = 0 \tag{e}$$

$$\frac{\partial \psi}{\partial x_2} = 2(x_2 - 1) + \frac{r}{(x_1 + x_2 - 2)^2} + \frac{-r}{(1 - x_2)^2} = 0 \tag{f}$$

From (e) and (f)

$$\frac{x_1 - 1}{x_2 - 1} = \frac{(x_2 - 1)^2}{(x_1 - 1)^2}; \quad x_1 = x_2 \tag{g}$$

Substituting (g) into (e) yields

$$2(x_1 - 1) + \frac{\frac{r}{4} - r}{(x_1 - 1)^2} = 0 \tag{h}$$

or

$$(x_1 - 1)^3 = r - \frac{r}{4}; \quad x_1 = 1 + \sqrt[3]{r - \frac{r}{4}} \tag{i}$$

Let r approach to 0, then $x_1 = 1.0$ and $x_2 = 1.0$. The optimal point is at

$$x^* = (1.0, 1.0) \tag{j}$$

From Eq. (a), the optimal solution is

$$F(x^*) = 0.0 \tag{k}$$

For the numerical solution, we usually do the following:

I) Select a monotonic decreasing sequence $\{r^{(k)}\}$ that approaches to 0 as $k \to \infty$; then find $x^{(0)}$ and set $k = 0$;

II) With $x^{(k)}$ as the starting point, minimize $\psi(x, r^{(k)})$ to find $x^{(k+1)} = x(r^{(k)})$; and

III) If the convergence criteria are not satisfied, set $k = k + 1$ and return to II).

The numerical-solution procedures are demonstrated in Examples 4.8.3 and 4.8.4.

Example 4.8.3 – The interior penalty function is used to find the optimum solution with the classical approach; it is then solved with the numerical procedure illustrated in the graph of Fig. 4.18. The given condition is

$$\text{Minimize} \quad F(x) = x_1^2 + x_2^2 + 3x_1x_2 \tag{a}$$

$$\text{Subject to} \quad g(x) = x_1 + x_2 - 2 \geq 0 \tag{b}$$

Solution:

The composite function is expressed as

$$\psi(x, r) = x_1^2 + x_2^2 + 3x_1x_2 - r\ln(x_1 + x_2 - 2) \tag{c}$$

for which we first find the close solution as

$$\frac{\partial \psi(x, r)}{\partial x_1} = 2x_1 + 3x_2 - \frac{r}{x_1 + x_2 - 2} = 0 \tag{d}$$

$$\frac{\partial \psi(x, r)}{\partial x_2} = 2x_2 + 3x_1 - \frac{r}{x_1 + x_2 - 2} = 0 \tag{e}$$

Eqs. (d) and (e) give

$$x_1 = x_2 = \frac{5 \pm \sqrt{25 + 10r}}{10} \tag{f}$$

Neglecting the negative root since it violates the constraint, we have

$$x_1^* = x_2^* = \lim_{r \to 0} \frac{5 + \sqrt{25 + 10r}}{10} = 1; \quad F(x^*) = 5 \tag{g}$$

For the numerical procedure, let us simplify the objective and constraint functions by using $x_1 = x_2 = x$, then

$$F(x) = 5x^2, \quad g(x) = 2x - 2 \geq 0 \tag{h}$$

The composite function becomes

$$\psi\,(x,r) = 5x^2 - r\ln(2x - 2) \tag{i}$$

The numerical results are illustrated in the graph as shown in Fig. 4.18 by using the procedures of I through III. From the numerical example we can observe that when $x\left(r^{(k)}\right)$ approaches to x^*, $\psi\left(x, r^{(k)}\right)$ is easy to minimize if $r^{(k)}$ remains reasonably large. But when $r^{(k)}$ becomes very small, the function becomes more and more difficult to minimize. Therefore, it is important to choose an appropriate sequence of $r^{(k)}$. $r^{(0)}$ is recommended to be chosen

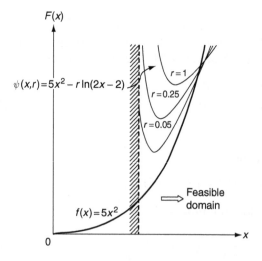

Figure 4.18 Convergence of interior-penalty-function algorithm of Example 4.8.3.

so that neither $F(x^{(0)})$ nor $r^{(0)}\phi(x^{(0)})$ dominates in the composite function $\psi(x^{(0)}, r^{(0)})$. The recommendation is

$$F(x^{(0)}) \approx r^{(0)}\phi(x^{(0)}) \tag{4.118}$$

The advantages that the method can offer are that it is relatively simple with great generality, and that it always yields a sequence of improving non-critical better designs. The method's disadvantages are that (a) it is unable to be applied to equality constraints but this can be overcome with the technique shown in Section 4.8.2, (b) the initial point should be feasible in feasible regions, and (c) when $r^{(k)}$ approaches to zero, it is difficult for a numerical process to achieve a solution.

Example 4.8.4 – Use the interior penalty function to optimize the truss shown in Fig. 4.12. The given data are: weight per unit volume $\rho = 1\,\text{lb in.}^{-3}$; allowable stress $\sigma \leq 20$ ksi (tension) and $\sigma \leq 15$ ksi (compression); the minimum cross-sectional area ≥ 0 and $l = 100$ in. Show the design results of

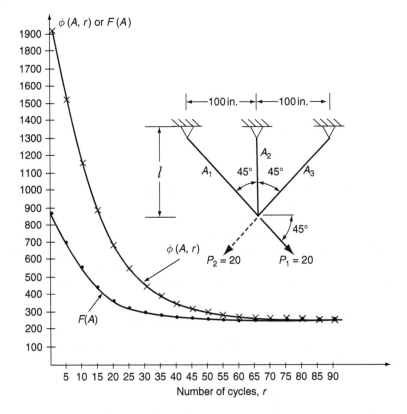

Figure 4.19 $w(A)$ and $\phi(A, r)$ for each optimization cycle, r.

the cross-sectional areas and the constraint values resulting from the optimization cycles. Also show graphically the values of the penalty-function value $\phi(A)$ and $F(A)$ at each iteration, r.

Solution:

Objective function: Since the structural configuration and loading are symmetric, we can assume that the two sloping members are identical. Thus,

$$F(A) = \rho l \left(2\sqrt{2}A_1 + A_2\right) = 100 \left(2\sqrt{2}A_1 + A_2\right) \tag{a}$$

The constraints

$$g_1(A) = 10\left(\frac{1}{A_1} + \frac{1}{A_1 + \sqrt{2}A_2}\right) - 20 \leq 0; \text{ (tension)} \tag{b}$$

$$g_2(A) = 10\left(\frac{2}{A_1 + \sqrt{2}A_2}\right) - 20 \leq 0; \text{ (tension)} \tag{c}$$

$$g_3(A) = -10\left(\frac{-1}{A_1} + \frac{1}{A_1 + \sqrt{2}A_2}\right) - 15 \leq 0; \text{ (compression)} \tag{d}$$

$$g_4(A) = -A_1 \leq 0; \text{ area} \tag{e}$$

$$g_5(A) = -A_2 \leq 0; \text{ area} \tag{f}$$

Penalty function:

$$\psi(A,r) = F(A) - r\sum_{i=1}^{5} g_i(x) \tag{g}$$

Let the initial design be $A_1 = 2.0$ in.^{2}, $A_2 = 3.0$ in.^{2}, and $r = 1,000$; for each iteration cycle r reduces by $1/10$, i.e. $r_{k+1} = 0.9r_k$.

The results of the first 90 cycles are tabulated in Table 4.1 for an interval of five cycles; $\phi(A)$ versus $F(A)$ are graphically given in Fig. 4.19 to reveal the convergent progress at the individual rth cycle. The optimal weight of the 90th cycle is $F(A) = 264.9583$ lbs. Note that this example only illustrates the fundamental procedure for the interior-penalty-function method; more sophisticated procedures will be used in Chapter 11 (see Appendix E).

4.8.2 *Extended interior-penalty-function techniques*

In order to employ $g_i(x) = 0$ in the method, the following *extended interior penalty function* can be used

$$\phi(x) = \begin{cases} \displaystyle\sum_{j=1}^{m} \frac{1}{g_j(x)} & \text{for } g_j(x) \geq \Delta \\[2ex] \displaystyle\sum_{j=1}^{m} \frac{2\Delta - g_j(x)}{\Delta^2} & \text{for } g_j(x) < \Delta \end{cases} \tag{4.119}$$

Table 4.1 Results at the individual optimization cycle of Example 4.8.4.

Cycle no. (r)	A_1	A_2	$g_1 (A)$	$g_2 (A)$	$g_3 (A)$
0 (1,000.0)	2.0	3.0	−13.398	−16.796	−11.602
5 (656.099)	1.63977	2.58634	−12.014	−16.225	−10.789
10 (387.419)	1.29163	1.99599	−9.827	−15.139	−9.688
15 (228.767)	1.04142	1.54910	−7.329	−13.812	−8.517
20 (135.084)	0.88471	1.22742	−4.825	−12.257	−7.568
25 (79.766)	0.80934	1.00939	−3.174	−11.059	−7.115
30 (47.100)	0.78024	0.85742	−2.165	−9.964	−7.201
35 (27.813)	0.77196	0.74308	−1.559	−9.028	−7.532
40 (16.423)	0.77183	0.65480	−1.154	−8.220	−7.934
45 (9.698)	0.77511	0.58689	−0.868	−7.539	−8.329
50 (5.726)	0.77977	0.53429	−0.663	−6.974	−8.688
55 (3.381)	0.78460	0.49413	−0.513	−6.518	−8.996
60 (1.997)	0.78760	0.46547	−0.387	−6.168	−9.219
65 (1.179)	0.7897	0.44595	−0.297	−5.919	−9.377
70 (0.696)	0.79103	0.43245	−0.229	−5.741	−9.488
75 (0.411)	0.79150	0.42373	−0.175	−5.619	−9.556
80 (0.243)	0.79150	0.41826	−0.135	−5.539	−9.596
85 (0.143)	0.79087	0.41596	−0.105	−5.498	−9.607
90 (0.0846)	0.79162	0.41253	−0.095	−5.455	−9.640

or

$$\phi (x) = \begin{cases} \sum_{j=1}^{m} \dfrac{1}{g_j(x)} & \text{for } g_j(x) \geq \Delta \\ \sum_{j=1}^{m} \dfrac{1}{\Delta \left[(g_j(x)/\Delta)^2 - 3(g_j(x)/\Delta) + 3 \right]} & \text{for } g_j(x) < \Delta \end{cases} \tag{4.120}$$

Therefore, the first derivative of the above functions is always continuous at $g_j (x) = \Delta$, where Δ can be a small quantity [8].

4.8.3 *Exterior-penalty-function method*

The popular penalty functions for this method are the quadratic function

$$\phi (x) = \frac{1}{r} \sum_{j=1}^{m} \left[g_j (x) \right]^2 \tag{4.121}$$

and the loss function

$$p\phi (x) = \frac{-1}{r} \sum_{j=1}^{m} \left[g_j (x) \right] \tag{4.122}$$

The exterior penalty function is obtained by using Eq. (4.121) or (4.122) to formulate a composite function for minimization of an unconstrained problem. For instance, using Eq. (4.121), the composite function is

$$\psi(x,r) = F(x) + \frac{1}{r}\sum_{j=1}^{m}[g_j(x)]^2 \tag{4.123}$$

where r is a decreasing sequence used to weigh the violation of constraints. As r approaches to zero then $1/r$ approaches to infinity. Consequently, $\phi(x)$ approaches to zero. The minimization process yields the solution of the unconstrained problem which naturally becomes a solution of the constrained problem. The numerical procedures are similar to that used for the interior-penalty-function method. The difference between these two methods is that the initial point $x^{(0)}$ should be in the feasible region for the interior-penalty-function method, but must be at the infeasible region for the exterior-penalty-function method. Thus, the method's disadvantage is that it generates a sequence of infeasible designs. Similar to the interior-penalty-function method, the method also has difficulty when $r^{(k)}$ approaches to zero as shown in the following numerical example.

Example 4.8.5 – Apply the quadratic function for the exterior-penalty-function method to solve the problem given in Example 4.8.3 using the quadratic function in Eq. (4.121).

Solution:

The composite function is

$$\psi(x,r) = x_1^2 + x_2^2 + 3x_1 x_2 + \frac{1}{r}(x_1 + x_2 - 2)^2 \tag{a}$$

for which the close-form solution can be obtained as

$$\frac{\partial\psi(x,r)}{\partial x_1} = 2x_1 + 3x_2 + \frac{2}{r}(x_1 + x_2 - 2) = 0 \tag{b}$$

$$\frac{\partial\psi(x,r)}{\partial x_2} = 2x_2 + 3x_1 + \frac{2}{r}(x_1 + x_2 - 2) = 0 \tag{c}$$

Eqs. (b) and (c) yield the optimal result as

$$x_1 = x_2 = \frac{4}{5r+4} \tag{d}$$

$$x_1^* = x_2^* = \lim_{r \to 0}\frac{4}{5r+4} = 1; \quad F(x^*) = 5 \tag{e}$$

The numerical solution is illustrated by simplifying Eq. (a) for $x_1 = x_2 = x$ as

$$\psi(x,r) = 5x^2 + \frac{1}{r}(2x - 2)^2 \tag{f}$$

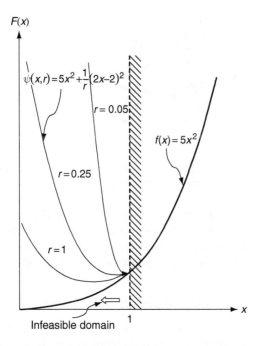

F(x)

$\psi(x,r)=5x^2+\frac{1}{r}(2x-2)^2$

r = 0.05

$f(x) = 5x^2$

r = 0.25

r = 1

Infeasible domain

x

1

Figure 4.20 Convergence of exterior-penalty-function algorithm of Example 4.8.5.

where

$$F(x) = 5x^2, \text{ and } g_j(x) = 2x - 2 \geq 0 \tag{g}$$

The numerical results based on Eq. (f) are shown in Fig. 4.20 which reveals that when $r = 1$, $\psi(x,1) = 5x^2 + (2x - 2)^2$. For $x = 0.5$, $\psi = 1.25 - (-1)^2 = 0.25$, then $F(x) = 1.25$ which is not optimal because $g(0.5) = -1$ violates the constraint in the feasible region. When r approaches to zero, then $\phi(x)$ is apparently difficult to handle. The method introduced in Section 4.8.4 can overcome the difficulty.

4.8.4 *Extended exterior-penalty-function techniques*

In order to overcome the ill condition of r approaching to zero, we can add the exterior penalty function with a Lagrangian function $\sum_{j=1}^{m} \lambda_j g_j(x)$, $j = 1, \ldots, m$, $\lambda_j \geq 0$ and $g_j \geq 0$ to form a composite function. For instance, using Eq. (4.121), we have

$$\psi(x, \lambda, r) = F(x) - \sum_{j=1}^{m} \lambda_j g_j(x) + \frac{1}{r} \sum_{j=1}^{m} [g_j(x)]^2 \tag{4.124}$$

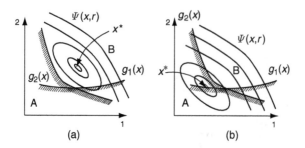

Figure 4.21 Convergence behaviour of Example 4.8.6: (a) interior-penalty-function method, (b) exterior-penalty-function method.

For given values of λ and r, the algorithm yields a point $x(\lambda,r)$; then adjust λ, r so that $x(\lambda,r)$ converges to x^* as the iteration proceeds. Using the technique, the convergence may occur without the need for r to be very small. The ill-conditioning associated with the penalty method can be avoided.

4.8.5 Comments on interior- and exterior-penalty-function methods

The convergence behaviours of the interior- and exterior-penalty-function methods are summarized in the simple diagram sketched in Fig. 4.21 based on two constraints, $g_1(x)$ and $g_2(x)$, as well as two design variables, x_1 and x_2; where x^* is the optimal solution of Example 4.8.6 as shown below.

Example 4.8.6 – Show the convergence behaviour of the problem given in Eqs. (a) and (b).

$$\text{Minimize} \quad x_1 + x_2 \tag{a}$$

$$\text{Subject to} \quad \left. \begin{array}{l} g_1 = 3 - x_1 < 0 \\ g_2 = 2 - x_2 < 0 \end{array} \right\} \tag{b}$$

Solution:

Using the interior-penalty-function method, we write

$$\psi(x,r) = x_1 + x_2 + r\left(\frac{1}{3-x_1} + \frac{1}{2-x_2}\right) \tag{c}$$

For simplicity, employing the closed-form-solution technique as $\partial\psi/\partial x_1 = 0$, $\partial\psi/\partial x_2 = 0$ which yields

$$\left. \begin{array}{l} x_1 = 3 - r^{1/2} \text{ or } 3 + r^{1/2} \\ x_2 = 2 - r^{1/2} \text{ or } 2 + r^{1/2} \end{array} \right\} \tag{d}$$

from which the values $x_1 = 3 + r^{1/2}$ and $2 + r^{1/2}$ are the feasible points for the desired solution. As r approaches to zero, the optimum result is 5 at $x_1^* = 3$ and $x_2^* = 2$ as shown in Fig. 4.21a. Observing Eq. (c) indicates that when the constraint approaches to the boundary, then the penalty blows up and r should be decreased to zero (see Fig. 4.18).

The solution based on the exterior-penalty-function method is sketched in Fig. 4.21b. The method is to search for the solution in the infeasible region (A); the advantage of which is that the search may be started from any infeasible point. The disadvantage is due to the fact that only the optimal solution can be used. However, to search for the solution in the feasible region (B) can offer the advantage that the solutions are always feasible. But the starting point should be carefully selected in order to be in the feasible region.

It is worthwhile noting that some optimization problems are difficult, if not impossible, to express derivatives of constraints with respect to design variables as described previously. This type of problem is relatively easier to solve by using the penalty-function methods.

A typical problem is the optimal design of nondeterministic structures subjected to a dead load, live load and dynamic forces with various considerations of probability concepts of variance formulation as well as loading models. Optimization of nondeterministic structures will be presented in Chapter 11 based on the interior-penalty-function method.

4.9 Concluding remarks

This chapter introduces nonlinear optimization with emphasis on the selected algorithms which will be used in later chapters. In general, the standards of selecting mathematical optimization methods should be: (a) robustness – the algorithms must be reliable for general design applications and must be theoretically guaranteed to converge to a solution point starting from an initial design estimate; (b) generality – the algorithms must be general in formulation and should not impose any restriction on the form of the cost and constraint functions; (c) efficiency – an efficient algorithm has a faster rate of convergence requiring a fewer number of iterations and consequently fewer system analyses, and least number of calculations within one design iteration; (d) ease of use – an algorithm that requires selection of parameters which are problem-dependent is difficult to use; and (e) accuracy – an accurate optimization algorithm is likely to have a sound mathematical basis and thus higher reliability.

References

1. Hadley, G., *Nonlinear and Dynamic Programming*, John Wiley and Sons, Hoboken, WJ, 1964.
2. Duffin, R.J., Peterson, E.L. and Zener, C.N., *Geometric Programming Theory and Application*, John Wiley and Sons, Hoboken, NJ, 1967.

3. Kuhn, H.W. and Tucker, A.W., Nonlinear programming, *Proceedings of the 2nd Berkeley Symposium on Mathematical Statistics and Probability*, Newyman, J. (ed.), University of California Press, Berkeley, CA, 1951.

4. Zangwill, W.I., *Nonlinear Programming – a Unified Approach*, Prentice Hall, Inc, NJ, 1969.

5. Morris, A.J. (ed.), *Foundation of Structural Optimization: A Unified Approach*, John Wiley and Sons, Hoboken, NJ, 1982.

6. Cheng, F.Y. and Bolkin, M.E., Nonlinear optimum design of dynamic damped frames, *Journal of Structural Engineering*, ASCE, Vol. 102, No. ST 3.

7. Reklaitis, G.V., Ravindran, A. and Ragsdell, K.M., *Engineering Optimization-Methods and Applications*, John Wiley and Sons, Hoboken, NJ, 1983.

8. Arora, J.S., *Introduction to Optimum Design*, 2nd edition, Elsevier, Amsterdam, 2004.

5 Optimization of rigid frames with $P - \Delta$ effects for static and dynamic loads

5.1 Introduction

This chapter is solely developed for rigid frames with emphasis on (1) matrix formulation of stiffness, mass and second-order matrices; (2) eigenvalue formulations; (3) calculations of various gradients; (4) step-by-step procedures for feasible direction optimization; and (5) numerical examples and design results. This chapter is intended to establish a solid foundation for dynamic analysis and mathematical optimization.

5.2 Characteristics of rigid frames

The structure considered in this chapter is assembled from members with standard wide flange sections. The joints are considered to be rigid and the member deformations are due to bending only. The positive forces and deformations of a typical member of rigid frames are sketched in Fig. 5.1 where the deformations are slopes θ_i, θ_j and deflections Y_i, Y_j measured from the original position $i\,j$ to the new position $i'j'$. The member forces are bending moments, M_i, M_j; and shears V_i, V_j. Based on Hooke's law, the stiffness coefficients to express the force–deformation relationship must be formulated in terms of material properties: Young's modulus, E, moment of inertia, I, and member length, L. Thus, the stiffness coefficients of a typical member, shown in Figs. 5.1a and b are

$$M_i = \frac{4EI}{L}\theta_i + \frac{2EI}{L}\theta_j - \frac{6EI}{L^2}Y_i - \frac{6EI}{L^2}Y_j \qquad (5.1a)$$

$$M_j = \frac{2EI}{L}\theta_i + \frac{2EI}{L}\theta_j - \frac{6EI}{L^2}Y_i - \frac{6EI}{L^2}Y_j \qquad (5.1b)$$

$$V_i = -\frac{6EI}{L^2}\theta_i - \frac{6EI}{L}\theta_j + \frac{12EI}{L^3}Y_i + \frac{12EI}{L^3}Y_j \qquad (5.1c)$$

$$V_j = -\frac{6EI}{L^2}\theta_i - \frac{6EI}{L}\theta_j + \frac{EI}{L^3}Y_i + \frac{12EI}{L^3}Y_j \qquad (5.1d)$$

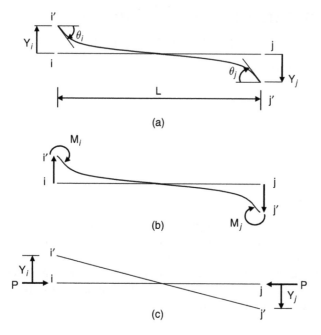

Figure 5.1 Typical deformed member: (a) end deformation, (b) end force, (c) P − Δ
effect on rigid bar.

Eq. (5.1) can be condensed as

$$\left\{\frac{M}{V}\right\} = \left[\begin{array}{c|c} S_{m\theta} & S_{my} \\ \hline S_{v\theta} & S_{vy} \end{array}\right]\left\{\frac{\theta}{Y}\right\} \tag{5.2a}$$

which may be rewritten for a typical element as

$$\{F\}_e = \left[\begin{array}{c|c} S_{m\theta} & S_{my} \\ \hline S_{v\theta} & S_{vy} \end{array}\right]_e \left\{\frac{\theta}{Y}\right\}_e = [S]_e \{e\}_e \tag{5.2b}$$

Note that $[S]_e$ is a symmetric matrix. Therefore $S_{v\theta} = S_{my}$. The force–
deformation relationship of a system composed of a number of members
may be modified from Eq. (5.2b) as shown in Eq. (5.2c)

$$\{F\} = \left[\begin{array}{c|c} \sum S_{m\theta} & \sum S_{my} \\ \hline \sum S_{my} & \sum S_{vy} \end{array}\right]\{e\} \tag{5.2c}$$

where $\sum$ represents a group of the constituent members of a system;
the stiffness coefficients of Eq. (5.2b) correspond to their respective
deformations.

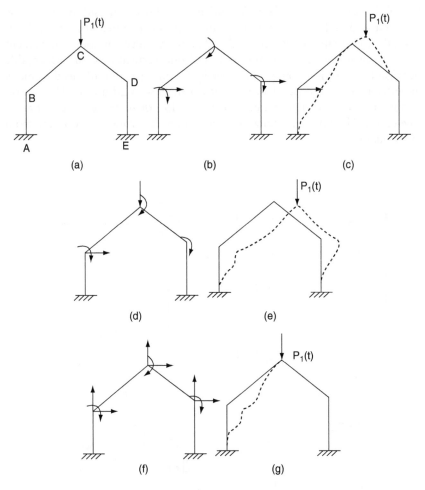

Figure 5.2 Rigid and elastic frame models: (a) gable frame, (b) possible d.o.f. for
   rigid frame model, (c) deformation due to sidesway at B, (d) possible
   d.o.f. for rigid frame model, (e) deformation due to sidesway at B,
   (f) possible d.o.f. for elastic frame model, (g) deformation due to
   displacement at B.

By considering member deformations, a structure can be generally classified as elastic frame or rigid frame. The main difference is the number
of d.o.f. assigned to the system. For instance, the gable frame of Fig. 5.2a
should have five d.o.f. for the rigid frame model as sketched in (b); but nine
d.o.f. for the elastic system model shown in (f). The difference is due to the
linear displacements. This rigid frame must have two lateral displacements
which are the independent sidesway of the structure. However, the elastic

frames should have two linear displacements at each node. Because the rigid frame's linear d.o.f. must be independent, two horizontal sideways can be assigned at B and D as shown in (b) but can't be assigned at B and C horizontally (horizontal movement at B will actually cause horizontal movement at C). However, one horizontal at B and one vertical at C are acceptable because B moves horizontally; the vertical d.o.f. at C doesn't need to move in order to keep independence as shown in Figs. 5.2d and e. The reason is obvious that the rigid frame's members do not have axial deformation (see Figs. 2.1b and c) and they must be in rigid-body motion resulting from individual sideways.

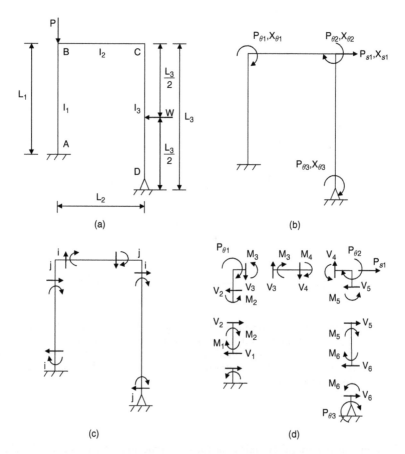

Figure 5.3 Diagrams for rigid frame analysis including the P − Δ effect: (a) given frame, (b) external d.o.f., (c) positive internal forces, (d) free-body diagrams, (e) compatibility between internal rotations due to $X_{\theta 1}$, (f) internal forces due to external loads, (g) compatibility between internal deformations due to X_{s1}, (h) internal forces resulting from the P − Δ effect.

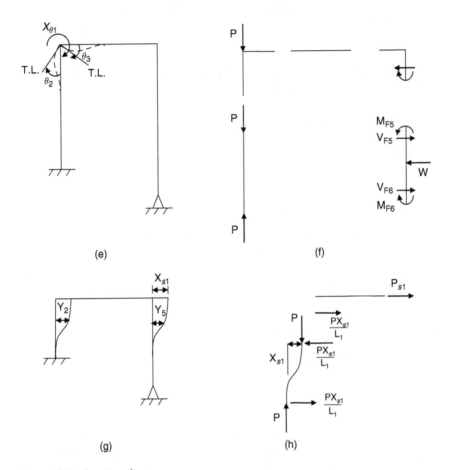

Figure 5.3 (Continued)

5.3 Formation of structural stiffness matrix, member forces and stresses

The stiffness matrix of a structure (regardless of rigid frame or elastic system) must be established based on (1) Hooke's law and (2) the equilibrium condition and (3) the compatibility condition [1]. The one-bay one-storey rigid frame shown in Fig. 5.3a is used to illustrate the fundamental concept. Our d.o.f. are assigned at Fig. 5.3b as $X_{\theta1}, X_{\theta2}, X_{\theta3}$, and X_{s1} and their associated possible forces (resulting from the externally applied loads) are positive in the clockwise direction. The internal deformations and the associated forces are sketched in Fig. 5.3c according to the typical member of Fig. 5.1. The presentation in this section includes seven parts: (A) equilibrium condition, (B) compatibility condition, (C) structural stiffness matrix, (D) loading matrix, (E) geometric matrix for the $P − Δ$ effect, (F) structural stiffness matrix with $P - \Delta$ effect, and (G) solution for displacements, internal forces and stresses.

(A) *Equilibrium condition*

From Fig. 5.3d, the equilibrium conditions of $\sum M = 0$ and $\sum F_x = 0$ give Eqs. (5.3a) and (5.4a), respectively

$$\begin{Bmatrix} P_{\theta 1} \\ P_{\theta 2} \\ P_{\theta 3} \end{Bmatrix} = \begin{bmatrix} 1 & 1 & 0 & 0 & 0 & 0 \\ 0 & 0 & 1 & 1 & 0 & 0 \\ 0 & 0 & 0 & 0 & 0 & 1 \end{bmatrix} \begin{bmatrix} M_1 & M_2 & M_3 & M_4 & M_5 & M_6 \end{bmatrix}^T$$

(5.3a)

$$\{P_{s1}\} = \begin{bmatrix} 0 & 1 & 0 & 0 & 1 & 0 \end{bmatrix} \begin{bmatrix} V_1 & V_2 & V_3 & V_4 & V_5 & V_6 \end{bmatrix}^T \qquad (5.4a)$$

Eqs. (5.3a) and (5.4a) may be written as symbolic expressions as shown in Eqs. (5.3b) and (5.4b) respectively

$$\{P_\theta\} = [A_\theta]\{M\} \tag{5.3b}$$

$$\{P_s\} = [A_s]\{V\} \tag{5.4b}$$

Combining Eqs. (5.3b) and (5.4b) yields

$$\{P\} = \begin{Bmatrix} P_\theta \\ P_s \end{Bmatrix} = \begin{bmatrix} A_\theta & 0 \\ 0 & A_s \end{bmatrix} \begin{Bmatrix} M \\ V \end{Bmatrix} = [A]\{F\} \tag{5.4c}$$

(B) *Compatibility condition*

Fig. 5.3e shows the compatibility of the member-end relations, θ_1 and θ_2, due to joint rotation $X_{\theta 1}$. When all the independent structural d.o.f. move one at a time, we can find the relationships between the member-end rotations, $\{\theta\}$, and nodal rotations, $\{X_\theta\}$, as

$$\begin{Bmatrix} \theta_1 \\ \theta_2 \\ \theta_3 \\ \theta_4 \\ \theta_5 \\ \theta_6 \end{Bmatrix} = \begin{bmatrix} 1 & 0 & 0 \\ 1 & 0 & 0 \\ 0 & 1 & 0 \\ 0 & 1 & 0 \\ 0 & 0 & 0 \\ 0 & 0 & 1 \end{bmatrix} \begin{Bmatrix} X_{\theta 1} \\ X_{\theta 2} \\ X_{\theta 3} \end{Bmatrix} \tag{5.5a}$$

as well as between member-end displacements, $\{Y\}$, and the system's sidesway, $\{X\}$, as

$$\begin{Bmatrix} Y_1 \\ Y_2 \\ Y_3 \\ Y_4 \\ Y_5 \\ Y_6 \end{Bmatrix} = \begin{bmatrix} 0 \\ 1 \\ 0 \\ 0 \\ 1 \\ 0 \end{bmatrix} X_{s1} \tag{5.5b}$$

Let Eqs. (5.5a) and (5.5b) be expressed in symbolic form as shown in Eqs. (5.5c) and (5.5d), respectively

$$\{\theta\} = [B_\theta]\{X_\theta\} \tag{5.5c}$$

$$\{Y\} = [B_s]\{X_s\} \tag{5.5d}$$

Combining Eqs. (5.5c) and (5.5d) yields a general equation expressing the relationship between internal deformations and external displacements as

$$\{e\} = \left\{\frac{\theta}{Y}\right\} = \left[\begin{array}{c|c} B_\theta & 0 \\ \hline 0 & B_s \end{array}\right]\left\{\frac{X_\theta}{X_s}\right\} = [B]\{X\} \tag{5.5e}$$

(C) *Structural stiffness matrix*

Combining Eqs. (5.2c), (5.4c) and (5.5e), we have

$$\{P\} = \left\{\frac{P_\theta}{P_s}\right\} = \left[\begin{array}{c|c} A_\theta & 0 \\ \hline 0 & A_s \end{array}\right]\left[\begin{array}{c|c} \sum S_{M\theta} & \sum S_{MY} \\ \hline \sum S_{MY} & \sum S_{VY} \end{array}\right]\left[\begin{array}{c|c} B_\theta & 0 \\ \hline 0 & B_s \end{array}\right]\left\{\frac{X_\theta}{X_s}\right\} \tag{5.6a}$$

from which the structural stiffness matrix may be represented by

$$[K] = \left[\begin{array}{c|c} A_\theta & 0 \\ \hline 0 & A_s \end{array}\right]\left[\begin{array}{c|c} \sum S_{M\theta} & \sum S_{MY} \\ \hline \sum S_{MY} & \sum S_{VY} \end{array}\right]\left[\begin{array}{c|c} B_\theta & 0 \\ \hline 0 & B_s \end{array}\right] = [A][S][B] \tag{5.6b}$$

Since the equilibrium matrix is the transpose of the compatibility matrix [ ], we then write

$$\left[\begin{array}{c|c} A_\theta & 0 \\ \hline 0 & A_s \end{array}\right] = \left[\begin{array}{c|c} B_\theta & 0 \\ \hline 0 & B_s \end{array}\right]^T = [A] = [B]^T \tag{5.6c}$$

Note that the submatrices of $[S]$ are composed of each of the constituent member's coefficients as shown in Eq. (5.2c). It must be pointed out that the above equations [Eqs. (5.6a)–(5.6c)] only serve to explain the mechanics of the stiffness matrix $[K]$ formulation. In practice, it is much more convenient to use either the *direct stiffness method* or *physical interpretation* [1]. A general equation to express the relationship between external loads $\{P\}$ and external displacements may be written as

$$\{P\} = [A][S][A]^T\{X\} \tag{5.6d}$$

(D) Loading matrix

The $\{P\}$ matrix in Eq. (5.4c) must be related to externally applied forces. Using Fig. 5.3f for the given external load, we first find fixed-end moments and fixed-end shears as

$$M_{F5} = M_{F6} = \frac{WL_3}{8} \tag{5.6e}$$

$$V_{F5} = \frac{W}{2} \tag{5.6f}$$

The load matrix $\{P\}$ is obtained by transferring the fixed-end moments to $\{P_\theta\}$ and fixed-end shear to $\{P_s\}$ as shown below

$$\begin{Bmatrix} P_{\theta 1} \\ P_{\theta 2} \\ P_{\theta 3} \end{Bmatrix} = \begin{Bmatrix} 0 \\ -\dfrac{WL}{8} \\ 0 \end{Bmatrix},$$

$$P_s = \begin{bmatrix} 0 & \dfrac{W}{2} & 0 & 0 & 0 & 0 \end{bmatrix} \begin{Bmatrix} V_{F1} & V_{F2} & V_{F3} & V_{F4} & V_{F5} & V_{F6} \end{Bmatrix}^T$$

$$\tag{5.7a,b}$$

Combining Eqs. (5.7a) and (5.7b) yields

$$\{P\} = \begin{Bmatrix} P_\theta \\ P_s \end{Bmatrix} = \pm \begin{Bmatrix} M_F \\ V_F \end{Bmatrix} \tag{5.7c}$$

It is worthwhile noting that the concentrated load P, applied vertically along member AB can't be formulated in the load matrix because there is no vertical d.o.f. at B. However, P can affect the structure's analysis in two ways: (a) combining shear at member end B (member BC) and the compression P to find the final axial stress of AB, (b) second-order $P − \Delta$ effect due to additional shear resulting from P times the sidesway displacement. This will be discussed in part (E).

(E) Geometric matrix (string stiffness) for the $P − \Delta$ effect

A compressive force can influence both deflection and rotation of a member. For lumped mass analysis of large building structures, the model may be simplified as a rigid bar such that an axial force will only influence deflection but not rotation, as shown in Fig. 5.1c (named string stiffness).

The string stiffness of a typical element is

$$[K_{eg}] = P \begin{bmatrix} \dfrac{1}{L} & \dfrac{1}{L} \\ \dfrac{1}{L} & \dfrac{1}{L} \end{bmatrix} \tag{5.8a}$$

For a structural system, the geometric stiffness matrix is composed of Eq. (5.8a) associated with the structural members subjected to compressive loads. Let Fig. 5.3a be used for illustration: since the axial force is applied vertically at B, only member AB induces the $P - \Delta$ effect. The geometric matrix $[K_g]$ of the frame may be established using Figs. 5.3g and 5.3h.

$$\{P\} = \left\{ \begin{array}{c} P_\theta \\ P_s \end{array} \right\} = \left[\begin{array}{c|c} 0 & 0 \\ \hline 0 & K_g \end{array} \right] \left\{ \begin{array}{c} X_\theta \\ X_s \end{array} \right\} = \left[\begin{array}{c|c} 0 & 0 \\ \hline 0 & K_g \end{array} \right] \{X\} \tag{5.8b}$$

in which

$$[K_g] = \frac{P}{L_1} \tag{5.8c}$$

(F) *Structural stiffness matrix with* $P - \Delta$ *effect*

The formulation of the structural stiffness matrix without the $P - \Delta$ effect is given in Eq. (5.6b). We can similarly establish the structural stiffness matrix with the $P - \Delta$ effect by including Eq. (5.8b) in Eqs. (5.6a) through (5.6c) to yield the following result [2, 3].

$$[K] = [A][S][A]^T - \left[\begin{array}{c|c} 0 & 0 \\ \hline 0 & K_g \end{array} \right] \tag{5.9}$$

Two items should be noted: (1) The axial force shown here is compression; if the axial force is tension, then the negative sign in Eq. (5.9) should be positive. Thus, compression can reduce the structural stiffness and consequently increase structural displacements and internal forces. (2) The axial force is static and does not have a time function; if it has a time function such as a harmonic axial force, the problem involves dynamic instability analysis [4].

(G) *Solution for displacements, internal forces and stresses*

For a system analysis, we first find nodal displacements then the member's deformations, and forces, finally the stresses. Substituting Eq. (5.6b) or (5.9) into Eq. (5.6a) where {P} is given in Eq. (5.7c), we find the nodal displacements

$$\{X\} = \left[[A][S][A]^T \right]^{-1} \{P\} \tag{5.10a}$$

For simplicity, the geometric matrix is not included in Eq. (5.10a). The member-end deformations are then obtained from Eq. (5.5e) as

$$\{e\} = [B]\{X\} \tag{5.10b}$$

The member-end moment is not only due to the deformation at that end but also due to the fixed-end moment, if there is any. Thus, the moments of a system can be expressed as

$$\{M\} = [S]\{e\} \pm \{M_F\} \qquad (5.10c)$$

The stresses of the constituent members are calculated using the member-end moments divided by their respective member section modulus as

$$\sigma = \frac{M}{\delta} \qquad (5.11a)$$

For the dynamic loading case, analysis can find the displacements resulting from vibrating modes, and then the stresses are found from the member deformations $\{e\}$. The bending stresses may be calculated only in terms of stress coefficients and deformations, thus

$$\sigma_i = \frac{M_i}{\delta_j} = [S]'\{e\} \qquad (5.11b)$$

where i = index corresponding to the force number, j = index corresponding to the member number, and δ_j = section modulus of member j, $I_j/(d_j/z)$, i and j can be related in any logical manner as

$$j = \frac{(i+1)}{2} + 1$$

which is an integer computation. In Eq. (5.10c), $[S]$ is of the form $\left[S_{M\theta} \, {}'S_{MY} \right]$ in which the elements are the bending stiffness coefficients given in Eqs. (5.1a) through (5.1d) as

$$[S_{M\theta}] = \frac{E\delta_j d_j}{2L_j} \begin{bmatrix} 4 & 2 \\ 2 & 4 \end{bmatrix}, \quad [S_{MY}] = \frac{E\delta_j d_j}{2L_j^2} \begin{bmatrix} -6 & -6 \\ -6 & -6 \end{bmatrix} \qquad (5.12a,b)$$

$$[S_{V\theta}] = [S_{MY}], \quad [S_{VY}] = \frac{E\delta_j d_j}{2L_j^3} \begin{bmatrix} 12 & 12 \\ 12 & 12 \end{bmatrix} \qquad (5.12c,d)$$

Thus, Eq. (5.11b) becomes

$$\{\sigma\} = [S]'\{e\} = \left[S'_{M\theta} \, {}'S'_{\theta V} \right]\{e\} \qquad (5.13)$$

where

$$[S_{M\theta}]'_j = \frac{Ed_j}{2L_j} \begin{bmatrix} 4 & 2 \\ 2 & 4 \end{bmatrix} \text{ and } [S_{MY}]'_j = \frac{Ed_j}{2L_j^2} \begin{bmatrix} -6 & -6 \\ -6 & -6 \end{bmatrix} \qquad (5.14)$$

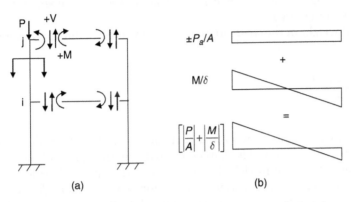

Figure 5.4 Axial stresses in columns due to girder shears: (a) assumed positive girder moments and shears, (b) stress distribution through section 1-1.

Note that bending stresses due to the static loads are superimposed upon bending stresses due to dynamic loads which involves vibrating modes and will be introduced in Section 5.7. The axial stress, however, can be obtained using the principle of superposition as shown in Fig. 5.4 after the moments due to dynamic forces are found.

5.4 Objective function and constraint

5.4.1 *Relationship between cross-sectional area and section modulus*

This chapter is devoted to steel structural design for minimum weight. Thus, the algorithm is to search for the lightest weight possible while not violating any of the constraints imposed upon it. However, structural weight must be based on the volume of each steel member as cross-sectional area times the member length. But the primary design variables for frameworks and continuous beams are the moment of inertia or section modulus, and there is no exact relationship between area and cross-sectional modulus (or moment of inertia) of the wide flange section; it was found that a very good approximation can be made if it is assumed that the section modulus is directly proportional to the product of the area and depth of the section. That is

$$\delta_j = q_j d_j A_j \tag{5.15}$$

where A_j = area of section, q_j = a constant between 0.34 and 0.36 for general sections as presented in Chapter 2. Eq. (5.15) now permits the mass or weight of the structure to be linearly related to the section modulus which will be the design variable for this proposed method. In equation form

$$W_j = \rho A_j L_j = \frac{\rho \delta_j L_j}{q_j d_j} \tag{5.16}$$

where W_j = weight of each member, L_j = length of each member, and ρ = density of material used.

The total weight of the structure is obtained by direct summation over all elements

$$W = \sum_{j=1}^{nm} \frac{\rho \delta_j L_j}{q_j d_j} \tag{5.17}$$

where W = total weight of the structure, and nm = number of members. The frameworks considered in this chapter are multi-storey and multi-bay rigid frames. The masses are modelled to be concentrated at the floor levels or nodal points, and only translational inertia forces will be considered. The total mass to be lumped at each nodal point will be the summation of floor slab and girder weights plus one-half the weight of each column above and below the floor level.

5.4.2 Objective function

For minimum weight design, the objective function may be expressed as

$$\text{Minimize} \quad W(\delta) = \rho \sum_{j=1}^{ndv} \frac{\delta_j L_j}{q_j d_j} \tag{5.18}$$

of which the variables have been defined already except ndv representing the number of design variables and introduced as follows: for practical engineering design sometimes it is desired to maintain two or more members identical. These members can be assigned the same variable. The member numbers corresponding to the design variables are stored in a *design variable correlation* matrix.

$$\overline{D} = \begin{Bmatrix} D_1 \\ D_2 \\ \cdot \\ \cdot \\ D_{ndv} \end{Bmatrix} = \begin{bmatrix} \delta_1 & \cdot & \cdot & \cdot & \delta_m \\ \delta_2 & \cdot & \cdot & \cdot & \delta_n \\ \cdot & & & \\ \cdot & & & \\ \cdot & & & \end{bmatrix}_{ndv \times m \text{ max}} \tag{5.19}$$

where m and n are subscripts relating to member numbers, ndv is the number of design variables and m max is the maximum number of members assigned to any one design variable. It should be noted, now, that if the summation in Eq. (5.18) is on the number of design variables, then Eq. (5.19) must be searched for each design variable, to obtain the total number of times for which the weight component due to that design variable must be added.

5.4.3 Description of constraints

(A) *Behaviour constraints*

The optimum design must be the lightest design possible subject to limitations on the displacements and stresses, and on the range permitted in which the frequencies can lie. These are termed *behaviour constraints* which will be discussed in detail according to static or dynamic cases shown in Sections 5.6 and 5.7, respectively.

(B) *Side constraints*

Limitations may also be imposed upon the size of the design variable and are called *side constraints*. Since, in general, the numbers tend to become smaller as the design proceeds, the lower limit is more likely to become active than the upper limit. It should be noted, however, that the stress in a given member will become bounded before the design variable for that member is reduced to zero. This eliminates the possibility of a member developing a zero or negative area, which is obviously physically impossible. In other words, the stress constraint automatically imposes some additional constraint upon the size of the members.

5.5 Outline of optimization algorithm

The optimization technique is based on the *steepest descent method* presented in Section 4.4 of Chapter 4 and briefly outlined here in three parts: taking a step, adjusting to the constraint, and finding a new feasible direction as follows.

5.5.1 Taking a step

A new set of design variables can be expressed (see Eq. (4.38c)) as

$$\{\delta\}^{(k+1)} = \{\delta\}^{(k)} + \alpha \{s\} \tag{5.20}$$

where $\{s\}$ is the direction of the search vector and α is a scaling factor related to a desired weight reduction (see Eq. (4.51)) as

$$\alpha = -\frac{\mu W_0}{\{s\}^T \{\nabla W\}} \tag{5.21}$$

where μ is the fractional weight reduction, W_0 is the initial weight, and $\{\nabla W\}$ is the gradient of the weight function. The general feasible direction vector is $\{s\}$; if the design, $\delta(k)$, is bounded then $\{s\}$ will be a new feasible direction vector. Note that the notation of the objective function, $F(x)$, used in Chapter 4, is replaced by $W(\delta)$ in this chapter. They are interchangeable because the structural weight, W, is a function of the design variable, δ, $W(\delta)$ is therefore used for convenience. The new set of design variables can now be compared with the limiting values, called side constraints, and linearly reduced, if necessary. At this time, a new analysis is made and another step is taken if the design is feasible.

5.5.2 Correcting the step

If the design results in one or more violated behaviour constraints, α will have to be reduced until the most violated constraint is bounded. A linear interpolation procedure was derived by assuming that α varies linearly between the last feasible design and the present violated one, as can be shown in Fig. 4.5. At α equal to zero, the constraint has a value of g_j (not violated). At the present value, say α_t, the constraint is violated and called g_{jt}. The curved surface represents the actual constraint. Since a zero constraint evaluation would be difficult to obtain in the finite arithmetic of the digital computer, some tolerance is permitted and is assumed to be $-\varepsilon/2$. The desired α can then be determined in terms of the constraint values as shown in Eq. (4.60) and represented here in Eq. (5.22)

$$\alpha_I = -\frac{g_{jo} - \frac{\varepsilon}{2}}{g_{jt} - g_{jo}}\alpha_t \qquad (5.22)$$

It should be noted that in some instances the linear interpolation fails to yield a positive α. A quadratic scheme was derived in Eq. (4.67).

5.5.3 Finding a direction and application of the Simplex method

By observing Fig. 4.7, the geometric interpolation of the feasible direction vector can be seen. Angle φ_1 must be greater than 90° so that the weight is decreased during the step, α, while at the same time φ_2 must be greater than 90° so that the tangent to the constraint is not followed. These angles can be related to the gradients in Eqs. (4.68) and (4.69) which is also repeated in Eqs. (5.22a) and (5.22b)

$$\cos\varphi_1 = \frac{\{s\}^T\{\nabla W\}}{|s||\nabla W|} \qquad (5.22a)$$

$$\cos\varphi_2 = \frac{\{s\}^T\{\nabla g_1\}}{|s||\nabla g_1|} \qquad (5.22b)$$

which were simplified to Eqs. (4.70) and (4.71) by normalizing $\{s\}^T\{s\} = 1$, $\{W\}^T\{W\} = 1$. For convenience, Eqs. (4.70) and (4.71) are given in Eqs. (5.23a) and (5.23b) as

$$\{s\}^T \{\nabla W\} < 0 \tag{5.23a}$$

$$\{s\}^T \{\nabla g_1\} < 0 \tag{5.23b}$$

In order to obtain the best possible feasible direction vector, Eqs. (4.72) and (4.73) are derived as

$$\{s\}^T \{\nabla g\}_j + \{\theta\}_j \beta \leq 0, \quad j = 1, \ldots, \text{nbc} \tag{5.24a}$$

$$\{s\}^T \{\nabla W\} + \beta \leq 0, \quad \text{where } \beta \to \beta_{\text{max}} \tag{5.24b}$$

in which nbc is the number of bounded constraints, $\{\theta\}_j$ is a vector of 'pushoff' factors and β is a positive scalar. Satisfaction of Eqs. (5.24a) and (5.24b) ensure the satisfaction of Eqs. (5.23a) and (5.23b) as long as β is finite and positive. The components of $\{\theta\}_j$ are usually unity for curved constraints. Eqs. (5.24a) and (5.24b) describe the classical linear programming problem where θ is the variable to be maximized. The result will be the very best feasible direction vector for the combination of constraint and weight gradients in the given stage of the problem. Note that some measure of the unknown, $\{s\}$, must be bounded so that β cannot be large without limit. The simplest form of limitation is

$$|s_i| \leq 1 \quad i = 1, \ldots, \text{ndv} \tag{5.25}$$

which forces all components of $\{s\}$ to lie between $+1$ and -1. Eqs. (5.24a), (5.24b) and (5.25) can be formulated as a linear programming problem by first redefining the independent variable. This is necessary so that the right-hand side of both inequalities obtained from Eq. (5.25) are non-negative. Let

$$s_i' = s_i + 1 \tag{5.26}$$

which when substituted into Eqs. (5.24a) and (5.24b) yields

$$\{s'\}^T \{\nabla g\}_j + \{\theta\}_j \beta \leq \sum_{i=1}^{\text{ndv}} \frac{\partial g_j}{\partial \delta_i} \quad j = 1, \ldots, \text{nbc} \tag{5.27a}$$

$$\{s'\}^T \{\nabla W\} + \beta \leq \sum_{i=1}^{\text{ndv}} \frac{\partial W}{\partial \delta_i} \tag{5.27b}$$

Eq. (5.25) now takes the form

$$s_i' \leq 2 \quad i = 1, \ldots, \text{ndv} \tag{5.28a}$$

$$s_i' \geq 0 \tag{5.28b}$$

If each row of Eqs. (5.27) through (5.28) is taken to be a row of a standard linear programming problem, such as

$$[A]\{Y\} = \{B\} \tag{5.29}$$

where $Y_i = s_i'\,(i=1,\ldots,\text{ndv})$, $Y_{n+1} = \beta$ and $Y_i\,(i=\text{ndv}+2,\ldots,t)$ are the slack variables. The total number of rows, t, equals $2 \times \text{ndv} + \text{nbc} + 1$. Also note that Y_i is always greater than or equal to zero.

As an example of the problem set-up, let $\text{nbc}=2$ and $\text{ndv}=2$. Matrix $[A]$ can be written

$$[A] = \begin{bmatrix} \dfrac{\partial g_1}{\partial \delta_1} & \dfrac{\partial g_1}{\partial \delta_2} & \theta_1 & 1 & 0 & 0 & 0 & 0 \\[2ex] \dfrac{\partial g_2}{\partial \delta_2} & \dfrac{\partial g_2}{\partial \delta_2} & \theta_2 & 0 & 1 & 0 & 0 & 0 \\[2ex] \dfrac{\partial W_1}{\partial \delta_2} & \dfrac{\partial W}{\partial \delta_2} & 1 & 0 & 0 & 1 & 0 & 0 \\[2ex] 1 & 0 & 0 & 0 & 0 & 0 & 1 & 0 \\[1ex] 0 & 1 & 0 & 0 & 0 & 0 & 0 & 1 \end{bmatrix} \tag{5.30}$$

and

$$\{B\} = \left[\, \dfrac{\partial g_1}{\partial \delta_1} + \dfrac{\partial g_1}{\partial \delta_2} \quad \dfrac{\partial g_2}{\partial \delta_1} + \dfrac{\partial g_2}{\partial \delta_2} \quad \dfrac{\partial W}{\partial \delta_1} + \dfrac{\partial W}{\partial \delta_2} \quad 2 \quad 2 \,\right]^T \tag{5.31}$$

and Eq. (5.28b) is satisfied by the requirements of the linear program, such as the Simplex method presented in Chapter 2. The problem is then stated: find $\{Y\}$ such that $F = -Y_{n+1} \to \min$; realizing that the variable Y_{n+1} corresponds to β; also remember that minimizing $-\beta$ is the same as maximizing β. There is a possibility that one or more terms of Eq. (5.31) may be negative. In this case, one or more artificials must be introduced into Eq. (5.30) to form a canonical basis and then form the basis in the usual manner.

5.6 Optimum design of rigid frames for static loads

5.6.1 *Response and constraints of displacements and stresses*

For a statically loaded structure, the displacement constraint including rotational and translational d.o.f. may be

$$\{y_{st}\} - \{Y_u\} \le 0 \tag{5.32}$$

where $\{Y_u\}$ is a vector of limiting displacements and $\{y_{st}\}$ is a vector of displacements obtained by using Eq. (5.10a) as

$$\{y_{st}\} = \left[[A][S][A]^T\right]^{-1}\{P\} \tag{5.33}$$

Internal deformations related to external displacements are obtained from Eq. (5.10b) as

$$\{e\} = [B]\{y_{st}\} \tag{5.34}$$

where $\{e\}$ = member and deformations, and

$$[B] = [A]^T \tag{5.35}$$

The moments at the constituent members' ends are then obtained using Eq. (5.10c) as

$$\{M\} = [S]\{e\} \pm \{M_F\} \tag{5.36}$$

The member stresses can be calculated from Eq. (5.36) and used to form a stress constraint as

$$\{\sigma_b\} - \{F_b\} \le 0 \tag{5.37}$$

where $\{F_b\}$ is a vector of the allowable bending stresses and $\{\sigma_b\}$ is a vector of actual bending stresses and may be calculated as

$$\sigma_{bi} = \frac{M_i}{\delta_j} \tag{5.38}$$

Eq. (5.37) is adequate to represent the stress constraint of continuous beams and frame's girders because these members mainly have bending deformations, but axial deformations are negligible. For columns, although axial deformations are sometimes negligible, axial stresses can be due to the axial force transmitted from girder shears as shown in Fig. 5.4a. Thus, the axial stress at the i end of member j may be obtained as

$$\sigma_{ai} = \frac{P_{aj}}{A_j} \tag{5.39}$$

where $P_{aj} = V + P$. The axial and bending stresses of a column may be combined as shown in Fig. 5.4b. Thus, the stress constraint may be expressed in the following interaction equation

$$\frac{\{\sigma_b\}}{F_b} + \frac{\{\sigma_a\}}{F_a} - 1 \le 0 \tag{5.40}$$

where F_a is the allowable axial stress.

The shears are obtained from the member-end deformations using Eq. (5.2a) as

$$\{V\} = [S_{V\theta}\ S_{VY}]\{e\} \tag{5.41}$$

where $[S_{V\theta}] = [S_{MY}]$ is given in Eqs. (5.12b and c), and

$$[S_{VY}] = \frac{E\delta_j d_j}{2L_j^3}\begin{bmatrix} 12 & 12 \\ 12 & 12 \end{bmatrix} \tag{5.42}$$

Note that the shears are usually not considered in the primary stages of design; they are employed to check the adequacy of the web area during the secondary design stage.

5.6.2   *Gradients of displacement and stress constraints*

The displacement gradients may be obtained by taking the derivative with respect to each design variable of Eq. (5.6b) as

$$\frac{\partial\{y_{st}\}}{\partial\delta_i}[K] + \{y_{st}\}\frac{\partial[K]}{\partial\delta_j} = \frac{\partial\{P\}}{\partial\delta_j} \tag{5.43}$$

where ∂/∂_j = the partial derivatives with respect to the variable j; and

$$[K] = [A][S][A]^T \tag{5.44}$$

The desired gradient may be solved for by recognizing that the vector of external static loads is invariant with respect to the change in design variable. Thus, Eq. (5.43) becomes

$$\frac{\partial\{y_{st}\}}{\partial\delta_j} = -[K]^{-1}\frac{\partial[K]}{\partial\delta_j}\{y_{st}\} \tag{5.45}$$

Using Eq. (5.44) yields

$$\frac{\partial[K]}{\partial\delta_j} = [A]\frac{\partial[S]}{\partial\delta_j}[A]^T \tag{5.46}$$

which is based on the fact that [A] or [B] is a function of the geometry of the structure only. It should be noted that Eq. (5.46) is relatively sparse since non-zero terms are only present for those stiffness coefficients related to variable δ_j.

The static stress gradient can be derived by taking the partial derivative of Eq. (5.38) with respect to the design variables. If Eq. (5.36) is first substituted into Eq. (5.38), then it is obvious that the design variable δ_j cancels out, we have

$$\{\sigma_b\} = [S]' \{e\} \tag{5.47}$$

where $[S]'$ is given in Eq. (5.13). The gradient of Eq. (5.47) then is

$$\frac{\partial \{\sigma_b\}}{\partial \delta_j} = [S]' \frac{\partial \{e\}}{\partial \delta_j} \tag{5.48}$$

in which

$$\frac{\partial \{e\}}{\partial \delta_j} = [B] \frac{\partial \{y_{st}\}}{\partial \delta_j} \tag{5.49}$$

and $\partial \{y_{st}\}/\partial \delta_j$ is given in Eq. (5.45). For the loads not directly applied at the nodes, the fixed-end moments must be included in the stress vector as

$$\{\sigma_f\} = \{\sigma_b\} + \left\{ \frac{M_f}{\delta} \right\} \tag{5.50}$$

where the second term is the vector of stresses due to the fixed-end moments $\{M_f\}$. Thus, the ith component of the final stress gradient

$$\frac{\partial \sigma_{f_i}}{\partial \delta_j} = \frac{\partial \sigma_{b_i}}{\partial \delta_j} + M_{f_i} \frac{\partial \left[\frac{1}{\delta_l} \right]}{\partial \delta_j} \tag{5.51}$$

where the subscript l refers to the member corresponding to the ith component. If l does not correspond to j, then the second term drops out.

5.6.3 *Gradient of weight function*

The optimum weight of a structural system was expressed in Eq. (5.18) as a linear function of the design variables, δ_j. The gradient may be obtained by taking the partial derivative with respect to the design variables as

$$\{\nabla W\} = \frac{\partial W}{\partial \delta_j} = \left\{ \frac{\rho l_j}{q_j d_j} \right\}, \quad j = 1, \ldots, \text{ndv} \tag{5.52}$$

in which ndv is the number of design variables. Eq. (5.52) can be not only for the static loading case in this section but also applicable to the dynamic case in Section 5.7.

144   *Optimization of rigid frames with* P −Δ

5.6.4   *Numerical illustrations*

Example 5.6.1 – To illustrate the design procedures described above, the continuous beam shown in Fig. 5.5a is used to demonstrate two design cycles. The initial-stage variables are based on a 10W45 of which the depth is 10 in. and the section modulus is 49.1 in.³. Assume the elastic modulus is 30,000 ksi and the allowable bending stress $F_b = 21.6$ ksi. No displacement constraint is considered for this problem.

Solution:

Based on the diagram shown in Figs. 5.5b and c, the equilibrium matrices of Eq. (5.4c) are

$$[A_\theta]=\begin{bmatrix} 0 & 1 & 1 & 0\end{bmatrix} \text{ and } [A_s]=\begin{bmatrix} 0 & 1 & -1 & 0\end{bmatrix} \tag{a,b}$$

The initial values are depth $= 10$ in., $\delta = 49.1$ in.³ and $E = 30,000$ ksi, for which the stiffness matrices of members 1 and 2 according to Eqs. (5.12a) through (5.12d) are

$$[S_{M\theta}]_1 = \frac{30,000(49.1)10}{2(144)}\begin{bmatrix} 4 & 2 \\ 2 & 4 \end{bmatrix} \text{ and}$$

$$[S_{M\theta}]_2 = \frac{30,000(49.1)10}{2(96)}\begin{bmatrix} 4 & 2 \\ 2 & 4 \end{bmatrix} \tag{c,d}$$

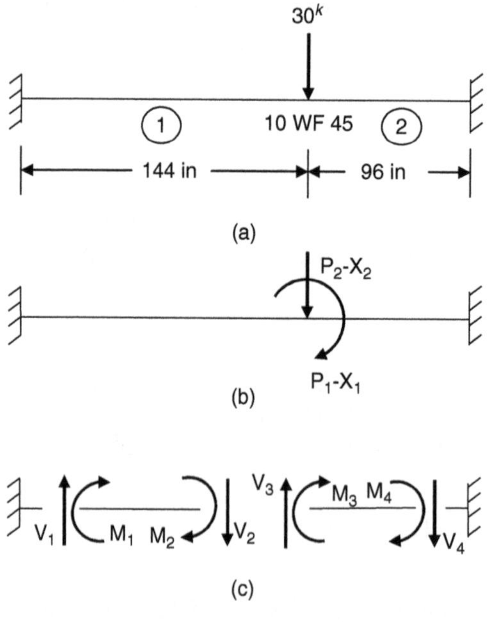

(a)

(b)

(c)

Figure 5.5 Beam with both ends fixed: (a) given beam, (b) external d.o.f., (c) internal forces.

$$[S_{MY}]_1 = \frac{30,000(49.1)10}{2(144)^2} \begin{bmatrix} -6 & -6 \\ -6 & -6 \end{bmatrix} \text{ and}$$

$$[S_{MY}]_2 = \frac{30,000(49.1)10}{2(96)^2} \begin{bmatrix} -6 & -6 \\ -6 & -6 \end{bmatrix} \tag{e,f}$$

$$[S_{VY}]_1 = \frac{30,000(49.1)10}{2(144)^3} \begin{bmatrix} 12 & 12 \\ 12 & 12 \end{bmatrix} \text{ and}$$

$$[S_{VY}]_2 = \frac{30,000(49.1)10}{2(96)^3} \begin{bmatrix} 12 & 12 \\ 12 & 12 \end{bmatrix} \tag{g,h}$$

Thus, the stiffness matrix of the structure using Eq. (5.6b) is obtained as

$$[K] = \begin{bmatrix} A_\theta S_{M\theta} A_\theta^T & A_\theta S_{MY} A_S^T \\ \text{sym} & A_S S_{VY} A_S^T \end{bmatrix} = \begin{bmatrix} 511,458 & 2,663.8 \\ 2,663.8 & 129.49 \end{bmatrix} \tag{i}$$

For the given external force the matrix is $\{P\} = \begin{bmatrix} 0 & 30 \end{bmatrix}$ kips and the nodal displacements are obtained using Eq. (5.33) as

$$\{y_{st}\} = [K]^{-1} \{P\} = \begin{Bmatrix} -1.35143 \times 10^{-3} \\ 2.59475 \times 10^{-1} \end{Bmatrix} \begin{matrix} \text{rad} \\ \text{in.} \end{matrix} \tag{j}$$

The moments are then obtained using Eq. (5.36) as

$$\{M\} = [S][A]^T \{y_{st}\} = \begin{Bmatrix} -57.6 \\ -69.12 \\ 69.12 \\ 86.4 \end{Bmatrix} \text{k-ft} \tag{k}$$

where the minus sign indicates that the moment direction is opposite to the positive direction assumed in Fig. 5.5c. Applying Eq. (5.38) to Eq. (k) yields the bending stresses as

$$\{\sigma_b\} = \begin{Bmatrix} -14.07 \\ -16.89 \\ 16.89 \\ 21.11 \end{Bmatrix} \text{ksi} \tag{l}$$

where the negative sign is associated with the moment direction in Eq. (k). The constraint shown in Eq. (5.37) is now evaluated

$$g = \left| \{\sigma_b\} \right| - \{F_b\} = \begin{Bmatrix} 14.07 \\ 16.89 \\ 16.89 \\ 21.11 \end{Bmatrix} - \begin{Bmatrix} 21.6 \\ 21.6 \\ 21.6 \\ 21.6 \end{Bmatrix} = \begin{Bmatrix} -7.53 \\ -4.71 \\ -4.71 \\ -0.49 \end{Bmatrix} \le 0 \tag{m}$$

where the negative sign indicates that all are feasible. If σ_4 is tested against some tolerance (say, $\epsilon = -1/10$) it should be bounded at 0.222 (i.e. $0.49/21.6 = 0.222$). It is now necessary to find the gradients of the stresses as presented in Section 5.6.2. First, find the stiffness gradients using Eq. (5.46). Note that the gradients of the individual member stiffness matrices are the same as shown in Eq. (i) except that the section modulus, δ, is left out in Eqs. (c) through (g).

$$\frac{\partial [K]}{\partial \delta_1} = \begin{bmatrix} 4,166.7 & -43.44 \\ -43.44 & 0.6028 \end{bmatrix} \text{ and } \frac{\partial [K]}{\partial \delta_2} = \begin{bmatrix} 6,250 & 97.65 \\ 97.65 & 2.034 \end{bmatrix} \quad \text{(n,o)}$$

Eq. (5.45) can now be used to obtain the displacement derivatives

$$\frac{\partial \{y_{st}\}}{\partial \delta_1} = \left\{ \begin{matrix} 4.668 \times 10^{-5} \\ -2.621 \times 10^{-3} \end{matrix} \right\} \text{ and } \frac{\partial \{y_{st}\}}{\partial \delta_2} = \left\{ \begin{matrix} -1.915 \times 10^{-5} \\ -2.663 \times 10^{-3} \end{matrix} \right\} \quad \text{(p,q)}$$

Combining Eqs. (5.48) and (5.49) yields the desired stress gradients as

$$\frac{\partial \{\sigma_b\}}{\partial \delta_1} = \left\{ \begin{matrix} 0.2110 \\ 0.3082 \\ 0.0357 \\ -0.1100 \end{matrix} \right\} \text{ and } \frac{\partial \{\sigma_b\}}{\partial \delta_2} = \left\{ \begin{matrix} 0.0756 \\ 0.0357 \\ -0.3798 \\ -0.3199 \end{matrix} \right\} \quad \text{(r,s)}$$

in which the only quantity of interest is relating to σ_4. Let the normalization procedure be used as

$$\left[\frac{\partial \sigma_4}{\partial \delta_1} \quad \frac{\partial \sigma_4}{\partial \delta_2} \right] = [-0.1100 \quad -0.3199]$$
$$= [-0.3253 \quad -0.9456] \text{ (normalized)} \quad \text{(t)}$$

and

$$\sum_{i=1}^{2} \frac{\partial \sigma_4}{\partial \delta_i} = -0.4299 = -1.2710 \text{ (normalized)} \quad \text{(u)}$$

Before the feasible direction can be found, the gradients of the weight function, Eq. (5.17), must be obtained. The structural weight is

$$W = 0.000283 \left(\frac{144(49.1)}{10(0.34)} + \frac{96(49.1)}{10(0.34)} \right) = 0.981 \text{ kips} \quad \text{(v)}$$

for which the weight gradient is calculated using Eq. (5.52) as

$$\frac{\partial W}{\partial \delta_i} = 0.000283 \left[\frac{144}{10(0.34)} \quad \frac{96}{10(0.34)} \right] = [0.01198 \quad 0.00799]$$

$$= [0.8320 \quad 0.5547] \text{ (normalized)} \tag{w}$$

and

$$\sum_{i=1}^{2} \frac{\partial W}{\partial \delta_i} = 0.01997 = 1.3870 \text{ (normalized)} \tag{x}$$

The feasible direction vector was found using linear programming (see Appendix A) and the relative directions of all the gradients are shown in Fig. 5.6 where

$$\{s\}_1 = s' - 1 = \left\{ \begin{array}{c} 0 \\ 1.77 \end{array} \right\} - \left\{ \begin{array}{c} 1 \\ 1 \end{array} \right\} = \left\{ \begin{array}{c} -1 \\ 0.77 \end{array} \right\} \tag{y}$$

Now it is necessary to reduce the weight in the direction of {s} using Eq. (5.21) as

$$\alpha = \frac{0.1\,(0.981)}{\{-1 \quad 0.77\} \left\{ \begin{array}{c} 0.01198 \\ 0.00799 \end{array} \right\}} = 16.9 \tag{z}$$

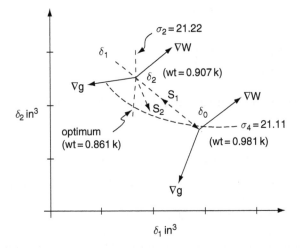

Figure 5.6 Graphical representation of static design example.

where it is desired to reduce the weight by 10 per cent ($\mu = 0.1$). The new design variables from Eq. (5.20) are then

$$\{\delta\}_1 = \left\{ \begin{array}{c} 49.1 \\ 49.1 \end{array} \right\} + 16.9 \left\{ \begin{array}{c} -1 \\ 0.77 \end{array} \right\} = \left\{ \begin{array}{c} 32.24 \\ 62.09 \end{array} \right\} \text{in.}^3 \qquad \text{(aa)}$$

which is sketched in the figure. It was found that σ_2 was violated making it necessary to adjust α according to Eq. (5.22) where

$$g_{jt} = \sigma_2 - 21.6 = 23.22 - 21.6 = 1.6235 \qquad \text{(bb)}$$

$$g_{j0} = 16.89 - 21.6 = -4.71 \qquad \text{(cc)}$$

then

$$\alpha_l = -\frac{-4.71 - 0.05}{1.6235 - (-4.71)}(16.9) = 12.55 \qquad \text{(dd)}$$

This new α should cause σ_2 to be bounded. The stresses at this new design (36.55, 58.77) are found to be

$$\{\sigma_b\} = \left\{ \begin{array}{c} -16.61 \\ -21.22 \\ 13.201 \\ 20.11 \end{array} \right\} \text{ksi} \qquad \text{(ee)}$$

And it is obvious that σ_2 is less than ε. The weight at this new design is 0.907 kips which is a reduction of 7.5 per cent. The confrontation with the constraint kept the desired weight reduction of 10 per cent from being attained. If the stress gradients are found at this new design, a new feasible direction becomes

$$\{s\}_2 = \left\{ \begin{array}{c} 1.377 \\ 0 \end{array} \right\} - \left\{ \begin{array}{c} 1 \\ 1 \end{array} \right\} = \left\{ \begin{array}{c} 0.377 \\ -1 \end{array} \right\} \qquad \text{(ff)}$$

This procedure can be repeated until it is no longer possible to find a feasible direction vector.

5.7   Optimum design of structures for dynamic forces

The optimum design of structures subjected to time-dependent forces such as dynamic loads or seismic excitations first involves how to find structural displacements, then the member forces due to these nodal displacements' so-called dynamic response. The response results are then employed in the constraint formulation for optimization procedures. The dynamic response analysis needs several fundamental formulations in order to understand why and how to conduct sensitivity analysis of constraints. The fundamental formulations comprise finding structural frequencies, vibrating mode shapes and response spectra for different types of dynamic excitations. These subjects are briefly introduced in Section 5.7.1, then the gradients of frequencies and vibrating mode shapes are formulated in Sections 5.7.2. and 5.7.3, respectively.

5.7.1 Displacement and stress response of single d.o.f. and multiple d.o.f. systems

Dynamic response analysis is based on vibrating mode shapes and the associated frequencies of a structural system subjected to external excitation expressed as a response spectrum (or shock spectrum). The methodology involves the response analysis of a single d.o.f. system which is then extended to a multiple d.o.f. system in general which are presented in Parts (A) and (B), respectively. The dynamic d.o.f. is different from the d.o.f. of a static structure; the former is associated with independent movement of structural mass, and the latter represents the independent movements of nodes of the structure. For instance, the static d.o.f. of Fig. 5.3 is four, the dynamic d.o.f. is one because the mass movement is considered for the girders vibrating in the horizontal direction only. However, the static d.o.f. is necessary for the dynamic system, the dynamic d.o.f. is condensed from the static d.o.f. which will be discussed later in Part (B).

(A) Single degree-of-freedom system

The differential equation of motion of the simple system shown in Fig. 5.7 may be obtained by summation of all horizontal forces acting on the mass, M.

$$M\ddot{y}(t) + C\dot{y}(t) + Ky(t) = Ff(t) \tag{5.53}$$

where M = system mass, C = damping coefficient, K = stiffness coefficient, $Ff(t)$ = time-dependent force, and $\ddot{y}(t)$, $\dot{y}(t)$, $y(t)$ = displacement, velocity and acceleration, respectively, where (t) indicates variation with respect to

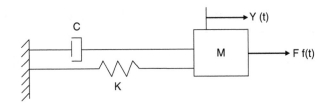

Figure 5.7 Mass-spring-damper system.

time. Eq. (5.53) can be modified for the case where the damping coefficient is a linear combination of the mass and stiffness as

$$C = \gamma M + \beta K \tag{5.54a}$$

in which γ and β are scalars dependent upon the frequency and the ratio of actual damping to critical damping, ξ. The scalars in Eq. (5.54a) can be derived and expressed as

$$\gamma = \xi p \tag{5.54b}$$

$$\beta = \frac{\xi}{p} \tag{5.54c}$$

where p is the natural frequency of the one-degree-of-freedom system [1]. Substituting Eqs. (5.54b) and (5.54c) into Eq. (5.54a) yields the resulting equation which is then substituted into Eq. (5.53), we have

$$\ddot{y}(t) + 2\xi p \dot{y}(t) + p^2 y(t) = \frac{F}{M} f(t) \tag{5.55}$$

in which

$$p^2 = \frac{K}{M} \tag{5.56}$$

Eq. (5.55) can be readily solved using various numerical integration techniques [1]. For a range of frequencies, and specified damping ratios, Eq. (5.55) is solved for the maximum displacements which are the ordinates of the *response spectrum* or *shock spectrum*.

The finite-difference-solution technique is based on Newton's backward differences for third-order polynomials [5] (see Appendix B). The acceleration and velocity at some time, t_n, can be expressed as a linear combination of displacements at earlier times

$$\ddot{y}(t_n) = \frac{1}{(\Delta t)^2} (2y(t_n) - A) \tag{5.57a}$$

$$\dot{y}(t_n) = \frac{1}{6\Delta t} (11y(t_n) - B) \tag{5.57b}$$

in which Δt is the time increment and A and B are the linear combinations of the displacements at previous times

$$A = 5y(t_{n-1}) - 4y(t_{n-2}) + y(t_{n-3}) \tag{5.57c}$$

and

$$B = 18y(t_{n-1}) - 9y(t_{n-2}) + 2y(t_{n-3})$$

where the displacements, $y(t_{n-1})$, $y(t_{n-2})$ and $y(t_{n-3})$ refer to those that occur in previous time increments, Δt, $2\Delta t$, $3\Delta t$, respectively. Eqs. (5.57a) through (5.57c) are derived in Eqs. (B.8) through (B.11). By substituting Eqs. (5.57a) and (5.57b) into Eq. (5.55), a recursion equation can be obtained which gives the displacement at successive, equal increments of time

$$y(t_n) = \frac{\Delta_{st} + A \cdot H + B \cdot G}{1 + 2H + 11G} \tag{5.58}$$

where $\Delta_{st} = \frac{F}{K} f(t_n)$; $H = 1/p^2 (\Delta t)^2$ and $G = \xi/3p\Delta t$. Δ_{st} is referred to as the static displacement and in actuality is the displacement at time t_n due to the force, $Ff(t_n)$, without considering the inertia force. It was found in a previous investigation that the best results are obtained using Δt equal to $T/40$, where T is the fundamental natural period of the system.

Eq. (5.58) was used to plot the shock spectrum for a forcing function approximated by a half sine wave

$$f(t) = \sin(\omega t) \tag{5.59}$$

where ω is the forcing frequency varying from 0 to π. Fig. 5.8 shows the period-type spectrum considering several damping ratios. A period spectrum is a plot of maximum displacements versus a range of natural periods. These displacements are normalized with respect to the static displacement. Note that Fig. 5.8 is divided into two regions. This is because the

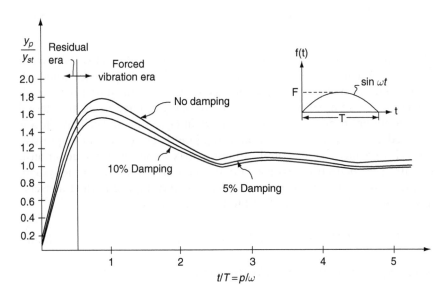

Figure 5.8 Shock spectrum for the half sine function.

maximum displacement for a given combination of natural and forcing frequencies may occur either before or after the force has gone to zero [1].
Since the curve in Fig. 5.8 was obtained numerically, only the points at
regular intervals are known. For displacements in between these points, a
central difference interpolation formula was employed. A similar equation
was derived to approximate the slope of the curve to be used in obtaining
the constraint gradients. Now it will be shown how this shock spectrum
can be used to predict the peak dynamic response of systems with several
degrees of freedom, in Part (B).

(B) *Multi-degree-of-freedom systems*

Similar to Eq. (5.53) the differential equation of motion for a system of
several degrees of freedom is [ ]

$$\begin{bmatrix} 0 & 0 \\ \hline 0 & M \end{bmatrix} \begin{Bmatrix} \ddot{x}_\theta \\ \ddot{x}_s \end{Bmatrix} + \begin{bmatrix} 0 & 0 \\ \hline 0 & C \end{bmatrix} \begin{Bmatrix} \ddot{x}_\theta \\ \ddot{x}_s \end{Bmatrix} + \begin{bmatrix} \begin{bmatrix} K_{11} & K_{12} \\ \hline K_{21} & K_{22} \end{bmatrix} - \begin{bmatrix} 0 & 0 \\ \hline 0 & K_g \end{bmatrix} \end{bmatrix} \begin{Bmatrix} x_\theta \\ x_s \end{Bmatrix} = \begin{Bmatrix} F_\theta(t) \\ F_s(t) \end{Bmatrix}$$

$$(5.60)$$

where $[M]$ = mass matrix corresponding to transverse displacements, $[C]$ =
viscous damping matrix, $\{x_\theta\}$ = nodal rotations of dynamic response, $\{x_S\}$ =
nodal translations of dynamic response, $\{F_\theta(t)\}$ = rotative applied dynamic
loads, $\{F_S(t)\}$ = transverse applied dynamic loads and

$$\begin{bmatrix} K_{11} & K_{12} \\ K_{12}^T & K_{22} \end{bmatrix} = [A][S][A]^T$$

$$(5.61)$$

$[K_g]$ is the second-order stiffness matrix due to the P–Δ effect shown in
Fig. 5.9a. When the columns of a multi-storey frame experience some relative lateral displacement, Δ, secondary moments transmitted may result
if there exists an axial force, P, either externally applied or as an internal
force resulting from girder shears as shown in Figs. 5.9b and c. This effect is
included in the equation of motion of Eq. (5.60). The effect of compressive
forces is to reduce the structural stiffness matrix components which, in turn,
will affect the deflections, stresses and the natural frequencies of the system.
    It was assumed that the dynamic forces are only applied in the transverse
directions which allows Eq. (5.60) to be condensed as

$$[M]\{\ddot{x}_s\} + [C]\{\dot{x}_s\} + [K_c - Kg]\{x_s\} = \{F_s(t)\}$$

$$(5.62)$$

where $[K_c]$ is the *condensed stiffness matrix*

$$[K_c] = [K_{22} - K_{12}^T K_{11}^{-1} K_{12}]$$

$$(5.63)$$

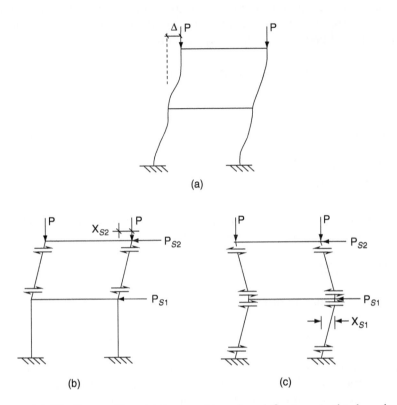

Figure 5.9 The $P - \Delta$ effect: (a) frame subjected to deflection, Δ, (b) shear forces
  required to resist displacement X_{s2}, (c) shear forces required to resist
  displacement X_{s1}.

For simplicity, we will not consider the P–Δ effect hereafter in the formulation. Note that the dynamic responses of $\{x_\theta\}$ are dependent on $\{x_S\}$ due to
the condensation as

$$\{x_\theta\} = -\begin{bmatrix} K_{11}^{-1} & K_{12} \end{bmatrix}\{x_s\} \tag{5.64}$$

Thus, the number of equations describing the system's vibration equals the
number of translational degrees of freedom.

Eq. (5.60) can be solved by using the superposition of eigenvectors as

$$\{x_s\} = \sum_{r=1}^{\text{nps}}\{X\}_r q_r = [X]\{q\} \tag{5.65}$$

where each column of $[X]$ is a modal matrix associated with normal
modes, $\{q\}$ represents the generalized response vector called the *generalized*

coordinates, and nps is the number of degrees of freedom in the translation. Eq. (5.62) can be transformed into generalized coordinates by introducing Eq. (5.65) and premultiplying by $[X]^T$ as shown in Eq. (5.66).

$$[I]\{\ddot{q}\} + [C_r]\{\dot{q}\} + [K_r]\{q\} = [X]^T\{F_s(t)\} \tag{5.66}$$

where

$$[I] = [X]^T[M][X] \tag{5.67a}$$

$$[C_r] = [X]^T[C][X] \tag{5.67b}$$

$$[K_r] = [X]^T[K][X] \tag{5.67c}$$

in which $[I]$ is the identity matrix, $[C_r]$ and $[K_r]$ are diagonal matrices. For the rth mode, Eq. (5.67a) may be written as

$$\{X\}_r^T[M][X]_r = 1 \tag{5.68}$$

Thus, the rth equation of Eq. (5.66) is

$$\ddot{q}_r + C_r\dot{q}_r + K_rq_r = \{X\}_r^T\{F_s(t)\} \tag{5.69}$$

which is identical to Eq. (5.53), the single-degree-of-freedom equation, with the exception that $\{X\}_r^T$ adds some contribution from the other modes. Eq. (5.69) can be expressed in the form of Eq. (5.55) as

$$\ddot{q}_r + 2p_r\dot{q}_r + p_r^2q_r = c_rf(t) \tag{5.70}$$

where, $c_r = \{X\}_r^T\{F_0\}$ (5.71a)

where $\{F_0\}$ is the vector of forcing function amplitudes; $f(t)$ is the time function which is usually considered to be the same for each force. The solution of Eq. (5.70) is then similar to Eq. (5.55)

$$q_r = c_ry(t) \tag{5.71b}$$

in which q_r is the rth component of the solution in normal coordinates and $y(t)$ is the single-degree-of-freedom solution, from Eq. (5.58), of a system having a frequency p_r and a dynamic force amplitude of unity. Now the peak response in normal coordinates of the multi-degree-of-freedom system can be obtained from the shock spectrum, which gives the peak response of the simple system, through Eq. (5.71b)

$$q_{rp} = c_ry_{pr} \tag{5.72}$$

where q_{rp} is the rth component of the peak response vector in normal coordinates and y_{pr} is obtained from the shock spectrum, Fig. 5.8, for a frequency of p_r. q_{rp} is termed the peak *modal participation coefficient.* Since all that has been obtained thus far is the solution in normal coordinates, Eq. (5.65) must be employed to get the actual peak modal response

$$Y_{pi} \cong \left| \lfloor X \rfloor^i \{q_r\} \right| \tag{5.73}$$

where Y_{pi} is the ith component of the peak response vector, $\{q_r\}$ is the r-component vector of normal mode peak responses from Eq. (5.72) and $\lfloor X \rfloor^i$ is the eigenvector in the ith row of the modal matrix.

Eq. (5.73) gives an upper bound to the maximum response since in actuality the maximum responses in each mode do not all occur at the same time. For this reason the resulting design will be conservative. Note that in dynamic analysis the *root-mean-square* technique can be used to overcome the conservative approach [1]. Combining Eqs. (5.72) and (5.73) and expressing in terms of each matrix component produces the final form of the peak response

$$Y_{pi} = \sum_{r=1}^{nps} \left| X_{ir} c_r y_{pr} \right| \tag{5.74}$$

in which X_{ir} is the ith component of the rth eigenvector.

The bending stresses due to dynamic loads can be expressed similar to the upper-bound displacements. These stresses are taken to be proportional to the peak modal participation coefficients. If $\{\sigma_b\}_r$ is a vector of bending stresses for a multi-degree-of-freedom system vibrating in its normal mode r, then an upper bound to the peak stress at member i can be written in component form as

$$\sigma_{pi} = \sum_{r=1}^{nps} \left| \sigma_{bir} c_r y_{pr} \right| \qquad i = 1, \ldots, 2m \tag{5.75}$$

where σ_{bir} is the ith component of $\{\sigma_b\}_r$ and m is the number of members.

To calculate the modal stress vector, the deformations in each mode, $\{e\}_r$, must be obtained as in Eq. (5.10b)

$$\{e\}_r = [A]^T \{X\}_r \tag{5.76}$$

in which the set of eigenvectors, $\{X\}_r$, must include all component directions. The rth mode bending stresses are then obtained by following Eq. (5.11b) as

$$\{\sigma_b\}_r = [S]' \{e\}_r \tag{5.77}$$

To form the constraints, the values from Eqs. (5.73) and (5.77) are used for the displacement and stress, respectively, which will be employed in Sections 5.7.6 and 5.7.7.

5.7.2   Gradient of frequencies

Note that the eigen-solution comprises eigenvalues (frequencies), eigenvectors (vibrating modes), changing design variables can significantly affect the solution. Since the dynamic characteristics of a structure are a function of structural stiffness and mass of the total system, the gradient of structural stiffness and mass must be involved.

The frequencies are usually obtained from vibration of a structure without damping for which the motion equation can be obtained from Eq. (5.62) as

$$[M]\{\ddot{x}_s\} + [K_c]\{x_s\} = 0 \tag{5.78}$$

Note that $[M]$ is a diagonal matrix and that, for simplicity, $[K_g]$ is not included here. The motion described in Eq. (5.78) is harmonic and can be assumed to be

$$\{x_s\} = \{X_r \cos(p_r t - \psi)\} \tag{5.79}$$

for which

$$\{\ddot{x}_s\} = \{-p_r^2 X_r \cos(p_r t - \psi)\} \tag{5.80}$$

where p_r is the natural frequency of the rth mode, and $\{X\}_r$ are the mode shapes. Substituting Eq. (5.79) into (5.80) yields the classical eigenvalue problem

$$[M\lambda_r]\{X\}_r - [K_c]\{X\}_r = 0 \tag{5.81a}$$

where

$$\lambda_r = p_r^2 \tag{5.81b}$$

Note that $\{X\}$ is associated with transverse displacement, for simplicity, the subscript s in Eq. (5.78) is dropped. For a condensed form Eq. (5.81a) becomes

$$[K_c - \lambda_r M]\{X\}_r = 0 \tag{5.81c}$$

from which the eigenvalues (natural frequencies) and eigenvectors (normal modes) can be obtained.

Let

$$[F_r] = [F(\lambda_r, \delta)] \equiv [K_c - \lambda_r M] \qquad (5.82)$$

upon substituting into Eq. (5.81), we have

$$[F_r]\{X\}_r = 0 \qquad (5.83)$$

Premultiplying by $\{X\}_r^T$, Eq. (5.83) becomes

$$\{X\}_r^T [F_r]\{X\}_r = 0 \qquad (5.84)$$

If the partial derivatives of Eq. (5.84) are taken, then

$$\frac{\partial \{X\}_r^T}{\partial \delta_j} [F_r]\{X\}_r + \{X\}_r^T \frac{\partial [F_r]}{\partial \delta_j} \{X\}_r + \{X\}_r^T [F_r] \frac{\partial \{X\}_r}{\partial \delta_j} = 0 \qquad (5.85)$$

Observing Eqs. (5.83) and (5.84) indicates that the first and third terms of Eq. (5.85) should be zero. However, the second term is not zero and can be derived by taking the partial derivatives of Eq. (5.82) with respect to δ_j as

$$\frac{\partial [F_r]}{\partial \delta_j} = \left[\frac{\partial [K_c]}{\partial \delta_j} - \frac{\partial \lambda_r}{\partial \delta_j} [M] - \lambda_r \frac{\partial [M]}{\partial \delta_j} \right] \qquad (5.86)$$

The gradients of stiffness and mass matrices will be presented in parts (C) and (D), respectively.

Introducing Eq. (5.86) into Eq. (5.85), and noting Eq. (5.68), the gradient of the eigenvalues can be expressed as

$$\frac{\partial \lambda_r}{\partial \delta_j} = \{X\}_r^T \left[\frac{\partial [K_c]}{\partial \delta_j} - \lambda_r \frac{\partial [M]}{\partial \delta_j} \right] \{X\}_r \qquad (5.87)$$

in which $\lambda_r = p_r^2$ as given in Eq. (5.81b). Eq. (5.87) needs the gradients of $[K_c]$ and $[M]$. It should be noted that only mode shapes related to lateral deflections are used in Eq. (5.87), and the resulting derivatives correspond only to the first nps eigenvalues.

5.7.3   *Gradient of normal modes*

The required derivatives can be expressed as a linear combination of the previously obtained mode shapes

$$\frac{\partial \{X\}_r}{\partial \delta_j} = \sum_{k=1}^{nps} a_{rjk} \{X\}_k \qquad (5.88)$$

where a_{rjk} is a scalar multiplier to be found. Differentiating Eq. (5.83) yields

$$\frac{\partial [F_r]}{\partial \delta_j} \{X\}_r + [F_r] \frac{\partial \{X\}}{\partial \delta_j} = 0 \tag{5.89}$$

Substituting Eq. (5.88) into the above gives

$$\frac{\partial [F_r]}{\partial \delta_j} \{X\}_r + [F_r] \sum_k a_{rjk} \{X\}_k = 0 \tag{5.90}$$

Premultiplying by $\{X\}^T$ for $l \neq r$ Eq. (5.90) becomes

$$\{X\}_l^T \frac{\partial [F_r]}{\partial \delta_j} \{X\}_r + \sum_k a_{rjk} \{X\}_l^T [F_r] \{X\}_k = 0 \tag{5.91}$$

in which a_{rjk} is a scalar and can be placed outside the matrix multiplication. Observing the second term of Eq. (5.91) and Eqs. (5.67a) through (5.67c), it can be seen that only when $l = k$ does there exist a non-zero value.

$$\{X\}_l^T [F_r] \{X\}_l = \lambda_l - \lambda_r \tag{5.92}$$

Upon substituting back into Eq. (5.91) a_{rjl} can be obtained

$$a_{rjl} = \frac{\{X\}_l^T \dfrac{\partial [F_r]}{\partial \delta_j} \{X\}_r}{(\lambda_r - \lambda_l)} \quad l \neq r \tag{5.93}$$

Obviously, if $l = r$, Eq. (5.93) approaches infinity and, therefore, is not valid. By differentiating Eq. (5.68) the scalar equation is obtained

$$2\{X\}_r^T [M] \frac{\partial \{X\}_r}{\partial \delta_j} + \{X\}_r^T \frac{\partial [M]}{\partial \delta_j} \{X\}_r = 0 \tag{5.94}$$

into which Eq. (5.88) is substituted, yielding

$$2\{X\}_r^T [M] \sum_k a_{rjk} \{X\}_k + \{X\}_r^T \frac{\partial [M]}{\partial \delta_j} \{X\}_r = 0 \tag{5.95}$$

Again, due to Eq. (5.67a) the first term drops out for $r \neq k$. And, due to Eq. (5.68) the first term reduces to $2 \sum a_{rjr}$. Solving for a_{rjr}

$$a_{rjr} = \frac{\{X\}_r^T \dfrac{\partial [M]}{\partial \delta_j} \{X\}_r}{2} \tag{5.96}$$

To obtain the gradients of the rotational eigenvectors, it is necessary to take the partial derivative of Eq. (5.64)

$$\frac{\partial \{X_\theta\}_r}{\partial \delta_j} = -\left[\frac{\partial [K_{11}]^{-1}}{\partial \delta_j} [K_{12}] + [K_{11}]^{-1} \frac{\partial [K_{12}]}{\partial \delta_j} \right] \{X\}_r - [K_{11}]^{-1} [K_{12}] \frac{\partial \{X\}_r}{\partial \delta_j}$$

$$(5.97)$$

in which the gradients of the stiffness matrices are presented in Part (C).

5.7.4   Gradient of condensed stiffness matrix

The gradient of the reduced stiffness matrix, $[K_c]$, may be derived by taking the partial derivative of Eq. (5.63) with respect to the design variables

$$\frac{\partial [K_c]}{\partial \delta_j} = \frac{\partial [K_{22}]}{\partial \delta_j} - \left[\frac{\partial [K_{12}]^T}{\partial \delta_j} K_{11}^{-1} K_{12} + K_{12}^T \frac{\partial [K_{11}^{-1}]}{\partial \delta_j} K_{12} + K_{12}^T K_{11}^{-1} \frac{\partial [K_{12}]}{\partial \delta_j} \right]$$

$$(5.98)$$

where $\partial [K_{22}]/\partial \delta_j$ and $\partial [K_{12}]/\partial \delta_j$ are submatrices of Eq. (5.61). The partial derivative of the inverse of the rotational sub-stiffness matrix, $\partial [K_{11}^{-1}]/\partial \delta_j$, is not obtainable in closed form, and can be found using a two-point central approximation

$$\frac{\partial [K_{11}^{-1} (D^{(j+)})]}{\partial \delta_j} = \frac{1}{2 \times h} [K_{11}^{-1} (D^{(j+)}) - K_{11}^{-1} (D^{(j-)})]$$

$$(5.99)$$

where h is some small change in the design variable δ_j. It was found that using h equal to $1/10\,\text{in}^3$ gives very accurate results. Therefore, the set of design variables at the $j+$ point is

$$D^{(j+)} \equiv \left\{ \delta_1, \dots, \delta_j + \frac{1}{10}, \dots, \delta_r \right\} \quad r = \text{ndv}$$

$$(5.100a)$$

and at the $j-$ point is

$$D^{(j-)} \equiv \left\{ \delta_1, \dots, \delta_j - \frac{1}{10}, \dots, \delta_r \right\} \quad r = \text{ndv}$$

$$(5.100b)$$

As an example of the accuracy obtainable from Eq. (5.98) a comparison was made (see Eq. (w)) for the Example 5.8.1 of Section 5.8. The following exact gradient was computed as

$$\frac{\partial K_{11}^{-1}}{\partial \delta_1} = \begin{bmatrix} -3.0333174 & \text{sym} \\ 1.5166589 & -0.7583290 \end{bmatrix} \times 10^{-7}$$

which compares closely with the approximation given by Eq. (5.99)

$$\frac{\partial K_{11}^{-1}}{\partial \delta_1} = \begin{bmatrix} -3.033322 & \text{sym} \\ 1.5166613 & -0.7583306 \end{bmatrix} \times 10^{-7}$$

5.7.5   Gradient of mass matrix

If the mass of the girder at any floor level and one-half the mass of the columns above and below the floor level are added to the superimposed mass so as to contribute to the internal force, the derivative of the mass matrix with respect to the design variables must also be considered. If Fig. 5.10 represents a typical floor level, i, and the subscripts, u, l and g refer to the dimensions of the upper column, lower column and girder, respectively, then the derivatives of the ith diagonal element of the mass matrix with respect to the jth design variable will be

$$\frac{\partial M_i}{\partial \delta_j} = \frac{\rho}{g} \left[\frac{l_g}{q_g d_g} \frac{\partial \delta_g}{\partial \delta_j} + \frac{l_u}{q_u d_u} \frac{\partial \delta_{cu}}{\partial \delta_j} + \frac{l_l}{q_l d_l} \frac{\partial \delta_{cl}}{\partial \delta_j} \right] \tag{5.101}$$

in which it is assumed that the columns on any given level are related to the same design variable due to the necessary symmetry in design. Therefore, each column mass actually represents both left and right sides of the columns. Note that various terms of Eq. (5.101) will drop out when j is not equal to g, u or l.

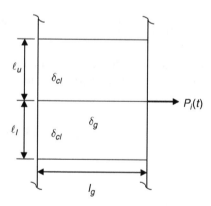

*Figure 5.10 i*th story of typical bay.

5.7.6   Gradient of dynamic displacements

A mathematically equivalent expression for Eq. (5.74) is

$$Y_{pi} = \sum_{r=1}^{nps} \text{sign}\left(X_{ir} c_r y_{pr}\right) \left[X_{ir} c_r y_{pr}\right] \tag{5.102}$$

Let $z = X_{ir}c_ry_{pr}$ and sign (z) is a multiplier used to replace the absolute value signs of Eq. (5.74); therefore

$$\text{sign}\,(z) = \left\{ \begin{array}{l} -1 \text{ if } z < 0 \\ 0 \text{ if } z = 0 \\ 1 \text{ if } z > 0 \end{array} \right\} \tag{5.103}$$

The response can be derived by taking the partial derivatives of Eq. (5.74). The gradient of the ith peak displacement can be expressed as

$$\frac{\partial Y_{pi}}{\partial \delta_j} = \sum_{r=1}^{\text{nps}} \text{sign}\,(z) \left[\frac{\partial X_{ir}}{\partial \delta_j}c_ry_{pr} + X_{ir}\frac{\partial c_r}{\partial \delta_j}y_{pr} + X_{ir}c_r\frac{\partial y_{pr}}{\partial \delta_j} \right] \quad j = 1, \ldots, \text{ndv} \tag{5.104}$$

where $\partial X_{ir}/\partial \delta_j$ is the rth component of Eq. (5.58). From Eq. (5.71a), the gradient of c_r can be obtained

$$\frac{\partial c_r}{\partial \delta_j} = \frac{\partial \{X\}_r^T}{\partial \delta_j}\{F_0\} \tag{5.105}$$

It is necessary to relate the gradient of y_{pr} to the slope of the shock spectrum (see Appendix C).

Let v be the ordinate of the spectrum, Fig. 5.8, where y_{st} is the so-called static displacement

$$y_{st} = \frac{1}{p_r^2} \tag{5.106}$$

where the actual forces will be introduced into the problem at a later step. Let u be the abscissa of the spectrum so the slope $\eta = dv/du$. By the chain rule

$$\frac{\partial y_{pr}}{\partial \delta_j} = \eta \times \frac{\partial u}{\partial p_r} \times \frac{\partial y_{pr}}{\partial v} \times \frac{\partial p_r}{\partial \delta_j} \tag{5.107}$$

where it should be remembered that partial derivatives are the same as total derivatives when only one variable is involved. The last multiplier of Eq. (5.107) is closely related to $\partial \lambda_r/\partial \delta_j$, since

$$p_r = \sqrt{\lambda_r} \tag{5.108a}$$

Then

$$\frac{\partial p_r}{\partial \delta_j} = \frac{\partial p_r}{\partial \lambda_r} \times \frac{\partial \lambda_r}{\partial \delta_j} = \frac{1}{2p_r}\frac{\partial \lambda_r}{\partial \delta_j} \tag{5.108b}$$

where $\partial\lambda_r/\partial\delta_j$ is the frequency gradient given in Eq. (5.87). Terms two and three of Eq. (5.107) are present because both the ordinate and abscissa of Fig. 5.8 are dependent upon p_r. Since $u=p_r/2\omega$, then

$$\frac{\partial u}{\partial p_r}=\frac{1}{2\omega} \tag{5.109}$$

Term three is somewhat more complicated and can be obtained by taking the differential of $v=y_{pr}/y_{str}=p_r^2 y_{pr}$ as

$$\partial v=p_r^2\partial y_{pr}+2p_r y_{pr}\partial p_r \tag{5.110}$$

Dividing through by ∂v and rearranging

$$\frac{\partial y_{pr}}{\partial v}=\frac{1}{p_r^2}\left(1-2p_r y_{pr}\frac{\partial p_r}{\partial v}\right)=\frac{1}{p_r^2}\left(1-\frac{4p_r\omega y_{pr}}{\eta}\right) \tag{5.111}$$

and noting Eq. (5.109), we have

$$\frac{\partial p_r}{\partial v}=\frac{\partial p_r}{\partial u}\times\frac{\partial u}{\partial v}=\frac{2\omega}{\eta} \tag{5.112}$$

Eq. (5.107) now takes this final form after substituting in Eqs. (5.108b), (5.109) and (5.111) as shown below

$$\frac{\partial y_{pr}}{\partial\delta_j}=\frac{y_{st}}{2p_r}\left(\frac{\eta}{2\omega}-2p_r y_{pr}\right)\frac{\partial\lambda_r}{\partial\delta_j} \tag{5.113}$$

where n is the spectrum's slope obtained by using the equations in Appendix C.

5.7.7 *Gradient of dynamic stresses*

Similar to Eq. (5.102), the stress can be expressed as

$$\sigma_{pi}=\sum_r \text{sign}\left(\sigma_{bir}c_r y_{pr}\right)\left[\sigma_{bir}c_r y_{pr}\right] \tag{5.114}$$

Using sign(z) in Eq. (5.114) and taking the derivative with respect to δ_j, we have

$$\frac{\partial\sigma_{pi}}{\partial\delta_j}=\sum_{r=1}^{nps}\text{sign}\,(z)\left[\frac{\partial\sigma_{bir}}{\partial\delta_j}c_r y_{pr}+\sigma_{bir}\frac{\partial c_r}{\partial\delta_j}y_{pr}+\sigma_{bir}c_r\frac{\partial y_{pr}}{\partial\delta_j}\right] \tag{5.115}$$

where $\partial \sigma_{bir}/\partial \delta_j$ is the ith component of the derivatives of the stress vector in the rth mode with respect to δ_j. From Eq. (5.77)

$$\frac{\partial \{\sigma_b\}_r}{\partial \delta_j} = [S]' \frac{\partial \{e\}_r}{\partial \delta_j} \qquad (5.116)$$

which can be obtained from Eq. (5.76) as

$$\frac{\partial \{e\}_r}{\partial \delta_j} = [A]^T \frac{\partial \{X\}_r}{\partial \delta_j} \qquad (5.117)$$

where $\partial \{X\}_r/\partial \delta_j$ represents the gradient of the total set of eigenvectors, lateral and rotational, in the rth mode, Eq. (5.88) and (5.97).

5.7.8 Gradient of axial stresses due to girder shears

When the girder shears are considered to contribute to the axial stresses in the columns, Eq. (5.39) must be modified as

$$\sigma_{aj} = \frac{P_{aj}}{A_j} = \frac{P_{aj} q_j d_j}{\delta_j} \qquad (5.118)$$

which should also be differentiated with respect to δ_j. Let us consider the axial stress in a member i which may not correspond to the design variable j. The general derivative expression may be written as

$$\frac{\partial \sigma_{ai}}{\partial \delta_j} = \frac{\partial P_{ai}}{\partial \delta_j}\frac{q_i d_i}{\delta_i} + P_{ai} q_i d_i \frac{\partial \left(\frac{1}{\delta_i}\right)}{\partial \delta_j} \qquad (5.119)$$

in which the second term drops out if the design variable j does not contain member i. If only static shears are considered, then the first term drops out also, since the axial load will be independent of the member sizes.

The dynamic girder shears can be written as

$$V_{pi} \cong \sum_{r=1}^{nps} |V_{ir} c_r y_{pr}| \quad i = 1, \ldots, 2m \qquad (5.120)$$

where V_{ir} refers to the ith component of $\{V\}_r$, which must be differentiated with respect to δ_j as

$$\frac{\partial V_{pi}}{\partial \delta_j} \cong \sum_r \text{sign}(z) \left[\frac{\partial V_{ir}}{\partial \delta_j} c_r y_{pr} + V_{ir} \frac{\partial c_r}{\partial \delta_j} y_{pr} + V_{ir} c_r \frac{\partial y_{pr}}{\partial \delta_j} \right] \qquad (5.121)$$

where $\partial V_{ir}/\partial \delta_j$ is the *i*th component of the gradient of the shear vector in the *r*th mode for which the shear is expressed as

$$\{V\}_r = \left[S_{v\theta} | S_{vy} \right] \{e\}_r \tag{5.122}$$

where $[S_{v\theta}]$ and $[S_{vy}]$ are defined in Eqs. (5.12c) and (5.12d), respectively, for a typical member. Then the gradient of shear associated with the *r*th mode is

$$\frac{\partial \{V\}_r}{\partial \delta_j} = \frac{\partial \left[S_{v\theta} | S_{vy} \right]}{\partial \delta_j} \{e\}_r + \left[S_{v\theta} | S_{vy} \right] \frac{\partial \{e\}_r}{\partial \delta_j} \tag{5.123}$$

Only the term $\partial \left[S_{v\theta} \,' S_{vy} \right]/\partial \delta_j$ in the above equation has not been derived and this can be obtained by differentiating Eqs. (5.12c) and (5.12d) with respect to δ_j. If the *i*th component does not correspond to the *j*th design variable, then the first term of Eq. (5.123) will be zero for that component.

5.7.9 *Combination of static and dynamic stresses*

The axial stresses in a column are due to the applied load acting on the column and the internal forces transferred from static and dynamic girder shears. Compressive axial stresses are assumed to be positive and the positive sign of the shears is shown in Fig. 5.4. The final stress is obtained by superimposing the bending and axial stresses of the column as

$$\{\sigma_t\} = \text{sign}\,(\sigma_a + \sigma_p)\left[|\{\sigma_a\}| + |\{\sigma_p\}|\right] \tag{5.124}$$

where $\text{sign}(z) = \left\{ \begin{matrix} +1 \text{ if } z > 0 \\ -1 \text{ if } z < 0 \end{matrix} \right\}$ is a multiplier to maintain the proper sign after superimposition.

Note that stresses of girders are due to bending only for which the static bending is combined with dynamic bending stress at each end of a girder for maximum magnitude. Note also that the dynamic displacement response is not combined with static displacement; it is because dynamic displacement is measured from the structural system's equilibrium position where the dynamic displacement constraints are required to be formulated.

5.8 Examples with numerical procedures and comparison of optimum solutions

Three examples are included in this section. The first is a two-lumped-mass cantilever beam to illustrate detailed numerical procedures associated with the equations derived. The second example is a one-bay and two-storey frame subjected to lateral dynamic forces for comparative studies of various

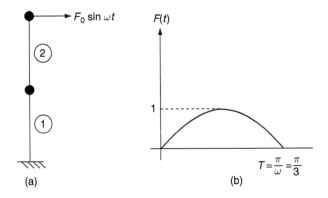

Figure 5.11 Characteristics of dynamic loading: (a) given model, (b) time function.

considerations of damping, the $P - \Delta$ effect, and the effect of dynamic stress combined with static stress. The third example is a one-bay four-storey frame subjected to seismic excitation for various design results.

Example 5.8.1 – The two-storey structure of the cantilever type shown in Fig. 5.11 is designed with the consideration of several sets of constraints for the purpose of showing (1) details of the numerical procedure, and (2) how the equations derived are used for the actual design. This example problem will be solved for one design cycle for the following set of constraints

$$Y_{pj} \leq \begin{bmatrix} 0.03 & 0.0715 \end{bmatrix} \text{ft}, \quad \sigma_{pi} \leq 30 \text{ksi} \quad i=1,\ldots,4; \quad j=1,2$$
$$1 \text{ rps} \leq p_1 \leq 10 \text{ rps}$$
$$\begin{bmatrix} 10 & 25 \end{bmatrix} \text{in.}^3 \leq \delta_l \leq 100 \text{ in.}^3, \quad l=1,2$$

Solution:

The numerical procedure is divided into four segments: (A) initial design variable step, (B) computation of constraint values, (C) computation of constraint gradient, and (D) feasible direction and the design results of this cycle.

(A) *Initial design variable step*

Let the initial design variable be

$$\delta_o = \begin{bmatrix} 69.1 & 31.7 \end{bmatrix} \text{in.}^3 \tag{a}$$

which is selected so that the design is unbounded and not in violation of any of the constraints. The initial weight can be determined from Eq. (5.17) as

$$W_o = 0.00283 \left[\frac{120(69.1)}{12(0.36)} + \frac{120(31.7)}{10(0.36)} \right] = 0.8417 \, \text{kips} \tag{b}$$

in which $\rho = 0.00283$ and $q = 0.36$. The gradient of the weight function is obtained from Eq. (5.52) as

$$\nabla W = \begin{bmatrix} 0.00786 & 0.00943 \end{bmatrix}^T \text{kips in.}^{-3} \tag{c}$$

The new design point can be expressed in terms of the old design point, the gradient of the weight equation and a scalar, α. If ΔW is 10 per cent of the initial weight, then from Eq. (5.21)

$$\alpha = -\frac{\mu W_o}{\{\nabla W\}^T \{\nabla W\}} = -\frac{0.0841}{0.0001506} = -558.4 \tag{d}$$

The new point is determined by Eq. (5.20)

$$\{\delta\}^{(1)} = \begin{Bmatrix} 69.1 \\ 31.7 \end{Bmatrix} - 558.4 \begin{Bmatrix} 0.00786 \\ 0.00943 \end{Bmatrix} = \begin{Bmatrix} 64.7 \\ 26.4 \end{Bmatrix} \text{in.}^3 \tag{e}$$

This new design point must now be analyzed to see if the constraints are bounded or violated.

(B) *Computation of constraint values*

Since the main objective of this example is the evaluation of the constraints and gradients and the carrying through of the design procedure, the determination of frequencies, mode shapes and the stiffness matrix will not be shown in detail. The frequencies and normal modes have been found as

$$p_1 = 4.6363 \, \text{rad s}^{-1}, \quad p_2 = 19.26 \, \text{rad s}^{-1} \tag{f}$$

$$\{X\}_1 = \begin{bmatrix} 0.2516 & 0.9345 \end{bmatrix}^T, \quad \{X\}_2 = \begin{bmatrix} 0.6608 & -0.3557 \end{bmatrix}^T \tag{g}$$

The elements in $\{X\}_1$ and $\{X\}_2$ correspond to x_{S_1} and x_{S_2} in Fig. 5.12a. The condensed stiffness matrix is

$$[K_c] = \begin{bmatrix} 653.4 & -164.3 \\ -164.3 & 65.7 \end{bmatrix} \text{kip ft}^{-1} \tag{h}$$

The maximum displacements can now be computed using the superimposition of normal modes shown in Eq. (5.74) for which individual elements are calculated as follows. First, c_r can be obtained from Eq. (5.71a) where

$$\{F_o\}^T = \begin{bmatrix} 0 & 1 \end{bmatrix} \text{kip} \tag{i}$$

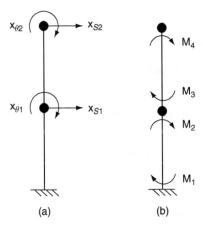

Figure 5.12 Internal and external actions on a two-member model: (a) external degrees of freedom, (b) internal forces.

and

$$c_1 = 0.9345, \quad c_2 = -0.3557 \tag{j}$$

Second, y_{pi} can be obtained from the shock spectrum, Fig. 5.8, for values of $p/2\omega$ based on the time function given in Fig. 5.11b,

$$\left\{\frac{p_i}{2\omega}\right\} = \begin{bmatrix} 0.772 & 3.21 \end{bmatrix} \tag{k}$$

where $\omega = \mathrm{rad\,s}^{-1}$. From Fig. 5.8, we have

$$v_i = \left(\frac{y_p}{y_{st}}\right)_i = \begin{bmatrix} 1.7393 & 1.1689 \end{bmatrix} \tag{l}$$

Since the eigenvectors are considered to be normalized with respect to the mass matrix, then

$$y_{st_i} = \frac{1}{p_i^2}, \quad i = 1, 2 \tag{m}$$

$$y_{pi} = \begin{bmatrix} \dfrac{1.7393}{(4.6363)^2} & \dfrac{1.1689}{(19.26)^2} \end{bmatrix} = \begin{bmatrix} 0.08091 & 0.00315 \end{bmatrix} \mathrm{ft} \tag{n}$$

Using the normal modes given in Eq. (g), the maximum displacements are finally calculated from Eq. (5.74) as

$$Y_{p1} = |(0.2516)(0.9345)(0.08091)$$
$$+ (0.6608)(-0.3557)(0.00315)| = 0.0197\,\text{ft} \tag{o}$$
$$Y_{p2} = |(0.9345)(0.9345)(0.08091)$$
$$+ (-0.3557)(-0.3557)(0.00315)| = 0.07104\,\text{ft} \tag{p}$$

The displacement constraints can now be evaluated

$$Y_{p1} - Y_{u1} = 0.0197 - 0.03 = -0.0103 \le 0 \tag{q}$$
$$Y_{p2} - Y_{u2} = 0.07104 - 0.0715 = -0.00046 \le 0 \tag{r}$$

Both constraints are satisfied, and no violation has occurred. Now it must be checked to see if either is less than $\varepsilon(=-0.01)$ which is the ratio of the actual constraint value to its limiting value

$$\frac{Y_{p1} - Y_{u1}}{Y_{u1}} = -0.343; \quad \frac{Y_{p2} - Y_{u2}}{Y_{u2}} = -0.006 \tag{s,t}$$

Eq. (s) is not bound, however, Eq. (t) is bound for it is less than (absolute value-wise) ε, hence Y_{p2} is bounded. The stresses can be calculated in a manner very similar to the displacements. It was found that the stresses were very small and, therefore, not bounded.

(C) *Computation of constraint gradient*

Now at this point a new direction called a feasible direction must be found so that another step can be taken toward a smaller weight. The gradient of the bounded constraint is shown in Eq. (5.104). Let us first calculate the gradient of the condensed stiffness matrix of Eq. (5.98) for which

$$\frac{\partial K_{22}}{\partial \delta_1} = \begin{bmatrix} 15 & 0 \\ 0 & 0 \end{bmatrix}; \quad \frac{\partial K_{21}}{\partial \delta_1} = \begin{bmatrix} -75 & 0 \\ 0 & 0 \end{bmatrix} \tag{u,v}$$

$$\frac{\partial K_{11}^{-1}}{\partial \delta_1} = \begin{bmatrix} -3.03 & 1.52 \\ 1.52 & -0.758 \end{bmatrix} \times 10^{-7} \tag{w}$$

$$\frac{\partial K_{22}}{\partial \delta_2} = \begin{bmatrix} 12.5 & -12.5 \\ -12.5 & 12.5 \end{bmatrix}; \quad \frac{\partial K_{21}}{\partial \delta_2} = \begin{bmatrix} 62.5 & -62.5 \\ 62.5 & -62.5 \end{bmatrix} \tag{x,y}$$

$$\frac{\partial K_{11}^{-1}}{\partial \delta_2} = \begin{bmatrix} -1.89 & 0.947 \\ 0.947 & -34.9 \end{bmatrix} \times 10^{-7} \tag{z}$$

Using the above in Eq. (5.98) yields

$$\frac{\partial K_c}{\partial \delta_1} = \begin{bmatrix} 5.04 & -0.516 \\ -0.516 & 0.206 \end{bmatrix}; \quad \frac{\partial K_c}{\partial \delta_2} = \begin{bmatrix} 12.4 & -4.96 \\ -4.96 & 2.00 \end{bmatrix} \frac{\text{kip}}{\text{ft-in.}^3} \quad \text{(aa,bb)}$$

Recognizing the given mass is constant, therefore $\partial [M]/\partial \delta_j = 0$; one may calculate the gradient of the eigenvalues from Eq. (5.87) as

$$\frac{\partial \lambda_1}{\partial \delta_1} = \begin{bmatrix} 0.2516 & 0.9345 \end{bmatrix} \begin{bmatrix} 5.04 & -0.516 \\ -0.516 & 0.206 \end{bmatrix} \begin{Bmatrix} 0.2516 \\ 0.9345 \end{Bmatrix} = 0.2565$$

$$\text{(cc)}$$

Similarly

$$\frac{\partial \lambda_1}{\partial \delta_2} = 0.185, \quad \frac{\partial \lambda_2}{\partial \delta_1} = 2.466, \quad \frac{\partial \lambda_2}{\partial \delta_2} = 7.99 \quad \text{(dd)}$$

The gradient of the eigenvectors can now be computed from Eq. (5.88) as

$$\frac{\partial \{X\}_1}{\partial \delta_1} = a_{111} \{X\}_1 + a_{112} \{X\}_2 = a_{112} \{X\}_2 \quad \text{(ee)}$$

in which a_{rjr} is shown in Eq. (5.96) and we have

$$a_{112} = \frac{\begin{bmatrix} 0.6608 & -0.3557 \end{bmatrix}}{(20.9 - 371.3)} \begin{bmatrix} 5.04 & -0.516 \\ -0.516 & 0.206 \end{bmatrix} \begin{Bmatrix} 0.2516 \\ 0.9345 \end{Bmatrix}$$

$$= -0.001422 \quad \text{(ff)}$$

Eq. (ee) yields

$$\frac{\partial \{X\}_1}{\partial \delta_1} = -0.001422 \begin{Bmatrix} 0.6608 \\ -0.3557 \end{Bmatrix} = \begin{Bmatrix} -0.9397 \\ 0.05058 \end{Bmatrix} \times 10^{-2}$$

Similarly the other gradients can be obtained as

$$\frac{\partial \{X\}_2}{\partial \delta_1} = \begin{Bmatrix} 0.3578 \\ 0.13290 \end{Bmatrix} \times 10^{-2} \quad \text{(gg)}$$

$$\frac{\partial \{X\}_1}{\partial \delta_2} = \begin{Bmatrix} 0.23004 \\ -0.12382 \end{Bmatrix} \times 10^{-2} \quad \text{(hh)}$$

$$\frac{\partial \{X\}_2}{\partial \delta_2} = \begin{Bmatrix} -0.08758 \\ -0.32534 \end{Bmatrix} \times 10^{-2} \quad \text{(ii)}$$

The gradient of the dynamic displacement is given in Eq. (5.113) for which elements are evaluated as follows.

First, the change in c with respect to δ_j can be obtained from Eq. (5.105) as

$$\frac{\partial c_1}{\partial \delta_1} = \begin{bmatrix} -0.09397 & 0.05058 \end{bmatrix} \times 10^{-2} \begin{Bmatrix} 0 \\ 1 \end{Bmatrix} = (0.05058) \times 10^{-2} \qquad \text{(jj)}$$

Similarly

$$\frac{\partial c_1}{\partial \delta_2} = (-0.12382) \times 10^{-2}; \quad \frac{\partial c_2}{\partial \delta_1} = (0.13290) \times 10^{-2};$$

$$\frac{\partial c_2}{\partial \delta_2} = (-0.32534) \times 10^{-2} \qquad \text{(kk)}$$

Next, the gradient term related to the slope of the shock spectrum can be obtained from Eq. (5.113). The slope of Fig. 5.8 can be computed using the finite-difference interpolation given in Appendix C, we have

$$n_1 = 0.1292 \text{ and } n_2 = 0.0112 \qquad \text{(ll)}$$

where $h = 0.05$, the parameters for n_1 are

$$u = 0.44; \ f_{-2} = 1.690; \ f_{-1} = 1.717; \ f_o = 1.735; \ f_1 = 1.741; \text{ and } f_2 = 1.742$$

The parameters for η_2 are

$$u = 0.20; \ f_{-2} = 1.165; \ f_{-1} = 1.167; \ f_o = 1.168; \ f_1 = 1.169; \text{ and } f_2 = 1.168$$

Then, the change in displacement for each mode with respect to each design variable change is obtained using Eq. (5.113) as

$$\frac{\partial y_{p1}}{\partial \delta_1} = \frac{1}{(2)(4.6363)^3} \left(\frac{0.1292}{(2)(3)} - 2(4.6363)(0.8091) \right)(0.2565)$$

$$= -0.9377 \times 10^{-3} \qquad \text{(mm)}$$

Similarly

$$\frac{\partial y_{p1}}{\partial \delta_2} = -0.6796 \times 10^{-3}; \quad \frac{\partial y_{p2}}{\partial \delta_1} = -0.2064 \times 10^{-4};$$

$$\frac{\partial y_{p2}}{\partial \delta_2} = -0.6686 \times 10^{-4} \qquad \text{(nn)}$$

The gradient of the response of the system's top floor can be found using Eq. (5.104). Note that only the bounded constraints need to be considered.

From Eq. (5.104)

$$\begin{aligned}
\frac{\partial Y_{p2}}{\partial \delta_1} =& \left[(0.0508 \times 10^{-2})\,(0.9345)\,(0.08091)\right.\\
&+ (0.9345)\,(0.0505 \times 10^{-2})\,(0.08091)\\
&+ (0.9345)\,(0.9345)\,(-0.9377 \times 10^{-3})\big]\\
&+ \big[0.1329 \times 10^{-2}\,(-0.3557)\,(0.00315)\\
&+ (-0.3557)\,(0.1329 \times 10^{-2})\,(0.00315)\\
&+ (-0.3557)\,(-0.3557)\,(-0.2064 \times 10^{-4})\big]\\
=& -0.7479 \times 10^{-3}\,\text{ft in.}^{-3}
\end{aligned}$$

(oo)

Similarly the other gradients can be computed as

$$\frac{\partial Y_{p2}}{\partial \delta_2} = -0.7820 \times 10^{-3}\,\text{ft in.}^{-3} \tag{pp}$$

(D) *The feasible direction vector*

Using linear programming to solve Eqs. (5.24a) and (5.24b), we can find the best feasible direction vector given the gradient of the weight function and of the bounded constraint. The results calculated by longhand are given in Appendix D in which Eq. (D-22) is

$$\{s\} = \begin{bmatrix} 1.0 & -0.893 \end{bmatrix} = 1.3406 \begin{bmatrix} 0.7458 & -0.6660 \end{bmatrix} \tag{qq}$$

Although this vector was chosen to be the best according to Eqs. (5.24a) and (5.24b), any vector will work as long as Eqs. (5.23a) and (5.23b) are satisfied; that is

$$\{s\}^T \{\nabla g\} < 0$$
$$\{s\}^T \{\nabla W\} < 0$$

If the gradients are normalized, i.e.

$$\{\nabla g\} = \begin{bmatrix} 0.7479 & 0.7820 \end{bmatrix} \times 10^{-3} = 1.082 \times 10^{-3} \begin{bmatrix} 0.6912 & 0.7226 \end{bmatrix} \tag{rr}$$

and

$$\{\nabla W\} = \begin{bmatrix} 0.00786 & 0.00943 \end{bmatrix} = 1.227 \times 10^{-2} \begin{bmatrix} 0.640 & 0.768 \end{bmatrix} \tag{ss}$$

then

$$\{s\}^T \{\nabla g\} = -0.0342 < 0; \quad \{s\}^T \{\nabla W\} = -0.0341 < 0 \tag{tt}$$

These two calculations ensure that for some step length, α, a new feasible design can be obtained. A new design was found for $\alpha = 1.44$

$$\{\delta\}_2 = \left\{\begin{array}{c} 64.7 \\ 26.4 \end{array}\right\} + 1.44 \left\{\begin{array}{c} 1.0 \\ -0.893 \end{array}\right\} = \left\{\begin{array}{c} 66.14 \\ 25.10 \end{array}\right\} \text{in.}^3 \qquad \text{(uu)}$$

which was found to be the optimum design for $\beta = 0.001$ from Eq. (5.24b). The optimum weight was found to be

$$W_{opt} = 0.7568 \text{ kip} \qquad \text{(vv)}$$

for a constraint tolerance of $\varepsilon = -1/10$. It was found that for a smaller tolerance, a similar weight could be obtained since it was possible to get closer to the constraint.

Example 5.8.2 – The two-storey planar frame of Fig. 5.13 consisting of wide flange members rigidly connected has eight degrees of freedom including four rotational d.o.f. $(X_{\theta 1} \ldots X_{\theta 4})$ and two sideways

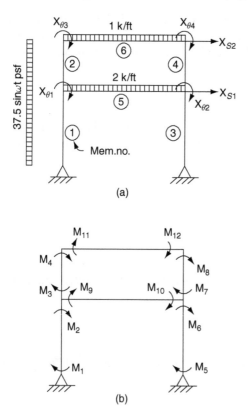

Figure 5.13 External and internal actions for a two-storey frame: (a) external degrees of freedom, (b) internal forces.

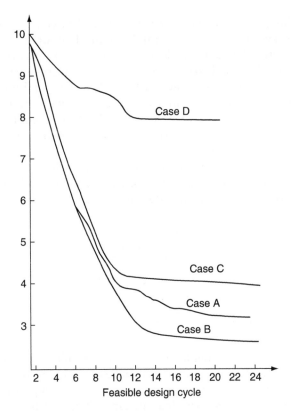

Figure 5.14 Weight versus cumulative feasible design cycle.

$(X_{s1}$ and $X_{s2})$. The system has four design variables, two for the columns and one for each of the girders which are identified in Fig. 5.13a where the design variables are the same for members (1) and (3); members (2) and (4) are identical. The member's length is: 15 ft of (1) and (3), 10 ft of (2) and (4), and 40 ft of (5) and (6). The two translational eigenvectors were used to approximate the dynamic analysis. The limits on the behaviour and side constraints were as follows

$$Y_{pj} \leq [\,0.20 \quad 0.25\,]\,\text{ft}, \quad \sigma_{pi} \leq 30\,\text{ksi} \quad i=1,\ldots,12; \quad j=1,2$$
$$1\,\text{rps} \leq p_1 \leq 100\,\text{rps}$$
$$10\,\text{in.}^3 \leq \delta_l \leq 100\,\text{in.}^3, \quad l=1,2,\ldots,4$$

Four cases were considered: (A) dynamic force only with no damping or axial load; (B) dynamic force with 10 per cent of critical damping; (C) 50 kips-per-column superimposed load considering the $P - \Delta$ effect, and (D) the same as case (A) but with consideration of static stresses due to uniform girder loading.

Solution:

In all cases the mass lumped at each girder resulted from the assumed slab weight resting on the girder equal to $2\,\text{kips ft}^{-1}$ for the first storey and $1\,\text{kip ft}^{-1}$ for the top storey. The dynamic loading at each storey was considered to be the result of a wind load of 37.5 psf with a time function approximated by half a sine wave as shown in Fig. 5.11b. The total force was computed on the basis of similar frames spaced 20 ft between frames. The design procedure attempts to decrease the weight by 10 per cent each step, but if a constraint is encountered during a step, the actual decrease may be somewhat less than 10 per cent.

A plot of the cumulative number of bounded designs versus the weight at the bounded design for the four cases is shown in Fig. 5.14. The optimum weight and active constraints are summarized in the table.

Case	Optimum weight (kips)	Active constraints
A	3.313	X_{s1}, X_{s2}
B	2.713	X_{s1}, X_{s2}, δ_4
C	4.032	y_2, σ_2, σ_3
D	8.016	$\sigma_6, \sigma_8, \sigma_{10}, \sigma_{12}$

It can be seen that the inclusion of damping in the analysis allows a much lighter weight design. It is seen that because of increased stresses due to secondary moments of the $P - \Delta$ effect as well as direct superimposition of axial upon bending stresses, the axial load design is bounded by σ_2 at a weight 20–30 per cent above the design in which the axial load is not present. The axial load also causes the columns to be at their maximum stresses in the final design, which results in columns somewhat larger than the other designs, while the girders are not affected as much.

All the designs terminated when β in Eqs. (5.24a) and (5.24b) became less than 0.001. This means that the constraints normal and the objective function normal are so related that no feasible direction can be found and therefore the minimum is encountered. This condition is an automatic satisfaction of the Kuhn–Tucker conditions for a minimum.

Example 5.8.3 – Find the maximum weight of the four-storey frame shown in Fig. 5.15 using the following constraints

$$Y_{pj} \le [0.1 \quad 0.15 \quad 0.2 \quad 0.25], \quad j = 1, \dots, 4$$
$$\sigma_{pi} \le 20\,\text{ksi}, \quad i = 1, \dots, 24$$
$$8\,\text{rps} \le p_1 \le 100\,\text{rps}$$
$$10\,\text{in.}^3 \le \delta_l \le 1{,}000\,\text{in.}^3, \quad l = 1, \dots, 8$$

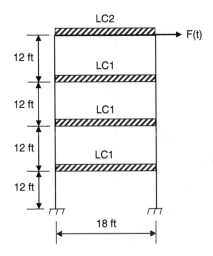

Figure 5.15 Four-storey shear building.

LC1 of loading condition number 1
   live load $ll = 0.6 \, \mathrm{k \, ft}^{-1}$
   dead load $dl = 1 \, \mathrm{k \, ft}^{-1}$
LC2 of loading condition number 2
   live load $ll = 1.2 \, \mathrm{k \, ft}^{-1}$
   dead load $dl = 2 \, \mathrm{k \, ft}^{-1}$

Design the structure for the base accelerations given in Fig. 5.16 which represented the recommended design loadings for nuclear power plant components. Similar spectral values could be used for any known earthquake acceleration time history. The maximum ground acceleration was assumed to be $0.15 \, g$ and there was no damping.

Solution:

Two cases were run to do the following:

(A) Test the method when frequency constraints are active.
(B) Test the effect of large column loads.

Curve (1) of Fig. 5.17 shows the weight of the frame versus the cumulative number of feasible designs for the case of no active frequency constraint and no column load. A feasible design is defined as one in which no constraints were violated. There were roughly twice as many analyses made as there were feasible designs.

   By comparison, Curve (2) shows the final design weight to be somewhat higher than that when the fundamental frequency is limited to a minimum of 11.7 rps. Curve (3) shows a most interesting result for the case of heavy

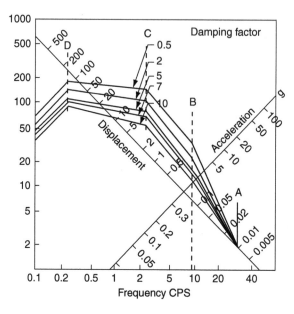

Figure 5.16 Design response spectrum.

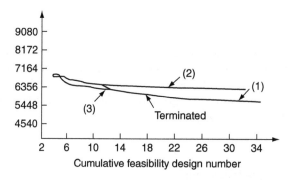

Figure 5.17 Weight versus feasible design number.

column loads of $P = 50$ kips. This type of loading causes lower frequencies and larger sidesway displacements because of the $P − \Delta$ effect and additional direct stress in the columns. The design can be seen to be terminated at the 18th iteration at a higher weight than the case in which there is no column load.

5.9 Concluding remarks

This chapter uses the nonlinear optimization algorithm based on the feasible direction search techniques presented in Chapter 4 for minimum weight design of rigid steel frames. The structures may be subjected to static loads, dynamic forces, and seismic excitations with $P - \Delta$ effects. The time-dependent forces and earthquake motions are expressed in the shock spectra of which the ordinates and slopes are obtained for a dynamic response by using the finite-difference technique. The formulations of finite difference and that of the stiffness-matrix method for rigid frames are reviewed with details. Two simple structures were designed by hand with the use of desk calculators. Therefore, the derived equations were inserted with various numerical values in order to make the design procedure easier to understand. Several selected example problems were given to show the significant effects of frequency, stress and displacement constraints and damping $P - \Delta$ effects on optimum design.

References

1. Cheng, F.Y., *Matrix Analysis of Structural Dynamics – Application and Earthquake Engineering*, (1005 pages), Marcel Dekkar, Inc., New York, and CRC Press, Boca Raton, FL, 2001.
2. Botkin, M. E., Shape optimization with buckling and stress constraints, *AIAA Journal*, Vol. 34, No. 2, pp. 423–425, 1996.
3. Cheng, F.Y. and Botkin, M. E., Second-order elasto-plastic analysis of tall buildings with damped dynamic excitations, *Proceedings of Finite Element Method in Civil Engineering*, Canadian Conference, Montreal, Canada, pp. 549–563, 1972.
4. Cheng, F.Y., ed., *Stability Under Seismic Loading*, ASCE, Reston, Virginia, 1986.
5. Cheng, F.Y. and Botkin, M. E., Nonlinear optimum design of dynamic damped frames, *Journal of Structural Engineering*, ASCE, Vol. 102, No. ST3, pp. 609–628, 1976.

6   Gradient-based search techniques

6.1   Introduction

Gradient-based optimization techniques have been a mainstay for unconstrained mathematical programming (optimization) procedures for decades. These techniques use gradients of the objective function in order to develop a path of sequential directions that lead to the optimal solution. For convex, well-behaved problems, the techniques essentially guarantee convergence making them extremely useful for select sets of problems. These unconstrained mathematical techniques typically have limited use within structural problems which often have constraints on responses such as displacements, stresses, natural frequencies or physical properties such as areas, moments of inertia or mass. The concepts of the gradient-based techniques coupled with their mathematical rigour are important and useful in certain constrained optimization algorithms. Within constrained optimization problems, adjustments to these ideas must be made to ensure that the algorithms will not allow the crossing or violation of a constraint as shown in Chapters 4 and 5. The following sections will provide the necessary theory as well as examples to illustrate the effectiveness of the different gradient-based, unconstrained optimization search techniques.

6.2   Unconstrained optimization

Unconstrained optimization with respect to structural systems has been successfully used in topology problems for material distribution to minimize volume or surface area, and geometric arrangements of members within structures to minimize potential energy. Although these types of problems have been solved with unconstrained optimization techniques, even topology and geometric structural optimization problems have constraints rendering unconstrained optimization as a minimally used tool for solving structural problems. In order to use these techniques, the structural problem will need to take the classical form of the unconstrained optimization problem.

The mathematical optimization problem for an unconstrained function with N variables can be defined as

$$\text{Minimize } W(\delta) \quad \text{where } \delta = \{\delta_i\} \quad \text{for } i = 1, ..., N \tag{6.1}$$

and δ is the set or vector of the independent (design) variables. The optimal solution to this problem requires every term of the gradient vector (first-order partial derivatives relative to the design variables) of the function to be zero, stated mathematically as

$$\nabla W(\delta^*) = 0 \tag{6.2}$$

where δ^* is the vector of independent variables that minimizes $W(\delta)$. In other words, δ^* is the optimal set of δ_i. The set of equations formed by setting the gradient to zero can rarely be solved directly due to implicit nonlinearities that exist between the function W and the independent variables δ. Therefore, these equations are best solved through an iterative process where logically based search directions, often based on the gradients, are used. This iterative process typically requires the use of an initial set of variables $\delta^{(0)}$ which is then modified through an equation of the form

$$\delta^{(j+1)} = \delta^{(j)} + \alpha^{(j)} d^{(j)} \tag{6.3}$$

where $d^{(j)}$ is the search direction found using the $\delta^{(j)}$ variables and $\alpha^{(j)}$ is the step size along this direction that will minimize or reduce $W(\delta^{(j+1)})$. Different techniques for determining the step size α are presented in Section 6.6, Line Search Methods. The search directions are developed such that $W(\delta^{(j+1)})$ is less than $W(\delta^{(j)})$ giving rise to the name, descent techniques, which is often used to describe these methods. The mathematical means of describing this is:

$$W(\delta^{(j+1)}) = W(\delta^{(j)} + \alpha^{(j)} d^{(j)}) < W(\delta^{(j)}) \tag{6.4}$$

Alternatively, this can also be stated as having the dot product of the search direction, $d^{(j)}$, and the gradient of the function to be minimized, $\nabla W(\delta^{(j)})$, being less than zero or

$$d^{(j)T} \nabla W(\delta^{(j)}) < 0 \tag{6.5}$$

Physically, this guarantees the search direction vector will have a component in the negative direction of the gradient of the objective function. Therefore, travelling along this direction (at least locally in the design space, small steps) will guarantee that the function W will be reduced since at least one component opposes the increasing direction (gradient) of W.

Algorithmically, these optimization processes are iterative and can involve two steps. The first step is to define the search direction and the second step is to choose a step size α. The selection of α is critical in guaranteeing that the function reduces from one iteration to the next. It can be a preselected constant (typically a very small step), it can be initially chosen and then modified from iteration to iteration based on an algorithm that guarantees a reduction in the objective or it can be optimized to take the optimal step along $d^{(j)}$ as discussed in Section 6.6, Line Search Methods. The first two methods of using a fixed step size or a non-optimized variable step size are less stable numerically with slower convergence rates for most problems, but simpler to implement. The third technique of optimizing the step length typically helps in guaranteeing that one will reach the optimal solution, and often helps convergence. Of course, one must first start with an initial set of design variables, $\delta^{(0)}$. The choice of the initial set of design variables can also slow or speed convergence depending on the mathematical characteristics of the problem (contour shapes) and the algorithm used to find the solution. At the end of each iteration, one must check for convergence. Convergence checks take many forms. Primarily, convergence is achieved when some norm measuring the change in the objective is within some prescribed value of precision. Convergence can also be related to the relative change in the design variables in a set of iterations, but this is typically not used by itself and is usually coupled with the relative change in the objective as well.

6.3   Steepest descent technique with a sample illustration

One of the first gradient-based, descent techniques developed was the method of steepest descent. This technique assumes that the best (most efficient) direction of descent is when the direction is exactly opposite to that of the gradient of the objective function. The method of steepest descent is classified as a first-order method since the negative of the gradient of the objective function at iteration j will be used to define the search direction. Mathematically stated,

$$d^{(j)} = -\nabla W(\delta^{(j)}) \tag{6.6}$$

The reason for choosing the negative of the gradient for the objective function is that it is diametrically opposed to the gradient and easily satisfies Eq. (6.5) as long as $\nabla W(\delta^{(j)})$ is not zero. These gradients are not always simple to find, but often explicit or finite-difference methods can be used to find these values.

As alluded to earlier, the value of α can be chosen as a constant or be modified during each iteration based on the direction chosen and the local value of the objective function. The most common procedure is to perform a one-dimensional search, as Section 6.6, to find the value of α which minimizes $\nabla W(\delta^{(j+1)})$ as a function of α, as shown in Eq. (6.4). It can be proven that if the correct value, α^*, is found, the successive search directions (gradients) must be orthogonal. Physically, this means that α^* would guarantee that the algorithm would travel along the search direction until it became tangent to a contour of the objective function at which point the next direction, $-\nabla W(\delta^{(j+1)})$ would have to be orthogonal to the tangent of W at the point $\delta^{(j+1)}$. The method of steepest descent is exceptionally efficient for problems that have smooth, convex contours, but can lose efficiency quickly when the contours become n-dimensionally elongated. This algorithm is lacking from the fact that it does not use the search direction histories to aid in finding the next search direction which can lead to slower convergence. Yet the steepest descent direction is often chosen as the initial search direction for the more robust algorithms which are discussed in the next sections.

Example 6.3.1 – Two-bar truss using steepest descent. Minimization of the potential energy function will be used to find the equilibrium-based deformed shape for the two-bar truss with the given properties, geometry and loading as shown in Fig. 6.1a. The final deformed shape and solution

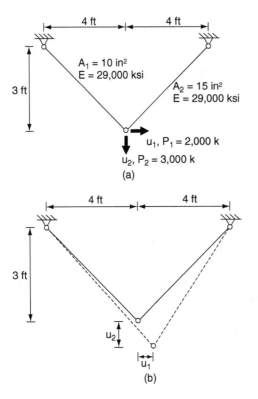

Figure 6.1 Two-bar truss: (a) undeformed, (b) deformed. (1 in. = 2.54 cm, 1 ft = 0.305 m, 1 kip = 4.45 kN).

are shown in Fig. 6.1b. The objective function which has been chosen as the potential energy function can be written as

$$\text{Minimize} \quad PE = \frac{1}{2}K_1(\Delta L_1)^2 + \frac{1}{2}K_2(\Delta L_2)^2 - P_1 u_1 - P_2 u_2 \tag{a}$$

where the change in relative length for each member is written as a function of the displacements, u_1 and u_2, as

$$\Delta L_1 = \sqrt{(3 + u_2)^2 + (4 + u_1)^2} - 5 \tag{b}$$

and

$$\Delta L_2 = \sqrt{(3 + u_2)^2 + (4 - u_1)^2} - 5 \tag{c}$$

These equations lead to the gradient equations (steepest descent direction) shown as

$$\frac{\partial PE}{\partial u_1} = K_1 \Delta L_1 \frac{\partial \Delta L_1}{\partial u_1} + K_2 \Delta L_2 \frac{\partial \Delta L_2}{\partial u_1} - P_1 \tag{d}$$

and

$$\frac{\partial PE}{\partial u_2} = K_1 \Delta L_1 \frac{\partial \Delta L_1}{\partial u_2} + K_2 \Delta L_2 \frac{\partial \Delta L_2}{\partial u_2} - P_2 \tag{e}$$

where the partial derivatives of the change in length are given as

$$\frac{\partial \Delta L_1}{\partial u_1} = \{(3 + u_2)^2 + (4 + u_1)^2\}^{-\frac{1}{2}} \times (4 + u_1) \tag{f}$$

$$\frac{\partial \Delta L_1}{\partial u_2} = \{(3 + u_2)^2 + (4 + u_1)^2\}^{-\frac{1}{2}} \times (3 + u_1) \tag{g}$$

$$\frac{\partial \Delta L_2}{\partial u_1} = -\{(3 + u_2)^2 + (4 - u_1)^2\}^{-\frac{1}{2}} \times (4 - u_1) \tag{h}$$

$$\frac{\partial \Delta L_2}{\partial u_2} = \{(3 + u_2)^2 + (4 - u_1)^2\}^{-\frac{1}{2}} \times (3 + u_2) \tag{i}$$

These gradients are used to define the steepest descent directions for use in predicting the iterative set of displacements as

$$u^{(k+1)} = u^{(k)} - \alpha(\nabla PE(u^{(k)})) \tag{j}$$

or in vector form

$$\begin{Bmatrix} u_1 \\ u_2 \end{Bmatrix}^{(k+1)} = \begin{Bmatrix} u_1 \\ u_2 \end{Bmatrix}^{(k)} - \alpha \begin{Bmatrix} \dfrac{\partial PE}{\partial u_1} \\ \dfrac{\partial PE}{\partial u_2} \end{Bmatrix} \tag{k}$$

In this example α is taken as a fixed value of 2×10^{-5}. If one chooses to use a fixed value, it usually requires several iterations of trial and error to pick an acceptable value. This choice is heavily dependent on the starting point and the shape of the contours associated with the objective function. For the two-bar truss in Fig. 6.1a with $A_1 = 10\,in.^2$ ($64.5\,cm^2$), $A_2 = 15\,in.^2$ ($96.8\,cm^2$) and $E = 29,000\,ksi$ ($20,010\,kN\,cm^{-2}$) the geometry and loading shown in Fig. 6.1a, the first iteration becomes ($1\,k\,in.^{-1} = 1.75\,kN\,cm^{-1}$)

$$K_1 = \frac{A_1 E}{L_1} = \frac{10 \times 29,000}{60} = 4,833\,k\,in.^{-1} \tag{l}$$

$$K_2 = \frac{A_2 E}{L_2} = \frac{15 \times 29,000}{60} = 7,250\,k\,in.^{-1} \tag{m}$$

coupled with the loading give initial displacements ($1\,in. = 2.54\,cm$) of

$$\begin{Bmatrix} u_1 \\ u_2 \end{Bmatrix}^{(0)} = \begin{Bmatrix} 2.0 \\ 4.0 \end{Bmatrix} in. \tag{n}$$

Substituting these displacements in Eqs. (a), (b) and (c) gives PE = $45,874\,k\text{-in.}$ ($518,514\,kN\text{-cm}$), $\Delta L_1 = 4.22\,in.$ ($10.7\,cm$) and $\Delta L_2 = 2.28\,in.$ ($5.79\,cm$). Eq. (f) becomes

$$\frac{\partial \Delta L_1}{\partial u_1} = \{(3+4)^2 + (4+2)^2\}^{-\frac{1}{2}} \times (4+2) = 0.651 \tag{o}$$

similarly, Eqs. (g), (h), and (i) give

$$\frac{\partial \Delta L_1}{\partial u_2} = \{(3+4)^2 + (4+2)^2\}^{-\frac{1}{2}} \times (3+4) = 0.759 \tag{p}$$

$$\frac{\partial \Delta L_2}{\partial u_1} = -\{(3+4)^2 + (4-2)^2\}^{-\frac{1}{2}} \times (4-2) = -0.275 \tag{q}$$

$$\frac{\partial \Delta L_2}{\partial u_2} = \{(3+4)^2 + (4-2)^2\}^{-\frac{1}{2}} \times (3+4) = 0.962 \tag{r}$$

The gradients for the potential energy which become the steepest descent directions, shown in Table 6.1 as d_{u1} and d_{u2}, are found using Eqs. (d) and (e) as

$$\frac{\partial PE}{\partial u_1} = d_{u1} = (4,833)(4.22)(0.651) + (7,250)(2.28)(-0.275)$$

$$-2,000 = 6,731.2 \tag{s}$$

$$\frac{\partial PE}{\partial u_2} = d_{u2} = (4,833)(4.22)(0.759) + (7,250)(2.28)(0.962)$$

$$-3,000 = 28,379 \tag{t}$$

The new displacements are found using Eqs. (j) or (k) (1 in. = 2.54 cm) as

$$\begin{Bmatrix} u_1 \\ u_2 \end{Bmatrix}^{(1)} = \begin{Bmatrix} 2 \\ 4 \end{Bmatrix} - (2 \times 10^{-5}) \begin{Bmatrix} 6,731.2 \\ 28,379 \end{Bmatrix} = \begin{Bmatrix} 1.87 \\ 3.43 \end{Bmatrix} \text{ in.} \tag{u}$$

The displacements in Eq. (u) become the initial point for the second iteration as shown in Table 6.1. The fixed value for the step size, α, causes the problem to converge slowly. The problem converges in approximately 75 iterations as shown in Table 6.1 and the path generated by the steepest descent method is shown in Fig. 6.2. From Table 6.1 one sees that the optimal solution for the potential energy should be zero (equilibrium), but the displacements have converged to an acceptable accuracy with minor improvements to be achieved with additional iterations. The rate of convergence can be improved significantly by using the techniques in Section 6.6 to optimize the step size, α, within each iteration.

Table 6.1 Two-bar truss results using the steepest descent method (1 in. = 2.54 cm).

Iteration	u_1 (in.)	u_2 (in.)	PE	d_{u1}	d_{u2}
1	2.00	4.00	45,874.00	6,731.20	28,379.00
2	1.87	3.43	30,599.00	6,007.60	22,463.00
3	1.75	2.98	20,840.00	5,466.50	17,944.00
4	1.64	2.62	14,454.00	5,041.20	14,472.00
5	1.54	2.33	10,167.00	4,686.40	11,785.00
.	.	.	.	.	.
.	.	.	.	.	.
.	.	.	.	.	.
74	0.40	0.67	−1,451.40	4.07	4.91
75	0.40	0.67	−1,451.40	3.66	4.43

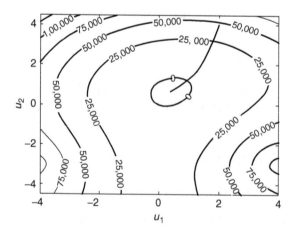

Figure 6.2 Two-bar truss using steepest descent method results, ($u^0 = [2,4]$) (1 in. = 2.54 cm).

6.4 Conjugate direction methods

6.4.1 *Quadratic form with sample illustrations*

Davidon in 1959 [2] and then Fletcher and Powell in 1963 [3] developed a method for solving for the minimum of quadratic functions in N or fewer iterations where N is the number of design variables. Unfortunately, most structural optimization problems are not quadratic in form and require a variation of this process to reach an optimal solution. Although this technique is not extremely useful for structural problems, the quadratic form is useful in understanding the concept of quadratic functions and for this reason it will be presented in theoretical form followed by a non-structural, quadratic function problem.

If A is an $N \times N$ symmetric, positive definite matrix, then the direction vectors $d^{(i)}$ and $d^{(j)}$, from linear algebra terminology, are said to be conjugate if and only if

$$d^{(i)T} A d^{(j)} = 0 \quad \text{where } i \neq j \tag{6.7}$$

Since A is positive definite and if Eq. (6.7) is true, it can be proven that $d^{(i)}$ and $d^{(j)}$ are linearly independent directions. This is important in that it guarantees that these directions form a basis of vectors (coordinate directions) that are linearly independent. If A is not positive definite, the directions are not linearly independent and the system is said to be A-conjugate.

The standard form of a quadratic problem would include the matrix A and vectors of coefficients B and C, written as

$$\left(\frac{1}{2}\right) \delta^T A \delta - B^T \delta + C \tag{6.8}$$

When the objective function takes this quadratic form and one uses conjugate directions with optimal step lengths, it can be proven (Bazaraa and Shetty, [1]) that finite convergence will occur in N or fewer iterations. A less rigorous proof than shown in [1] is shown here.

If A is symmetric positive definite, the minimization of Eq. (6.8) is equivalent to solving the linear problem in Eq. (6.9) which is found by taking the derivative of Eq. (6.8) and setting it equal to zero.

$$A \delta = B \tag{6.9}$$

If δ^* is the optimal solution of Eq. (6.8) and (6.9) and knowing that δ^* can be written as a vector sum of the conjugate (independent or basis) directions as

$$\delta^* = \sum_{j=0}^{N-1} \alpha^{(j)} d^{(j)} \tag{6.10}$$

then the coefficients can be found by substituting Eq. (6.10) into Eq. (6.8), taking the derivative with respect to $\alpha^{(j)}$ and solving for $\alpha^{(j)}$ as

$$\alpha^{(j)} = \frac{d^{(j)T} A \delta^*}{d^{(j)T} A d^{(j)}} = \frac{d^{(j)T} B}{d^{(j)T} A d^{(j)}} \tag{6.11}$$

The fact that $\alpha^{(j)}$ is independent of δ^* clearly indicates that the linear equations can be solved by using N iterative steps. Beginning with Eq. (6.10),

$$\delta^{(j+1)} = \delta^{(0)} + \sum_{i=0}^{j} \alpha^{(i)} d^{(i)} \tag{6.12}$$

and knowing that $\alpha^{(j)}$ minimizes Eq. (6.8) with respect to the direction $d^{(j)}$ we know that the direction $d^{(j)}$ is orthogonal to the gradient, g, of $W(\delta)$ giving

$$d^{(j)T} g^{(j+1)} = 0 \tag{6.13}$$

where $g^{(j+1)}$ can be found as the derivative of Eq. (6.8) and substituting Eq. (6.12) in this gradient gives

$$g^{(j+1)} = A\delta^{(j+1)} - B = A\delta^{(0)} - B + \sum_{i=0}^{j} \alpha^{(i)} A d^{(i)} \tag{6.14}$$

Using Eqs. (6.13) and (6.14) and, premultiplying Eq. (6.14) by $d^{(j)T}$ and recognizing Eq. (6.7) and that this quantity must be zero according to Eq. (6.13), $\alpha^{(j)}$ can be found to be

$$\alpha^{(j)} = -\frac{d^{(j)T} \left\{ A\delta^{(0)} - B \right\}}{d^{(j)T} A d^{(j)}} \tag{6.15}$$

and for $j = N - 1$ the solution for $\delta^{(N)}$ becomes

$$\delta^{(N)} = \delta^{(0)} - \sum_{j=0}^{N-1} \left\{ \frac{d^{(j)T} \left\{ A \delta^{(0)} - B \right\}}{d^{(j)T} A d^{(j)}} \right\} d^{(j)} = A^{-1} B \tag{6.16}$$

Note that Eq. (6.16) shows that the optimal solution for a quadratic function can be found by simply finding the N conjugate directions and substituting them into Eq. (6.16). This will be illustrated in the following example.

Example 6.4.1 – Conjugate direction method. The given example is a quadratic function; therefore, it will be shown that the problem can easily be solved either in two iterations of the conjugate direction method or using Eq. (6.16) directly.

$$W(\delta) = 6\delta_1^2 + 6\delta_2^2 - 6\delta_1\delta_2 - 18\delta_2$$

$$= \left(\frac{1}{2}\right)\begin{Bmatrix} \delta_1 \\ \delta_2 \end{Bmatrix}^T \begin{bmatrix} 12 & -6 \\ -6 & 12 \end{bmatrix} \begin{Bmatrix} \delta_1 \\ \delta_2 \end{Bmatrix} + \begin{Bmatrix} 0 \\ 18 \end{Bmatrix}^T \begin{Bmatrix} \delta_1 \\ \delta_2 \end{Bmatrix} + 0 \tag{a}$$

Choosing $d^{(0)T} = \{1\ 0\}$ the second direction $d^{(1)}$ must be found as the conjugate direction relative to A or mathematically stated

$$\{1\ \ 0\}\begin{bmatrix} 12 & -6 \\ -6 & 12 \end{bmatrix}\begin{Bmatrix} d_1^{(1)} \\ d_2^{(1)} \end{Bmatrix} = 0 \tag{b}$$

giving $d^{(1)} = \{1\ 2\}$. Using Eq. (6.15) $\alpha^{(1)}$ must be found to minimize the function along $d^{(0)}$ first, then a second $\alpha^{(2)}$ must be found for $d^{(1)}$ which will then result in the solution for δ. The values for α can be found using Eq. (6.15) or derived from Eq. (6.16) as the term in the large brackets of Eq. (c). Arbitrarily choosing $\{-1/2\ -1/2\}$ as the starting point $\delta^{(0)}$ and using Eq. (6.16) as shown

$$\delta^* = \begin{Bmatrix} -1/2 \\ -1/2 \end{Bmatrix} - \left(\frac{\{1\ \ 0\}\left\{ \begin{bmatrix} 12 & -6 \\ -6 & 12 \end{bmatrix}\begin{Bmatrix} -1/2 \\ -1/2 \end{Bmatrix} - \begin{Bmatrix} 0 \\ 18 \end{Bmatrix} \right\}}{\{1\ \ 0\}\begin{bmatrix} 12 & -6 \\ -6 & 12 \end{bmatrix}\begin{Bmatrix} 1 \\ 0 \end{Bmatrix}} \right) \begin{Bmatrix} 1 \\ 0 \end{Bmatrix}$$

$$- \left(\frac{\{1\ \ 2\}\left\{ \begin{bmatrix} 12 & -6 \\ -6 & 12 \end{bmatrix}\begin{Bmatrix} -1/2 \\ -1/2 \end{Bmatrix} - \begin{Bmatrix} 0 \\ 18 \end{Bmatrix} \right\}}{\{1\ \ 2\}\begin{bmatrix} 12 & -6 \\ -6 & 12 \end{bmatrix}\begin{Bmatrix} 1 \\ 2 \end{Bmatrix}} \right) \begin{Bmatrix} 1 \\ 2 \end{Bmatrix} \tag{c}$$

$$= \begin{Bmatrix} -1/2 \\ -1/2 \end{Bmatrix} + 1/4\begin{Bmatrix} 1 \\ 0 \end{Bmatrix} + 5/4\begin{Bmatrix} 1 \\ 2 \end{Bmatrix} = \begin{Bmatrix} 1 \\ 2 \end{Bmatrix}$$

the solution is shown to be $\delta^{*T} = \{1\ 2\}$ which is equivalent to $[A]^{-1}B$ (for a quadratic problem) as shown here

$$\delta^* = \begin{bmatrix} 1/9 & 1/18 \\ 1/18 & 1/9 \end{bmatrix}\begin{Bmatrix} 0 \\ 18 \end{Bmatrix} = \begin{Bmatrix} 1 \\ 2 \end{Bmatrix} \tag{d}$$

From Eq. (c) one realizes that $\alpha^{(1)}$ is $-1/4$ and $\alpha^{(2)}$ is $-5/4$.

A second solution using a different initial point and the numerical approach for conjugate directions coupled with line searches yielded the

Table 6.2 Results for the quadratic function using the conjugate direction method.

Iteration	δ_1	δ_2	$F(\delta)$	$d_{\delta 1}$	$d_{\delta 2}$	α
0	6.00	20.00	1,536.00	48.00	−186.00	0.067099
1	9.22	7.52	298.03	−59.57	−40.00	0.13799
2	1.00	2.00	−18.00	0.00	0.00	–

values in Table 6.2 and the numerical solution shown in Fig. 6.3. The details are shown here for each cycle:

$$W(\delta) = 6\delta_1^2 + 6\delta_2^2 - 6\delta_1\delta_2 - 18\delta_2$$

$$= \frac{1}{2} \begin{Bmatrix} \delta_1 \\ \delta_2 \end{Bmatrix}^T \begin{bmatrix} 12 & -6 \\ -6 & 12 \end{bmatrix} \begin{Bmatrix} \delta_1 \\ \delta_2 \end{Bmatrix} + \begin{Bmatrix} 0 \\ 18 \end{Bmatrix}^T \begin{Bmatrix} \delta_1 \\ \delta_2 \end{Bmatrix} + 0 \qquad \text{(e)}$$

producing the gradient

$$\nabla W = \begin{Bmatrix} 12\delta_1 - 6\delta_2 \\ 12\delta_2 - 6\delta_1 - 18 \end{Bmatrix} \qquad \text{(f)}$$

and Hessian matrix

$$A = \nabla^2 W = \begin{bmatrix} 12 & -6 \\ -6 & 12 \end{bmatrix} \qquad \text{(g)}$$

Therefore, cycle 1 begins with the arbitrarily chosen starting point of {6 20} providing a new set of directions and step lengths, yet converging in two cycles.

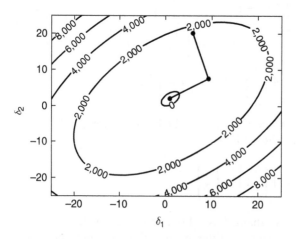

Figure 6.3 Conjugate gradient method results, $(\delta^0 = [6, 20])$.

$$\delta^{(0)} = \begin{Bmatrix} 6 \\ 20 \end{Bmatrix} \tag{h}$$

which gives a starting objective value of

$$W(\delta^{(0)}) = 6(6)^2 + 6(20)^2 - 6(6)(20) - 18(20) = 1,536 \tag{i}$$

Using Eq. (f) the negative gradient of W becomes the first direction as

$$d^{(1)} = -\nabla W^{(1)T} = \{48 \quad -186\} \tag{j}$$

and the step size can be found as

$$\alpha^{(1)} = \frac{\{d^{(1)}\}\{d^{(1)}\}^T}{\{d^{(1)}\}[A]\{d^{(1)}\}^T} = 0.0671 \tag{k}$$

giving a new set of variables as

$$\delta^{(1)} = \{\delta^{(0)}\} + \alpha^{(1)}\{d^{(1)}\}^T = \begin{Bmatrix} 6 \\ 20 \end{Bmatrix} + 0.0671 \begin{Bmatrix} 48 \\ -186 \end{Bmatrix} = \begin{Bmatrix} 9.22 \\ 7.52 \end{Bmatrix} \tag{l}$$

Cycle 2 begins with the new starting point found in Eq. (l)

$$\delta^{(1)} = \begin{Bmatrix} 9.22 \\ 7.52 \end{Bmatrix} \tag{m}$$

which gives a significantly reduced objective value compared to the initial starting point

$$W(\delta^{(0)}) = 6(9.22)^2 + 6(7.52)^2 - 6(9.22)(7.52) - 18(7.52) = 298.03 \tag{n}$$

The gradient of the objective can be found in two ways, one directly using Eq. (f)

$$\nabla W^{(2)T} = \{65.533 \quad 16.911\} \tag{o}$$

or using the equation developed in Eq. (6.14)

$$g^{(2)} = [A]\{\delta^{(0)}\} - \alpha^{(1)}[A]\{d^{(1)}\}^T = \begin{Bmatrix} -65.532 \\ -16.911 \end{Bmatrix} \tag{p}$$

Note that both Eqs. (o) and (p) produce the same results. The parameter β, a numerical constant derived from Eq. (6.14),

$$\beta^{(2)} = \frac{\{d^{(1)}\}[A]\{g^{(2)}\}^T}{\{d^{(1)}\}[A]\{d^{(1)}\}^T} = -0.124 \tag{q}$$

is used to develop the next direction vector

$$\{d^{(2)}\}^T = \{g^{(2)}\}^T - \beta^{(2)}\{d^{(1)}\}^T = \{-59.57 \quad -40.00\} \tag{r}$$

Once this new direction is defined, the optimal step size is found using Eq. (6.15) as

$$\alpha^{(2)} = \frac{\{d^{(2)}\}\{g^{(2)}\}^T}{\{d^{(2)}\}[A]\{d^{(2)}\}^T} = 0.138 \tag{s}$$

With the new direction vector and step size the optimal set of variables is found as

$$\delta^{(2)} = \{\delta^{(1)}\} + \alpha^{(2)}\{d^{(2)}\}^T = \left\{\begin{array}{c} 9.22 \\ 7.52 \end{array}\right\} + 0.138 \left\{\begin{array}{c} -59.57 \\ -40.00 \end{array}\right\} = \left\{\begin{array}{c} 1.00 \\ 2.00 \end{array}\right\} \tag{t}$$

The intermediate values and final results for this example, Eqs. (a)–(t), are summarized in Table 6.2. For simple problems, finding the conjugate directions is as easy as shown in the example, but the difficulty of this technique increases due to the need to define the conjugate directions for N directions where N is large. This is where the conjugate optimization procedures differ as shown in the following sections.

6.4.2   Conjugate gradient method with a sample illustration

The conjugate gradient method developed by Fletcher and Reeves [4] begins by using the steepest descent direction for the first iteration. Thereafter a new direction, conjugate in the case of a quadratic function, is formed by adding to the steepest descent a scaled version of the previous step's direction. This is the same as the generalized conjugate direction method where the conjugate directions are formed by orthogonalizing the successive directions, in this case the gradients.

From Eq. (6.13) it was shown for a quadratic function that

$$\nabla W^{(i+1)T} d^{(i)} = 0 \quad \text{for} \quad i = 1, \ldots, N \tag{6.17}$$

which can be shown to be equivalent to

$$\nabla W^{(i+1)T} d^{(i)} = 0 \tag{6.18}$$

due to orthogonality of the directions to the A matrix, Eq. (6.7). This leads to the main notion of the conjugate gradient approach which is

$$\nabla W^{(i+1)T} \nabla W^{(i)} = 0 \quad \text{for} \quad i = 1, \ldots, N \tag{6.19}$$

This relationship holds by the fact that each $d^{(i)}$ is formed by a successive orthogonalization of the gradients and therefore are defined within the same subspace. Each successive direction is found through the equation

$$d^{(i+1)} = -\nabla W^{(i+1)} + \sum_{j=1}^{i} \beta^{(j)} d^{(j)} \tag{6.20}$$

where $\beta^{(j)}$ is derived in a manner to ensure orthogonality. By premultiplying Eq. (6.20) by $d^{(i)T}A$ and using the concept of conjugate directions, Eq. (6.7), the equation for $\beta^{(j)}$ can be written as

$$\beta^{(j)} = \frac{\nabla W^{(j+1)T} A d^{(j)}}{d^{(j)T} A d^{(j)}} \tag{6.21}$$

and knowing that

$$A d^{(j)} = \nabla W^{(j+1)} - \nabla W^{(j)} \tag{6.22}$$

$\beta^{(j)}$ must be zero for all j prior to the ith iteration so Eq. (6.20) becomes

$$d^{(i+1)} = -\nabla W^{(i+1)} + \beta^{(i)} d^{(i)} \tag{6.23}$$

And by premultiplying by $d^{(i)T}A$, using Eq. (6.22) and conjugacy Eq. (6.7), $\beta^{(i)}$ can be found as

$$\beta^{(i)} = \frac{\nabla W^{(i+1)T}(\nabla W^{(i+1)} - \nabla W^{(i)})}{d^{(i)T}(\nabla W^{(i+1)} - \nabla W^{(i)})} \tag{6.24}$$

If the problem is quadratic it can be shown that

$$d^{(i)T}\nabla W^{(i+1)} = 0 \tag{6.25}$$

and that

$$\nabla W^{(i+1)}\nabla W^{(i)} = 0 \tag{6.26}$$

so that

$$d^{(i)T}\nabla W^{(i)} = -\nabla W^{(i)T}\nabla W^{(i)} \tag{6.27}$$

giving rise to the Fletcher–Reeves conjugate gradient formula for β as

$$\beta^{(j)} = \frac{\nabla W^{(j+1)T}\nabla W^{(j+1)}}{\nabla W^{(j)T}\nabla W^{(j)}} \tag{6.28}$$

If the system is not quadratic but Eqs. (6.25) and (6.27) hold true, Polak and Ribiere [7] show that

$$\beta^{(j)} = \frac{\nabla W^{(j+1)T}(\nabla W^{(j+1)} - \nabla W^{(j)})}{\nabla W^{(j)T} \nabla W^{(j)}} \tag{6.29}$$

and if Eqs. (6.25) and (6.26) hold true, Lenard [6] shows that

$$\beta^{(j)} = \frac{\nabla W^{(j+1)} \nabla W^{(j+1)}}{d^{(j)T}(\nabla W^{(j+1)} - \nabla W^{(j)})} \tag{6.30}$$

All three cases can be considered by implementing Eq. (6.24) in the algorithm since those conditions in Eqs. (6.25)–(6.27) will create by default the Eqs. (6.28)–(6.30).

Although the concept of conjugate directions and conjugate gradient techniques is based on quadratic forms, non-quadratic problems can be solved using these techniques. As one approaches the local minimum of a problem, typically the problem can be closely approximated with a quadratic form which leads to a finite convergent solution. If one assumes that the $\delta^{(n)}$ solution is closer to the optimal solution than $\delta^{(0)}$, then the problem could be restarted at $\delta^{(0)} = \delta^{(n)}$ giving a series of solutions that approach the global solution [7]. In addition to restarting the algorithm every n iterations, a non-quadratic problem can become ill-conditioned by violating Eq. (6.5) or mathematically stated

$$d^{(j)T} \nabla W^{(j+1)} > 0 \tag{6.31}$$

and if the one-dimensional search does not improve the objective, the algorithm should be restarted (if the steepest descent direction used in the restart fails to improve the objective, the minimum has been reached) (Vanderplatts, [8]).

Example 6.4.2 – Conjugate gradient method – Fletcher–Reeves technique. The given function is to be minimized by using the Fletcher–Reeves technique

$$\text{Minimize} \quad W(\delta) = (\delta_2 - 1)^4 + (\delta_2 - \delta_1)^2 \tag{a}$$

where the gradients can be calculated as

$$\nabla W = \left\{ \begin{array}{c} -2(\delta_2 - \delta_1) \\ 4(\delta_2 - 1)^3 + 2(\delta_2 - \delta_1) \end{array} \right\} \tag{b}$$

and the Hessian matrix (second derivatives) is given as

$$A = \nabla^2 W = \begin{bmatrix} 2 & -2 \\ -2 & 12(\delta_2 - 1)^2 + 2 \end{bmatrix} \tag{c}$$

Here are the detailed calculations for the first two cycles of this example.

Cycle 1

Starting with an initial point of

$$\{\delta^{(0)}\} = \begin{Bmatrix} 10 \\ 10 \end{Bmatrix} \tag{d}$$

gives the initial value for the objective function as

$$W = (10 - 1)^4 + (10 - 10)^2 = 6,561 \tag{e}$$

and the numerically generated gradient from Eq. (b) of

$$\nabla W = \begin{Bmatrix} 0 \\ 2,916 \end{Bmatrix} \tag{f}$$

As noted in the text, the initial value for β is taken as zero

$$\beta^{(1)} = 0 \tag{g}$$

providing the first direction as just the negative of the gradients found in Eq. (f)

$$d^{(1)} = \{-\nabla W(\delta^{(0)})\}^T = \{0 \quad -2,916\} \tag{h}$$

Using the quadratic line search techniques discussed in Section 6.6, $\alpha^{(1)}$ is found as 0.00253 which in turn provides the new set of variables as

$$\{\delta^{(1)}\} = \{\delta^{(0)}\} + \alpha^{(1)}\{d^{(1)}\}^T = \begin{Bmatrix} 10 \\ 10 \end{Bmatrix} + 0.00253 \begin{Bmatrix} 0 \\ -2,916 \end{Bmatrix} = \begin{Bmatrix} 10 \\ 2.63 \end{Bmatrix} \tag{i}$$

Cycle 2

Cycle 2 begins with the initial values from Eq. (i)

$$\{\delta^{(1)}\} = \begin{Bmatrix} 10 \\ 2.63 \end{Bmatrix} \tag{j}$$

and gives the significantly reduced weight of

$$W = (2.63 - 1)^4 + (2.63 - 10)^2 = 61.38 \tag{k}$$

The gradient for this new point becomes

$$\nabla W^{(2)} = \begin{Bmatrix} 14.74 \\ 2.61 \end{Bmatrix} \tag{l}$$

Using Eq. (6.28) the value for β becomes

$$\beta^{(2)} = \frac{\{\nabla W^{(2)}\}\{\nabla W^{(2)}\}^T}{\{\nabla W^{(1)}\}\{\nabla W^{(1)}\}^T} = 2.63 \times 10^{-5} \tag{m}$$

giving the new direction shown in Eq. (n).

$$\{d^{(2)}\} = \{-\nabla W(\delta^{(1)})\}^T - \beta^{(2)}\{d^{(1)}\} = \{-14.74 \quad -2.53\} \tag{n}$$

Finding the optimal step size using the quadratic line search technique gives $\alpha^{(2)} = 0.60$ and the new set of design variables can be found as

$$\{\delta^{(2)}\} = \{\delta^{(2)}\} + \alpha^{(2)}\{d^{(2)}\}^T = \begin{Bmatrix} 10 \\ 2.63 \end{Bmatrix} + 0.60 \begin{Bmatrix} -14.74 \\ -2.53 \end{Bmatrix} = \begin{Bmatrix} 1.15 \\ 1.11 \end{Bmatrix} \tag{o}$$

This new set of variables greatly reduces the objective function again as seen in Table 6.3, iteration 3. The first and second iteration results, from Eqs. (a)–(o), are shown in the final numerical results in Table 6.3. (The final path to the optimal solution can be seen in Fig. 6.4.) The objective function is not actually zero in iterations 2 through 8, but is sufficiently small to round to 0.00 in each iteration.

Table 6.3 Results using the Fletcher–Reeves conjugate gradient method.

Iteration	δ_1	δ_2	$W(\delta)$	$d_{\delta 1}$	$d_{\delta 2}$	A
0	10.00	10.00	6,561.00	0.000	−2,916	0.0025
1	10.00	2.63	61.38	−14.74	−2.53	0.60
2	1.15	1.11	0.00	−0.079	0.075	0.24
3	1.13	1.13	0.00	−0.004	−0.004	10.38
4	1.09	1.09	0.00	−0.001	−0.001	2.98
5	1.08	1.08	0.00	0.011	0.000	0.10
6	1.08	1.08	0.00	−0.002	−0.002	11.42
7	1.06	1.06	0.00	0.018	0.009	0.15
8	1.06	1.06	0.00	−0.001	−0.003	0.73

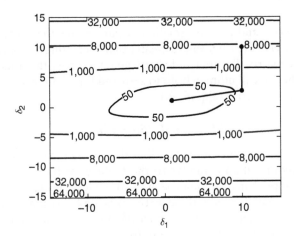

Figure 6.4 Fletcher–Reeves method results, $(\delta^0 = [10, 10])$.

6.4.3   *Newton method*

Newton methods are based on a Taylor series expansion of a function in order to approximate it with a quadratic form. Using this expansion, one can derive an iterative process for changing the variables (finding a search direction for the approximate function) using the Hessian matrix. The Taylor series expansion (approximate equation) becomes

$$Q(\delta) = W(\delta^{(i)}) + \nabla W(\delta^{(i)})^T(\delta - \delta^{(i)}) + \frac{1}{2}(\delta - \delta^{(i)})^T H(\delta^{(i)})(\delta - \delta^{(i)}) \quad (6.32)$$

where H is the Hessian matrix of W at $\delta^{(i)}$. At the optimal solution $\nabla Q(\delta)$ must equal 0 giving

$$\nabla W(\delta^{(i)}) + H(\delta^{(i)})(\delta - \delta^{(i)}) = 0 \quad (6.33)$$

If the inverse of the Hessian exists, the new variables for δ can be found as

$$\delta^{(i+1)} = \delta^{(i)} - H(\delta^{(i)})^{-1} \nabla W(\delta^{(i)}) \quad (6.34)$$

The difficulty with using the Newton method is the necessity for finding the Hessian matrix in every iteration and the fact that the Hessian must be inverted, if possible, in every iteration. There is no guarantee that the Newton method will converge to a solution unless one is sufficiently close to the optimal solution and H is of full rank. Bazaraa and Shetty [1] proposed a modification to the Newton method and proved that it will converge to an optimal solution. They proposed using

$$\delta^{(i+1)} = \delta^{(i)} - B(\delta^{(i)}) \nabla W(\delta^{(i)}) \tag{6.35}$$

where $B(\delta^{(i)}) = (\varepsilon I + H)^{-1}$ and ε is the smallest scalar greater than 0 that makes the matrix $(\varepsilon I + H)$ have eigenvalues greater than 0 creating a positive definite and invertible matrix. Bazaraa and Shetty go on to prove that this is a descent function where ε will ultimately go to zero and Eq. (6.35) becomes equivalent to the Newton method Eq. (6.34).

6.4.4   Quasi-Newton method with a sample illustration

Quasi-Newton methods make use of the fact that the Hessian matrix or its inverse can be approximated and, similar to the findings of Bazaraa and Shetty [1], that these approximations approach the true Hessian as the optimal solution is approached. Typically these methods begin by assuming the Hessian to be the identity matrix, I, which relegates the initial iteration to using the steepest descent direction given in Eq. (6.7).

From Vanderplaats [10], the search directions for the Davidon–Fletcher–Powell method and the Broydon–Fletcher–Goldfarb–Shanno method can be written in this form:

$$d^{(j)} = -H \nabla W(\delta^{(j)}) \tag{6.36}$$

where H is an approximation of the inverse of the Hessian matrix (the initial value used for H is the identity matrix) as given by this equation

$$H^{(j+1)} = H^{(j)} + h^{(j)} \tag{6.37}$$

where h is a symmetric matrix that updates H and is given as

$$h^{(j)} = \left(\frac{\gamma + \Gamma \varsigma}{\gamma^2} \right) r r^T + \left(\frac{\Gamma - 1}{\varsigma} \right) H^{(j)} q \left(H^{(j)} q \right)^T - \frac{\Gamma}{\gamma} \left(H^{(j)} q r^T + r \left(H^{(j)} q \right)^T \right) \tag{6.38}$$

and the vectors r and q and the scalars γ and ς are defined as

$$r = \delta^{(j)} - \delta^{(j-1)} \tag{6.39}$$

$$q = \nabla W(\delta^{(j)}) - \nabla W(\delta^{(j-1)}) \tag{6.40}$$

$$\gamma = r^T q \tag{6.41}$$

$$\varsigma = q^T H^{(j)} q \tag{6.42}$$

The Davidon–Fletcher–Powell method requires Γ to be 0 and the Broydon–Fletcher–Goldfarb–Shanno method requires Γ to be 1.

Example 6.4.3 – Newton method example. Using the same function as in Example 6.4.2

$$\text{Minimize} \quad W(\delta) = (\delta_2 - 1)^4 + (\delta_2 - \delta_1)^2 \tag{a}$$

and comparing Figs. 6.4 and 6.5, one can see the benefits of using the Newton method. Even though the contours are elongated and the same initial point is used, the Newton method converges in half the number of iterations. The iterative solution can be seen in Table 6.4 and the path for the solution can be seen in Fig. 6.5.

Table 6.4 Results using the Newton method.

Iteration	δ_1	δ_2	$W(\delta)$	$d_{\delta 1}$	$d_{\delta 2}$	A
1	2.10	2.10	1.48	−0.37	−0.37	2.63
2	1.13	1.13	0.00	−0.05	−0.05	2.63
3	1.02	1.02	0	−0.01	−0.01	4.71
4	0.9906	0.9906	0.00	0.00	0.00	4.71

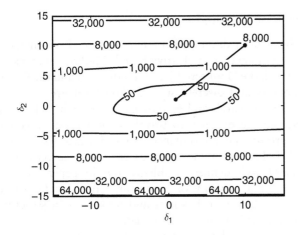

Figure 6.5 Newton method results, $(\delta^0 = [10, 10])$.

Numerically, the steps are

$$\{\delta^{(0)}\} = \begin{Bmatrix} 10 \\ 10 \end{Bmatrix} \tag{b}$$

and

$$\nabla W = \begin{Bmatrix} 0 \\ 2,916 \end{Bmatrix} \tag{c}$$

and

$$\nabla^2 W = \begin{bmatrix} 2 & -2 \\ -2 & 12(\delta_2 - 1)^2 + 2 \end{bmatrix} = \begin{bmatrix} 2 & -2 \\ -2 & 974 \end{bmatrix} \qquad \text{(d)}$$

Using Eqs. (b)–(d) the new direction can be found as

$$d^{(0)} = -(\nabla^2 W^{-1})\, \nabla W = \begin{Bmatrix} -3 \\ -3 \end{Bmatrix} \qquad \text{(e)}$$

Using the quadratic approximation line search technique discussed in Section 6.6, $\alpha = 2.6327$ and

$$\{\delta^{(1)}\} = \begin{Bmatrix} 10 \\ 10 \end{Bmatrix} + 2.633 \begin{Bmatrix} -3 \\ -3 \end{Bmatrix} = \begin{Bmatrix} 2.1 \\ 2.1 \end{Bmatrix} \qquad \text{(f)}$$

The results for the remaining three iterations are given in Table 6.4 and shown in Fig. 6.5

6.5 Non-derivative-based methods

Numerous methods have been developed for unconstrained optimization techniques that are not gradient- or derivative-based. Methods such as the Cyclic Coordinate method, the Hooke and Jeeves method [5] and the Rosenbrock method (continuous and discrete step methods) [9] are based on using a defined direction of search and employing line search techniques, Section 6.6, to determine the distance to travel along a given direction. Detailed discussions, proofs and convergence properties are described in Bazaraa and Shetty [1]. Each of these methods can produce reasonable results for select sets of well defined problems, but in general cannot guarantee convergence for a majority of nonlinear, large variable objective functions. Due to the lack of robustness and usefulness in structural problems, these techniques will be briefly discussed.

The Cyclic Coordinate method searches along the coordinate axes, one at a time. Therefore, each direction vector is populated with zeros for all but one component which is one for the given coordinate axis being used as the search direction. This is equivalent to changing one variable at a time, searching for the optimal step length and iterating through all of the coordinate axes (often multiple times) until the optimal solution is reached.

The Hooke and Jeeves method uses a two-step process. First, it uses the Cyclic Coordinate method (one pass through all coordinate directions) to generate a new set of variables. Then a search along the vector formed by subtracting the initial point from the new point is performed; this is called a pattern search. This generates another starting point which is then used to begin the two-step process over again. For many problems this greatly improves the rate of convergence relative to the Cyclic Coordinate method.

The Rosenbrock method searches along n linearly independent and orthogonal directions. Once the new point is found, a new set of n orthogonal directions is generated and the process continues until convergence is achieved. The original Rosenbrock method used discrete, predetermined steps along these orthogonal search directions, but line search techniques can be used to find optimal step lengths making the process more robust.

6.6   Line search methods

Most methods using descent directions for updating the design variables use a uni-variable optimization of the distance to travel (step size) along the descent direction. The step size is found such that it will minimize the original objective along the chosen direction. For a given updating scheme with an updating direction $d^{(j)}$ the problem becomes

$$\text{Minimize}\quad W(\delta^{(j+1)}) = W(\delta^{(j)} + \alpha\, d^{(j)}) = W(\alpha) \tag{6.43}$$

where the function W is approximated using the newly updated design variables, $\delta^{(i)}$, with α as the only unknown variable. Therefore, an internal optimization needs to be performed to find α^*, the optimal step size, which minimizes the approximate function. These line search techniques are dependent upon the availability of derivatives and the mathematical form of W.

6.6.1   One-point method – Newton–Raphson

If the first and second derivatives of W are available, the function W can be approximated by a second-order Taylor series expansion as

$$Q(\alpha) = W(\alpha^{(1)}) + W'(\alpha^{(1)})(\alpha - \alpha^{(1)}) + \frac{1}{2} W''(\alpha^{(1)})(\alpha - \alpha^{(1)})^2 \tag{6.44}$$

By taking the derivative of Q and setting it equal to zero, the value of α that minimizes the approximate function is

$$\alpha^{(2)} = \alpha^{(1)} - \frac{W'(\alpha^{(1)})}{W''(\alpha^{(1)})} \tag{6.45}$$

Then $\alpha^{(2)}$ can be used to generate $\alpha^{(3)}$ until the minimum of Q is reached with the updated α's. This is equivalent to using the Newton–Raphson technique to find $Q' = 0$.

6.6.2   *Two-point method*

Two-point methods use the first derivatives of W to generate an iterative process for finding α^*. These methods are based on the premise that the second derivatives of W are unavailable or prohibitively difficult to calculate. These techniques need a bounded region that contains α^* in order to be used. This bounded region is called the interval of uncertainty. A simple check to see if α^* lies in this interval is to calculate the derivatives of the end points of the interval and check to make sure that the derivatives have different signs.

Most of the two-point methods are based on decreasing this interval until the end points of the interval are sufficiently close to one another with α^* being considered the midpoint of the interval. This can be written as

$$\alpha^{(3)} = \alpha^{(2)} - c(\alpha^{(2)} - \alpha^{(1)}) \tag{6.46}$$

where the original interval is given as $[\alpha^{(1)}, \alpha^{(2)}]$ and c must be in the interval $[0,1]$. The derivative of the objective function will be evaluated using $\alpha^{(3)}$ and depending on the value of this derivative a new interval will be generated as one of these two $[\alpha^{(1)}, \alpha^{(3)}]$ or $[\alpha^{(3)}, \alpha^{(2)}]$. If the derivative of $Q(\alpha)$ is positive, the minimum lies in $[\alpha^{(1)}, \alpha^{(3)}]$, and if the derivative is negative, the minimum lies in $[\alpha^{(3)}, \alpha^{(2)}]$. If the function is convex, α^* is the global minimum, but if the function is not convex, α^* could be a global or local minimum. This technique is often called the bisecting search method, often using a constant value of $c = 1/2$. This technique is crude, but effective. The efficiency of this algorithm can be improved by developing methods for finding the value c based on polynomial approximations of the function and using the first derivatives of the polynomial approximating function.

One way to find a better value for c is to use a quadratic fit to the function if the first derivatives at two local points can be found. This fit is only locally, near $\alpha^{(1)}$, valid which gives rise to the need for an iterative process to find the actual α^*. This concept leads to the approximate function

$$Q(\alpha) = W(\alpha^{(1)}) + W'(\alpha^{(1)})(\alpha - \alpha^{(1)}) + \frac{1}{2}\left(\frac{(W'(\alpha^{(2)}) - W'(\alpha^{(1)}))}{(\alpha^{(2)} - \alpha^{(1)})}\right)(\alpha - \alpha^{(1)})^2 \tag{6.47}$$

By taking the derivative of this function and setting it equal to zero, an optimal value for α can be found which will be the new value $\alpha^{(3)}$ used in the next iteration.

$$\alpha^{(3)} = \alpha^{(2)} - \frac{W'(\alpha^{(2)})(\alpha^{(2)} - \alpha^{(1)})}{(W'(\alpha^{(2)}) - W'(\alpha^{(1)}))} \tag{6.48}$$

therefore, c in Eq. (6.46) can be written as

$$c = \frac{W'(\alpha^{(2)})}{(W'(\alpha^{(2)}) - W'(\alpha^{(1)}))} \tag{6.49}$$

Note that c is only dependent on the derivatives. Essentially it is finding the location where the derivative of the approximate quadratic function goes to zero. This gives rise to a new interval, either $[\alpha^{(1)}, \alpha^{(3)}]$ or $[\alpha^{(3)}, \alpha^{(2)}]$ which can be used to generate the next quadratic Q. This process is continued until the interval is small enough to produce an accurate α^*. This method is often called the method of false position or the secant method.

If the second derivatives of the function are known, Newton's method can be used to provide a quadratic approximation. The quadratic approximation is

$$Q(\alpha) = W(\alpha^{(1)}) + W'(\alpha^{(1)})(\alpha - \alpha^{(1)}) + \frac{1}{2} W''(\alpha^{(1)})(\alpha - \alpha^{(1)})^2 \tag{6.50}$$

Similar to the method of false position, the solution for α^* can be easily found by taking the derivative of the approximate function, setting it equal to 0 and solving for α.

$$\alpha^{(2)} = \alpha^{(1)} - \left(\frac{W'(\alpha^{(1)})}{W''(\alpha^{(1)})} \right) \tag{6.51}$$

Once successive iterations yield α's that are sufficiently close, the process ends. Note that the method of false position is the same as the Newton method except that the second derivative of the function is approximated by taking the difference of the first derivatives at two points divided by the difference of the two points.

6.6.3 Cubic interpolation method

The objective function, $W(\delta)$, can be represented as a cubic approximation, Q, about a given point, δ, leaving α as the single variable

$$Q(\alpha) = A(\alpha)^3 + B(\alpha)^2 + C(\alpha) + D \tag{6.52}$$

when the values of $W(\alpha^{(1)})$, $W(\alpha^{(2)})$ and their derivatives at these points are known and the derivatives have opposite signs (indicating that a minimum lies between the two points). Using these values, the coefficients A, B, C and D can be found as formulated in many numerical methods texts. The derivatives of Eq. (6.52) become

$$Q'(\alpha) = 3A(\alpha)^2 + 2B(\alpha) + C \tag{6.53}$$

and

$$Q''(\alpha) = 6A(\alpha) + 2B \tag{6.54}$$

and provide the necessary information to find the minimum of Q by solving the quadratic equation Eq. (6.53).

$$\alpha^* = \frac{-2B + (4B^2 - 12AC)^{\frac{1}{2}}}{6A} \tag{6.55}$$

This is valid when A is not equal to zero. When A is zero, the solution to Eq. (6.53) becomes simply $(-C/2B)$. Using this value coupled with the appropriate value of $\alpha^{(1)}$ or $\alpha^{(2)}$, the process can be repeated until convergence is achieved for α^*.

6.6.4 *Three-point method – quadratic interpolation*

There is another family of approximations, three-point or quadratic interpolation methods that do not require any derivatives, but one must have the value of the function to be minimized at three points. The benefit of such a method is that the objective function is the only necessary function. These techniques use the three points to develop a quadratic approximation to the original function. Once this function is formed, the derivative is found and set to zero to find the optimal step size. This is easily implemented in many algorithms and is also very efficient. These techniques will not be presented here, but can be explored further in [2–4, 8, 9].

6.7 Concluding remarks

Unconstrained optimization with gradient-based search directions has limited uses within the field of structural design or optimization, but the concepts used are germane to a complete understanding of many constrained structural optimization algorithms. Many constrained optimization algorithms use gradient-based search directions and line search techniques coupled with techniques that check for or prevent violation of the constraints such as displacements, drift, stresses and natural frequencies. Many topology optimization algorithms make use of unconstrained optimization by including the constraints in the objective function.

In effect, optimality-criteria techniques that use the Lagrangian function, Chapter 8, could be considered a completely different set of unconstrained optimization problems that do not use gradient-based search directions. In reality these techniques eventually require additional constraints that require a modified optimality-criteria approach that accounts for these constraints.

As seen from the examples, these unconstrained techniques can be very efficient for well-behaved functions. In many cases the efficiency is also a function of the uni-variable line search technique implemented in the algorithm.

References

1. Davidon, W.C., Variable metric methods for minimization, *AEC Research Development Report*, Argonne National Laboratories, ANL-5990, 1959.
2. Fletcher, R. and Powell, M.J.D., A rapidly convergent descent method for minimization, *British Computer Journal*, Vol. 6, pp. 163–168, 1963.
3. Bazaraa, M.S. and Shetty, C. M., *Nonlinear Programming: Theory and Algorithms*, John Wiley & Sons, New York, 1979.
4. Fletcher, R. and Reeves, C.M., Function minimization by conjugate gradients, *Computer Journal*, Vol. 7, pp. 149–154, 1964.
5. Polak, E. and Ribiere, G., Note sur la convergence de methods de directions conjuges, *Revue Francaise Informat, Recherche Operationelle*, Vol. 16, pp. 35–43, 1969.
6. Lenard, M.L., Accelerated conjugate direction methods for unconstrained optimization, *Journal of Optimization Theory and Applications*, Vol. 25, No. 1, pp. 11–31, 1978.
7. Powell, M.J.D., Restart procedures for the conjugate gradient method, *Mathematical Programming*, Vol. 12, pp. 241–254, 1977.
8. Vanderplaats, G.N., *Numerical Optimization Techniques for Engineering Design: with Applications*, McGraw-Hill, New York, 1984.
9. Hooke, R. and Jeeves, T.A., Direct search solution of numerical and statistical problems, *Journal of Association of Computer Machinery*, Vol. 8, pp. 212–229, 1961.
10. Rosenbrock, H.H., An automated method for finding the greatest or least value of a function, *Computer Journal*, Vol. 3, pp. 175–184, 1960.

7 Energy distribution algorithm for optimality-criteria method and optimization of 2-D seismic resistant frames

7.1 Introduction

Several optimization strategies presented previously are focused on classical linear and nonlinear mathematical algorithms. Now the modern optimization algorithms of various optimality-criteria methods are introduced in Chapters 7 through 10, with extensive applications for 2-D and 3-D structures subjected to static and seismic forces. This chapter starts with the following items: (1) formulation of a structural model suitable for finite element analysis and energy distribution, with consideration of coupling ground motions and the $P - \Delta$ effect, (2) derivation of a primary recursion formula based on energy distribution, (3) derivation of a secondary recursion formula based on constraint gradients, (4) development of minimum cost design, (5) derivation and selection of seismic response spectra, (6) observation of the influence of interacting earthquake motions and building code provision on structural stiffness requirements, and (7) comparison of the optimum solutions based on minimum weight and minimum cost design.

7.2 Formulation of structural model and dynamic excitations

7.2.1 Stiffness, geometric, and mass matrices for consistent mass model

General structural systems of plane frameworks with and without bracing members are formulated on the basis of the displacement method and the consistent-mass technique. The constituent members are prismatic between nodes and may have bending and axial deformations. The *second-order effect* of the axial loads on the columns is included [1]. The coordinates, positive forces and their associated deformations for typical constituent members of a structural system are shown in Fig. 7.1. The beam-column elements are for either girders or columns and two-force bar elements are for bracings. The matrices of stiffness, $[K_s]_i$, mass, $[M_s]_i$, and geometric stiffness, $[K_g]_i$, of a girder or column are

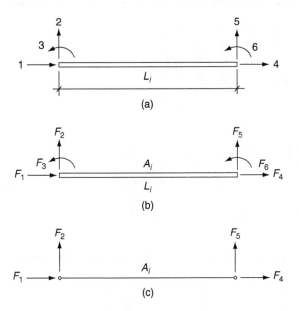

Figure 7.1 Typical member for consistent mass formulation: (a) coordinates of a planar member, (b) positive forces for both girders and columns, (c) positive forces for bracings.

$$[K_s]_i = \frac{EI_i}{L_i} \begin{bmatrix} A_i/I_i & 0 & 0 & -A_i/I_i & 0 & 0 \\ & 12/L_i^2 & 6/L_i & 0 & -12/L_i^2 & 6/L_i \\ & & 4 & 0 & -6/L_i & 2 \\ & & & A_i/I_i & 0 & 0 \\ & \text{sym} & & & 12/L_i^2 & -6/L_i \\ & & & & & 4 \end{bmatrix} \quad (7.1)$$

$$[M_s]_i = \frac{\rho_i A_i L_i}{g} \begin{bmatrix} 1/3 & 0 & 0 & 1/6 & 0 & 0 \\ & 13/35 & 11L_i/210 & 0 & 9/70 & -13L_i/420 \\ & & L_i/105 & 0 & 13L_i/420 & -L_i/140 \\ & & & 1/3 & 0 & 0 \\ & \text{sym} & & & 13/35 & -11L_i/210 \\ & & & & & L_i^2/105 \end{bmatrix} \quad (7.2)$$

$$[K_g]_i = \begin{bmatrix} 0 & 0 & 0 & 0 & 0 & 0 \\ & 6/(5L_i) & 1/10 & 0 & -6/(5L_i) & 1/10 \\ & & 2L_i/15 & 0 & -1/10 & -L_i/130 \\ & & & 0 & 0 & 0 \\ & \text{sym} & & & 6/(5L_i) & -1/10 \\ & & & & & 2L_i/15 \end{bmatrix} \quad (7.3)$$

The matrices of stiffness, $[K_s]_i$, and mass, $[M_s]_i$ of a bracing member are

$$[K_s]_i = \frac{EA_i}{L_i} \begin{bmatrix} 1 & 0 & 0 & -1 & 0 & 0 \\ & 0 & 0 & 0 & 0 & 0 \\ & & 0 & 0 & 0 & 0 \\ & \text{sym} & & 1 & 0 & 0 \\ & & & & 0 & 0 \\ & & & & & 0 \end{bmatrix} \tag{7.4}$$

$$[M_s]_i = \frac{\rho_i A_i L_i}{g} \begin{bmatrix} 1/3 & 0 & 0 & 1/6 & 0 & 0 \\ & 1/3 & 0 & 0 & 1/6 & 0 \\ & & 0 & 0 & 0 & 0 \\ & & & 1/3 & 0 & 0 \\ & \text{sym} & & & 1/3 & 0 \\ & & & & & 0 \end{bmatrix} \tag{7.5}$$

The cross-sectional property matrix, $[b]_i$, of a beam-column element is

$$[b]_i = \begin{bmatrix} 1/A_i & 0 & -1/S_i & 0 & 0 & 0 \\ 0 & 1/v_i & 0 & 0 & 0 & 0 \\ 1/A_i & 0 & 1/S_i & 0 & 0 & 0 \\ 0 & 0 & 0 & 1/A_i & 0 & -1/S_i \\ 0 & 0 & 0 & 0 & 1/v_i & 0 \\ 0 & 0 & 0 & 1/A_i & 0 & 1/S_i \end{bmatrix} \tag{7.6}$$

For bracings, the cross-sectional property matrix becomes

$$[b]_i = \begin{bmatrix} 1/A_i & 0 & 0 & 0 & 0 & 0 \\ 0 & 0 & 0 & 0 & 0 & 0 \\ 0 & 0 & 0 & 0 & 0 & 0 \\ 0 & 0 & 0 & 1/A_i & 0 & 0 \\ 0 & 0 & 0 & 0 & 0 & 0 \\ 0 & 0 & 0 & 0 & 0 & 0 \end{bmatrix} \tag{7.7}$$

The notations in Eqs. (7.1) through (7.7) are: A_i = cross-sectional area of member i, E = modulus of elasticity, g = gravity acceleration, I_i = moment of inertia of member i, L_i = length of member i, S_i = section modulus of member i, v_i = shear flow of member i and ρ_i = mass density of member i.

7.2.2   Equations of motion for static loads and multi-component seismic excitations

The equations of motion for a multi-degree-of-freedom structural system that is subjected to both static load and multi-component ground motions can be expressed as

$$([M_n]+[M_s])\{\ddot{r}(x,t)\}+[C]\{\dot{r}(x,t)\}+([K_s]-[K_g])(\{r(x,t)\}+\{r(x)\})$$
$$=-([M_n]+[M_s])\{\ddot{r}_g(t)\}+\{R\} \qquad (7.8)$$

in which $[M_n]=$ non-structural mass matrix resulting from the superimposed weight. $[M_s]=$ structural mass matrix resulting from the weight of the structural members. $[C]=$ viscous damping matrix. $[K_s]=$ structural stiffness matrix. $[K_g]=$ geometric stiffness matrix. $\{\ddot{r}(x,t)\}$, $\{\dot{r}(x,t)\}$, $\{r(x,t)\}=$ acceleration, velocity, and displacement vectors, respectively, of system coordinates measured from the equilibrium position after deformation caused by static loads. $\{r(x)\}=$ static displacement vector measured from the undeformed position to the equilibrium position. $\{\ddot{r}_g(t)\}=$ vector of ground accelerations. And $\{R\}=$ vector of fixed-end moments due to the static load applied between two structural nodes of a beam or girder.

Let the generalized coordinates of the ith member and the structure to be related be

$$\{v_i(x)\}=[a]_i\{r(x)\} \qquad (7.9)$$

in which $v_i(x)$ is the displacement vector of the ith member, and $[a]_i$ is the compatibility matrix connecting the generalized coordinates of the ith member and those of the structural system. Then the system matrices in Eq. (7.8) can be obtained by assembling the member matrices as follows [1]

$$[M_s]=\sum_{i=1}^{m}[a]_i^T[M_s]_i[a]_i \qquad (7.10)$$

$$[K_s]=\sum_{i=1}^{m}[a]_i^T[K_s]_i[a]_i \qquad (7.11)$$

$$[K_g]=\sum_{i=1}^{m}P_i[a]_i^T[K_g]_i[a]_i \qquad (7.12)$$

in which m is the total number of the constituent members of the system, and $[M_s]_i$, $[K_s]_i$, and $[K_g]_i$ are the mass, structural stiffness, and geometric stiffness matrices of the ith member, respectively, and

$$P_i=N_i+\sum_{k=1}^{CN_i}\rho_k A_k l_k \qquad (7.13)$$

in which P_i is the total axial force, N_i is the weight of non-structural masses acting axially on the ith member, CN_i is the total number of structural masses lumped on the ith member, and l_k is the length of member k. A typical structure is sketched in Fig. 7.2 for which typical P's are shown below

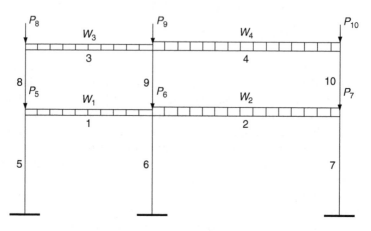

Figure 7.2 Typical diagram for the axial forces of the $P - \Delta$ effect.

$$P_5 = \frac{1}{2}\left[w_1 l_1 + w_3 l_3\right] + \frac{1}{2}\rho\left[A_1 l_1 + A_5 l_5 + A_8 l_8 + A_3 l_3 + A_8 l_8\right], \quad CN_5 = 5$$

$$P_6 = \frac{1}{2}\left[w_1 l_1 + w_2 l_2 + w_3 l_3 + w_4 l_4\right]$$

$$\qquad + \frac{1}{2}\rho\left[A_1 l_1 + A_2 l_2 + A_6 l_6 + A_9 l_9 + A_3 l_3 + A_4 l_4 + A_9 l_9\right], \quad CN_6 = 7$$

$$P_7 = \frac{1}{2}\left[w_2 l_2 + w_4 l_4\right] + \frac{1}{2}\rho\left[A_2 l_2 + A_7 l_7 + A_{10} l_{10} + A_4 l_4 + A_{10} l_{10}\right],$$

$$\qquad CN_7 = 5$$

$$P_8 = \frac{1}{2}\left[w_3 l_3\right] + \frac{1}{2}\rho\left[A_3 l_3 + A_5 l_5\right], \quad CN_8 = 2$$

$$P_9 = \frac{1}{2}\left[w_3 l_3 + w_4 l_4\right] + \frac{1}{2}\rho\left[A_3 l_3 + A_4 l_4 + A_9 l_9\right], \quad CN_9 = 3$$

$$P_{10} = \frac{1}{2}\left[w_4 l_4\right] + \frac{1}{2}\rho\left[A_4 l_4 + A_{10} l_{10}\right], \quad CN_{10} = 2$$

$[M_n]$ can be similarly established as $[M_s]$ for uniformly distributed non-structural masses and may include lumped masses associated with the concentrated superimposed weight.

The static displacement, $\{r(x)\}$, can be obtained directly from the static equilibrium equation

$$\{r(x)\} = [K]^{-1}\{R\} \qquad (7.14)$$

in which

$$[K] = [K_s] - \left[K_g \right] \tag{7.15}$$

In Eq. (7.15), the negative sign indicates that the geometric stiffness matrix is due to the compressive forces only. The dynamic displacements, $\{r(x,t)\}$, must be solved by employing the dynamic equilibrium equations, which can be derived from Eq. (7.8) by using

$$[M]\{\ddot{r}(x,t)\} + [C]\{\dot{r}(x,t)\} + [K]\{r(x,t)\} = -[M]\{\ddot{r}_g(t)\} \tag{7.16}$$

in which

$$[M] = [M_n] + [M_s] \tag{7.17}$$

Let the system coordinates in Eq. (7.16) be expressed separately in terms of horizontal, vertical and rotational directions, then

$$[M]\begin{Bmatrix}\ddot{r}_h(x,t) \\ \ddot{r}_v(x,t) \\ \ddot{r}_\theta(x,t)\end{Bmatrix} + [C]\begin{Bmatrix}\dot{r}_h(x,t) \\ \dot{r}_v(x,t) \\ \dot{r}_\theta(x,t)\end{Bmatrix} + [K]\begin{Bmatrix}r_h(x,t) \\ r_v(x,t) \\ r_\theta(x,t)\end{Bmatrix} = -[M]\begin{Bmatrix}\ddot{r}_{gh}(t) \\ \ddot{r}_{gv}(t) \\ \ddot{r}_{g\theta}(t)\end{Bmatrix} \tag{7.18}$$

in which the subscripts h, v and θ indicate the degree of freedom corresponding to horizontal, vertical and rotational movements, respectively.

By neglecting the effect of rotational ground motion, one may rewrite the horizontal and vertical ground accelerations in the following forms

$$\{\ddot{r}_{gh}(t)\} = \{e\}x_h(t) \tag{7.19}$$

and

$$\{\ddot{r}_{gv}(t)\} = \{e\}x_v(t) \tag{7.20}$$

in which $x_h(t)$ and $x_v(t)$ are the functions of the specific horizontal and vertical ground accelerations, respectively and $\{e\}$ is a unit vector. The substitution of Eqs. (7.19) and (7.20) for the right side of Eq. (7.18) yields

$$[M]\{\ddot{r}(x,t)\} + [C]\{\dot{r}(x,t)\} + [K]\{r(x,t)\} = -[M]\begin{Bmatrix}e \\ 0 \\ 0\end{Bmatrix}x_h(t)$$

$$-[M]\begin{Bmatrix}0 \\ e \\ 0\end{Bmatrix}x_v(t) \tag{7.21}$$

Eq. (7.21) represents a set of coupling second-order differential equations. These equations can be uncoupled by using the modal superposition technique, for which the damping matrix [C] can be expressed in terms of the mass and stiffness matrices as

$$[C] = a_o [M] + b_o [K] \tag{7.22}$$

in which a_o and b_o are determined by employing the damping coefficient and the natural frequency at each mode to be used [1].

The uncoupling procedures require coordinate transformations. This can be done by changing the system coordinates to normal coordinates and the eigen-solutions of the eigenvalues and eigenvectors. The eigen-solutions can be obtained first by employing Eq. (7.23) as

$$[K]\{\phi_i\} = \omega_i^2 [M]\{\phi_i\} \tag{7.23}$$

in which ω_i is the *i*th undamped natural frequency associated with the mode shape $\{\phi_i\}$. Then the coordinate transformation can be expressed as

$$\{r(x,t)\} = [\Phi]\{q(t)\} \tag{7.24}$$

in which $[\Phi]$ represents the eigenvector of each mode arranged in columns, and $\{q(t)\}$ denotes the displacement vector associated with the normal coordinates. Thus, Eq. (7.21) becomes

$$[m]\{\ddot{q}(t)\} + [c]\{\dot{q}(t)\} + [k]\{q(t)\} = -[\Phi]^T [M] \begin{Bmatrix} e \\ 0 \\ 0 \end{Bmatrix} x_h(t)$$

$$-[\Phi]^T [M] \begin{Bmatrix} 0 \\ e \\ 0 \end{Bmatrix} x_v(t) \tag{7.25}$$

In Eq. (7.25), $[m], [c]$ and $[k]$ are diagonal matrices and are represented by

$$[m] = [\Phi]^T [M] [\Phi] \tag{7.26}$$
$$[c] = [\Phi]^T [C] [\Phi] \tag{7.27}$$

and

$$[k] = [\Phi]^T [K] [\Phi] \tag{7.28}$$

By employing Eqs. (7.26) and (7.28), one can rewrite Eq. (7.23) for all the modes as

$$[k] = [m] [\Omega] \tag{7.29}$$

in which $[\Omega] = [\omega_i^2]$ is a diagonal matrix containing the square of the natural frequencies.

The ith equation of the uncoupled system of Eq. (7.25) can now be expressed as

$$m_i \ddot{q}_i(t) + c_i \dot{q}_i(t) + k_i q_i(t) = -\{\Phi_i\}^T [M] \begin{Bmatrix} e \\ 0 \\ 0 \end{Bmatrix} x_h(t)$$

$$- \{\Phi_i\}^T [M] \begin{Bmatrix} 0 \\ e \\ 0 \end{Bmatrix} x_v(t) \qquad (7.30)$$

From Eq. (7.29)

$$k_i = m_i \omega_i^2 \qquad (7.31)$$

Also, let

$$\beta_i = \frac{c_i}{2 m_i \omega_i} \qquad (7.32)$$

Then Eq. (7.30) becomes

$$\ddot{q}_i(t) + 2\beta_i \omega_i \dot{q}_i(t) + \omega_i^2 q_i(t) = -\frac{1}{m_i}\{\Phi_i\}^T [M] \begin{Bmatrix} e \\ 0 \\ 0 \end{Bmatrix} x_h(t)$$

$$- \frac{1}{m_i}\{\Phi_i\}^T [M] \begin{Bmatrix} 0 \\ e \\ 0 \end{Bmatrix} x_v(t) \qquad (7.33)$$

Eq. (7.33) represents the motion equation for a single-degree-of-freedom system having a frequency, ω_i, a mass, m_i, and a relative viscous damping ratio, β_i, and which is subjected to multi-component horizontal and vertical ground motions.

7.2.3 Eigenvalue comparison due to the $P - \Delta$ effect and for various structural models

Example 7.2.1 – For the one-storey shear building shown in Fig. 7.3, first find the buckling load, P_{cr}, then find the influence of various ratios of P_{cr} on the natural frequency. Let $E = 10 \times 10^6$ psi, $I = 100$ in.^{4}, $l = 1,000$ in., $W = 1,159.2$ lbs.

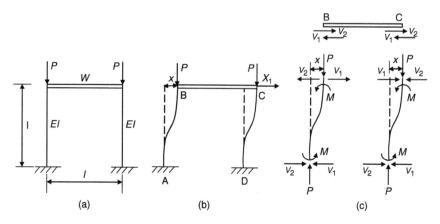

Figure 7.3 Example 7.2.1: (a) given structure, (b) movement due to the d.o.f. X_1, (c) free-body diagrams.

Solution:

(A) Find the buckling load. The structure shown in Fig. 7.3a has one degree of freedom, X_1, for which the possible displacement x is sketched in Fig. 7.3b. Due to the movement x, the column moment is $M = 6EI/l^2x$ which is balanced by the shear $V_1 = 2M/l = 12EI/l^3x$; and the secondary moment, Px, balanced by the shear $V_2 = Px/l$ as shown in Fig. 7.3c. Transfer V_1 and V_2 to the girder BC for which the equilibrium condition yields

$$2\left(\frac{12EI}{l^3}\right)x - \frac{2Px}{l} = 0 \tag{a}$$

The determinant of Eq. (a) is

$$\left|\frac{24EI}{l^3} - \frac{2P}{l}\right| = 0 \tag{b}$$

which yields the buckling load as

$$P_{cr} = \frac{12EI}{l^2} = \frac{12\left(10^7\right)(100)}{10^6} = 12,000\,\text{lbs} \tag{c}$$

(B) The frequency equation for the structure may be generally expressed as

$$\left([K_s] - \omega^2[M]\right)\{X\} - [K_g]\{X\} = 0 \tag{d}$$

For this one-storey shear building, we have

$$\omega^2 = \frac{1}{m}[K_s - K_g] = \frac{1}{m}\left[\frac{24EI}{l^3} - \frac{2\left(rP_{cr}\right)}{l}\right] \tag{e}$$

The influence of the $P - \Delta$ effect on ω^2 is

$$
\left.
\begin{array}{ll}
\text{when} \quad r = 0, & \omega^2 = 8 \\[4pt]
\qquad\quad r = 0.5, & \omega^2 = \dfrac{1}{m}\left[\dfrac{24EI}{l^3} - 2\,(0.5)\,\dfrac{12EI}{l^3}\right] = 4 \\[6pt]
\qquad\quad r = 1, & \omega^2 = \infty
\end{array}
\right\}
\tag{f}
$$

Example 7.2.2 – For the rigid frame of Fig. 7.4 find the eigenvalues due to different mathematical structural models as (1) lumped mass, (2) consistent mass, and (3) dynamic stiffness. Assume that all members are identical with $m = 0.04837\,\text{kg s}^2\,\text{m}^{-2}$, $I = 0.00286\,\text{cm}^4$, $E = 20,684.27\,\text{kN cm}^{-2}$ and $L = 0.2413\,\text{m}$.

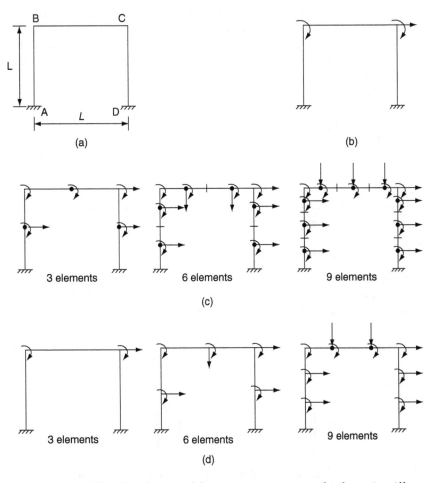

Figure 7.4 Modelling for three models: (a) given structure, (b) dynamic stiffness model, (c) lumped mass model, (d) consistent mass model.

Table 7.1 Comparison of eigenvalues by lumped mass, consistent mass and dynamic stiffness.

Methods	Number of elements	First mode (rad s⁻¹)	Second mode (rad s⁻¹)	Third mode (rad s⁻¹)
Lumped mass	3	210.5	531.2	840.0
	6	195.8	762.5	1,323.6
	9	194.5	765.7	1,265.0
Consistent mass	3	194.5	891.7	1,988.6
	6	194.3	771.7	1,264.9
	9	194.3	767.8	1,254.2
Dynamic stiffness	3	194.3	766.8	1,250.7

Solution:

The constituent members are divided into three groups of elements as shown in Fig. 7.4b with three elements for the dynamic stiffness model; Fig. 7.4c with three, six and nine elements for the lumped mass model; and Fig. 7.4d with three, six and nine elements for the consistent mass model. The results are given in Table 7.1 [1].

Note that the consistent mass model and dynamic stiffness model (exact solution) can yield almost the same fundamental frequency for three elements but the lumped mass model needs nine elements to yield similar results. In general, the consistent mass model requires a fewer number of elements than the lumped mass model to achieve an accurate eigenvalue solution.

7.3 Seismic response based on direct procedures, response spectra, and building provisional recommendations

7.3.1 Closed formulation for seismic response

The solution of the modal-motion equation shown in Eq. (7.33) for multi-component ground motions can be obtained by using the *Laplace transformation*. Let Eq. (7.33) be written as

$$\ddot{q}_i(t) + 2\beta_i\omega_i\dot{q}_i(t) + \omega_i^2 q_i(t) = -\frac{\Upsilon_{hi}}{\bar{m}_i}x_{gh}(t) - \frac{\Upsilon_{vi}}{\bar{m}_i}x_{gv}(t) \qquad (7.34)$$

in which

$$\Upsilon_{hi} = \{\Phi_i\}^T[M]\begin{Bmatrix} e \\ 0 \\ 0 \end{Bmatrix}; \quad \Upsilon_{vi} = \{\Phi_i\}^T[M]\begin{Bmatrix} 0 \\ e \\ 0 \end{Bmatrix} \qquad (7.35a, b)$$

The terms Υ_{hi} and Υ_{vi} are called *modal participation factors* and are associated with the *i*th mode. Using the *Laplace transformations* with the initial conditions of $q_i(0) = 0$ and $\dot{q}_i(0) = 0$, one may write

$$\ddot{q}_i(t) = s^2 \bar{q}_i(s) - s q_i(0) - \dot{q}_i(0) = s^2 \bar{q}_i(s) \tag{7.36}$$

$$\dot{q}_i(t) = s \bar{q}_i(s) - q_i(0) = s \bar{q}_i(s) \tag{7.37}$$

$$q_i(t) = \bar{q}_i(s) \tag{7.38}$$

and

$$c_1 x_h(t) + c_2 x_v(t) = c_1 F_h(s) + c_2 F_v(s) \tag{7.39}$$

in which c_1 and c_2 are constants and the F's are functions of the Laplace transformation coordinates. Thus, Eq. (7.34) becomes

$$\left(s^2 + 2\beta_i \omega_i s + \omega_i^2\right) \bar{q}_i(s) = -\frac{\Upsilon_{hi}}{\bar{m}_i} F_h(s) - \frac{\Upsilon_{vi}}{\bar{m}_i} F_v(s) \tag{7.40}$$

Let

$$a = -\beta_i \omega_i - \omega_i \sqrt{1 - \beta_i^2 \bar{i}} \tag{7.41}$$

and

$$b = -\beta_i \omega_i + \omega_i \sqrt{1 - \beta_i^2 \bar{i}} \tag{7.42}$$

in which $\bar{i}$ is an imaginary number. Substitute Eqs. (7.41) and (7.42) for their equivalents in Eq. (7.40) and solve for $\bar{q}_i(s)$ as follows

$$\bar{q}_i(s) = \frac{1}{(s-a)(s-b)} \left[-\frac{\Upsilon_{hi}}{\bar{m}_i} F_h(s) - \frac{\Upsilon_{vi}}{\bar{m}_i} F_v(s) \right] \tag{7.43}$$

By employing

$$\frac{1}{(s-a)(s-b)} = \frac{e^{bt} - e^{at}}{b-a} \quad a \neq b \tag{7.44}$$

and

$$F(s)G(s) = \int_0^t F(\tau)G(t-\tau)d\tau \tag{7.45}$$

in which F and G are functions and τ is the time variable for integration, Eq. (7.43) can now be transformed back to its former condition as follows

$$q_i(t) = \frac{1}{(b-a)} \left[-\frac{\Upsilon_{hi}}{\tilde{m}_i} \left[\left(\int_0^t x_{gh}(\tau)e^{b(t-\tau)}d\tau - \int_0^t x_{gh}(\tau)e^{a(t-\tau)}d\tau \right) \right. \right.$$

$$\left. \left. -\frac{\Upsilon_{vi}}{\tilde{m}_i} \left(\int_0^t x_{gv}(\tau)e^{b(t-\tau)}d\tau - \int_0^t x_{gv}(\tau)e^{a(t-\tau)}d\tau \right) \right] \right] \tag{7.46}$$

By substituting a and b of Eqs. (7.41) and (7.42) for their equivalents in Eq. (7.46) and using

$$\omega_{di} = \omega_i \sqrt{1 - \beta_i^2} \tag{7.47}$$

$$e^{\bar{i}\omega_{di}(t-\tau)} = \cos\omega_{di}(t-\tau) + \bar{i}\sin\omega_{di}(t-\tau) \tag{7.48}$$

and

$$e^{-\bar{i}\omega_{di}(t-\tau)} = \cos\omega_{di}(t-\tau) + \bar{i}\sin\omega_{di}(t-\tau) \tag{7.49}$$

One then finds the closed form solution of Eq. (7.34) as

$$q_i(t) = -\frac{\Upsilon_{hi}}{\tilde{m}_i\omega_{di}} \int_0^t x_{gh}(\tau)e^{-\beta_i\omega_i(t-\tau)} \sin\omega_{di}(t-\tau)d\tau$$

$$-\frac{\Upsilon_{vi}}{\tilde{m}_i\omega_{di}} \int_0^t x_{gv}(\tau)e^{-\beta_i\omega_i(t-\tau)} \sin\omega_{di}(t-\tau)d\tau \tag{7.50}$$

This shows that the dynamic response of a structural system subject to multi-component earthquake motions can be obtained by using the super-position of the response associated with each individual component. If the structure is subjected to both static loads, $\{R\}$, and dynamic forces, such as wind excitations, $\{R(t)\}$, the motion equations can be similarly established as

$$[M]\{\ddot{r}(x,t)\} + [C]\{\dot{r}(x,t)\} + [K]\left(\{r(x,t)\} + \{r(x)\}\right) = \{R(t)\} + \{R\} \tag{7.51}$$

in which

$$\{R(t)\} = \{R_0\}f(t) \tag{7.52}$$

R_0 is the dynamic force's magnitude and $f(t)$ is a time function. The solution associated with the dynamic response of Eq. (7.51) becomes

$$q_i(t) = \frac{\Upsilon_i}{\tilde{m}_i\omega_{di}} \int_0^t f(\tau)e^{-\beta_i\omega_i(t-\tau)} \sin\omega_{di}(t-\tau)d\tau \tag{7.53}$$

For this case, the modal participation factor in Eq. (7.53) is

$$\Upsilon_i = \{\Phi_i\}^T \{R_0\} \tag{7.54}$$

Note that ω_{di} in Eq. (7.47) includes a damping ratio, β_i, and that the ratio is small, in magnitude less than 5 per cent for steel structures. The damping effect can be neglected by using ω_i instead of ω_{di} in Eq. (7.50) [1].

7.3.2   *Direct time-history response*

It is apparent that the integration of Eqs. (7.50) and (7.53) can be evaluated exactly if the functions, $x_h(\tau)$, $x_v(\tau)$ and $f(\tau)$, are mathematical expressions suitable for integration. For seismic accelerations, such as El Centro Earthquake 1940 shown in Fig. 7.5, these functions are not continuous expressions, so the integration must be performed numerically. Let us consider a typical discontinuous forcing function of an earthquake record. For a small portion of a time interval, $\Delta t = t_2 - t_1$, the seismic excitation can be represented linearly as shown in Fig. 7.6. For simplicity, let Eq. (7.34) consist of one earthquake component, $x_g(t)$, (for simplicity,

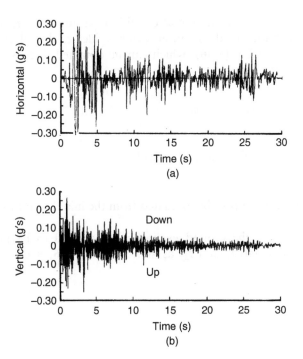

Figure 7.5 El Centro 1940 earthquake: (a) N-S component, (b) vertical component.

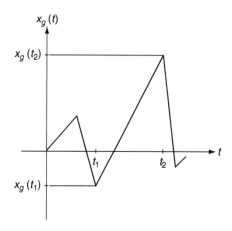

Figure 7.6 Arbitrary forcing function for dynamic and seismic excitations.

the double dot is omitted in $\ddot{x}_g(t)$) and be expressed as

$$\ddot{q}_i(t) + 2\beta_i\omega_i\dot{q}_i(t) + \omega_i^2 q_i(t) = -\frac{\Upsilon_i}{m_i}x_g(t) \qquad (7.55)$$

In order to include the effect of the initial conditions at any time, t, the solution associated with the initial conditions can be added to the solution of Eq. (7.50) as shown in Eq. (7.56), which can be obtained directly by using the Laplace transformations

$$q_i(t) = -\frac{\Upsilon_{hi}}{m_i}[e^{-\beta_i\omega_i t}(c_1\cos\omega_{di}t + c_2\sin\omega_{di}t)$$

$$+\frac{1}{\omega_{di}}\int_0^t x_g(\tau)e^{-\beta_i\omega_i(t-\tau)}\sin\omega_{di}(t-\tau)d\tau] \qquad (7.56)$$

in which c_1 and c_2 are constants to be evaluated from the initial conditions at a given time.

Because the function of the ground excitation is assumed to be linear during any time interval, t_1 and t_2, as shown in Fig. 7.6, it can be expressed as

$$x_g(t) = A + Bt \qquad (7.57)$$

in which

$$A = x_g(t_1) \qquad (7.58)$$

and

$$B = \frac{x_g(t_2) - x_g(t_1)}{t_2 - t_1} \tag{7.59}$$

By substituting Eq. (7.57) for its equivalent in Eq. (7.56) and considering only the second term in brackets on the right side of the equation, one can obtain

$$\frac{1}{\omega_{di}} \int_0^t x_g(\tau) e^{-\beta_i \omega_i (t-\tau)} \sin \omega_{di}(t-\tau) d\tau$$

$$= \frac{e^{-\beta_i \omega_i t} \sin \omega_{di} t}{\omega_{di}} \int_0^t (A + B\tau) e^{\beta_i \omega_i \tau} \cos \omega_{di} \tau \, d\tau \tag{7.60}$$

$$- \frac{e^{-\beta_i \omega_i t} \cos \omega_{di} t}{\omega_{di}} \int_0^t (A + B\tau) e^{\beta_i \omega_i \tau} \sin \omega_{di} \tau \, d\tau$$

The integrals of Eq. (7.60) become

$$\frac{1}{\omega_{di}} \int_0^t x_g(\tau) e^{-\beta_i \omega_i (t-\tau)} \sin \omega_{di}(t-\tau) d\tau = \frac{A + Bt}{\omega_i^2} - \frac{2\beta_i B}{\omega_i^3} \tag{7.61}$$

Thus, the solution of Eq. (7.56) is

$$q_i(t) = -\frac{\Upsilon_{hi}}{m_i} \left[e^{-\beta_i \omega_i t} (c_1 \cos \omega_{di} t + c_2 \sin \omega_{di} t) + \frac{A + Bt}{\omega_i^2} - \frac{2\beta_i B}{\omega_i^3} \right] \tag{7.62}$$

Differentiating Eq. (7.62) with respect to time yields the velocity, $\dot{q}_i(t)$, and acceleration, $\ddot{q}_i(t)$, at any time, t, during the linear forcing function; $\dot{q}_i(t)$ and $\ddot{q}_i(t)$ are then used to evaluate c_1 and c_2 by employing the following initial conditions at time t_1

$$q_i(0) = q_i(t_1) \tag{7.63}$$

and

$$\dot{q}_i(0) = \dot{q}_i(t_1) \tag{7.64}$$

By substituting Eqs. (7.63) and (7.64) respectively into Eq. (7.62) and its associated velocity equation and then solving for c_1 and c_2, we have

$$c_1 = q_i(t_1) - \frac{A}{\omega_i^2} + \frac{2\beta_i B}{\omega_i^3} \tag{7.65}$$

and

$$c_2 = \frac{1}{\omega_{di}} \left(\dot{q}_i(t_1) + \beta_i \omega_i q_i(t_1) - \frac{\beta_i A}{\omega_i} + \frac{(2\beta_i^2 - 1)B}{\omega_i^2} \right) \tag{7.66}$$

Consequently, the modal response at any time, t, within the interval of time, t_1 and t_2, can be computed by substituting c_1 and c_2 in Eqs. (7.65) and (7.66) for their equivalents in Eq. (7.62) as follows

$$
\begin{aligned}
q_i(t) = {} & -\frac{\Upsilon_i e^{-\beta_i \omega_i t}}{m_i} \left[\left(\dot{q}_i(t_1) - \frac{A}{\omega_i^2} + \frac{2\beta_i B}{\omega_i^3} \right) \cos \omega_{di} t \right. \\
& \left. + \frac{1}{\omega_{di}} \left(\dot{q}_i(t_1) + \beta_i \omega_i q_i(t_1) - \frac{\beta_i A}{\omega_i} + \frac{(2\beta_i^2 - 1)B}{\omega_i^2} \right) \sin \omega_{di} t \right] \\
& + \frac{\Upsilon_i}{m_i} \left[\frac{A + Bt}{\omega_i^2} - \frac{2\beta_i}{\omega_i^3} \right]
\end{aligned}
\tag{7.67}
$$

The velocity and the acceleration associated with Eq. (7.67) are

$$
\begin{aligned}
\dot{q}_i(t) = {} & -\frac{\Upsilon_i e^{-\beta_i \omega_i t}}{m_i} \left[\left(\dot{q}_i(t_1) - \frac{B}{\omega_i^2} \right) \cos \omega_{di} t \right. \\
& \left. + \frac{1}{\omega_{di}} \left(A - \omega_i^2 q_i(t_1) - \beta_i \omega_i \left(\dot{q}_i(t_1) + \frac{B}{\omega_i^2} \right) \right) \sin \omega_{di} t \right] \\
& + \frac{\Upsilon_i B}{m_i \omega_i^2}
\end{aligned}
\tag{7.68}
$$

and

$$
\begin{aligned}
\ddot{q}_i(t) = {} & \frac{\Upsilon_i e^{-\beta_i \omega_i t}}{m_i} \left[\left(A - \omega_i^2 q_i(t_1) - 2\beta_i \omega_i \dot{q}_i(t_1) \right) \cos \omega_{di} t \right. \\
& \left. + \frac{1}{\omega_{di}} \left(-\beta_i \omega_i A + B + \beta_i \omega_i^3 q_i(t_1) + \omega_i^2 (2\beta_i^2 - 1) \dot{q}_i(t_1) \right) \sin \omega_{di} t \right]
\end{aligned}
\tag{7.69}
$$

Eqs. (7.67) and (7.68) are used to compute the end values, which become the initial conditions for the next linear portion. The complete solution of the modal response over the required time-history can be obtained by repeating these equations.

7.3.3 Structural displacements and internal stresses for static and dynamic loads

Once $q_i(t)$ is computed for all significant modes, the displacements associated with the system coordinates can be obtained by means of Eq. (7.24) as follows

$$\{r(x,t)\} = [\Phi][\psi][D][\Phi]^T\{R_0\} \tag{7.70}$$

In this equation, $\{\psi\}$ and $\{D\}$ are diagonal matrices, and their elements are given by

$$\psi_i = \frac{1}{m_i\omega_i} \tag{7.71}$$

and

$$D_i = \int_0^t f(\tau)e^{-\beta_i\omega_i(t-\tau)}\sin\omega_{di}(t-\tau)d\tau \tag{7.72}$$

The total displacements, $\{r\}$, resulting from both the static loading and dynamic loading can be computed as follows:

$$\{r\} = \{r(x)\} \pm \{r(x,t)\} \tag{7.73}$$

in which $\{r(x)\}$ and $\{r(x,t)\}$ are taken from Eqs. (7.14) and (7.70), respectively. The nodal stresses of the constituent members can be computed in a manner similar to that used for the displacement response. On the basis of Eq. (7.9), the internal forces, $\{F\}_i$, of the ith member can be determined by using

$$\{S\}_i = [F]_i\{v(x)\}_i \tag{7.74}$$

in which $[K]_i = [K_s]_i - [K_g]_i$ for the member. Then the stresses, $\{\sigma\}_i$, at the nodes of any member, i, can be computed from

$$\{\sigma\}_i = [sp]_i\{S\}_i \tag{7.75}$$

in which $[sp]_i$ contains the cross-sectional properties of member i as given in Eqs. (7.6) and (7.7).

Substitution of Eqs. (7.9), (7.73) and (7.74) for their equivalents in Eq. (7.75) yields Eq. (7.76), in which the first term is due to static loading, and the second to dynamic loading

$$\{\sigma\}_i = [sp]_i[K]_i[b]_i\{r(x)\} \pm [sp]_i[K]_i[b]_i\{r(x,t)\} \tag{7.76}$$

It is apparent that Eq. (7.76) gives the upper-bound solutions for the individual modes. An adjustment of the stress and displacement responses on the basis of root-mean-square [1] is employed in the design. It is also apparent that a considerable amount of computer time is needed for the analysis of the eigenvalue problems. It has been found, however, that only certain eigenvalues and their associated eigenvectors are significant in the optimum design. A method particularly suitable for finding a limited number of eigen-solutions can also be found [1]. The method is based on the *Sturm sequence* property in conjunction with a simple bisection procedure with which any eigenvalue can be determined without having to find any of the other eigenvalues.

7.3.4 Response spectra for horizontal and vertical components of El Centro 1940 earthquake

The response of a structure subject to seismic excitation is parametric with time. To avoid using excessive computing time to solve for the dynamic response, the time parameter is eliminated by using three methods: the response spectrum, the average response spectrum, and the equivalent static lateral forces recommended in building code provisions. These three methods are briefly introduced here for they will be used later for optimum design and parameter assessments.

As discussed previously, the response of a multi-degree-of-freedom system can be simplified to that of a single-degree-of-freedom system for simultaneous horizontal and vertical ground motions, for which the result is given in Eq. (7.50). The solution of Eq. (7.50), which requires integration at different time intervals, demands a considerable amount of computational effort. In order to avoid the time dependency, one can define the spectral displacement as

$$s_d(\omega, \beta) = \left| \frac{1}{\omega_d} \int_0^t x_g(\tau) e^{-\beta \omega(t-\tau)} \sin \omega_d(t-\tau) d\tau \right|_{max} \qquad (7.77)$$

in which $x_g(t)$ is a ground excitation as a function of time. By using the integration technique shown in Eq. (7.67), one can eliminate the time dependency by finding the maximum values of the integral of Eq. (7.77) for various values of the natural frequencies, ω, and the damping ratio, β. Therefore, the maximum response as expressed in Eq. (7.50) can be found as follows

$$|q_i(t)|_{max} = -\frac{\Upsilon_{hi}}{m_i} s_{dh}(\omega_i, \beta_i) - \frac{\Upsilon_{vi}}{m_i} s_{dv}(\omega_i, \beta_i) \qquad (7.78)$$

in which

$$s_{dh}(\omega_i, \beta_i) = \left| \frac{1}{\omega_{di}} \int_0^t x_h(\tau) e^{-\beta_i \omega_i(t-\tau)} \sin \omega_{di}(t-\tau) d\tau \right|_{max} \qquad (7.79)$$

and

$$s_{dv}(\omega_i, \beta_i) = \left| \frac{1}{\omega_{di}} \int_0^t x_v(\tau) e^{-\beta_i \omega_i(t-\tau)} \sin \omega_{di}(t-\tau) d\tau \right|_{max} \tag{7.80}$$

The spectral accelerations are

$$s_{ah}(\omega_i, \beta_i) = \omega_{di}^2 s_{dh}(\omega_i, \beta_i) \tag{7.81}$$

and

$$s_{av}(\omega_i, \beta_i) = \omega_{di}^2 s_{dv}(\omega_i, \beta_i) \tag{7.82}$$

The spectral accelerations of El Centro 1940 earthquake are shown in Fig. 7.7. As pointed out in the previous section, Eq. (7.50) always gives the conservative value of the response, because the maximum values for the individual modes do not occur at the same time. A technique of levelling off the maximum responses such as root-mean-square is recommended [1].

7.3.5 Housner's average spectra, Newmark's inelastic design spectra, and equivalent lateral force procedure

The spectra shown in Fig 7.7 are developed for one particular earthquake which apparently cannot be representative for different construction sites

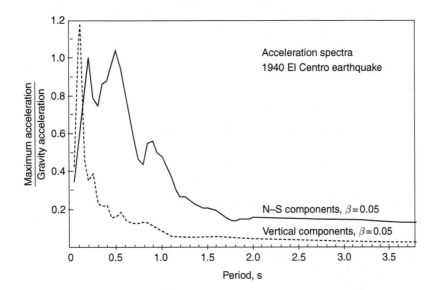

Figure 7.7 Acceleration spectrum of El Centro earthquake, May 18, 1940.

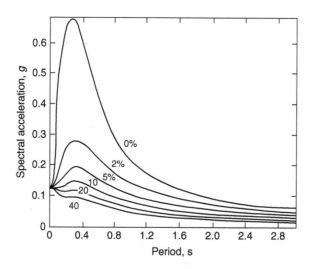

Figure 7.8 Housner's average response spectra. (1 in. = 2.54 cm).

and for a structure's lifetime. Housner thus developed *average response spectra* as shown in Fig. 7.8 where spectral shapes were obtained by averaging the normalized response spectra obtained in four earthquakes (El Centro 1934, El Centro 1940, Olympia 1949 and Tehachapi 1952). The ordinates of the figure should be multiplied by factors representative of the spectrum intensities to bring them into agreement with the different recorded ground motions. The scale factor of each different recorded ground motion is presented in Table 7.2 [2]. For convenience in computer programming, the spectral shape corresponding to 5 per cent damping is generated by using least-square, curve-fitting polynomial expressions as

$$(S_a)_i = 0.12250006 + 0.23919934 \, T_i \quad \text{for } T_i \leq 0.3 \, s \tag{7.83}$$

and

$$(S_a)_i = 0.25845146 - 0.22122407 \, T_i + 0.08188504 \, T_i^2$$
$$- 0.0108108 \, T_i^3 \text{ for } 0.3 \, s < T_i \leq 3.0 \, s \tag{7.84}$$

Table 7.2 The scaling factors of different earthquake records used in Housner's average response spectra.

Earthquake	Factor
El Centro, 18 May 1940	2.7
El Centro, 30 December 1940	1.9
Olympia, 13 April 1949	1.9
Taft, 21 July 1952	1.6
Vernon, 10 March, 1933	1.5

Newmark and Hall [3] considered inelastic responses and developed design spectra with various ductility factors. The method and various spectra are given in Appendix E. Note that all the spectra require dynamic modal analysis for eigen-solutions which need a significant amount of computation and are not necessarily needed for all types of structural design. The building code provisions are based on selection of structural analysis procedures such as the lateral force method, given in Appendix F.

7.4 Optimality-criteria method

7.4.1 *Modified Kuhn–Tucker conditions and recurrence formulation*

Let us assume the objective function is $W(x\text{'s})$. This represents either the structural weight or the structural cost, and the x's are the primary variables. Although the optimization formulation has been extensively discussed in previous chapters some fundamentals are somewhat repeated here in order to smoothly derive the algorithm. Let the behaviour constraint requirements be y_i $(x\text{'s})$, such as allowable stresses, allowable deflections, and lower bounds of natural frequency of any particular mode. In the mathematical expressions, we will minimize

$$W(x_1, x_2, \ldots, x_m) \tag{7.85}$$

subject to

$$y_i(x_1, x_2, \ldots, x_m) \le b_i \qquad i = 1, 2, \ldots, n \tag{7.86}$$

$$y_{n+j}(x_1, x_2, \ldots, x_m) \le -x_j^0 \quad j = 1, 2, \ldots, m \quad j = 1, 2, \ldots, m \tag{7.87}$$

in which the b's are the behaviour constraints, x°'s are the side constraints for lower limits of the member sizes, and $y_{n+j}(x_1, x_2, \ldots, x_m)$ is equal to $-x_j$. The necessary requirement for a local minimum is to satisfy the Kuhn–Tucker condition as

$$\frac{\partial}{\partial x_i}(W(x_1, x_2, \ldots, x_m)) + \sum_{j=1}^{n+m} \lambda_j \frac{\partial}{\partial x_i}(y_j(x_1, x_2, \ldots, x_m)) = 0 \quad i = 1, 2, \ldots, m$$

$$\tag{7.88}$$

with

$$\lambda_i(y_i(x_1, x_2, \ldots, x_m) - b_i) = 0 \quad i = 1, 2, \ldots, n \tag{7.89}$$

$$\lambda_{n+j}(y_{n+j}(x_1, x_2, \ldots, x_m) + x_j^{\circ}) = 0 \quad j = 1, 2, \ldots, m \tag{7.90}$$

$$\lambda_i \ge 0 \quad i = 1, 2, \ldots, n \tag{7.91}$$

$$\lambda_{n+j} \ge 0 \quad j = 1, 2, \ldots, m \tag{7.92}$$

By substituting Eq. (7.90) into Eq. (7.88), one obtains

$$\frac{\partial}{\partial x_i}(W(x_1, x_2, \ldots, x_m)) + \sum_{j=1}^{n} \lambda_j \frac{\partial}{\partial x_i}(y_j(x_1, x_2, \ldots, x_m))$$

$$- \lambda_{n+i} = 0 \quad i = 1, 2, \ldots, m \tag{7.93}$$

Let us consider two sets of design variables in Eq. (7.87) as J, a set of $\{x_i\}$ satisfying an inequality, and J_o, a set of $\{x_i\}$ satisfying an equality. Then, for x_i in J, λ_{n+i} is equal to zero in Eq. (7.93). Therefore, for such i, Eq. (7.93) becomes

$$\frac{\partial}{\partial x_i}(W(x_1, x_2, \ldots, x_m)) + \sum_{j=1}^{n} \lambda_j \frac{\partial}{\partial x_i}(y_j(x_1, x_2, \ldots, x_m)) = 0 \tag{7.94}$$

For x_i in J_o, Eq. (7.93) becomes

$$\frac{\partial}{\partial x_i}(W(x_1, x_2, \ldots, x_m)) + \sum_{j=1}^{n} \lambda_j \frac{\partial}{\partial x_i}(y_j(x_1, x_2, \ldots, x_m)) - \lambda_{n+i} = 0 \tag{7.95}$$

Eqs. (7.94) and (7.95) are the modified Kuhn–Tucker conditions. From these two equations, one obtains

$$\frac{-\sum_{j=1}^{n} \lambda_j \dfrac{\partial}{\partial x_i}(y_j(x_1, x_2, \ldots, x_m))}{\dfrac{\partial}{\partial x_i}(W(x_1, x_2, \ldots, x_m))} = 1 \quad \text{for } x_i \in J \tag{7.96}$$

and

$$\frac{-\sum_{j=1}^{n} \lambda_j \dfrac{\partial}{\partial x_i}(y_j(x_1, x_2, \ldots, x_m))}{\dfrac{\partial}{\partial x_i}(W(x_1, x_2, \ldots, x_m))} \leq 1 \quad \text{for } x_i \in J_o \tag{7.97}$$

If any behaviour constraint, y_t, is not active, then from Eq. (7.86),

$$y_t(x_1, x_2, \ldots, x_m) - b_t < 0 \tag{7.98}$$

and from Eq. (7.89)

$$\lambda_t = 0 \tag{7.99}$$

Let the number of active constraints be N, then the number of non-active constraints can be given by $n - N$. In Eqs. (7.96) and (7.97), we may pick up the potentially non-zero λ_j's, which correspond to active behaviour constraints, and rearrange these λ_j's in Eqs. (7.96) and (7.97), which may now be expressed as

$$\frac{-\sum_{j=1}^{N} \lambda_j \frac{\partial}{\partial x_i}(y_j(x_1,x_2,\ldots,x_m))}{\frac{\partial}{\partial x_i}(W(x_1,x_2,\ldots,x_m))} = 1 \quad \text{for } x_i \in J \tag{7.100}$$

and

$$\frac{-\sum_{j=1}^{N} \lambda_j \frac{\partial}{\partial x_i}(y_j(x_1,x_2,\ldots,x_m))}{\frac{\partial}{\partial x_i}(W(x_1,x_2,\ldots,x_m))} \leq 1 \quad \text{for } x_i \in J_o \tag{7.101}$$

We consider all x_i's belonging to set J. By multiplying x_i^2 on both sides of Eq. (7.100), and then taking the square root as follows

$$x_i = \left(\frac{-\sum_{j=1}^{N} \lambda_j \frac{\partial}{\partial x_i}(y_j(x_1,x_2,\ldots,x_m))}{\frac{\partial}{\partial x_i}(W(x_1,x_2,\ldots,x_m))}\right)^{1/2} x_i \quad \text{for } x_i \in J \tag{7.102}$$

According to Eq. (7.102), a recurrence relation on the basis of the Kuhn–Tucker condition can be obtained as

$$x_i^{(v+1)} = \left(\frac{-\sum_{j=1}^{N} \lambda_j \frac{\partial}{\partial x_i}(y_j(x_1,x_2,\ldots,x_m))}{\frac{\partial}{\partial x_i}(W(x_1,x_2,\ldots,x_m))}\right)^{\frac{1}{2}} x_i^{(v)} \tag{7.103}$$

where v is the cycle number. If $x_i^{(v+1)}$ converges to $x_i^{(v)}$, then x_i, which is a solution of Eq. (7.102), also satisfies Eq. (7.100).

7.4.2 Optimum criterion based on static stiffness for stress constraints

The strain energy of a structure is expressed [4, 23] by

$$U = \frac{1}{2}\{R\}^T\{r\} \tag{7.104}$$

in which $\{R\}$ and $\{r\}$ represent the load and displacement vectors, respectively. The total strain energy, however, should have some limited value

$$\frac{1}{2}\{R\}^T\{r\} \le \text{given value} \tag{7.105}$$

which is a measurement of structural stiffness. According to Eqs. (7.94) and (7.95), one of the Kuhn–Tucker conditions may be expressed as

$$\frac{\partial W}{\partial x_i} + \lambda \frac{\partial}{\partial x_i}\left(\frac{1}{2}\{R\}^T\{r\}\right) = 0 \quad i = 1,2,\ldots,m \tag{7.106}$$

The force–displacement relationship can be expressed in terms of the structural stiffness matrix, $[K]$, as

$$\{R\} = [K]\{r\} = \sum_{i=1}^{m}[K]_i\{r\} \tag{7.107}$$

in which $[K]_i$ is the element stiffness matrix in global coordinates. In Eq. (7.106)

$$\frac{\partial}{\partial x_i}\left(\frac{1}{2}\{R\}^T\{r\}\right) = \frac{1}{2}\{R\}^T\frac{\partial\{r\}}{\partial x_i} \tag{7.108}$$

By differentiating Eq. (7.107) with respect to x_i, one obtains

$$\begin{aligned}
0 &= \left([K]_1\frac{\partial\{r\}}{\partial x_i} + \cdots + [K]_m\frac{\partial\{r\}}{\partial x_i}\right) + \frac{\partial[K]_i}{\partial x_i}\{r\} \\
&= [K]\frac{\partial\{r\}}{\partial x_i} + \frac{\partial[K]_i}{\partial x_i}\{r\}
\end{aligned} \tag{7.109}$$

Since the compatibility matrix $[b]$ for formulating the stiffness matrix of a structural system (see Eq. (7.11) or (7.12)) is independent of the design variables, we can, for simplicity, examine $\partial[K]_i/\partial x_i$ without $[b]$. Thus, for a truss element, the design variable is the cross-sectional area, therefore the stiffness matrix in local coordinates is

$$[K]_i = A_i\begin{pmatrix} E/l_i & -E/l_i \\ -E/l_i & E/l_i \end{pmatrix} = A_i[K]_i' \tag{7.110}$$

in which $[K]_i' = [K]_i/A_i$. For a column element of a bending deformation combined with axial deformation, or a beam element with bending deformation only, the design variable is the moment of inertia of the cross-section,

I_i. For simplicity, the ratio of the cross-sectional area to the moment of inertia is assumed to be constant, η_i. Thus

$$[K]_i = \begin{pmatrix} EA_i/l_i & 0 & 0 & -EA_i/l_i & 0 & 0 \\ & 12EI_i/l_i^3 & 6EI_i/l_i^2 & 0 & -12EI_i/l_i^3 & 6EI_i/l_i^2 \\ & & 4EI_i/l_i & 0 & -6EI_i/l_i^2 & 2EI_i/l_i \\ & & & EA_i/l_i & 0 & 0 \\ & \text{sym} & & & 12EI_i/l_i^3 & -6EI_i/l_i^2 \\ & & & & & 4EI_i/l_i \end{pmatrix}$$

$$= \begin{pmatrix} E\eta_i/l_i & 0 & 0 & -E\eta_i/l_i & 0 & 0 \\ & 12E/l_i^3 & 6E/l_i^2 & 0 & -12E/l_i^3 & 6E/l_i^2 \\ & & 4E/l_i & 0 & -6E/l_i^2 & 2E/l_i \\ & & & E\eta_i/l_i & 0 & 0 \\ & \text{sym} & & & 12E/l_i^3 & -6E/l_i^2 \\ & & & & & 4E/l_i \end{pmatrix}$$

$$= I_i [K]_i' \tag{7.111}$$

Since I_i is a design variable and is replaced by x_i, we may write

$$[K]_i = x_i [K]_i' \tag{7.112}$$

from which

$$[K]_i' = \frac{\partial [K]_i}{\partial x_i} \tag{7.113}$$

Note that Eq. (7.111) is in local coordinates as the truss element discussed previously. In order to use Eq. (7.109), Eq. (7.113) should be expressed in global coordinates. Now rewrite Eq. (7.109) as

$$0 = [K] \frac{\partial \{r\}}{\partial x_i} + \frac{1}{x_i} [K]_i \{r\} \tag{7.114}$$

from which

$$\frac{\partial \{r\}}{\partial x_i} = -\frac{1}{x_i} [K]^{-1} [K]_i \{r\} \tag{7.115}$$

By combining Eqs. (7.107), (7.108) and (7.115), one obtains

$$\frac{\partial}{\partial x_i} \left(\frac{1}{2} \{R\}^T \{r\} \right) = -\frac{1}{2x_i} \{r\}^T [K]_i \{r\} \tag{7.116}$$

The symbolic form of the objective function is

$$W = \sum_{i=1}^{m} \rho_i A_i l_i = \sum_{i=1}^{m} \rho_i \eta_i x_i l_i \tag{7.117}$$

in which l_i is the length of an element, i, ρ_i the unit weight of an element, i, and η_i the ratio of the cross-sectional area, A_i, to the design variable, x_i. Then by substituting Eqs. (7.116) and (7.117) into Eq. (7.106), we have

$$\rho_i \eta_i l_i - \frac{\lambda}{2x_i} \{r\}^T [K]_i \{r\} = 0 \tag{7.118}$$

from which

$$1 = \frac{\lambda \{r\}^T [K]_i \{r\}}{2\rho_i \eta_i l_i x_i} \tag{7.119}$$

The strain energy stored in an element, i, is

$$u_i = \frac{1}{2} \{r\}^T [K]_i \{r\} \tag{7.120}$$

Therefore

$$\sum_{i=1}^{m} u_i = \sum_{i=1}^{m} \frac{1}{2} \{r\}^T [K]_i \{r\} = \frac{1}{2} \{r\}^T [K] \{r\} \tag{7.121}$$

which may also be expressed as

$$\sum_{i=1}^{m} u_i = \frac{1}{2} \{r\}^T \{R\} = U \tag{7.122}$$

From Eq. (7.119)

$$\frac{1}{\lambda} \sum_{i=1}^{m} \rho_i \eta_i x_i l_i = \sum_{i=1}^{m} \frac{1}{2} \{r\}^T [K]_i \{r\} = U \tag{7.123}$$

we have

$$\lambda = \frac{W}{U} \tag{7.124}$$

By multiplying by xi^2 on both sides of Eq. (7.119) and then taking the square root, one obtains

$$x_i = (\lambda)^{\frac{1}{2}} \left(\frac{\frac{1}{2} \{r\}^T [K]_i \{r\}}{\rho_i \eta_i l_i} \right)^{\frac{1}{2}} (x_i)^{\frac{1}{2}} \tag{7.125}$$

By substituting Eqs. (7.123) and (7.124) into the above equation

$$x_i = \left(\frac{W}{U} \right)^{\frac{1}{2}} \left(\frac{u_i}{\rho_i \eta_i l_i} \right)^{\frac{1}{2}} (x_i)^{\frac{1}{2}} \tag{7.126}$$

in which x_i must satisfy at the optimum state. Let us now consider the following recursion relationship at v and $v+1$ cycles,

$$x_i^{(v+1)} = \left(\left(\frac{W}{U} \right)^{\frac{1}{2}} \left(\frac{u_i}{\rho_i \eta_i l_i} \right)^{\frac{1}{2}} \right)^{(v)} \left((x_i)^{(v)} \right)^{\frac{1}{2}} \tag{7.127}$$

When $\lim_{v \to \infty} x_i^{(v)} = x_i$, $\lim_{v \to \infty} U^{(v)} = U$, $\lim_{v \to \infty} W^{(v)} = W$, $\lim_{v \to \infty} u_i^{(v)} = u_i$, then Eq. (7.127) converges to Eq. (7.126), which indicates that $\lim_{v \to \infty} x_i^{(v)} = x_i$ satisfies both the recursion relationship and the necessary conditions of optimality. By introducing $x_i = \Lambda \alpha_i$ into Eq. (7.127) one obtains

$$(\Lambda \alpha_i)^{(v+1)} = \left(\left(\frac{W}{U} \right)^{1/2} \left(\frac{u_i \Lambda}{\alpha_i \eta_i \rho_i l_i} \right)^{1/2} \alpha_i \right)^{(v)} \tag{7.128}$$

in which α_i is the relative design variable, and Λ the *scaling factor* whose value is equal to the maximum moment of inertia among all members of a framework or the maximum cross-sectional area among all members of a truss. If $u_i' = u_i \Lambda$, and $\zeta_i' = \alpha_i \eta_i \rho_i l_i$, Eq. (7.128) becomes

$$(\Lambda \alpha_i)^{(v+1)} = \left(\left(\frac{W}{U} \right)^{1/2} \left(\frac{u_i'}{\zeta_i'} \right)^{1/2} \alpha_i \right)^{(v)} \tag{7.129}$$

If one considers the $P - \Delta$ effect resulting from the axial force on a member, Eq. (7.123) may be rewritten as

$$\frac{1}{\lambda} \sum_{i=1}^{m} \rho_i \eta_i x_i l_i = \sum_{i=1}^{m} \frac{1}{2} \left(\{r\}^T [K]_i \{r\} - P_i' \{r\} [K_g]_i \{r\} \right) \tag{7.130}$$

in which P_i' is equal to $\rho_i \eta_i l_i / 2$ which is half the weight of the member differentiated with respect to the design variable x_i, and $[K_g]_i$ is the geometric stiffness matrix in global coordinates associated with the $P - \Delta$ effect. Because the structural weight is to be minimized, sufficient strain energy is required to bring the structure to stay in the feasible region. For this purpose, u_i'/ζ_i' in Eq. (7.129) needs to be selected as large as possible, that is, λ should be selected as small as possible. For programming convenience, the Lagrange multiplier in Eq. (7.129) associated with W/U may be replaced by

$$
\lambda_{\min} = \min \left(\left(\frac{\zeta_i'}{u_i'} \Lambda^2 \right) \left(\frac{R_{x_j}}{R_{x\max}} \right)^2 \right)
$$

$$
= \min \left(\left(\frac{\eta_i x_i \rho_i l_i}{\frac{1}{2} \left(\{r\}^T [K]_i \{r\} - P_i' \{r\}^T [K_g]_i \{r\} \right)} \right) \left(\frac{R_{x_j}}{R_{x\max}} \right)^2 \right) \tag{7.131}
$$

This means that $\lambda_{\min}$ is the minimum value chosen from all the members under each loading. R_{x_j} is the maximum ratio of the actual stress to the allowable stress among all members at the loading condition j, and $R_{x\max}$ is the maximum value of all the ratios based on the constraints of stresses, displacements, and frequencies. $R_{x_j}/R_{x\max}$ is used as an approximate approach to adjust the original scaling factor, Λ, which is based on the maximum ratio for all constraints. Note that *the stiffness constraint is treated as a stress constraint because stresses in a structure are directly affected by the system's stiffness*. For NLC loading conditions, one should obtain NLC $\lambda_{\min}$ and may use either Eq. (7.124) or (7.131) to evaluate the Lagrange multiplier.

7.4.3   *Optimum criterion based on static displacement constraints*

For convenience of formulation and calculation, let us consider one of several displacement constraints. If the jth degree of freedom is active, one may express this displacement as

$$
u_j(x_1, x_2, \ldots, x_m) \leq \text{given value} \tag{7.132}
$$

in which the x's are primary variables. One of the Kuhn–Tucker conditions of Eqs. (7.88) through (7.92) is

$$
\frac{\partial W}{\partial x_i} + \lambda \frac{\partial u_j}{\partial x_i} = 0 \qquad i = 1, 2, \ldots, m \tag{7.133}
$$

The term u_j can be expressed as

$$
u_j = \{Q_j\}^T \{r\} \tag{7.134}
$$

in which $\{r\}$ is the actual displacement vector, and $\{Q_j\}$ the *virtual load vector* in the following form

$$\{Q_j\} = [0\,0\ldots 0\,1\,0\ldots 0\,0]^T$$

↑
jth column

(7.135)

In Eq. (7.133)

$$\frac{\partial u_j}{\partial x_i} = \{Q_j\}^T \frac{\partial \{r\}}{\partial x_i}$$

(7.136)

If one uses the same process of derivation for the energy constraints of Eq. (7.115), then

$$\frac{\partial \{r\}}{\partial x_i} = -\frac{1}{x_i} [K]^{-1} [K]_i \{r\}$$

(7.137)

If we consider $\{Q_j\}$ of Eq. (7.135) as a load vector that is virtually applied to the structure, the force–displacement relationship can be expressed as

$$\{Q_j\} = [K]\{q_j\}$$

(7.138)

in which $\{q_j\}$ is the displacement resulting from the virtual load vector $\{Q_j\}$. By substituting Eqs. (7.137) and (7.138) into Eq. (7.136), one then obtains

$$\frac{\partial u_j}{\partial x_i} = -\frac{1}{x_i} \{q_j\}^T [K]_i \{r\}$$

(7.139)

If we introduce

$$\left\{\bar{Q}_j\right\}_{gi} = [K]_i \{q_j\}$$

(7.140)

then $\left\{\bar{Q}_j\right\}_{gi}$ becomes the nodal forces of element i in global coordinates due to the virtual load, $\{Q_j\}$. From Eqs. (7.139) and (7.140),

$$\frac{\partial u_j}{\partial x_i} = -\frac{1}{x_i} \left\{\bar{Q}_j\right\}_{gi}^T \{r\}$$

(7.141)

Remove those degrees-of-freedom that are not related to the element i in $\{q_j\}$, which is modified as $\{\bar{q}_j\}$. Then Eq. (7.140) becomes

$$\left\{\bar{Q}_j\right\}_{gi} = [K]_i \{\bar{q}_j\}$$

(7.142)

Let $\{\bar{r}\}_{gi}$ be the actual displacements at the nodes of element i in global coordinates; then Eq. (7.141) yields

$$\frac{\partial u_j}{\partial x_i} = -\frac{1}{x_i} \left\{\bar{Q}_j\right\}_{gi}^T \{\bar{r}\}_{gi} \tag{7.143}$$

By using the transformation matrix, $[T]$, from the local coordinates to the global coordinates, we have

$$\left\{\bar{Q}_j\right\}_{gi} = [T] \left\{\bar{Q}_j\right\}_{ei} \tag{7.144}$$

and

$$\{\bar{r}\}_{gi} = [T] \{\bar{r}\}_{ei} \tag{7.145}$$

in which the subscript ei represents the element i in local coordinates. Substituting Eqs. (7.144) and (7.145) into Eq. (7.143) yields

$$\frac{\partial u_j}{\partial x_i} = -\frac{1}{x_i} \left\{\bar{Q}_j\right\}_{ei}^T \{\bar{r}\}_{ei} \tag{7.146}$$

Consider the objective function

$$W = \sum_{i=1}^{m} \rho_i \eta_i x_i l_i \tag{7.147}$$

then substituting Eqs. (7.146) and (7.147) into Eq. (7.145) gives

$$\eta_i \rho_i l_i = \frac{\lambda}{x_i} \left\{\bar{Q}_j\right\}_{ei}^T \{\bar{r}\}_{ei} \tag{7.148}$$

or

$$\lambda = \frac{\eta_i \rho_i l_i x_i}{\left\{Q_j\right\}_{ei}^T \{r\}_{ei}} \tag{7.149}$$

Multiplying both sides by x_i^2 and then taking the square root yields

$$x_i = (\lambda)^{\frac{1}{2}} \left(\frac{\left\{\bar{Q}_j\right\}_{ei}^T \{\bar{r}\}_{ei}}{\eta_i \rho_i l_i} \right)^{\frac{1}{2}} (x_i)^{\frac{1}{2}} \tag{7.150}$$

By using $x_i = \Lambda\alpha_i$, and then considering the recursion relationship, we will have

$$
(\Lambda\alpha_i)^{(v+1)} = (\lambda_{\min})^{\frac{1}{2}} \left(\frac{\{Q_j\}_{ei}^{T} \{r\}_{ei} \Lambda}{\eta_i \alpha_i \rho_i l_i} \right)^{\frac{1}{2}(v)} \alpha_i^{(v)}
\tag{7.151}
$$

In the computer analysis, $\lambda_{\min}$ may be obtained as follows

$$
\lambda_{\min} = \min\left(\left(\frac{\zeta_i'}{u_i'} \Lambda^2 \right) \left(\frac{R_{x_j}}{R_{x\max}} \right)^2 \right)
$$

$$
= \min\left(\left(\frac{\eta_i x_i \rho_i \ell_i}{\{\bar{Q}_j\}_{ei}^{T} \{\bar{r}\}_{ei}} \right) \left(\frac{R_{x_j}}{R_{x\max}} \right)^2 \right)
\tag{7.152}
$$

in which ζ_i' is equal to $\eta_i \alpha_i \rho_i l_i$, u_i' is equal to $\{\bar{Q}_j\}_{ei}^{T} \{\bar{r}\}_{ei} \Lambda$, and R_{x_j} is the ratio of the actual displacements (i.e., only those displacements associated with the j active displacement constraints) to the allowable displacement for each loading condition. $R_{x\max}$ is the maximum value of all the ratios for all the constraints and loading conditions considered. Thus, $\lambda_{\min}$ is sorted out as the minimum value for all members under each active displacement due to each loading condition. For NLC loading conditions with total nJ active displacements, we can have $nJ\lambda_{\min}$. $R_{x_j}/R_{x\max}$ is used to adjust the original scaling factor, Λ, which is based on the maximum value of the ratios. By considering the $P - \Delta$ effect on the stiffness, one may write Eq. (7.151) as

$$
(\Lambda\alpha_i)^{(v+1)} = \left((\lambda_{\min})^{\frac{1}{2}} \left(\frac{\left(\{\bar{Q}_j\}_{ei}^{T} \{\bar{r}\}_{ei} - P_i' \{\bar{Q}_{gj}\}_{ei}^{T} \{\bar{r}\}_{ei} \right) \Lambda}{\eta_i \alpha_i \rho_i l_i} \right)^{\frac{1}{2}} \alpha_i \right)^{(v)}
\tag{7.153}
$$

in which $\{\bar{Q}_{gj}\}_{ei}$ is the geometric force vector of the element i occasioned by the displacement resulting from the load vector $\{Q_j\}$, and $\lambda_{\min}$ should be obtained in a similar manner as shown in Eq. (7.152) with the inclusion of the $P - \Delta$ effect.

7.4.4 Optimality criterion based on dynamic stiffness stress constraints

The derivation of the dynamic stress constraints is similar to that of the static stress constraints expressed in terms of structural stiffness. The recursion equation may be expressed as

$$(\Lambda\alpha_i)^{(\nu+1)} = \left(\left(\frac{W}{U}\right)^{\frac{1}{2}} \begin{array}{c} [(\{r(x,t)\}^T [K]_i \{r(x,t)\} - P'_i \{r(x,t)\}^T [K_g]_i \{r(x,t)\} \\ -p^2 \{r(x,t)\}^T [M_s]_i \{r(x,t)\}) \Lambda / 2\eta_i\alpha_i\rho_i l_i]^{1/2} \alpha_i \end{array} \right)^{(\nu)}$$

(7.154)

in which p is the frequency obtained on the basis of the *Rayleigh quotient*, $[M_s]_i$ is the structural mass matrix of member i in global coordinates, and the variables, x and t, represent the design variable and time, respectively.

$$U = \sum_{i=1}^{m} u_i = \sum_{i=1}^{m} \frac{1}{2} (\{r(x,t)\}^T [K]_i \{r(x,t)\} - P'_i \{r(x,t)\}^T [K_g]_i \{r(x,t)\}$$

$$-p^2 \{r(x,t)\}^T [M_s]_i \{r(x,t)\})$$

(7.155)

and

$$W = \sum_{i=1}^{m} w_i = \sum_{i=1}^{m} \eta_i x_i \rho_i l_i$$

(7.156)

However, W/U may be replaced by $\lambda_{\min}$ as in

$$\lambda_{\min} = \min\left(\left(\frac{\zeta'_i}{u'_i}\Lambda^2\right)\left(\frac{R_{x_i}}{R_{x_{\max}}}\right)^2\right) = \min\left(\left(\frac{w_i}{u'_i}\right)\left(\frac{R_{x_i}}{R_{x_{\max}}}\right)^2\right)$$

(7.157)

in which u_i and w_i are expressed in Eqs. (7.155) and (7.156), respectively, ζ'_i is equal to w_i/Λ, u'_i is equal to $u_i\Lambda$ and the other terms have already been explained for Eq. (7.131).

7.4.5 Optimality criterion based on dynamic displacement constraints

Similar to the derivation of the recursion for static displacement constraints, the recursion expression for the dynamic case can be expressed as

$$(\Lambda\alpha_i)^{(\nu+1)} = \left((\lambda_{\min})^{1/2} \left(\begin{array}{c} \alpha_i \left(\{\bar{Q}_i\}^T_{ei} \{r(x,t)\}_{ei} - P'_i \{\bar{Q}_{gi}\}^T_{ei} \{r(x,t)\}_{ei} \right. \\ \left. -p^2 \{\bar{\bar{Q}}_i\}^T_{ei} \{r(x,t)\}_{ei}\right) \Lambda / \eta_i\alpha_i\rho_i l_i \end{array} \right)^{1/2} \right)^{(\nu)}$$

(7.158)

in which $\{\bar{\bar{Q}}_i\}_{ei}$ is the inertia force vector of the element i due to displacement resulting from the load vector, $\{Q_j\}$, and

$$\lambda_{\min} = \min\left(\left(\frac{\zeta_i'}{u_i'}\Lambda^2\right)\left(\frac{R_{x_j}}{R_{x\max}}\right)^2\right)$$

$$= \min\left(\left(\eta_i x_i \rho_i l_i \Big/ \left(\{\bar{Q}_j\}_{ei}^T \{r(x,t)\}_{ei} - P_i'\{\bar{Q}_{gj}\}_{ei}^T \{r(x,t)\}_{ei}\right.\right.\right.$$

$$\left.\left.\left. - p^2\{\bar{\bar{Q}}_j\}_{ei}^T \{r(x,t)\}_{ei}\right)\right)\left(\frac{R_{x_j}}{R_{x\max}}\right)^2\right) \tag{7.159}$$

in which R_{x_j} and $R_{x\max}$ are for the dynamic case, that are similarly defined for Eq. (7.152), ζ_i' is equal to $\eta_i\alpha_i\rho_i l_i$, and u_i' is equal to

$$\Lambda\left(\{\bar{Q}_j\}_{ei}^T\{r(x,t)\}_{ei} - P_i'\{\bar{Q}_{gj}\}_{ei}^T\{r(x,t)\}_{ei} - p^2\{\bar{\bar{Q}}_j\}_{ei}^T\{r(x,t)\}_{ei}\right)$$

7.4.6 Optimal criterion based on dynamic frequency constraints

For the constraint on any natural frequency, ω_j, the recursion equation may be written as

$$(\Lambda\alpha_i)^{(v+1)} = \left((\lambda_{\min})^{1/2}\alpha_i\left(\frac{\{\phi_j\}^T[K]_i\{\phi_j\} - P_i'\{\phi_j\}^T[K_g]_i\{\phi_j\}}{-\omega_j^2\{\phi_j\}^T[M_s]_i\{\phi_j\})\Lambda/\eta_i\alpha_i\rho_i l_i}\right)^{1/2}\right)^{(v)} \tag{7.160}$$

in which $\{\phi_j\}$ is the jth normal mode [1] and

$$\lambda_{\min} = \min\left(\left(\frac{\zeta_i'}{u_i'}\Lambda^2\right)\left(\frac{R_{x_j}}{R_{x\max}}\right)^2\right)$$

$$= \min\left(\left(\eta_i x_i \rho_i l_i/(\{\phi_j\}^T[K]_i\{\phi_j\} - P_i'\{\phi_j\}^T[K_g]_i\{\phi_j\}\right.\right.$$

$$\left.\left. -\omega_j^2\{\phi_j\}^T[M_s]_i\{\phi_j\})\right)\left(\frac{R_{x_j}}{R_{x\max}}\right)^2\right) \tag{7.161}$$

where R_{x_j} is the ratio of the allowable frequency to the actual frequency of the jth mode and $R_{x\max}$ is the maximum value of all the ratios for various constraints and loading conditions including static and dynamic loadings, if any. It is apparent that the maximum is searched for in each member i corresponding to the particular mode being investigated. ζ_i' is equal to $\eta_i \alpha_i \rho_i l_i$, and u_i' is equal to $\Lambda \left(\{\phi_i\}^T [K]_i \{\phi_i\} - P_i' \{\phi_i\}^T [K_g]_i \{\phi_i\} - \omega_j^2 \{\phi_i\}^T [M_s]_i \{\phi_i\} \right)$.

7.4.7   *Comments on energy density for dynamic design*

As discussed previously, *the strain energy virtual strain energy* and *kinetic energy* are included in the calculation of *energy density* for each individual member i. However, in designing high-rise buildings or buildings with a low value of the fundamental natural frequency, a negative value of energy density may be obtained for members on the upper storeys. Thus, the algorithm may not yield reasonable results. The numerical procedures are modified such that the members with negative values of energy density should be redesigned as passive elements. To prevent an incorrect redistribution in the sizes of the members, Eqs. (7.154), (7.158) and (7.160) can be rewritten as Eqs. (7.162), (7.163) and (7.164), respectively,

$$(\Lambda \alpha_i)^{(v+1)} = \left(\left(\frac{W}{U} \right)^{1/2} \begin{array}{l} (\{r(x,t)\}^T [K]_i \{r(x,t)\} \\ -P_i' \{r(x,t)\}^T [K_g]_i \{r(x,t)\}) \, \Lambda / 2\eta_i \alpha_i \rho_i l_i)^{1/2} \, \alpha_i \end{array} \right)^{(v)} \tag{7.162}$$

$$(\Lambda \alpha_i)^{(v+1)} = \left((\lambda_{\min})^{1/2} \left(\left(\{\bar{Q}_i\}_{ei}^T \{r(x,t)\}_{ei} - P_i' \{\bar{Q}_{gi}\}_{ei}^T \{r(x,t)\}_{ei} \right) \right. \right.$$
$$\left. \left. \Lambda / \eta_i \alpha_i \rho_i l_i \right)^{1/2} \alpha_i \right)^{(v)} \tag{7.163}$$

$$(\Lambda \alpha_i)^{(v+1)} = \left((\lambda_{\min})^{1/2} \left(\left(\{\phi_i\}^T [K]_i \{\phi_i\} - P_i' \{\phi_i\}^T [K_g]_i \{\phi_i\} \right) \right. \right.$$
$$\left. \left. \Lambda / \eta_i \alpha_i \rho_i l_i \right)^{1/2} \alpha_i \right)^{(v)} \tag{7.164}$$

in which the total energy, U, should not include the kinetic energy term [21, 23].

7.5   Multiple active constraints for optimality-criteria method

7.5.1   *Recursion relation based on multiple active constraints*

In any design, it is possible to have more than one active constraint when the restraints on displacements, stresses and frequencies are imposed simultaneously. Therefore, it is necessary to find Lagrange multipliers corresponding to the active constraints of the current design variables. However, because of the difficulty in making a numerical calculation, the minimum value of λ for all active constraints of member i is adopted in the recursion relation, which is given as

$$(\alpha_i)^{(v+1)} = (\alpha_i)^{(v)} \left(\frac{\max (u'_{ij} \lambda_J)}{\zeta'_{ij}} \right)^{1/2} \tag{7.165}$$

in which λ_J is the minimum value of λ determined from all active constraints for an individual member i and expressed as $\min (\lambda_{ij})$. The maximum value of u'_{ij} is also obtained from all active constraints of member i. Thus, $\max (u'_{ij} \lambda_J)$ is the upper bound of the member size requirement for all active constraints for all loading conditions. Note that the static displacements are combined with dynamic displacement from which the stresses are calculated.

7.5.2 Calculation of constraint gradients

The recursion relation based on the strain energy criterion in conjunction with the scaling procedure presented in the previous sections is sufficient for designing optimum structures with various constraints. The design, however, can be further improved by using an iterative algorithm based on constraint gradients for which the algorithm is of the following form [4, 6].

$$\alpha_i^{(v+1)} = \alpha_i^{(v)} + s(\Delta \alpha_i) \tag{7.166}$$

in which v and $v+1$ refer to the cycle of iteration, α_i is the relative design variable of member i, and s is the step size determining the rate of approach to the optimum solution. The value $\Delta \alpha_i$ is determined by the influence of the design variable of member i on the active constraint. The procedures for determining $\Delta \alpha_i$, which is based on the active displacement constraint and active stress constraint, are presented below.

(A) *Determination of $\Delta \alpha_i$ based on displacement constraints.* Let us consider that a structure has been optimized to a local minimum. If we increase a member size and reanalyze it, two possibilities can be found in the behaviour of the structure: a) the displacements for active degrees-of-freedom are increased, and b) the displacements for active degrees-of-freedom are decreased. The former is called a negative influence because the structure is less stiff when the member size is increased, and the latter is called a positive influence. If the procedure is repeated by changing member sizes one by one, one can discover which member can be reduced in size. However, this information does not provide the magnitude of change. An overshooting problem could occur in the optimal process if the magnitude of change in the member size is not suitably chosen. To overcome this problem, the following procedures are derived.

Let us reduce the size of all members in a structure by a certain percentage. This change will reduce the structural stiffness, and, consequently, the active displacement, r_j, is increased beyond the constrained surface by Δr_j. In order to bring it back to the constrained surface, one can increase the size of members which have positive influence. The magnitude of the increment of member size, Δx_i, is determined on the basis of the assumption that the change in displacement is directly proportional to the change of member size, and the change of member size is inversely proportional to the member length because an increase in size of a member with a larger length causes a larger increase in weight. Thus, the increment in member size, Δx_i, can be expressed as follows

$$\Delta x_i = \Omega \frac{dr_{ji}}{l_i} \tag{7.167}$$

in which dr_{ji} is the change of displacement in the active degree of freedom j due to the unit change in the size of member i, and Ω is a constant of proportionality, which is determined by evaluating the average influence on the change of the displacement in the active degree of freedom occasioned by the unit change in the size of members that have positive influence. In order to determine Ω, let us assume that dr_{ji} is directly proportional to the member size, x_i, and the change in r_j due to Δx_i is Δr_{ji}, then

$$\frac{\Delta r_{ji}}{\Delta x_i} = \frac{dr_{ji}}{x_i} \tag{7.168}$$

from which

$$\Delta r_{ji} = \frac{dr_{ji}}{x_i} \Delta x_i \tag{7.169}$$

By introducing $x_i = \Lambda \alpha_i$ into the above equation, one has

$$\Delta r_{ji} = \frac{dr_{ji}}{\alpha_i} \Delta \alpha_i \tag{7.170}$$

Thus, the total change in r_j due to the change in member size, Δx_i, is

$$\Delta r_j = \sum_{i=1}^{m} \Delta r_{ji} = \sum_{i=1}^{m} \frac{dr_{ji}}{\alpha_i} \Delta \alpha_i \tag{7.171}$$

in which m is the number of members that have a positive influence on r_j. By substituting Eq. (7.167) into Eq. (7.171), one has

$$\Delta r_j = \bar{\Omega} \sum_{i=1}^{m} \frac{1}{\alpha_i l_i} \left(dr_{ji} \right)^2 \tag{7.172}$$

in which $\bar{\Omega} = \Omega/\Lambda$. From Eq. (7.172), one finds that

$$\bar{\Omega} = \frac{\Delta r_j}{\sum\limits_{k=1}^{m} \frac{1}{\alpha_k l_k}(dr_{jk})^2}$$
(7.173)

Substitution of Eq. (7.173) into Eq. (7.167) gives

$$\Delta\alpha_i = \frac{\Delta r_j}{\sum\limits_{k=1}^{m} \frac{1}{\alpha_k l_k}(dr_{jk})^2}\left(\frac{dr_{ji}}{l_i}\right)$$
(7.174)

$\Delta\alpha_i$ is assumed to be zero if the value computed from Eq. (7.173) is negative. It is possible that more than one displacement exceeds the limit. Under these circumstances then, it is necessary to determine the change in each element size separately for each constraint. The largest value of $\Delta\alpha_i$ shall be used for the actual change in the size of the *i*th element.

(B) Determination of $\Delta\alpha_i$ based on stress constraints. The determination of $\Delta\alpha_i$ based on stress constraints is similar to that discussed in (A). If '*j*' is the member at which the stress constraint is active, and $d\sigma_{ji}$ is the change in σ_j occasioned by a unit change in the size of element *i*, then the required change in the size of the *i*th element is determined by the following expression:

$$\Delta\alpha_i = \bar{\Omega}_1 \frac{d\sigma_{ji}}{l_i}$$
(7.175)

in which $d\sigma_{ji}$ is calculated on the basis of dr_{ji}, and the constant of proportionality is

$$\bar{\Omega} = \frac{\Delta\sigma_j}{\sum\limits_{k=1}^{m} \frac{1}{\alpha_k l_k}(d\sigma_{jk})^2}$$
(7.176)

in which $\Delta\sigma_j = \sum\limits_{i=1}^{m}\Delta\sigma_{ji}$, which is determined from the summation of the differential stress (the difference between the actual stress and the allowable stress) associated with the active stress constraint of member *j*, occasioned by the size changes of member *i*. The required change in relative design variable for Eq. (7.166) is then given as

$$\Delta\alpha_i = \frac{\Delta\sigma_j}{\sum\limits_{k=1}^{m} \frac{1}{\alpha_k l_k}(d\sigma_{jk})^2}\left(\frac{d\sigma_{ji}}{l_i}\right)$$
(7.177)

7.5.3 *Numerical procedures for calculating constraint gradients*

The detailed numerical procedures for calculating the constraint gradients are given in the following 10 steps.

Step 1. Determine the most active constraint based on the design variables that satisfy the optimum criterion. At this step, the system stiffness matrix, $[K]$, and the displacement vector, $\{r\}$, are stored as well.

Step 2. Calculate the stiffness matrices, $[K]_i$ and $[K']_i$, of each individual element i. Here, $[K]_i$ is the element stiffness matrix in global coordinates and corresponds to the design variable x_i determined in Step 1, and $[K']_i$ is the element stiffness matrix in global coordinates and corresponds to the design variable with the unit change $x_i + 1$.

Step 3. Calculate $\{dr\}_i$ according to Eq. (7.166). If the displacement constraint is the most active constraint, then proceed to Step 5.

Step 4. Calculate $\{d\sigma\}_i$ by using the result of Step 3.

Step 5. Reduce the design variable x_i of each element i by a reduction factor β. (Assume that β is 10–20 per cent.) Then check the reduced design variable $x'_i = (1 - \beta)x_i$, with the size constraint. If x'_i is less than the lower bound of the member size, $x_{\min}$, then assume that x'_i is equal to $x_{\min}$.

Step 6. Analyze the structure by using the reduced design variable to obtain the new displacement vector $\{r\}$. If the displacement constraint is the most active constraint, which is determined in Step 1, then proceed to Step 8.

Step 7. Calculate the new stress vector, $\{\sigma\}$, by using the displacement vector, $\{r\}$.

Step 8. Compute Δr_j or $\Delta \sigma_j$ by using the result of either Step 6 or 7: a) if the displacement constraint controls, then $\Delta r_j = |r_j| - r_{ja}$ in which r_j is the displacement of $\{r\}$ in the active constraint direction j, and r_{ja} is the allowable displacement, and b) if the stress constraint is the most active constraint, then $\Delta \sigma_j$ is determined according to $\Delta \sigma_j = |\sigma_j| - \sigma_{ja}$ in which σ_j is the active stress member j in $\{\sigma\}$, and σ_{ja} is the allowable value of the stress.

Step 9. Calculate the required change in the size of each element i, α_i, by using either Eq. (7.174) or (7.177). If a displacement constraint is active, Eq. (7.174) is used. For the value of dr_{ji} in Eq. (7.174), when $r_j < 0$ and $dr_{ji} > 0$ or when $r_j > 0$ and $dr_{ji} < 0$ then let dr_{ji} be equal to zero. If the stress constraint is active, then Eq. (7.177) is used. For the value of $d\sigma_{ji}$ in Eq. (7.176), when $\sigma_j < 0$ and $d\sigma_{ji} > 0$ then let $d\sigma_{ji}$ be equal to zero.

Step 10. Redistribute the member sizes according to Eq. (7.166).

7.5.4 Numerical illustration of optimality-criteria method

Example 7.5.1 – The truss given in Fig. 7.9a has the material properties: $A_{min} = 0.1$ in.^{2}, $E = 1.0 \times 10^7$ psi, $\rho = 0.1$ lbs in^{-3}. Optimize the structure for minimum weight considering the following two cases: (A) Stress constraints of $\bar{\sigma}_1 = 10,000$ psi, $\bar{\sigma}_2 = 40,000$ psi, $\bar{\sigma}_3 = 10,000$ psi and (B) Displacement constraints of $\bar{c}_1 = 0.05$ in., $\bar{c}_2 = 0.05$ in. Illustrate (A) and (B) with detailed calculations.

Solution:

Let the initial design variables be

$$[A_1 \, A_2 \, A_3] = [1 \; 1 \; 1] \text{ in.}^2 = \Lambda \, [\alpha_1 \; \alpha_2 \; \alpha_3] \quad \text{where } \Lambda = 1.0, \; \alpha_1 = 1.0,$$
$$\alpha_2 = 1.0, \; \alpha_3 = 1.0.$$

The free-body diagram is sketched in Fig. 7.9 where internal forces, $\{F\}$, and external load, $\{P\}$, are symbolically assigned in the positive direction. The equilibrium condition at the node, the relationship between member forces $\{F\}$ and deformations $\{e\}$, as well as compatibility conditions between

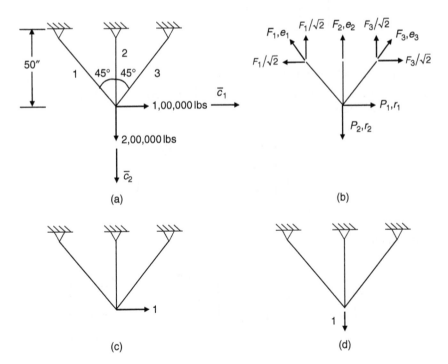

Figure 7.9 Example 7.5.1: (a) given structure, (b) free-body diagram, (c) virtual load in r_1 direction, (d) virtual load in r_2 direction.

external displacements $\{r\}$ and internal deformations $\{e\}$ are formulated in Eqs. (a), (b) and (c), respectively [1].

$$\{P\}=[A]\{F\}=\begin{pmatrix} \dfrac{1}{\sqrt{2}} & 0 & -\dfrac{1}{\sqrt{2}} \\[2mm] \dfrac{1}{\sqrt{2}} & 1 & \dfrac{1}{\sqrt{2}} \end{pmatrix}\begin{Bmatrix} F_1 \\ F_2 \end{Bmatrix} \tag{a}$$

$$\{F\}=[S]\{e\}=\begin{pmatrix} \dfrac{A_1E}{l_1} & 0 & 0 \\[2mm] 0 & \dfrac{A_2E}{l_2} & 0 \\[2mm] 0 & 0 & \dfrac{A_3E}{l_3} \end{pmatrix}=\dfrac{AE}{50}\begin{pmatrix} \dfrac{\alpha_1}{\sqrt{2}} & 0 & 0 \\[2mm] 0 & \alpha_2 & 0 \\[2mm] 0 & 0 & \dfrac{\alpha_3}{\sqrt{2}} \end{pmatrix} \tag{b}$$

$$\{e\}=[B]\{r\}=\begin{pmatrix} \dfrac{1}{\sqrt{2}} & \dfrac{1}{\sqrt{2}} \\[2mm] 0 & 1 \\[2mm] -\dfrac{1}{\sqrt{2}} & \dfrac{1}{\sqrt{2}} \end{pmatrix}\begin{Bmatrix} r_1 \\ r_2 \end{Bmatrix} \tag{c}$$

Note that the $[B]$ compatibility matrix of a system (similar to $[b]_i$ in Eq. (7.9) for an individual member) is always equal to $[A]^T$. Thus, the structural stiffness matrix is

$$\{K\}=[A][S][A]^T=\dfrac{AE}{50}\begin{Bmatrix} \dfrac{\alpha_1+\alpha_3}{2\sqrt{2}} & \dfrac{\alpha_1-\alpha_3}{2\sqrt{2}} \\[2mm] \text{sym} & \dfrac{\alpha_1+\alpha_3}{2\sqrt{2}}+\alpha_2 \end{Bmatrix} \tag{d}$$

from which

$$[K]^{-1}=\dfrac{50}{AE}\dfrac{2\sqrt{2}}{\alpha_1\alpha_3\sqrt{2}+\alpha_1\alpha_2+\alpha_2\alpha_3}\begin{Bmatrix} \dfrac{\alpha_1+\alpha_3}{2\sqrt{2}}+\alpha_2 & -\dfrac{\alpha_1-\alpha_3}{2\sqrt{2}} \\[2mm] \text{sym} & \dfrac{\alpha_1+\alpha_3}{2\sqrt{2}} \end{Bmatrix} \tag{e}$$

Thus, the displacements, member forces and stresses can be expressed by Eqs. (f), (g) and (h), respectively.

$$\begin{Bmatrix} r_1 \\ r_2 \end{Bmatrix} = [K]^{-1} \begin{Bmatrix} P_1 \\ P_2 \end{Bmatrix} = \frac{50}{\Lambda E} \frac{2\sqrt{2}}{\alpha_1\alpha_3\sqrt{2}+\alpha_1\alpha_2+\alpha_2\alpha_3}$$

$$\begin{Bmatrix} \left(\dfrac{\alpha_1+\alpha_3}{2\sqrt{2}}+\alpha_2\right)P_1 - \dfrac{\alpha_1-\alpha_3}{2\sqrt{2}}P_2 \\[4mm] -\dfrac{\alpha_1-\alpha_3}{2\sqrt{2}}P_1 + \dfrac{\alpha_1+\alpha_3}{2\sqrt{2}}P_2 \end{Bmatrix} \tag{f}$$

$$\begin{Bmatrix} F_1 \\ F_2 \\ F_3 \end{Bmatrix} = [S][A]^T \begin{Bmatrix} r_1 \\ r_2 \end{Bmatrix} = \frac{2\sqrt{2}}{\alpha_1\alpha_3\sqrt{2}+\alpha_1\alpha_2+\alpha_2\alpha_3}$$

$$\begin{Bmatrix} \left(\dfrac{\alpha_1\alpha_2}{2\sqrt{2}}+\dfrac{\alpha_1\alpha_2}{2}\right)P_1 + \dfrac{\alpha_1\alpha_2}{2\sqrt{2}}P_2 \\[4mm] -\dfrac{\alpha_1\alpha_2-\alpha_2\alpha_3}{2\sqrt{2}}P_1 + \dfrac{\alpha_1\alpha_2+\alpha_2\alpha_3}{2\sqrt{2}}P_2 \\[4mm] \left(-\dfrac{\alpha_1\alpha_2}{2\sqrt{2}}-\dfrac{-\alpha_2\alpha_3}{2}\right)P_1 + \dfrac{\alpha_1\alpha_3}{2\sqrt{2}}P_2 \end{Bmatrix} \tag{g}$$

$$[\sigma_1 \ \sigma_2 \ \sigma_3]^T = \begin{bmatrix} \dfrac{F_1}{A_1} & \dfrac{F_2}{A_2} & \dfrac{F_3}{A_3} \end{bmatrix}^T \tag{h}$$

(A) Stress constraints only

Cycle 1

Substituting $\alpha_1 = \alpha_2 = \alpha_3 = 1.0$, $\Lambda = 1.0$, $P_1 = 1,000,000\,\text{lbs}$, $P_2 = 2,000,000\,\text{lbs}$ into Eqs. (f), (g) and (h) yields Eqs. (i), (j) and (k), respectively.

$$\begin{Bmatrix} r_1 \\ r_2 \end{Bmatrix} = \frac{50}{1.0 \times 10^7} \frac{2\sqrt{2}}{\sqrt{2}+1+1} \begin{Bmatrix} \left(\dfrac{2}{2\sqrt{2}}+1\right)100,000 - (0)\,200,000 \\[4mm] -(0)\,100,000 + \left(\dfrac{2}{2\sqrt{2}}\right)200,000 \end{Bmatrix}$$

$$= \begin{Bmatrix} \dfrac{2}{\sqrt{2}} \\[4mm] \dfrac{\sqrt{2}}{1+\sqrt{2}} \end{Bmatrix} \text{in.} \tag{i}$$

$$
\begin{Bmatrix} F_1 \\ F_2 \\ F_3 \end{Bmatrix} = \frac{2\sqrt{2}}{\sqrt{2}+1+1} \begin{Bmatrix} \left(\dfrac{1}{2\sqrt{2}}+\dfrac{1}{2}\right)100,000 + \left(\dfrac{1}{2\sqrt{2}}\right)200,000 \\[2mm] \left(\dfrac{2}{2\sqrt{2}}\right)200,000 \\[2mm] \left(-\dfrac{1}{2\sqrt{2}}+\dfrac{1}{2}\right)100,000 + \left(\dfrac{1}{2\sqrt{2}}\right)200,000 \end{Bmatrix}
$$

$$
= \begin{Bmatrix} 129,289.32 \\ 117,157.29 \\ -12,132.03 \end{Bmatrix} \text{lbs} \tag{j}
$$

$$
\begin{Bmatrix} \sigma_1 \\ \sigma_2 \\ \sigma_3 \end{Bmatrix} = \begin{Bmatrix} F_1/A_1 \\ F_2/A_2 \\ F_3/A_3 \end{Bmatrix} = \begin{Bmatrix} 129,289.32 \\ 116,157.29 \\ -12,132.03 \end{Bmatrix} \text{psi} \tag{k}
$$

Comparing with the given allowable stresses gives

$$
\left. \begin{aligned} \left|\frac{\sigma_1}{\bar{\sigma}_1}\right| &= \frac{129,289.32}{10,000} = 12.928932 \\[2mm] \left|\frac{\sigma_2}{\bar{\sigma}_2}\right| &= \frac{117,157.29}{10,000} = 2.92893 \\[2mm] \left|\frac{\sigma_3}{\bar{\sigma}_3}\right| &= \frac{12,132.03}{10,000} = 1.213203 \end{aligned} \right\} \quad \max\left(\left|\frac{\sigma_i}{\bar{\sigma}_i}\right|\right) = 12.928932 \tag{l}
$$

Now, $(\Lambda)_{\text{new}} = (\Lambda)_{\text{old}} \max\left(\left|\dfrac{\sigma_i}{\bar{\sigma}_i}\right|\right) = (1.0)\,12.928932 = 12.928932$. Because every member is increased with the same area, the displacements in Eq. (i) may be modified as

$$
\begin{Bmatrix} r_1 \\ r_2 \end{Bmatrix} = \frac{1}{12.928932} \begin{Bmatrix} \dfrac{\sqrt{2}}{2} \\[3mm] \dfrac{\sqrt{2}}{1+2} \end{Bmatrix} = \begin{Bmatrix} 0.05469 \\ 0.04531 \end{Bmatrix} \text{in.} \tag{m}
$$

we then have

$$
\begin{Bmatrix} \sigma_1 \\ \sigma_2 \\ \sigma_3 \end{Bmatrix} = \begin{Bmatrix} 10,000 \\ 9,061.637 \\ -938.363 \end{Bmatrix} \text{psi;} \quad \begin{Bmatrix} |\sigma_1/\bar{\sigma}_1| \\ |\sigma_2/\bar{\sigma}_2| \\ |\sigma_3/\bar{\sigma}_3| \end{Bmatrix} = \begin{Bmatrix} 1.0 \\ 0.22654 \\ 0.93836 \end{Bmatrix}
$$

The structure's weight is

$$W^{(1)} = \sum_{i=1}^{3} \rho_i A_i l_i = 0.1(12.928932)(50\sqrt{2} + 50 + 50\sqrt{2}) = 247.4874 \, \text{lbs}$$

(n)

Cycle 2

Since $\sigma_1/\bar{\sigma}_1 = 1$, one of the members is fully stressed, we start λ according to Eq. (7.124) based on the strain energy $[F]^T \{e\}$. From Eqs. (c) and (m),

$$\begin{Bmatrix} e_1 \\ e_2 \\ e_3 \end{Bmatrix} = \begin{pmatrix} \dfrac{1}{\sqrt{2}} & \dfrac{1}{\sqrt{2}} \\ 0 & 1 \\ -\dfrac{1}{\sqrt{2}} & \dfrac{1}{\sqrt{2}} \end{pmatrix} \begin{Bmatrix} 0.05469 \\ 0.04531 \end{Bmatrix} = \begin{Bmatrix} 0.07071 \\ 0.04531 \\ -0.006633 \end{Bmatrix} \, \text{in.}$$

(o)

Using Eqs. (j) and (o), we find the total strain energy as

$$U^{(1)} = \frac{1}{2} [e_1 F_1 + e_2 F_2 + e_3 F_3]$$

$$= \frac{1}{2}[(0.07071)129,289.32 + (0.04531)117,157.29$$

(p)

$$+ (0.006633)12,132.03]$$

$$= 7,065.458$$

$$\lambda = \frac{W^{(1)}}{U^{(1)}} = \frac{247.4874}{7,065.458} = 0.0340636$$

(q)

From Eq. (7.129)

$$(\Lambda\alpha_i)^{(v+1)} = \left[\left(\frac{W}{U}\right)^{\frac{1}{2}} \left(\frac{u_i}{\rho_i A_i l_i}\right)^{\frac{1}{2}} (\Lambda\alpha_i) \right]^{(v)}$$

$$(\Lambda\alpha_1)^{(2)} = \left[(0.0340636)^{\frac{1}{2}} \left(\frac{\dfrac{1}{2}(0.07071)\,129,289.32}{0.1\,(12.928932)\,50\sqrt{2}} \right)^{\frac{1}{2}} (12.928932) \right]$$

$$= 16.873015$$

(r)

$$(\Lambda\alpha_2)^{(2)} = \left[(0.0340636)^{\frac{1}{2}} \left(\frac{\dfrac{1}{2}(0.04531)\,117,157.29}{0.1\,(12.928932)\,50} \right)^{\frac{1}{2}} (12.928932) \right]$$

$$= 15.29002 \tag{s}$$

$$(\Lambda\alpha_3)^{(2)} = \left[(0.0340636)^{\frac{1}{2}} \left(\frac{\dfrac{1}{2}(0.006633)\,12,132.03}{0.1\,(12.928932)\,50\sqrt{2}} \right)^{\frac{1}{2}} (12.928932) \right]$$

$$= 1.583035 \tag{t}$$

Now, the new factors are

$$\Lambda = 16.873015 \tag{u}$$

$$\alpha_1 = 1.0, \quad \alpha_2 = \frac{\Lambda\alpha_2}{\Lambda} = \frac{15.29002}{16.873015} = 0.906182 \tag{v}$$

$$\alpha_3 = \frac{\Lambda\alpha_3}{\Lambda} = \frac{1.583035}{16.873015} = 0.938205 \tag{w}$$

Following Eqs. (f), (g) and (h) yields

$$\begin{Bmatrix} r_1 \\ r_2 \end{Bmatrix} = \begin{Bmatrix} 0.0486343 \\ 0.033788 \end{Bmatrix} \text{in.} \tag{x}$$

$$\begin{Bmatrix} F_1 \\ F_2 \\ F_3 \end{Bmatrix} = \begin{Bmatrix} 139,071.27 \\ 103,323.856 \\ -2,350.221 \end{Bmatrix} \tag{y}$$

$$\begin{Bmatrix} \sigma_1 \\ \sigma_2 \\ \sigma_3 \end{Bmatrix} = \begin{Bmatrix} 8,242.23 \\ 6,757.60 \\ -1,484.63 \end{Bmatrix} \text{psi} \tag{z}$$

$$\begin{Bmatrix} |\sigma_1/\bar{\sigma}_1| \\ |\sigma_2/\bar{\sigma}_2| \\ |\sigma_3/\bar{\sigma}_3| \end{Bmatrix} = \begin{Bmatrix} 0.824223 \\ 0.16894 \\ 0.148463 \end{Bmatrix} \to \max\left(\left| \frac{\sigma_i}{\bar{\sigma}_i} \right| \right) = 0.824223 \tag{a1}$$

The new areas are

$$(\Lambda)_{\text{new}} = (\Lambda)_{\text{old}} \times \left(\max \left| \frac{\sigma_i}{\bar{\sigma}_i} \right| \right) = (16.873015)\,0.824223 = 13.907127 \tag{a2}$$

$$[A_1 \ A_2 \ A_3] = [13.907127 \ 12.602388 \ 1.304774] \text{ in.}^2 \tag{a3}$$

for which

$$W^{(2)} = \sum_{i=1}^{n} \left(\rho_i \Lambda_i \alpha_i l_i \right) = 0.1 \left[(13.907127)\, 50\sqrt{2} + (12.602388)\, 50 \right.$$

$$\left. + (1.304774)\, 50\sqrt{2} \right] = 170.57632 \text{ lbs } < W^{(1)} = 247.4874 \text{ lbs}$$

(a4)

Repeating the previous procedures for

Cycle 3

$$W^{(3)} = 0.1 \left[(14.13114)\, 50\sqrt{2} + (10.49887)\, 50 + (0.23881)\, 50\sqrt{2} \right]$$

$$= 154.105 \text{ lbs } < W^{(2)} = 170.57632 \text{ lbs}$$

(a5)

Cycle 4

This cycle gives that the area of member 3 is less than A_{min} as

$$(\Lambda\alpha_1)^{(4)} = 14.4344, \quad (\Lambda\alpha_2)^{(4)} = 10.230436, \quad (\Lambda\alpha_3)^{(4)} = 0.01123 \text{ in.}^2$$

(a6)

Now, the factor becomes

$$\Lambda = 14.4344, \quad \alpha_1 = 1.0, \quad \alpha_2 = 0.708754,$$

$$\alpha_3 = 0.1/14.4344 = 0.006928$$

(a7)

which are used to find the stress to compare with the allowable

$$\left\{ \begin{array}{c} |\sigma_1/\bar{\sigma}_1| \\ |\sigma_2/\bar{\sigma}_2| \\ |\sigma_3/\bar{\sigma}_3| \end{array} \right\} = \left\{ \begin{array}{c} 0.9797 \\ 0.24438 \\ 0.00223 \end{array} \right\}^{\rightarrow\max.}$$

(a8)

Because the most stressed member is less than the allowable, the scaling factor can be decreased as

$$(\Lambda)_{\text{new}} = (\Lambda)_{\text{old}} \times \left(\max \left| \frac{\sigma_i}{\bar{\sigma}_i} \right| \right) = (14.4344)\, 0.9797 = 14.1419$$

(a9)

from which

$$[A_1\ A_2\ A_3] = [14.1419\ 10.0231\ 0.098], \text{ thus } 0.1 \text{ in.}^2, \text{ thus } A_3 = 0.1 \text{ in.}^2$$

$$[\alpha_1\ \alpha_2\ \alpha_3] = [1.0\quad 0.70854\quad 0.0070712]$$

(a10)

For the above member sizes, we have

$$\left\{\begin{array}{c}\sigma_1\\\sigma_2\\\sigma_3\end{array}\right\} = \left\{\begin{array}{c}\dfrac{141,419.08}{14.1419}\\[2mm]\dfrac{100,003.219}{10.02314}\\[2mm]\dfrac{-2.2759}{0.1}\end{array}\right\} = \left\{\begin{array}{c}10,000\\9,977.23\\-22.759\end{array}\right\}\text{psi} \qquad \text{(a11)}$$

$$W^{(4)} = 0.1\left[(14.1419)\,50\sqrt{2}+(10.02314)\,50+(0.1)\,50\sqrt{2}\right]$$

$$= 150.821\,\text{lbs} < W^{(3)} = 154.105\,\text{lbs} \qquad \text{(a12)}$$

Cycle 5

The comparison of the stresses is

$$\left\{\begin{array}{c}|\sigma_1/\bar{\sigma}_1|\\|\sigma_2/\bar{\sigma}_2|\\|\sigma_3/\bar{\sigma}_3|\end{array}\right\} = \left\{\begin{array}{c}9,969.125/10,000\\9,969.623/40,000\\0.5/10,000\end{array}\right\} = \left\{\begin{array}{c}0.99691\\0.24924\\5\times10^{-5}\end{array}\right\}^{\to\text{max.}} \qquad \text{(a13)}$$

$$(\Lambda)_{\text{new}} = (\Lambda)_{\text{old}}\times\left(\max\left|\dfrac{\sigma_i}{\bar{\sigma}_i}\right|\right) = (14.1419)\,0.99691 = 14.14213 \qquad \text{(a14)}$$

$$[A_1\ A_2\ A_3] = [14.14213\ 10.0005\ 0.09969], \qquad \text{(a15)}$$

use $0.1\,\text{in.}^2$ for member 3, then

$$[\alpha_1\ \alpha_2\ \alpha_3] = [1.0\quad 0.70714\quad 0.0070711] \qquad \text{(a16)}$$

$$\left\{\begin{array}{c}\sigma_1\\\sigma_2\\\sigma_3\end{array}\right\} = \left\{\begin{array}{c}141,424.128/14.14213\\100,004.658/10.0005\\0.02/0.1\end{array}\right\} = \left\{\begin{array}{c}10,000\\10,000\\0.2\end{array}\right\}\text{psi} \qquad \text{(a17)}$$

$$W^{(5)} = 0.1\left[(14.14213)\,50\sqrt{2}+(100,005)\,50+(0.1)\,50\sqrt{2}\right]$$

$$= 150.71\,\text{lbs} \cong W^{(4)} = 150.82\,\text{lbs} \qquad \text{(a18)}$$

Because $W^{(5)} \cong W^{(4)}$, we say the structural weight has converged to a local minimum, the constraint-gradient method is employed.

Cycle 6

Apply the constraint-gradient method given in (B) of Section 7.5.2 for the next cycle as

Step 1. Change each member size by $1.0\,\text{in.}^2$ and find $\Delta\sigma_{ij}$

Change A_1 to $A_1 + 1 = 15.14213\,\text{in.}^2$ $(\Delta A_1 = 1\,\text{in.}^2)$, then $\Lambda = 15.14213$ and $[\alpha_1\ \alpha_2\ \alpha_3] = [1.0\ 0.66044\ 0.006604]$

From Eqs. (7.114) and (7.115), we find

$$[K]\{\Delta r\} = -[\Delta K]\{r\}, \quad \{\Delta r\} = -[K]^{-1}[\Delta K]\{r\}$$

Using Eq. (a) through (c) yields

$$[\Delta K]_1 = \begin{pmatrix} \dfrac{1}{\sqrt{2}} & 0 & -\dfrac{1}{\sqrt{2}} \\ \dfrac{1}{\sqrt{2}} & 1 & \dfrac{1}{\sqrt{2}} \end{pmatrix} \begin{pmatrix} \dfrac{\Delta A_1 E}{l_1} & 0 & 0 \\ & 0 & 0 \\ \text{sym} & & 0 \end{pmatrix} \begin{pmatrix} \dfrac{1}{\sqrt{2}} & \dfrac{1}{\sqrt{2}} \\ 0 & 1 \\ -\dfrac{1}{\sqrt{2}} & \dfrac{1}{\sqrt{2}} \end{pmatrix}$$

$$= \dfrac{E}{50} \begin{pmatrix} \dfrac{1}{2\sqrt{2}} & \dfrac{1}{2\sqrt{2}} \\ \text{sym} & \dfrac{1}{2\sqrt{2}} \end{pmatrix} \tag{a19}$$

According to Eq. (f)

$$\begin{Bmatrix} \Delta r_{11} \\ \Delta r_{12} \end{Bmatrix} = \dfrac{50}{14.14213E} \dfrac{2\sqrt{2}}{0.0070711\sqrt{2} + 0.70714 + (0.70714)0.0070711}$$

$$\times \begin{pmatrix} \dfrac{1.0070711}{2\sqrt{2}} + 0.70714 & \dfrac{1 - 0.0070711}{2\sqrt{2}} \\ -\dfrac{1 - 0.0070711}{2\sqrt{2}} & \dfrac{1.0070711}{2\sqrt{2}} \end{pmatrix}$$

$$\dfrac{E}{50} \begin{pmatrix} \dfrac{1}{2\sqrt{2}} & \dfrac{1}{2\sqrt{2}} \\ \dfrac{1}{2\sqrt{2}} & \dfrac{1}{2\sqrt{2}} \end{pmatrix} \begin{Bmatrix} 0.050002 \\ 0.05 \end{Bmatrix}$$

$$= \begin{Bmatrix} -6.97329 \times 10^{-3} \\ -4.89603 \times 10^{-5} \end{Bmatrix} \tag{a20}$$

$$\left\{ \begin{array}{c} \Delta e_{11} \\ \Delta e_{21} \\ \Delta e_{31} \end{array} \right\} = \begin{pmatrix} \dfrac{1}{\sqrt{2}} & 0 & -\dfrac{1}{\sqrt{2}} \\ \dfrac{1}{\sqrt{2}} & 1 & \dfrac{1}{\sqrt{2}} \end{pmatrix} \left\{ \begin{array}{c} -6.97329 \times 10^{-3} \\ -4.89603 \times 10^{-5} \end{array} \right\}$$

$$= \left\{ \begin{array}{c} -4.96548 \times 10^{-3} \\ -4.89603 \times 10^{-5} \\ -4.89624 \times 10^{-4} \end{array} \right\} \text{ in.} \tag{a21}$$

$$\left\{ \begin{array}{c} \Delta\sigma_{11} \\ \Delta\sigma_{21} \\ \Delta\sigma_{31} \end{array} \right\} = \left\{ \begin{array}{c} \dfrac{E\Delta e_{11}}{l_1} \\ \dfrac{E\Delta e_{21}}{l_2} \\ \dfrac{E\Delta e_{31}}{l_3} \end{array} \right\} = \left\{ \begin{array}{c} \dfrac{10^7 \left(-4.96548 \times 10^{-3} \right)}{50\sqrt{2}} \\ \dfrac{10^7 \left(-4.89603 \times 10^{-5} \right)}{50} \\ \dfrac{10^7 \left(-4.89624 \times 10^{-3} \right)}{50\sqrt{2}} \end{array} \right\}$$

$$= \left\{ \begin{array}{c} -702.22492 \\ -9.79206 \\ 692.433 \end{array} \right. \begin{array}{l} \leftarrow \text{ positive influence} \\ \leftarrow \text{ positive influence} \\ \leftarrow \text{ negative influence} \end{array} \tag{a22}$$

Change A_2 to $A_2 + 1 = 11.0005 \text{ in.}^2 \ (\Delta A_2 = 1 \text{ in.}^2)$

$$[\Delta K]_2 = \begin{pmatrix} \dfrac{1}{\sqrt{2}} & 0 & -\dfrac{1}{\sqrt{2}} \\ \dfrac{1}{\sqrt{2}} & 1 & \dfrac{1}{\sqrt{2}} \end{pmatrix} \begin{pmatrix} 0 & 0 & 0 \\ & \dfrac{\Delta A_2 E}{l_2} & 0 \\ \text{sym} & & 0 \end{pmatrix} \begin{pmatrix} \dfrac{1}{\sqrt{2}} & \dfrac{1}{\sqrt{2}} \\ 0 & 1 \\ -\dfrac{1}{\sqrt{2}} & \dfrac{1}{\sqrt{2}} \end{pmatrix}$$

$$= \dfrac{E}{50} \begin{pmatrix} 0 & 0 \\ 0 & 1 \end{pmatrix} \tag{a23}$$

$$\left\{ \begin{matrix} \Delta r_{12} \\ \Delta r_{22} \end{matrix} \right\} = \frac{50}{14.14213E} \frac{2\sqrt{2}}{0.0070711\sqrt{2} + 0.70714 + (0.70714)0.0070711}$$

$$\times \left(\begin{matrix} \dfrac{1.0070711}{2\sqrt{2}} + 0.70714 & \dfrac{1 - 0.0070711}{2\sqrt{2}} \\[3mm] -\dfrac{1 - 0.0070711}{2\sqrt{2}} & \dfrac{1.0070711}{2\sqrt{2}} \end{matrix} \right)$$

$$\frac{E}{50} \begin{pmatrix} 0 & 0 \\ 0 & 1 \end{pmatrix} \left\{ \begin{matrix} 0.050002 \\ 0.05 \end{matrix} \right\}$$

$$= \left\{ \begin{matrix} 4.861292 \times 10^{-3} \\ -4.93053 \times 1410^{-3} \end{matrix} \right\} \text{in.} \tag{a24}$$

$$\left\{ \begin{matrix} \Delta e_{12} \\ \Delta e_{22} \\ \Delta e_{32} \end{matrix} \right\} = \left(\begin{matrix} \dfrac{1}{\sqrt{2}} & 0 & -\dfrac{1}{\sqrt{2}} \\[3mm] \dfrac{1}{\sqrt{2}} & 1 & \dfrac{1}{\sqrt{2}} \end{matrix} \right)^{T} \left\{ \begin{matrix} 4.861292 \times 10^{-3} \\ -4.9305314 \times 10^{-3} \end{matrix} \right\}$$

$$= \left\{ \begin{matrix} -4.895965 \times 10^{-3} \\ -4.9305314 \times 10^{-3} \\ -6.9238674 \times 10^{-3} \end{matrix} \right\} \text{in.} \tag{a25}$$

$$\left\{ \begin{matrix} \Delta \sigma_{12} \\ \Delta \sigma_{22} \\ \Delta \sigma_{32} \end{matrix} \right\} = \left\{ \begin{matrix} \dfrac{E \Delta e_{12}}{l_1} \\[3mm] \dfrac{E \Delta e_{22}}{l_2} \\[3mm] \dfrac{E \Delta e_{32}}{l_3} \end{matrix} \right\} = \left\{ \begin{matrix} \dfrac{10^7 \left(-4.895965 \times 10^{-5} \right)}{50\sqrt{2}} \\[3mm] \dfrac{10^7 \left(-4.9305314 \times 10^{-5} \right)}{50} \\[3mm] \dfrac{10^7 \left(-6.9238674 \times 10^{-3} \right)}{50\sqrt{2}} \end{matrix} \right\}$$

$$= \left\{ \begin{matrix} -6.92394 \\ -986.106 \\ -979.18234 \end{matrix} \right\} \begin{matrix} \leftarrow \\ \leftarrow \text{ positive influence} \\ \leftarrow \end{matrix} \tag{a26}$$

Change A_3 to $A_3 + 1 = 1.1$ in.^{2} ($\Delta A_3 = 1$ in.^{2}), we can similarly obtain

$$[\Delta K]_3 = \frac{E}{50}\begin{pmatrix} \dfrac{1}{2\sqrt{2}} & -\dfrac{1}{2\sqrt{2}} \\ \text{sym} & \dfrac{1}{2\sqrt{2}} \end{pmatrix}\begin{Bmatrix} \Delta r_{13} \\ \Delta r_{23} \end{Bmatrix} = \begin{Bmatrix} -2.76961 \times 10^{-7} \\ 1.384773 \times 10^{-7} \end{Bmatrix} \quad (a27)$$

$$\begin{Bmatrix} \Delta e_{13} \\ \Delta e_{23} \\ \Delta e_{33} \end{Bmatrix} = \begin{Bmatrix} -9.79228 \times 10^{-8} \\ 1.384773 \times 10^{-7} \\ 9.79228 \times 10^{-8} \end{Bmatrix}$$

$$\begin{Bmatrix} \Delta\sigma_{12} \\ \Delta\sigma_{22} \\ \Delta\sigma_{32} \end{Bmatrix} = \begin{Bmatrix} -0.013848 \\ 0.027695 \\ 0.013848 \end{Bmatrix} \begin{matrix} \leftarrow \text{positive influence} \\ \leftarrow \text{negative influence} \\ \leftarrow \text{negative influence} \end{matrix} \quad (a28)$$

Step 2. Reduce each member size by 10 per cent

$[A_1 \quad A_2 \quad A_3] = [12.729 \quad 9.00045 \quad 0.09]$ in.^{2}; then

$$\Lambda = 12.7279, \quad [\alpha_1 \, \alpha_2 \, \alpha_3] = [1.0 \quad 0.70714 \quad 0.0070711]$$

$$\begin{Bmatrix} \sigma_1 \\ \sigma_2 \\ \sigma_3 \end{Bmatrix} = \begin{Bmatrix} 141,424.128/12.7279 \\ 100,004.658/9.00045 \\ 0.02/0.09 \end{Bmatrix} = \begin{Bmatrix} 11,111.348 \\ 11,111.073 \\ 0.22222 \end{Bmatrix} \text{psi}$$

The most violated stress constraint is the stress of member 1

$$\Delta\alpha_i = 11,111.348 - 10,000 = 1,111.348 \, \text{psi}$$

From Eq. (7.177)

$$\Delta\alpha_i = \frac{\Delta\sigma_j}{\displaystyle\sum_{k=1}^{3} \frac{1}{\alpha_k l_k}(\Delta\sigma_{jk})^2}\left(\frac{\Delta\sigma_{ji}}{l_i}\right)$$

$$\Delta\alpha_1 = \frac{\Delta\sigma_1}{\displaystyle\sum_{k=1}^{3} \frac{1}{\alpha_k l_k}(\Delta\sigma_{1k})^2}\left(\frac{\Delta\sigma_{11}}{l_1}\right)$$

$$= \frac{1,111.348}{\dfrac{(702.22492)^2}{1.0 \times 50\sqrt{2}} + \dfrac{(6.92394)^2}{0.70714 \times 50} + \dfrac{(0.013848)^2}{0.0070711 \times 50\sqrt{2}}}$$

$$\left(\frac{702.22492}{50\sqrt{2}}\right) = 1.5823 \quad (a29)$$

$$\Delta\alpha_2 = \frac{1,111.348}{\dfrac{(702.22492)^2}{1.0 \times 50\sqrt{2}} + \dfrac{(6.92394)^2}{0.70714 \times 50} + \dfrac{(0.013848)^2}{0.0070711 \times 50\sqrt{2}}}$$

$$\left(\frac{6.92394}{50}\right) = 0.02206 \tag{a30}$$

$$\Delta\alpha_3 = \frac{1,111.348}{\dfrac{(702.22492)^2}{1.0 \times 50\sqrt{2}} + \dfrac{(6.92394)^2}{0.70714 \times 50} + \dfrac{(0.013848)^2}{0.0070711 \times 50\sqrt{2}}}$$

$$\left(\frac{0.013838}{50\sqrt{2}}\right) = 3.12033 \times 10^{-5} \tag{a31}$$

Use step size $= 0.5$, then $\alpha_1 = 1 + \dfrac{1}{2}(1.5823) = 1.79115$, and

$$[\alpha_1\ \alpha_2\ \alpha_3] = [1.79115\ 0.71817\ 0.00700867],$$
$$[A_1\ A_2\ A_3] = [22.7976\ 9.14096\ 0.09].$$

Use $0.1\,\text{in.}^2$ for member 3. Thus

$$\Lambda = 22.7976,\ [\alpha_1\ \alpha_2\ \alpha_3]_{new} = [1\ 0.40095\ 0.00438642]$$

$$\left\{\begin{matrix} r_1 \\ r_2 \end{matrix}\right\} = \left\{\begin{matrix} 0.007896634 \\ 0.0543406 \end{matrix}\right\}\text{in.} \qquad \left\{\begin{matrix} e_1 \\ e_2 \\ e_3 \end{matrix}\right\} = \left\{\begin{matrix} 0.0440083 \\ 0.0543406 \\ 0.032841 \end{matrix}\right\}\text{in.}$$

$$\tag{a32}$$

$$\left\{\begin{matrix} |\sigma_1/\bar{\sigma}_1| \\ |\sigma_2/\bar{\sigma}_2| \\ |\sigma_3/\bar{\sigma}_3| \end{matrix}\right\} = \left\{\begin{matrix} 6,223.7118/10,000 \\ 10,868.111/40,000 \\ 4,644.39/10,000 \end{matrix}\right\}$$

$$= \left\{\begin{matrix} 0.6224 \\ 0.2717 \\ 0.4644 \end{matrix}\right\} \to \text{max but not on constraint surface} \tag{a33}$$

Reduce the member sizes as

$$\Lambda = 22.7979(0.6224) = 14.1886,$$

$$[A_1\ A_2\ A_3] = [14.1886\ 5.68896\ 0.0662]\ \text{in.}^2.\text{Use } 0.1 \text{ for member 3},$$

then $[\alpha_1\ \alpha_2\ \alpha_3] = [1.0\ 0.40095\ 0.0070479]$

$$\left\{\begin{matrix} r_1 \\ r_2 \end{matrix}\right\} = \left\{\begin{matrix} 0.0132198 \\ 0.0869727 \end{matrix}\right\} \text{in.} \qquad \left\{\begin{matrix} e_1 \\ e_2 \\ e_3 \end{matrix}\right\} = \left\{\begin{matrix} 0.070847 \\ 0.0869727 \\ 0.052151 \end{matrix}\right\} \text{in.} \quad \text{(a34)}$$

$$\left\{\begin{matrix} |\sigma_1/\bar{\sigma}_1| \\ |\sigma_2/\bar{\sigma}_2| \\ |\sigma_3/\bar{\sigma}_3| \end{matrix}\right\} = \left\{\begin{matrix} 10,019.26/10,000 \\ 17,394.55/40,000 \\ 7,375.29/10,000 \end{matrix}\right\}$$

$$= \left\{\begin{matrix} 1.001925 \\ 0.43486 \\ 0.7375 \end{matrix}\right\} \rightarrow \text{max and over stressed} \qquad \text{(a35)}$$

Increase the member sizes as $\Lambda = 14.1886(1.001926) = 14.21593$; then

$$[A_1 \ A_2 \ A_3] = [14.21593 \ 5.69993 \ 0.10019],$$
$$[\alpha_1 \ \alpha_2 \ \alpha_3] = [1.0 \ 0.40095 \ 0.00704793]$$
$$\left\{\begin{matrix} r_1 \\ r_2 \end{matrix}\right\} = \left\{\begin{matrix} 0.013194 \\ 0.086806 \end{matrix}\right\}$$

The stresses are compared with the allowable as

$$\left\{\begin{matrix} |\sigma_1/\bar{\sigma}_1| \\ |\sigma_2/\bar{\sigma}_2| \\ |\sigma_3/\bar{\sigma}_3| \end{matrix}\right\} = \left\{\begin{matrix} 1.00 \\ 0.43403 \\ 0.7361 \end{matrix}\right\} \qquad \text{(a36)}$$

$$W^{(6)} = 0.1(14.21593 \times 50\sqrt{2} + 5.69993 \times 50 + 0.10019 \times 50\sqrt{2})$$
$$= 129.73 < W^{(5)} = 150.71 \text{ lbs} \qquad \text{(a37)}$$

Now, use the energy distribution as shown in the first five cycles for more cycles without detailed calculations.

Cycle 7

At this cycle, we have

$$\left\{\begin{matrix} |\sigma_1/\bar{\sigma}_1| \\ |\sigma_2/\bar{\sigma}_2| \\ |\sigma_3/\bar{\sigma}_3| \end{matrix}\right\} = \left\{\begin{matrix} 11,939.139/10,000 \\ 12,123.03/40,000 \\ 183.944/10,000 \end{matrix}\right\}$$

$$= \left\{\begin{matrix} 1.1939 \\ 0.303077 \\ 0.01839 \end{matrix}\right\} \rightarrow \text{max and over stressed} \qquad \text{(a38)}$$

Reduce the member sizes as

$$\Lambda = 11.8467(1.1939) = 14.1438, \quad [A_1 \ A_2 \ A_3] = [14.1438 \ 9.84559 \ 0.119]$$

$$W^{(7)} = 0.1(14.1438 \times 50\sqrt{2} + 9.84559 \times 50 + 0.119 \times 50\sqrt{2})$$

$$= 150.08 > W^{(6)} = 129.73 \,\text{lbs} \tag{a39}$$

Apply the *constraint-gradient method* again and use the result of Cycle 6.
Increase size of member 1 by $\Delta A_1 = 1$ in.^{2}

$$\begin{Bmatrix} \Delta\sigma_{11} \\ \Delta\sigma_{21} \\ \Delta\sigma_{31} \end{Bmatrix} = \begin{Bmatrix} -685.0589 \\ -17.030 \\ 702.089 \end{Bmatrix} \begin{matrix} \leftarrow \\ \leftarrow \\ \leftarrow \end{matrix} \begin{matrix} \text{positive influence} \\ \\ \text{negative influence} \end{matrix}$$

Increase size of member 2 by $\Delta A_2 = 1$ in.^{2}

$$\begin{Bmatrix} \Delta\sigma_{12} \\ \Delta\sigma_{22} \\ \Delta\sigma_{32} \end{Bmatrix} = \begin{Bmatrix} -21.9964 \\ -2,987.2224 \\ -2,966.311 \end{Bmatrix} \begin{matrix} \leftarrow \\ \leftarrow \\ \leftarrow \end{matrix} \text{all positive influence} \tag{a40}$$

Increase size of member 3 by $\Delta A_3 = 1$ in.^{2}

$$\begin{Bmatrix} \Delta\sigma_{13} \\ \Delta\sigma_{23} \\ \Delta\sigma_{33} \end{Bmatrix} = \begin{Bmatrix} 504.2855 \\ -1,778.694 \\ -2,282.9798 \end{Bmatrix} \tag{a41}$$

Reduce each member size by 10 per cent as $\Lambda = 12.731337$ and the stresses are

$$\begin{Bmatrix} \sigma_1 \\ \sigma_2 \\ \sigma_3 \end{Bmatrix} = \begin{Bmatrix} 10,000/0.9 \\ 17,361.11/0.9 \\ 7,361.1145/0.9 \end{Bmatrix} = \begin{Bmatrix} 11,111.11 \\ 19,290.12 \\ 8,179.016 \end{Bmatrix} \text{psi} \tag{a42}$$

The most violated stress constraint is member 1, therefore

$$\Delta\alpha_1 = \frac{1111.11}{\dfrac{(685.0589)^2}{1.0 \times 50\sqrt{2}} + \dfrac{(21.9964)^2}{0.40095 \times 50} + \dfrac{(0)^2}{0.00704793}} \left(\frac{685.0589}{50\sqrt{2}} \right)$$

$$= 1.61604 \tag{a43}$$

$$\Delta\alpha_2 = \frac{1111.11}{\dfrac{(685.0589)^2}{1.0 \times 50\sqrt{2}} + \dfrac{(21.9964)^2}{0.40095 \times 50} + 0} \left(\frac{21.9964}{50}\right) = 0.073382$$

(a44)

$$\Delta\alpha_3 = 0$$

Use step size 0.08, then

$$[\alpha_1\ \alpha_2\ \alpha_3]_{new} = [\alpha_1 + 0.08\Delta\alpha_1\ \ \alpha_2 + 0.08\Delta\alpha_2\ \ \alpha_3]$$
$$= [1.1272832\ 0.40682\ 0.00704793]$$
$$[A_1\ A_2\ A_3] = [12.731337 \times 1.1292832\ \ 12.731337$$
$$\times 0.40682\ \ 12.731337 \times 0.00704793]$$
$$= [14.37728\ 5.17937\ 0.0897]\ in.^2$$

Use $0.1\,in.^2$ for member 3; then

$$\Lambda = 14.78877, \quad [\alpha_1\ \alpha_2\ \alpha_3] = [1.0\ 0.360247\ 0.0069554]$$

for which

$$\left\{\begin{array}{l} |\sigma_1/\bar{\sigma}_1| \\ |\sigma_2/\bar{\sigma}_2| \\ |\sigma_3/\bar{\sigma}_3| \end{array}\right\} = \left\{\begin{array}{l} 9,900.14/10,000 \\ 19,057.33/40,000 \\ 9,157.198/10,000 \end{array}\right\} = \left\{\begin{array}{l} 0.99001 \\ 0.47643 \\ 0.9157 \end{array}\right\} \rightarrow max$$

(a45)

Reduce the member sizes as

$$\Lambda = 14.78877 \times 0.99001 = 14.2337,$$
$$[A_1\ A_2\ A_3] = [14.2337\ 5.1276\ 0.99]\,in.^2, \text{ replace } 0.99 \text{ by } 0.1, \text{ then}$$
$$[\alpha_1\ \alpha_2\ \alpha_3] = [1.0\ 0.360247\ 0.0070256]$$

for which the comparison of the actual stress with the allowable is

$$\left\{\begin{array}{l} |\sigma_1/\bar{\sigma}_1| \\ |\sigma_2/\bar{\sigma}_2| \\ |\sigma_3/\bar{\sigma}_3| \end{array}\right\} = \left\{\begin{array}{l} 1.000 \\ 0.481178 \\ 0.92465 \end{array}\right\}$$

(a46)

$$W^{(7)} = 0.1\left[(14.2337)\,50\sqrt{2} + (5.12765)\,50 + (0.1)\,50\sqrt{2}\right]$$
$$= 126.993 < W^{(6)} = 129.73\,lbs$$

(a47)

We could somewhat further reduce the weight, but we stop here. Thus, the optimum weight at cycle 7 is $W^{(7)} = 126.993\,lbs$. The optimization cycles are shown in Fig. 7.10.

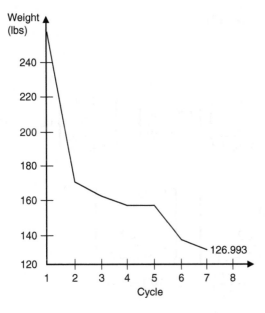

Figure 7.10 Example 7.5.1 with stress constraint.

(B) Displacement constraint only

As shown in the solution procedure of (A)

$$[K] = [A][S][A]^T = \begin{pmatrix} \dfrac{1}{\sqrt{2}} & 0 & -\dfrac{1}{\sqrt{2}} \\ \dfrac{1}{\sqrt{2}} & 1 & \dfrac{1}{\sqrt{2}} \end{pmatrix}$$

$$\left(\dfrac{\Lambda E}{50} \begin{pmatrix} \dfrac{\alpha_1}{\sqrt{2}} & 0 & 0 \\ & \alpha_2 & 0 \\ \text{sym} & & \dfrac{\alpha_3}{\sqrt{2}} \end{pmatrix} \right) \begin{pmatrix} \dfrac{1}{\sqrt{2}} & -\dfrac{1}{\sqrt{2}} \\ 0 & 1 \\ -\dfrac{1}{\sqrt{2}} & \dfrac{1}{\sqrt{2}} \end{pmatrix}$$

$$= \dfrac{\Lambda E}{50} \begin{pmatrix} \dfrac{\alpha_1 + \alpha_3}{2\sqrt{2}} & 0 \\ \text{sym} & \dfrac{\alpha_1}{2\sqrt{2}} + \alpha_2 + \dfrac{\alpha_3}{2\sqrt{2}} \end{pmatrix} \qquad \text{(a48)}$$

Cycle 1

(For initial design, $\alpha_1 = \alpha_2 = \alpha_3 = 1$)

$$\left\{ \begin{matrix} r_1 \\ r_2 \end{matrix} \right\} = [K]^{-1} \left\{ \begin{matrix} P_1 \\ P_2 \end{matrix} \right\}$$

$$= \frac{AE}{50} \begin{pmatrix} \sqrt{2} & 0 \\ 0 & \frac{\sqrt{2}}{1+2\sqrt{2}} \end{pmatrix} \left\{ \begin{matrix} 100,000 \\ 200,000 \end{matrix} \right\} = \left\{ \begin{matrix} 0.5\sqrt{2} \\ \frac{\sqrt{2}}{1+\sqrt{2}} \end{matrix} \right\} \text{ in.} \qquad (a49)$$

$$\left\{ \begin{matrix} \frac{r_1}{\bar{c}_1} \\ \frac{r_2}{\bar{c}_2} \end{matrix} \right\} = \left\{ \begin{matrix} \frac{0.5\sqrt{2}}{0.05} \\ \frac{\sqrt{2}}{0.05(1+\sqrt{2})} \end{matrix} \right\} = \left\{ \begin{matrix} 10\sqrt{2} & \leftarrow\text{max} \\ 20\frac{\sqrt{2}}{(1+\sqrt{2})} \end{matrix} \right\} \qquad (a50)$$

$$(\Lambda)_{new} = (\Lambda)_{old} \left(\max \left(\frac{r}{\bar{c}} \right) \right) = (1)10\sqrt{2} = 10\sqrt{2}$$

Use $(\Lambda)_{new}$, then

$$\left\{ \begin{matrix} r_1 \\ r_2 \end{matrix} \right\} = \frac{50}{\left(10\sqrt{2}\right)1.0 \times 10^7} \left\{ \begin{matrix} 100,000\sqrt{2} \\ 200,000\left(\frac{\sqrt{2}}{1+\sqrt{2}}\right) \end{matrix} \right\} = \left\{ \begin{matrix} 0.05 \\ 0.04142 \end{matrix} \right\} \text{ in.}$$

$$(a51)$$

which shows that displacement r_1 is active, then

$$[A_1 \ A_2 \ A_3] = \Lambda [\alpha_1 \ \alpha_2 \ \alpha_3] = \left[10\sqrt{2} \ 10\sqrt{2} \ 10\sqrt{2}\right] \qquad (a52)$$

$$W^{(1)} = \sum_{i=1}^{3} \rho_i A_i l_i = 0.1 \left(10\sqrt{2}\right)\left(50\sqrt{2}+50+50\sqrt{2}\right) = 270.71 \text{ lbs} \qquad (a53)$$

Cycle 2

According to Eq. (7.135), the energy distribution is carried as follows:
Apply the unit virtual load shown in Fig. 7.9c and the force vector is $\left\{ \begin{matrix} 1 \\ 0 \end{matrix} \right\}$,
then the virtual displacements are

$$\begin{Bmatrix} q_1 \\ q_2 \end{Bmatrix} = \frac{50}{AE} \begin{pmatrix} \frac{\sqrt{2}}{2} & 0 \\ 0 & \frac{\sqrt{2}}{1+\sqrt{2}} \end{pmatrix} \begin{Bmatrix} 1 \\ 0 \end{Bmatrix}$$

$$= \frac{50}{\left(10\sqrt{2}\right) 1 \times 10^7} \begin{Bmatrix} \frac{\sqrt{2}}{2} \\ 0 \end{Bmatrix} = \begin{Bmatrix} (5)10^{-7} \\ 0 \end{Bmatrix} \tag{a54}$$

The virtual internal forces are calculated as $\{t\} = [S][A]^T \{q\}$ which yields

$$\begin{Bmatrix} t_{11} \\ t_{21} \\ t_{31} \end{Bmatrix} = \frac{AE}{50} \begin{pmatrix} \frac{\alpha_1}{2} & \frac{\alpha_1}{2} \\ 0 & \alpha_2 \\ -\frac{\alpha_3}{2} & \frac{\alpha_3}{2} \end{pmatrix} \begin{Bmatrix} 5 \times 10^{-7} \\ 0 \end{Bmatrix} = \begin{Bmatrix} \frac{\sqrt{2}}{2} \\ 0 \\ -\frac{\sqrt{2}}{2} \end{Bmatrix} \tag{a55}$$

The actual member deformations are $\{e\} = [A]^T \{r\}$ as

$$\begin{Bmatrix} e_1 \\ e_2 \\ e_3 \end{Bmatrix} = \begin{pmatrix} \frac{1}{\sqrt{2}} & -\frac{1}{\sqrt{2}} \\ 0 & 1 \\ -\frac{1}{\sqrt{2}} & \frac{1}{\sqrt{2}} \end{pmatrix} \begin{Bmatrix} 0.05 \\ 0.04142 \end{Bmatrix} = \begin{Bmatrix} 0.06464 \\ 0.04142 \\ -0.006067 \end{Bmatrix} \tag{a56}$$

The virtual strain energy due to the virtual load in the r_1 direction is

$$U_{11} = t_{11}e_1 = \frac{\sqrt{2}}{2} (0.06464) = 0.04571$$

$$U_{21} = t_{21}e_2 = 0 \tag{a57}$$

$$U_{31} = t_{31}e_3 = \left(-\frac{\sqrt{2}}{2}\right)(-0.06464) = 0.00429$$

Similarly, for the virtual load shown in Fig. 7.9d, we have

$$\begin{Bmatrix} q_1 \\ q_2 \end{Bmatrix} = \frac{50}{AE} \begin{pmatrix} \frac{\sqrt{2}}{2} & 0 \\ 0 & \frac{\sqrt{2}}{1+\sqrt{2}} \end{pmatrix} \begin{Bmatrix} 1 \\ 0 \end{Bmatrix} = \begin{Bmatrix} 0 \\ 2.0711 \times 10^{-7} \end{Bmatrix} \tag{a58}$$

$$\begin{Bmatrix} t_{12} \\ t_{22} \\ t_{32} \end{Bmatrix} = \frac{AE}{50} \begin{pmatrix} \frac{\alpha_1}{2} & \frac{\alpha_1}{2} \\ 0 & \alpha_2 \\ -\frac{\alpha_3}{2} & \frac{\alpha_3}{2} \end{pmatrix} \begin{Bmatrix} 0 \\ 2.0711 \times 10^{-7} \end{Bmatrix} = \begin{Bmatrix} 0.2929 \\ 0.5858 \\ 0.2929 \end{Bmatrix} \tag{a59}$$

The virtual strain energy due to the virtual load in the r_2 direction is

$$\left.\begin{array}{l} U_{12} = 0.2929\,(0.06464) = 0.018933 \\ U_{22} = 0.5858\,(0.04142) = 0.02464 \\ U_{32} = 0.2929\,(-0.06067) = -1.777 \times 10^{-3} \end{array}\right\} \tag{a60}$$

Based on Eq. (7.149), determine the Lagrange multiplier as

$$\left.\begin{array}{l} U_{11} = \dfrac{0.04571}{0.1(10\sqrt{2})(50\sqrt{2})} = 4.571 \times 10^{-4} \\[3mm] U_{21} = \dfrac{0}{0.1(10\sqrt{2})(50)} = 0 \\[3mm] U_{31} = \dfrac{0.00429}{0.1(10\sqrt{2})(50\sqrt{2})} = 4.29 \times 10^{-5} \end{array}\right\} \tag{a61}$$

which yields $\lambda_{\min}$ as $1/\lambda_1$ since U_{11}/W_1 is the maximum.

$$\left.\begin{array}{l} \dfrac{U_{12}}{W_1} = \dfrac{0.018933}{0.1(10\sqrt{2})(50\sqrt{2})}\left(\dfrac{0.05}{0.04142}\right)^2 = 2.7587 \times 10^{-4} \\[3mm] \dfrac{U_{22}}{W_2} = \dfrac{0.024164}{0.1(10\sqrt{2})(50)}\left(\dfrac{0.05}{0.04142}\right)^2 = 5.0 \times 10^{-4} \\[3mm] \dfrac{U_{32}}{W_3} = \dfrac{-0.001777}{0.1(10\sqrt{2})(50\sqrt{2})}\left(\dfrac{0.05}{0.04142}\right)^2 = 2.2893 \times 10^{-5} \end{array}\right\} \tag{a62}$$

where $(0.05/0.04142)$ represents $\left(R_{x_{\max}}/R_{x_j}\right)$ which is used to scale the design variable as the r_2 direction is active and λ_2 is the minimum. Use Eq. (7.151) and let $X = \alpha A$, we have

$$\left.\begin{array}{l} (X_1)^{(2)} = \sqrt{\lambda_1 \dfrac{U_{11}}{W_1}}\,(X_1)^{(1)} = \sqrt{\dfrac{(4.571)\,10^{-4}}{(4.571)\,10^{-4}}}\left(10\sqrt{2}\right) = 10\sqrt{2} = 14.142 \\ \text{or} \\ (X_1)^{(2)} = \sqrt{\lambda_2 \dfrac{U_{22}}{W_1}}\,(X_1)^{(1)} = \sqrt{\dfrac{1}{(2.7587)\,10^{-4}}\dfrac{0.018933}{(0.1)\left(10\sqrt{2}\right)\left(50\sqrt{2}\right)}} \\ \qquad\qquad \left(10\sqrt{2}\right) = 11.7158 \end{array}\right\} \tag{a63}$$

$$(X_2)^{(2)} = \sqrt{\lambda_1 \frac{U_{21}}{W_2}} (X_2)^{(1)} = 0$$

or

$$(X_2)^{(2)} = \sqrt{\lambda_2 \frac{U_{22}}{W_2}} (X_2)^{(1)} = \sqrt{\frac{1}{(5.0)\,10^{-4}} \frac{0.024264}{(0.1)\left(10\sqrt{2}\right)(50)}} \left(10\sqrt{2}\right) = 10\sqrt{2} = 11.7157 \tag{a64}$$

$$(X_3)^{(2)} = \sqrt{\lambda_1 \frac{U_{31}}{W_3}} (X_3)^{(1)} = \sqrt{\frac{1}{(4.57)\,10^{-4}}(4.29 \times 10^{-5})} \left(10\sqrt{2}\right) = 4.333$$

or

$$(X_3)^{(2)} = \sqrt{\lambda_2 \frac{U_{32}}{W_3}} (X_3)^{(1)} = \sqrt{\left(\frac{1}{(5.0)\,10^{-4}}\right)\frac{0.001777}{(0.1)\left(10\sqrt{2}\right)\left(50\sqrt{2}\right)}} \left(10\sqrt{2}\right) = 2.666 \tag{a65}$$

Use $\Lambda = 14.142$, then we have $[\alpha_1 \; \alpha_2 \; \alpha_3] = [1.0 \; 0.8284 \; 0.30639]$, we have

$$\begin{Bmatrix} r_1 \\ r_2 \end{Bmatrix} = \frac{50}{14.142 \times 10^7} \frac{2\sqrt{2}}{0.3069\sqrt{2} + 0.8284 + (0.8284)0.30639}$$

$$\times \left\{ \begin{matrix} \left(\dfrac{1.30639}{2\sqrt{2}} + 0.8284\right)100{,}000 - \dfrac{1 - 0.30639}{2\sqrt{2}}(200{,}000) \\[2mm] -\dfrac{1 - 0.30639}{2\sqrt{2}}(100{,}000) + \dfrac{1.30639}{2\sqrt{2}}(200{,}000) \end{matrix} \right\} \tag{a66}$$

$$= \begin{Bmatrix} 0.05278 \\ 0.044773 \end{Bmatrix} \text{in.}$$

$$\begin{Bmatrix} \dfrac{r_1}{\bar{c}_1} \\[2mm] \dfrac{r_2}{\bar{c}_2} \end{Bmatrix} = \begin{Bmatrix} \dfrac{0.05278}{0.05} \\[2mm] \dfrac{0.044773}{0.05} \end{Bmatrix} = \begin{Bmatrix} 1.0555 \\ 0.89545 \end{Bmatrix} \tag{a67}$$

For $\Lambda = 14.142 \times 1.0555 = 14.927$,

$$[A_1 \; A_2 \; A_3] = [14.927 \; 12.3657 \; 4.5736] \tag{a68}$$

$$\begin{Bmatrix} r_1 \\ r_2 \end{Bmatrix} = \begin{Bmatrix} 0.05278/1.0555 \\ 0.044773/1.0555 \end{Bmatrix} = \begin{Bmatrix} 0.05 \\ 0.04242 \end{Bmatrix} \text{in.} \tag{a69}$$

$$W^{(2)} = 0.1 \left[14.927(50\sqrt{2}) + 12.3657(50) + 4.5736(50\sqrt{2}) \right]$$

$$= 199.72\,\text{lbs} \tag{a70}$$

Repeating the previous procedures, one may find the solution of the next three cycles as

Cycle 3

$$\begin{Bmatrix} \dfrac{r_1}{\bar{c}_1} \\ \dfrac{r_2}{\bar{c}_2} \end{Bmatrix} = \begin{Bmatrix} \dfrac{0.047206}{0.05} \\ \dfrac{0.047597}{0.05} \end{Bmatrix} = \begin{Bmatrix} 0.9441 \\ 0.95194 \end{Bmatrix} \leftarrow \text{max} \tag{a71}$$

$$\Lambda = 14.927 \times 0.95194 = 14.2099,$$

$$[A_1 \quad A_2 \quad A_3] = [14.2099 \quad 9.9869 \quad 2.2529]\,\text{in.}^2 \tag{a72}$$

$$\begin{Bmatrix} r_1 \\ r_2 \end{Bmatrix} = \begin{Bmatrix} 0.04959 \\ 0.05 \end{Bmatrix} \text{in.} \tag{a73}$$

$$W^{(3)} = 0.1 \left[14.2099(50\sqrt{2}) + 9.9869(50) + 2.2529(50\sqrt{2}) \right]$$

$$= 166.344\,\text{lbs} < W^{(2)} \tag{a74}$$

Cycle 4

$$\begin{Bmatrix} \dfrac{r_1}{\bar{c}_1} \\ \dfrac{r_2}{\bar{c}_2} \end{Bmatrix} = \begin{Bmatrix} 1.00546 \\ 1.00142 \end{Bmatrix} \leftarrow \text{max} \tag{a75}$$

$$\Lambda = 14.0929 \times 1.00546 = 14.17,$$

$$[A_1 \; A_2 \; A_3] = [14.17 \; 10.0414 \; 0.3627]\,\text{in.}^2$$

$$\begin{Bmatrix} r_1 \\ r_2 \end{Bmatrix} = \begin{Bmatrix} 0.05 \\ 0.04980 \end{Bmatrix} \text{in.} \tag{a76}$$

$$W^{(4)} = 0.1 \left[14.17(50\sqrt{2}) + 10.0414(50) + 0.3627(50\sqrt{2}) \right]$$

$$= 152.968\,\text{lbs} < W^{(3)} \tag{a77}$$

Cycle 5

$$\begin{Bmatrix} \dfrac{r_1}{\bar{c}_1} \\ \dfrac{r_2}{\bar{c}_2} \end{Bmatrix} = \begin{Bmatrix} 0.9962 \\ 0.99987 \end{Bmatrix} \leftarrow \text{max} \tag{a78}$$

$\Lambda = 14.17 \times 0.99987 = 14.168,$

$[A_1 \; A_2 \; A_3] = [14.168 \; 9.9997 \; 0.10179] \; \text{in.}^2$

$$\left\{ \begin{matrix} r_1 \\ r_2 \end{matrix} \right\} = \left\{ \begin{matrix} 0.04982 \\ 0.05 \end{matrix} \right\} \; \text{in.} \tag{a79}$$

$$W^{(5)} = 0.1 \left[14.168 \left(50\sqrt{2} \right) + 9.9997(50) + 0.10179 \left(50\sqrt{2} \right) \right] \tag{a80}$$

$$= 150.902 \; \text{lbs} < W^{(4)} = 152.968 \; \text{lbs}$$

Cycle 6

$$\left\{ \begin{matrix} \dfrac{r_1}{\bar{c}_1} \\ \dfrac{r_2}{\bar{c}_2} \end{matrix} \right\} = \left\{ \begin{matrix} 1.00355 \\ 1.00005 \end{matrix} \right\} \tag{a81}$$

Because the displacements are over the allowable, reduce the member sizes as

$$\Lambda = (14.11165)1.00355 = 14.1666,$$

$$[A_1 \; A_2 \; A_3] = [14.1666 \; 10.035 \; 0.10035] \; \text{in.}^2$$

The displacements are now

$$\left\{ \begin{matrix} r_1 \\ r_2 \end{matrix} \right\} = \left\{ \begin{matrix} 0.05 \\ 0.04983 \end{matrix} \right\} \tag{a82}$$

$$W^{(6)} = 0.1 \left[14.1666(50\sqrt{2}) + 10.035(50) + 0.10035(50\sqrt{2}) \right]$$

$$= 151.058 \; \text{lbs} > W^{(5)} \tag{a83}$$

Since $W^{(6)} > W^{(5)}$, the constraint-gradient method will be used in the next cycle.

Cycle 7

(Constraint-gradient method)
Increase A_1 to $A_1 + 1 = 15.168 \; \text{in.}^2$

$$\left\{ \begin{array}{c} \Delta r_{11} \\ \Delta r_{21} \end{array} \right\} = \frac{50}{14.168 \times 10^7} \frac{2\sqrt{2}}{0.0071842\sqrt{2} + 0.70579 + (0.70579)0.0071842}$$

$$\times \left(\begin{array}{cc} \dfrac{1.0071842}{2\sqrt{2}} + 0.70579 & -\dfrac{1 - 0.0071842}{2\sqrt{2}} \\[3mm] -\dfrac{1 - 0.0071842}{2\sqrt{2}} & \dfrac{1 + 0.0071842}{2\sqrt{2}} \end{array} \right)$$

$$\frac{10^7}{50} \left(\begin{array}{cc} \dfrac{1}{2\sqrt{2}} & \dfrac{1}{2\sqrt{2}} \\[3mm] \dfrac{1}{2\sqrt{2}} & \dfrac{1}{2\sqrt{2}} \end{array} \right) \left\{ \begin{array}{c} 0.04982 \\ 0.05 \end{array} \right\}$$

$$= \left\{ \begin{array}{c} -6.9463 \times 10^{-3} \\ -4.96392 \times 10^{-5} \end{array} \right\} \begin{array}{l} \text{positive influence} \\ \text{in.} \end{array} \tag{a84}$$

Increase A_2 to $A_2 + 1 = 10.997 \, \text{in.}^2$

$$\left\{ \begin{array}{c} \Delta r_{12} \\ \Delta r_{22} \end{array} \right\} = \frac{50}{14.168 \times 10^7} \frac{2\sqrt{2}}{0.0071842\sqrt{2} + 0.70579 + (0.70579)0.0071842}$$

$$\times \left(\begin{array}{cc} \dfrac{1.0071842}{2\sqrt{2}} + 0.70579 & -\dfrac{1 - 0.0071842}{2\sqrt{2}} \\[3mm] -\dfrac{1 - 0.0071842}{2\sqrt{2}} & \dfrac{1 + 0.0071842}{2\sqrt{2}} \end{array} \right)$$

$$\frac{10^7}{50} \left(\begin{array}{cc} 0 & 0 \\ 0 & 1 \end{array} \right) \left\{ \begin{array}{c} 0.04982 \\ 0.05 \end{array} \right\}$$

$$= \left\{ \begin{array}{c} 4.8593 \times 10^{-3} \\ -4.9297 \times 10^{-3} \end{array} \right\} \begin{array}{l} \leftarrow \text{negative influence} \\ \leftarrow \text{positive influence} \end{array} \tag{a85}$$

Similarly for $A_3 + 1 = 1.10179 \, \text{in.}^2$

$$\left\{ \begin{array}{c} \Delta r_{12} \\ \Delta r_{22} \end{array} \right\} = \left\{ \begin{array}{c} 2.48958 \times 10^{-5} \\ -1.245952 \times 10^{-5} \end{array} \right\} \begin{array}{l} \leftarrow \text{negative influence} \\ \leftarrow \text{positive influence} \end{array} \tag{a86}$$

Reduce each member size by 10 per cent, then $[A_1 \, A_2 \, A_3] =$ [12.7512 8.997 0.0916], we have

$$\left\{ \begin{array}{c} r_1 \\ r_2 \end{array} \right\} = \left\{ \begin{array}{c} 0.04982/0.9 \\ 0.05/0.9 \end{array} \right\} = \left\{ \begin{array}{c} 0.05536 \\ 0.05556 \end{array} \right\} \text{in.} \tag{a87}$$

$$\left\{ \begin{array}{c} \Delta r_1 \\ \Delta r_2 \end{array} \right\} = \left\{ \begin{array}{c} 0.05536 - 0.05 \\ 0.05556 - 0.05 \end{array} \right\} = \left\{ \begin{array}{c} 0.00536 \\ 0.00556 \end{array} \right\} \tag{a88}$$

From Eq. (7.174)

$$\Delta\alpha_1 = \frac{\Delta r_1}{\displaystyle\sum_{k=1}^{3} \frac{1}{\alpha_k l_k} (\Delta r_{1k})^2} \left(\frac{\Delta r_{11}}{l_1} \right)$$

$$= \frac{0.00536}{\dfrac{(6.9463 \times 10^{-3})^2}{(1.0)50\sqrt{2}}} \frac{6.9463(10^{-3})}{50\sqrt{2}} = 0.77163$$

or

$$\Delta\alpha_1 = \frac{0.0556}{\dfrac{(4.96392 \times 10^{-5})^2}{(1.0)50\sqrt{2}} + \dfrac{(4.9297 \times 10^{-3})^2}{(0.70579)50} + \dfrac{(1.2459502 \times 10^{-5})^2}{(0.007184)50\sqrt{2}}}$$

$$\frac{4.9639(10^{-5})}{50\sqrt{2}} = 0.005565$$

$\left. \vphantom{\begin{array}{c} a \\ b \\ c \\ d \\ e \\ f \\ g \\ h \\ i \end{array}} \right\} \tag{a89}$

$$\Delta\alpha_2 = \frac{0.00536}{\dfrac{(6.9463 \times 10^{-3})^2}{(1.0)50\sqrt{2}}} \frac{0}{50} = 0$$

or

$$\Delta\alpha_2 = \frac{0.00556}{\dfrac{(4.96392 \times 10^{-5})^2}{(1.0)50\sqrt{2}} + \dfrac{(6.9297 \times 10^{-3})^2}{(0.70579)50} + \dfrac{(1.245952 \times 10^{-5})^2}{(0.007184)50\sqrt{2}}}$$

$$\frac{4.9297(10^{-3})}{50} = 0.79564$$

$\left. \vphantom{\begin{array}{c} a \\ b \\ c \\ d \\ e \\ f \\ g \\ h \end{array}} \right\} \tag{a90}$

$$\Delta\alpha_3 = 0$$

or

$$\left.\begin{array}{c} \Delta\alpha_3 = \dfrac{0.00556}{\dfrac{\left(4.96392 \times 10^{-5}\right)^2}{(1.0)50\sqrt{2}} + \dfrac{\left(4.9297 \times 10^{-3}\right)^2}{(0.70579)50} + \dfrac{\left(1.245952 \times 10^{-5}\right)^2}{(0.007184)50\sqrt{2}}} \\[3em] \dfrac{1.245952(10^{-5})}{50\sqrt{2}} = 0.001422 \end{array}\right\}$$

(a91)

Using $\Delta\alpha_1 = 0.77163$, $\Delta\alpha_2 = 0.79564$, and $\Delta\alpha_3 = 0.001422$ yields

$$\alpha_1 = 1.0 + (0.01)0.77163 = 1.0077163$$
$$\alpha_2 = 0.70579 + (0.01)0.79564 = 0.713746$$
$$\alpha_3 = 0.0071842 + (0.01)0.001422 = 0.00719842$$
$$[A_1\ A_2\ A_3] = [12.8496\ 9.10112\ 0.0918]$$

(a92)

Use $A_3 = 0.1$, then $[\alpha_1\ \alpha_2\ \alpha_3] = [1.0\ 0.70828\ 0.0077283]$ for which

$$\left\{\begin{array}{c} r_1 \\ r_2 \end{array}\right\} = \left\{\begin{array}{c} 0.055118 \\ 0.05494 \end{array}\right\}; \qquad \left\{\begin{array}{c} \dfrac{r_1}{\bar{c}_1} \\ \dfrac{r_2}{\bar{c}_2} \end{array}\right\} = \left\{\begin{array}{c} 1.10236 \\ 1.09879 \end{array}\right\}$$

(a93, a94)

Increase member sizes in order to reduce the displacements; we have

$$\Lambda = (12.8496)\,1.10236 = 14.165,$$
$$[A_1\ A_2\ A_3] = [14.165\ 10.033\ 0.1102]\ \text{in.}^2$$
$$\left\{\begin{array}{c} r_1 \\ r_2 \end{array}\right\} = \left\{\begin{array}{c} 0.05 \\ 0.04984 \end{array}\right\}\ \text{in.}$$

(a95)

The structural weight at this cycle is

$$W^{(7)} = 0.1\left[14.165\left(50\sqrt{2}\right) + 10.033(50) + 0.1102\left(50\sqrt{2}\right)\right]$$
$$= 151.10\,\text{lbs} > W^{(6)} > W^{(5)}$$

(a96)

Thus, the minimum weight is achieved at cycle 5 as $W^{(5)} = 150.902$ lbs. The solutions of the 6 cycles are shown in Fig. 7.11.

Example 7.5.2 – The cantilevered ten-bar truss is shown in Fig. 7.12 with its ten design variables and eight degrees-of-freedom subjected to static loads of 100 kips (444.8 kN) applied at nodes 2 and 4. The truss modulus of elasticity, $E = 10 \times 10^6$ psi (6,895 kN cm^{-2}), its mass density, $\rho = 0.10$ lbs in.$^{-3}$

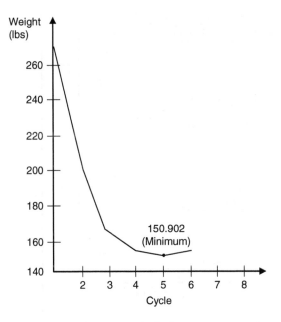

Figure 7.11 Example 7.5.1 with displacement constraint.

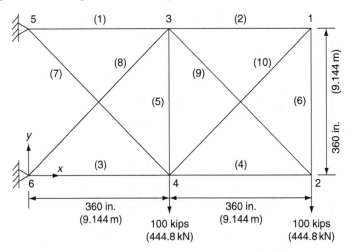

Figure 7.12 Example 7.5.2 ten-bar truss.

$(0.0271\,\mathrm{N\,cm^{-2}})$, and allowable stress, $\sigma_{\mathrm{all}} = \pm 25,000$ psi $(17.236\,\mathrm{kN\,cm^{-2}})$. Its allowable displacements were ± 2.0 in. $(5.08\,\mathrm{cm})$ in both the x and y directions at nodes 1, 2, 3 and 4. The truss was designed for four different cases: Case (1) stress constraint without using the constraint gradients; Case (2) stress constraint and the constraint gradients; Case (3) both stress

Table 7.3a The final design results of the ten-bar truss for constraint gradient studies. (1 in. = 2.54 cm, 1 lb = 4.448 N)

Member No.	Case (1)		Case (2)		Case (3)		Case (4)	
	Area (in.^{2})	Stress (lbs in.$^{-2}$)	Area (in.^{2})	Stress (lbs in.$^{-2}$)	Area (in.^{2})	Stress (lbs in.$^{-2}$)	Area (in.^{2})	Stress (lbs in.$^{-2}$)
1	7.9379	24,999.9	7.9379	25,000.0	30.1512	6,621.0	30.720	6,598.0
2	0.1000	15,533.0	0.1000	15,533.0	0.100	52.0	0.100	1,198.0
3	8.0622	24,999.9	8.0622	25,000.0	21.901	9,040.0	22.390	8,814.0
4	3.9379	25,000.0	3.9379	25,000.0	15.154	6,599.0	15.140	6,612.0
5	0.1000	0.078	0.1000	0.039	0.101	20,047.0	0.100	24,645.0
6	0.1000	15,533.1	0.1000	15,533.1	0.100	52.0	0.530	223.0
7	5.7447	25,000.0	5.7447	25,000.0	8.924	15,526.0	7.660	17,959.0
8	5.5690	24,999.9	5.5690	25,000.0	21.496	6,712.0	21.530	6,744.0
9	5.5690	25,000.0	5.5690	25,000.0	21.431	6,599.0	21.420	6,612.0
10	0.1000	21,966.9	0.1000	21,966.9	0.100	73.0	0.100	1,695.0
Final Wt.(lbs)	1,593.18		1,593.18		5,088.2		5,065.0	
No. of Iterations	19		19		20		11	

and displacement constraints without using the constraint gradients; Case (4) both stress and displacement constraints and the constraint gradient.

Solution:

The problem was solved by the computer program ODSEWS-2D-II [7]. The optimum solutions, which include the sectional areas, stresses and optimum weight, are shown in Table 7.3a. The plot of weight versus cycles of iteration is shown in Fig. 7.13. The optimum weight of Case (3), as shown in Fig. 7.13, can be reduced by applying the constraint gradient at the 10th cycle; as shown in Table 7.3b, the stresses of the passive elements are increased. However, the result of Case (1) is not improved significantly by applying the constraint gradient method. The ten-bar cantilevered truss has been studied by many researchers. Comparison of final results with previous results is made in Table 7.3b.

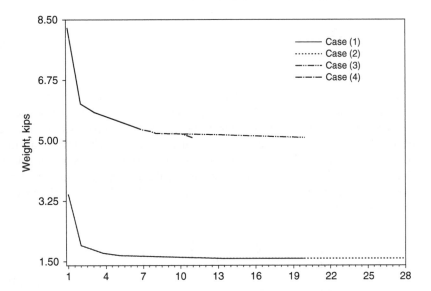

Figure 7.13 Weight versus cycles of iteration plot of the ten-bar truss. (1 kip = 4.448 kN).

Table 7.3b Comparison of final weights for the ten-bar cantilevered truss (1 lbs = 4.448 N) (I – Stress displacement constraints), (II – Stress constraints only).

	Ref. 8	*Ref. 9*	*Ref. 10*	*Ref. 11*	*Ref. 12*	*Present*
I, lbs (No. of cycles)	5,076.85 (13)	5,089.0 (23)	5,066.98 (16)	5,080.0 (15)	5,076.66 (12)	5,065.0 (11)
II, lbs (No. of cycles)	1,593.23 (15)	1,593.2 (20)		1,622.11 (11)	1,593.18 (15)	1,593.18 (19)

7.6   Cost functions and design variables

The objective function has been focused on structural weight. This section will be centred on minimum cost. The objective function of such a minimum cost design includes the costs of structural members, painting, connections and damage [4, 13, 14] which are presented individually in the following sections.

7.6.1   Structural member costs

The steel costs may be divided into two parts: first, the basic charge, which is estimated on the basis of the weight of the purchased steel members, and second, the extra-size charge, which is appraised on the basis of the shapes of the members. An evaluation of the basic charge can be easily calculated in accordance with the following equation

$$C_{TB} = C_s \sum_{i=1}^{m} \rho_i A_i l_i \tag{7.178}$$

in which C_{TB} is the total basic charge of the structural members (\$), C_s the unit price of steel (\$/lbs), m the total number of members of the structure, ρ_i the mass density of a steel member, i (lbs in.$^{-3}$), A_i the cross-sectional area of member, i (in.^{2}), and l_i the length of a member, i (in.). The design variable of the girders and columns is the moment of inertia, I. By replacing A_i in Eq. (7.178) with $A_i I_i / I_i$ for the girders and columns and assuming that the mass density of all the members is the same, Eq. (7.178) may be changed to the following form

$$
\begin{aligned}
C_{TB} &= C_s \rho \left(\sum_{i=1}^{m_g} \frac{A_i}{I_i} I_i l_i + \sum_{i=1}^{m_c} \frac{A_i}{I_i} I_i l_i + \sum_{i=1}^{m_b} \frac{A_i}{I_i} I_i l_i \right) \\
&= C_s \rho \left(\sum_{i=1}^{m_g} \eta_i I_i l_i + \sum_{i=1}^{m_c} \eta_i I_i l_i + \sum_{i=1}^{m_b} \eta_i I_i l_i \right)
\end{aligned} \tag{7.179}
$$

in which η_i is equal to A_i/I_i, m_g is the total number of girders, m_c the total number of columns, and m_b the total number of bracings.

An appraisal of the extra-size charge is based on the shape of the steel members. According to industrial practice, there is a higher extra cost per pound for smaller shapes than for larger ones as indicated in Fig. 7.14 which may be represented by

$$C_{ex} = 0.00916 \, A^{-0.21} \tag{7.180}$$

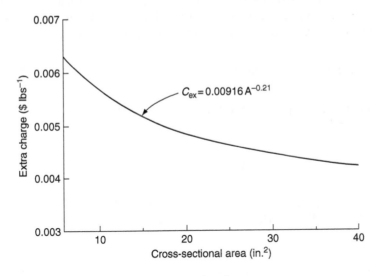

Figure 7.14 Extra charge versus cross-sectional area.

in which C_{ex} is quoted in dollars per pound. The extra-size charge of each individual member, $(C_{EX})_i$, is then obtained as

$$(C_{EX})_i = 0.00916\, \rho_i A_i^{0.79} l_i \tag{7.181}$$

By assuming that the extra-size charges of bracings have a similar model to that of the members with wide flange sections, one can obtain the total extra-size charge of the system with the following formula

$$(C_{EX})_i = 0.00916 \sum_{i=1}^{m} \rho_i A_i^{0.79} l_i$$

$$= 0.00916\, \rho \left(\sum_{i=1}^{m_g} \eta_i^{0.79} I_i^{0.79} l_i + \sum_{i=1}^{m_c} \eta_i^{0.79} I_i^{0.79} l_i + \sum_{i=1}^{m_b} A_i^{0.79} l_i \right) \tag{7.182}$$

7.6.2 Painting costs

The amount of painting is measured according to the surface area of the members. For simplicity, the cross-section of a bracing member is assumed to be a square. Thus, the surface area of each individual bracing member, i, is $4/A_i l_i$. For wide flange sections, the surface area is $(4b + 2d)l$ for which b and d are the flange width and the depth of the section, respectively. In order to model the cost of painting the wide flange sections, the relationship between the flange width and the design variable, I, and the relationship between the depth of the section and I must be developed. Unfortunately, there is no direct relationship that can be used for this development. For the selected economic sections from the AISC Manual, the relationship developed between the radius of the gyration, r_x, and depth, d, is

$$r_x \simeq 0.52 \, d^{0.92} \quad \text{(for beams)} \tag{7.183}$$

in which the appropriate moment of inertia is between 180 in.[4] and 2,500 in.[4] and

$$r_x \simeq 0.39 \, d^{1.04} \quad \text{(for columns)} \tag{7.184}$$

where the appropriate moment of inertia is between 200 in.[4] and 1,500 in.[4]. Because $I = A r_x^2$, Eqs. (7.183) and (7.184) become

$$d = \left(\frac{I}{0.2704 \, A} \right)^{\frac{1}{1.84}} \quad \text{(for beams)} \tag{7.185}$$

and

$$d = \left(\frac{I}{0.1521 \, A} \right)^{\frac{1}{2.08}} \quad \text{(for columns)} \tag{7.186}$$

The development of the relationship between the flange width and moment of inertia is based on the assumption that the web thickness, t_w, and the flange thickness, t_f, can be expressed as $t_w \simeq t_f$ for most compact sections. By using this relationship and the cross-sectional depth, the approximate equations for moment of inertia and cross-sectional area can be established from which the flange width can be roughly expressed by

$$b = \frac{2.35 \left(\dfrac{I}{Z_x} \right)}{2.09 - 0.812 \left(\dfrac{Z_x^4}{I^3} \right)} \tag{7.187}$$

in which Z_x is the plastic modulus of the section that is expressed as $Z_x = 0.953 \sqrt{AI}$. The mean value computation of Z_x for the selected sections results in

$$Z_x \simeq 2.25 \frac{I}{d} \tag{7.188}$$

By substituting Eq. (7.188) into (7.187), one obtains

$$b = \frac{2.35 \dfrac{d}{2.25}}{2.09 - 0.812 \dfrac{\left(2.25 \dfrac{I}{d}\right)^4}{I^3}} \tag{7.189}$$

By using Eqs. (7.185), (7.186) and (7.189), the surface area of wide flange sections is obtained with the following formula:

$$(4b_i + 2d_i)\, l_i = \eta_i^{-0.54328}\left(\frac{8.5043}{2.09 - 1.212\,\eta_i^{2.17391} I_i} + 4.0712\right) l_i \quad \text{(for beams)} \tag{7.190}$$

and

$$(4b_i + 2d_i)\, l_i = \eta_i^{-0.4808}\left(\frac{10.3312}{2.09 - 0.5565\,\eta_i^{1.9231} I_i} + 4.946\right) l_i \quad \text{(for columns)} \tag{7.191}$$

Let us suppose that the unit price of painting is C_{pt}, then the total painting cost of the structure, C_p, is

$$C_p = C_{pt}\left(\sum_{i=1}^{m_g} \eta_i^{-0.54328}\left(\frac{8.5043}{2.09 - 1.212\,\eta_i^{2.17391} I_i} + 4.0712\right) l_i\right.$$
$$\left. + \sum_{i=1}^{m_c} \eta_i^{-0.4808}\left(\frac{10.3312}{2.09 - 0.5565\,\eta_i^{1.9231} I_i} + 4.946\right) l_i + 4\sum_{i=1}^{m_b} A_i^{0.5} l_i\right) \tag{7.192}$$

7.6.3 Connection costs

The discussion of connection costs is primarily concerned with welded plate beam-column connections, which include the costs of steel plate and welding. Fig. 7.15 shows a typical model of a steel plate, which is to be welded to the top and bottom flanges of beams at each beam-column connection. The connections are required to be able to develop the full moment capacity of beams. That is, the plastic moment of the connection, M_p', must be larger than the plastic moment of the beam M_p. For simplicity, M_p' is taken to be equal to M_p. The plastic moment of the connections is given as

$$M_p' = \sigma_y' d\, A' \tag{7.193}$$

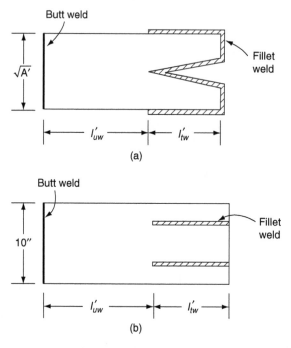

Figure 7.15 Model of plates at beam-column connections: (a) plate on the top flange
of the beam, (b) plate on the bottom flange of the beam.

in which σ'_y is the yielding stress of the plate (psi), A' is the cross-sectional
area of the plate (in.²), and d the depth of the beam (in.). By rearranging
Eq. (7.193) and assuming that the plates are A514 steel and the beams are
A36 steel, it can be said that the cross-sectional area of the plate is

$$A' = \frac{M'_p}{\sigma'_y d} = \frac{M_p}{\sigma'_y d} = 0.36 \frac{Z_x}{d} \qquad (7.194)$$

Equation (7.193) can be also rearranged as follows:

$$\sigma'_y = \frac{M'_p}{A'd} \qquad (7.195)$$

Then, from the stress–strain relationship

$$\sigma'_y = E\varepsilon'_y \qquad (7.196)$$

Let δ' be the deformation of the plate at the unwelded part, then the strain becomes

$$\varepsilon_y' = \frac{\delta'}{l_{uw}'} \tag{7.197}$$

From Eqs. (7.195) through (7.197)

$$\delta' = \frac{M_p'}{A'dE} l_{uw}' \tag{7.198}$$

Thus, the rotation, θ, at the connection is found to be

$$\theta = \delta' \frac{2}{d} = \frac{2M_p' l_{uw}'}{A'd^2 E} \tag{7.199}$$

Let us suppose that k' is the initial stiffness of the connection, which is defined as the ratio of the plastic moment to the rotation of the connection, then

$$k' = \frac{M_p'}{\theta} = \frac{A'd^2 E}{2l_{uw}'} \tag{7.200}$$

The unwelded length of the plate, l_{uw}', is then given as

$$l_{uw}' = \frac{A'd^2 E}{2k'} \tag{7.201}$$

By substituting Eq. (7.194) into Eq. (7.201), we will then have

$$l_{uw}' = \frac{0.36 Z_x E d}{2k'} \tag{7.202}$$

If the modulus of elasticity, E, is 30×10^6 psi, then

$$l_{uw}' = \frac{54 \times 10^5 Z_x d}{k'} \tag{7.203}$$

The welding length of the plate is necessarily designed to develop a plastic moment of the section. Because the plate is A514 steel, the luxury of an E110 electrode with a 3/8 in. fillet weld is assumed, and a strength of 8,900 lbs in.$^{-1}$ is obtained. Suppose that the total length of the fillet weld is l_w' then the moment capacity developed by the weld is $8900 l_w' d$ in.–lbs which must be larger than the plastic moment of the beam, i.e.,

$$8,900 l_w' d \geq M_p \tag{7.204}$$

Thus, the minimum welding length is given as

$$l'_w = \frac{\sigma_y Z_x}{8,900\,d} = 4.04 \frac{Z_x}{d} \tag{7.205}$$

If the length of the butt weld in Fig. 7.15 is neglected, and the four primary legs of the fillet weld on the top plate are equal in length, then l'_{tw} would be equal to $l'_w/4$. The bottom plate in Fig. 7.15 is welded only along the outside edges of the bottom flange, thus l'_{bw} is equal to $l'_w/2$. The lengths of the plates are then found to be $l'_{uw} + l'_{tw}$ for the top plate and $l'_{uw} + l'_{bw}$ for the bottom plate. Because there are two beam-column connections on each beam, combining the previous equations with Eq. (7.188) yields the cost of the steel plates for each beam as

$$\left(C_{pc}\right)_i = 2C_{pl}\rho A' \left(2l'_{uw} + l'_{tw} + l'_{bw}\right)$$

$$= 0.72 C_{pl}\rho Z_x^2 \left(\frac{10,800,000}{k'} + 0.6 \left(\frac{Z_x}{I}\right)^2\right) \tag{7.206}$$

in which C_{pl} is the unit price, \$ in.$^{-3}$, of the steel plate. By substituting Eqs. (7.196) and (7.188) into Eq. (7.206), one obtains

$$\left(C_{pc}\right)_i = 0.8796\, C_{pl}\rho \eta_i^{1.087} I_i^2 \left(\frac{10,800,000}{k'_i} + 0.73301\eta_i^{1.087}\right) \tag{7.207}$$

In order to determine the cost of the welding, the total amount of metal contained in the weld should be estimated. First, it is assumed that the cross-section of the top plate is square, that the root gap between the ends of the plates and the column is 3/16 in. wide, and that the plates have been bevelled to 45°. Then the metal volume in the butt weld for the top plate is $A'\left(\sqrt{A'}/2 + 3/16\right)$ in.^{3}. If the width of the bottom plate is assumed to be 10 in., then the volume of the metal in the butt weld is $A'\,(A'/20 + 3/16)$ in.^{3}. The metal volume of a 3/8 in. fillet weld for a beam is $0.3403l'_w$ in.^{3}. This is 10 per cent over the specification to provide an additional factor of safety.

The total volume of metal used for the welding of each beam is then found by summing up the metal volume of the butt weld and of the fillet weld as follows:

$$\left(V_{TW}\right)_i = 2A'_i \left(\frac{\sqrt{A'_i}}{2} + \frac{3}{16}\right) + 2A'_i \left(\frac{A'_i}{20} + \frac{3}{16}\right) + 0.3403 \left(l'_w\right)_i$$

$$= A'_i \left(0.75 + 0.1A'_i + \sqrt{A'_i}\right) + 0.3403 \left(l'_w\right)_i \tag{7.208}$$

By substituting Eqs. (7.188), (7.194) and (7.205) into Eq. (7.208) and assuming that the unit price of welding is C_w, then the welding cost of each individual beam, i, is

$$(C_{wc})_i = C_w \rho \left(0.8919 \, \eta_i^{1.087} \, I_i + 3.881 \times 10^{-3} \, \eta_i^{2.174} \, I_i^2 + 0.08463 \, \eta_i^{1.6304} \, I_i^{1.5}\right)$$

(7.209)

Then, by adding the steel plate cost obtained with Eq. (7.207) and the welding cost obtained with Eq. (7.209) for all the beams, one obtains the connection cost, which is given as

$$C_{\mathrm{con}} = \sum_{i=1}^{mg} \left((C_{pc})_i + (C_{wc})_i\right)$$

(7.210)

7.6.4 Damage costs

Several factors influence the appraisal damage costs: these include the magnitude of seismic excitations, site condition, properties of construction materials and others. The difficulty in damage modelling is that two seismic excitations of equal magnitude will yield totally different damage results. The absence of data, which could be used to develop models of damage costs, causes further difficulties in modelling. Consequently, the development of damage models herein is limited to the costs of repairing non-structural damage. This type of damage can be simply modelled by using a function of storey drift, which according to most investigators is the best indicator for damage [4, 13]. Let the damage ratio, ξ, be the ratio of non-structural damage repair cost per storey to the construction cost of damaged items on that storey. The damaged items indicated here are partitions and glass. The relationship between the damage ratio and the storey drift, Δ, can therefore be obtained on the basis of the previous research data and is given by

$$\xi = 8.52 \Delta$$

(7.211)

If the construction costs of the damaged items on the ith storey is $(C_{nc})_i$, then the repair costs on that storey will be

$$(C_r)_i = (C_{nc})_i \, (\xi)_i = 8.52 \, (C_{nc})_i \, \Delta_i$$

(7.212)

The storey drift is a function of ground acceleration and member properties; however, as soon as the size of members is determined, the storey drift may be simply expressed as a function of ground acceleration such that

$$\Delta_i = c_i a$$

(7.213)

where the subscript i represents the ith storey. According to Eq. (7.213), if the structure is designed on the basis of ground acceleration, a^*, and the resulting storey drift of the ith story is Δ_i^*, then c_i is given as

$$c_i = \frac{\Delta_i^*}{a^*} \tag{7.214}$$

By substituting Eq. (7.213) into Eq. (7.212) one obtains

$$(C_r)_i = 8.52 \, (C_{nc})_i \, c_i a \tag{7.215}$$

Equation (7.215) represents the relationship between damage repair costs and ground acceleration.

In its lifetime, a structure may be subjected to earthquake excitations of different magnitudes. To estimate the non-structural damage repair costs in the lifetime of the structure, the repair costs for all expected earthquake damage must be evaluated. The number of seismic shocks may be reasonably estimated by using n_o to designate earthquake frequency. This term is defined as

$$n_o = -\frac{dN}{dM} = \frac{1}{B} A N_o e^{\frac{-M}{B}} \tag{7.216}$$

in which N is the annual number of shallow earthquakes having magnitudes equal to or greater than M, in area A, M the Richter magnitude, A the amount of area, B the distribution parameter that describes seismic severity, and N_o the annual number of seismic shocks per unit area. The value of n_o then represents the number of shocks having magnitude between M and $M + dM$ in area A. To develop a similar formula in terms of ground acceleration, an idealized relationship between area, ground acceleration and magnitude, as shown in Fig. 7.16 [4], is used. According to this figure, the area over which a certain range of ground acceleration will exist for a given magnitude of earthquake can be obtained. For an example, let us assume that the parameters N_o and B are known, then if we substitute each individual covered area listed in Table 7.5, which is established on the basis of Fig. 7.16, into Eq. (7.216), we can estimate the annual number of seismic shocks within the specified range of ground acceleration for a given seismic intensity, because the influenced area in Fig. 7.16 decreases when the ground acceleration increases for any given magnitude. If we assume that the influenced area is linearly decreased, then it is clear that the average number of shocks for any given magnitude within the specified range of ground acceleration could be obtained by dividing by a factor of 2. The results of an estimation of the number of shocks are given in Table 7.6 for Southern California, where N_o was 1.7 mile in.$^{-2}$ and B was 0.48.

To estimate the number of shocks for any given ground acceleration, the mean values in Table 7.6 were obtained by dividing the total number of

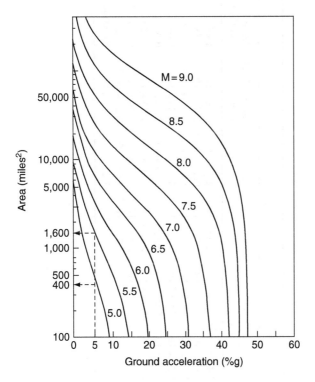

Figure 7.16 Affected area curves.

Table 7.5 Covered area in 1,000 miles².

Acceleration (% g)	M							
	5	5.5	6	6.5	7	7.5	8	8.5
≥ 5	0.4	1.6	3.6	6.8	13	28	56	130
≥ 10		0.6	1.6	3.6	7.6	14	32	72
≥ 15			0.6	2	4.4	9.6	21	47
≥ 20				0.9	2.5	6	14	33
≥ 25					1.3	4	10	24
≥ 30					0.25	2	6.4	17
≥ 35						0.6	4	12
≥ 40							1.2	5.8
≥ 45								0.2

shocks by the increment of ground acceleration of each specified range of acceleration, 0.05. These mean values are plotted in Fig 7.17. A least square curve fitted function is obtained by using

$$n_o = 2.9335e^{-14.6118a} \tag{7.217}$$

Table 7.6 Calculations of earthquake frequencies.

Acceleration (%g)	M								Total = Σ of Row	Mean = Total/0.05
	5	5.5	6	6.5	7	7.5	8	8.5		
0.05–0.10	2.12×10^{-2}	1.87×10^{-2}	1.32×10^{-2}	7.45×10^{-3}	4.44×10^{-3}	4.06×10^{-3}	2.45×10^{-3}	2.09×10^{-3}	7.36×10^{-2}	1.47
0.10–0.15		1.12×10^{-2}	6.60×10^{-3}	3.72×10^{-3}	2.63×10^{-3}	1.28×10^{-3}	1.12×10^{-3}	9.02×10^{-4}	2.75×10^{-2}	0.55
0.15–0.20			3.96×10^{-3}	2.56×10^{-3}	1.56×10^{-3}	1.04×10^{-3}	7.16×10^{-4}	5.05×10^{-4}	1.03×10^{-2}	0.206
0.20–0.25				2.09×10^{-3}	9.86×10^{-4}	5.80×10^{-4}	4.09×10^{-4}	3.25×10^{-4}	4.39×10^{-3}	0.0878
0.25–0.30					8.62×10^{-4}	5.80×10^{-4}	3.68×10^{-4}	2.53×10^{-4}	2.06×10^{-3}	0.0412
0.30–0.35					2.05×10^{-4}	4.06×10^{-4}	2.45×10^{-4}	1.80×10^{-4}	1.04×10^{-3}	0.0208
0.35–0.40						1.74×10^{-4}	2.86×10^{-4}	2.24×10^{-4}	6.84×10^{-4}	0.0137
0.40–0.45							1.23×10^{-4}	2.02×10^{-4}	3.25×10^{-4}	0.0065
0.45–0.50								7.22×10^{-6}	7.22×10^{-6}	0.000144

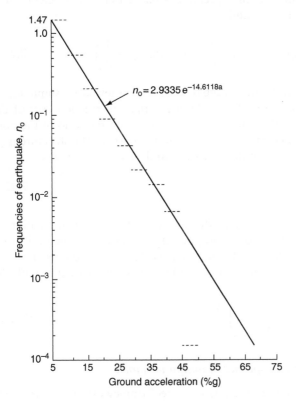

Figure 7.17 Annual number of earthquake shocks for the specified ground accelerations.

in which a is the ground acceleration normalized by the gravity acceleration, g and $n_o da$ represent the annual number of shocks having a normalized ground acceleration between a and $a + da$. As indicated in Fig. 7.17, the number of shocks drops off the curve at a ground acceleration $0.45\,g$. Equation (7.217) may be modified to the following general form as

$$n_o = \gamma e^{-\vartheta a} \tag{7.218}$$

Let us suppose that the lifetime of the structure is N_ℓ years, then the number of shocks with ground accelerations between a and $a + da$ in the lifetime will be $N_\ell n_o da$. Then, by using Eqs. (7.215) and (7.218), we can determine that the damage repair cost for the ith storey is

$$(C_D)_i = 8.52\,N_\ell\,(C_{nc})_i\,c_i\gamma \int^{a_{max}} a e^{-\vartheta a}\,da \tag{7.219}$$

$$= 8.52\,N_\ell\,(C_{nc})_i\,c_i\gamma\,f^o$$

in which a_{max} is the maximum expected normalized ground acceleration in the lifetime of the structure, and

$$\pounds = -\frac{1}{\vartheta}\left(a_{max}\exp(-\vartheta a_{max}) + \frac{1}{\vartheta}\exp(-\vartheta a_{max}) - \frac{1}{\vartheta}\right) \tag{7.220}$$

The total non-structural damage repair cost is then levied by summing up the damage repair costs of all floor levels. If the construction cost of the non-structural items is the same for every storey, and the ratio of the non-structural construction costs to the construction costs of the structure is χ, then the construction cost of the non-structural items per storey is

$$(C_{nc})_i = \frac{\chi C_{cl}}{n_s} \tag{7.221}$$

in which the construction cost of the structure, C_{cl}, is the summation of the steel costs, connection costs and painting costs of the structural members. By substituting Eq. (7.221) into Eq. (7.219), one obtains the total non-structural damage repair cost in the lifetime of the structure with the following formula

$$C_D = 8.25 N_\ell \chi C_{cl} \gamma \pounds \left(\sum_{i=1}^{n_s} c_i\right)\frac{1}{n_s} \tag{7.222}$$

7.6.5 Cross-sectional properties of design variables

In general the girders and columns of a steel framework are wide flange sections. The wide flange sections can be either selected from the AISC Manual [15] or designed for built-up wide flange sections which are presented in (A) and (B) as follows:

(A) In the AISC Manual, the sectional properties of WF steel sections are given. The interesting sectional properties in structural optimum design are cross-sectional area, A, sectional modulus, S_x, and moment of inertia, I_x. Unfortunately, the direct relationships are not available for these three values. Therefore, approximate relationships for the most economical WF sections need to be developed on the basis of curve fitting with selected algebraic expressions [16, 4]. These relationships are given as

(a) for $0 \le I_x \le 9,000$ in.[4]

$$S_x = \sqrt{60.61\,I_x + 84,100} - 290 \tag{7.223}$$

and

$$A = 0.465\sqrt{I_x} \tag{7.224}$$

(b) for $9{,}000 \text{ in.}^4 \le I_x \le 20{,}300 \text{ in.}^4$

$$S_x = \frac{I_x - 8{,}056.3}{1.876} \tag{7.225}$$

and

$$A = \frac{I_x + 2{,}300}{256} \tag{7.226}$$

(B) The AISC WF sections given above may not be adequate for the design of tall and heavy frameworks. The built-up sections are needed to allow for this inadequacy. The cross-sectional area, A, the moment of inertia, I_x, the sectional modulus, S_x, and the shear flow, v, are respectively given as follows

$$A = d^2 \left(\frac{t_w}{d} + 2\frac{t_f}{d}\left(\frac{b}{d} - \frac{t_w}{d}\right) \right) \tag{7.227}$$

$$I_x = d^4 \left(\frac{b}{2d}\left(\frac{t_f}{d}\right)\left(1 - \frac{t_f}{d}\right)^2 + \frac{1}{12}\frac{t_w}{d}\left(1 - \frac{2t_f}{d}\right)^3 \right) \tag{7.228}$$

$$S_x = d^3 \left(\frac{b}{d}\left(\frac{t_f}{d}\right)\left(1 - \frac{t_f}{d}\right)^2 + \frac{1}{6}\frac{t_w}{d}\left(1 - \frac{2t_f}{d}\right)^3 \right) \tag{7.229}$$

and

$$v = \frac{d^2 \left(\frac{b}{2d}\left(\frac{t_f}{d}\right)\left(1 - \frac{t_f}{d}\right)^2 + \frac{1}{12}\frac{t_w}{d}\left(1 - \frac{2t_f}{d}\right)^3 \right)\frac{t_w}{d}}{\left(\frac{b}{2d}\left(\frac{t_f}{d}\right)\left(1 - \frac{t_f}{d}\right) + \frac{1}{8}\frac{t_w}{d}\left(1 - \frac{2t_f}{d}\right)^2 \right)} \tag{7.230}$$

in which b is the flange width, d the depth of the cross-section, t_f the flange thickness, and t_w the web thickness.

7.7 Optimization of 15-storey frames and assessments of design results

7.7.1 Design comparisons due to the effect of mass models, $P - \Delta$ effect, horizontal and vertical El Centro 1940 earthquake

In this chapter, the design results are for a 15-storey unbraced building which is subjected to the horizontal and vertical ground motions of the 1940 El Centro earthquake for which the acceleration spectra with 5 per cent damping are given in Fig. 7.5. In addition to the consideration of verti-cal component of the ground motion, the second-order $P - \Delta$ effect on the design is also included. The $P - \Delta$ effect is a result of all the axial forces exerted on the columns. These forces are composed of the dead loads of the structural and non-structural masses and the associated inertial forces occasioned by vertical acceleration. The structures are assumed to have two models: Model I has two nodes at both ends of a girder, and Model II has three nodes at both ends and at the mid-span of a girder [17, 18].

The 15-storey building has the following structural dimensions, weights, member properties, and allowable stresses: the span length and floor height are 21 ft (6.405 m) and 12 ft (3.66 m), respectively. The dead load (non-structural mass) on each floor is 180 lbs in.$^{-1}$ (3,210.2 kg m^{-1}). The mass density and the modulus of elasticity of the construction material are 0.283 lbs in.$^{-3}$ (7.823 g cm^{-3}) and 29,000 ksi (200.1 GN m^{-2}), and the allowable shear stress, σ_v, is less than or equal to 0.65σ. The allowable axial stress of the bracings is assumed to be 20 ksi (138 MN m^{-2}).

Although different allowable deflections may be imposed at any particu-lar node, the allowable deflection of each floor is limited to 0.005 times the height of that floor from the ground level. The columns and girders are made of the built-up sections described in 7.6.5 and have the following properties: $b = 25$ in. (63.50 cm), $d_{max} = 75$ in. (190.50 cm), $d_{min} = 15$ in. (38.10 cm), $(t_f/d)_{max} = 0.045$, $(t_f/d)_{min} = 0.023$, and $t_w/d = 0.020$. The framework is designed for the following two groups of constraints listed as (A) and (B) as follows:

Group (A) *Stress constraints only* The 15-storey unbraced frame shown in Fig. 7.18a is used to study the significant effect of multi-component ground motion and of various constraints on the optimal design results. The opti-mal design results of the structure are observed by considering the stress constraints and the constraints on the relative stiffness of the constituent members. Four design cases are considered by using Models I and II as shown in Fig. 7.18b. Case (a) is designed for horizontal ground motion only, Case (b) is designed for horizontal ground motion and the $P - \Delta$ effect of the dead load associated with the structural masses, Case (c) is designed for hor-izontal and vertical earthquake components but no $P - \Delta$ effect, and Case (d) is designed for horizontal and vertical earthquake components as well as the $P - \Delta$ effect occasioned by the dead load and the vertical inertial forces associated with the structural and non-structural masses. The design results are identified by $H, H + P\Delta\,(DL), H + V, H + V + P\Delta\,(DL + V)$, which cor-respond to Cases (a) through (d), respectively. The stresses caused by gravity loads on the girders are not included in the design. Thus, the design results are based on the dynamic excitations only.

The plot of the total weights versus the number of iterations for Case (a) through Case (d) is shown in Fig. 7.19. Table 7.7 lists the final weights of these four cases, which are identified as Group A. The table also includes the final displacements at the top floor, the number of modes, and the associated

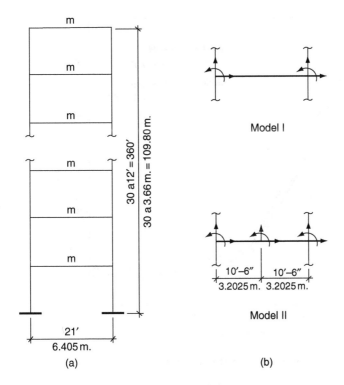

Figure 7.18 15-storey, single-bay unbraced frame: (a) given structure, (b) structural models.

natural periods used in the design. It is apparent that the multi-component ground motion when combined with the $P - \Delta$ effect of the structural and non-structural masses can yield nearly 3 per cent of an increase in the structural weight over that which would be required for one horizontal component only. Fig 7.20 shows the ratio of the energy (kinematic and strain), W, of the individual modes to the total energy, W_T, associated with the total number of natural modes included in the design. This plot signifies that the first mode is the most significant for all cases. The modes beyond the third have little effect on a structure subject to a horizontal motion in Cases (a) and (b) of Model I. However, when a structure is subjected to combined horizontal and vertical ground motions as in Cases (c) and (d) of Model II, the first five modes are needed for an adequate design. Figs 7.21 and 7.22 show the distribution of the amount of inertia of the columns and girders, respectively. The moment of inertia for the columns is the smallest at the top floor and increases almost linearly from the top floor to the second floor and abruptly increases at the first floor. The moment of inertia for the girders is relatively small at the first floor and increases suddenly at the second floor.

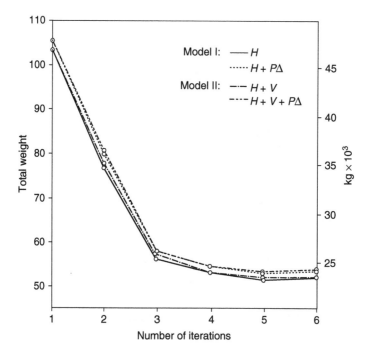

Figure 7.19 Optimum weights of 15-storey frame.

Table 7.7 Final weights, natural periods, and displacements of 15-story frames. (1 kip = 453 kg, 1 in. = 2.54 cm)

Group	Case	Final Weight (kips)	Natural period (s)					Disp. at top floor (in.)
			1	2	3	4	5	
A	a	51.88	2.409	0.796	0.462	0.327	0.245	13.19
	b	53.03	2.412	0.792	0.461	0.323	0.243	13.21
	c	52.15	2.413	0.792	0.461	0.326	0.245	13.22
	d	53.32	2.409	0.786	0.458	0.322	0.242	13.22
B	a	60.44	2.152	0.694	0.404	0.282	0.212	10.8
	b	62.21	2.179	0.671	0.386	0.268	0.202	10.8
	c	60.77	2.158	0.692	0.402	0.281	0.217	10.8
	d	62.21	2.152	0.677	0.395	0.276	0.212	10.8

It decreases almost linearly from the second to the 13th floor and abruptly decreases from the 14th to the top.

Group (B) *Stress and displacement constraint* In addition to the constraints discussed in Group (A), the displacement constraint is included as the allowable displacement of 0.005 times the storey height from the ground imposed

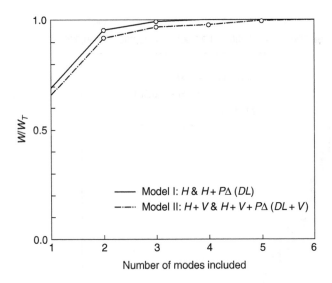

Figure 7.20 Contribution of energy of individual modes to the design.

on each floor. The final weights, periods and displacements at the top storey for all cases are also listed in Table 7.7 for Group B. The design result similarly reveals that the structural weight, which is needed to withstand the effects of a vertical ground motion and a $P - \Delta$ effect, is about 3.0 per cent greater than the structural weight needed for a horizontal ground motion alone. It is apparent that the displacement constraints cause the structure of Group B to be more rigid than those in Group A; thus, the final structural weight of Group B is much heavier than that of Group A. The distribution shapes of the moments of inertia of the columns and girders are similar to those of Group A and are therefore not revealed here.

7.7.2 Influence of effect of peak ground acceleration and velocity with various soil types on design results

The 15-storey, one-bay, unbraced frame shown in Fig. 7.23 was designed according to ATC-3-06 [19, 20], equivalent lateral force procedures. The design was based on a response modification factor of $R = 8$, a deflection amplification factor of $C_d = 5.5$, and an allowable storey drift of $\Delta_a = 0.05h_{sx}/C_d$ in which h_{sx} is the storey height below level x. The stress constraint was not taken into consideration, and the non-structural dead load on each floor was set at 50,000 lbs (222.4 kN). The AISC wide-flange sections were used in the design. Thirty-six cases were designed on the basis of different combinations of the effective peak ground acceleration, A_a, the

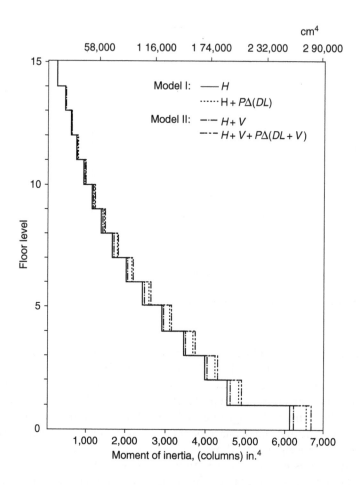

Figure 7.21 Distribution of moment of inertia of columns.

effective peak ground velocity, A_v, and the soil profile type. The coefficients of A_a were changed to 1, 3, 5 and 7, and those of A_v to 3, 5 and 7 for three soil profile types, S_1, S_2 and S_3. The plots of the fundamental periods, the seismic design coefficients, C_s, and the optimum weights of the 36 design cases are shown in Figs. 7.24, 7.25 and 7.26, respectively. It is worth noting that all the fundamental periods are larger than the approximate fundamental period, T_a, as well as $1.2T_a$. The seismic design coefficients of Fig. 7.25 varied only when A_a and the soil profile types were changed, because $1.2T_a$ controlled the design. In Fig. 7.25, points marked by an asterisk were computed in accordance with the formula $C_s = 2.5A_a/R$ in which the natural period is not needed. Based on the optimum design results, it was

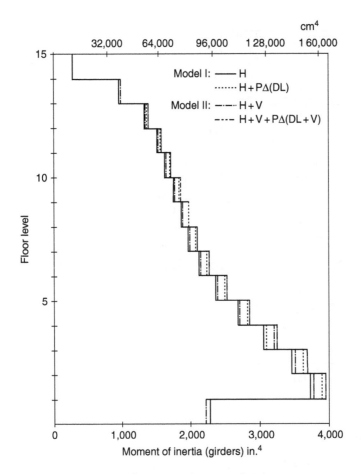

Figure 7.22 Distribution of moment of inertia of girders.

not necessary to use mechanical methods to determine the natural period, especially for those areas having low values of A_a and A_v. Because $1.2T_a$ controls all the design results, some of the 36 design cases have an identical optimal weight. The optimum solutions may be classified into 11 groups. In Fig. 7.26, the number in parentheses indicates the group that the design case belongs to. The plot of weights versus cycles of iteration for these 11 groups is shown in Fig. 7.27. The stability coefficients, that are determined according to Eq. (F.12), are shown in Fig. 7.28 in which all of the values are less than the upper bound 0.1 except for the case corresponding to $A_a = 1$, $A_v = 3$ and soil profile type S_1. This case has a maximum value of 0.117 at the bottom storey. Among these 36 design cases, the case corresponding

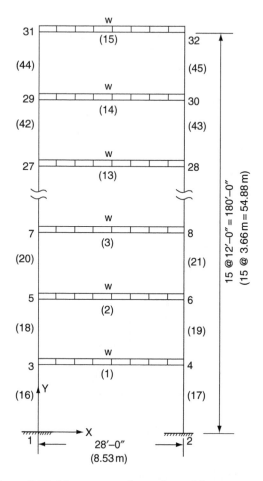

Figure 7.23 15-storey, one-bay unbraced frame.

to $A_a = 7$, $A_v = 7$ and soil profile type S_3 has a maximum eccentricity of 5.66 ft (1.73 m) as measured from the centre of the bay to the outside of the bay and induced by the resultant of the seismic forces as well as the vertical loads at the foundation–soil interface.

The moment of inertia of the girder on the first floor for all 11 groups, as shown in Fig. 7.29, is much less than that of the girder on the second and third floors. The moment of inertia of the girder on either the second or third floors is the largest. The decrement of the moment of inertia decreases gradually from either the second or the third floor to the 15th floor. As illustrated in Fig. 7.30, in descending order, the moment of inertia of a column on the top storey is the smallest but increases at the 14th storey. The distribution from the 14th storey to the second storey is almost linearly increased, and at the first storey, it abruptly increases.

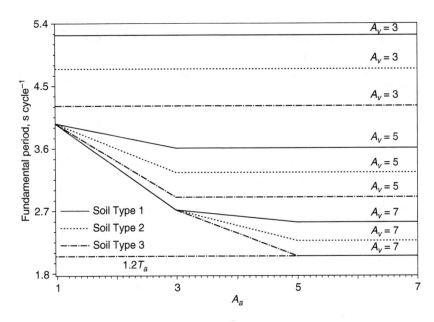

Figure 7.24 Fundamental periods of 15-storey frame using equivalent lateral force design.

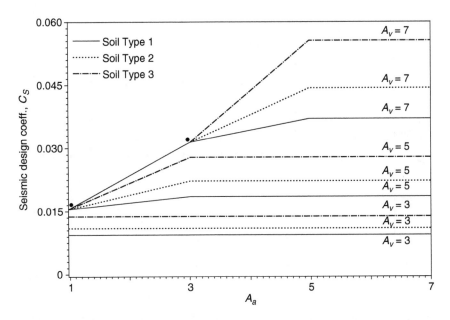

Figure 7.25 Seismic design coefficients, C_S, of the 15-storey frame.

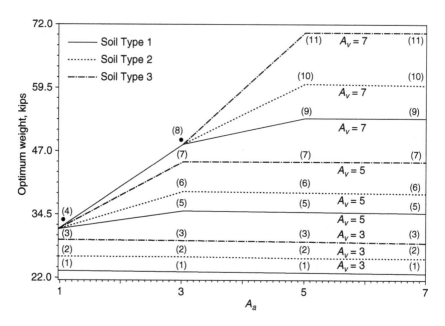

Figure 7.26 Optimum weights of the 15-storey frame. (1 kip = 4.448 kN).

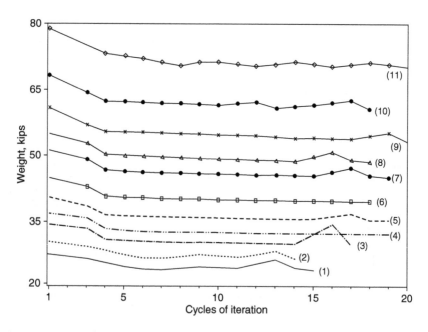

Figure 7.27 Weights versus cycles of iteration plot of the 15-storey frame. (1 kip = 4.448 kN).

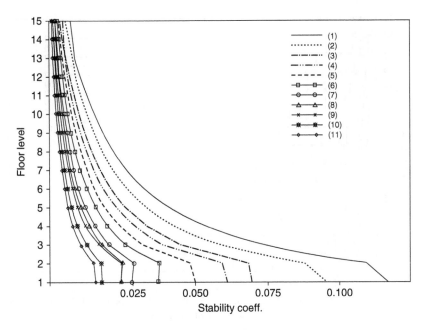

Figure 7.28 Stability coefficients of the 15-storey frame.

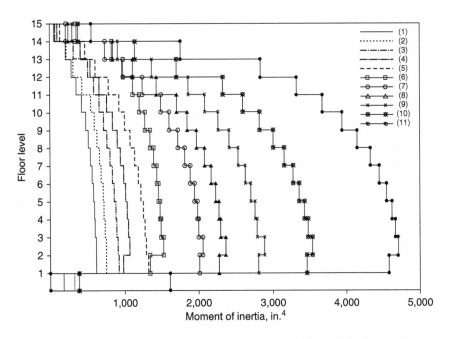

Figure 7.29 Distribution of moment of inertia of the girders of the frame (1 in. = 2.54 cm).

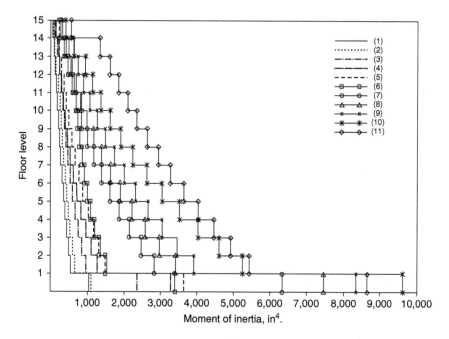

Figure 7.30 Distribution of moment of inertia of the columns of the 15-storey frame (1 in. = 2.54 cm).

7.7.3 *Comparison of design results based on minimum weight and minimum cost*

The structure shown in Fig. 7.31 was designed to resist a non-structural weight of 100,000 lbs (444.8 kN) per floor and seismic design forces. The design was based on a response modification factor of $R = 8$, a deflection amplification factor of $C_d = 5.5$, and an allowable storey drift of $\Delta_a = 0.015\,h_{sx}/C_d$ in which h_{sx} is the storey height below level x according to the ATC-06 equivalent lateral force procedure outlined in Appendix F. The stress constraint was not taken into consideration, and wide-flange sections were used. The minimum cost function has the unit prices of the steel, painting, steel at the connections, and welding metal being 0.24 $ lbs^{-1} (0.054 $ N^{-1}), 0.796×10^3 (0.124 $\times 10^3$ $ cm^{-2}), 0.3 $ lbs^{-1} (0.0674 $ N^{-1}), and 5.5 $ lbs^{-1} (1.24 $ N^{-1}), respectively. The design was based on a ground acceleration of 0.4 g. The maximum expected intensity of seismic excitations for 50 years, the lifetime of the structure, was assumed to be 0.5 g. In addition, rigid connections were assumed, therefore a large value of 9×10^9 was used as the initial stiffness connection, k. The input parameters of the equivalent lateral method were taken to be the same as those of the minimum weight design.

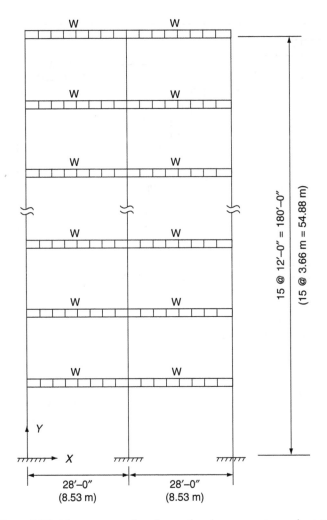

Figure 7.31 15-storey, two-bay frame for the comparison of minimum weight and minimum cost designs.

The moments of inertia of the girders and exterior columns from the bottom to the top of the building have a distribution that is similar to that of the one-bay frame. The moments of inertia of the girders are as shown in Fig. 7.32. The distribution of the moments of inertia of the interior columns, however, decreases from the third floor to the second floor and then to the first floor as sketched in Fig. 7.33. The decrement from the second floor to the first floor is almost equal to the decrement from the third floor to the second floor. The stability coefficients in Fig. 7.34 reveal that θ is 0.0194

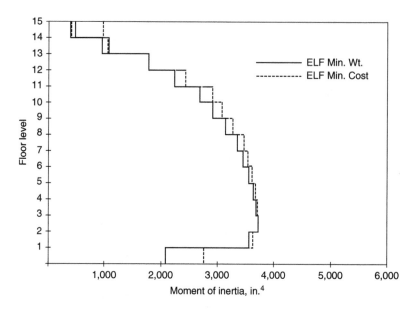

Figure 7.32 Moment of inertia of girders of the 15-storey, two-bay frame for minimum weight and minimum cost designs (1 in. = 2.54 cm).

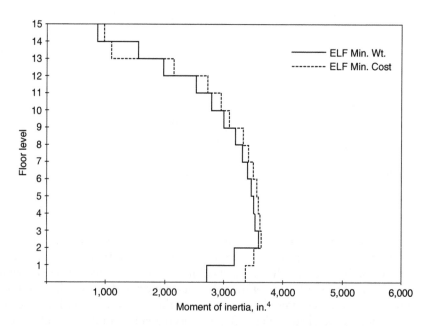

Figure 7.33 Moment of inertia of the interior columns of the 15-storey, two-bay frame for minimum weight and minimum cost designs (1 in. = 2.54 cm).

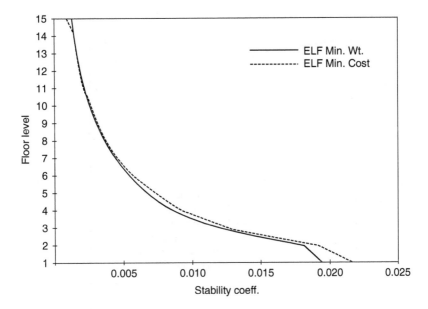

Figure 7.34 Stability coefficient of the 15-storey frame for minimum weight and minimum cost designs.

and 0.0217 corresponding to minimum weight and minimum cost, respectively. Fig. 7.35 illustrates the plot of the costs versus cycles of iteration for each type of cost. As indicated, the total structural cost is governed by the base charge, which is the cost of the structural members. All other kinds of costs are just a small percentage of the total cost. Therefore, we may reasonably suggest that the minimum weight design can be a reasonable objective function for steel structures. It is also worth noting that the damage cost increases whenever the total cost decreases. It is obvious that the decrease in total cost will increase the flexibility of the structure and hence enlarge the storey drifts. The damage cost is therefore raised because of the increase of the storey drifts.

7.8 Concluding remarks

The optimality-criteria method based on energy distribution is presented for designing various two-dimensional steel structures subjected to multi-component inputs of static, dynamic and earthquake forces. The seismic input can be one-dimensional or two-dimensional; one-dimensional is horizontal, two-dimensional is horizontal coupled with vertical. The seismic forces may be excitations at the base or equivalent lateral forces at the structural nodes. The base excitations include actual earthquake records and response spectra. The structural formulation is derived on the basis of the

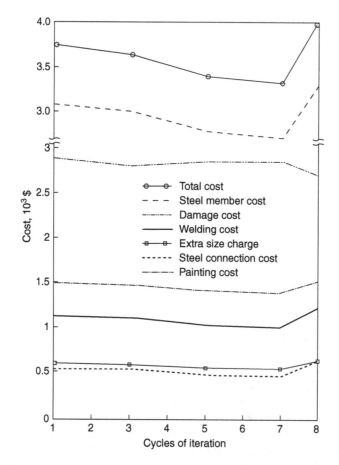

Figure 7.35 Costs versus cycles of iteration plot of the 15-storey frame minimum cost design.

matrix displacement method and the consistent mass method with consideration of the second-order $P - \Delta$ forces. The constituent members of a system are made of either built-up sections or AISC WF sections. The constraints include stresses, displacements, storey drifts, natural frequencies, maximum differences between relative stiffnesses, and lower bound of cross-sections. The objective function can be either minimum weight or minimum cost. Extensive numerical examples are provided to illustrate the advantages of using the proposed design method, the assessment of the ATC-3-06 parameters, the effect of soil–structure interaction on the ATC-3-06 provisions, the effect of the $P - \Delta$ forces and the vertical ground excitations on the optimum design, the stiffness distribution of high-rise buildings, the comparison of the minimum weight and minimum cost design, and the influence of storey drift constraint and displacement constraint on optimum design. The presented optimization method with energy distribution and gradient search is the foundation of the optimality-criteria techniques which will be further discussed in later chapters including Chapters 8, 9, 10 and 11 with Uniform Building Code [22].

References

1. Cheng, F.Y. *Matrix Analysis of Structural Dynamics-Applications and Earthquake Engineering*, Marcel Dekker, Inc., New York, (1005 pages), 2001.
2. Housner, G.W. *Design Spectrum, Chapter 5, Earthquake Engineering*, Wiegel, R.L., ed., Prentice-Hall, Upper Saddle River, New Jersey, 1970.
3. Newmark, N.M. and Hall, W.J. *Earthquake Spectra and Design*, Earthquake Engineering Research Institute, Oakland, CA, 1982.
4. Cheng, F.Y. and Juang, D.S. *Optimum Design of Braced and Unbraced Frames Subjected to Static, Seismic and Wind Forces with UBC, ATC-3 and TJ-11*, NSF Report, U.S. Department of Commerce, National Technical Information Service, Virginia, NTIS No. PB87-162988/AS (376 pages), 1985.
5. Venkayya, V.B. and Cheng, F.Y. Resizing of frames subjected to ground motion, *Proceedings of the International Symposium on Earthquake Structural Engineering*, University of Missouri– Rolla, August 1976.
6. Cheng, F.Y., Srifuengfung, D. and Sheng, L.H., *ODSEWS – Optimum Design of Static, Earthquake, and Wind Steel Structures*, NSF Report, U.S. Department of Commerce, National Technical Information Service, Virginia, NTIS, PB81-232738, 1981.
7. Cheng, F.Y. and Juang, D.S. *ODSEWS-2D-II: User's Manual: A Computer Program for Optimum Design of 2-D Structures for Static, Earthquake, and Wind Forces-Version II*, NSF Report, U.S. Department of Commerce, National Technical Information Service, Virginia, NTIS No. PB87-163093/AS (134 pages), 1985.
8. Schmit, L.A. and Miura, M. *Approximation Concepts for Efficient Structural Synthesis*, NASA CR-2552, 1976.
9. Schmit, L.A. and Farshi, B. Some approximation concepts for structural synthesis, *AIAA Journal*, Vol. 12, No. 5, pp. 692–699, 1974.
10. Kan, M.R., Willmert, K.D. and Thorntown, W.A. A new optimality criterion method for large scale structures, *AIAA/ASME 19th Structures, Structural Dynamics, and Materials Conference*, Bethesda, MD, April 1978, pp. 47–58.
11. Dobbs, M.W. and Nelson, R.B. Application of optimality criteria to automated structural design, *AIAA Journal*, Vol. 14, No. 10, pp. 1436–1443, 1976.
12. Rizzi, P. Optimization of multiconstrained structures based on optimality criteria, *AIAA/ASME/SAE 17th Structures, Structural Dynamics, and Materials Conference*, King of Prussia, PA, May 1976, pp. 448–462.
13. Walker, N.D. *Automated Design of Earthquake Resistant Multi-story Steel Building Frames*, Report EERC 77-12, University of California, Berkeley, 1977.
14. Walker, N.D. and Pister, K.G. *Study of a Method of Feasible Directions for Optimal Elastic Design of Framed Structures Subjected to Earthquake Loading*, Report EERC 75-39, University of California, Berkeley, 1975.
15. AISC *Manual of Steel Construction*, 8th Edition, American Institute of Steel Construction, Chicago, Illinois, 1980.

16. Brown, D.M. and Ang, A.H. Structural optimization by nonlinear programming, *Journal of the Structural Division*, ASCE, Vol. 92, ST 6, pp. 319–340, 1966.

17. Cheng, F.Y. and Srifuengfung, D. Optimum design for simultaneous multicomponent static and dynamic input, *International Journal for Numerical Methods in Engineering*, Vol. 13, No. 2, pp. 353–372, 1978.

18. Cheng, F.Y., Venkayya, V.B. and Khachaturian, N., *Computer Methods of Optimum Structural Design*, 5th Annual Short Course Notes, Vols. 1 and 2, University of Missouri–Rolla, MO, 1976.

19. Applied Technology Council. *Tentative Provisions for the Development of Seismic Regulations for Buildings*, ATC-3-06, National Bureau of Standards, Washington, DC, 1978.

20. Building Seismic Safety Council. *Amendments to ATC-3-06 Tentative Provisions for the Development of Seismic Regulations for Buildings for Use in Trial Designs*, NBSIR 82-2626, BSSC 82-2, 1982 Foundation, 1985.

21. Cheng, F.Y. and Juang, D.S. Recursive optimization for seismic steel frames, *Journal of Structural Engineering*, ASCE, Vol. 115, No. 2, pp. 445–466, 1989.

22. Uniform Building Code. *International Conference of Building Officials*, 1984 Edition.

23. Venkayya, V.B., Khot, N.S. and Berke, L. Application of optimality criteria approaches to automated design of large practical structures, *Preprint of Paper Presented at the AGARD Second Symposium on Structural Optimization*, Milan, Italy, April, 1973.

8 Generalized optimality-criteria approach

8.1 Introduction

Generalized optimality-criteria (GOC) approaches are based on the mathematical satisfaction of a set of the necessary conditions, not necessarily all of the conditions, for an optimal solution. The Kuhn–Tucker conditions for an optimal solution are used in a customized procedure that will ensure that the prescribed set of optimality conditions is met [1–11]. Unlike mathematical programming that is typically based on defining a set of directions that will move each iteration to a 'better' solution using prescribed methods of choosing how far to travel in those directions, GOC uses the gradients of the objective and constraints, combined with Lagrange multipliers and a recursion relationship to iteratively reach the optimal solution through the satisfaction of the Kuhn–Tucker conditions for optimality.

The development and use of GOC can be traced back to the early 1970s as introduced in Chapter 7 [7, 11, 12]. The major improvement in using a GOC approach is the significant reduction in the number of iterations required to reach a solution. Unlike mathematical programming, the number of iterations required to reach a solution is largely independent of the number of design variables.

The GOC approach is heavily dependent on finding accurate gradients of the constraints and the objectives. Many of the structurally related gradients can be found numerically by direct differentiation of the structural-response equations as shown in subsequent sections.

8.2 Lagrangian formulation

Beginning with the general form for an optimization problem which is stated as

$$\text{Minimize } W(\delta) \quad \text{where } \delta = \{\delta_i\} \quad \text{for } i = 1, \ldots, N \tag{8.1}$$

$$\text{Subject to } g_j(\delta) \leq 0 \quad \text{for } j = 1, \ldots, l \tag{8.2}$$

$$\underline{\delta}_i \leq \delta_i \leq \overline{\delta}_i \quad \text{for } i = 1, \cdots, N \tag{8.3}$$

where $W(\delta)$ is the objective function (weight, cost, reliability, etc.), $g_j(\delta)$ are the structural constraints (displacements, drifts, frequencies, stresses, etc.) and δ_i are the design variables. The upper and lower bars define the lower and upper limits, minimum or maximum size, for each design variable. N is the number of primary design variables, l is the number of constraints, and δ_i is the ith elemental design variable.

8.2.1   Lagrangian function

In order to define the optimality criteria for a structural problem, the problem must be rewritten as a Lagrangian function incorporating the constraints and the objective into a single function. The Lagrangian can be written as

$$L(\delta, \lambda) = W(\delta) + \sum_{j=1}^{l} \lambda_j g_j(\delta) \qquad (8.4)$$

where λ_j is the Lagrange multiplier associated with the jth constraint. Mathematically, this formulation requires the constraints to be equality constraints, and any constraint that is contained in the Lagrangian will be forced to take on its limiting value. Therefore, extra care must be taken to choose the correct constraints for the optimal solution. This will be discussed in Section 8.4 on finding the Lagrange multipliers. In order to define the 'correct' constraints to be included in Eq. (8.4), two terms are commonly used, active and inactive constraints. Active constraints are constraints that are considered to be equality constraints and must be included in Eq. (8.4); inactive constraints are constraints that do not take on the prescribed constraint limit and are not included in Eq. (8.4). For example, if the prescribed upper limit for a displacement is 1 in. (2.54 cm) and the calculated value for the displacement is 0.9 in. (2.29 cm), this constraint could be considered active due to the proximity of the calculated value to the prescribed upper limit and would be included within the Lagrangian. If the calculated displacement was 0.3 in. (0.76 cm), this constraint would most likely be considered inactive since it is not 'close' to the prescribed limit of 1 in. (2.54 cm) and would not be included in the Lagrangian for that specific iteration. Since the structural design changes with each iteration, one must check each constraint for its level of mathematical participation in each iteration as described in Section 8.5.

8.2.2   Kuhn–Tucker necessary conditions for optimality

The Lagrangian in Eq. (8.4) is used to derive the Kuhn–Tucker necessary conditions for an optimal solution. The Kuhn–Tucker conditions for optimality are

$$\frac{\partial L}{\partial \delta_i}(\delta^*, \lambda^*) = 0 \quad i = 1, \ldots, N \tag{8.5}$$

$$\lambda_j^* \geq 0 \quad j = 1, \ldots, l \tag{8.6}$$

$$\lambda_j^* g_j(\delta^*) = 0 \quad j = 1, \ldots, l \tag{8.7}$$

where the δ^* refers to a set of design variables and Lagrange multipliers that are the values for the optimal solution. These conditions are necessary, but not sufficient conditions to guarantee a globally optimal solution. The sufficient conditions of optimality which are needed in addition to the necessary conditions can be found in numerous references that present mathematical optimization theory such as [1, 2, 12].

The GOC to be used within the customized algorithm is derived using the Kuhn–Tucker necessary conditions. Substituting Eq. (8.4) into Eqs. (8.5), (8.6) and (8.7) provides the optimality criteria as

$$\frac{\partial L}{\partial \delta_i} = \frac{\partial W}{\partial \delta_i} + \sum_{j=1}^{l} \lambda_j \frac{\partial g_j}{\partial \delta_i} = 0 \quad i = 1, \ldots, N \tag{8.8}$$

with

$$\lambda_j \geq 0 \quad j = 1, \ldots, l \tag{8.9}$$

$$\lambda_j g_j = 0 \quad j = 1, \ldots, l \tag{8.10}$$

Rearranging Eq. (8.8) gives

$$T_i = -\sum_{j=1}^{l} \lambda_j \left(\frac{\partial g_j}{\partial \delta_i}\right) \bigg/ \left(\frac{\partial W}{\partial \delta_i}\right) = 1 \quad i = 1, \ldots, N \tag{8.11}$$

which along with Eqs. (8.9) and (8.10) must be true in order for a local or global optimum to be possible. (For a single load case and a single displacement constraint for a truss, it can be shown that Eq. (8.11) implies that the virtual strain energy density of each element in the optimal design must be equivalent.)

Eq. (8.11) is the basis for the optimality-criteria-based algorithms. The algorithms are based on finding ways of forcing the T_i values for each 'active' element to converge to 1 as quickly as possible. Active elements are elements that are not controlled by an upper or lower bound as shown in Eq. (8.3). Elements controlled by the upper or lower limits are called passive elements and are controlled by these limits (called side constraints) rather than the optimality criteria specified in Eq. (8.11). Therefore, Eq. (8.11) becomes

$$T_i = -\sum_{j=1}^{l} \lambda_j \left(\frac{\partial g_j}{\partial \delta_i}\right) \bigg/ \left(\frac{\partial W}{\partial \delta_i}\right) = 1 \quad i = 1, \ldots, N_1 \tag{8.12}$$

where N_1 is the number of active elements or elements not controlled by the side constraints, and $(N - N_1)$ is the number of passive elements controlled by their upper or lower limits. The optimality criteria are only valid in the case where elements are free to take on any value in a continuous design space. Therefore, those elements that have reached a lower or upper bound are no longer controlled by Eq. (8.11). These passive elements are forced to take on their fixed (discrete) maximum or minimum value, therefore, the mathematical continuity required by the optimality criteria is no longer valid for these design variables.

Within the algorithms, the value of T_i provides a measure of the current iteration's position relative to a global solution. It is easy to see that T_i can be used in resizing the elements through the use of recurrence relationships. Note that the real difficulty is in the fact that T_i is dependent on the unknown values of the Lagrange multipliers and only those constraints that are truly active. Once a reasonable method for determining the Lagrange multipliers can be found (Section 8.4), they can be used to sort the active and inactive constraints by using Eq. (8.9).

8.3   Recurrence relationships

Recurrence relationships generally take one of two forms: exponential or linear. The linear form is derived from the exponential form. For any given active design variable, an exponential recurrence relationship can be derived by multiplying δ_i taken to the rth root to give

$$\delta_i^{(k+1)} = \delta_i^{(k)} (T_i)_k^{1/r} \quad i = 1, \ldots, N_1 \tag{8.13}$$

where k represents the kth iteration. Eq. (8.13) is often written as

$$\delta_i^{(k+1)} = \delta_i^{(k)} (1 + (T_i - 1))_k^{1/r} \quad i = 1, \ldots, N_1 \tag{8.14}$$

From Eq. (8.14), the second term in the parentheses is clearly a measure of how close $\delta^{(k)}$ is to the optimal solution. The parameter r is called a convergence control parameter (similar to move limits or step size in mathematical programming) and is used to control how large a change can occur for each element in each iteration. A value of two for r works well for most structural problems. For highly nonlinear problems, larger values of

r cause smaller incremental changes which helps stabilize the numerically based algorithms.

The linear form for recurrence is found by multiplying the optimality criteria, $T_i = 1$, by $\delta_i (1 - \alpha)$ where α is often called a relaxation parameter to give

$$\delta_i^{(k+1)} (1 - \alpha) = \delta_i^{(k)} (1 - \alpha) T_i \quad i = 1, \ldots, N_1 \tag{8.15}$$

which ignores the fact that $\delta_i^{(k+1)}$ and $\delta_i^{(k)}$ are not the same (as the algorithm converges, the difference becomes minor). Rearranging Eq. (8.15) and ignoring the difference in values between iterations gives

$$\delta_i^{(k+1)} = \delta_i^{(k)} (\alpha + (1 - \alpha) T_i) \quad i = 1, \ldots, N_1 \tag{8.16}$$

The most widely used approach for a linear recurrence relationship is based on the Taylor series expansion of Eqs. (8.13) and (8.14). If only the first two terms of the expansion are retained (binomial expansion), the linear recurrence relationship becomes

$$\delta_i^{(k+1)} = \delta_i^{(k)} \left(1 + \frac{1}{r} (T_i - 1) \right) \quad i = 1, \ldots, N_1 \tag{8.17}$$

Note that Eqs. (8.16) and (8.17) are related by

$$\alpha = \left(1 - \frac{1}{r} \right) \tag{8.18}$$

The recurrence relationships presented need several critical pieces of information prior to their use. The convergence control parameter, r, or the relaxation parameter, α, must be chosen; the constraint gradients, objective-function gradients and the Lagrange multipliers must be determined. As mentioned, a convergence-control parameter of two or a relaxation parameter of $1/2$ are typically adequate for structural-optimization problems. The gradients of the constraints, the gradients of the objective functions and the determination of the Lagrange multipliers will be discussed in Sections 8.10 and 8.4, respectively.

8.4 Lagrange multipliers

Prior to using Eqs. (8.13) through (8.17) to resize the design variables, the Lagrange multipliers needed for the optimal solution must be provided. Except for the simplest of cases, the optimal Lagrange multipliers can only be approximated. Therefore, the recursive relationships generally require an initial estimate which can be very difficult to assess. Instead of 'guessing' an initial set of Lagrange multipliers, other techniques can be devised to find a set of Lagrange multipliers that will satisfy the optimality criteria as long as a set of probable active constraints have been chosen. The Lagrange multiplier estimations can be divided into two categories based on which type of recurrence relationship is used: (1) exponential recurrence or (2) linear recurrence relationships. The accuracy of this estimation of the Lagrange multipliers is very important to the convergence, stability and accuracy of the derived optimal solution.

The exponential recurrence relationship for Lagrange-multiplier determination is based on the same equations as seen in Eqs. (8.13) through (8.17). Since the Lagrange multipliers are related to the satisfaction of constraints, it makes sense to use the associated constraint instead of the optimality criterion, T_i. If the constraints in Eq. (8.2) are rewritten as two separate constraints as

$$g_j = (u_j - \bar{u}_j) \le 0 \quad j = 1, \dots, l_1 \tag{8.19}$$

and

$$g_j = (\underline{u}_j - u_j) \le 0 \quad j = 1, \dots, l_2 \tag{8.20}$$

where the upper and lower bars represent the maximum and minimum values for the constrained structural response u_j, and l_1 and l_2 are the number of upper and lower constraints, respectively. These two relationships can be rewritten as

$$D_j = \frac{u_j}{\bar{u}_j} \le 1 \quad j = 1, \dots, l_1 \tag{8.21}$$

and

$$D_j = \frac{u_j}{\underline{u}_j} \le 1 \quad j = 1, \dots, l_2 \tag{8.22}$$

If only active constraints are considered, the optimal solution provides equality constraints and D_j will be unity, similar to the optimality criteria, T_i, leading one to use the same recurrence equations except substituting D_j for T_i giving

$$\lambda_j^{(k+1)} = \lambda_j^{(k)} \left(1 + (D_j - 1)\right)^{\frac{1}{z}} \quad j = 1, \dots, l \tag{8.23}$$

or

$$\lambda_j^{(k+1)} = \lambda_j^{(k)} \left(1 + \frac{1}{z}(D_j - 1)\right) \quad j = 1, \dots, l \tag{8.24}$$

In both cases, an initial estimate of the Lagrange multiplier, $\lambda^{(1)}$, is required. This is nearly impossible for large-scale problems with different types of constraints since Lagrange multipliers can range from near zero to numbers as high as or higher than 10^{12}.

The second category of methods for finding the Lagrange multipliers is based on using the linear recurrence relationships given in Eqs. (8.16) and (8.17). With appropriate assumptions and algebraic manipulation, Eq. (8.17) in particular can be used to produce a set of linear equations which can be solved for the Lagrange multipliers, eliminating the need to have an initial approximation as required for Eqs. (8.23) and (8.24). Approximating the change in a constraint as

$$\Delta g_j = g_j(\delta + \Delta \delta) - g_j(\delta) = \sum_{i=1}^{N} \frac{\partial g_j}{\partial \delta_i}(\Delta \delta_i) \quad j=1,\dots,l \tag{8.25}$$

and assuming that if g_j is an active constraint, the change, $\Delta \delta$, should force $g_j(\delta + \Delta \delta)$ to become zero giving

$$-g_j(\delta) = \sum_{i=1}^{N} \frac{\partial g_j}{\partial \delta_i}(\Delta \delta_i) \quad j=1,\dots,l \tag{8.26}$$

Note that the change in the design variable, $\Delta \delta_i$, can be written using Eq. (8.17) as

$$\Delta \delta_i = \delta_i^{(k+1)} - \delta_i^{(k)} = \frac{1}{r}(T_i - 1)\delta_i^{(k)} \quad i=1,\dots,N_1 \tag{8.27}$$

Substituting Eq. (8.27) into Eq. (8.26) gives

$$-g_j(\delta) = \sum_{i=1}^{N} \frac{\partial g_j}{\partial \delta_i} \frac{1}{r}(T_i - 1)\left(\delta_i^{(k)}\right) \quad j=1,\dots,l \tag{8.28}$$

and substituting Eq. (8.11) into (8.28)

$$-g_j(\delta) = \sum_{i=1}^{N} \frac{\partial g_j}{\partial \delta_i} \frac{1}{r}\left(-\sum_{t=1}^{l}\lambda_t\left(\frac{\partial g_t}{\partial \delta_i} \Big/ \frac{\partial W}{\partial \delta_i}\right) - 1\right)\delta_i^{(k)} \quad j=1,\dots,l \tag{8.29}$$

where t represents constraints that are within the set of active constraints. Rearranging Eq. (8.29) gives

$$rg_j(\delta) - \sum_{i=1}^{N} \frac{\partial g_j}{\partial \delta_i}\delta_i^{(k)} = \sum_{t=1}^{l}\lambda_t\left(\sum_{i=1}^{N}\left(\frac{\partial g_j}{\partial \delta_i}\frac{\partial g_t}{\partial \delta_i} \Big/ \frac{\partial W}{\partial \delta_i}\right)\delta_i^{(k)}\right) \quad j=1,\dots,l$$

$$\tag{8.30}$$

Eq. (8.30) can be converted to equations using the relaxation parameter, α, by replacing r with $1/(1-\alpha)$. Eq. (8.30) generates a series of l linear equations that can be used to find the l Lagrange multipliers, removing the need to have an initial estimate for the Lagrange multipliers. It is important to have a fairly accurate set of active constraints in order to reduce the number of simultaneous equations and to reduce the number of times these equations need to be solved in each iteration (to be explained further in Section 8.5). Eq. (8.30) also takes into account the dependence of one constraint on another and their combined effect on the solution where as the exponential recurrence relationships do not. If only the diagonal terms associated with $(\partial g_i/\partial \delta_i)$ $(\partial g_t/\partial \delta_i)$ are considered, the recurrence techniques Eqs. (8.24) and (8.30) can be shown to be equivalent as long as $1/z$ is equivalent to r.

It should be noted that the recurrence methods are not affected by side constraints (design variables reaching their upper and lower limits), but the linear techniques (Eq. (8.30)) are affected. These effects must be added discretely to the linear equations since the T_i's do not consider side constraints. Rewriting Eq. (8.28) gives

$$-g_j(\delta) = \sum_{i=1}^{N_1} \frac{\partial g_j}{\partial \delta_i}(\Delta \delta_i) + \sum_{i=N_1+1}^{N} \frac{\partial g_j}{\partial \delta_i}\left(\delta_i^P - \delta_i^{(k)}\right) \quad j=1,\ldots,l \qquad (8.31)$$

where δ_i^P represents an element that becomes passive during the kth iteration and N_1 is the number of active (non-passive) elements. Using Eqs. (8.27) and (8.12), Eq. (8.30) can be adjusted to incorporate passive elements and Eq. (8.31) becomes

$$rg_j(\delta) - \sum_{i=1}^{N_1} \frac{\partial g_j}{\partial \delta_i}\delta_i^{(k)} + r\sum_{i=N_1+1}^{N} \frac{\partial g_j}{\partial \delta_i}\left(\delta_i^P - \delta_i^{(k)}\right)$$

$$= \sum_{t=1}^{l} \lambda_t \left(\sum_{i=1}^{N_1} \left(\frac{\partial g_j}{\partial \delta_i} \frac{\partial g_t}{\partial \delta_i} \Big/ \frac{\partial W}{\partial \delta_i} \right) \delta_i^{(k)} \right) \quad j=1,\ldots,l \qquad (8.32)$$

Unfortunately, these passive elements are generally not known until the recursive relationships are used to find the new design variables which are then checked for violation of the upper and lower side constraints. Therefore, an iterative process must be used within the kth iteration for determining the Lagrange multipliers.

Eq. (8.32) is a general equation that can be used to find the Lagrange multipliers associated with the l active constraints. In order for these Lagrange multipliers to be valid they must satisfy Eq. (8.6) by having a non-zero, positive value. If the equations yield a negative Lagrange multiplier, this constraint should be removed from the active set according to the Kuhn–Tucker

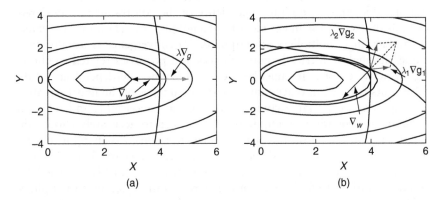

Figure 8.1 Effects of Lagrange multipliers with (a) single constraints and (b) multiple constraints.

conditions. In effect, the negative Lagrange multiplier changes the direction of the constraint gradient vector indicating that enforcing this constraint will move the solution to a higher value of the objective.

Eq. (8.8) can be explained using Fig. 8.1. Fig. 8.1a shows the necessary conditions with only one constraint and the optimal solution. Fig. 8.1b shows the necessary conditions if there are two constraints. From Fig. 8.1b it is obvious that the Lagrange multipliers are scalars that are necessary to create a vector sum of the gradients that directly opposes the gradient of the objective function. Therefore, if a negative Lagrange multiplier is found, this constraint and hence the Lagrange multiplier should be removed from the set of equations, and the new set of equations should be solved. This iterative process must continue until a set of non-zero, positive Lagrange multipliers is found. There are several different approaches to this iterative process when there is more than one negative Lagrange multiplier. Since there is interdependency of constraints within Eq. (8.32), removal of one Lagrange multiplier can change the solution, so one can choose to eliminate the negative Lagrange multipliers one at a time or one can choose to eliminate a set of negative Lagrange multipliers or one can choose to remove all of the negative Lagrange multipliers simultaneously. All methods seem to work well, and with most structural problems, removing all of the negative Lagrange multipliers at one time produces identical results to the other methods.

Each of the methods for finding the Lagrange multipliers has its advantages and disadvantages. The advantages associated with the straight-recurrence techniques are: (1) there is no need to predict the active set of constraints since Eq. (8.24) will force the Lagrange multipliers associated with passive constraints to become small, (2) very little computational effort is required, and (3) these techniques are unaffected by side constraints. The disadvantages of the recurrence techniques are: (1) the initial values of the Lagrange multipliers must be given, (2) convergence can be slow and unstable, and (3) no interdependence of the constraints is considered. The advantages of the linear-equation techniques are: (1) no initial values of the Lagrange multipliers are required, (2) convergence is usually more stable although numerically good choices and processes related to choosing the active constraints must be used, and (3) the interdependence of the constraints is taken into account. The disadvantages of the linear-equation techniques are: (1) a large computational effort to assemble and solve the simultaneous equations is required, (2) the need to accurately choose the initial set of active constraints to minimize iterations, (3) an algorithm for eliminating the equations and coefficients related to negative Lagrange multipliers (passive constraints) must be implemented, and (4) the need for iterative solutions to find the set of non-zero, positive Lagrange multipliers.

The linear-equation techniques are typically the best option. They are based on logical assumptions (expectations of the optimal solution), and capture the interdependence of the constraints with the mixed-gradient terms in Eq. (8.32). This technique has proved to be more stable and provides a logical means for assessing the set of active constraints. A possible technique could be to use the linear equations for the first few iterations in order to find the set of active constraints and a reasonable set of Lagrange multipliers, and to use these as the initial set of Lagrange multipliers for the simpler recurrence techniques.

8.5　Active constraints

Active constraints are considered to be any constraints that are close to the upper or lower limit constraint surfaces. In order to save computational effort, it is important to choose a reasonably accurate set of active constraints. A constraint that needs to be considered active but is not included in the active set when determining the Lagrange multipliers will not be controlled numerically and it could be violated with the newly generated set of design variables. This can be corrected by going back and including it in the set of active constraints or it can be included in the next iteration of active constraints.

One method of determining active constraints is to use Eqs. (8.21) and (8.22). When these equations are close to unity, the structural response is close to one of the extreme limits. One technique for choosing the initial set of active constraints and for adding constraints in subsequent iterations is to check Eqs. (8.21) and (8.22) for each constraint and set an acceptable range surrounding unity for inclusion of the constraint. These equations are one possible approach:

$$(1 - Q_1) \le \frac{u_j}{\bar{u}_j} \le (1 + Q_2) \quad j = 1, \ldots, l_1 \tag{8.33}$$

for upper-bound limits (constraints) and

$$(1-Q_1) \le \frac{u_j}{u_j} \le (1+Q_2) \quad j=1,\ldots,l_2 \tag{8.34}$$

for lower-bound limits (constraints). These equations allow the flexibility of establishing a region along the constraint surface which can be as large or as small as desired for the purpose of determining active constraints. Note that if a constraint is highly violating an upper or lower limit (non-feasible region) beyond the prescribed value Q_2, it needs to be included in the active set of constraints in order to be numerically controlled and/or satisfied. Reasonable values for Q are between 0.5 and 10 per cent or 0.05 and 0.1.

8.6 Scaling of the design

If the linear-equation approach for finding the Lagrange multipliers is to be used, a set of active constraints must be identified before the equations for finding the Lagrange multipliers can be generated. Generally, an initial set of design variables (preliminary design) will be either conservative (feasible with no active constraints, none satisfying Eqs. (8.33) or (8.34)) or non-conservative (infeasible with one or more violated constraints). Therefore, scaling is often used to adjust the preliminary design variables such that a set of active constraints as identified by Eqs. (8.33) and (8.34) will be satisfied. (In most cases, the initial set of active constraints will be small, possibly only one constraint.) For linear structures, scaling is often a one-step process, by finding a factor that is the maximum value of either of these two equations

$$f_j = \frac{u_j}{u_j} \quad j=1,\ldots,l_1 \tag{8.35a}$$

for the lower-limit constraint or

$$f_j = \frac{u_j}{u_j} \quad j=1,\ldots,l_2 \tag{8.35b}$$

for the upper-limit constraint where l_1 and l_2 are the number of possible lower- and upper-level constraints, respectively. For a stiffness matrix which is linear with respect to the design variables, the response is adjusted by a simple factor as well, $1/f_j$, which would force u_j to become either the upper or lower limit which is the definition of an active constraint.

The use of scaling for nonlinear (in terms of the design variable) stiffness and response becomes an iterative process. A simple example would be natural frequency. If the stiffness and mass are scaled by the same factor (design variables are only cross-sectional area), the frequency will not change, and if the primary design variable is the moment of inertia, the stiffness and mass will be scaled differently depending on the relationship between the cross-sectional area relative to the moment of inertia. Scaling can be problematic if displacement and frequency constraints are used together as frequency is typically affected by the direct scaling factor while the displacements are affected by the inverse of the factor which can cause oscillations in the sets of active constraints.

Scaling is important for many problems using GOC for two different reasons. The first has been described as finding the set of initial active constraints by using the maximum scaling factor described in Eqs. (8.35a) and (8.35b). The second reason is to force a subsequent design in the ith iteration back to the feasible domain (no violated constraints). It is very possible that a 'new' constraint that was not part of the original set of active constraints used in the linear equations to find the Lagrange multipliers will be violated, therefore the optimality criteria were not mathematically controlling the design relative to this 'new' constraint. Since it becomes violated, it needs to be added to the set of active constraints in the next iteration. Although scaling may not be numerically necessary, it could be used to move the design back to the feasible region giving a feasible design prior to the next iteration. As the optimization progresses, the set of active constraints typically becomes stable and often scaling is only necessary in a few iterations.

8.7   Termination criteria

Due to the iterative nature of structural optimization, in particular, GOC, the algorithm needs termination criteria. A primary condition is to check for convergence (or divergence) of the objective function. The actual validity of the optimization is measured with respect to T_i as defined in Eq. (8.11) or (8.12); unfortunately, most structural engineers would find this inappropriate, being difficult to assess relative to reality. Typically, in the latter stages of the optimization, there are small changes made to the design variables which lead to small changes in the objective function. Typically, a mathematician would consider convergence to have occurred if all of the active design variables had T_i values very close to unity, whereas a structural designer would be satisfied if the design was within a certain range of the optimal solution. Therefore, convergence can be considered by comparing the T_i to unity and the values of the objective function for successive optimization cycles to a specified percentage of change, Q_3, which can be written as

$$\frac{W^{(k-1)} - W^{(k)}}{W^{(k)}} \leq Q_3 \tag{8.36}$$

and

$$Q_4 \leq T_i \leq Q_5 \quad \text{for all } i = 1, \ldots, N_1 \tag{8.37}$$

where Q_4 and Q_5 bracket unity by a prescribed amount. If Eqs. (8.36) and (8.37) are satisfied, the algorithm is terminated. In many 'practical' cases only Eq. (8.36) is used with the assumption that a small change in the objective indicates that an optimal solution has been achieved.

8.8 Sample algorithm

All structural-optimization algorithms have four major components: the analysis, active-constraint determination and sensitivity analysis (gradient determination), numerical optimization procedure, and termination determination. Within each of these major categories there is a range of techniques and procedures that may be used as outlined in this and other chapters. The algorithm presented and shown in Fig. 8.2 is one used by Truman and Cheng et al. [3, 4, 6, 8, 9] for numerous types of problems, some of which will be described in Chapters 9 and 10. The first step is to define the structural geometry, loading, material properties and initial element sizes. The second step is to define the primary design variables. The third step is to define the optimization criteria for termination, for active-constraint determination, and the convergence-control parameter. The fourth step is to perform the analysis. Any type of analysis, static, dynamic time-history, frequency, response-spectra, or pseudo-dynamic can be used as long as the sensitivity analysis is modified to match the type of analysis. (Different analysis combinations can be used together if the structure has to satisfy simultaneously static and dynamic constraints.) The fifth step is to check whether any constraints have been compromised or are not close to being active. If either is the case, scaling may help the algorithm converge faster. (Often scaling can be eliminated from the sequence, but the solution may oscillate and not necessarily produce a feasible solution at each iteration until the set of active constraints becomes well defined.) If scaling is needed, a re-analysis (step 4) is required and may require several iterations when using a nonlinear analysis or frequency constraints as mentioned in Section 8.6. If scaling is not needed, the sixth step is to define a set of active constraints. The seventh step is to calculate the gradients of the active constraints relative to the design variables. The eighth step is to generate the linear Lagrange-multiplier equations and solve for the Lagrange multipliers. If any Lagrange multipliers are negative, it implies that the associated constraint can be removed from the active set and the equations must be resolved with the inactive-constraint-related equation and coefficients removed. (In matrix terminology, the associated row and column are removed and the equations resolved.) Once all of the Lagrange multipliers are positive, they are used in step nine to solve for the optimality criteria for each design element. These values are then used in step ten to resize the design variables. If any of the design variables violate the minimum or maximum size (side constraints), they are considered passive and the Lagrange multiplier equations must be regenerated using the

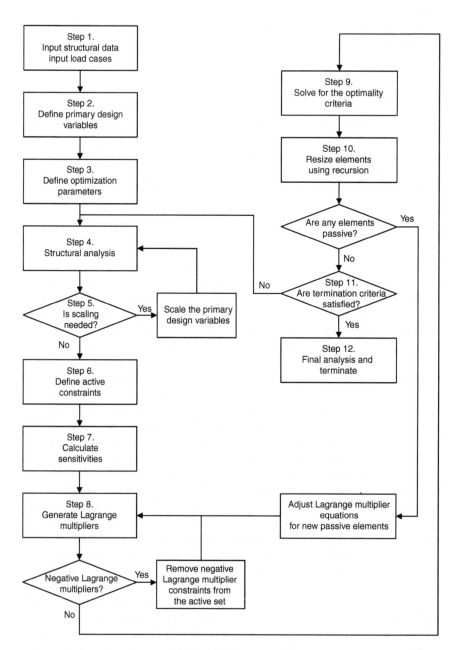

Figure 8.2 Generalized optimality-criteria algorithm flow chart.

passive-element formulation in step eight. Once again the new Lagrange multipliers must be checked for any negative values and once they are all positive, the optimality criteria are regenerated and the design variables are resized. If none of the remaining active design variables violate the side constraints, step eleven, checking the termination criteria, is performed. If the termination criteria are not satisfied, the algorithm returns the new design values to step four and a new analysis and optimization iteration begins. If the termination criteria are met, the last set of design variables is used in step twelve to perform the final analysis in order to obtain the final structural responses for the optimal design at which point the algorithm terminates.

8.9 Linking of design variables

In real systems, it is often required to have certain elements all remain the same size. For instance, practically speaking, it may be necessary to specify that all columns on the first two floors of a building be the same size (or some subset of these columns to be exactly the same). This can be handled in the optimization algorithm through a technique called linking. The linked variables can all be represented by one design variable by making these adjustments to the critical algorithmic equations. Linked variables are nothing more than a mapping of the original design variables δ on to the set Λ through this equation:

$$\{\delta\} = [R]\{\Lambda\} \tag{8.38}$$

in which $\{\Lambda\}$ is the vector of m global, linked, design variables required to represent the structure, $[R]$ is a matrix of zeros and ones which relate each δ_i to the appropriate global design variable. This translates into adjusting Eq. (8.8) to accommodate linking as

$$\frac{\partial L}{\Lambda_v} = \sum_{i=1}^{s} \frac{\partial L}{\partial \delta_i} = \sum_{i=1}^{s} \left(\frac{\partial W}{\partial \delta_i} + \sum_{j=1}^{l} \lambda_j \frac{\partial g_j}{\partial \delta_i} \right) = 0 \quad v = 1, \cdots, N_2 \tag{8.39}$$

where N_2 is the total number of active global design variables and s is the number of elemental design variables, δ_i, linked to the design variable Λ_v. The total effect of the linked variable is nothing more than a summing of the effects of each elemental design variable associated with the global design variable. (Note that if $s = 1$, a one-to-one relationship between the global and elemental design variables causes Eq. (8.39) to degenerate to Eq. (8.8) for the non-linking case.)

The need for Eq. (8.39) is to adjust the optimality criteria such that a single T_v can represent the effect of all of the linked variables simultaneously in finding the Lagrange multipliers and to create a recurrence relationship that can give a single change to the set of linked variables. Therefore, the optimality criteria T_v becomes

$$T_v = -\frac{\sum_{i=1}^{s}\sum_{j=1}^{l}\lambda_j\frac{\partial g_j}{\partial \delta_i}}{\sum_{i=1}^{s}\frac{\partial W}{\partial \delta_i}} = 0 \quad v = 1,\ldots,N_2 \tag{8.40}$$

Including linking and side constraints coupled with Eqs. (8.28) and (8.40), the linear equation for finding the Lagrange multipliers becomes

$$rg_j - \sum_{i=1}^{N_1}\frac{\partial g_j}{\partial \delta_i}\delta_i^{(k)} + r\sum_{i=N_1}^{N}\frac{\partial g_j}{\partial \delta_i}\left(\delta_i^P - \delta_i^{(k)}\right)$$

$$= \sum_{c=1}^{l}\lambda_c\left(\sum_{i=1}^{N_2}\left(\sum_{q=1}^{s}\left(\frac{\partial g_j}{\partial \delta_q}\frac{\partial g_c}{\partial \delta_q}\right)\bigg/\sum_{q=1}^{s}\frac{\partial W}{\partial \delta_q}\right)_i \Lambda_i^{(k)}\right) \quad j = 1,\ldots,l \tag{8.41}$$

which is the most general equation for finding the Lagrange multipliers and becomes Eq. (8.32) if s is one, or no linking of a variable occurs. The subscript i for the parenthetical expression refers to the fact that only those gradients associated with Λ_i are included in the summation over q.

The concept of linking is very important if realistic designs are to be achieved using optimization techniques. Without linking each element can take on its own value providing an optimal, yet often impractical, solution. Linked solutions will always provide less optimal but more practical solutions.

8.10   Sensitivity analyses

As seen by Eqs. (8.40) and (8.41), an important aspect of the optimization procedure is determining accurate gradients of the constraints with respect to the design variables. This portion of the algorithm, step seven, in Fig. 8.2, is typically the most time-consuming portion of the numerical optimization.

8.10.1   *Static displacement gradients*

The gradients for the static displacements are easily derived by taking the derivative of the stiffness formulation $[K]\{u\} = \{P\}$ and algebraically rearranging the terms. The equation becomes

$$\frac{\partial \{u\}}{\partial \delta_i} = \frac{\partial \{P\}}{\partial \delta_i} - [K]^{-1}\frac{\partial [K]}{\partial \delta_i}\{u\} \tag{8.42}$$

where $\{P\}$ is the vector of loads, $[K]$ is the stiffness matrix, and $\{u\}$ is the vector of displacements. In most static cases, the structural load is independent of the element sizes (if the weight of the structure is negligible) making the first term zero. Also note that the partial derivative of the stiffness matrix typically provides a sparse matrix for large structures since each element can only affect a limited number of internal degrees of freedom.

8.10.2 Static stress gradients

The gradients for stress can be related to the displacement gradients through a technique that has been referred to as a virtual-load technique. This technique is based on the premise that the stresses can be written as a linear combination of the structural displacements. In equation form this becomes

$$\sigma_j = \{b\}_j^T \{u\} \tag{8.43}$$

giving a stress gradient of

$$\frac{\partial \sigma_j}{\partial \delta_i} = \frac{\partial \{b\}_j^T}{\partial \delta_i} \{u\} + \{b\}_j^T \frac{\partial \{u\}}{\partial \delta_i} \tag{8.44}$$

Once the vector b for the jth stress component has been defined, the gradients for the vector b are usually very straightforward and the gradients for the displacements are given in Eq. (8.42). For a truss or brace element, b is

$$\{b\}_j^T = \left(\frac{E}{L} \quad \frac{-E}{L} \right) \tag{8.45}$$

relative to the two end displacements of the member (the actual vector is filled with zeros for the other degrees of freedom), L is the length of the brace and E is the modulus of elasticity. The b vector must be developed for each type of element. These vectors can be found in references [2, 4, 6]. In many cases $\{b\}$ will be independent of the design variable and the first term in Eq. (8.44) will be zero. (If the area of the brace is the design variable used to describe each brace, Eq. (8.45) is independent of the area, leaving only the second term of Eq. (8.44) and therefore only requires multiplying the vector b with the gradient of the displacements.)

8.10.3 Natural frequency gradients

Frequency constraint gradients are found by direct differentiation of the free-vibration equation used to find the natural frequencies [1, 3–5, 8, 9, 12].

$$[[K] - \omega_j^2 [M]] \{\phi\}_j = 0 \tag{8.46}$$

where $[M]$ is the mass matrix, ω_j is the jth natural frequency, and $\{\phi\}_j$ is the jth mode shape. Taking the derivative of this equation, pre multiplying by $\{\phi\}_j^T$ and rearranging the terms algebraically gives

$$\frac{\partial \omega_j^2}{\partial \delta_i} = \frac{\{\phi\}_j^T \left[\dfrac{\partial [K]}{\partial \delta_i} - \omega_j^2 \dfrac{\partial [M]}{\partial \delta_i} \right] \{\phi\}_j}{\{\phi\}_j^T [M] \{\phi\}_j} = 0 \qquad (8.47)$$

Although not derived here, a very similar expression can be derived for buckling loads since they too are governed by a similar eigenvector-solution process [1, 5, 10].

8.10.4 Mode shape gradients

For many of the dynamic responses found using modal superposition or response-spectrum approaches, the gradient of the modes will be required. The gradients of the mode shape can be found by direct differentiation of Eq. (8.46) and rearranging the terms algebraically gives

$$\frac{\partial \{\phi\}_j}{\partial \delta_i} = - \left[[K] - \omega_j^2 [M] \right]^{-1} \left[\frac{\partial [K]}{\partial \delta_i} - \frac{\partial \omega_j^2}{\delta_i} [M] - \omega_j^2 \frac{\partial [M]}{\partial \delta_i} \right] \{\phi\}_j \qquad (8.48)$$

The problem with this equation is that the matrix to be inverted is singular and therefore cannot be used to solve for the gradient. This can easily be overcome by understanding that the terms within the eigenvector are only multiples of one another. Therefore, forcing the change in one term of the gradient of the eigenvector to be zero is the same as always normalizing (setting to unity) the eigenvector relative to this same component in each iteration. This provides a boundary condition for elimination of a row and column of the matrix to be inverted which then becomes non-singular and can be inverted giving a solution of this form:

$$\frac{\partial \{\phi\}_j^T}{\partial \delta_i} = \left\{ \frac{\partial \phi_{1j}}{\partial \delta_i}, \ldots, 0, \ldots, \frac{\partial \phi_{nj}}{\partial \delta_i} \right\} \qquad (8.49)$$

where the zero term reflects the component associated with the degree of freedom which has the normalized component. If the eigenvector is normalized with respect to the mass or some other quantity, the change in the jth component will not be zero from one iteration to the next and a correction term as defined in reference [4] must be applied to the solution. Solving for these gradients can require large computational effort, but if each eigenvector is normalized to the same component, the inverse need only be calculated one time and utilized for each design variable for which the gradients are needed.

8.10.5 Dynamic and pseudo-dynamic displacement and stress gradients

These gradients are fairly straightforward once the displacement, frequency and mode shape gradients have been defined. Depending on how the dynamics' problems are solved and loaded, the formulations are different. Therefore, the appropriate gradients will be shown in the specific applications given in Chapter 9.

8.11 Illustrative examples

8.11.1 Three-bar truss

In order to understand the algorithm presented in Fig. 8.2, a simple three-bar truss as shown in Fig. 8.3 is to be optimized for its weight subject to one constraint. Therefore, the objective function, structural weight, can be written as

$$\text{Minimize} \quad W(A_i) = \gamma \, (A_1 L_1 + A_2 L_2 + A_3 L_3) \tag{a}$$

$$\text{Subject to} \quad g(A_i) = u_1 - 0.20 \, \text{in.} \tag{b}$$

where u_1 is the horizontal deflection and the elements are numbered from left to right, respectively. The optimization was performed using a convergence-control parameter of two and no scaling. (If scaling had been

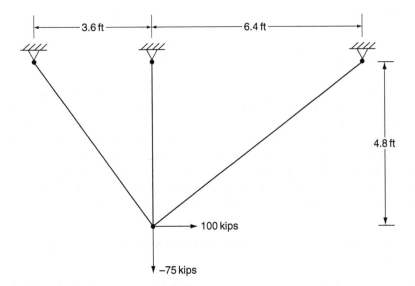

Figure 8.3 Three-bar truss example.

Table 8.1 Optimality-criteria history for a three-bar truss (1 in. = 2.54 cm, 1 k = 4.45 kN).

Cycle	A_1 (in.^{2})	A_2 (in.^{2})	A_3 (in.^{2})	W (k)	u (in.)	λ	T_1	T_2	T_3
0	1.440	2.000	2.000	116.51	0.16325	261.78	0.7107	0.0046	0.3984
1	1.232	1.005	1.398	79.62	0.21006	298.24	1.3462	−0.0497	0.7777
2	1.445	0.477	1.243	71.14	0.19936	286.79	1.1265	−0.1892	0.7511
3	1.536	0.194	1.088	64.15	0.19968	287.54	1.0526	−0.3670	0.8627
4	1.577	0.061	1.014	60.79	0.19986	292.02	1.0089	−0.5121	0.9567
5	1.584	0.015	0.992	59.57	0.19995	294.74	0.9989	−0.5806	0.9919
6	1.583	0.003	0.988	59.25	0.19999	295.61	0.9995	−0.6004	0.9987
7	1.583	0.001	0.990	59.19	0.19999	295.80	0.9999	−0.6047	0.9997
8	1.582	0.000	0.987	59.17	0.20000	295.86	1.0000	−0.6103	1.000

used, the number of iterations would have been reduced by at least three cycles.)

The history for the optimization is shown in Table 8.1. Note that the optimality criteria reach unity for the left and right bars, but do not reach unity for the centre bar. The optimality criteria are trying to eliminate this member (make it passive at its minimum value of 0) since it is not required to constrain the horizontal deflection. (Adding a second constraint for the vertical deflection could cause the centre bar to become active in the solution.) This is reflected in the gradients of the displacements as well. The small gradient for u_1 with respect to A_2 (−0.0003) indicates that this element has a minor effect on the horizontal deflection. In reality this should be zero, but numerically due to rounding it takes on a very small value.

$$\left[\frac{\partial \{u\}}{\partial A_1} \quad \frac{\partial \{u\}}{\partial A_2} \quad \frac{\partial \{u\}}{\partial A_3} \right] = \left[\begin{array}{ccc} -0.5543 & -0.0003 & -0.0414 \\ 0.0277 & 0.0147 & -0.0114 \end{array} \right] \qquad \text{(c)}$$

Note that the displacements in Table 8.1 reach the constraint value (active constraint) and that all of the solutions with the exception of the first cycle are feasible solutions with cycle eight producing the least weight. If the design had been scaled in the initial sizing, the first cycle would have produced a solution that was closer to the constraint value. Also note the convergence of the Lagrange multiplier, although not as stable as the optimality criteria or the weight, it quickly settles into a value near 295. The solution was terminated when the weight changed less than 0.05 per cent.

8.11.2 Two-dimensional frame; primary versus secondary design variables

The simple frame, shown in Fig. 8.4, consisting of two bending elements with rectangular cross-sections and a brace, is used to illustrate primary versus secondary design variables and the use of multiple constraints. Since

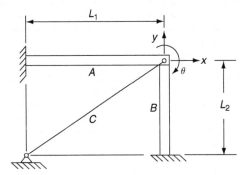

Figure 8.4 Two-dimensional frame with two bending elements and one brace.

the objective function, the weight, is linear in terms of the cross-sectional areas, and the dominant stiffness parameters for the bending elements are linear with respect to the major-axis moments of inertia, either of these quantities could be used as the primary design variables for the bending elements, whereas the area is the only logical choice for the brace.

The frame consists of two rectangular cross-sectional beam-column elements and a brace. All elements were made of steel with a modulus of elasticity of 30,000 ksi (20,700 kN cm^{-2}). The beam-column cross-sections were forced to have a depth-to-width ratio of 2.0 which provides the major-axis moment of inertia, I_{xi}, in terms of the cross-sectional area, A_i, as

$$I_{xi} = \frac{1}{6}A_i^2 \tag{a}$$

The use of this explicit relationship allows a member to be represented by one variable, either I_x or A, which will be called the primary design variable. The other variable is termed the secondary design variable. (In Chapter 9, three-dimensional systems will be developed that require multiple secondary-design variables that must be related to a primary-design variable.) The lengths L_1 and L_2 are 15.0 ft (4.58 m). Each element starts with an initial area of 30.0 in.^{2} (194 cm^2). The analysis is considered similar to a steady-state problem such that the load vector is 50.0 kip (2,223 kN) in the x-direction, -100 kip (445 kN) in the y-direction, and 1,000 k-in (113 kN-m) in the θ-direction. It is similar to a steady-state situation in that the solution was based upon this equation:

$$[[K] - \omega^2 [M]] \{u\} = \{P\} \tag{b}$$

where $[K]$ is the stiffness matrix, ω is the frequency of the applied load, $[M]$ is the mass matrix, $\{u\}$ is the displacement vector, and $\{P\}$ is the constant-load vector. The fact that a constant-load vector was used creates a situation which is not purely steady state. The value for ω was 100 rad s^{-1}.

Table 8.2 Results for a two-dimensional frame using different primary-design variables after three cycles of optimization (1 in. = 2.54 cm, 1 k = 4.45 kN).

	Primary design variables			
	2-cycle results		3-cycle results	
	Case 1 – Area	Case 2 – Moment of inertia	Case 1 – Area	Case 2 – Moment of inertia
A_A (in.^{2})	26.1	25.8	25.5	25.2
A_B (in.^{2})	26.3	26.1	27.8	27.4
A_C (in.^{2})	13.1	11.7	6.7	5.9
Wt. (k)	3.62	3.52	3.21	3.11
No. of scalings	5	2	6	2
No. of analyses	7	4	9	5

Each element was constrained to have its stress below 22 ksi (15 kN cm^{-2}); while the bending elements were to have cross-sectional areas above 0.1 in.^{2} (0.6 cm^2) and the brace was to have a cross-sectional area above 2.0 in.^{2} (12.9 cm^2). The x-displacement was to remain below 0.020 in. (0.05 cm), and the y-displacement was to remain below 0.026 in. (0.066 cm). Small displacement constraints were chosen in order to have the displacements close to their active values concurrently with the stress constraints.

The results are presented in Table 8.2. The results in Table 8.2 correspond to results after two and three cycles of optimization. The optimization problem was solved using two different sets of primary-design variables. In case 1 the area was used as the primary-design variable for each element, and in case 2 the moments of inertia were used as the primary-design variable for the bending elements and the area was used for the brace. Each design was controlled by the stress in element B. Within the first cycle of each problem the stress in each bending element was within 95 per cent of the active value, therefore, both of these stresses were considered as active constraints. In each case the Lagrange multipliers for the stress in element A was negative (approximately -4.0×10^{-3}) while the Lagrange multiplier for the stress in element B was positive (approximately 4.2×10^{-3}). Since the constraint for element A provided a negative Lagrange multiplier, it was dropped from the active set of constraints, and the Lagrange-multiplier equations were resolved providing a value for the remaining constraint, the stress in element B, of approximately 1.7×10^{-4}. Note that there is considerable difference in the values for the Lagrange multipliers after one constraint is removed which supports the notion that constraint interaction in these equations is important. The negative Lagrange multiplier indicates that by controlling the stress in element A, the weight (objective function) would have increased and therefore should be removed from the possibilities for active constraints.

GOC approaches are sensitive to the set of constraints chosen to be active. Making the correct choices for this active set is still an area being studied by the optimality-criteria-based structural-optimization researchers.

Using the major-axis moment of inertia for the bending elements and the cross-sectional area for the brace provides the lowest weight with the least amount of computational effort. The cross-sectional area produced similar results to those produced by the major-axis moment of inertia problem except that the number of analyses required was considerably larger. As seen from Table 8.2, this is due to the number of scalings required to reach a set of active constraints. A constraint is not considered an active constraint unless the response is 'close' to the limiting value. The large number of scalings is due to the nonlinear factoring associated with the moment of inertia when using the areas as the primary-design variable. From Eq. (a) it is seen that a linear scaling of the area provides a moment of inertia factor which is one-sixth of the square of the area. Therefore, it is seen that the major-axis moment of inertia is the best choice for the primary-design variable for the bending elements, and the cross-sectional area is the best choice for brace elements.

8.11.3 A three-dimensional frame illustrated step-by-step

A one-storey, one-bay, three-dimensional frame, as shown in Fig. 8.5, will be used to illustrate the steps used to develop the overall understanding of the algorithm. An in-depth look at the first cycle will be presented along with the results of the entire design. The specifics of the element types and the gradient calculations are presented in detail in Chapter 9. This example is only used to clarify the steps within the algorithm for static stress and displacement constraints.

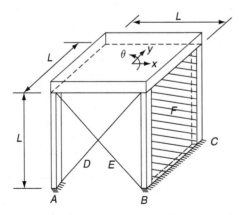

Figure 8.5 Three-dimensional system with beam-columns, braces and panels.

The three-dimensional structure is being used to show how systems with mixed element types and mixed constraints can be optimized using the algorithm presented in Section 8.8. The frame consists of a square, rigid-slab which was 10.0 ft by 10.0 ft (3.05 m by 3.05 m) which was supported by three 10.0 ft (3.05 m) rectangular, steel columns (A, B and C). The rectangular cross-sections are assumed to have a depth-to-width ratio of 1.5 (which can be used as in the previous example to create a direct relationship between the area and the moment of inertia), and the concrete panel is assumed to have a fixed depth which spans between columns B and C leaving the thickness as the primary-design variable. The column depths are parallel to the y-direction. Steel X-bracings (D and E) are in one vertical plane as shown in Fig. 8.5 and use their areas as their primary-design variables. The concrete panel does not include steel in this example. The loading consists of 300 kips (1,335 kN) in the x-direction, 100 kips (445 kN) in the y-direction, and 3,000 kip-in (339 kN-m) in the θ-direction.

The constraints consist of static displacement and stress constraints. The maximum allowable deflection is 0.5 in. (1.27 cm) for both the x- and y-directions. The maximum allowable stresses are 30 ksi (20.7 kN cm^{-2}) for the steel elements, and 3.0 ksi (2.1 kN cm^{-2}) for the concrete panel. Three side constraints are used: 865,000 in.^{4} (0.359 m^4) is the lower limit for the concrete panel's moment of inertia, 35.0 in.^{4} (1,451 cm^4) is the lower limit for the steel columns, and 5.0 in.^{2} (32.3 cm^2) is the lower limit for the steel brace areas. Moduli of elasticity of 30,000 ksi (20,700 kN cm^{-2}) and 3,000 ksi (2,070 kN cm^{-2}) were used for the steel and concrete elements, respectively. The final results of each iteration are shown in Table 8.3.

Table 8.3 Results for the one-bay, one-storey, structural system with stress constraints (1 in. = 2.54 cm, 1 k = 4.45 kN).

Cycle	A(in.^{4})	B(in.^{4})	C(in.^{4})	D, E(in.^{2})	F(in.^{4})	Weight (k)
0	1,500	1,500	1,500	100	2,000,000	—
0[a]	1,418	1,418	1,418	95	1,891,180	36.39
1[b]	1,683	788	1,205	50	910,334	22.72
2	2,010	437	988	27	921,852	19.99
3	2,492	233	722	15	958,700	18.79
4	2,988	128	604	12	926,560	17.87
5	3,344	67	438	10.6	893,500	17.14
6	3,553	36	295	10.5	877,810	16.59
7	3,688	35	188	10.4	872,890	16.32
8	3,771	35	113	10.4	868,810	16.06
9	3,811	35	65	10.5	864,000	15.80

[a] These values are found by scaling the initial values by 0.947.
[b] Element F has become passive in this step, but takes on the value of 910,334 due to a scaling factor of 1.053. (Scaling is performed after each iteration, but is relatively close to 1.0 in. each cycle.)

This example has two exceptions with respect to the algorithm presented in Section 8.8. A pseudo-scaling was used. A scaling based upon the cross-sections and constraints was developed which used the largest ratio of the actual response to the constraint limit raised to the 4/3 power. This type of power scaling can be useful in well-defined problems where the relationships between the primary-design variables and the constraints can be shown to be nonlinear. In most problems, especially in two-dimensions, a standard scaling using a power of 1.0 is adequate. Secondly, a scaling was performed at the end of each optimization cycle in order to be very close to the constraint value for the beginning of the next iteration. Typically, a scaling is not necessary except when the constraints violate the upper or lower limit or when there is no violation, and all constraints are below the lower active limits by a significant amount.

The step-by-step procedure is described, with these two exceptions from the algorithm presented in Section 8.8, for the first cycle as:

Step 1 – The initial values for the element sizes were chosen as 1,500 in.^{4} (62,400 cm^4) for elements A, B and C, 100 in.^{2} (645 cm^2) for elements D and E, and 2,000,000 in.^{4} (0.832 m^4) for the concrete panel.

Step 2 – The static displacements were determined by using a stiffness-based structural analysis as $x = 0.19$ in. (0.49 cm), $y = 0.17$ in. (0143 m), and $\theta = -0.003$ rad.

Step 3 – The maximum normal stress was determined for each element as $\sigma_A = 28.8$ ksi (19.9 kN cm^{-2}), $\sigma_B = 0.76$ ksi (0.52 kN cm^{-2}), $\sigma_C = 19.4$ ksi (13.4 kN cm^{-2}), $\sigma_{D,E} = 1.74$ ksi (1.20 kN cm^{-2}), and $\sigma_F = -0.02$ ksi (0.014 kN cm^{-2}).

Step 4 – The largest ratio of structural response to their respective constraint value was determined (it was the stress in element A of 28.8 ksi relative to the maximum value of 30 ksi). This ratio was then used to find the scaling factor as:

$$f = \left(\frac{28.8}{30}\right)^{\frac{4}{3}} = 0.947 \tag{a}$$

In the normal algorithm this factor would have been taken as the ratio raised to a power of 1. If this were used, the solution would be slightly different.

Step 5 – The initial sizes were adjusted by the scaling factor of 0.947. A lower limit of 95 per cent of the constraint limit was set as the acceptable limit. Within the current algorithm the ratio within the parenthesis of the scaling factor would have provided an acceptable factor (above .95) and a scaling would not have taken place. The new values are 1,418 in.^{4} (59,000 cm^4) for elements A, B and C, 94.7 in.^{2} (610 cm^2) for elements D and E, 1,891,180 in.⁴ (0.787 m⁴) for element F. Using these values and the relationship between the major-axis moment of inertia, the areas, and the minor-axis moment of inertia, the secondary-design variables were determined from the new primary-design variables.

Step 6 – Solve the new problem for the new displacements. None of the displacements are close to the active value. Therefore, no displacement constraints are considered potentially active.

Step 7 – Solve the new problem for the new stresses. Only the stress for element A has a stress which is near the active value. Element A had a maximum stress of 30.0 ksi (20.7 kN cm⁻²).

Step 8 – All constraints are checked for violation (was the scaling factor above $(1 + Q_2 = 1.05)$?). None of the constraints were violated. All constraints were then checked to see if any constraints were active (was the scaling factor between $(1 - Q_1 = 0.95)$ and $(1 + Q_2 = 1.05)$?). The stress for element A was the only constraint within this range and was chosen as an active constraint.

Step 9 – The gradient of the stress for element A was determined using the virtual-load technique which will be described thoroughly in Chapter 9, but is mathematically described here as:

$$\frac{\partial \sigma_A}{\partial \delta_i} = -\{v\}_A^T \frac{\partial [K]}{\partial \delta_i} \{u\} \quad i = A, \ldots, F \tag{b}$$

where δ_i represents the primary-design variable for each element A through F, $\{u\}$ is the vector of displacements, and $\{v\}_A^T$ is a vector of virtual displacements determined from

$$\{v\}_A = [K]^{-1} \{b\}_A \tag{c}$$

where $\sigma_A = \{b_A^T\}\{u\}$, significant details are provided in Chapter 9 on how to find the vectors, $\{b\}$, for each type of constraint and element type.

Step 10 – The linear equation for the determination of the Lagrange multiplier was developed. Since there is only one active constraint, there is only one Lagrange multiplier to be found and one equation to be formed. The Lagrange multiplier was found to be 0.118. If more than one active constraint was found, a set of Lagrange multipliers would be found. All of the Lagrange multipliers that were not positive would be removed from the active set. Then the new set of equations would be resolved, and this process would iteratively continue until all of the Lagrange multipliers were positive.

Step 11 – The optimality criteria are determined for each element as shown in Eq. (8.12). The optimality criteria for this cycle were $T_A = 1.25$, $T_B = 0.06$, $T_C = 0.61$, $T_{D,E} = 0.006$, and T_F was a negative quantity which was nearly zero. An optimal solution is obtained when all active elements (elements not at their upper or lower limits (passive elements)) have optimality criteria near unity. From the optimality criteria, small and negative values, it is seen that elements D, E and F are trying to rapidly reduce their size.

Step 12 – The optimality criteria are used within the linear recurrence relationship given as

$$\delta_i^{(k+1)} = \left(1 + \frac{1}{r}(T_i - 1)\right)\delta_i^{(k)} \quad i = A,\ldots,F \qquad (d)$$

with a convergence-control parameter, r, of 2.0 to produce the second-cycle element sizes. The new sizes are $I_{xA} = 1,598$ in.^{4} (66,500 cm^4), $I_{xB} = 748.8$ in.^{4} (31,100 cm^4), $I_{xC} = 1,143$ in.^{4} (47,600 cm^4), $A_{D,E} = 47.55$ in.^{2} (307 cm^2), and $I_{xF} = 864,000$ in.^{4} (0.360 m^4). Since the optimality criterion for element F was negative, element F was reduced to its lower limit (passive value) without the use of the recurrence relationship.

Step 13 – The new structural responses are determined and checked for constraint violation. If violation occurs, the design is scaled. If there is no violation, the termination criteria of percentage weight change, the number of optimization cycles and number of analyses are checked. The percentage weight change for this cycle was approximately 37.5 per cent. This is largely due to the large decrease in size associated with the concrete panel. If none of the termination criteria are satisfied, the problem is shifted back to step 4 if scaling is necessary. If scaling is unnecessary, the algorithm goes to step 6 for the next iteration.

The final results are shown in Table 8.3. This problem was terminated due to the number of cycles of optimization which had a maximum of ten cycles. Realistically, this problem could have been terminated earlier due to the small change in weight after the sixth or seventh cycle. Note that elements B and F reached their passive values and the braces stabilized at a value of 10.5 in.^{2} (67.7 cm^2). Since only the stress for element A remained an active constraint throughout this problem, the global optimum result would most likely force many of the elements except A to become passive.

This problem is only used to provide insight into the GOC algorithm presented in Section 8.8. This algorithm has tremendous flexibility to be customized relative to scaling, active-constraint selection, convergence-control parameters, explicit or implicit or approximate gradient calculations. This flexibility allows one to customize the algorithm to the problem to be solved in order to be efficient both in computation as well as in cycles of optimization.

8.12   Concluding remarks

The algorithm presented in Fig. 8.2 has been used to solve numerous types of structural problems [3, 4, 6, 8, 9]. Within this algorithm, the different techniques for recursion, scaling, Lagrange-multiplier determination, and termination can be used in their respective steps to best suit the problem to be solved. This has always been the advantage of GOC approaches: the flexibility to customize the process and gain the related numerical benefits of such. The authors have had great success in using the linear recurrence relationship Eq. (8.17), the linear-equation technique for solving for the Lagrange multipliers (with side constraints and linking) Eq. (8.41) coupled with accurate gradient calculations to solve these problems.

Chapters 9 and 10 will provide examples of the many different problems that have been solved using this algorithm based on GOC. Chapter 9 will provide additional information used in the calculation of displacement, stress and frequency constraint gradients coupled with illustrative examples. Chapter 10 will provide an assortment of examples that reflect the versatility of the GOC approach. It is important that the algorithms have exact or good approximations to the gradients in order for the optimization algorithm to maintain the constraint values while reducing the objective function from one cycle to the next. If these constraint values are not maintained, scalings must be performed between each cycle, which can significantly increase the total number of analyses and computational effort. As illustrated in the provided examples in Chapters 8, 9 and 10, it will be seen how efficient this approach can be for large-scale systems with many design variables, load cases and constraints.

References

1. Arora, J.S., *Introduction to Optimum Design*, McGraw-Hill, New York, 1989.
2. Bazaraa, M.S. and Shetty, C.M., *Nonlinear Programming: Theory and Algorithms*, John Wiley & Sons, New York, 1979.
3. Cheng, F.Y. and Srifuengfung, D., Earthquake structural design based on optimality criterion, in *Proceedings of the 6th World Conference on Earthquake Engineering*, Vol. 5, New Delhi, India, 1977.
4. Cheng, F.Y. and Truman, K.Z., *Optimal Design of 3-D Reinforced Concrete and Steel Buildings Subjected to Static and Seismic Loads Including Code Provisions*, Structural Series 85–20, University of Missouri-Rolla, (Final Report for NSF), Rolla, MO, 1985.
5. Haug, E.J. and Arora, J.S., *Applied Optimal Design*, John Wiley and Sons, New York, 1979.
6. Hoback, A.S. and Truman, K.Z., A new method for finding global and discrete optimums for structural systems, *Computers and Structures*, Vol. 52, No. 1, pp. 127–134, 1994.
7. Khot, N.S., Generalized optimality criteria, *Foundations of Structural Optimization: A Unified Approach*, Morris, A.J. (ed.), John Wiley & Sons, New York, 1982.

8. Truman, K.Z. and Cheng, F.Y., How to optimization for seismic loads, Chapter 12, *Guide to Structural Optimization*, Arora, J.S. (ed.), ASCE Manuals and Reports on Engineering Practice, No.90, New York, pp. 237–262, 1997.

9. Truman, K.Z. and Cheng, F.Y., Optimum assessment of irregular 3-D seismic buildings, *Journal of Structural Engineering*, ASCE, Vol. 116, No. 12, pp. 3324–3337, 1990.

10. Vanderplaats, G.N., *Numerical Optimization Techniques for Engineering Design: With Applications*, McGraw-Hill, New York, 1984.

11. Venkayya, V.B., Khot, N.S. and Berke, L., Application of optimality criteria approaches to automated design of large practical structures, in *Proceedings for the 2nd Symposium on Structural Optimization, AGARD*, No. 123, Milan, Italy, 1973.

12. Powell, M.J.D., Algorithms for nonlinear constraints that use Lagrangian functions, *Mathematical Programming*, Vol. 14, pp. 224–248, 1978.

9 Generalized optimality criteria applied to statically and dynamically loaded structural systems

9.1 Introduction

Many different optimization approaches have been and can be used to optimize structural systems subjected to dynamic loads. This chapter focuses on the use of generalized optimality criteria (GOC) as a means of optimizing both three-dimensional and two-dimensional structural systems subject to static and dynamic, essentially seismic, loads. The seismic loads in this chapter are reflected in code-based static equivalent (lateral) loads, code-based modal loads, response-spectra-based loads and time-history accelerograms. The optimality-criteria approach presented is an amalgamation of work performed by Truman and Cheng *et al.* [1–7] and closely parallels algorithms independently developed by Grierson and Chan [8, 9]. Many others have provided algorithms that optimize structural systems with respect to dynamic loads, but most use mathematical programming or genetic algorithm approaches instead of generalized optimality-criteria techniques [10–18].

Several of the examples use a (computer) program developed by Truman and Cheng, ODRESB-3D, which is based on the use of three-dimensional elements and structures, but it can also be used to simulate two-dimensional systems [1]. In the structural optimization of three-dimensional systems, there are two common approaches used for building systems: (1) each geometric property for every element is used as its own independent design variable, and (2) a primary design variable (geometric property such as the major-axis moment of inertia) is used for each structural element with developed relationships between the primary design variable and the secondary design variables (remaining geometric properties such as area, minor-axis moment of inertia, torsional moment of inertia and section moduli). These are not the only approaches but they are two of the most common and most other approaches are a combination of these two approaches. An original three-dimensional, structural-optimization method for steel and concrete buildings based on optimality criteria was developed by Cheng and Truman *et al.* in the early 1980s [1–7]. At basically the same time a similar approach was being developed by Grierson and Chang (1983–1990) for three-dimensional steel structures [8, 9]. Both approaches used optimality criteria, Cheng's work resulted in a public domain (NSF-sponsored) program called ODRESB-3D (Optimal Design of 3-Dimensional Reinforced Concrete and Steel Buildings) [1] and Grierson's work resulted in a commercial program called SODA (Structural Optimization, Design and Analysis) [8]. The primary structure of the optimization algorithms for both programs is based on the theory presented in Chapter 8. This chapter will provide detail in the application of the theory from Chapter 8 in order to apply it to structures subjected to static and seismic loadings. The primary enhancements from Chapter 8 are in the development of the structural elements, the constraint gradients, and the application for different dynamic and pseudo-dynamic loads.

9.2 ODRESB-3D optimal design of 3-dimensional reinforced concrete and steel buildings

The internal workings and several of the examples generated by ODRESB-3D will be presented in this chapter along with the theory from Terlaje and Truman [19, 20] which was modified to include the International Building Code loads as well as the theory and results from other two-dimensional frames as developed by Jan [6] and Petruska [7]. ODRESB-3D used five different types of structural elements while the recent two-dimensional programs use traditional beam or beam-column elements within their structures. Each type of element in ODRESB-3D is characterized by its local degrees of freedom, primary and secondary design variables, and construction material. The five types of elements can be classified as steel beam-columns, beams, braces as well as reinforced concrete beam-columns (shear walls) and flexural panels. These elements were developed originally for a public domain program called INRESB-3D developed by Cheng and Kitipitayangkul [21]. This approach was used in order to provide a simple, quick-analysis approach for three-dimensional mixed-building systems with rigid in-plane yet flexible out-of-plane diaphragms. The presentation of ODRESB-3D analysis procedures and elements provides the necessary insight for the customization of the optimality-criteria approach to any analysis and elemental system.

9.2.1 Steel elements

The steel element cross-sections can be of any shape: rectangular, tubular, I-shape, etc. as long as the element has secondary design variables that can be geometrically (analytically) or statistically linked to the primary design variable.

The ODRESB-3D beam-columns have twelve degrees of freedom and are represented by six geometric properties. Each element has three translational and three rotational degrees of freedom at each node as shown in

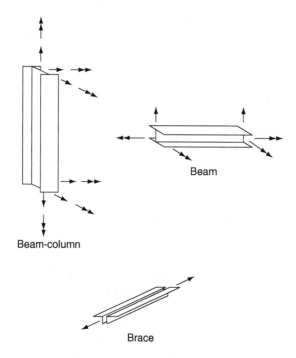

Beam-column

Beam

Brace

Figure 9.1 Steel elemental names and degrees of freedom.

Fig. 9.1. Each element requires these geometric properties: the major-axis, minor-axis and torsional moment of inertias, the major-axis and minor-axis section moduli and the cross-sectional area. The section moduli are only necessary if stress constraints are used in the optimization of the building system. The major-axis moment of inertia is used as the primary design variable with the other five variables being represented as secondary design variables.

The beams use six degrees of freedom. Each element has one degree of translation and two degrees of rotation at each node as shown in Fig. 9.1. These degrees of freedom are consistent with the notion that the floor diaphragms are rigid in their plane and flexible out-of-plane. Therefore, the analysis requires each beam to be represented by three geometric properties: the major-axis moment of inertia, major-axis section modulus (if stress constraints are used) and the torsional moment of inertia. As in the case of the beam-columns, the major-axis moment of inertia is used as the primary design variable and the remaining two properties are treated as secondary design variables.

The steel braces have two degrees of freedom. Each elemental node is allowed to displace along the member axis as shown in Fig. 9.1. Therefore, the cross-sectional area is the only necessary geometric property required to represent a brace which becomes the primary design variable.

9.2.2 Reinforced concrete elements

The two different concrete elements, beam-columns and panels (walls), are developed based on the following assumptions. The elements must be rectangular with a fixed depth, h. The reinforcing steel must be equally distributed along the major and minor axes with the amount of steel based upon the chosen value of ρ, the percentage of steel area per the gross cross-sectional area. Also the cracking depth is based upon the theory of elastic systems or working stress for bending about a single axis.

Both the concrete panels and the beam-columns use the same working stress theory in order to determine their cross-sectional properties based on a cracked section where no tension stresses are allowed. The panels have six degrees of freedom while the beam-columns use twelve degrees of freedom as shown in Fig. 9.2. Each corner of a panel is allowed to translate in the vertical direction, while the upper and lower faces of the panel are allowed to displace in the horizontal direction as shown in Fig. 9.2. This requires that each panel be represented by three geometric properties: the major-axis moment of inertia, the major-axis section modulus (stress constraints only), and the cross-sectional area. The reinforced-concrete beam-columns have the same degrees of freedom as the steel beam-columns and require the same six geometric properties in order to represent the element.

The working stress model is based on the transformed cross-sections as shown in Fig. 9.3. The transformed cross-sectional properties can be derived as:

$$I_x = \frac{1}{3}b(kd)^3 + (n-1)A_s\left(kd - d'\right)^2 + nA_s\left(d - kd\right)^2 \tag{9.1}$$

$$I_y = \frac{1}{3}h(kb')^3 + (n-1)A_s\left(kb' - b''\right)^2 + nA_s\left(b' - kd'\right)^2 \tag{9.2}$$

$$A_t = b(kd) + (n-1)A_s + nA_s \tag{9.3}$$

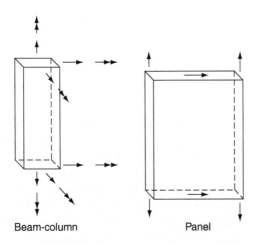

Beam-column   Panel

Figure 9.2 Reinforced concrete element names and degrees of freedom.

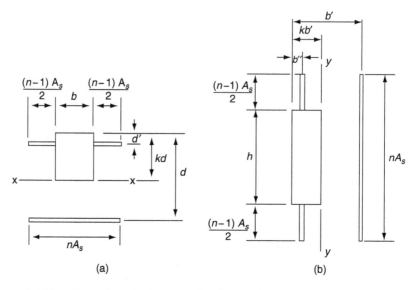

Figure 9.3 Transformed cracked sections for the reinforced concrete elements with respect to (a) the major axis and (b) the minor axis.

where $A_s = \rho bd$, $d = Ph$, $d' = (1 - P)h$, $k = f_c/(f_c + f_s)$ and $n = E_s/E_c$ in which E_s is Young's modulus for steel. E_c is Young's modulus for concrete, f_s is the working stress for steel, f_c is the working stress for concrete. Eqs. (9.1)–(9.3) can be rewritten in terms of ρ, the steel ratio, P, the percentage of the depth to the lumped steel, k, the percentage of the effective depth for the cracked section based on the position of the lumped steel, n, the modular ratio, b, the variable width, and h, the fixed depth as:

$$I_x = bh^3 \left[\frac{1}{3}(kP)^3 + (n-1)\rho P\left(P\left(k+1\right) - 1\right)^2 + n\rho P^3 \left(1 - k\right)^2 \right] \quad (9.4)$$

$$I_y = bh^3 \left[\frac{1}{3}(kP)^3 + (n-1)\rho P\left(P\left(k+1\right) - 1\right)^2 + n\rho P^3 \left(1 - k\right)^2 \right] \quad (9.5)$$

$$A_t = Pbh\left[k + 2n\rho - \rho\right] \quad (9.6)$$

Note that the terms in the brackets are independent of the dimensions of the cross-section, thereby greatly simplifying the equations to a constant times the relationships between the depth and the width. These properties are clearly defined by the assumptions for this formulation where (1) there is a uniform distribution of steel with respect to the major and minor axes, (2) there is no interaction with respect to bending about both axes, (3) the depth is fixed coupled with a variable width, and (4) no tensile strength is allowed within the concrete. These assumptions are slightly restrictive, but are reasonable in developing a system for generating reasonable optimal designs.

9.3   Primary and secondary design variables

Pure mathematical optimization or optimality-criteria approaches could use each geometric property as a single design variable. This would provide the most efficient set of geometric properties for each individual element which would provide the most efficient structural system while maintaining the structural responses (stress, displacements, drifts and frequencies) within the specified limits. Although this might be the most efficient system for the given objective function, the set of optimal geometric properties (six for a beam-column, etc.) for a given element will most likely provide an impractical element, one that cannot be made or purchased. For example, if the optimal structure needed a beam-column that needed very little axial strength yet large bending resistance, the optimal shape would have a small cross-sectional area and a large moment of inertia. As the optimization pushes these properties to their extreme, it would be difficult or impossible to find an appropriate steel or reinforced-concrete cross-section that could satisfy both, or all, of the optimal conditions. Also, the use of each geometric property as a separate design variable would create a large increase in computational effort. Because of these reasons, a model was developed for both steel and concrete elements that allows each element to be represented by one geometric property called the primary design variable. All other geometric properties other than the primary design variable are defined as secondary design variables or sometimes called pseudo-design variables. The model used provides a continuous (pseudo-continuous) relationship between the primary and secondary design variables. The pseudo or secondary relationships relative to the primary design variables for regular geometric shapes can produce exact relationships between the primary and secondary design variables as was shown in Section 8.11.2 for rectangular cross-sections while irregular shapes can be represented using polynomials or statistically derived equations as pseudo-discrete approximations.

The model developed produces an exact relationship for geometrically regular steel shapes and for the reinforced-concrete elements, while requiring an approximate relationship for steel wide-flange sections, double angles, T-sections or any other manufactured shapes. All element types except the braces use the major-axis moment of inertia as their primary design variable. Whereas the braces use their cross-sectional area as their primary design variable. Each secondary design variable is represented in this form:

$$S_{ij} = C_{1j}\delta_i^{C_{2j}} + C_{3j} \qquad (9.7)$$

where S_{ij} is the jth secondary design variable (for example the minor-axis moment of inertia) for the ith element and C_{1j}, C_{2j} and C_{3j} are the appropriate constants (mathematically or statistically derived) for the δ_i, the ith element primary design variable relative to the jth secondary design variable.

9.3.1 Regular cross-sections

For most regular cross-sections such as rectangles, circles and tubes, the constants in Eq. (9.7) can be derived explicitly. For example, a rectangular cross-section with a fixed ratio of depth to width of R provides a set of equations for the minor-axis moment of inertia and the cross-sectional area as:

$$I_y = \left(\frac{1}{R^2}\right) I_x \tag{9.7a}$$

$$A = \left(\frac{1}{R}\right)^{\frac{1}{2}} I_x^{\frac{1}{2}} \tag{9.7b}$$

Note that in each of Eqs. (9.7a) and (9.7b) the coefficient $C_{3j} = 0$. Similar equations can be developed for circular sections, tubes with fixed ratios of inner and outer radii, as well as other regular shapes. Illustrative examples will be shown where rectangular elements are used not necessarily because they are practical, but because they are simple.

9.3.2 Steel wide-flange sections

In order to represent a large range of I-shaped steel sections, a numerically derived set of pseudo-discrete equations was developed in the form of Eq. (9.7). Although the actual, available manufactured sizes are discrete, the elements are treated as a continuous spectrum of sizes.

The equations presented were developed in a manner that provides an upper bound for each of the secondary design variables. It is important to note that these equations do not provide a one-to-one correspondence for the primary and secondary design variables with respect to a specific wide-flange cross-section in the American Institute of Steel Construction Manual (AISCM) [22]. Reasonable judgement coupled with the optimal results must be used to select the appropriate wide-flange cross-section for each element. (There are several discrete optimization techniques that can be used as a final step in the optimization process in order to make 'good' decisions for the final discrete sizes of the elements. See Chapter 10 for a description of the branch and bound techniques used in pile optimization.) Reasonable equations that were statistically derived from the AISCM for wide-flange shapes using I_x as the primary design variable δ in Eq. (9.7) are shown in Table 9.1.

Table 9.1 Steel wide-flange coefficients for primary versus secondary design variables (1 in. = 2.54 cm).

J	S_{ij}	C_{1j}	C_{2j}	C_{3j}
		$I_x < 1,550 \text{ in.}^4$		
1	I_y	0.0389	0.925	0
2	J	0.0221	0.958	0
3	A	0.5008	0.487	0
4	S_x	0.4531	0.774	0
5	S_y	0.0423	0.732	0
		$1,550 < I_x < 12,100 \text{ in.}^4$		
1	I_y	0.0265	1.00	20.47
2	J	0.0124	0.905	0
3	A	0.5008	0.487	0
4	S_x	0.0462	1.00	78.46
5	S_y	0.0041	1.00	7.64
		$12,100 \text{ in.}^4 < I_x$		
1	I_y	0.0518	1.00	159.1
2	J	0.0124	0.905	0
3	A	0.5008	0.487	0
4	S_x	0.0520	1.00	56.00
5	S_y	0.0076	1.00	0.566

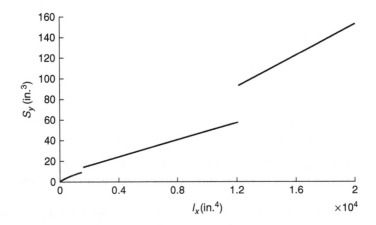

Figure 9.4 Minor-axis section modulus versus major-axis moment of inertia (1 in. = 2.54 cm).

These functions are shown in Figs. 9.4–9.8. It is easy to see that these are pseudo-continuous and can cause some numerical difficulties when an element is in the vicinity of the discrete changes in equations representing the secondary design variables. These elements are typically optimized in a

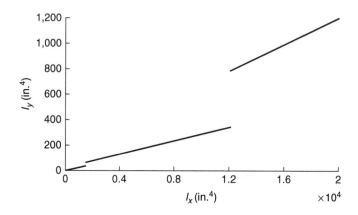

Figure 9.5 Minor-axis moment of inertia versus major-axis moment of inertia (1 in. = 2.54 cm).

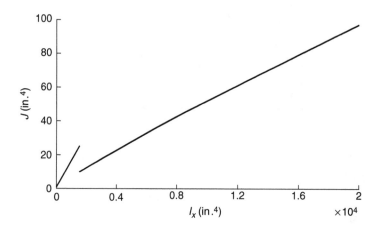

Figure 9.6 Torsional moment of inertia versus major-axis moment of inertia (1 in. = 2.54 cm).

manner that will place them on one curve or the other after one cycle causing short-lived numerical sensitivities to the discrete, discontinuous regions of the curves.

These values were determined with C_{2j} as the slope of a log-log plot of each secondary variable versus the primary design variable. The other two constants are found by choosing two points on the log-log line and solving two simultaneous equations for the two remaining constants. Other statistical approaches and curve forms can be used, but these work well for the shapes being used in ODRESB-3D. These equations cover a wide range of I-shapes and could be refined for specific shapes (depths) such as W14's or any other standard W shapes. These coefficients are reasonable for a wide spectrum of W shapes.

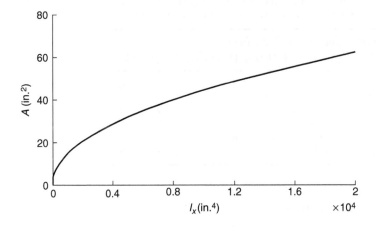

Figure 9.7 Cross-sectional area versus major-axis moment of inertia (1 in. = 2.54 cm).

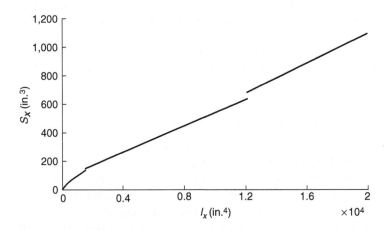

Figure 9.8 Major-axis section modulus versus major-axis moment of inertia (1 in. = 2.54 cm).

9.3.3 Reinforced concrete sections

The reinforced-concrete element equations are based on the working stress model described earlier. This approach is a good method for finding reasonable preliminary sizes. The form of the concrete equations is that of Eq. (9.7), but the constants can be explicitly derived as shown in Table 9.2.

These constants are derived using Eqs. (9.1)–(9.6). The value of b is derived from Eq. (9.4) in terms of I_x and substituted into Eqs. (9.5) and (9.6) to give the values in Table 9.2 which uses I_x as the primary design variable.

Table 9.2 Exact coefficients for rectangular reinforced-concrete sections with a given value of h and D.

For a given value of h and D where
$$D = (1/3(Pk)^3) + \rho P(n-1)(P(k+1)-1)^2 + nP^3\rho(1-k)^2$$

J	S_{ij}	C_{1j}	C_{2j}	C_{3j}
1	I_y	$1/(h^8 D^2)$	3.00	0
2	J	$1/(h^8 D^2)$	3.00	I_x
3	A	$(P(k+2n\rho-\rho)/h^2 D)$	1.00	0
4	A_N	$1/(h^2 D)$	1.00	0

9.4   Structural analysis

The structural elements and modelling used in ODRESB-3D were developed in order to reduce the overall number of degrees of freedom and to alleviate the heavy computational effort of finding the gradients of the displacements and mode shapes. These gradients are key to the optimization algorithm where dynamic displacement and stresses are being constrained. As shown in later sections, the inverse of the stiffness matrix is required in finding these gradients for each and every cycle (and possibly at several different time steps within a cycle). The elimination of degrees of freedom can significantly speed the optimization process.

9.4.1   *Global degrees of freedom for ODRESB-3D*

As previously discussed, the elements and modelling used in several of the examples are based on that of ODRESB-3D. This analysis model is based on representing each floor with a rigid, in-plane floor, which is flexible out-of-plane. This allows each floor to be represented with two translational degrees of freedom and one vertical rotational degree of freedom at the centre of mass of the floor coupled with one vertical translational degree of freedom and two rotational degrees of freedom at each column (node) for each floor as shown in Fig. 9.9. (The two rotational degrees of freedom at each column are eventually removed using static condensation since no rotational loading is applied at these nodes.) The final set of global degrees of freedom is shown in Fig. 9.10. As mentioned, this decreases the number of degrees of freedom making for a more efficient structural analysis as well as an efficient optimization tool.

9.4.2   *Second-order effects, P–Δ*

Second-order effects are included using a separate geometric stiffness matrix. The geometric matrix is developed using the string-stiffness

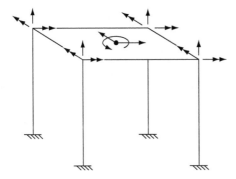

Figure 9.9 Global degrees of freedom per floor before condensation.

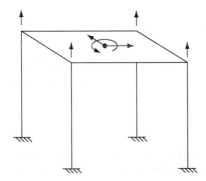

Figure 9.10 Global degrees of freedom per floor after condensation.

approach as shown in Fig. 9.11. The string-stiffness technique assumes that the given column with axial force P' has a second-order moment equivalent to the axial force multiplied by the storey drift, Δ. In order to enforce equilibrium, an additional shear of P'/L is required where L is the length of the flexible portion of the column. This term, P'/L, is used to reduce the lateral stiffness of the structure, therefore increasing the lateral deflections and increasing the internal moments. The elemental geometric stiffness becomes:

$$\begin{Bmatrix} V_{xi} \\ V_{yi} \\ V_{xj} \\ V_{yj} \end{Bmatrix} = \begin{bmatrix} -P'/L & 0 & P'/L & 0 \\ 0 & -P'/L & 0 & P'/L \\ P'/L & 0 & -P'/L & 0 \\ 0 & P'/L & 0 & -P'/L \end{bmatrix} \begin{Bmatrix} x_i \\ y_i \\ x_j \\ y_j \end{Bmatrix} \tag{9.8}$$

Note that D_T and D_B, shown in Fig. 9.11, are rigid zones at the top and bottom of the columns that can be used to represent the beam-column-slab regions. The geometric stiffness is transformed and added directly to the global stiffness and provides displacements that include first-order $P-\Delta$

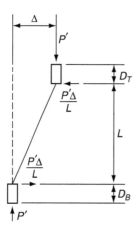

Figure 9.11 String stiffness approach for P–Δ effects.

effects. The inclusion of these geometric effects was used to study the effects of code based equations used to increase deflections as a means of including second-order stability effects [1].

9.4.3   External stiffness

External stiffness, essentially external springs, can be added to any one or a combination of the floor degrees of freedom which act in the horizontal plane of the floors as shown in Fig. 9.12. Therefore, three-dimensional structures coupled with external springs can be used to simulate two-dimensional structures by eliminating the rotational effects due to non-symmetry or by eliminating the orthogonal translational motion along with the rotational response.

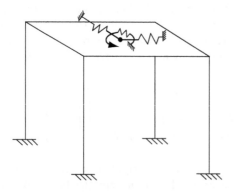

Figure 9.12 External springs used to simulate two-dimensional structures with the ODRESB-3D system.

9.4.4 *Structural and non-structural mass*

For dynamic analyses, the structural mass must be recalculated for each cycle of the optimization since the structural elements undergo a resizing in each cycle. A lumped mass including both structural and non-structural masses is used in the problems presented although a consistent mass approach could easily be used. The non-structural mass is considered to be a fixed, constant value while the distributed structural masses change with each cycle of the optimization and must be accounted for in order to correctly produce the dynamic and pseudo-dynamic loads which are based on mass (inertia). If the non-structural mass is large compared to the structural mass, the structural mass can be ignored with little effect on the optimal solution.

The structural mass is found by summing the tributary mass for each element at a given node coupled with the non-structural mass associated with that node as shown in Fig. 9.13. Both translational and rotational

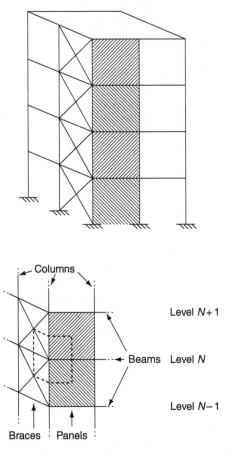

Figure 9.13 Elemental contribution to the nodal lumped mass.

inertia are included in the structural analyses. It is important to note that for the dynamic calculations or the pseudo-dynamic calculations the mass is involved as an inertia driver or used in the pseudo-dynamic forces. Since the mass is not constant, it must be included in the gradient calculations and optimization algorithm.

9.5   ODRESB-3D-based structural optimization

Structural optimization is best defined as the application of optimization techniques for improving designs with respect to distinct objectives while staying within well-defined constraints. The objective could be reduced weight or cost or increased reliability or any combination of these, see Cheng *et al.* [1, 2, 13–16]. The constraints generally represent the structural responses: displacements, drifts, stresses and natural frequencies along with limitations on member dimensions (side constraints) as described in Chapter 8. The optimization approach used in these examples will be the generalized optimality-criteria technique and algorithm described in Chapter 8. In order to use this approach, quality estimates or preferably exact constraint sensitivities (constraint gradients) are necessary. Unlike the statically loaded systems, seismically loaded systems and their analyses, both pseudo-dynamic and time history, are dependent upon the structural mass. Therefore, the gradients of the loadings are no longer zero, but are dependent on the changes in the structural mass, both in value and distribution, within each cycle of the optimization. Therefore, additional information beyond that presented in Chapter 8 regarding the constraint gradients needs to be developed, but the algorithm from Chapter 8 remains essentially unchanged as long as the appropriate gradients are used.

9.5.1   Objective function

The objective function to be used in the following examples will be the structural weight. The inclusion of non-structural mass or weight would change the solution and is often dominant in seismic design. The objective function must be written in a form that is a function of the primary design variables, δ_i, which represent the major-axis moment of inertia in these problems. The objective function for the structural weight is written as:

$$\text{Minimize} \quad W(\delta) = \sum_{i=1}^{n} \gamma_i \left(\text{Vol}\right)_i = \sum_{i=1}^{n} \gamma_i \left(l_i \left(C_{1A}\delta_i^{C_{2A}} + C_{3A}\right)_i\right) \tag{9.9}$$

where $\left(\text{Vol}\right)_i$ is the volume, γ_i is the specific weight, l_i is the length, and C_{1A}, C_{2A} and C_{3A} are the appropriate constants for the area, A_i, relative to the primary design variable, δ_i.

The gradient of this function is needed within the optimality-criteria algorithm in order to generate the optimality criterion and to find the necessary Lagrange multipliers. The gradient is easily found by taking the derivative of Eq. (9.9) which gives:

$$\frac{\partial W(\delta)}{\partial \delta_i} = \gamma_i \frac{\partial (\text{Vol})_i}{\partial \delta_i} = \gamma_i l_i \frac{\partial A_i}{\partial \delta_i} = \gamma_i l_i C_{1A} C_{2A} \delta_i^{(C_{2A}-1)} \tag{9.10}$$

Eq. (9.10) can be substituted into each of the necessary equations in Chapter 8 where the gradient of the objective function is required, primarily the optimality criterion Eq. (8.12) and the equations for determining the Lagrange multipliers, Eqs. (8.32) or (8.41).

9.5.2 Sensitivity analyses

In addition to finding the gradient of the objective function, it is necessary to find the gradients of the active constraints. Once these gradients are determined they can be substituted into the equations for the optimality criterion Eq. (8.12) and the equations for finding the Lagrange multipliers, Eqs. (8.32) or (8.41). Each type of constraint requires its own derivation and methods for determining its gradients. The constraints can be written in a general form as:

$$g_j = (u_j - \bar{u}_j) \le 0 \tag{9.11}$$

for an upper-bound constraint and

$$g_j = (\underline{u}_j - u_j) \le 0 \tag{9.12}$$

for a lower-bound constraint where the upper and lower bars represent the upper and lower bounds, respectively, on the structural response u_j which could be displacements, stresses, frequencies or storey drift. Therefore, the gradients of the jth constraint can be written as:

$$\frac{\partial g_j}{\partial \delta_i} = \frac{\partial u_j}{\partial \delta_i} \quad i = 1, \dots, n \tag{9.13}$$

for an upper-bound constraint and

$$\frac{\partial g_j}{\partial \delta_i} = -\frac{\partial u_j}{\partial \delta_i} \quad i = 1, \dots, n \tag{9.14}$$

for a lower-bound constraint since the upper- and lower-bound values are constants and n is the total number of elements. Keep in mind that many structural problems have multiple load cases associated with each response which requires Eqs. (9.11)–(9.14) to have an additional subscript, u_{jl} and g_{jl} where l represents the lth load case. For clarity in the development of the gradients in this section, the equations will be derived for a single load case.

9.5.3 Stiffness gradients

In the examples to be presented, the stiffness and mass are directly differentiable with respect to the primary and secondary design variables. Using the chain rule and Eq. (9.7), the gradients of the stiffness and mass matrices with respect to the primary design variables can be derived. Note that the stiffness and mass matrices are nonlinear with respect to the primary design variable due to the relationships with the secondary design variables as seen in Eq. (9.9).

The stiffness can be broken into the elemental stiffness matrices which can then be broken into the contributions from the different geometric properties. This allows the ith elemental stiffness to be written as:

$$[K]_i = \sum_{j=1}^{t} [K]_{ij} = [K]_{iA} + [K]_{il_x} + [K]_{il_y} + [K]_{ij} \tag{9.15}$$

where t represents the necessary number of geometric properties to represent element i's stiffness. Eq. (9.15) represents the number of secondary design variables necessary to represent a steel or concrete beam-column. Therefore, the gradient of the stiffness with respect to the primary design variable I_x can be written as:

$$\frac{\partial [K]_T}{\partial \delta_i} = \frac{\partial [K]_T}{\partial I_{xi}} = \frac{\partial [K]_i}{\partial I_{xi}} = \sum_{j=1}^{t} \frac{\partial [K]_{ij}}{\partial I_x} = \frac{\partial \left[[K]_{iA} + [K]_{il_x} + [K]_{il_y} + [K]_{ij} \right]}{\partial I_x}$$

$$\tag{9.16}$$

where

$$\frac{\partial [K]_{ij}}{\partial I_x} = \frac{\partial [K]_{ij}}{\partial S_{ij}} \frac{\partial S_{ij}}{\partial I_x} = \frac{\partial [K]_{ij}}{\partial S_{ij}} \left(C_{1j} C_{2j} I_x^{(C_{2j}-1)} \right) \tag{9.17}$$

Note that the stiffness $[K]_{ij}$ is linear in terms of the secondary design variable S_{ij} so that the derivative of $[K]_{ij}$ relative to S_{ij} is nothing more than $[K]_{ij}/S_{ij}$. For the most complicated element, the beam-column, this can be written as:

$$\frac{\partial [K]_T}{\partial I_{xi}} = \frac{[K]_{iA}}{A_i} \left(C_{1A} C_{2A} I_{xi}^{(C_{2A}-1)} \right) + \frac{[K]_{il_x}}{I_{x_i}} + \frac{[K]_{ij}}{J_i} \left(C_{1J} C_{2J} I_{xi}^{(C_{2J}-1)} \right)$$

$$+ \frac{[K]_{il_y}}{I_{y_i}} \left(C_{1l_y} C_{2l_y} I_{xi}^{(C_{2l_y}-1)} \right) \tag{9.18}$$

Note that $[K]_T$ is the total stiffness and that the gradient for all terms other than those related to element i is zero. This is a very specific form based on the method used to relate primary and secondary design variables and would have to be modified accordingly for a different strategy. It can be seen that this approach is fairly simple and reasonable once the coefficients given in Tables 9.1 and 9.2 are generated. Every other type of element is a subset of the beam-column Eq. (9.18). For example, the braces would only require the first term of Eq. (9.18).

9.5.4 Mass gradients

The gradients for the mass are easily found in the same manner as the stiffness. Since the structural mass is dependent only upon the cross-sectional areas of the elements, and the non-structural mass is independent of the elemental geometric properties, the gradient of the mass for the ith element becomes:

$$\frac{\partial\,[M]_T}{\partial \delta_i} = \frac{\partial\,[M]_T}{\partial I_{xi}} = \frac{\partial\,[M]_i}{\partial I_{xi}} = \frac{\partial\,[M]_i}{\partial A_i}\frac{\partial A_i}{\partial I_{xi}} = \frac{[M]_i}{A_i}\left(C_{1A}C_{2A}I_{xi}^{(C_{2A}-1)}\right) \quad (9.19)$$

Note that the elemental mass matrix $[M]_i$ is linearly dependent upon A_i. Therefore, the first term of Eq. (9.19) can be found as the elemental mass matrix divided by A_i. The second term is based on the relationship between the primary design variable I_x and the elemental area. Also note that for braces the term in the parentheses becomes unity and the gradients are simply found as the elemental mass matrix divided by the braces' cross-sectional area.

9.5.5 Static displacement and stress gradients

The virtual-load technique described in Section 8.10.2 is used to find the gradients for the static displacements, drifts and stresses. This technique is based on the premise that the static displacements, drifts and stresses can be written as a linear combination of the structural displacements. In equation form it becomes:

$$u_j = \{b\}_j^T\{U\} \quad (9.20)$$

where $\{b\}_j^T$ is the appropriate vector to enforce this relationship, u_j is the jth global displacement, drift or elemental stress and $\{U\}$ is the vector of global displacements. The general form of the gradients will be discussed here with the specific $\{b\}_j^T$ vectors discussed in detail for the specific gradients and element types in subsequent sections. Using Eq. (9.20), the gradient for the jth displacement, drift or stress can be written as:

$$\frac{\partial u_j}{\partial I_{xi}} = \frac{\partial \{b\}_j^T}{\partial I_{xi}} \{U\} + \{b\}_j^T \frac{\partial \{U\}}{\partial I_{xi}} \tag{9.21}$$

In several cases, such as static displacements and drifts as well as stresses in brace elements, the first term in Eq. (9.21) goes to zero as the vector $\{b\}_j^T$ is a vector of constants. In most of the remaining cases the vector is not independent of the major-axis moment of inertia (the primary design variable). In either case the gradients of the global displacements $\{U\}$ are needed to solve Eq. (9.21).

The gradient of the static global displacements is easily found by taking the derivatives of the standard static relationship between stiffness, displacements and load, $[K]\{u\} = \{P\}$ where $\{P\}$ represents the static loading. Taking the partial derivatives of this relationship gives:

$$\frac{\partial [K]_T}{\partial I_{xi}} \{U\} + [K]_T \frac{\partial \{U\}}{\partial I_{xi}} = \frac{\partial \{P\}}{\partial I_{xi}} \tag{9.22}$$

Since the static loads are independent of the primary design variable, Eq. (9.22) is equal to zero and can be rearranged to give:

$$\frac{\partial \{U\}}{\partial I_{xi}} = -[K]_T^{-1} \frac{\partial [K]_T}{\partial I_{xi}} \{U\} \tag{9.23}$$

where the derivative of the stiffness is given in Eqs. (9.16) or (9.18). Therefore, the gradient for a specific displacement or stress component becomes:

$$\frac{\partial u_j}{\partial I_{xi}} = \frac{\partial \{b\}_j^T}{\partial I_{xi}} \{U\} + \{b\}_j^T \frac{\partial \{U\}}{\partial I_{xi}} = \frac{\partial \{b\}_j^T}{\partial I_{xi}} \{U\} - \{b\}_j^T [K]_T^{-1} \frac{\partial [K]_T}{\partial I_{xi}} \{U\} \tag{9.24}$$

where $\{b\}_j^T [K]_T^{-1}$ is often described as a virtual displacement, $\{v\}_j$, and $\{b\}$ is considered the virtual load. When the first term of Eq. (9.24) is zero (or can be approximated as zero), Eq. (9.24) can be written in the virtual-load form as:

$$\frac{\partial u_j}{\partial I_{xi}} = -\{v\}_j^T \frac{\partial [K]_T}{\partial I_{xi}} \{U\} = -\{v\}_j^T \frac{\partial [K]_i}{\partial I_{xi}} \{U\} \tag{9.25}$$

Eqs. (9.24) and (9.25) provide the component of the gradient for the jth static displacement, drift or stress constraint. Every term in Eqs. (9.24) and (9.25) is known with the exception of $\{b\}_j$ which can be used to find $\{v\}_j$. Therefore, $\{b\}_j$ must be derived for each type of constraint.

9.5.6 Displacement and drift constraints

The virtual load vector $\{b\}_j$ for a displacement constraint is independent of the primary design variable and consists of a vector of zeros with one single value of 1 in the jth location.

$$\{b\}_j^T = \{0,\ldots,0,1,0,\ldots,0\} \tag{9.26}$$

For displacement constraints, the first term of Eq. (9.21) is zero and Eq. (9.24) can be used to find the jth component of the gradient of the displacement. For storey drifts, the $\{b\}$ vector has an additional term of -1 in order to capture the difference between two storey displacements.

9.5.7 Stress constraints

The virtual-load vectors for the stresses must be developed for each type of element. Each virtual-load vector must satisfy the equation:

$$\sigma_j = \{b\}_j^T \{U\} \tag{9.27}$$

where σ_j represents the jth stress for the ith element. This was described in Chapter 8 for brace elements. Each element type must have a separate $\{b\}$ vector that relates to their specific displacements.

9.5.8 Beam-column stress constraints

The beam-column stress constraint virtual-load vectors, one for each end of the beam-column, are based upon a biaxial-bending and axial stress combination. Starting with the typical equation for this combination:

$$\sigma = \frac{P'}{A} \pm \frac{M_x c}{I_x} \pm \frac{M_y d}{I_y} = \frac{P'}{A} \pm \frac{M_x}{S_x} \pm \frac{M_y}{S_y} \tag{9.28}$$

where P' is the axial load, M_x and M_y are the moments about the x- and y-axes, respectively, c and d are the appropriate distances from the principal axes to the outermost fibres, and A, I_x, I_y, S_x and S_y are the previously defined geometric properties which are related to the primary design variables through Eq. (9.7). The mixed signs are needed to represent the stress in each quadrant of a given cross-section within the beam-column element. Torsional and shear stresses were assumed to be negligible and were not considered within the development of the beam-column stress constraints. Eq. (9.28) is the basis for developing the virtual-load vector using the beam-column stiffness coefficients along with the local displacements shown in

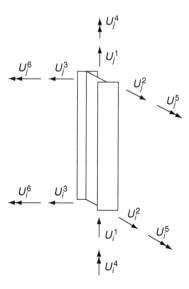

Figure 9.14 Beam-column local degrees of freedom for determination of the beam-column stress vector.

Fig. 9.14. The values for the internal forces at end i can be written as:

$$P' = \frac{EA}{L}\left(U_i^1 - U_j^1\right) \tag{9.29}$$

$$M_x = \frac{6EI_x}{L^2}\left(U_i^2 - U_j^2\right) + \frac{4EI_x}{L}\left(U_i^6 + \frac{1}{2}U_j^6\right) \tag{9.30}$$

$$M_y = \frac{6EI_y}{L^2}\left(U_i^3 - U_j^3\right) + \frac{4EI_y}{L}\left(U_i^5 + \frac{1}{2}U_j^5\right) \tag{9.31}$$

where E is Young's modulus and L is the length of the elastic portion of the beam-column. Substituting Eqs. (9.29)–(9.31) into Eq. (9.28), the stress at node i of the beam-column element can be written as:

$$\sigma_i = \left\{ \frac{E}{L} \quad -\frac{E}{L} \right\} \left\{ \begin{array}{c} U_i^1 \\ U_j^1 \end{array} \right\} \pm \left\{ \frac{6Ec}{L^2} \quad \frac{4Ec}{L} \quad -\frac{6Ec}{L^2} \quad \frac{2Ec}{L} \right\} \left\{ \begin{array}{c} U_i^2 \\ U_i^6 \\ U_j^2 \\ U_j^6 \end{array} \right\}$$

$$\pm \left\{ -\frac{6Ed}{L^2} \quad \frac{4Ed}{L} \quad \frac{6Ed}{L^2} \quad \frac{2Ed}{L} \right\} \left\{ \begin{array}{c} U_i^3 \\ U_i^5 \\ U_j^3 \\ U_j^5 \end{array} \right\} \tag{9.32}$$

Comparing Eqs. (9.32) and (9.27), one can derive the virtual load vector $\{b\}_j$ for the stress at the ith end of the beam as:

$$\{b\}_j^T = \left\{ \begin{matrix} \dfrac{E}{L} & \pm\dfrac{6Ec}{L^2} & \pm\dfrac{-6Ed}{L^2} & 0 & \pm\dfrac{4Ed}{L} & \pm\dfrac{4Ec}{L} \\[2ex] -\dfrac{E}{L} & \pm\dfrac{-6Ec}{L^2} & \pm\dfrac{6Ed}{L^2} & 0 & \pm\dfrac{2Ed}{L} & \pm\dfrac{2Ec}{L} \end{matrix} \right\}$$

(9.33)

where the signs are chosen according to the active stress found by Eq. (9.28) and j represents the jth active constraint. The values for c and d are found as the ratio of I_x/S_x and I_y/S_y where the section moduli are defined as functions of the major-axis moment of inertia through Eq. (9.7). Therefore, the gradient of the $\{b\}_j$ vector is not independent of the primary design variable, the major-axis moment of inertia, and should not be considered zero unless these terms are insignificant compared to the second terms in Eq. (9.21).

9.5.9 Beam stress constraints

The beam stresses are based on the model shown in Fig. 9.15 and are strictly purely bending stresses. The stresses at end i can be written as:

$$\sigma_i = \pm\frac{M_{xi}c}{I_x} = \pm\frac{M_{xi}}{S_x}$$

(9.34)

where M_{xi} is the moment about the major axis at end i and c is the distance from the major axis to the outermost fibre, I_x is the major-axis moment of inertia and S_x is the major-axis section modulus. Using the beam stiffness coefficients with the local degrees of freedom shown in Fig. 9.15, the moment M_{xi} can be written as:

$$M_{xi} = \frac{EI_x}{L^2}\{\, 6 \quad 4L \quad -6 \quad 2L \,\} \left\{ \begin{matrix} U_i \\ \theta_i \\ U_j \\ \theta_j \end{matrix} \right\}$$

(9.35)

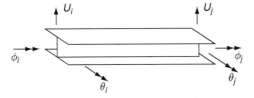

Figure 9.15 Beam local degrees of freedom for determination of the beam stress vector.

where E is Young's modulus and L is the length of the elastic portion of the beam. Substituting Eq. (9.35) into Eq. (9.34), the stresses can be written in terms of the displacements as:

$$\sigma_i = \pm \left\{ \begin{array}{cccc} \dfrac{6Ec}{L^2} & \dfrac{4Ec}{L} & -\dfrac{6Ec}{L^2} & \dfrac{2Ec}{L} \end{array} \right\} \left\{ \begin{array}{c} U_i \\ \theta_i \\ U_j \\ \theta_j \end{array} \right\} \qquad (9.36)$$

which leads to a virtual-load vector for the kth beam stress constraint at end i as:

$$\{b\}_k^T = \left\{ \begin{array}{cccc} \dfrac{6Ec}{L^2} & \dfrac{4Ec}{L} & -\dfrac{6Ec}{L^2} & \dfrac{2Ec}{L} \end{array} \right\} \qquad (9.37)$$

Note that only the positive sense is needed since the same level of stress is obtained for both tension and compression within the steel beam element. Also note that just like the beam-column, the values for c and d are found as the ratio of I_x/S_x and I_y/S_y where the section moduli are defined as functions of the major-axis moment of inertia through Eq. (9.7). Therefore, the gradient of the $\{b\}_k$ vector is not independent of the primary design variable, the major-axis moment of inertia, and should not be considered zero unless these terms are insignificant compared to the second terms in Eq. (9.21).

9.5.10   Brace stress constraint

The brace stresses are derived from axial loadings only and the local degrees of freedom are as shown in Fig. 9.16. The stress can be written as:

$$\sigma = \frac{P'}{A} = \frac{1}{A} \left\{ \begin{array}{cc} \dfrac{EA}{L} & -\dfrac{EA}{L} \end{array} \right\} \left\{ \begin{array}{c} U_1 \\ U_2 \end{array} \right\} \qquad (9.38)$$

where P' is the axial load, A is the cross-sectional area, E is Young's modulus and L is the length. Using Eq. (9.38), the virtual-load vector can easily be written as:

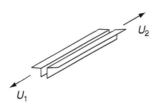

Figure 9.16 Brace local degrees of freedom for determination of the brace stress vector.

$$\{b\}_i^T = \left\{ \begin{array}{cc} \dfrac{E}{L} & -\dfrac{E}{L} \end{array} \right\} \qquad (9.39)$$

and the tensile or compressive nature of the jth stress is determined strictly by the elemental displacements. Note that the gradient of this vector relative to the primary design variable of the area does produce a zero vector and Eq. (9.25) can be used.

9.5.11 Panel stress constraints

The panel stress constraint virtual-load vectors are based upon an axial stress combined with bending about one axis. The stresses considered are:

$$\sigma_i = \frac{P'}{A} \pm \frac{M_{xi}c}{I_x} = \frac{P'}{A} \pm \frac{M_{xi}}{S_x} \qquad (9.40)$$

where i represents the ith end of the panel. Using the panel stiffness coefficients and Fig. 9.17, the stress can be written as:

$$\sigma_i = \left\{ \begin{array}{cc} \dfrac{E}{L} & -\dfrac{E}{L} \end{array} \right\} \left\{ \begin{array}{c} U_1 \\ U_2 \end{array} \right\} \pm \left\{ \begin{array}{cccc} \dfrac{6Ec}{L^2} & \dfrac{4Ec}{L} & -\dfrac{6Ec}{L^2} & \dfrac{2Ec}{L} \end{array} \right\} \left\{ \begin{array}{c} U_3 \\ U_4 \\ U_5 \\ U_6 \end{array} \right\} \qquad (9.41)$$

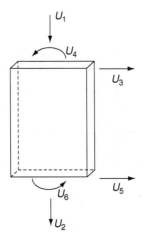

Figure 9.17 Panel local degrees of freedom for determination of the panel stress vector.

giving rise to a virtual load vector of:

$$\{b\}_k^T = \left\{ \begin{array}{ccccccc} \dfrac{E}{L} & -\dfrac{E}{L} & \pm\dfrac{6Ec}{L^2} & \pm\dfrac{4Ec}{L} & \pm\dfrac{-6Ec}{L^2} & \pm\dfrac{2Ec}{L} \end{array} \right\} \qquad (9.42)$$

and the signs are chosen according to the active constraint being tensile or compressive. Note that for a panel the value of c is constant as half of the depth of the system [1], therefore $\{b\}$ is a vector of constants and Eq. (9.25) can be used to find the gradients.

9.5.12 Coordinate transformations

All of the virtual-load vectors presented are coordinate system dependent. The displacement response and its virtual load vector are given in the global coordinate system. The stress virtual-load vectors are developed in the local or elemental coordinate system. The virtual displacements are found in the global coordinate system, therefore, the stress virtual-load vectors need to be transformed from the local to the global coordinates. ODRESB-3D uses a reference coordinate system and a global coordinate system as shown in Fig. 9.18. Therefore, the stress virtual-load vectors must be transformed to the reference system as:

$$\{b\}_j^{REF} = [T]^T \{b\}_j^E \qquad (9.43)$$

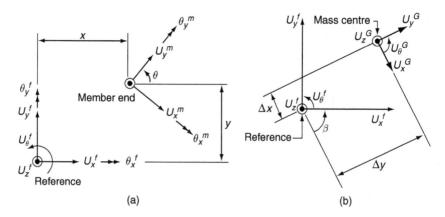

Figure 9.18 Coordinate systems for transformations between element, reference and global coordinates.

where

$$[T]\{U^f\} = \begin{bmatrix} s & -c & a & 0 & 0 & 0 \\ c & s & b & 0 & 0 & 0 \\ 0 & 0 & 1 & 0 & 0 & 0 \\ 0 & 0 & 0 & s & -c & 0 \\ 0 & 0 & 0 & c & s & 1 \\ 0 & 0 & 0 & 0 & 0 & 1 \end{bmatrix} \begin{Bmatrix} U_x^f \\ U_y^f \\ U_\theta^f \\ \theta_x^f \\ \theta_y^f \\ U_z^f \end{Bmatrix} \tag{9.43a}$$

and $s = \sin\theta$, $c = \cos\theta$, $a = -ys - xc$ and $b = -yc + xs$. The vector can then be transformed to the global system using:

$$\{b\}_j^G = [T']^T \{b\}_j^{REF} = [T']^T [T]^T \{b\}_j^E \tag{9.44}$$

where $\{b\}_j^G$, $\{b\}_j^{REF}$ and $\{b\}_j^E$ are the virtual-load vectors in the global, reference and elemental coordinate systems, respectively, and $[T]$ is given in Eq. (9.43a) and $[T']$ is:

$$[T'] = \begin{bmatrix} A_1 & & & \\ & A_2 & & \\ & & \ddots & \\ & & & A_n \end{bmatrix} \tag{9.44a}$$

where $[A]_n$ is:

$$[A]_n = \begin{bmatrix} \cos\beta & \sin\beta & -\Delta y\cos\beta + \Delta x\sin\beta \\ -\sin\beta & \cos\beta & \Delta x\cos\beta + \Delta y\sin\beta \\ 0 & 0 & 1 \end{bmatrix}_n \tag{9.44b}$$

and n represents the number of levels.

Although the virtual displacements are found in the global system, the static response gradients are best found using the elemental or local coordinates. As long as $\{U\}$, $[K]_i$ and $\{b\}_i$ are all transformed to the same reference system, the gradients will be calculated correctly. Therefore, any coordinate system can be used, as long as each component of the calculation is consistent with that frame of reference.

9.5.13 Reduction effects

Prior to using Eq. (9.33), the components of $\{b\}_i$ associated with the rotational degrees of freedom must be condensed. (The global stiffness matrix was used to condense these degrees of freedom.) This requires that the virtual-load vector be modified and reduced as:

$$\{b\}_j^{RED} = \{b_2\}_j - [K_{21}][K_{11}]^{-1}\{b_1\}_j, \tag{9.45}$$

where $\{b_2\}_j$ is the portion of the $\{b\}_j$ vector which corresponds to the vertical and translational degrees of freedom, and $\{b_1\}_j$ is the portion corresponding to the rotational, global degrees of freedom that were originally condensed from the global stiffness matrix.

Using Eq. (9.33) and $\{b\}_j^{RED}$, the $\{v_2\}_j$ terms corresponding to the vertical and translational, global virtual displacements can be found. Eq. (9.46) can be used to find $\{v_1\}_j$, the corresponding rotational, global virtual displacements as:

$$\{v_1\}_j = [K_{11}]^{-1}\left[\{b_1\}_j - [K_{12}]\{b_2\}_j\right] \tag{9.46}$$

These values are needed when the stress is evaluated at the elemental level, as shown by Eq. (9.33), since it requires the rotational degrees of freedom. Therefore, the virtual displacements, including the rotational degrees of freedom, are transferred into the local, virtual displacements and used to find the gradients. The local level is easiest to use since the derivative of the stiffness matrix requires only that portion of the stiffness supplied by element i as seen in Eqs. (9.16) and (9.18).

9.5.14 Effects of primary versus secondary design variables

Eq. (9.25) was derived as if the virtual-load vector was independent of the primary design variable which is not the case except as noted for the displacements and the brace elements. Each of the elements that have bending (rotational) stresses has a relationship that is related to the depth or width of the element and, to be exact, should use Eq. (9.24). The approximate values for c and d, the depths from the neutral axes to the extreme fibres, are found as noted earlier using these equations:

$$c = \frac{I_x}{S_x} = \frac{I_x}{\left(C_{1S_x}I_x^{C_{2S_x}} + C_{3S_x}\right)} \tag{9.47}$$

and

$$d = \frac{I_y}{S_y} = \frac{\left(C_{1I_y}I_x^{C_{2I_y}} + C_{3I_y}\right)}{\left(C_{1S_y}I_x^{C_{2S_y}} + C_{3S_y}\right)} \tag{9.48}$$

The algorithm used in the examples that follow assumes that the virtual-load vector is independent of the primary design variable and uses Eq. (9.25) instead of the more accurate version of Eq. (9.24). This can cause some violations of the stress constraints in the early cycles, but has very little effect after the first few cycles. This assumption has proved to be reasonable, and reduces the computational effort [1]. Using the exact equations that reflect the dependence of the $\{b\}$ vectors on the primary design variables, using Eq. (9.24), does help to better control the constraints and can cause convergence to occur a few cycles sooner.

9.6 Static displacement and stress constraint examples

Several examples will be used to illustrate the effectiveness of using generalized optimality-criteria techniques for structural problems. The examples will be used to illustrate the use of static and stress constraints for the generation of designs that satisfy a given set of design constraints.

9.6.1 Two-storey, steel setback structure

A two-storey all-steel, setback structure subjected to static loads is used to illustrate a three-dimensional, unsymmetrical optimization problem. The structure is shown in Fig. 9.19. The maximum allowable displacements are chosen as 0.25 in. (0.64 cm) for the first floor and 0.50 in. (1.27 cm) for the second-level displacements. These displacements are to be constrained with respect to both the x and y directions at each mass centre as shown in Fig. 9.19. Stress constraints were given as 36 ksi (25 kN cm^{-2}) for all elements, columns, beams and braces. The termination criteria were to have less than a 2 percent change in weight from one cycle to the next or a maximum of 20 cycles. A set of lateral loads of 400 kips (1,780 kN) and 450 kips (2,003 kN) was applied at the upper and lower mass centres, respectively, and was applied in both the x and y directions. Along with the lateral loads, a set of 5 kip (22 kN) downward vertical nodal forces was applied at each column node. The initial sizes used for the columns were 56.6 in.^{4} (2,356 cm^4), for the beams 146.0 in.^{4} (6,077 cm^4), and for the braces 3.63 in.^{2} (23.4 cm^2). After the preliminary analysis, the initial sizes (moments of inertia for the bending elements and cross-sectional area for the braces) were scaled by a factor of 121 to provide responses that were close to at least one of the prescribed constraints, displacements or stresses. This gives element sizes used for the columns of 6,849 in.^{4} (285,100 cm^4), for the beams 17,667 in.^{4} (735,400 cm^4), and for the braces 439 in.^{2} (2,832 cm^2). Linking of elements was not used, therefore, each element was free to take on its own optimal size. Linking was not used for the design variables in order to illustrate the effectiveness of the optimization algorithm, not necessarily to develop a practical design.

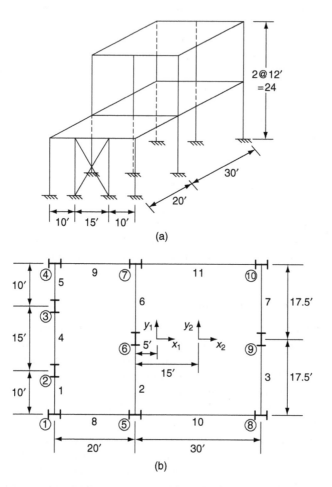

Figure 9.19 Two-storey, steel setback structure subjected to static loads (1 ft = 0.305 m).

The optimization required 10 cycles, 12 analyses (two scaling analyses were required), and was terminated due to a 1.8 percent change in weight between the last two cycles. The initial weight after the first scaling was 589 kips (267 Mg) and the final weight was 322 kips (146 Mg), a reduction of approximately 45 percent based on the initial weight. (This uses the primary to secondary design variable relationships given in Eq. (9.7).) Therefore, this weight is based on the relationship between the area and the major-axis moment of inertia for the beams and the columns. The initial set of active constraints included only the second-level x displacement, $X_2 = 0.461$ in. (1.17 cm). The next cycle included the x displacements on both levels with $X_1 = 0.228$ in. (0.579 cm) and $X_2 = 0.507$ in. (1.29 cm), but the first-floor displacement constraint was eliminated from the active set since its Lagrange multiplier was −375.2 and the negative Lagrange multiplier would violate the Kuhn–Tucker conditions for optimality. After three cycles, the active set of constraints stabilized and consisted of the x and y displacements on the second level. The final results for the displacements, member sizes and optimality criteria are given in Tables 9.3 through 9.6. Note that the optimality criteria have a large range of values. The values which are less than 0.5 typically indicate that the element would eventually reach its minimum value while those above 0.5 are members that would eventually converge to an optimality criterion of one. Once a large portion of the dominating elements have optimality criteria close to unity, the objective function shows little change and the design is terminated.

Table 9.3 Final displacements for the all-steel, two-storey setback structure (1 in. = 2.54 cm).

Responses	Level 1	Level 2
X (in.)	0.120	0.501[a]
Y (in.)	0.200	0.499[a]
Φ (rad)	3.90×10^{-4}	3.80×10^{-4}

[a]These displacements reached the maximum allowed values.

Table 9.4 Optimal column sizes and their final optimality-criteria values in parentheses (1 in. = 2.54 cm).

Level	Columns (in.⁴)									
	1	2	3	4	5	6	7	8	9	10
2	–	–	–	–	11,789	7,624	8,708	6,235	6,202	6,161
					(1.23)	(1.44)	(0.84)	(0.93)	(1.09)	(0.89)
1	45,079	702	555	1,270	27,284	689	893	4,755	2,195	1,205
	(1.41)	(0.21)	(0.20)	(0.17)	(1.15)	(0.22)	(0.37)	(0.73)	(1.04)	(0.70)

Table 9.5 Optimal beam sizes and their final optimality-criteria values in parentheses (1 in. = 2.54 cm).

Level	Beams (in.⁴)										
	1	2	3	4	5	6	7	8	9	10	11
2	–	2,060	1,809	–	–	2,081	1,734	–	–	6,814	6,155
		(0.94)	(0.68)			(0.90)	(0.64)			(0.91)	(0.85)
1	894	2,383	2,986	758	919	2,212	2,096	21,257	2,447	5,446	3,636
	(0.10)	(1.02)	(0.96)	(0.01)	(0.05)	(0.85)	(0.69)	(1.22)	(0.36)	(0.52)	(0.64)

Table 9.6 Optimal brace sizes and their final optimality-
criteria values in parentheses (1 in. = 2.54 cm).

Upper column line	Lower column line	Area (in.^{2})
3	2	47.2 (0.87)
2	3	48.2 (0.89)

This solution is instructive as it illustrates the tendency for the optimal solution, without linking of elements, to select critical bays or frames in order to provide the most efficient stiffness with the smallest amount of material. On the lower level, columns 1 and 5 along with beam 8 provide most of the stiffness required to limit the lower-level x displacement which helps reduce the second-level x displacement as well. This is the reason for the column sizes on the second floor to be considerably smaller than the first floor columns 1 and 5. In other words, this allows the intra-storey drift for level 2 to be larger than 0.25 in. (0.64 cm) thereby using less stiffness. Drift is also a constrainable quantity as described previously and is often necessary to produce reasonable designs that will avoid 'soft' storey scenarios. The second-level y displacements are mostly controlled by columns 5, 6 and 7 and beams 2 and 6.

This example also shows the need for several different types of constraints, linking of elements, side constraints and drift constraints in order to provide realistic optimal designs. Side constraints are used to place minimum and maximum allowable sizes on the elements. Linking of elements allows one to force certain columns or beams to have the same sizes leading to a more practical solution for design and construction. In the case of this structure, it would be very likely, in real life, that all of the columns would be forced to be the same size for ease of fabrication and economies of scale. A drift constraint would not allow the second storey to become 'soft' in order to reach its maximum displacement, but would most likely force both the first- and second-floor displacements (and drifts) to become active and better distribute the stiffness vertically.

9.6.2   *Two-storey, steel and concrete structure*

A two-storey structure comprising every element type, beams, beam-columns, braces and panels, will be used to illustrate a statically constrained problem with both steel and concrete elements. The structure is shown in Fig. 9.20. Due to the large amount of stiffness supplied by the panels and the shear walls (concrete beam-columns), the deflection constraints were set at small values of 0.60 in. (1.52 cm) for the upper level and 0.20 in. (0.51 cm) for the lower level. These values were used for both the x and y directions. The maximum allowable stresses for the steel columns were 20 ksi (13.8 kN cm^{-2}), for the steel beams 16.5 ksi (11.4 kN cm^{-2}), for the

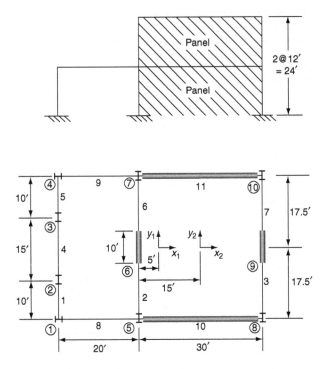

Figure 9.20 Two-storey setback structure with steel and concrete elements subject to static constraints (1 ft = 0.305 m).

steel braces 15.0 ksi (10.4 kN cm^{-2}), for the concrete columns (shear walls) 3 ksi (2.10 kN cm^{-2}) in compression and 50 ksi (34.5 kN cm^{-2}) in tension (steel reinforcing), and for the concrete panels 3 ksi (2.10 kN cm^{-2}) in compression and 3 ksi (2.10 kN cm^{-2}) in tension (no reinforcing). The concrete beam-column elements were assumed to have a steel ratio of 0.025 and a modular ratio of 10 for the columns and 1 for the panels. The steel columns were constrained to be between 7.0 and 50.0 in^4 (29.1 and 2,080 cm^4), the beams between 5.0 and 50.0 in.^{4} (20.8 and 208 cm^4), the braces between 1.0 and 15.0 in.^{2} (6.5 and 96.8 cm^2), the concrete columns' thicknesses between 5.0 and 20.0 in. (12.5 and 50.8 cm) and the panels' thicknesses between 4.0 and 10.0 in. (10.2 and 25.4 cm). The loading consisted of 400 kips (1,780 kN) concentrated lateral loads on the top level in both the x and y directions at the centre of mass, 450 kip (2,003 kN) concentrated, lateral loads on the lower level in both the x and y directions at the centre of mass, and 5 kips (22 kN) of downward vertical concentrated load at each column node. The convergence control parameter was set at 4.0 due to the many different types of elements coupled with displacement and stress constraints. No linking of elements was used. The termination criteria were to stay below 15 cycles or a 2 percent weight change between cycles.

Table 9.7 Initial and final sizes and maximum stresses for the two-storey steel–concrete setback structure columns (1 in. = 2.54 cm, 1 k = 4.45 kN).

Level	Columns			
	Column line	Initial size (in.⁴)	Final size (in.⁴)	Max. stress (ksi) (+ tension)
2	5	56.6	15.1	14.9
	6	8.0	5.0 in.ª	28.3 (−1.7)
	7	56.6	29.0	14.2
	8	56.6	15.3	−18.5
	9	8.0	5.0 in.ª	27.3 (−1.6)
	10	56.6	19.7	−13.6
1	1	56.6	12.9	5.2
	2	56.6	28.1	12.6
	3	56.6	27.4	−12.0
	4	56.6	12.9	4.6
	5	56.6	7.8	17.0
	6	8.0	8.8 in.	39.0 (−2.3)
	7	56.6	49.2	8.4
	8	56.6	7.0ª	−17.1
	9	8.0	7.8 in.	42.8 (−2.5)
	10	56.6	25.2	−8.9

ªMinimum allowable sizes.

The optimization of this structure required six cycles to reach a less than 2 percent change in weight. The initial weight was 391 kips (177 Mg) and the final weight was 131 kips (59.3 Mg). Initial and final design sizes are given in Table 9.7. The final solution produces optimality criteria which range in values from 0.4 to 1.4. The reason for the variation is due to the fact that the concrete elements have stabilized close to their minimum values providing little change in the weight which in turn terminated the process. If a true optimal solution was wanted, a smaller percentage of weight change for termination would be required. When the structure contains both steel and concrete elements, it may be more realistic to use cost as an objective function. Cheng *et al.* developed several cost functions that accounted for material fabrication, concrete forming, painting, volume, etc. which are very useful with mixed-material structures [13–16, 21]. The trend for this optimal solution is to reduce columns 5 and 8 to small values and use concrete columns 6 and 9 along with steel columns 8 and 10 and beams 6 and 7 to resist the *y* displacements. Along with these frames, columns 2 and 3 on the first floor coupled with the *x* bracing provide a system which helps resist the torsion of the first floor which in effect reduces the *y* displacement on the second level due to the offset of the first- and second-level centres of mass.

The final set of active constraints includes stresses, displacements and side constraints. The second-level *y* displacement is active with a value of 0.58 in. (1.47 cm). The stresses for the beams on level 2, bays 3, 6 and 7, are active as well as the stress in the panel on the lower level between columns 5 and 8. The stresses are shown in Tables 9.7–9.10. Note that there is a slight overstress of approximately 5 percent for the beam on level 2, bay 3, which was allowed by the fact that the feasible range for an active constraint was set at 5 percent above or below the maximum value. All stresses, displacements and sizes are within the acceptable ranges, thereby producing a feasible design, although a wide range of member sizes (an impractical design) was obtained since linking was not used. The panels are so rigid that the x displacements are approximately 10 percent of the constraint

Table 9.8 Initial and final sizes and maximum stresses for the two-storey steel–concrete setback structure beams (1 in. = 2.54 cm, 1 k = 4.45 kN).

Level	Beams			
	Bay	Initial (in.⁴)	Final (in.⁴)	Max. Stress (ksi)
2	2	146.0	24.5	13.4
	3	146.0	25.3	17.4[a]
	6	146.0	34.8	16.4
	7	146.0	22.9	15.8
	10	146.0	12.6	0.0
	11	146.0	12.6	0.1
1	1	146.0	12.8	4.9
	2	146.0	17.9	10.6
	3	146.0	18.5	15.0
	4	146.0	12.6	1.4
	5	146.0	12.8	4.9
	6	146.0	21.7	14.0
	7	146.0	17.5	13.4
	8	146.0	12.6	1.0
	9	146.0	12.6	0.5
	10	146.0	12.6	0.3
	11	146.0	12.6	0.0

[a]Maximum allowable stresses.

Table 9.9 Initial and final sizes and maximum stresses for the two-storey steel–concrete setback structure panels (1 in. = 2.54 cm, 1 k = 4.45 kN).

Level	Panels			
	Bay	Initial (in.)	Final (in.)	Max. stress (ksi)
2	10	12.0	4.0[a]	0.9
	11	12.0	4.0[a]	0.6
1	10	12.0	4.0[a]	3.0[b]
	11	12.0	4.0[a]	−0.9

[a]Minimum allowable sizes.
[b]Maximum allowable stresses.

Table 9.10 Initial and final sizes and maximum stresses for the two-storey steel–concrete setback structure braces (1 in. = 2.54 cm, 1 k = 4.45 kN).

Top column line	Braces Bottom column line	Initial (in.²)	Final (in.²)	Max. stress (ksi)
3	2	3.63	15.0[a]	12.7
2	3	3.63	15.0[a]	−12.5

[a] Maximum allowable sizes.

values, but it is interesting to note that the optimal solution tries to reduce the original structure to a set of one-bay frames parallel to the direction of the active displacement in order to control this displacement. This is a common trend for a pure optimization problem where every element is free to reach its own specific value at the optimal solution. For the simplest cases of one loading condition where there are no side constraints, the problems often reach a statically determinate or simplified frame solution.

9.7 Frequency constrained systems with a sample illustration

The natural frequencies of a structure are of the utmost importance when considering dynamic loads. By controlling one or multiple frequencies of a structure, it may be possible to avoid resonance, thereby reducing the dynamically generated displacements and stresses. Controlling more than one frequency along with displacements and stresses can cause difficulties due to competing optimal structural stiffness and mass distributions for the different constraints. The structural optimization will try to find the optimal stiffness and mass distributions to control multiple frequencies (and mode shapes), but these can numerically compete with the optimal control of the displacement and constraints. Typically, controlling one frequency coupled with displacements and stresses is not a problem. In structural systems subjected to seismic loads, which is the case for the presented examples, the first mode is a significant contributor to the response of the structure and is the most important to control, from both a code-compliance issue and the dynamic responses.

Frequency constraint gradients are found by a direct differentiation of the free-vibration equation used to find the undamped natural frequencies. The equation used to find a single frequency ω_j is:

$$[[K]_T - \omega_j^2 [M]_T] \{\phi\}_j = 0 \tag{9.49}$$

where the subscript T represents the total global systems, and $\{\phi\}_j$ is the mode shape associated with the jth natural frequency. Differentiating Eq. (9.49) with respect to the primary design variable, δ_i, yields:

$$\left[\frac{\partial [K]_T}{\partial \delta_i} - \frac{\partial \omega_j^2}{\partial \delta_i}[M]_T - \omega_j^2 \frac{\partial [M]_T}{\partial \delta_i}\right]\{\phi\}_j + \left[[K]_T - \omega_j^2 [M]_T\right]\frac{\partial \{\phi\}_j}{\partial \delta_i} = 0$$

(9.50)

Since the stiffness and mass are symmetric, pre multiplying Eq. (9.50) by $\{\phi\}_j^T$ causes the second term to become zero as seen in Eq. (9.49), leaving this equation:

$$\{\phi\}_j^T \left[\frac{\partial [K]_T}{\partial \delta_i} - \frac{\partial \omega_j^2}{\partial \delta_i}[M]_T - \omega_j^2 \frac{\partial [M]_T}{\partial \delta_i}\right]\{\phi\}_j = 0$$

(9.51)

Rearranging this equation gives the gradient of the square of the natural frequency as:

$$\frac{\partial \omega_j^2}{\partial \delta_i} = \frac{\{\phi\}_j^T \left[\dfrac{\partial [K]_T}{\partial \delta_i} - \omega_j^2 \dfrac{\partial [M]_T}{\partial \delta_i}\right]\{\phi\}_j}{\{\phi\}_j^T [M]_T \{\phi\}_j}$$

(9.52a)

In many instances the eigenvectors (mode shapes) are normalized relative to the mass matrix giving:

$$\frac{\partial \omega_j^2}{\partial \delta_i} = \{\phi\}_j^T \left[\frac{\partial [K]_T}{\partial \delta_i} - \omega_j^2 \frac{\partial [M]_T}{\partial \delta_i}\right]\{\phi\}_j$$

(9.52b)

Using Eqs. (9.52a) and (9.52b), a direct solution for the gradient of the square of the natural frequency can easily be obtained.

Example 9.7.1 – Five-storey, L-plan Structure with frequency constraints. A five-storey, L-shaped structure, shown in Fig. 9.21, with seven column lines and seven beams, all made of steel wide-flange sections, will be used to illustrate the use of multiple-frequency constraints. This example does not use linking of elements. Each storey is 12 ft (3.6 m) tall. Each level has a translational mass of 0.31 k-s² in.$^{-1}$ (54.3 Mg) and a rotational mass of 16,403 k-s²-in. (1,854 Mg-m²). Initially the columns have major-axis moments of inertia of 9,500 in.^{4} (395,400 cm^4). This produces an initial weight of 206 kips (93.3 Mg). The optimization termination criteria are 20 cycles of optimization or less than a 5 percent weight change between cycles. The constraints consisted of keeping the first period between the values of 0.75 and 1.0 s, the second period below 0.50 s and the third period below 0.40 s while keeping each element size between 10.0 in.^{4} (416 cm^4) and 20,000 in.^{4} (832,400 cm^4).

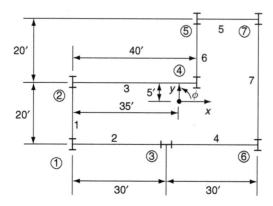

Figure 9.21 Five-storey L-shaped building plan (1 ft = 0.305 m).

Table 9.11 Results for the columns of the five-storey, L-shaped building subjected to frequency (period) constraints (1 in. = 2.54 cm).

Level	Columns (in.⁴)						
	1	2	3	4	5	6	7
5	1,158	944	1,235	1,161	1,261	1,471	1,251
4	2,710	1,898	2,036	2,576	3,147	3,075	2,656
3	4,492	2,873	3,009	4,298	5,459	4,790	3,889
2	5,420	3,591	4,324	5,625	7,262	5,342	4,508
1	5,370	4,493	18,127	6,309	7,274	9,522	9,092

Table 9.12 Results for the beams of the five-storey, L-shaped building subjected to frequency (period) constraints (1 in. = 2.54 cm).

Level	Beams (in.⁴)						
	1	2	3	4	5	6	7
5	661	543	348	592	338	857	846
4	1,932	1,139	506	1,246	571	2,821	2,212
3	3,608	1,939	699	2,041	904	5,898	3,896
2	4,888	2,488	813	2,588	1,115	8,364	4,613
1	4,061	1,702	737	1,805	982	5,747	2,649

The optimization was terminated in four cycles due to less than a 5 percent change in weight between cycles. The final weight is 104 kips (47.1 Mg) or nearly a 50 percent reduction of the initial weight. All three period constraints became active with a first period of 1.02 s, a second period of 0.50 s, and a third period of 0.41 s. None of the side constraints became active as they were the extreme practical limits for sizes of normal wide-flange systems. The final column and beam sizes are shown in Tables 9.11 and 9.12.

There are a few items worth noting with regard to this example. The beams increase in size from bottom to top with the exception of the first level. This is a direct reflection of the added resistance provided to the bottom of the columns by having a fixed base and is common in optimal designs which do not have stress constraints. It is important to note that the modes are coupled since the centre of mass does not coincide with the centre of rigidity. Coupling just implies that the mode shapes will have x, y and ϕ components for each mode. The large value of moments of inertia for the first-level, third-column line is due to the need to increase the stiffness in the x direction in order to keep the first mode period under one second. Since column 3 is the only column oriented with its major-axis moment of inertia in the x direction, the optimization provides the most resistance with the least area with this element. The y direction modes require producing a strong frame with columns 4 and 5 and beam 6. This is not as evident as the reason for column 3, but the fact that both columns are oriented in the y direction and they are coupled together through beam action provides the optimal means for controlling the y modes. The ability to maintain certain frequencies or periods is very important when trying to force a structure to remain in a certain region of a response spectrum or if one is trying to control a code-based minimum or maximum specified period during a design.

9.8 Dynamic displacement and stress gradients

The dynamic response gradients are derived by direct differentiation of the equations used to solve for the displacements. Once the gradients of the dynamic displacements are known, the virtual-load vectors derived for the static stresses can be multiplied by the dynamic-displacement gradients to generate the dynamic-stress gradients.

Dynamic-related displacements can be found in several ways. Older codes such as UBC [23], BOCA [24] and the ATC-03 (not technically a code) [25] as well as the new IBC [26] use equivalent lateral forces that are loosely based on first-mode response-spectra techniques. A second method is either code developed [23–26] or actual modal response-spectra analyses and a third technique would be time-history analyses using an actual time-based loading [27–29]. Equivalent lateral forces will be covered in later examples while the classical techniques for dynamic analysis and the development of the necessary gradients will be presented.

9.8.1 *Direct modal superposition displacement gradients*

Although the direct modal superposition method for finding dynamic displacements is rarely used due to the overestimation of responses that occur with this technique, it is simple and useful for understanding how to find dynamic displacement gradients based on modal-solution techniques. The direct modal superposition method for finding dynamic displacements is based on the equation:

$$\{U\} = [\phi]\{P\}_{max} \tag{9.53}$$

where $[\phi]$ is the matrix of eigenvectors and $\{P\}_{max}$ is the generalized displacement found using a response spectrum. This can also be written in component form as:

$$U_j = \sum_{l=1}^{t} \phi_{jl} P_l = \sum_{l=1}^{t} X_l \tag{9.54}$$

where t is the number of modes to be used. The gradient of the jth displacement can then be written as:

$$\frac{\partial U_j}{\partial \delta_i} = \sum_{l=1}^{t} \left(\frac{\partial \phi_{jl}}{\partial \delta_i} P_l + \phi_{jl} \frac{\partial P_l}{\partial \delta_i} \right) = \sum_{l=1}^{t} \frac{\partial X_l}{\partial \delta_i} \tag{9.55}$$

Note that Eq. (9.55) requires the gradients of both the eigenvectors and the generalized displacements.

As noted earlier, the generalized displacement is a function of the response spectrum generated for the prescribed dynamic loading. These response spectra can be generated for a specific dynamic loading or they can be an amalgamation of several different dynamic loads or even pseudo-dynamic loads. The last two scenarios are very common for the development of code-based response spectra that are used to reflect potential seismic loads for large regions. The generalized displacement gradients can be found for the jth component from the equation of the generalized displacement.

$$P_j^{max} = \frac{\{\phi\}_j^T [M]}{m_j \omega_j^2} \{S_a\}_j \tag{9.56}$$

where $\{S_a\}_j$ is the value for the spectral acceleration for the jth natural frequency (or period) and

$$m_j = \{\phi\}_j^T [M] \{\phi\}_j \tag{9.57}$$

Taking the partial derivative of Eq. (9.56) gives:

$$\begin{aligned}
\frac{\partial P_j^{max}}{\partial \delta_i} &= \frac{\partial \{\phi\}_j^T}{\partial \delta_i} \frac{[M]}{m_j \omega_j^2} \{S_a\}_j + \{\phi\}_j^T \frac{\partial [M]}{\partial \delta_i} \frac{\{S_a\}_j}{m_j \omega_j^2} + \frac{\{\phi\}_j^T [M]}{m_j \omega_j^2} \frac{\partial \{S_a\}_j}{\partial \delta_i} \\
&+ \frac{\{\phi\}_j^T [M]}{\omega_j^2} \{S_a\}_j \frac{\partial \left(\frac{1}{m_j} \right)}{\partial \delta_i} + \frac{\{\phi\}_j^T [M]}{m_j} \{S_a\}_j \frac{\partial \left(\frac{1}{\omega_j^2} \right)}{\partial \delta_i}
\end{aligned} \tag{9.58}$$

The solution of Eq. (9.58) requires the gradients of the eigenvectors, as well as the gradients for the reciprocal generalized mass, the spectral accelerations, the total mass and the reciprocal of the squared natural frequency.

Using the chain rule, the gradient for the reciprocal generalized mass can be defined as:

$$\frac{\partial \left(\dfrac{1}{m_j}\right)}{\partial \delta_i} = \frac{\partial \left(\dfrac{1}{m_j}\right)}{\partial m_j} \frac{\partial m_j}{\partial \delta_i} = -m_j^{-2} \frac{\partial m_j}{\partial \delta_i} \tag{9.59a}$$

where

$$\frac{\partial m_j}{\partial \delta_i} = 2 \frac{\partial \{\phi\}_j^T}{\partial \delta_i} [M] \{\phi\}_j + \{\phi\}_j^T \frac{\partial [M]}{\partial \delta_i} \{\phi\}_j \tag{9.59b}$$

giving

$$\frac{\partial \left(\dfrac{1}{m_j}\right)}{\partial \delta_i} = -m_j^{-2} \left[2 \frac{\partial \{\phi\}_j^T}{\partial \delta_i} [M] \{\phi\}_j + \{\phi\}_j^T \frac{\partial [M]}{\partial \delta_i} \{\phi\}_j \right] \tag{9.59c}$$

which requires the gradients of the eigenvectors as well.

The spectral acceleration gradients can be found using the chain rule. Typically the spectral accelerations in design codes are provided in terms of the period. Using the chain rule gives:

$$\frac{\partial \{S_a\}_j}{\partial \delta_i} = \frac{\partial \{S_a\}_j}{\partial T_j} \frac{\partial T_j}{\partial \omega_j} \frac{\partial \omega_j}{\partial \omega_j^2} \frac{\partial \omega_j^2}{\partial \delta_i} \tag{9.60}$$

where the period can be written as $T_j = (2\pi)/\omega_j$ giving:

$$\frac{\partial T_j}{\partial \omega_j} = -\frac{2\pi}{\omega_j^2} \tag{9.61}$$

and

$$\frac{\partial \omega_j}{\partial \omega_j^2} = \frac{\partial (\omega_j^2)^{\frac{1}{2}}}{\partial \omega_j^2} = \frac{1}{2} \frac{1}{\omega_j} \tag{9.62}$$

Substituting Eqs. (9.61) and (9.62) into Eq. (9.60) gives

$$\frac{\partial \{S_a\}_j}{\partial \delta_i} = -\frac{\partial \{S_a\}_j}{\partial T_j} \frac{\pi}{\omega_j^3} \frac{\partial \omega_j^2}{\partial \delta_i} = -\{\eta\}_j \frac{\pi}{\omega_j^3} \frac{\partial \omega_j^2}{\partial \delta_i} \tag{9.63}$$

where $\{\eta\}_j$ represents the slope of the response spectrum for period T_j. If the response spectrum is represented with an equation, the slope can be replaced by the derivative of that equation. Often in code-based modal analyses, the response spectrum is represented as a coefficient times a function of the period such as the reciprocal of the square root of the period. This concept will be shown in the examples.

The gradient of the reciprocal, square of the natural frequency can be developed similarly to that of the reciprocal of the generalized mass. The chain rule can be used to produce:

$$\frac{\partial\left(\dfrac{1}{\omega_j^2}\right)}{\partial\delta_i} = \frac{\partial\left(\dfrac{1}{\omega_j^2}\right)}{\partial\omega_j^2}\frac{\partial\omega_j^2}{\partial\delta_i} = -\frac{1}{\omega_j^4}\frac{\partial\omega_j^2}{\partial\delta_i} \tag{9.64}$$

Using the information in Eqs. (9.59b), (9.63) and (9.64), the jth generalized-displacement gradient can be written as:

$$\begin{aligned}
\frac{\partial P_j^{\max}}{\partial\delta_i} =\ & \frac{\partial\{\phi\}_j^T}{\partial\delta_i}\frac{[M]_T}{m_j\omega_j^2}\{S_a\}_j + \{\phi\}_j^T\frac{\partial[M]_T}{\partial\delta_i}\frac{\{S_a\}_j}{m_j\omega_j^2} \\[2mm]
& -\{\phi\}_j^T\frac{[M]_T}{m_j^2\omega_j^2}\left[2\frac{\partial\{\phi\}_j^T}{\partial\delta_i}[M]_T\{\phi\}_j + \{\phi\}_j^T\frac{\partial[M]_T}{\partial\delta_i}\{\phi\}_j\right]\{S_a\}_j \\[2mm]
& -\{\phi\}_j^T\frac{[M]_T}{m_j\omega_j^2}\left[\frac{\pi}{\omega_j^3}\{\eta\}_j + \frac{\{S_a\}_j}{\omega_j^2}\right]\frac{\partial\omega_j^2}{\partial\delta_i}
\end{aligned} \tag{9.65}$$

All of the gradients have been derived with the exception of those for the eigenvectors. The gradients of the eigenvectors will be discussed in Section 9.8.2. Note that by orthogonalizing the eigenvectors relative to the mass, the generalized mass term will always be unity. Therefore, the gradient of the reciprocal generalized mass, Eq. (9.59b), will be zero and $m_j = 1$ and Eq. (9.65) becomes:

$$\begin{aligned}
\frac{\partial P_j^{\max}}{\partial\delta_i} =\ & \frac{\partial\{\phi\}_j^T}{\partial\delta_i}\frac{[M]_T}{\omega_j^2}\{S_a\}_j + \{\phi\}_j^T\frac{\partial[M]_T}{\partial\delta_i}\frac{\{S_a\}_j}{\omega_j^2} \\[2mm]
& -\{\phi\}_j^T\frac{[M]_T}{\omega_j^2}\left[\frac{\pi}{\omega_j^3}\{\eta\}_j + \frac{\{S_a\}_j}{\omega_j^2}\right]\frac{\partial\omega_j^2}{\partial\delta_i}
\end{aligned} \tag{9.66}$$

The third term of Eq. (9.65) becomes zero and these effects are absorbed into the first term which provides the effects of the changes in the eigenvectors. If the eigenvectors are not normalized relative to the mass, Eq. (9.65) must be used in its entirety.

9.8.2   Eigenvector gradients

As can be seen by Eqs. (9.65) and (9.66), the gradients of the generalized displacements hinge on finding a numerical means for finding the gradients of the eigenvectors. A direct approach to finding the gradients of the eigenvectors can be implemented by differentiating the free-vibration equation, Eq. (9.49). This differentiation gives:

$$\left[\frac{\partial [K]_T}{\partial \delta_i} - \frac{\partial \omega_j^2}{\partial \delta_i}[M]_T - \omega_j^2 \frac{\partial [M]_T}{\partial \delta_i}\right]\{\phi\}_j + \left[[K]_T - \omega_j^2[M]_T\right]\frac{\partial \{\phi\}_j}{\partial \delta_i} = 0$$

(9.67)

Rearranging Eq. (9.67) gives:

$$\left[[K]_T - \omega_j^2[M]_T\right]\frac{\partial \{\phi\}_j}{\partial \delta_i} = -\left[\frac{\partial [K]_T}{\partial \delta_i} - \frac{\partial \omega_j^2}{\partial \delta_i}[M]_T - \omega_j^2 \frac{\partial [M]_T}{\partial \delta_i}\right]\{\phi\}_j$$

(9.68)

Since the matrix $\left[[K]_T - \omega_j^2[M]_T\right]$ is singular, a simple inversion cannot be performed in order to solve for the gradient of the eigenvector. The fact that the rank of this matrix is less than the number of degrees of freedom implies that the eigenvector is only valid within a multiple of itself. This realization is the key to solving Eq. (9.68). By forcing one component of the gradient of the eigenvector to be zero, it is equivalent to forcing that component of the eigenvector to take on a specific, unchanging value. This value could be unity or any other unchanging value. The specified value of zero for one component in the gradient of the eigenvector provides the necessary boundary condition (elimination of a row and column, in effect the singularity) needed to provide a matrix that is non-singular and which can be inverted. Therefore, the eigenvector gradient will take the form:

$$\frac{\partial \{\phi'\}_j^T}{\partial \delta_i} = \left\{\frac{\partial \phi'_{1j}}{\partial \delta_i}, \ldots, 0, \ldots, \frac{\partial \phi'_{nj}}{\partial \delta_i}\right\}$$

(9.69)

where one term has been chosen to be zero. Any term can be chosen to be zero. The prime indicates that this gradient is enforcing a term to be zero. Therefore, the eigenvector gradients can be written as:

$$\frac{\partial \{\phi'\}_j}{\partial \delta_i} = -\left[[K]_T - \omega_j^2[M]_T\right]_R^{-1}\left[\frac{\partial [K]_T}{\partial \delta_i} - \frac{\partial \omega_j^2}{\partial \delta_i}[M]_T - \omega_j^2 \frac{\partial [M]_T}{\partial \delta_i}\right]_R\{\phi'\}_j$$

(9.70)

where R represents the reduced stiffness and mass matrices (the zeroed degree of freedom row and column removed) and the prime stands for the eigenvector with the one component removed.

If the eigenvector is normalized with respect to the mass or some other quantity, the change in the jth component will not be zero from one iteration to the next, and a correction term must be applied. Assuming that the eigenvector is normalized with respect to the mass, the correction term is found by writing the eigenvector which is normalized with respect to the mass in terms of the eigenvector with the jth component equal to unity as:

$$\{\phi''\}_j = \frac{\{\phi'\}_j}{\left[\{\phi'\}_j^T [M]_T \{\phi'\}_j\right]^{\frac{1}{2}}} = \frac{\{\phi'\}_j}{m_j^{1/2}} \tag{9.71}$$

The gradient of the normalized eigenvector becomes:

$$\frac{\partial \{\phi''\}_j}{\partial \delta_i} = \left(1/m_j^{1/2}\right) \frac{\partial \{\phi'\}_j}{\partial \delta_i} + \{\phi'\}_j^T \frac{\partial \left(1/m_j^{1/2}\right)}{\partial \delta_i} \tag{9.72}$$

Using the chain rule provides:

$$\frac{\partial \{\phi''\}_j}{\partial \delta_i} = \frac{1}{m_j^{1/2}} \frac{\partial \{\phi'\}_j}{\partial \delta_i}$$

$$- \frac{1}{2} \{\phi'\}_j^T \frac{1}{m_j^{3/2}} \left[2\frac{\partial \{\phi'\}_j^T}{\partial \delta_i} [M]_T \{\phi'\}_j + \{\phi'\}_j^T \frac{\partial [M]_T}{\partial \delta_i} \{\phi'\}_j \right] \tag{9.73}$$

where the prime indicates the reduced unnormalized eigenvector and the double prime represents the reduced normalized eigenvector.

Solving for the gradients of the eigenvectors is computationally intense, but is necessary for accurate gradients and optimization of pseudo-dynamic (modal) analysis problems. The beauty of this approach is that the gradients of the eigenvectors needed or used in the analysis can be found without including all other eigenvectors. If the same component is held constant for each eigenvector, the matrix in Eq. (9.70) only needs to be inverted once per optimization cycle. If different components are held constant in different eigenvectors, the matrix must be inverted for each eigenvector. Now that the gradients of the eigenvector are found, the gradients of the direct modal superposition displacements in Eq. (9.55) can be found.

9.8.3 Eigenvector gradient numerical example

The single-bay, single-storey example used in Chapter 8 to illustrate the optimality-criteria algorithm will be used to numerically illustrate the accuracy of the eigenvector-gradient calculations. The three-dimensional structure consists of a square, rigid-slab which is 10 ft by 10 ft (3.05 m by 3.05 m) and is supported by three 10 ft (3.05 m) rectangular steel columns (A, B and C). The rectangular cross-sections are assumed to have a depth to width ratio of 1.5, and the concrete panel is assumed to have a fixed depth which spans between columns B and C. The column depths are parallel to the y direction. The steel x-bracings (D and E) are in one vertical plane with a concrete flexural panel (F) in a vertical plane as shown in Fig. 9.22. In this example, the concrete panel does not include steel. The loading consists of 300 kips (1,335 kN) in the x direction, 100 kips (445 kN) in the y direction and 3,000 kip-in (339 kN-m) in the θ direction. The mass used in the example is the structural mass associated with the elements using the specific weight of steel as 490 pcf (7,820 kg m^{-3}) and the specific weight of concrete of 150 pcf (2,395 kg m^{-3}). The eigenvector, natural frequency and subsequent gradient calculations are generated for the element sizes of $I_{Xa} = 1,576$ in.^{4}, $I_{xB} = 743$ in.^{4}, $I_{xC} = 1,121$ in.^{4}, $A_{D,E} = 47.1$ in.^{2}, and $I_F = 955,130$ in.^{4} (1 in.$^4 = 41.6$ cm^4, 1 in.$^2 = 6.45$ cm^2).

The first natural frequency is found to be 92.14 rad s^{-1}, and the first eigenmode, normalized to the x-degree of freedom, is $\{\phi\} = \{\ x\quad y\quad \theta\ \} = \{\ 1.0\quad 0.951\quad -0.016\ \}$. The actual gradients of the natural frequency and the eigenvector calculations will be compared to numerically generated values found by making a small change in each variable.

Several quantities are needed to find both the gradients for the natural frequency and the eigenvectors with respect to element A. From Eq. (9.52), the important quantities are:

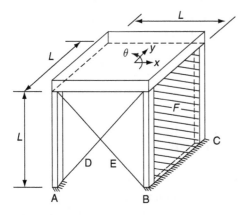

Figure 9.22 One-storey one-bay with all possible elements.

$$\frac{\partial [K]}{\partial I_A} \{\phi\} = \left\{ \begin{array}{l} 4.32 \times 10^{-3} \\ 3.96 \times 10^{-1} \\ -2.57 \times 10^1 \end{array} \right\} \tag{a}$$

and

$$\frac{\partial [M]}{\partial I_A} \{\phi\} = \left\{ \begin{array}{l} 1.57 \times 10^{-6} \\ 1.49 \times 10^{-6} \\ -1.79 \times 10^{-4} \end{array} \right\} \tag{b}$$

and

$$[M] \{\phi\} = \left\{ \begin{array}{l} 7.08 \times 10^{-2} \\ 6.73 \times 10^{-2} \\ -4.30 \times 10^0 \end{array} \right\} \tag{c}$$

From Eq. (9.52)

$$\frac{\partial \omega^2}{\partial I_A} \{\phi\} = \frac{7.90 \times 10^{-1} - 8,490 \left(5.84 \times 10^{-6}\right)}{0.203} = 3.647 \tag{d}$$

and the numerically calculated value for a 0.01 change in the moment of inertia for element A gives $\Delta \omega^2 / \Delta I_A = 3.648$ which is very close to the numerically generated value. Using Eqs. (a)–(d) and substituting these values into Eq. (9.68) gives

$$[[K]_T - \omega^2 [M]_T] \frac{\partial \{\phi\}}{\partial I_A} = -\left\{ \begin{array}{l} 4.32 \times 10^{-3} \\ 3.96 \times 10^{-1} \\ -2.57 \times 10^1 \end{array} \right\}$$

$$+ 8,490 \left\{ \begin{array}{l} 1.57 \times 10^{-6} \\ 1.49 \times 10^{-6} \\ -1.79 \times 10^{-4} \end{array} \right\}$$

$$+ 3.647 \left\{ \begin{array}{l} 7.08 \times 10^{-2} \\ 6.73 \times 10^{-2} \\ -4.31 \times 10^0 \end{array} \right\} \tag{e}$$

$$= \left\{ \begin{array}{l} 2.67 \times 10^{-1} \\ -1.39 \times 10^{-1} \\ 8.51 \times 10^0 \end{array} \right\}$$

Once the right side of Eq. (9.68) is generated as shown in Eq. (e), a boundary condition is applied to generate a solution for the eigenvector gradients. Since the left matrix is singular, the eigenvector only provides a direction vector where the magnitude of the vector is meaningless. By setting the first term to be one and scaling the other terms in proportion, the direction remains unchanged. If this first term is always set to be unity in each analysis, there will never be a change in this term and the partial derivative of the component of the eigenvector that has been defined to be unity will be zero, giving the necessary boundary condition for Eq. (9.68) to be solved. This allows the reduced system to become:

$$[[K]_T - \omega^2 [M]_T] \left\{ \begin{array}{c} 0 \\ \partial\phi_2/\partial I_A \\ \partial\phi_3/\partial I_A \end{array} \right\} = \left\{ \begin{array}{c} 0 \\ -1.39 \times 10^{-1} \\ 8.51 \times 10^0 \end{array} \right\} \tag{f}$$

giving:

$$\left\{ \begin{array}{c} \partial\phi_2/\partial I_A \\ \partial\phi_3/\partial I_A \end{array} \right\} = \left\{ \begin{array}{c} -3.85 \times 10^{-5} \\ 5.28 \times 10^{-7} \end{array} \right\} \tag{g}$$

The numerically generated gradient quantities for a 0.01 change in the moment of inertia for element A are -3.85×10^{-5} and 5.32×10^{-7} showing the validity of the numerical calculations. Note that, as explained in Section 9.8.2, a correction term can be applied to these gradients if the eigenvectors are normalized with respect to the mass. These quantities would be used in the generation of the gradients for the dynamic displacements with respect to element A. This type of calculation must be performed for each element, therefore, it is easily seen that the gradient calculations are time-consuming, but the accuracy of the numerically generated values provides the information for accurate and quick convergence of the optimality-criteria algorithm.

9.8.4 Root-sum-of-the-squares displacement gradients

The previous derivations were for the case where direct modal superposition was used to find the displacements. Typically, the square-root-of-the-sum-of-the-squares (SRSS) or the complete-quadratic-combination (CQC) techniques are used as a means of minimizing the effects of adding the maximum contributions from each mode which is the case in the direct modal superposition approach. Only the SRSS technique will be presented, but a similar derivation could be used for the CQC approach. The square root of the sum of the squares can be written as:

$$U_j = \left[\sum_{k=1}^{t} \left(\phi_{jk} P_j^{max} \right)^2 \right]^{1/2} \tag{9.74}$$

where k represents the kth eigenvector, the term in the parentheses represents the kth modal component of U_j and t is the total number of eigenvectors used in the modal analysis. The gradient can be written as:

$$\frac{\partial U_j}{\partial \delta_i} = \frac{1}{2} \left[\sum_{k=1}^{t} \left(\phi_{jk} P_j^{\max} \right)^2 \right]^{-1/2} \frac{\partial \left[\sum_{k=1}^{t} \left(\phi_{jk} P_j^{\max} \right)^2 \right]}{\partial \delta_i}$$

$$= \frac{1}{2} \left[\sum_{k=1}^{t} \left(\phi_{jk} P_j^{\max} \right)^2 \right]^{-1/2} 2 \left[\sum_{k=1}^{t} \left(\phi_{jk} P_j^{\max} \right) \right]$$

(9.75)

$$\left[\sum_{k=1}^{t} \left(\frac{\partial \phi_{jk}}{\partial \delta_i} P_j^{\max} + \phi_{jk} \frac{\partial P_j^{\max}}{\partial \delta_i} \right) \right]$$

$$= \frac{1}{U_j} \left[\sum_{k=1}^{t} \left(\phi_{jk} P_j^{\max} \right) \right] \left[\sum_{k=1}^{t} \left(\frac{\partial \phi_{jk}}{\partial \delta_i} P_j^{\max} + \phi_{jk} \frac{\partial P_j^{\max}}{\partial \delta_i} \right) \right]$$

in which the gradient of the eigenvector as derived in Eq. (9.70) is required as well as the gradient of the generalized displacement as shown in Eqs. (9.65) and (9.66).

9.8.5   Dynamic-stress gradients

The dynamic-stress gradients are found by using Eqs. (9.24) and (9.25). Since Eq. (9.75) produces all of the dynamic displacement gradients with respect to the primary design variables, the stress gradients can be written as:

$$\frac{\partial \sigma_j}{\partial \delta_i} = \frac{\partial \{b\}_j^T}{\partial \delta_i} \{U\} + \{b\}_j^T \frac{\partial \{U\}}{\partial \delta_i}$$

(9.76)

where the vectors $\{b\}_j$ are identical to those created for the static stresses. Just as in the static case, the transformations have no effect as long as both $\{b\}_j$ and $\{U\}$ are in the same coordinate system. As noted in the static examples, the first term of Eq. (9.76) can be neglected, but it will affect the accuracy of the gradients and therefore the rate (and stability) of the convergence.

The gradient calculations are generally the most time-consuming portion of an optimality-criteria algorithm. Yet, accurate gradients are essential to having an algorithm that converges quickly. The gradients found using the developed equations are exact as long as the first term in Eq. (9.76) is not ignored.

9.9 Seismic loads and associated gradients

This section will develop and describe how to use the optimality-criteria optimization procedures for the design of structures subject to earthquake loads. Examples of how to optimize for code-based equivalent lateral forces, modal response-spectra analyses and time-history analysis procedures are presented. Examples including results for each analysis procedure are presented in order to clarify how structural optimization can be used to better the understanding of seismic-resistant designs.

Many times the word 'optimization' is taken literally as finding the globally optimal design. Realistically, structural optimization for earthquake loads should be viewed as an automated design process. A major advantage of using structural optimization for seismically designed structures is the capability of simultaneously optimizing a structure for multiple load cases, i.e. multiple earthquakes. Traditionally, seismic designers design for the anticipated worst earthquake and check other earthquake loads for constraint or design violations. This requires an iterative procedure which can be avoided by using structural-optimization procedures that can simultaneously design for multiple earthquake loadings. The optimization procedure will resize members in a fashion providing the most efficient dynamic properties for all of the predetermined earthquake events, whereas a traditional design focuses on only one event resulting in an overdesigned structure with respect to a suite of different earthquakes.

Structural optimization for seismic loads has several unique features compared to those for static loads. Although probabilistic seismic loads [13, 14] can be used in seismic optimization, this section will focus on deterministic systems where the seismic loads have specific load equations or accelerograms which are used to reflect the possible seismic events. A major difference in optimizing a dynamically excited structure versus a statically loaded structure is that the inertia-generated earthquake loads and responses change each time the structure is modified. Therefore, each iteration of the optimization procedure requires the regeneration of the seismic-related loads and may require the gradients of the loads relative to the primary design variables. A statically loaded structure changes with each iteration of the optimization procedure, but the loading remains constant allowing certain simplifications within the optimization algorithms, as noted earlier, that cannot be made for seismic optimization.

Deterministic, seismically designed structures are typically loaded either through equivalent static forces based on design-code formulas, a response spectrum or a time-dependent accelerogram. Equivalent static-force procedures can be a single set of forces or multiple sets of modal forces used as a means of approximating a full modal response-spectra analysis. The equivalent static-force method is a common approach for designers when actual earthquake data is scarce or its use is unwarranted (non-critical structures). Modal response-spectrum and time-history analysis procedures are very common when earthquake data for a specific site is known or when the structure is deemed a critical facility. Many times these procedures are used for structures located in a region of frequent seismic activity or a region of potentially high-magnitude seismic events.

9.9.1   Equivalent static-force analyses

The Uniform Building Code (UBC) [23] equivalent static-design forces are based on finding a total base shear and then distributing this base shear to each floor according to the mass distribution and distance above the ground. The UBC equivalent static-base shear equation is of the form:

$$V = \frac{1.25ZISW}{R_w T^{2/3}} \tag{9.77}$$

where Z is a seismic zone factor, I is an importance factor, S is a soil factor, R_w is a structural framing-related ductility factor, T is the fundamental period of the structure and W is the seismic dead load. Note that both the weight, W, and the period, T, will change after each cycle of optimization which clearly shows the difference between seismic and static loadings. (Also note that the term $1/T^{2/3}$ is a means of approximating the shape of a seismic design response spectrum.) Typically, the lateral forces are developed as a percentage of the base shear distributed according to height and mass for a certain level. The UBC lateral-force distribution is of the form:

$$F_x = \frac{(V - F_t)\,w_x h_x}{\sum\limits_{i=1}^{n} w_i h_i} = (V - F_t)\,\rho_x \tag{9.78}$$

where F_x is the xth level horizontal force, $F_t = 0.07TV$ and is the top-storey horizontal lateral force, w_x is the xth level portion of the weight W used in Eq. (9.77), h_x is the height above the base to level x and ρ_x is the xth storey distribution percentage. Note that the base shear and lateral forces in Eqs. (9.77) and (9.78) are dependent upon the natural period of the structure and the weight of each floor which will change in each cycle of the optimization.

For the UBC equivalent static case the gradient of load vector $\{P\}$ from Eq. (9.78) becomes:

$$\frac{\partial \{P\}}{\partial \delta_i} = (1 - 0.07T)\,\frac{1.25ZIS}{R_w}\left[T^{-2/3}\frac{\partial W}{\partial \delta_i} + W\frac{\partial T^{-2/3}}{\partial \delta_i}\right]\{\rho\} - 0.07V\frac{\partial T}{\partial \delta_i} \tag{9.79}$$

where $\{\rho\}$ is the vector of storey percentages for the distribution of the base shear according to storey heights and masses. The gradient of the period can be found by using the chain rule (Cheng and Truman [2]) and the well-known natural frequency gradient as shown in Section 9.7.

9.9.2 Modal response-spectra analyses

The modal response-spectrum procedure as suggested by the NEHRP Provisions (1992) [25] is also based on finding a set of equivalent static forces based on modal base shear forces for a prescribed number of modes. The displacements, stresses and drifts resulting from individual modes are then combined through the use of a statistically based procedure such as the square-root-of-the-sum-of-the squares (SRSS) or complete-quadratic-combination (CQC). The jth modal base shear for a response-spectrum analysis is of the form:

$$V_j = \frac{\overline{W}_j}{g} S_{Aj} \tag{9.80}$$

where the effective weight $\overline{W}_j$ is defined as:

$$\overline{W}_j = \frac{\left[\{\phi\}_j^T [M] \{I\}\right]^2}{\{\phi\}_j^T [M] \{\phi\}_j} g \tag{9.81}$$

where $\{\phi\}_j$ is the jth mode shape, $[M]$ is the mass matrix, S_{Aj} is the jth mode spectral acceleration, g is the gravitational constant and $\{I\}$ is a unity vector. Note that the mode shapes and the spectral acceleration are dependent upon the natural period of the structure. A classical response-spectrum procedure where the displacements are found directly from the response-spectrum analysis coupled with optimization procedures was presented in Section 9.8.1.

The gradients for the modal response-spectra displacements, stresses and drifts are also derived by direct differentiation and application of the chain rule by Cheng and Truman [1–4]. Therefore, the gradient for the jth modal base shear can be written as:

$$\frac{\partial V_j}{\partial \delta_i} = \frac{1}{g} \left[\frac{\partial \overline{W}_j}{\partial \delta_i} S_{Aj} + \overline{W}_j \frac{\partial S_{Aj}}{\partial \delta_i} \right] \tag{9.82}$$

where the partial derivative of the effective weight from Eq. (9.81) requires the partial derivatives of the mode shapes, the mass and the spectral acceleration as developed in Section 9.8.1. The gradient of the load vector $\{P\}$ is found by using the appropriate lateral-force-distribution equations which would be similar to Eq. (9.78).

9.9.3 Ground motion accelerogram – time-history analyses

Time-history analyses are based on the solution of the classical differential equation for seismic analysis using a ground-motion accelerogram. The differential equation can be solved by numerous techniques as described in Bathe and Wilson [27] or Cheng [28]. Truman and Petruska [7] have successfully used Newmark integration procedures within the optimization algorithms. The classical differential equations are of the form:

$$[M]\{\ddot{u}\} + [C]\{\dot{u}\} + [K]\{u\} = [M]\{I\}\ddot{x} = \{R\} \tag{9.83}$$

where $[C]$ is the damping matrix, $[K]$ is the stiffness matrix, $\{u\}$ is the relative displacement vector, $\{R\}$ is the load vector, $\ddot{x}$ is the ground acceleration and each dot is one derivative with respect to time. Note that the inertia force on the right side of Eq. (9.83) is mass- and time-dependent, therefore, the maximum response for each iterative design typically occurs at different times within a single earthquake loading. Using a Newmark integration scheme allows direct formulation of the gradients for the velocity and acceleration in terms of the displacement gradients (Truman and Petruska, [7]).

The algorithm for using the Newmark integration method [27, 28] can be broken down into two parts: initial calculations and calculations for each time step. The initial calculations involve forming the mass, damping and stiffness matrices, initial displacement, velocity and acceleration at time zero, selecting a suitable time step, Δt, calculating the integration constants a_{0-7} and forming the effective stiffness matrix as:

$$\left[\hat{K}\right] = [K] + a_0[M] + a_1[C] \tag{9.84}$$

The integration constants for this method have long been established and can be found in [27, 28]. For each time step, the effective loads at time $t + \Delta t$ are calculated as:

$$\begin{aligned}\left\{\hat{R}_{t+\Delta t}\right\} = \{R_{t+\Delta t}\} &+ [M]\left(a_0\{U_t\} + a_2\left\{\dot{U}_t\right\} + a_3\left\{\ddot{U}_t\right\}\right) \\ &+ [C]\left(a_1\{U_t\} + a_4\left\{\dot{U}_t\right\} + a_5\left\{\ddot{U}_t\right\}\right)\end{aligned} \tag{9.85}$$

thus, the displacements at time $t + \Delta t$ can be solved using:

$$\left[\hat{K}\right]\{U_{t+\Delta t}\} = \left\{\hat{R}_{t+\Delta t}\right\} \tag{9.86}$$

and the accelerations and velocities at time $t + \Delta t$ are given as:

$$\left\{\ddot{U}_{t+\Delta t}\right\} = a_0\left(\{U_{t+\Delta t}\} - \{U\}_t\right) - a_2\left\{\dot{U}_t\right\} - a_3\left\{\ddot{U}_t\right\} \tag{9.87}$$

$$\left\{\dot{U}_{t+\Delta t}\right\} = \left\{\dot{U}_t\right\} + a_6\left\{\ddot{U}_t\right\} + a_7\left\{\ddot{U}_{t+\Delta t}\right\} \tag{9.88}$$

Eq. (9.83) indicates that the solution to the differential equation is highly dependent on $[M]$ and $[K]$. This is equivalent to forcing the period and resonance effects of the structure to change with respect to the inertia loads, $\{R\}$, for each iteration of the structural optimization.

The equilibrium equations of motion are approximately solved by using Newmark's numerical integration method outlined in Eqs. (9.84)–(9.88). Differentiating Eq. (9.86) with respect to the design variables, and noting that the derivatives of the integration constants with respect to the design variables δ are zero, gives the gradient of the dynamic displacement for each time step as:

$$\frac{\partial \{U_{t+\Delta t}\}}{\partial \delta_i} = [K]^{-1} \left(\frac{\partial \{\hat{R}_{t+\Delta t}\}}{\partial \delta_i} - \left[\frac{\partial [K]}{\partial \delta_i} + a_0 \frac{\partial [M]}{\partial \delta_i} + a_3 \frac{\partial [C]}{\partial \delta_i} \right] \{U_{t+\Delta t}\} \right)$$

(9.89)

where

$$
\begin{aligned}
\frac{\partial \{\hat{R}_{t+\Delta t}\}}{\partial \delta_i} = {} & \frac{\partial R_{t+\Delta t}}{\partial \delta_i} + \frac{\partial [M]}{\partial \delta_i} \left(a_0 \{U_t\} + a_2 \{\dot{U}_t\} + a_3 \{\ddot{U}_t\} \right) \\[6pt]
& + [M] \left(a_0 \frac{\partial \{U_t\}}{\partial \delta_i} + a_2 \frac{\partial \{\dot{U}_t\}}{\partial \delta_i} + a_3 \frac{\partial \{\ddot{U}_t\}}{\partial \delta_i} \right) \\[6pt]
& + [C] \left(a_1 \frac{\partial \{U_t\}}{\partial \delta_i} + a_4 \frac{\partial \{\dot{U}_t\}}{\partial \delta_i} + a_5 \frac{\partial \{\ddot{U}_t\}}{\partial \delta_i} \right) \\[6pt]
& + \frac{\partial [C]}{\partial \delta_i} \left(a_1 \{U_t\} + a_4 \{\dot{U}_t\} + a_5 \{\ddot{U}_t\} \right)
\end{aligned}
$$

(9.90)

and the derivatives of the acceleration and velocity can be found by directly differentiating Eqs. (9.87) and (9.88). With the displacement gradients for each time step calculated, the storey-drift gradients can be obtained and so can the stress gradients as shown in Section 9.8.5 or Cheng and Truman [1].

9.9.4 Optimization procedures for seismic loads and analyses

Any structural optimization algorithm (computer program) that can account for the gradient of the loads with respect to the design variables can be used for structural optimization for earthquake or inertia-driven dynamic loads. The gradients of these loads with respect to the design variables can be found explicitly or implicitly, but are typically necessary for stable convergence of the optimization algorithms. The need for the gradients of the loads can easily be seen by exploring the use of an optimality-criteria approach which is generally described but specifically modified for earthquake loads.

From a practical point of view, the constraints used to produce a feasible design for seismic design include drift, displacements, frequencies, stresses, stability, reliability and maximum and minimum member sizes. Note that each constraint must have quantifiable gradients, as produced in the preceding sections, for use in the optimality-criteria algorithm. Once these gradients are derived and quantifiable, the structural optimization, using the Kuhn–Tucker relationship, is able to control these responses while finding an optimal structural system as shown in Chapter 8. To be able to control these responses simultaneously for multiple earthquake loads is extremely useful for seismic-resistant design.

9.10   Seismic design examples

Four earthquake (or dynamic) examples are presented. Three use code-based quasi-dynamic (equivalent lateral) or modal analysis while the fourth uses time-history analysis. These techniques, although illustrated using seismic-design code-based processes, can be used for any dynamic excitations. The examples are used to show how structural optimization can provide automated designs and information regarding the behaviour of prescribed structures. The path of optimization can often provide insight that can be used to enhance the designer's understanding of the structure and its interaction with the seismic loading. This enhanced understanding can lead to better seismic-resistant structures.

9.10.1   *Non-linked, wide-flange steel system subject to multi-component modal response-spectrum loading*

A five-storey, L-shaped, rigid diaphragm, moment-resistant frame structure was optimized with respect to two separate load cases in order to study the effects of multi-component modal response-spectra loadings. Material constants are $E = 200,100\,\text{MPA}$ (29,000 ksi) and $\gamma = 7,820\,\text{kg}\,\text{m}^{-3}$ (490 lbs ft^{-3}). The beams and columns (beam-columns) are wide-flange cross-sections as shown in Figs. 9.14, 9.15 and 9.21 with the primary design variable as the major-axis moment of inertia (Truman and Cheng [5]), and each storey height was 3.05 m (10 ft). The initial size of the columns and beams was 395,400 cm^4 (9,500 in.^{4}) for all levels. Each member was free (non-linked) to be any size between 416 cm^4 (10 in.^{4}) and 832,000 cm^4 (20,000 in.^{4}). The constraints consisted of displacement constraints of 1.14 cm (0.45 in.) per storey. The response spectrum in equation form is:

$$\frac{S_a}{a_{\text{max}}} = -26.14T^2 + 13.94T + 0.935 \tag{9.91}$$

for $T \leq 0.4$ s and

$$\frac{S_a}{a_{max}} = -0.1606\,(T-0.4)^4 - 1.141\,(T-0.4)^3 + 2.996\,(T-0.4)^2$$
$$- 3.618\,(T-0.4) + 2.229 \tag{9.92}$$

for $T > 0.4$ s where a_{max} is the maximum ground acceleration.

The first load case used 0.4 g as the maximum ground acceleration in the x direction only and the second load case was 0.4 g in the x direction and 0.267g in the y direction. The third load case includes a third component of 0.267 g in the z direction as well. The modal analyses were performed using three modes and the square-root-sum-of-the-squares (SRSS) method of modal superposition.

Initially, each load case started with active x-displacement constraints on levels 2 through 4. Finally, each loading produced active constraints as shown in Table 9.13. Note that the multi-component load cases have additional y-displacement constraints in the active set. Initially, the multi-component load case requires a smaller weight as shown in Fig. 9.23. This is due to the fact that, initially, the rigidity centre is located within the third quadrant of the global coordinate system. Therefore, the positive y-loading coupled with the rigid diaphragm causes a reduction in the x-displacements. Finally, the multi-component excited structure produces a final design which is 25 percent larger than the single component excited structure. It is important to note that the centre of rigidity for the non-linked case is free to relocate on each level for each cycle of optimization. The increase in weight, as seen in Fig. 9.23, for the last cycle of the x-excitation case is due to a change in the active set of constraints. The final distributions of stiffness are shown in Tables 9.14–9.16. The x-only loading produces a strong frame using columns 1, 3 and 6 with beams 2 and 4. For the multi-component cases, the dominant columns are again 1, 3 and 6 with beams 1, 2, 4 and 6. Columns 1, 3 and 6 with beams 2 and 4 are primarily used to control the x-deflections, while columns 1, 2, 4 and 5 with beams 1 and 6 are used to control the y-deflections. Column 1 is critical since it can be used effectively

Table 9.13 Loads and results for the non-linked, L-shaped steel structure (1 k = 4.45 kN).

Load case	Excitation			Initial Wt.(k)	Final Wt.(k)	Active constraints
	X(g)	Y(g)	Z(g)			
1	0.400	–	–	274.1	108.2	X_4, X_5, Y_4, Y_5
2	0.400	0.267	–	261.7	135.6	$X_4, X_5, Y_2\text{–}Y_5$
3	0.400	0.267	0.267	261.5	135.5	$X_3\text{–}X_5, Y_2\text{–}Y_5$

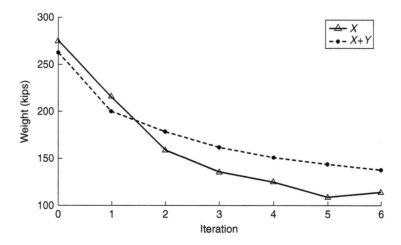

Figure 9.23 Optimal weights for the non-linked five-storey L-shaped structure subjected to multi-component excitations (1 kip = 4.45 kN).

Table 9.14 Final stiffness distribution for the five-storey, L-shaped structure subjected to a single component of excitation, load case 1 (1 in.^{4} = 41.6 cm^4).

Level\ID	Columns (in.^{4})						
	1	2	3	4	5	6	7
5	4,039	1,186	5,778	1,168	2,339	4,244	2,455
4	7,876	1,228	6,854	2,188	4,384	10,864	3,326
3	16,631	1,331	9,106	1,467	3,489	4,627	2,979
2	17,070	1,389	8,045	2,966	6,372	19,220	6,147
1	2,621	984	20,000	1,043	2,070	2,380	1,573
	Beams (in.^{4})						
5	281	1,564	217	1,501	361	462	182
4	116	4,367	401	4,692	891	2,164	199
3	10	7,737	433	4,827	928	2,862	149
2	10	9,715	478	6,717	1,174	400	96
1	50	9,381	450	7,300	914	154	141

as a part of the frame used to resist both sets of orthogonal deflections and at the same time to force the rigidity centre to be within the second quadrant which allows the rotation to help control the y-displacements.

9.10.2 *Linked, wide-flange steel system subject to multi-component modal response-spectrum loads*

This structural system is identical, geometrically and its loading, to that used in the preceding example with the exception that linking is used. The initial

Table 9.15 Final stiffness distribution for the five-storey, L-shaped structure subjected to two components of excitation, load case 2 ($1 \, in.^4 = 41.6 \, cm^4$).

Level\ID	Columns (in.⁴)						
	1	2	3	4	5	6	7
5	9,235	2,860	6,736	3,288	3,556	6,789	1,898
4	10,305	3,913	8,790	4,068	4,789	7,697	2,019
3	20,000	4,981	6,472	4,753	4,929	14,887	1,612
2	20,000	7,503	12,865	5,487	5,232	20,000	2,073
1	20,000	20,000	20,000	6,020	4,547	4,296	2,417
	Beams (in.⁴)						
5	1,833	2,113	297	2,395	314	2,709	1,095
4	5,541	5,760	543	7,172	619	6,895	894
3	8,639	6,697	541	7,406	610	9,612	879
2	12,529	9,065	621	9,724	607	8,184	664
1	9,860	11,601	790	7,833	557	4,323	727

Table 9.16 Final stiffness distribution for the five-storey, L-shaped structure subjected to three components of excitation, load case 3 ($1 \, in.^4 = 41.6 \, cm^4$).

Level\ID	Columns (in.⁴)						
	1	2	3	4	5	6	7
5	9,282	2,866	6,849	3,227	3,469	7,320	1,919
4	10,247	3,884	9,195	3,775	4,100	7,697	1,840
3	20,000	4,836	6,466	4,750	4,830	15,916	1,546
2	20,000	6,942	12,846	5,722	5,023	20,000	2,243
1	20,000	18,544	20,000	8,807	5,360	2,254	1,723
	Beams (in.⁴)						
5	1,856	2,156	298	2,448	313	2,750	1,172
4	5,805	5,953	531	7,396	581	6,000	835
3	8,864	6,749	524	7,629	571	9,046	847
2	12,382	9,008	584	9,840	614	8,247	680
1	9,860	11,692	776	7,873	586	4,586	739

data is exactly the same except that each level is required to have the same size columns throughout that level and the same beam size throughout each level. The linking was used to stabilize the location of the rigidity centre and to provide a more realistic design. Due to the linking and orientation of the columns, the rigidity centre is permanently located within the third quadrant of the global coordinate system for each cycle and level. An even more realistic design could have been achieved by linking multiple levels of columns and/or beams. Truman and Cheng have multiple examples of this scenario [1–3, 5].

The final results for the given linked systems are given in Figs. 9.24 and 9.25 and Table 9.17. The linked system with the fixed location of the rigidity centre allows the multi-component excited structure to continually produce results with less weight than the single component excitation as shown in Fig. 9.24. The optimization path of both systems is nearly identical which is a result of the column orientation. Most of the major axes are oriented to resist translation in the *y* direction. This fact coupled with the linking of columns only produces active constraints in the

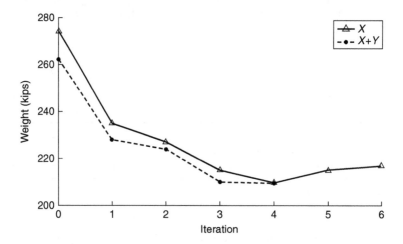

Figure 9.24 Optimal weights for the linked five-storey L-shaped structure subjected to single and two-component excitations (1 kip = 4.45 kN).

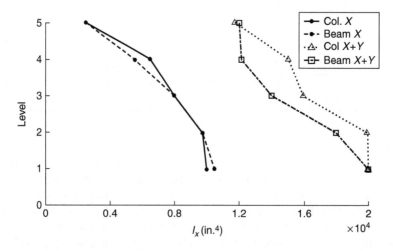

Figure 9.25 Stiffness distributions for the linked five-storey L-shaped structure subjected to single and two-component excitations (1 in. = 2.54 cm).

Table 9.17 Comparison of the linked and non-linked five-storey, L-shaped structure
subjected to multi-component seismic loads (1 k = 4.45 kN).

Excitation		Linked	Initial weight (k)	Final weight (k)	Active constraints
X(g)	Y(g)				
0.40	–	No	274.1	108.2	X_4, X_5, Y_4, Y_5
0.40	–	Yes	274.1	208.2	X_2–X_5
0.40	0.27	No	261.7	135.6	X_4, X_5, Y_2–Y_5
0.40	0.27	Yes	261.7	206.5	X_2–X_5

x direction. Therefore, the only effect of the y direction loading is to reduce
the x-displacements of the mass centre due to the rotation about the centre
of rigidity causing a slightly lighter design. This highlights the need for multiple load cases in order to produce designs that capture the worst-case
scenarios.

The final results and comparison to the non-linked cases are given in
Table 9.17. The single-component, linked case gives a 92 percent increase
in weight over the non-linked case, whereas the multi-component, linked
case provides a 52 percent increase over the non-linked case. The additional weight is accumulated since the y-displacements for the linked case
were only seven to ten percent of the allowable values, and the non-linked
case allowed each element to be designed causing several y-displacement
constraints to become active.

These results suggest that a new design is in order. A lighter structure
could be obtained by reorienting several of the columns such that their
major-axis moment of inertia would help to resist the x-displacements. This
reorientation would allow the y-displacement criteria to play a more significant role in the final design, therefore providing a better balance in
resistance.

9.10.3 *Two-storey, two-bay steel frames subjected to equivalent lateral forces – international building code criteria*

This example uses the International Building Code [26] equivalent lateral-
force equations which are slightly different from the UBC equations and
derivations used in Section 9.9.1. Most of the equivalent lateral-force procedures and equations from the different codes from the last two decades
are very similar. The procedures typically find a base shear based on several constants that are used to reflect site conditions, building ductility and
importance coupled with the inertia-based weight (structural, dead and limited live loads) and the structural period. The base shear equation for this
problem is:

$$V = C_s W \tag{9.93}$$

where C_s is the seismic response coefficient and W is the total dead load of the structure. The value for C_s is dependent on the site location and geotechnical conditions, S_{DS}, the structural ductility factor, R_w, and the structure's importance factor, Π, and can be written as:

$$C_s = \frac{S_{DS}}{R_w / \Pi} \qquad (9.94)$$

where C_s has upper and lower limits as described in these equations.

$$C_{s,max} = \frac{S_{DS}}{T(R_w / \Pi)} \qquad (9.95)$$

$$C_{s,min} = 0.044 S_{DS} \Pi \qquad (9.96)$$

The vertical distribution of forces is similar to the equation used for the UBC [23] as presented in Section 9.9.1. The major exception is that a force of F_t is not required as it was in Eq. (9.78). For the IBC, the forces are distributed using the equation:

$$F_x = C_{vx} V \qquad (9.97)$$

$$C_{vx} = \frac{w_x h_x^k}{\sum\limits_{i=1}^{n} w_i h_i^k} \qquad (9.98)$$

where n is the number of storeys, C_{vx} is the vertical distribution factor, V is the total seismic base shear, w_i and w_x are portions of the gravity load of the structure located or assigned to level i or x, h_i and h_x are the height from the base to level i or x, k is an exponent related to the structure period as follows: for structures having a period of 0.5 s or less, $k = 1$, for structures having a period of 2.5 s or more, $k = 2$, and for structures having a period between 0.5 and 2.5 s, k shall be interpolated between 1 and 2. (The k factor is used to simulate a first-period mode shape between linear and parabolic depending on the period of the structure.)

Many seismic codes have limits on the storey drifts. These limits are used to make sure that a structure will not have a 'soft' storey even though the overall building displacements are within acceptable limits. The code-based drift limits for storey x are controlled by this equation:

$$\delta_x = \frac{C_d \delta_{xe}}{\Pi} \qquad (9.99)$$

where C_d is the deflection amplification factor (used to predict inelastic deflections relative to the elastic deflections), δ_{xe} is the drift determined by an elastic analysis. For the structure in Figs. 9.26 and 9.27, the first-storey

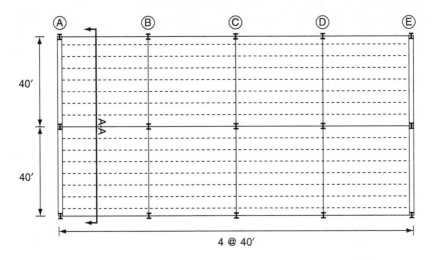

Figure 9.26 Plan view of two-storey office building with moment frames along lines
A and E (1 ft = 0.305 m).

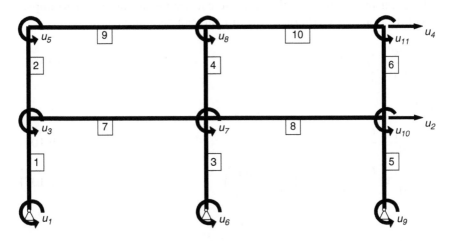

Figure 9.27 Elevation view A-A of the moment frame from Fig. 9.26.

drift is $\delta_{1e} = u_2$, and the second-storey drift is $\delta_{2e} = u_4 - u_2$. The code-based
allowable drifts typically take a form such as $\delta_{x,\text{lim}} = 0.025h_{sx}$ where h_{sx} is
the storey height below level x.

The building system is shown in Fig. 9.26 with the lateral-force-resisting
frame to be optimized shown in Fig. 9.27. The floor dead load is taken to
be 60 psf (2.87 kN m^{-2}) and the roof dead load is 40 psf (1.92 kN m^{-2}). The
building is located in Seattle, WA, on a site that has spectral-response accelerations of $S_s = 1.435$ and $S_1 = 0.485$ and Site Class B. The importance factor

Π is taken as 1.25 and the response modification factor R is 3.5 for an ordinary steel moment frame. Therefore, $F_a = 1.0$, $F_v = 1.0$, $S_{ms} = F_a \times S_s = 1.503$, $S_{m1} = F_v \times S_1 = 0.534$, $S_{DS} = 2/3 \times S_{ms} = 1.002$, and $S_{D1} = 2/3 \times S_{m1} = 0.356$. Assuming that the building is in Seismic Design Category D, $C_u = 1.4$, $C_t = 0.028$, $x = 0.8$, $h_n = 30\,\text{ft}$ (9.15 m), $C_d = 1.45$ and the building is in Seismic Use Group I.

The objective is to minimize the weight while controlling the drift. This can be stated mathematically as:

$$\text{Minimize } W = \sum_{k=1}^{n} w_k \tag{9.100}$$

subject to the two constraints

$$g_1 = \frac{C_d \delta_{xe1}}{\Pi} - 0.025 h_1 \tag{9.101}$$

$$g_2 = \frac{C_d \delta_{xe2}}{\Pi} - 0.025 h_2 \tag{9.102}$$

The gradients of the objective and the constraints are necessary to implement the optimality-criteria algorithm and can be written as:

$$\frac{\partial W}{\partial I_i} = \sum_{k=1}^{n} \frac{\partial w_k}{\partial I_i} \tag{9.103}$$

and

$$\frac{\partial g_1}{\partial I_i} = \frac{C_d}{\Pi} \frac{\partial \delta_{xe1}}{\partial I_i} = \frac{C_d}{\Pi} \frac{\partial u_2}{\partial I_i} \tag{9.104}$$

$$\frac{\partial g_2}{\partial I_i} = \frac{C_d}{\Pi} \frac{\partial \delta_{xe2}}{\partial I_i} = \frac{C_d}{\Pi} \left(\frac{\partial u_4}{\partial I_i} - \frac{\partial u_2}{\partial I_i} \right) \tag{9.105}$$

where

$$\frac{\partial \{u\}}{\partial I_i} = -[K]^{-1} \left[\frac{\partial K}{\partial I_i} \right] \{u\} + [K]^{-1} \left\{ \frac{\partial F}{\partial I_i} \right\} \tag{9.106}$$

as described in Section 9.5.5. As noted in this section, the second term is typically zero in statically loaded systems where the force vector $\{F\}$ is independent of the weight and period of the system. This is not the case in most seismic equivalent lateral-force systems. The gradient of the IBC equivalent lateral forces becomes:

$$\frac{\partial F_x}{\partial I_i} = \frac{\partial C_{vx}}{\partial I_i} V + C_{vx} \frac{\partial V}{\partial I_i} \tag{9.107}$$

where

$$C_{vx} = \frac{w_x h_x^k}{\sum\limits_{j=1}^{n} w_j h_j^k} \tag{9.108}$$

and

$$\frac{\partial C_{vx}}{\partial I_i} = \frac{\left(\sum\limits_{j=1}^{N} w_j h_j^k\right)\left(\frac{\partial w_x}{\partial I_i} h_x^k\right) + w_x h_x^k \ln(h_x)\frac{\partial k}{\partial I_i} - w_x h_x^k \left(\sum\limits_{j=1}^{N} \frac{\partial w_j}{\partial I_i} h_j^k\right) + \sum\limits_{j=1}^{n} w_j h_j^k \ln(h_j)\frac{\partial k}{\partial I_i}}{\left(\sum\limits_{j=1}^{N} w_j h_j^l\right)^2}$$

$$\tag{9.109}$$

and $w_j = \sum\limits_{m} A_m L_m \gamma$ where m represents the member numbers comprising level j and

$$\frac{\partial w_j}{\partial I_i} = \sum\limits_{m} \frac{\partial A_m}{\partial I_i} L_m \gamma \tag{9.110}$$

Note that the primary and secondary design variable relationship in Eq. (9.7) of $A_k = 0.5008 \cdot I_k^{0.487}$ for rolled steel sections is used. Since the base shear V is a function of the period and the weight as shown above, the gradient becomes:

$$\frac{\partial V}{\partial I_i} = \frac{\partial C_s}{\partial I_i} W + C_s \frac{\partial W}{\partial I_i} \tag{9.111}$$

where

$$\frac{\partial C_s}{\partial I_i} = 0 \quad \text{if } C_s \text{ is greater than } C_{s,\max} \tag{9.112}$$

or

$$\frac{\partial C_s}{\partial I_i} = -\frac{S_{D1}}{\left(\frac{R}{\Pi}\right)T^2}\frac{\partial T}{\partial I_i} \quad \text{if } C_s \text{ is less than } C_{s,\max} \tag{9.113}$$

and the gradient of the weight is just the gradient of the objective function as previously derived.

To aid in producing a useable design, useable in terms of constructability and fabrication, it is desired that the final member sizes contain cross-sectional properties similar to members readily available in the AISC Manual of Steel Construction [22]. In this case, it is required that all column members have a depth of at least 20 in. (50.8 cm) (W21 or larger) and all beam members have a minimum depth of 14 in. (35.6 cm) (W14 or larger). To ensure that these requirements are met, the lower stiffness limit is set to 843 in.[4] (35,090 cm[4]) for all columns and 200 in.[4] (8,325 cm[4]) for all beam members. These lower limits correspond to the smallest moment-of-inertia value for W21 and W14, respectively. Therefore, the side constraints become (1 in. = 2.54 cm):

$$I_{1,2,3,4,5,6} \geq 843 \text{ in.}^4 \tag{a}$$

$$I_{7,8,9,10} \geq 200 \text{ in.}^4 \tag{b}$$

As seen in Figs. 9.28 and 9.29, the system reaches its optimal solution in 14 cycles. The iteration data is shown in Table 9.18. The stiffness terms were initialized at a value of 2,000 in.[4] (83,250 cm[4]) to provide the algorithm

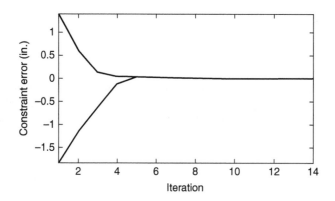

Figure 9.28 Constraint optimization of the two-storey moment frame using equivalent lateral-force procedures (1 in. = 2.54 cm).

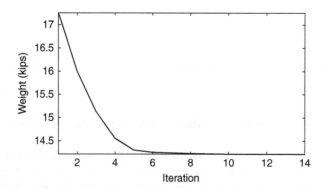

Figure 9.29 Objective function convergence of the two-storey moment frame using equivalent lateral-force procedures (1 kip = 4.45 kN).

Table 9.18 IBC equivalent lateral-force example results (in.^{4}) (1 in. = 2.54 cm, 1 k = 4.45 kN).

Iteration	I_1	I_2	I_3	I_4	I_5	I_6	I_7	I_8	I_9	I_{10}	Weight (k)	g_1	g_2
1	2,000	2,000	2,000	2,000	2,000	2,000	2,000	2,000	2,000	2,000	17.26	1.38	−1.80
2	2,716	1,278	3,767	1,325	2,716	1,278	1,995	1,995	1,108	1,108	15.99	0.62	−1.16
3	2,646	843	5,010	859	2,646	843	2,316	2,316	640	640	15.14	0.14	−0.61
4	2,378	843	5,566	843	2,378	843	2,487	2,487	368	368	14.56	0.05	−0.13
5	2,021	843	5,597	843	2,021	843	2,680	2,680	276	276	14.31	0.03	0.05
6	1,713	843	5,564	843	1,713	843	2,865	2,865	251	251	14.26	0.03	0.01
7	1,554	843	5,747	843	1,554	843	2,973	2,973	234	234	14.24	0.01	0.00
8	1,445	843	5,913	843	1,445	843	3,040	3,040	223	223	14.23	0.00	0.00
9	1,370	843	6,059	843	1,370	843	3,086	3,086	216	216	14.23	0.00	0.00
10	1,321	851	6,188	843	1,321	851	3,114	3,114	209	209	14.23	0.00	0.00
11	1,288	870	6,301	843	1,288	870	3,130	3,130	202	202	14.22	0.00	0.00
12	1,275	864	6,420	843	1,275	864	3,119	3,119	200	200	14.20	0.00	0.02
13	1,261	875	6,470	843	1,261	875	3,127	3,127	200	200	14.22	0.00	0.00
14	1,253	874	6,522	843	1,253	874	3,125	3,125	200	200	14.22	0.00	0.00
	W21X62	W21X55	W30X148	W21X44	W21X62	W21X55	W18X175	W18X175	W14X22	W14X22	21.15		

an initial structure with realistic response values. Had the algorithm been initialized with unrealistic stiffness values (either very large or very small), the possibility of instability becomes much more likely as an ill-conditioned problem may result. The first-floor drift controls the optimization process for the first three iterations, which is to be expected due to the pinned bases. However, both drift constraints become and remain active for the final eleven iterations. It can be seen that elements 2, 4, 6, 9 and 10 reached the lower stiffness limits as specified in the side constraints. Essentially the algorithm has attempted to create a stiff lower level and minimize the weight of the upper floor. As a final step of the design process, the calculated stiffness values can be used to choose a suitable set of wide-flange sections that satisfies the final design requirements. The final row of Table 9.18 provides member sizes that have stiffness properties similar to those calculated by the algorithm. The structural weight of the moment frame using the specified section is somewhat larger than that calculated by the algorithm and so the loads using the equivalent lateral-force method would increase; however, the weight of the moment frame contributes very little to the weight of the entire structure and so the increased force would be minimal. Therefore, the member sizes listed produce a structure close to the optimal design calculated by the algorithm.

9.10.4   Three-storey, two-bay frame subject to ground accelerations – time-history analysis

The three-storey, two-bay frame shown in Fig. 9.30 was analyzed to show the capability of a time-history optimization algorithm. The constraints

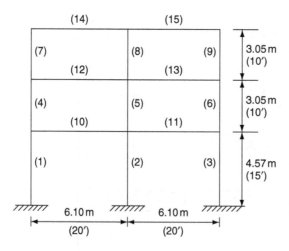

Figure 9.30 Three-storey two-bay frame subjected to seismic accelerograms.

for this design problem are displacement, and minimum and maximum member sizes. The absolute value of the horizontal displacement at each floor was limited to 1.00, 1.65 and 2.35 in. (2.54, 4.19 and 5.97 cm) for the first, second and third floors, respectively. The lower limit on the moment of inertia is 290 in.^{4} (12,070 cm^4). Initial values for the member sizes were 1,500 in.^{4} (62,400 cm^4). Material constants are $E = 29,000$ ksi (200,100 MPa) and $\gamma = 490$ lbs ft.$^{-3}$ (7,820 kg m^{-3}). A uniformly distributed non-structural weight totalling 33.33 kips (148.3 kN) was applied to each beam member of the frame. Damping of the structure was neglected. Three different loadings were examined. At time $t = 0$, the structure is at rest and the base undergoes a transient acceleration. The first base acceleration is defined as:

$$\ddot{a}_g(t) = 343 \sin(2\pi t) \quad 0 \le t \le 1.0 \text{s} \tag{9.114}$$

and is zero for t greater than 1.0 s. The second base acceleration is defined by the accelerogram for the north–south component of the El Centro, California earthquake on 18 May 1940 and is shown in Fig. 9.31. The third base acceleration is for the north–south component of the San Fernando, California earthquake on 9 February 1971 taken at the Caltech Seismological Laboratory and is shown in Fig. 9.32. The constant of 343 used in the sine function was chosen to match the peak acceleration of the El Centro earthquake record. The San Fernando earthquake was also multiplied by

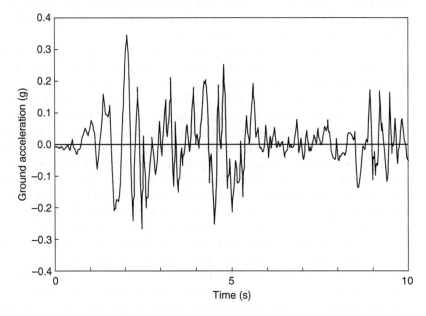

Figure 9.31 1940 El Centro earthquake acceleration record.

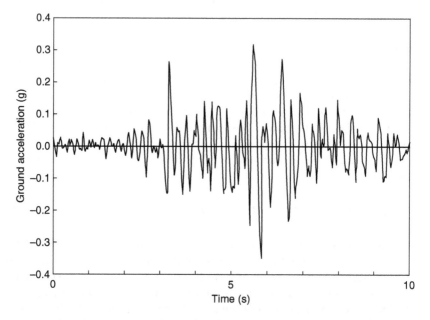

Figure 9.32 1971 San Fernando earthquake acceleration record.

a constant of 1.81 so that both earthquakes would have the same peak acceleration. The time step selected was 0.01 s. The examples use different linking schemes as indicated in Table 9.19.

For the sine load, the time frame examined was 2.0 s and for the El Centro and San Fernando earthquakes, the time frame was 10.0 s. Table 9.19 shows the optimum designs and Fig. 9.33 shows the weight versus iteration.

Comparing Case A to Case B, and Case C to Case D (defined in Table 9.19) shows that the linking examples produce heavier structures since each member is not able to take on its optimum size. It is important to note that the El Centro earthquake results in a slightly lighter structure than the structures simultaneously optimized for the San Fernando and El Centro earthquakes. It is worth noting that both earthquakes in Case E, simultaneous earthquakes, produce active constraints. This indicates that a design based on only one earthquake would most likely violate the constraints when loaded by the other earthquake. Fig. 9.33 shows for Case C a zigzag pattern between iteration 5 through 9 because at iteration 4, the process violates the displacement constraint, and the optimization process is returning to the constraint in iterations 5 through 9. Case D shows a large decrease in weight at iteration 6 due to a change in the active constraints thus allowing the optimization process to use a new path to produce a better solution.

Table 9.19 Optimum solutions for ground accelerations example (1 in. = 2.54 cm, 1 k = 4.45 kN).

Member	Moment of inertia (in.^{4})				
	Case A	Case B	Case C	Case D	Case E
1	1,222.1	1,631.8	2,775.0	1,503.0	1,076.3
2	2,274.1	1,631.8	2,775.0	5,694.3	9,706.3
3	1,222.1	1,631.8	2,775.0	1,501.2	1,308.7
4	381.6	639.8	944.0	621.5	1,004.8
5	1,072.0	639.8	944.0	1,827.3	1,719.0
6	381.6	639.8	944.0	623.7	1,007.9
7	290.0	351.7	657.5	355.0	397.1
8	543.7	351.7	657.5	1,050.7	1,251.8
9	290.0	351.7	657.5	358.4	389.3
10	1,250.8	1,287.2	1,713.9	1,887.0	2,394.9
11	1,250.8	1,287.2	1,716.7	1,887.1	2,632.8
12	740.8	782.4	1,076.8	1,035.0	1,769.0
13	740.8	782.4	1,058.0	1,042.1	1,745.8
14	290.0	290.0	403.9	304.4	356.2
15	290.0	290.0	393.0	307.5	357.0
W^a_{Total}(k)	9.74	9.95	12.14	11.87	13.52
Iteration	4	5	11	10	12
			Active constraints		
d.o.f.	X_3	X_3	X_2	X_2	X_3(El-C)
Time (s)	$t=0.69$	$t=0.69$	$t=8.00$	$t=3.58$	$t=4.74$
Value (in.)	2.35	2.35	1.67	1.63	−2.36
d.o.f.			X_3	X_3	X_3(SF)
Time (s)			$t=8.02$	$t=3.59$	$t=6.81$
Value (in.)			2.40	2.38	2.35
d.o.f.				X_3	
Time (s)				$t=4.22$	
Value (in.)				−2.33	

^{a}Only reflects the optimal weight.
Case A sine function (no linking of members)
Case B sine function (linking of members)
Case C El Centro earthquake (linking of columns)
Case D El Centro earthquake (no linking of members)
Case E El Centro and San Fernando earthquakes as separate load cases (no linking)

9.11 Concluding remarks

Several observations can be made with regard to the use of optimization procedures for earthquake loads. Structural optimization is a viable means of producing seismic-resistant designs. The optimization procedures can produce a series of feasible designs which can be used for the final seismic-resistant design. This series of feasible structures can be extremely useful

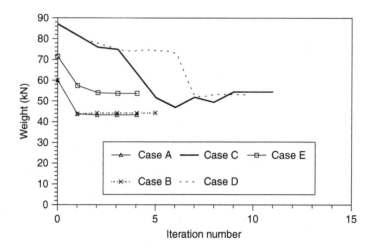

Figure 9.33 Weight convergence for the three-storey two-bay frame subjected to seismic accelerograms [4, 7] (1 kip = 4.45 kN).

in enhancing the designer's understanding of the interaction between the structure and the seismic loads. The use of structural-optimization procedures for full modal response-spectra and time-history analyses provides an opportunity to design structures for multiple earthquakes simultaneously. Being able to design a structure for a family of varied seismic accelerograms is of tremendous benefit. Objective functions can be related to cost, weight, damage, energy absorption or any other quantity that can be related to the actual design of the structure. Nondeterministic procedures are available in order to explore seismic-resistant design from a probabilistic point of view. These techniques are useful in exploring potential structural costs associated with probable events and failures. The use of structural optimization for earthquake loads is not as easily applied as that for static loads, but it is possible. Structural optimization can be an extremely useful and beneficial tool for seismic-resistant-system designers.

References

1. Cheng, F.Y. and Truman, K.Z., *Optimal Design of 3-D Reinforced Concrete and Steel Buildings Subjected to Static and Dynamic Loads Including Code Provisions*, Final Report Series 85-20 for National Science Foundation, U.S. Department of Commerce, NTIS Access No. PB87-168564/AS, Washington, DC, 1985.
2. Cheng, F.Y. and Truman, K.Z., Optimization of 3-D building systems for static and seismic loadings, *Modeling and Simulation in Engineering*, III, North-Holland Publishing Co., New York, NY, pp. 315–326, 1983.

3. Truman, K.Z., *Optimal Design of 3-D Reinforced Concrete and Steel Buildings Subjected to Static and Seismic Loads*, Ph.D. Dissertation, Advisor: F.Y. Cheng, University of Missouri-Rolla, Rolla, MO, May 1985.
4. Truman, K.Z. and Cheng, F.Y., How to optimize for seismic loads, Chapter 12, *Guide to Structural Optimizations,* Arora, J.S. (ed.), ASCE Manuals and Reports on Engineering Practice, No. 90, pp. 237–262, 1997.
5. Truman, K.Z. and Cheng, F.Y., Optimum assessment of irregular 3-D buildings, *Journal of Structural Engineering*, ASCE, Vol. 116, No. 12, pp. 3324–3337, 1990.
6. Truman, K.Z. and Jan, C.T., Optimal bracing schemes for structural systems subject to the ATC 3-06, UBC, and BOCA seismic provisions, in *Proceedings of the 9th World Conference on Earthquake Engineering*, Tokyo and Kyoto, Japan, Vol. V, pp. 1149–1155, 1988.
7. Truman, K.Z. and Petruska, D.J., Optimum design of dynamically excited structural systems using time history analysis, in *Proceedings for OPTI 91 – International Conference for Computer Aided Optimum Design of Structures*, Boston, MA, 1991, (*Optimization of Structural Systems and Industrial Applications*, Elsevier Applied Science, London, Hernandez, S. and Brebbia, C.A. (eds.), pp. 197–207, 1992).
8. Grierson, D.E., *SODA – Structural Optimal Design Algorithm*, SODA Inc., Waterloo, Canada, 1991.
9. Grierson, D.E. and Chan, C.-M., Design optimization of tall steel buildings, in *Proceedings for OPTI 91 – International Conference for Computer Aided Optimum Design of Structures*, Boston, MA, 25–27 June 1991, (*Optimization of Structural Systems and Industrial Applications*, Elsevier Applied Science, London, Hernandez, S. and Brebbia, C.A. (eds.), pp. 541–551, 1991).
10. Austin, M.A., Pister, K.S. and Mahin, S.A., Probabilistic design of moment-resistant frames under seismic loading, *Journal of Structural Engineering*, ASCE, Vol. 118, No. 8, pp. 1660–1677, 1978.
11. Balling, R.J., Pister, K.S. and Ciampi, V., Optimal seismic-resistant design of a planner steel frame, *Earthquake Engineering Structural Dynamics*, Vol. 11, pp. 541–556, 1983.
12. Bhatti, M.A. and Pister, K.S., A dual criteria approach for optimal design of earthquake-resistant structural systems, *Earthquake Engineering Structural Dynamics*, Vol. 9, pp. 557–572, 1981.
13. Chang, C.C., Ger, J.F. and Cheng, F.Y., Reliability-based optimum design for UBC and nondeterministic seismic spectra, *Journal of Structural Engineering*, ASCE, Vol. 120, No. 1, pp. 139–160, 1994.
14. Cheng, F.Y. and Chang, C.C., *Safety-based Optimum Design of Nondeterministic Structures Subjected to Various Types of Seismic Loads*, NSF Report, U.S. Department of Commerce, Virginia, NTIS No. PB90-133489/AS, 1988.
15. Cheng, F.Y. and Juang, D.S., Assessment of various code provisions based on optimum design of steel structures, *Journal of Earthquake Engineering and Structural Dynamics*, Vol. 16, pp. 46–61, 1988.
16. Cheng, F.Y. and Juang, D.S., Recursive optimization for seismic steel frames, *Journal of Structural Engineering*, ASCE, Vol. 115, No. 2, pp. 445–466, 1989.
17. Feng, T.T., Arora, J.S. and Haug, E.J., Optimal structural design under dynamic loads, *International Journal for Numerical Methods in Engineering*, Vol. 11, No. 1, pp. 39–52, 1977.

18. Yamakawa, H., *Optimum Structural Designs for Dynamic Response, New Directions in Optimum Structural Design*, John Wiley, New York, pp. 249–266, 1984.
19. Terlaje, A. and Truman, K.Z., Parameter identification and damage detection using structural optimization and static response data, *Advances in Structural Engineering – An International Journal*, Vol. 10, No. 6, pp. 607–621, 2007.
20. Terlaje, A. and Truman, K.Z., Damage detection of 2-D frames using static response data and optimality criterion, in *Proceedings of The Tenth East Asia-Pacific Conference on Structural Engineering*, Asian Institute of Technology, Bangkok, Thailand, Vol. 5, pp. 449–454, 2006.
21. Cheng, F.Y. and Kitipitayangkul, P., *INRESB-3D, A Computer Program for Inelastic Analysis of Reinforced-concrete Steel Buildings Subjected to 3-Dimensional Ground Motions*, Report No. 2, Civil Engineering Study Structural Series 79-11, University of Missouri-Rolla, Rolla, Missouri, 1979.
22. American Institute of Steel Construction Manual (AISCM), *Manual of Steel Construction*, 8th–13th Editions, Chicago, IL, 1980–2006.
23. UBC, *Uniform Building Code*, International Conference of Building Officials, Whittier, CA, 1991.
24. BOCA, *Building Officials Code Association*, Chicago, IL, 1980–1996.
25. ATC-03 NEHRP, *Recommended Provisions for the Development of Seismic Regulations for New Buildings*, Building Seismic Safety Council (for FEMA), Washington, DC, 1992.
26. IBC, *International Building Code*, International Code Council (ICC), Washington, DC, 2006.
27. Bathe, K. and Wilson, E., *Numerical Methods in Finite Element Analysis*, Prentice Hall, New Jersey, 1976.
28. Cheng, F.Y., *Matrix Analysis of Structural Dynamics – Applications and Earthquake Engineering*, Marcel Decker Inc., New York, 2001; CRC Press Taylor & Francis Group, Boca Raton, London, New York.
29. Newmark, N.M. and Hall, W.J., *Earthquake Spectra and Design*, Earthquake Engineering Research Institute, Berkeley, California, 1982.

10 Generalized optimality-criteria application to topological design, pile foundations, damage detection and structural identification

10.1 Introduction

Chapter 8 presented the theory and the associated algorithm for the generalized optimality-criteria method for use in traditional elemental or sizing optimization for building systems. Chapter 9 showed how to implement the algorithm from Chapter 8 for structures subjected to both static and dynamic loads. In this chapter an amalgamation of examples, where the generalized optimality-criteria techniques derived in Chapter 8 have been modified to solve several non-traditional structural problems, is presented. The unique quality of using optimality-criteria techniques is that they can be modified to accommodate the types of relationships between the design variables, the analysis methods, the constraints and the objectives for the optimization. As seen in Chapter 8 and illustrated in Chapter 9, as long as the sensitivities (gradients) of the objective functions and constraints relative to the design variables can be determined, the algorithm in Chapter 8 can be used as a means for optimizing the system relative to almost any objective function and well-defined constraints. An additional benefit is that the customization of the algorithm for specific problems often proves to be numerically effective in reducing the computational effort in reaching an optimal solution.

The generalized optimality-criteria technique presented in Chapter 8 has to be modified when a design variable is independent of the objective function and/or the constraints. For example, if the weight of a structure is to be optimized, rotating a structural element along its longitudinal axis will have no bearing on the weight. It could affect the constraints such as a displacement, but the gradient of the weight relative to this rotation would be zero which would create an ill-defined optimality criterion for that variable.

The examples in this chapter will show how optimality criteria can be modified and used for structural topology optimization, pile layout (geometric/topology) and size optimization, structural identification (non-weight based) and structural damage detection (non-weight based). The objective of this chapter is to show the versatility of the generalized optimality-criteria technique. The modifications, as necessary, will be briefly explained and will show how the basic concept of optimality criterion can be used for non-standard structural-optimization problems. The attractiveness of optimality-criteria techniques comes from their ability to be adapted to the specific problem to be solved, but this typically requires creativity and development by the users.

10.2   Topology optimization

This section demonstrates the application of the optimality-criteria method to topology-based optimization problems. The techniques and examples presented are based on the work of Hendel and Truman [1, 2]. The methods presented allow designers to find the most efficient material layout for a structure while meeting specific design requirements (constraints). Beginning with a finite element mesh and a set of known external loads and boundary conditions, classical topology-optimization methods often create the optimal structure by treating material density or thickness of each element as the design variable to be modified to optimize the system [3–6]. Additionally, many classical topology-optimization methods use mathematical programming approaches to perform the optimization which can require significant computational effort.

Most topology-optimization problems differ from sizing-optimization problems because they encourage or require solutions with the presence of internal material voids rather than solutions with large areas of minimal element thicknesses. In contrast, sizing-optimization problems are typically initialized by predefining where elements will be located (predefinition of void locations). These types of problems are capable of eliminating elements but typically are incapable of creating new elements to fill these predefined voids; therefore, they will only provide the optimal solution for the given configuration of elements. Topology-optimization techniques help define the necessary elements and their sizes for the optimal solution. Unfortunately, the technique is prohibitive for large-scale structures and for many structures that requires initially large voids such as 3-D building frames.

The next three examples show the application of the optimality-criteria method to the optimization of a cantilever, modelled using 2-D, four node, rectangular finite elements. The first two examples are based on minimization of weight with and without voids for a single-load case and the third example uses multiple load cases which is often a problem in classical topology-optimization algorithms. The problem is first approached as a sizing-optimization problem and then as a topology-optimization problem.

10.2.1   Weight-based sizing-optimization example

In this example the topology-optimization problem is first approached as a sizing-optimization problem. Fig. 10.1 shows a 1,200-element cantilever under a single point load at the free end. The element thicknesses, t_i, are treated as design variables and optimized with the goal of minimizing the

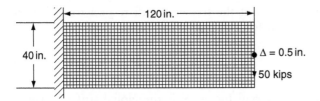

Figure 10.1 A 1,200-element cantilever with a single load case and a single deflection constraint [1] (1 in. = 2.54 cm, 1 kip = 4.45 kN).

weight of the structure. A deflection limit, Δ, of 0.5 in. (1.27 cm) at the point where the load is applied represents the single design constraint. The side constraints for the design variables are a maximum thickness, t_{max}, of 3.0 in. (22.9 cm) and a minimum thickness, t_{min}, of 0.001 in. (0.0025 cm), essentially zero or non-existent. The initial uniform thickness for this example, 0.9 in. (2.29 cm), was selected so that the deflection at the point of application was close to the maximum allowable value of 0.5 in. (1.27 cm). The material has a Young's modulus of 10,000 ksi (68,940 MPa) and a Poisson's ratio of 0.3.

The objective function is given by:

$$\text{Minimize } W(t) = \sum_{i=1}^{n} t_i A_i \tag{10.1}$$

In this equation, t_i represents the variable element thickness for the *i*th element, n represents the number of elements and A_i represents the constant cross-sectional area for element *i*. The single constraint for this problem is calculated as the difference between the actual deflection at the free end, $\delta(t)$, and the maximum allowable deflection at the free end, Δ.

$$g(t) = \delta(t) - \Delta \le 0 \tag{10.2}$$

Side constraints, as described in Chapter 8, are used to ensure that the element thicknesses stay within the allowable design domain.

$$t_{min} \le t_i \le t_{max} \quad \text{for} \quad i = 1, 2, \ldots, n \tag{10.3}$$

To find the optimality criterion, the sensitivities of the objective function and design constraint are required as shown in Chapter 8. For the case of the weight-based objective function, the derivative with respect to the thickness of any element will be constant.

$$\frac{\partial W(t)}{\partial t_i} = \frac{\partial}{\partial t_i} \sum_{i=1}^{n} t_i A_i = A_i \quad \text{for} \quad i = 1, 2, \ldots, n \tag{10.4}$$

To determine the sensitivity of the constraint, it is necessary to find the sensitivity of the deflection at the free end with respect to each element thickness. Using the linear force displacement relationship:

$$\{P\} = [K]\{u\} \tag{10.5}$$

where P is the set of external forces, K is the global stiffness matrix of the structure, and u is the set of nodal deflections, the derivative of deflection with respect to thickness can be found by applying the product rule, setting the derivative of the force with respect to thickness equal to zero (independent of t), and solving algebraically to find the familiar form for stiffness-based displacements.

$$\frac{\partial \{u\}}{\partial t_i} = -[K]^{-1} \frac{\partial [K]}{\partial t_i} \cdot \{u\} \quad \text{for} \quad i = 1, 2, \ldots, n \tag{10.6}$$

Since the stiffness matrix, $[K]$, is linear with respect to the thickness of each element, taking the derivative of $[K]$ with respect to t_i will result in any cell in $[K]$ not dependent upon t_i becoming a sub-matrix of zeros, and t_i being set to one for the calculation of all values where t_i is a factor. This is equivalent to solving the stiffness matrix for a set of thicknesses where all t_i are set to zero except for the thickness of the element of interest, which is set to one.

Using the optimality-criteria algorithm presented in Chapter 8 with the sensitivities as derived in Eqs. (10.4) and (10.6), Fig. 10.2 shows the optimal solution. A grayscale legend, shown to the right of the optimal solution, is used to represent element thickness. The darkest elements reached the maximum thickness of 3 in. while the lightest elements reached the minimum thickness of 0.001 in. and have been essentially removed from the mesh.

The optimal solution shown in Fig. 10.2 is as expected. The optimal solution at the supports strongly resembles a flanged beam with hinged supports at the corners, a thin web in the middle portion and thicker, tapered flange sections on the top and bottom. The upper and lower portions of the free end are eliminated (minimum thickness) since they provide minimal or no resistance for the prescribed loading.

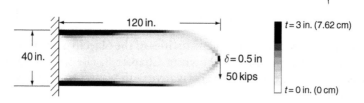

Figure 10.2 Optimal topology of the 1,200-element cantilever with a single-load case and a single deflection constraint [1] (1 in. = 2.54 cm, 1 kip = 4.45 kN).

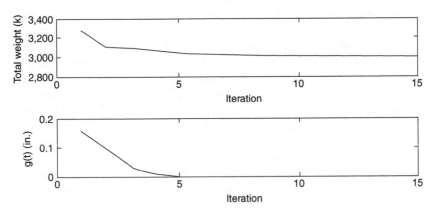

Figure 10.3 Objective function and constraint history for the 1,200-element cantilever with a single-load case and single deflection constraint [1] (1 kip = 4.45 kN).

Fig. 10.3 shows the data for convergence of the objective function and constraint to the optimal solution. The convergence plots shown in Fig. 10.3 once again demonstrate the efficiency of the generalized optimality-criteria technique presented in Chapter 8. A near optimal solution was achieved in seven iterations, while the remainder of the iterations was used to continue to refine the solution. Once many of the elements have reached their maximum or minimum sizes (passive elements as described in Chapter 8), small changes in the weight are to be expected. Note that the displacement constraint is controlled in all of the iterations after iteration 5.

The disadvantage of this approach is that it does not allow for elements to fully vanish or for the creation of voids. The weight-based sizing techniques will typically produce the standard beam shapes when bending is the primary loading for the system. Therefore, a modification to the weight-based sizing technique can be used to generate a design with voids. The following adjustments create an generalized optimality-criteria approach that is more in line with the classical approaches to topology optimization.

10.2.2 Void creation in topology optimization

While the solution shown in Fig. 10.2 meets the single design constraint and minimizes the weight of the structure, it may not be a practical or cost-effective solution from a fabrication standpoint. The flanges are tapered in thickness, and the web is very thin. To address this problem, topology-optimization methods can be used to alter the problem statement in a manner that favours solutions with the presence of internal voids rather than solutions with large areas of thin material as seen in Fig. 10.2.

This example approaches the void-creation problem through the use of a compound objective function that minimizes surface area as well as weight.

The minimization of surface areas causes structurally unnecessary, thin, elements to be eliminated. To do this, a function for calculating surface area based on element thickness is necessary. The actual functional relationship between surface area and element thickness is a Dirac delta function. Essentially, when the thickness has value, the element has a surface area, and when the thickness is zero, the surface area should be zero. However, because the gradient of the objective function is used in the calculation of the Lagrange multiplier and the optimality criteria, see Chapter 8, it is simpler if the objective function is a continuous function for all $t_i \geq 0$. In order to meet this requirement, the continuous function developed by Hendel [1] and given by Eq. (10.7) is one possibility that can be used to calculate an idealized surface area (pseudo-surface area) that mimics a Dirac delta function.

$$(SA)_i = t_i^v A_i$$

$$\text{where } v = \frac{1}{\ln(1,000t_i + 1)} \quad \text{for} \quad i = 1, 2, \ldots, n \tag{10.7}$$

When plotted (setting $A_i = 1$), Eq. (10.7) provides the curve seen in Fig. 10.4.

The modified objective function becomes the sum of the weight and the surface area times a weighting factor ζ. Combining Eqs. (10.1) and (10.7) yields:

$$O(t_i) = \sum_{i=1}^{n} (t_i A_i + \zeta \, t_i^v A_i) \quad \text{for} \quad i = 1, 2, \ldots, n \tag{10.8}$$

It is important to note that the weighting factor, ζ, which appears in Eq. (10.8), must be included with the pseudo-surface area portion of the objective function to ensure that the pseudo-surface area totals have similar magnitudes as the weight totals. This weighting factor can be swept through a series of values to find the proper balance of weight minimization and surface area minimization for each problem. When the combined pseudo-surface area and weight-based objective function is applied to the previous problem using the generalized optimality-criteria techniques of Chapter 8, the topology shown in Fig. 10.5 is obtained. As seen in Fig. 10.5,

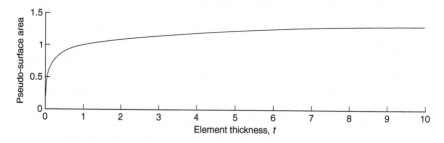

Figure 10.4 Pseudo-surface area function with respect to element thickness, t, using an area of unity.

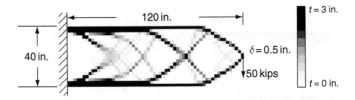

Figure 10.5 Chequerboard pattern resulting from the combined pseudo-surface area/weight-based objective function [1] (1 in. = 2.54 cm, 1 kip = 4.45 kN).

the solution now becomes essentially a truss where geometric locations of nodes and elements replace the concept of a beam with tapered flanges and a variable-thickness web. This solution has several issues, primarily chequerboarding and hinging of the elements.

The chequerboard patterns, lesser defined or extraneous elements, and the hinging effect, two or more elements connected at one node, as seen in Fig. 10.5, are commonly encountered in topology-optimization problems. The alternating solid/empty patterns are a result of the formulation of the stiffness of the four node elements [7, 8]. (Other elements could be used at the expense of additional analytical computations with mixed results.) While the chequerboard pattern coupled with hinging is unrealistic for design due to the lack of connectivity between many of the solid elements, these patterns can also suffer numerically by producing artificially high stiffness values [7–9].

To resolve the chequerboarding problem when using the combined objective function, mesh connectivity filters can be utilized. One such exemplary filter was created by Sigmund [10] and is easily utilized. This connectivity filter makes the sensitivity of each element thickness with regard to the constraints partially dependent on the values of the sensitivities of the surrounding elements. The new sensitivities are calculated using Eq. (10.9). In this equation, the hat on the constraint indicates the adjusted sensitivity, $\text{dist}(i,j)$ is the distance between the centre points of elements i and j, and $r_{\min}$ is a measure of the proximity of elements that will be included in the averaging of the sensitivities. (Typically $r_{\min}$ is taken as the distance (or a slightly larger distance) from centre to centre of diagonally adjacent elements.)

$$\frac{\partial \hat{g}}{\partial t_i} = \frac{1}{t_i \sum\limits_{j=1}^{n} \hat{H}_j} \sum\limits_{j=1}^{n} \hat{H}_j t_j \frac{\partial g}{\partial t_j}$$

(10.9)

$$\hat{H}_j = \begin{cases} r_{\min} - \text{dist}(i,j) & \text{when } r_{\min} - \text{dist}(i,j) \geq 0 \\ 0 & \text{when } r_{\min} - \text{dist}(i,j) \leq 0 \end{cases}$$

As seen in this equation, the sensitivities and thicknesses of the elements nearest the element in question have the greatest weighting in the adjustment

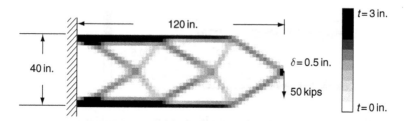

Figure 10.6 Optimal design for the 1,200-element cantilever using the combined
        pseudo-surface area/weight-based objective function with the connec-
        tivity filter [1] (1 in. = 2.54 cm, 1 kip = 4.45 kN).

of the sensitivity. Elements whose centres fall outside a radius of r_{min} from
the centre of the element in question have no effect on the sensitivity. When
using this filter, the sensitivity value for solid elements that are completely
surrounded by empty elements is reduced. Likewise, the sensitivity of an
empty element surrounded by solid elements is increased. Fig. 10.6 shows
the optimal cantilever design generated by the algorithm when the calculated sensitivities are filtered using Eq. (10.9). Plots of the convergence of the
total weight, pseudo-surface area, total objective function with a weighting
factor of 2.5 (weight + 2.5 × pseudo-surface area), and constraint are shown
in Fig. 10.7.

As seen in Fig. 10.6, the connectivity filter completely eliminated the
occurrence of chequerboard patterns and 'hinging'. Comparing these results
to those of the identical problem that was optimized using solely the weight-
based objective function shows that the addition of the more complicated
objective function resulted in a large increase in the number of iterations necessary for convergence, but with arguably more realistic results.
As expected, some weight minimization was sacrificed by minimizing the
pseudo-surface area as well. Although it is a bit unrealistic to compare
the weights of the two systems, the final weight of the structure optimized
using the combined objective function was 12 per cent greater than that of
the structure optimized using solely the weight-based objective function. In
effect, you are comparing a customized beam design to a truss design. The
surface area function, its weighting factor and the mesh connectivity filter
have been explored in detail and can affect the final solution, but mostly in
the rate of convergence and not the geometry and sizes [1].

10.2.3 Topology optimization using multiple load cases

One of the strengths of the generalized optimality-criteria method applied to
topology-optimization problems is its ability to handle multiple load cases
and control deflections at multiple points through the use of additional
inequality constraints. Many topology solutions are interested only in the
geometric shape and not the actual design (controlled displacements and

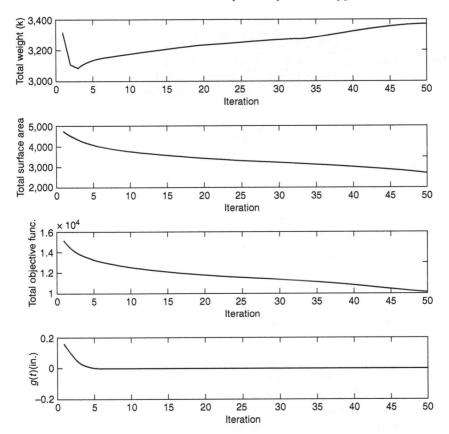

Figure 10.7 Objective and constraint histories for the 1,200-element cantilever
using the combined pseudo-surface area/weight-based objective function with the connectivity filter [1] (1 in. = 2.54 cm, 1 kip = 4.45 kN).

stresses). The optimality-criteria method can incorporate both. The 1,200-
element cantilever from the previous example is shown in Fig. 10.8, but
with an additional load case and point of controlled deflection added at
mid-span.

To form constraints from the given load case and deflection information, each load case is paired with each deflection limit to create four
constraints, as shown in Table 10.1. Using the deflections from Table 10.1
and the allowable deflections at points A and B, $\Delta_{A,B}$, the four constraints
are obtained:

$$g_1(t) = \delta_1(t) - \Delta_A \leq 0$$
$$g_2(t) = \delta_2(t) - \Delta_B \leq 0$$
$$g_3(t) = \delta_3(t) - \Delta_A \leq 0 \qquad (10.10)$$
$$g_4(t) = \delta_4(t) - \Delta_B \leq 0$$

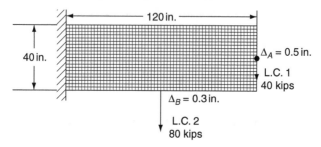

Figure 10.8 1,200-element cantilever with multiple load cases and multiple deflection constraints [1] (1 in. = 2.54 cm, 1 kip = 4.45 kN).

Table 10.1 Deflections used for constraints.

Constraint	Load case	Point	Deflection
1	1	A	δ_1
2	1	B	δ_2
3	2	A	δ_3
4	2	B	δ_4

Notice that the four deflection constraints are inequality constraints where the previous examples had one equality constraint. It is likely that not all deflections will reach their maximum allowable values for every load case in the final solution. Therefore, it is necessary to select the active (violated) and passive (not violated) constraints at the beginning of each iteration as outlined in Chapter 8. Using the filter and combined objective function as outlined in the previous example, the resulting topology is shown in Fig. 10.9. The convergence plots of the weight, pseudo-surface area, total objective function and constraints are shown in Figs. 10.10 and 10.11.

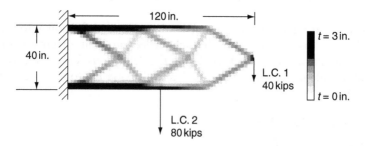

Figure 10.9 Optimal topology for a 1,200-element cantilever subjected to two load cases with two deflection constraints [1] (1 in. = 2.54 cm, 1 kip = 4.45 kN).

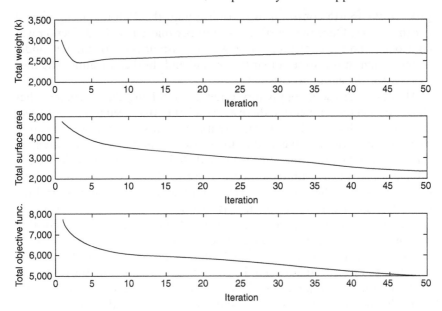

Figure 10.10 Objective function history for the 1,200-element cantilever subjected to two load cases with two deflection constraints [1] (1 kip = 4.45 kN).

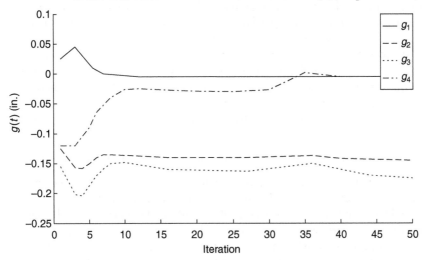

Figure 10.11 Constraint histories for the 1,200-element cantilever subjected to two load cases with two deflection constraints (1 in. = 2.54 cm).

Examination of the constraint-convergence plot, Fig. 10.11, shows that two load cases were active in the final solution: deflection of point A under load case 1, g_1, and deflection of point B under load case 2, g_4, which could be expected. While the first constraint was active through the entire process, the fourth constraint was passive until the 35th iteration, when it became active. Once the fourth constraint became active, its deflection was controlled and the constraint value was a constant zero for the remainder of the design. It is also seen that the second and third constraints remained inactive for the entire process.

Although the use of optimality criteria for solving these problems has shown tremendous potential, there are still issues that need to be addressed. The solution shown in Fig. 10.9 is essentially a truss, and one would expect that the joint closest to the load in load case 2 would be located exactly at the load. The fact that the joint is located slightly to the left of the load is a function of the algorithmic process. The truss had been, essentially, fully developed within the first seven cycles in a manner that would control the tip deflection or the first constraint as seen in Fig. 10.11. The fourth constraint did not become active until after the 35th iteration and the algorithm does not have the flexibility to make major adjustments by adding or rearranging the elements to make the final solution more realistic. If the first and fourth constraints were forced to be equality constraints from the beginning of the algorithm, this joint would have been located directly above the load. Also, the inclusion of stress constraints would enhance the ability to produce very realistic designs using generalized optimality criteria. A controlled stress constraint in the vicinity of the second load would have forced more material to be located near this load. The inclusion of stress constraints is more difficult due to the use of finite elements instead of dedicated elements such as braces, beams and beam-columns as were used in Chapter 9.

As seen from these examples, generalized optimality-criteria techniques such as that presented in Chapter 8 can be used to perform topology-based optimization. Adjustments such as restructuring the objective function to encourage void creation and adding a connectivity filter are necessary to reach solutions that are classically thought of as topology-based solutions.

10.3   Pile optimization – topological layouts and elemental sizes

At first glance, optimization of pile foundations appears simple, but in reality it is very complex due to the many geometric and constructability variables coupled with the sizing of the piles. The examples to be shown in this section are from work done by Hoback, Hurd and Truman [11–19]. This collection of work consists of optimizing pile layouts and sizes for piles that support rigid or flexible concrete slabs (structures) using the generalized optimality-criteria method described in Chapter 8. These slabs could be supporting buildings, bridges, locks, dams or any similar structure that has a concrete base (pile cap) with supporting piles. The original work was created for use by the U.S. Army Corps of Engineers to optimize large pile foundations under locks and dams using rigid slabs [20, 21]. This work was eventually modified to include flexible slabs which can greatly influence the optimal pile layouts and sizes. As with the topology optimization discussed in Section 10.2, the algorithm from Chapter 8 must be adjusted due to the geometric (topological) design variables and the discrete nature of pile sizes. Hoback's research [11–15, 18, 19] details many of the issues and possible modifications for achieving optimal geometric and discrete solutions for large-scale pile foundations. The examples here are to show the possibilities for using generalized optimality-criteria approaches in optimal pile design and not necessarily to provide all of the possible techniques and the specific details that have been developed.

In the case of large pile foundations, the beauty of the generalized optimality-criteria-based optimization algorithm lies in its ability to produce a reasonable design in a very short period of time. In a design office without an optimization tool, the iterative analysis and design of these large foundations are extremely time-consuming and difficult due to the large number of piles, 3-D geometries and large number of loading conditions. The structural-optimization algorithm presented with the use of the correct constraints produces a series of designs that satisfy the constraints and could be used as final designs. (At any iteration, the constraints will be satisfied creating a series of feasible designs.) As with many large-scale projects, constructability and practicality must also play a significant role in the final solution. All of these issues will be addressed in this section. The examples presented here were produced by the USACE sponsored program OPTPILE which also has a finite-element program developed by Hoback and Truman [11, 12, 14, 15, 21] called FEMPILE embedded in it to perform the rigid and flexible slab–pile analyses. The specifics of the pile stiffness coefficients, analysis procedures and finite-element analyses can be found in [11, 22–24]. The pile model is based on representing each pile as a set of springs based on the work of Dawkins [22] and presented in the USACE document for basic pile behaviour [23, 24]. The examples presented here will use steel HP piles, but the algorithm can easily be applied to timber piles or cast-in-place concrete piles.

10.3.1 Pile layouts (topology) and cross-sections (discrete sizes)

Pile optimization is complicated by the fact that it is a combination of topology and discrete-size optimization. Most often pile optimization is based on saving weight (tonnage of steel) or construction costs which are essentially related to the number of required piles, their slopes and their sizes. The topology optimization is used to locate piles under the foundation, to ensure that the piles will not interfere with each other and to geometrically orient the piles. The geometric orientation includes the batter (slope), the rotation, φ, of the sloped pile (for a given slope it can be located anywhere on the surface of the cone that has the same slope), and rotation, θ, of the cross-section (orientation of the flanges of an HP pile) as shown in Fig. 10.12. The rotational variables have no bearing on the weight of the piles and the batter will not affect the weight if the length of the pile is fixed. (Most cases

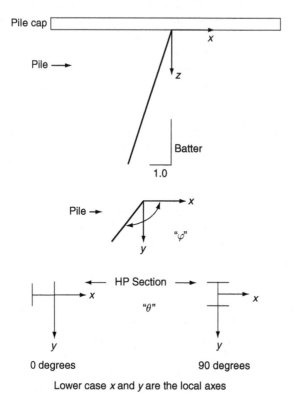

Figure 10.12 Pile coordinate system for batters, planar rotations, φ, and cross-sectional rotations, θ [11].

have a fixed depth and in this case the batter will make a difference in the weight.) The variables that have no effect on the objective function must be handled in a slightly different manner than the weight-based variables.

Initially this pile optimization algorithm uses the continuous optimization procedures using primary and secondary variables to reach the optimal sizes for each pile or pile group as described in Chapters 8 and 9. Often practical solutions rely on having groups of piles that have the same cross-sections and geometric orientation for economies related to constructability and purchasing. Once the continuous optimization process finishes, the algorithm switches to a discrete optimization, branch and bound, technique to find 'real' cross-sections that can be purchased and placed in the foundation.

Using the primary and secondary design variable concept presented in Chapter 8 and demonstrated in Chapter 9, each pile or pile group can be represented with one design variable, the major-axis moment of inertia, I_x. Since most steel piles are of a very specific type, called HP shapes, the coefficients are different from those given in Chapter 8. The coefficients presented in Chapter 8 cover a wide range (nearly all) of the AISC-based steel wide-flange sections [25]. The required secondary-design variables for piles are the minor-axis moment of inertia, I_y, the area, A, and the extreme fibre distances from the major and minor axes, $d/2$ and $b_f/2$ (stress calculations). The coefficients are found through linear and power law regressions for AISC HP shapes [25] giving these approximations for the pile examples:

$$I_y = 0.371I_x - 9.22 \tag{10.11}$$

$$I_y = 0.049I_x^{0.922} \tag{10.12}$$

$$d = 1.211 \times 10^{-3}I_x + 12.719 \tag{10.13}$$

$$b_f = 6.110 \times 10^{-4}I_x + 14.139 \tag{10.14}$$

These equations are used to approximate the secondary properties while allowing the gradients related to the primary-design variables to be continuous and easily implemented in the algorithm as presented in Chapter 8.

10.3.2 *Pile design variables and pile-related constraints*

The optimization algorithm uses all of the design variables, size as well as topological, to develop a series of designs that will ultimately be considered the optimal solution while satisfying all of the applied constraints. The algorithm is capable of having multiple load cases (in the case of a lock monolith, it could be as high as 10–20 load cases), multiple design constraints (displacements and stresses), side (size) constraints and construction-related constraints.

The design variables used to represent each pile or pile group are in two categories, topological (geometric) and structural element sizes. The topological variables include the batter (slope), the angles of orientation φ and θ, the pile head fixity (pinned through fixed) and pile location. The structural sizes are based on the primary-design variable, the major-axis moment of inertia, and Eqs. (10.11)–(10.14) representing the secondary-design variables.

Additionally, the pile structures are subjected to a variety of design constraints. As described in Chapter 8, these design constraints are incorporated into the optimality criteria through the Lagrangian. For each of these constraints the gradients will be needed as shown in Chapters 8 and 9. The pile-related constraints are based on placing limits (upper and/or lower) on:

1. Pile stresses: The steel piles have a limit on the maximum tensile and compressive stresses that can be allowed. These stresses are a combination of axial stress and bending stress. The allowable stresses for each of the pile foundation problems are from the USACE *Basic Pile Group Behavior* manual [23]. They are: 10 ksi (68.9 MPa) for compression at the pile tip, 17 ksi for axial compression (117.1 MPa) in the upper

region of the pile, and 18 ksi (124.9 MPa) for bending. These stresses are combined in several interaction equations defined in [23].

2. Bearing: The axial force in a bearing pile cannot exceed the strength of the bearing strata. In each of the examples it was assumed that the allowable compressive stress of the strata matched or exceeded that of the pile tip compression stress.

3. Displacements: These structures will typically have limitations on their displacements in order to provide clearances or minimize the potential for contact with adjacent structures. These displacements were held to 0.5 in. (1.27 cm) unless stated otherwise in the example.

4. Sizing: The piles (HP-14) typically have maximum and minimum available sizes for the major-axis moment of inertia, $I_{min} = 729$ in.^{4} (30,340 cm^4) and $I_{max} = 1,220$ in.^{4} (50,780 cm^4).

5. Batter: There are limits on the constructable batters (slopes). Too large a slope (small batter) makes the piles impractical to place. The batters were held to nothing less than 2.

6. Interference: Piles cannot intersect. For constructability, they typically need a cone of clearance around each pile since they can never be perfectly placed. This is forced to occur automatically as side constraints in OPTPILE [11, 21].

7. Edges and Borders: The pile groups must have pile heads that stay within the edges of the slab while the borders of the pile groups (could be a single pile) must not occupy the same space. This is forced to occur automatically as side constraints in OPTPILE [11, 21].

8. Obstructions: The pile tips may be restrained from extending under adjacent structures. Additionally, the piles cannot pass through objects such as sheet piles (although they can pass under these objects). This was not enforced in the following examples.

9. Slab stresses: The stresses in the slab must be prevented from exceeding their maximum allowable tensile or compressive stresses. The slab stresses were not considered in the following examples.

Any combination of these constraints can be considered simultaneously in the optimization algorithm to produce a practical and constructable final design.

10.3.3   Zero-weight gradient (topological) variables

From the generalized optimality criteria presented in Chapter 8 and applied in Chapter 9 and the previously shown topology examples, it is clear that these criteria cannot be used for design variables that do not affect the objective function. These examples will be using weight as the objective function; therefore, zero-weight gradients are not permissible since the optimality-criteria-based recurrence equations become unbounded for these variables. Typical weightless variables such as pile head fixity, pinning or fixing the pile head to the pile cap (foundation); the rotation, φ, of a pile about a vertical axis, the rotation, θ, of the pile about its longitudinal axis and the batter have no direct effect on the weight of the design. These topological variables do have an indirect effect on the weight, but one that is not readily quantifiable in the weight-based objective function. For example, if the pile head is fixed to the pile cap, a smaller pile will be required to control a displacement constraint, but it may increase the stresses in the pile. Both will affect the size of the pile through control of the constraints, but pile head fixity is not directly included in the objective function as a weight-dependent variable.

Accordingly, new weightless optimality criteria had to be developed for these weightless variables. The standard formulation given in Chapter 8 is still used for all weight-related variables. The approach used here is to make the weightless optimality criteria only a function of the constraint gradients. Essentially, this provides Kuhn–Tucker conditions leading to an equation where the sum of the constraint gradients is zero at the optimum solution. Therefore, the weightless optimality criteria are written as:

$$T_i = \sum_{j=1}^{m} \lambda_j \frac{\partial g_j}{\partial \delta_i} = 0 \tag{10.15}$$

where λ_j is the Lagrange multiplier for constraint j. The optimality criteria in Eq. (10.15) can be thought of as the efficiency of the relevant design variable, δ_i. At the optimum the efficiency is one, therefore the weightless optimality criteria can be modified to be:

$$T_i = -\sum_{j=1}^{m} \frac{\partial g_j}{\partial \delta_i} \frac{\lambda_j}{w_i} + 1 = 1 \tag{10.16}$$

where w_i is a scaling factor.

The recurrence equations have to be modified to account for these weightless variables. Therefore, the recurrence relationship from Chapter 8 becomes:

$$\delta_i^{(k+1)} = \delta_i^{(k)} \left(1 + \frac{1}{r} \left(-\sum_{j=1}^{m} \frac{\partial g_j}{\partial \delta_i} \frac{\lambda_j}{w_i} \right) \right) \tag{10.17}$$

and the equations for finding the Lagrange multipliers from Chapter 8 must be modified to reflect these weightless variables as:

$$r g_j^{(k)} - \sum_{i=1}^{n_2} \frac{\partial g_j}{\partial \delta_i} \delta_j^{(k)} = \sum_{s=1}^{m} \lambda_s \sum_{i=1}^{n_2} \frac{\partial g_s}{\partial \delta_i} \frac{\partial g_j}{\partial \delta_i} \left/ \frac{\partial O}{\partial \delta_i} \frac{\partial O}{\partial \delta_i} + \sum_{s=1}^{m} \lambda_s \sum_{i=n_2+1}^{n} \frac{\partial g_s}{\partial \delta_i} \frac{\partial g_j}{\partial \delta_i} \frac{\delta_i}{w_i} \right. \tag{10.18}$$

where n_2 is the number of variables with non-zero weight gradients, and $n - n_2$ is the number of weightless variables.

The parameter w_i is a factor that is used to prevent ill-conditioning or instability in the optimization process. This factor replaces the weight-gradient term which is zero for the weightless variables. Therefore, it can be considered a pseudo-weight gradient. A reasonable value of w_i is taken as an estimate of the weight gradient which is discussed in [11] and is automatically adjusted in OPTPILE [21].

Another alteration in the algorithm that must be considered when weight-less variables are considered is the fact that their change is zero when these variables are passive or unconstrained. Eq. (10.15) indicates that there is no direct benefit derived by increasing or decreasing these weightless variables (such as pile fixity). Therefore, they must be considered (forced to be) active if the designer wants them to be considered as design variables. If this is not forced to occur, the variables will be allowed to remain unchanged or stationary throughout the optimization process.

10.3.4 *Discrete member selection*

In the pile optimization process used by Hoback [11, 13, 14], a branch-and-bound technique was used to change the continuous variable designs into discrete, practical designs. This branch-and-bound technique takes the final continuously found variables and converts them to realistic discrete values while preserving as many of the optimal gains as possible. For example, a pile that is determined to have a moment of inertia that is between two commercially available HP cross-sections will be forced through the branch-and-bound technique to take on the actual value of one of these commercially available sections. Another example would be pile fixity. Pile fixity is assumed to be continuous during the optimization process, but in reality the pile will be constructed as either pinned (limited embedment in the slab) or fixed (3–4 ft (0.92–1.22 m) of embedment in the slab) and the branch-and-bound technique is used to find the optimal choice, fixed or pinned.

The branch-and-bound procedure begins by selecting a continuous variable which is to be converted to a discrete value. Two acceptable discrete values (scenarios) are chosen such that they 'bound' the continuous variable being considered. Two 'branches' or optimization sub-problems are created; the upper branch places a lower limit on the variable at the higher discrete value and the lower branch places an upper limit on the lower value at the lower discrete value. Each sub-problem is then optimized and the minimal solution is chosen and that discrete variable becomes fixed at that value. From all the branched sub-problems, the one with the lowest weight is used in the next branched sub-problem for the next discrete variable. This is repeated for all discrete variables until a solution is obtained. The conversion to a discrete solution increases the original continuous variable-based computational effort by as much as a factor of 15–20 times for large pile optimization problems, but it does produce a realistic set of results. This process is thoroughly discussed in Hoback [14].

An alternative that can be used to significantly reduce this effort but not necessarily achieve an optimal solution is to find those continuous variables that have values that are significantly close to an acceptable discrete value and set those design variables to the significantly close discrete value. Then perform the branch-and-bound procedure on the remaining continuously determined variables that are not significantly close to acceptable discrete values. For example, if in the continuous optimal solution a pile is indicated to have 90 per cent fixity, it could be set to have 100 per cent fixity before the branch-and-bound technique is applied. An additional way of handling certain discrete variables in a continuous fashion is presented in Section 10.3.6 which can alleviate some of the branch-and-bound sub-problems.

10.3.5 Pile group optimization

In the examples being presented, group pile optimization is performed by using the stiffness-multiplier method developed by Hoback [11, 13]. This method was developed in order to eliminate piles from groups of piles since the continuous variable, size-based optimization algorithm presented is incapable of element removal. Once individual piles reach their lower limits, they cannot become any smaller nor can they be physically removed from the design using the given algorithm (also discussed in Section 10.2). Most large-scale pile foundations have multiple groups of piles of similar size and orientation for ease of construction. The optimization algorithm presented can optimize these groups of piles through linking as described in Chapter 8, but when optimized as a group using linking, it is still not possible to eliminate single piles within the group without modifying the procedure. Using a method Hoback termed the stiffness-multiplier method, single piles within a group can be eliminated.

The stiffness-multiplier method allows the number of piles in a given group to be a continuous design variable. Of course, the final solution requires the number of piles to be an integer, a discrete value. The optimal integer can be found using the branch-and-bound technique for discrete variables or it may be obvious if the real value is close to an integer value. The stiffness-multiplier method is based on assembling the 'effective' pile stiffness over the region of the pile group. The simplest method is to smear the total stiffness uniformly over the pile group region. The stiffness multiplier, S_p, is the ratio of the real number of piles to the integer number used in the assembly:

$$K_{real} = S_p (K_{individual}) = \frac{N_{real}}{N_{assembled}} (K_{individual})$$ (10.19)

where K_{real} is the total stiffness for $N_{assembled}$ individual piles. For example, if 5.5 piles exist in the group then 5 piles are assembled in a symmetrical grid and the stiffness multiplier is 5.5/5 or 1.1 applied to the stiffness of each pile. (For purposes of stress constraints and flexible slabs, the piles are assumed to be uniformly spaced in a region located under the slab.)

The stress constraints for the piles in a group are handled on an individual basis. The stresses of the piles are evaluated by applying the appropriate displacements to each individual pile in a group, and then calculating the maximum combined stresses similar to those for a beam-column in Chapter 9, but altered slightly to reflect USACE criteria [23]. The highest stressed pile is then used to form the stress constraint for the entire group. Each pile in the group must remain the same size with the same orientation; yet any pile can be eliminated using the stiffness multiplier as described previously.

This method relies on the assumption that the exact layout of the piles within a group is insignificant. The exact placement of the piles does not significantly affect the structural displacements, but it can affect the local concrete stresses and the individual pile stresses. The pile stresses are significantly affected by the exact pile layout only when the slab is very flexible. Finding the optimal stiffness multiplier, the non-integer number of piles, is a very efficient method for pile elimination.

10.3.6   *Percentage-selection method*

Another technique that has been used for finding optimal solutions of discrete and continuous problems was developed by Hoback [13, 14] and is called the percentage-selection method. This method can be very effective in the elimination of piles and optimizing discrete designs. This technique is best suited for the selection of two or more discrete options, such as pinned versus fixed pile heads, for the final values of a design variable. This method allows the use of the continuous gradient-based approach to choose between two or more discrete options.

The percentage-selection method uses averaged properties to perform the analyses and the optimization. This allows the discrete problem to become continuous. Each discrete variable's options have a percentage associated with them which measures the efficiency of each option. The sum of all of these percentages must total 100 per cent giving:

$$\sum_{j=1}^{nopt_i} p_{ij} = 1.0 \tag{10.20}$$

where nopt is the number of discrete options for member i and p_{ij} is the percentage of option j for member i. These percentages are used to develop an equivalent member stiffness matrix which is found by using the per-cent-weighted average of the different discrete options:

$$K_i = \sum_{j=1}^{nopt_i} p_{ij} K_j \tag{10.21}$$

where K_i is the total effective stiffness for element i and K_j is the stiffness matrix with terms that are dependent upon discrete option j. The constraints that are elemental stiffness-dependent, such as stresses, are handled similarly by finding an equivalent stress constraint as:

$$g_i = \sum_{j=1}^{\text{nopt}_i} p_{ij} g_{ij} \qquad (10.22)$$

where g_{ij} is the ith constraint for the jth discrete option.

The percentage variable of each discrete option is allowed to take on a value between 0 and 1. If it reaches one of the extremes, it indicates that one of the discrete options is optimal. If it falls between 0 and 1 and is not significantly close to one of these values, a branch-and-bound procedure as described previously can be used to make the final selection. Of course, the use of these percentage variables requires additional constraints in order to keep the sum of the values for each element between 0 and 1.

10.3.7 Pile-founded wall example

A retaining wall pile foundation is optimized for two different load cases. Three different optimal pile foundations will be found; the first solution provides the optimal foundation using individual piles, the second solution provides the optimal foundation using two pile groups, and the third solution uses four pile groups (individual piles as groups). The initial design is shown in Fig. 10.13. The two load cases are shown in Table 10.2. This example is used to optimize the number, size and orientation of the piles while the pile-head locations are to remain fixed. This solution uses the percentage-selection method as described in Section 10.3.6.

The initial design has five piles in the retaining wall segment as shown in Fig. 10.13. This represents a repetitive slice (geometry and loads) of a longer wall. The initial design sizes, orientations and fixity are given in Table 10.3. The individual pile optimization is explored first. For this example the low number of piles in a repetitive pattern (along the length of the wall) would suggest the use of individual piles if not in size at least in orientation.

The optimal solution reduced the weight of the piles from 12,214 lbs (54.35 kN) to 10,950 lbs (48.73 kN) in seven iterations. The convergence of the weight is shown in Fig. 10.14. The final optimal solution using continuous variables gives the results in Table 10.4.

Note that the optimal moments of inertia were controlled by the lower bound of the useable HP cross-sections, 729 in.⁴ (30,340 cm⁴), and that the most efficient design uses partial fixity for all of the piles. The use of partial fixity in a real design is not typically allowed. Therefore, this continuous optimization was followed by a branch-and-bound procedure to produce piles with a pinned head or fully fixed head making it a useable, discrete, least-weight design.

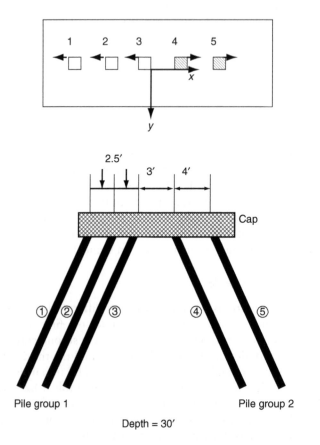

Figure 10.13 Initial pile layouts for the retaining wall; plan and elevation views [11] (1 ft = 0.305 m).

Table 10.2 Retaining wall load cases (1 k = 4.45 kN, 1 k-ft = 1.357 kN-m).

Load case	P_x (k)	P_z (k)	M_y (k-ft)
1	−220	440	500
2	250	250	0

Table 10.3 Initial retaining wall properties (1 in.^{4} = 41.62 cm^4).

Pile	I_x (in.^{4})	Batter	φ (deg)	θ (%)	% Fixity
1	729	2	180	50	50
2	729	2	180	50	50
3	729	2	180	50	50
4	729	2	0	50	50
5	729	2	0	50	50

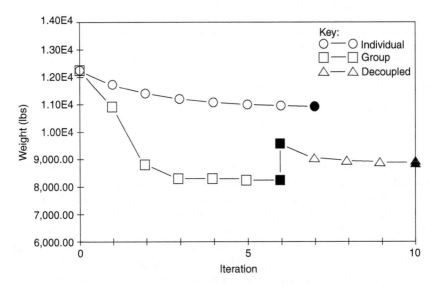

Figure 10.14 Objective function histories showing convergence to the optimal solution for the retaining wall example for three different scenarios [11] (1 lb = 4.45 N).

Table 10.4 Continuous optimized pile properties for individual piles (1 in.^{4} = 41.62 cm^4).

Pile	I_x (in.^{4})	Batter	φ (deg)	θ (%)	% Fixity
1	729	35.4	180	96.2	51.0
2	729	24.2	180	100	57.6
3	729	11.8	180	87.2	45.4
4	729	8.65	0	90.5	22.0
5	729	35.1	0	100	49.5

The final, useable design, after the branch-and-bound procedure was used, is shown in Table 10.5. The moments of inertia remain at 729 in.^{4} (30,340 cm^4), all piles are 0 percent fixed (pinned) and all piles are 100 percent (0 degrees of rotation).

The final branch-and-bound weight is 10,929 lbs (48.63 kN) which is almost identical to the initial continuously-optimized weight. This is to be expected since the moment of inertia and cross-sectional areas did not change. The geometric property of the batter is the only remaining variable that affects the weight (fixed depth of pile requires increased length with smaller batters). Note that any batter over 10 to one is nearly vertical; therefore, in this case the piles are essentially vertical. Therefore, no figure for the final results is presented. The only active constraints in this solution

Table 10.5 Branch-and-bound derived properties ($1 \text{in.}^4 =$ 41.62 cm^4).

Pile	I_x (in.^{4})	Batter	φ (deg)	θ (%)	% Fixity
1	729	53	0	100	0
2	729	100	0	100	0
3	729	100	180	100	0
4	729	17	0	100	0
5	729	31	0	100	0

were the minimum pile sizes for all piles. Therefore, this pile foundation is over-designed, and the number of piles could be reduced.

The group-optimization method was used in the second case. The group-optimization method allows the number of piles to be varied as a real number through the stiffness multiplier, S_p, and it thereby can allow the number of piles to be reduced. (A branch-and-bound method is used at the end to find an integer number of piles.) The initial properties for this example are the same as those given in Table 10.3, but the piles have now been grouped into two groups. Group 1 consists of piles 1, 2 and 3 while group 2 has piles 4 and 5.

The group-optimization method reduced the weight of steel from 12,214 lbs (54.35 kN) to 8,218 lbs (36,570 kN) in six iterations. The continuously-optimized results are given in Table 10.6. Since this solution provided a non-integer number of piles, a branch-and-bound procedure was used to produce a discrete, useable design. The branch-and-bound results increased the optimal weight to 9,489 lbs (42.23 kN) as shown in Fig. 10.14. The major change from the initial, individual optimization is the reduction of one pile, pile number 5. The final design values are shown in Table 10.7 and the final pile layout is shown in Fig 10.15. The final design

Table 10.6 Continuous variable, group optimized pile results ($1 \text{in.}^4 =$ 41.62 cm^4).

Pile	# of piles	I_x (in.^{4})	Batter	φ (deg)	θ (%)	% Fixity
1	2.34	729	4.51	180	100	50
2	1.22	729	2.00	0	100	32

Table 10.7 Branch-and-bound group optimized pile results ($1 \text{in.}^4 =$ 41.62 cm^4).

Pile	# of piles	I_x (in.^{4})	Batter	φ (deg)	θ (%)	% Fixity
1	3	729	11.3	180	100	100
2	1	904	2.00	0	100	100

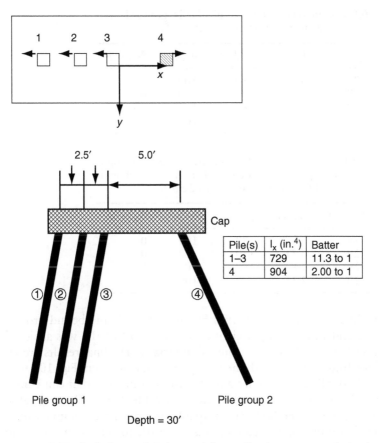

Figure 10.15 Final pile layout for the retaining wall using group optimization [11] (1 ft = 0.305 m).

has several active constraints. The stresses in pile 4 are active, and the piles in group 1 are at their minimum size with stresses near their maximum values.

From Tables 10.6 and 10.7 one can see that the removal of one pile from group 2 and forcing group 2 (pile 4) to take on the next available size of pile has changed the design significantly. It has affected the batter of group 1 and now each group is required to be fully fixed rather than pinned at the pile cap.

As a final comparison, the group-optimization technique can be used to optimize the individual pile problem by treating each individual pile as its own group, decoupled solution. This will allow the original problem to remove any unnecessary piles from the initial design. Using the final design for the group optimization as the initial design, the optimal solution decreased to 8,881 lbs (39.52 kN). Since the final design was not discrete,

Table 10.8 Continuous variable, individually grouped optimized pile results (1 in.^{4} = 41.62 cm^4).

Pile	I_x (in.^{4})	Batter	φ (deg)	θ (%)	% Fixity
1	729	37.5	180	100	43
2	729	9.10	180	100	64
3	729	5.17	180	100	30
4	729	3.49	0	100	22

Table 10.9 Continuous variable, individually grouped optimized pile results after applying the branch-and-bound technique (1 in.^{4} = 41.62 cm^4).

Pile	I_x (in.^{4})	Batter	φ (deg)	θ (%)	% Fixity
1	729	14.3	180	100	0
2	729	6.51	180	100	0
3	729	9.83	180	100	0
4	729	2.35	0	100	0

in that the final design had partial fixities, a branch-and-bound procedure was used and a final design weight of 8,971 lbs (39.92 kN) was achieved as shown in Fig. 10.14 as the decoupled approach. The final results are given in Tables 10.8 and 10.9 with the final pile layout shown in Fig. 10.16.

This decoupling of the piles and the use of a method that can eliminate individual piles has produced the lightest, useable design. The original individual, continuous method was incapable of reducing the number of piles and the first grouping placed unnecessary restrictions on some piles within the group. In terms of construction, it may still be more acceptable to use the original group solution since the batters for group 1 remain constant. Note that the individual piles found using the grouping process with each pile being its own group, the decoupled approach, produces a different batter for each and every pile which would be unacceptable to many contractors for construction purposes.

10.3.8 *Gravity dam pile foundation example*

The pile foundation for an existing structure called the Old River Low Sill Control Structure as analyzed and shown in the CPGA User's Guide [24] is used as the initial design. The only alteration to the original design is for all of the piles to have a final tip depth of 90 ft (27.5 m) where the original design used 90 ft (27.5 m) regardless of their batter. The group method and percentage optimization methods are used to initially optimize the pile foundation. This example is used to illustrate the effectiveness of the pile optimization procedures for large pile foundations.

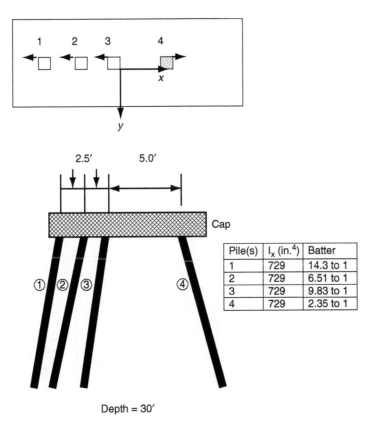

Pile(s)	I_x (in.⁴)	Batter
1	729	14.3 to 1
2	729	6.51 to 1
3	729	9.83 to 1
4	729	2.35 to 1

Figure 10.16 Final pile layout for the decoupled (individual group for each pile) optimization [11] (1 ft = 0.305 m).

The structure is shown in Fig. 10.17 [24]. The single-load case is given in Table 10.10 and the initial pile layouts and groups are given in Table 10.11. Initially, the three groups have 60, 52 and 128 piles for a total of 240 piles. Note that groups B and C are installed in the same area which is allowed as long as the piles do not intersect.

The optimal number of piles is 120 or half of the initial number. A direct comparison to the original design is not appropriate since this optimal design is based on only one load case whereas the original design was subject to numerous load cases. What can be seen is that given an initial design, for a given loading(s) and constraints, the optimization procedures described can be very effective. The final results are shown in Tables 10.12 and 10.13. Note that a branch-and-bound procedure was not used as the percentage method produced values that are all optimized to reasonably close discrete values. For example, all of the moments of inertia are significantly close to the minimum value of 729 in.⁴ (30,340 cm⁴) and can be assumed to be

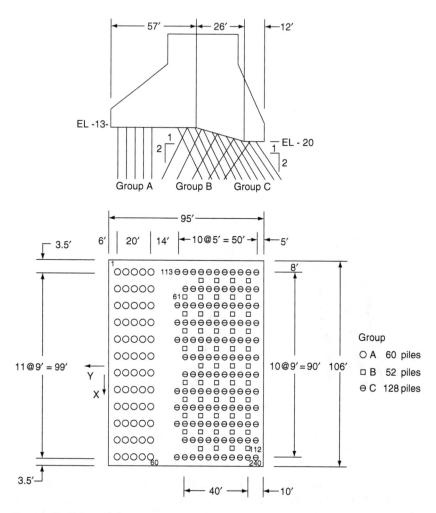

Figure 10.17 Actual dam structure in elevation view and its pile layout in elevation and plan views [24] (1 ft = 0.305 m).

Table 10.10 Loads for the dam example (1 k = 4.45 kN, 1 k-ft = 1.357 kN-m).

Load case	P_x(k)	P_y (k)	P_z (k)	M_x (k-ft)	M_y (k-ft)
1	0	−12,000	20,000	-1.0×10^6	0

this value for a useable design. Also note that in Table 10.13 the location of each pile group is given in order to eliminate any pile interference. The idealized initial and final pile layouts are shown in Fig. 10.18. Note that the bold arrows indicate the direction of batter (slope from top to bottom of the

Table 10.11 Initial values for the dam example (1 in.^{4} = 41.62 cm^4).

Group	# of piles	I_x (in.^{4})	Batter	φ (deg)	θ (%)	% Fixity
A	60	729	100	0	50	50
B	52	729	2	90	50	50
C	128	729	2	270	50	50

Table 10.12 Optimized pile results for the dam example (1 in.^{4} = 41.62 cm^4).

Group	# of piles	I_x (in.^{4})	Batter	φ (deg)	θ (%)	% Fixity
A	23	750	2.23	270	0	0
B	19	730	2.09	270	0	0
C	78	732	2.01	270	0	0

Table 10.13 Optimal pile group positions for the dam example (1 ft = 0.305 m).

Group	X_{low}	X_{high}	Y_{low}	Y_{high}
A	−46.8	49.5	−15.7	−10.9
B	−48.9	39.5	−31.6	−24.8
C	−45.3	43.8	−89.5	−45.3

pile(s) in the indicated direction) for groups B and C, while group A has vertical piles in the initial design. The final design has all of the piles battered in the same direction (nearly parallel). Also note that in the original design the piles in groups B and C were intertwined whereas the final design separates the two groups. (The major change in design is due to using only one load case; therefore, direct comparisons are strictly to show the versatility of the generalized optimality-criteria method.)

The weight of the steel was initially 1,667 kips (7,418 kN). The optimal weight is 870 kips (3,872 kN). Fig. 10.19 shows the convergence of the weight to this value. The initial spike in the weight for the first cycle is due to the fact that the stresses were exceeded when the 50 per cent fixity was used to initialize the percentage method and the sizes were scaled (as described in Chapter 8) to keep the stresses in the feasible range. The pile alignments for each pile group are very similar which is due to the use of a single-load case. If multiple load cases were used, many of the variables for each pile group as well as their locations could be significantly different. The final design has nearly all of the piles reaching their maximum stress levels along with the batter limits for groups B and C. This example shows that large pile foundations can be optimized in a relatively low number of cycles while controlling displacements and stresses, as well as geometric properties and locations.

10.3.9 Lock and dam gate monolith example

The gate monolith is representative of an actual monolith from the Melvin Price Lock and Dam. The initial pile layout was one of the original, feasible designs for this monolith. Several views of this monolith are given in Figs. 10.20 and 10.21. Note that AL-3 stands for auxiliary lock monolith 3, one of 11 monoliths. This monolith is abutted by AL-2 and AL-4 as indicated. Note that D-1 stands for dam monolith 1, and it abuts AL-3 as well. The other side of the monolith is abutted by the embankment (saturated and dry soils). As shown in the isometric view, Fig. 10.20, and in the elevation view, Fig. 10.21, the monolith is U-shaped with thick wall sections and a thinner (yet thick) slab of approximately 17 ft (5.19 m). The walls have voids (culverts) which are used to fill and empty the lock structure. Walkways and machinery voids are unsymmetrical in their placement throughout

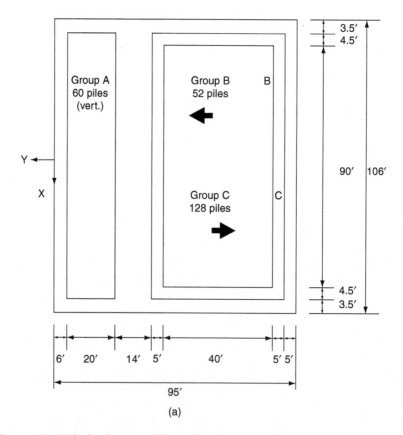

(a)

Figure 10.18 Idealized (a) initial and (b) final optimal pile layouts for the dam example [11] (1 ft = 0.305 m).

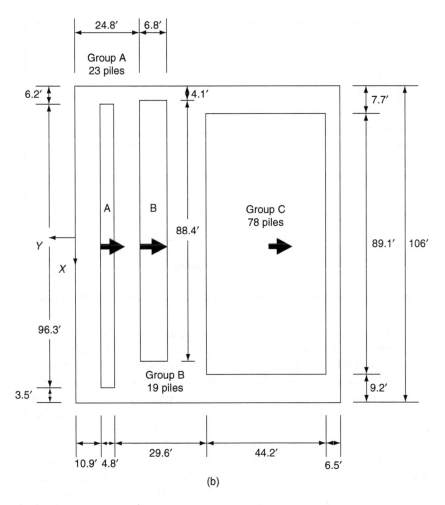

Figure 10.18 (Continued)

the walls. Typical voids and their locations are shown in Fig 10.21. The actual design has 710 piles where the idealized and initial pile designs have 734 piles as shown in Figs. 10.22, 10.23 and 10.24, respectively. The additional row of 24 piles was added to alleviate a slight overstress of the piles in group A actual, preliminary design. (The pile groups and layout in Fig. 10.24 were used as the initial layout for the optimization procedures since each group has to be confined to one region which was not the case in Fig. 10.23.) All of the piles are originally the maximum size of HP-14 piles, 1,220 in.[4] (50,780 cm[4]), that are initially in five groups. Initially, group A has 168 vertical piles, group B has 38 piles, battered at a slope of 2.5 in the y-direction, group C has 24 vertical piles, group D has 260 piles, battered at a slope of 2.5 in the negative x-direction, and group E has 28 piles,

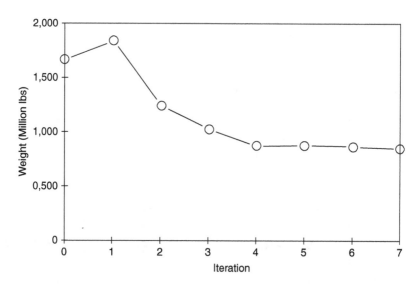

Figure 10.19 Objective function history for the dam pile optimization example [11]
(1 lb = 4.45 N).

battered at a slope of 2.5 in the y-direction, as seen in Figure 10.24. The
large number of battered piles is necessary to resist the large lateral forces
from the backfill as well as the water when the gates are closed. Over 100
actual load cases are used to design such a structure, many of which do not
control the design but must be checked. The designers narrowed this list to
20 significant load cases and for the purposes of this optimization the list
was narrowed to eight cases. The load cases and their descriptions are listed
in Table 10.14 with specific details provided in [20].

The complexity of these loads precludes providing all of the components
and their interaction with the 3-D model, but can be found in the USACE
documents [20, 21]. The largest components of the load cases generally are
the dead weight, the lateral and vertical down-drag forces from the backfill,
and the effects of the water on the sides, top and underside of the lock monolith (uplift). The water causes seven different loadings: chamber loads, side
pressures, uplift pressures, mitre gate forces, bulkhead forces, culvert pressures, and pressures in the joints between monoliths. Along with these loads
are the self-weight of the concrete elements and the steel gates. The different
load cases also require different allowable stresses or load factors based on
probability of occurrence and consequences of failure. The extreme cases
are typically of short duration and low probability requiring a lesser factor of safety than a daily, normal operating load with high probability and
frequent occurrence. The displacements are limited to very small values of
0.3 in. (0.76 cm) and 0.5 in. (1.27 cm) in the plane of the slab for the normal
and extreme load cases, respectively.

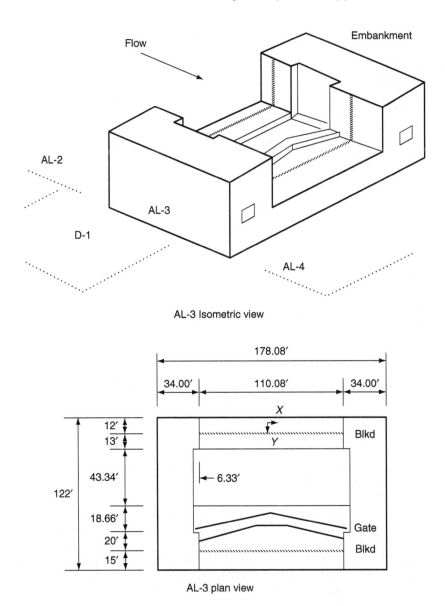

Figure 10.20 Lock monolith shown in isometric and plan views [20] (1 ft = 0.305 m).

The goal of this optimization was to reduce the cost of the structure while maintaining stresses and displacements within reasonable limits and maintaining reasonable constructability. The optimization process was allowed to vary pile size, number and batter of piles, pile fixity, geometric orientation and the location of the pile groups.

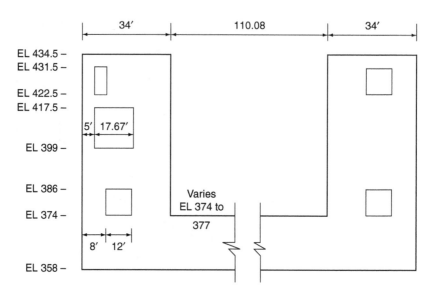

Figure 10.21 A cross section of the lock monolith showing void locations [20]
(1 ft = 0.305 m).

The optimization was initialized using the idealized, original design including the same number and orientation of piles, but their positions were altered slightly to create five pile groups. The initial pile group positions used in the optimization are shown in Fig. 10.24, and the initial pile data is shown in Tables 10.15 and 10.16.

After optimizing, the number of piles decreased from 734 to 501 and the weight of the steel reduced from 6.13×10^3 kips (27,280 kN) to 2.97×10^3 kips (13,220 kN) in ten iterations as shown in Fig. 10.25. The solution was terminated due to the number of cycles, therefore minor additional weight savings may have been realized if the optimization had been continued. The optimal pile layout is shown in Fig. 10.26 and quantified in Tables 10.17 and 10.18. This optimization was allowed to find the optimal solution independent of conventionally constructed systems. For example, the piles were allowed to choose an angle φ which would optimize the system whereas most constructed projects choose to align φ with one of the coordinate axes as in the initial design. (If the designers and contractors would loosen these types of constructability constraints, great savings may be seen in many projects.) Most of the piles are battered towards the negative x direction in order to resist the embankment loads. It is also important to note that the piles do not intersect each other, and they do not extend beyond the edges of the slab which is controlled by geometric side constraints.

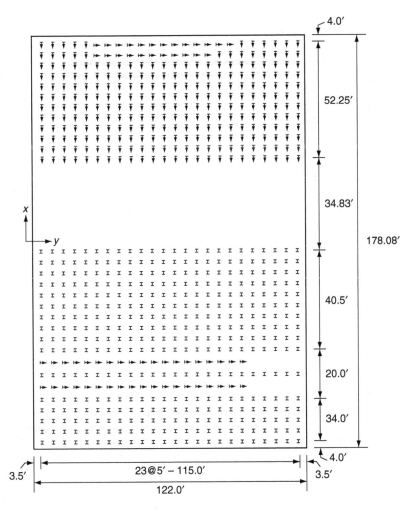

Figure 10.22 Actual pile design for monolith AL-3. (The arrows indicate the direction of batter for each pile. Piles without arrows are vertical.) [11, 20] (1 ft = 0.305 m).

In order to be a realistic solution for construction, this solution will be further processed using a branch-and-bound analysis. Even though the continuous variable results may not produce a useful design, these results can be used to illustrate the possible savings if other cross sections and more flexible construction layouts were allowed.

After the branch-and-bound process was completed, the weight increased to 3.26×10^3 kips (14,510 kN) (290 kips (1,291 kN) increase) and the number of piles increased from 501 to 592 as shown in Fig 10.27. All of the constraints were controlled except one pile was overstressed in load case

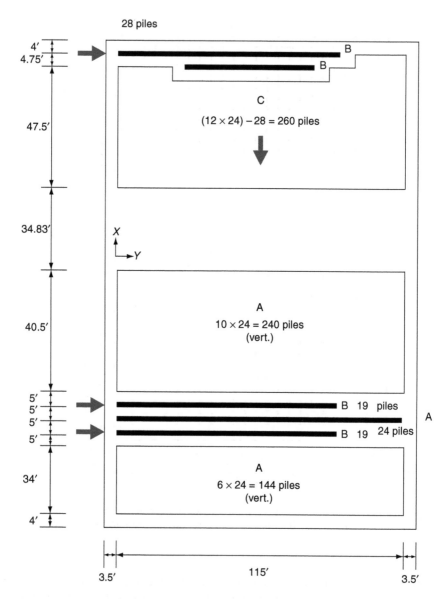

Figure 10.23 Idealized initial pile layout for monolith AL-3 (reflects actual design in Fig. 10.22) [11] (1 ft = 0.305 m).

four by less than 1 percent. The final discrete design is very similar to the previous design with the exception of changing the fixity of group B, and forcing the piles to take on actual HP-14 cross-sections. The results shown in Fig. 10.27 were used as an initial design and reoptimized using realistic

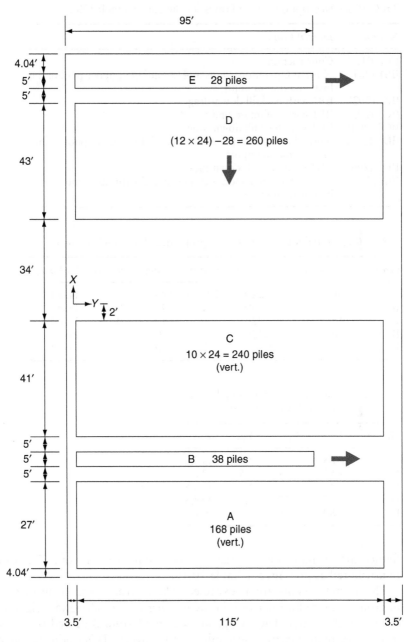

Figure 10.24 Initial grouping and pile layout used for the optimization procedure [11] (1 ft = 0.305 m).

Table 10.14 Descriptions of load cases for the dam example [20].

Number	Description
IA (101)	Construction
IIB1 (201)	Normal Operation, Pools at 419 and 395', upper pool in the lock, maximum uplift
IIB2 (202)	IIB1 with backfill down-drag
IIC1 (203)	IIB1 with minimum uplift
IIC2 (204)	IIC1 with backfill down-drag
IID1 (205)	Normal operation, pools at 419 and 395', lower pool in the lock, minimum uplift
IID2 (206)	IID1 with backfill down-drag
IIIA (301)	Extreme maintenance, lock dewatered with upstream bulkhead in place

Table 10.15 Initial lock monolith pile properties ($1 \text{ in.}^4 = 41.62 \text{ cm}^4$).

Group	# of piles	I_x (in.^{4})	Batter	φ (deg)	θ (%)	% Fixity
A	168	1,220	100	180	0	Fixed
B	38	1,220	2.5	90	0	Fixed
C	240	1,220	100	180	0	Fixed
D	260	1,220	2.5	180	0	Fixed
E	28	1,220	2.5	90	0	Fixed

Table 10.16 Initial lock monolith pile locations ($1 \text{ ft} = 0.305 \text{ m}$).

Group	X_{low}	X_{high}	Y_{low}	Y_{high}
A	−85	−58	3.5	118.5
B	−53	−48	3.5	95
C	−43	−2	3.5	118.5
D	32	75	3.5	118.5
E	80	85	3.5	95

fixities and available discrete sizes to give the final results of 557 piles but redistributed as presented in Table 10.19.

The discrete results are as expected. The increase in moment of inertia in group E caused a reduction in the number of the piles and a significant change in the batter. The change in fixity for Group B caused the stresses to increase requiring five more piles in this group. This optimization was performed simultaneously using eight load cases, and it is always prudent to check the optimal design using the other load cases. It is possible that some of the other 100 or more load cases could have become active once the design was significantly trimmed and rearranged.

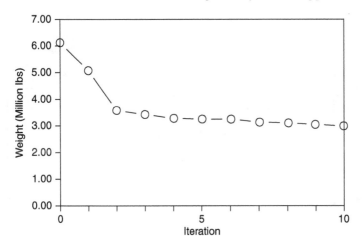

Figure 10.25 Objective function history for the lock monolith example [11] (1 lb = 4.45 N).

In any case, the optimization dramatically reduced the weight of steel and it accurately controlled the design limits. The example demonstrates how efficient and effective generalized optimality-criteria approaches can be for solving large-scale pile optimization problems. USACE documents [20, 21] have additional optimized designs where φ and θ are forced to take on conventional construction values which cause an increase in the optimal weight and number of piles.

10.4 Damage detection and structural identification in structural systems

The previous sections in Chapter 10 were all focused on finding the optimal design for a set of loads and constraints using generalized optimality criteria. Section 10.4 will change this focus and will explore the use of optimality-criterion-based optimization as a tool to identify the elemental stiffnesses for a given structure using limited loadings and measured response data. This is essentially the definition of structural identification of actual stiffness parameters in real structures. This technique can be used to find damage by comparing the calculated stiffness obtained from the structural-identification process to the actual undamaged structure's elemental stiffnesses.

Structures will inevitably sustain damage throughout their service life. The sources of this damage are many: wind, earthquakes, service loads, fatigue, etc. Predicting the location within a structure where damage will occur or has occurred can be challenging. For example, following a seismic event, a structure may sustain one type or level of damage and another structure, just feet away, will sustain a completely different type of damage.

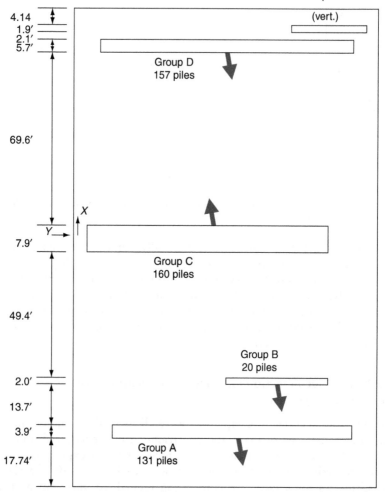

Figure 10.26 Optimal pile layout for the lock monolith example [11] (1 ft = 0.305 m).

Table 10.17 Optimized lock monolith pile properties (1 in.^{4} = 41.62 cm^4).

Group	# of piles	I_x (in.^{4})	Batter	φ (deg)	θ (%)	% Fixity
A	131	930	3.50	148	0	Fixed
B	20	936	2.00	146	90	51%
C	160	729	5.38	336	0	Fixed
D	157	729	2.00	169	0	Fixed
E	33	1,161	10.9	155	0	Pinned

Table 10.18 Optimized lock monolith pile locations (1 ft = 0.305 m).

Group	X_{low}	X_{high}	Y_{low}	Y_{high}
A	−71.3	−67.4	14.6	110.6
B	−53.7	−51.7	70.7	101.6
C	−2.3	5.6	6.2	99.8
D	75.2	80.9	11.6	114.2
E	83.0	84.9	87.0	118.3

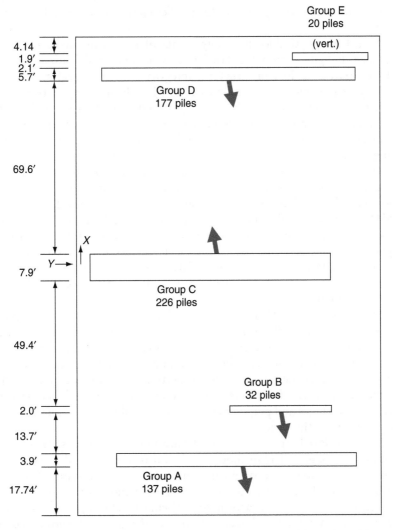

Figure 10.27 Pile layout used for the final optimization and branch-and-bound procedures [11] (1 ft = 0.305 m).

Table 10.19 Final discrete lock monolith pile properties ($1\,\text{in.}^4 = 41.62\,\text{cm}^4$).

Group	# of piles	I_x (in.^{4})	Batter	φ (deg)	θ (%)	(%) Fixity
A	152	904	3.54	149	0	Fixed
B	25	904	2.15	143	90	Fixed
C	180	729	5.12	337	0	Fixed
D	171	729	2.00	169	0	Fixed
E	29	1,220	6.51	165	0	Pinned

Structures such as gravity dams are subject to large hydrostatic pressures that create unique complications for predicting the locations and severity of damage.

Aside from physical loads, simply the age of a structure can be a source of damage. Fatigue, material deterioration, structural modifications, etc. can all be sources of loss of structural integrity in a structural system. The variety of possible damage sources coupled with the complexity of many structural systems make it challenging to predict the location and severity of damage. Therefore, a single damage-detection process that can detect damage regardless of its source is highly beneficial to critical structures in all geographic areas.

The most common method of detecting such damage is through visual inspection. The many drawbacks of damage detection using visual inspection have spurred research in the area of structural health monitoring (SHM). The main advantage SHM has over visual inspection methods is that first and foremost it is non-invasive. The disadvantage is that some types of physical response such as displacements, strains or frequencies must be collected. Depending on the type of damage-detection algorithm or SHM method being used, the responses may or may not be associated with a known loading.

There are several SHM methods that have been researched over recent years. First, there is SHM using dynamic response data, which is typically broken down into two sub-categories, one involving ambient vibration testing, and the second, forced vibration testing. Both methods detect damage by monitoring changes in dynamic properties, such as mode shapes and/or natural frequencies of elements within the structure. The goal then is to analyze these changes and pinpoint either stiffness reductions or changes in the mass matrix of the structure.

A less explored area of structural health monitoring uses static response data of a structure. The method, which will be presented, detects areas of stiffness reduction based on changes in displacements resulting from statically applied loads. The advantage of this method is that the damage-detection procedure is practical to implement. The input loads can be altered in magnitude and location to suit the test and output requirements. The main disadvantages of this method are the equipment and personnel costs required to apply the external loads and instrument the structure. Additionally, due to the fact that loads are applied inside the structure, obstructions such as desks, tables and partitions can complicate the procedure.

The generalized optimality-criteria technique presented uses static loads coupled with displacement measurements to identify potentially damaged elements and is an example of the work performed by Terlaje, Johnson and Truman [26–32]. This algorithm can be easily extended to use dynamic loads or dynamic measurements by using time-history-response optimization as presented in Chapter 9, but for bridges and many building systems using static loads, it provides simple, repeatable, accessible and assessable sets of data. This measured data is then used to form the constraints and the optimality criterion finds the optimal set of elements to create the structure that complies with the measured data. By comparing the actual (initial/real) design with the optimal design for the given measured displacements, one can estimate which elements are damaged. In many cases this can be done with relatively few load cases and measurements.

Similarly, structural identification can be performed by having sets of load and measured displacement data for a given structure. By setting the measured displacements as equality constraints for the given load(s) and formulating an appropriate objective function, usually potential or strain energy, the optimality-criteria algorithm will provide a structure for the given geometry with structural elements (sizes) that will produce the measured displacements. By having several loadings and their displacements, one can use all of this data in one optimization or it can be used separately or in other combinations to provide a suite of possibilities for the damaged structural system. If these different scenarios all produce similar results, one can be assured that the structural-identification solution is reasonably accurate.

10.4.1 Damage detection in a one-bay, one-storey frame

To illustrate the damage-detection procedure, damage will be located in the one-bay, one-storey frame, with dimensions and members labelled as shown in Fig. 10.28.

Assume that the healthy moment of inertia values are known, arbitrarily chosen as $I_1 = 1,500\,\text{in.}^4$, $I_2 = 1,000\,\text{in.}^4$, $I_3 = 1,500\,\text{in.}^4$ $(1\,\text{in.}^4 = 41.62\,\text{cm}^4)$.

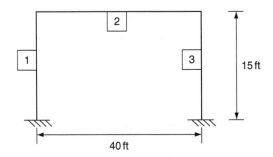

Figure 10.28 One-bay, one-storey moment frame $(1\,\text{ft} = 0.305\,\text{m})$.

Now assume that some time has elapsed since the structure was first constructed and that the structure has experienced damage, assume a 15 percent reduction in the moment of inertia of member 2.

The first step of the damage-detection procedure is to apply static point loads to the existing structure. In this case, three load cases as shown in Fig. 10.29 will be applied. A 10 kip (44.5 kN) load will be applied, first to the left-quarter point of the beam, then a 10 kip (44.5 kN) load at the mid-span of the beam, and finally a 10 kip (44.5 kN) load will be applied to the right-quarter point of the beam. For both load cases 1 and 2, the mid-span deflections are measured, and for load case 3, the mid-span deflection and the lateral displacement of the first storey are measured, with displacement measurements also shown in Fig. 10.29.

After applying the three different load cases to the damaged structure and obtaining the measured displacements, a mathematical model of the frame must be created. Using two-dimensional bending elements, the frame is created using two beam-columns and four beam elements resulting in a two-dimensional moment frame consisting of six elements and nine total degrees of freedom, as shown in Fig. 10.30. To give the optimization algorithm a starting point from which to begin adjusting moment of inertia values, the healthy moment of inertia values are used as the initial values.

For damage-detection problems a variety of objective functions can and have been used. This example minimizes strain energy making the assumption that the minimal strain energy should provide the equilibrium-based structure that produces the measured displacements.

$$O = \sum_{a=1}^{q} \frac{1}{2} u_a^T K u_a \tag{10.23}$$

where u_a is the displacement found from load case 'a', and q is the number of load cases which is three for this problem. The constraints in damage-detection problems are equality constraints that are used to find an analytically derived optimal structure that replicates the actual measured displacements. In other words, the calculated values from the optimal solution must eventually take on the values of the measured displacements. This is why the optimality-criteria approach can be used; it essentially forces the structure to assign elemental stiffness values such that the structure will conform to the applied loads and their measured displacements. For this problem after the first iteration the constraint error is

$$g_1 = u_{1,6} - u_{1,6\text{measured}} = -0.121 - (-0.125) = 0.004$$
$$g_2 = u_{2,6} - u_{2,6\text{measured}} = -0.207 - (-0.213) = 0.006$$
$$g_3 = u_{3,1} - u_{3,1\text{measured}} = -0.020 - (-0.018) = -0.002$$
$$g_4 = u_{3,6} - u_{3,6\text{measured}} = -0.124 - (-.129) = 0.005$$

where for u_{ij}, i = load case, and j = degree of freedom number.

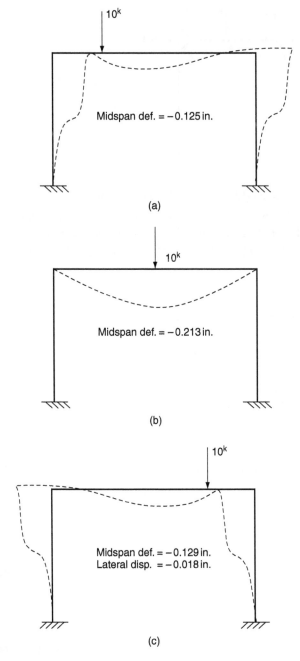

Figure 10.29 Load cases applied and measurements taken on the actual damaged structure (1 in. = 2.54 cm, 1k = 4.45 kN).

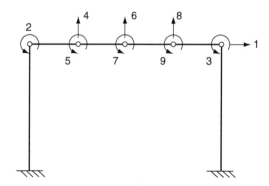

Figure 10.30 Mathematical model of the actual moment frame.

For this example it has been assumed that members 3 and 4 are linked and that members 5 and 6 are linked where the beam elements are numbered 3 to 6 from left to right in Fig. 10.30. Therefore, each optimality-criterion equation is of the form:

$$T_l = \frac{\sum_{j=1}^{4} \lambda_j \times \sum_{v=1}^{s} \frac{\partial g_j}{\partial I_v}}{\sum_{v=1}^{s} \frac{\partial O}{\partial I_v}} = 1 \quad \text{for} \quad l=1,2,\ldots,6 \tag{10.24}$$

where

$$\frac{\partial g_j}{\partial I_v} = \frac{\partial u_j}{\partial I_v} = -K^{-1} \frac{\partial K}{\partial I_v} u_j \tag{10.24a}$$

where

$$\frac{\partial O}{\partial I_v} = \frac{1}{2} \sum_{a=1}^{3} \frac{\partial u_a}{\partial I_v} K u_a + u_a \frac{\partial K}{\partial I_v} u_a + u_a^T K \frac{\partial u_a}{\partial I_v} \tag{10.24b}$$

giving

$$T_1 = -\frac{\sum_{j=1}^{4} \lambda_j \frac{\partial g_j}{\partial I_1}}{\frac{\partial O}{\partial I_1}} \tag{10.24c}$$

$$T_2 = -\frac{\sum_{j=1}^{4} \lambda_j \frac{\partial g_j}{\partial I_2}}{\frac{\partial O}{\partial I_2}} \tag{10.24d}$$

$$T_3 = T_4 = -\frac{\sum_{j=1}^{4} \lambda_j \times \left(\frac{\partial g_j}{\partial I_3} + \frac{\partial g_j}{\partial I_4}\right)}{\left(\frac{\partial O}{\partial I_3} + \frac{\partial O}{\partial I_4}\right)} \tag{10.24e}$$

$$T_5 = T_6 = -\frac{\sum_{j=1}^{4} \lambda_j \times \left(\frac{\partial g_j}{\partial I_5} + \frac{\partial g_j}{\partial I_6}\right)}{\left(\frac{\partial O}{\partial I_5} + \frac{\partial O}{\partial I_6}\right)} \tag{10.24f}$$

All values are defined in the above equations except for the Lagrange multiplier values. Using the equations derived in Chapter 8 for finding the Lagrange multipliers with linking, the Lagrange multipliers are found to be

$$\lambda_1 = 5.1898$$
$$\lambda_2 = -1.4855$$
$$\lambda_3 = 11.6737$$
$$\lambda_4 = -0.3623$$

Using these Lagrange multiplier values in Eqs. (10.24c), (10.24d), (10.24e) and (10.24f) gives optimality-criteria values of

$$T_1 = 0.99878$$
$$T_2 = 0.66013$$
$$T_3 = T_4 = 0.998599$$
$$T_5 = T_6 = 1.001443$$

As can be seen, members 1, 3, 4, 5 and 6 are nearly optimal as expected, however, the moment of inertia of member 2 must be reduced to better satisfy optimality. Using the optimality-criteria values in the recurrence relationship given in Chapter 8, the set of design variables for the next iteration (iteration 2) can be found as

$$I_1^{(2)} = 1,500 + \frac{1}{2}(0.9988 - 1) \times 1,500 = 1,499.1$$

$$I_2^{(2)} = 1,000 + \frac{1}{2}(0.6601 - 1) \times 1,000 = 830.1$$

$$I_3^{(2)} = I_4^{(2)} = 1,500 + \frac{1}{2}(0.9986 - 1) \times 1,500 = 1,498.95$$

$$I_5^{(2)} = I_6^{(2)} = 1,500 + \frac{1}{2}(1.001 - 1) \times 1,500 = 1,500.75$$

Given this set of design variables, the process is repeated until optimality is satisfied. The results of the iterative process are shown in Table 10.20. Plots of the constraint error, objective function, and actual and calculated damage states are shown in Fig. 10.31.

The algorithm reached an optimal set of design values in four iterations as shown in Table 10.20. All optimality-criteria values reached unity, meaning that an optimal solution was reached. Fig. 10.31 shows that the constraint error converged to zero, meaning that the optimization algorithm reached a feasible solution. Fig. 10.31 also shows the variation of the objective function for each iteration. After a large increase between iterations 1 and 2, the algorithm reduced the objective function as it reached a feasible solution. Finally, as seen in Fig. 10.31, the correct damage state was found for the structure considered.

The steps performed in this section outline the procedure used to complete the more complicated examples that follow. Although not relevant in this example, other examples may have side constraints that become active between iterations which require using the Lagrange multiplier generating equations that include active and passive elements as well as linking from Chapter 8.

10.4.2   Damage detection for a two-bay, two-storey moment frame

This example uses a two-bay, two-storey moment frame. Each bay is 40 ft (12.19 m) wide and each storey is 15 ft (4.57 m) tall. A total of 22 beam elements are used in the analysis (6 columns, 16 beams). As in the previous example, the healthy moment of inertia values will be used as the initial design values to seed the optimization algorithm. The structure and healthy moment of inertia values are shown in Fig. 10.32.

The mathematical model constructed assumes that members 7 and 8 are linked. Similarly, each pair of members 9 and 10, members 11 and 12, members 13 and 14, members 15 and 16, members 17 and 18, members 19 and 20, and members 21 and 22 are linked. This creates a total of 14 independent design variables to be used for this structure. The model contains a total of 32 d.o.f., as shown in Fig. 10.33. The load cases applied to the structure apply 10 kip (44.5 kN) loads in only the vertical direction along the span of each bay. (Vertical loads are assumed to be more easily applied than horizontal loads.) The load cases that will be utilized are shown in Table 10.21.

The load cases provided use 14 measured displacements. The generalized optimality-criteria algorithm was tested using six different damage states as

Table 10.20 Optimization results for the one-bay, one-storey damage-detection example ($1\,in.^4 = 41.62\,cm^4$).

Iterations	I_1	I_2	I_3	I_4	I_5	I_6	λ_1	λ_2	λ_3	λ_4	T_1	T_2	T_3	T_4	T_5
1	1,500	1,000	1,500	1,500	1,500	1,500	5.2	−1.49	11.67	−0.36	0.999	0.66	0.999	0.999	1.001
2	1,499	830.1	1,499	1,499	1,501	1,501	0.1	2.868	6.68	3.553	1.001	1.047	1.001	1.001	0.999
3	1,500	849.6	1,500	1,500	1,500	1,500	0.7	2.351	7.272	3.105	1	1.001	1	1	1
4	1,500	850	1,500	1,500	1,500	1,500	0.7	2.342	7.283	3.098	1	1	1	1	1

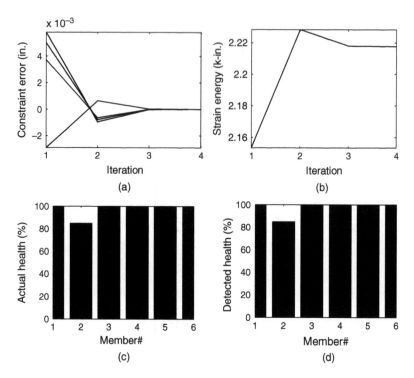

Figure 10.31 Damage-detection results: (a) constraint histories, (b) objective function history and (c) actual damage versus (d) calculated member damage, for the one-bay one-storey structure (1 in. = 2.54 cm, 1k = 4.45 kN).

shown in Table 10.22. Damage state 1 concentrates damage in the left and right columns of the first storey. This damage state is used to assess the algorithm's ability to discern damage in multiple columns on a single floor level. Damage state 2 reflects damages of the moment frame in the left column on the first floor and the centre column on the second floor. Damage state 3 assumes damage in the right column on the second storey and the middle column on the second storey. This damage state shows the ability to detect localized damage to the correct storey and correct column elements. Damage state 4 assumes damaged beams on first and second storeys. The damage locations are positioned far apart to show the ability of the algorithm to find damage in different regions. Damage state 5 assumes damage to a combination of column and beam elements, and damage state 6 assumes damage in all elements.

The same procedures used for the one-bay, one-storey moment frame example are used here. Each load case combination is solved using the lower side constraints set at 5 in.⁴ (208.1 cm⁴) while upper side constraints are not used.

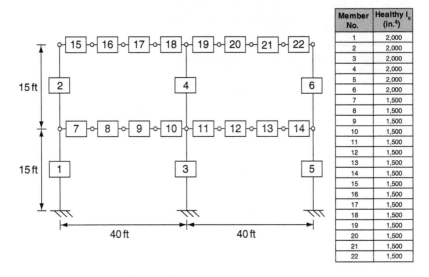

Member No.	Healthy I_x (in.⁴)
1	2,000
2	2,000
3	2,000
4	2,000
5	2,000
6	2,000
7	1,500
8	1,500
9	1,500
10	1,500
11	1,500
12	1,500
13	1,500
14	1,500
15	1,500
16	1,500
17	1,500
18	1,500
19	1,500
20	1,500
21	1,500
22	1,500

Figure 10.32 Two-bay, two-storey moment frame and the healthy moment of inertia values (1 in.⁴ = 41.62 cm⁴, 1 ft = 0.305 m).

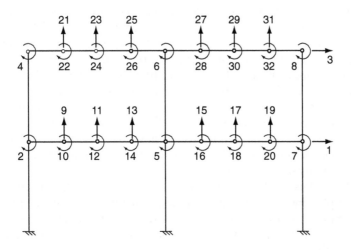

Figure 10.33 Two-bay, two-storey moment frame degrees of freedom.

Using the algorithm presented, the correct damage state was found in each case. Terlaje has shown that the large number of measured degrees of freedom coupled with not having upper side constraints on the element sizes is necessary for all six damage states to be correctly identified [27, 29]. The results are shown in Fig. 10.34.

Table 10.21 Load case schedule for the two-bay, two-storey damage-detection example (1 k = 4.45 kN).

Load case	d.o.f. loaded[a]	d.o.f. measured
1	9	11
2	11	11
3	13	1, 11
4	15	17
5	17	17
6	19	17
7	21	23
8	23	23
9	25	3, 23
10	27	29
11	29	29
12	31	29

[a]Each d.o.f. is loaded with a 10 k vertical load.

The algorithm remained stable for all damage states and the correct design values were calculated for all cases. The reason that the upper side constraints cannot be used is due to the fact that the algorithm wants to follow a path where several design values are increased above the healthy values but are then slowly reduced to the correct damaged values. Numerical anomalies such as this are being studied by Terlaje and Truman [29, 31]. For example, the iteration results for damage state 5 are shown in Table 10.23. As can be seen, the moment of inertia values for members 2, 5, 9, 10, 11 and 12 are increased above the healthy values after the first iteration. However, by the fourth iteration all design values are reduced to less than or equal to the healthy values. If upper side constraints were enforced, these members would have become inactive after the first iteration and then the number of active design variables would have been less than the number of displacement constraints. In some cases this situation can result in numerical instability of the algorithm.

Terlaje [27], Johnson [26] and Sanayei [33] have shown that when the number of displacement constraints is equal to the number of independent design variables, the optimality-criteria algorithm is able to detect the correct damage state in all cases. When this is not true, the damage state may still be detectable.

10.4.3 *Structural identification of a two-bay, two-storey moment resistant frame*

The presented optimality-criteria optimization algorithm can also be used as a structural identification tool. Without knowing the original member sizes, but knowing the element arrangements, the measured displacements can be used to find the 'actual' elemental sizes. Structural identification is a useful

Table 10.22 Tested damage states for the two-bay, two-storey damage-detection example (1 in.^{4} = 41.62 cm^4).

Damage state	Member No.	Actual damage ΔI (in.^{4})	Damage state	Member No.	Actual damage ΔI (in.^{4})	Damage state	Member no.	Actual damage ΔI (in.^{4})
1	1	−500	2	1	−500	3	1	0
	2	0		2	0		2	0
	3	0		3	0		3	0
	4	0		4	−500		4	−500
	5	−500		5	0		5	0
	6	0		6	0		6	−500
	7	0		7	0		7	0
	8	0		8	0		8	0
	9	0		9	0		9	0
	10	0		10	0		10	0
	11	0		11	0		11	0
	12	0		12	0		12	0
	13	0		13	0		13	0
	14	0		14	0		14	0
	15	0		15	0		15	0
	16	0		16	0		16	0
	17	0		17	0		17	0
	18	0		18	0		18	0
	19	0		19	0		19	0
	20	0		20	0		20	0
	21	0		21	0		21	0
	22	0		22	0		22	0
4	1	0	5	1	−500	6	1	−500
	2	0		2	0		2	−500
	3	0		3	−500		3	−500
	4	0		4	0		4	−500
	5	0		5	0		5	−500
	6	0		6	−500		6	−500
	7	−500		7	0		7	−500
	8	−500		8	0		8	−500
	9	0		9	0		9	−500
	10	0		10	0		10	−500
	11	0		11	0		11	−500
	12	0		12	0		12	−500
	13	0		13	0		13	−500
	14	0		14	0		14	−500
	15	0		15	0		15	−500
	16	0		16	0		16	−500
	17	0		17	−500		17	−500
	18	0		18	−500		18	−500
	19	0		19	−500		19	−500
	20	0		20	−500		20	−500
	21	−500		21	0		21	−500
	22	−500		22	0		22	−500

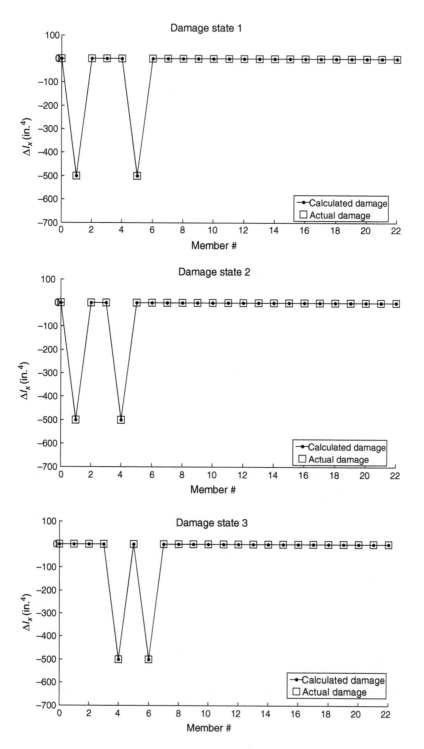

Figure 10.34 Damage-detection results for the two-bay, two-storey problem without upper side constraints for load cases 1–12 and damage states 1–6 (1 in.^{4} = 41.62 cm^4).

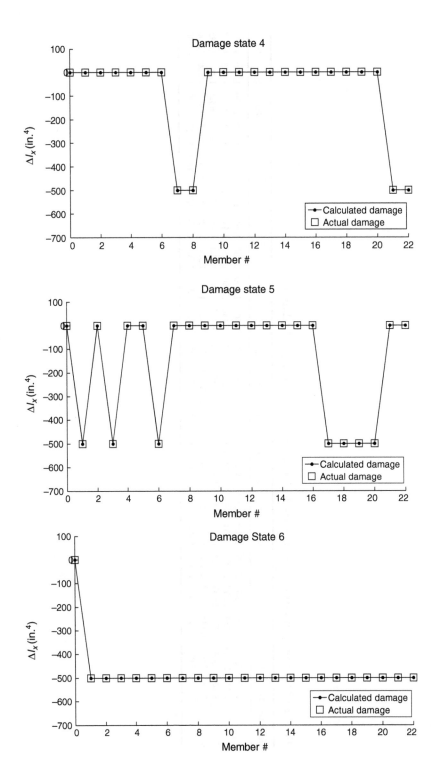

Figure 10.34 (Continued)

Table 10.23 Two-bay, two-storey iteration results for damage state 5 ($1\,\text{in.}^4 = 41.62\,\text{cm}^4$).

Iterations	I_1	I_2	I_3	I_4	I_5	I_6	I_7	I_8	I_9	I_{10}	I_{11}	I_{12}	I_{13}	I_{14}	I_{15}	I_{16}	I_{17}	I_{18}	I_{19}	I_{20}	I_{21}	I_{22}
1	2,000	2,000	2,000	2,000	2,000	2,000	1,500	1,500	1,500	1,500	1,500	1,500	1,500	1,500	1,500	1,500	1,500	1,500	1,500	1,500	1,500	1,500
2	1,474	2,145	1,334	1,858	2,100	1,451	1,477	1,477	1,522	1,522	1,519	1,519	1,480	1,480	1,454	1,454	813.5	813.5	793.4	793.4	1,463	1,463
3	1,504	2,048	1,474	1,948	2,024	1,521	1,493	1,493	1,507	1,507	1,507	1,507	1,493	1,493	1,482	1,482	972.7	972.7	963.2	963.2	1,485	1,485
4	1,500	2,000	1,499	1,999	2,000	1,500	1,500	1,500	1,500	1,500	1,500	1,500	1,500	1,500	1,500	1,500	999.3	999.3	998.7	998.7	1,500	1,500
5	1,500	2,000	1,500	2,000	2,000	1,500	1,500	1,500	1,500	1,500	1,500	1,500	1,500	1,500	1,500	1,500	1,000	1,000	1,000	1,000	1,500	1,500

tool for determining the current stiffness distribution within a structure, but it is incapable of discerning whether a structure is damaged unless the original stiffness distribution is known.

To show the robustness of this model, all initial design values were set just above the lower side constraint at a value of 6 in.[4] ($250\,\text{cm}^4$) (grossly undersized). Next, a FEM model using these grossly inaccurate member sizes is created and analyzed using the previous load combination. The 14 displacement constraints are generated using the displacements of the healthy structure as the measured displacements (in other words, the optimization algorithm will try to find a structure that has the 'real/healthy' displacements). These constraints are similar to those used in the optimality-criteria optimization algorithm presented for damage detection. The results are shown in Table 10.24.

The algorithm found the correct initial healthy state in 13 iterations. These results show that the damage-detection procedure is also useful as a structural identification tool. The only information used for this procedure was the displacement measurements taken on the actual structure. However, the number of iterations required to determine the current state of the structure was nearly three times that of the damage-detection example. This is due to the fact that the initial design values for the structural identification problem were much further from the actual solution.

Researchers [26, 27, 33] have found that the number of independent measurements, NIM, provides a strong indication of when the algorithm will be able to detect the correct damage state or be capable of performing a meaningful structural identification. In damage-detection problems, there is no unique solution unless the NIM is greater than or equal to the number of unknown parameters. NIM is calculated by:

$$\text{NIM} = (m_1 + m_2)(m_2 + m_3) - \frac{1}{2}m_2(m_2 - 1) \tag{10.25}$$

where m_1 is the number of measured degrees of freedom with a force applied, m_2 is the number of measured degrees of freedom where no force is applied, and m_3 is the number of unmeasured degrees of freedom where a force is applied.

The two-bay, two-storey structure contained a total of 14 independent design variables which is why the load combination had 14 measured displacements. With the number of measured displacements equal to the NIM, one is almost assured of a successful outcome. If the NIM is less than the number of design variables, one may still have success if the displacements are measured in 'good' locations. Much research is being performed in determining the optimal number of measurements and their location in order to guarantee success for damage-detection and structural-identification algorithms.

Table 10.24 Structural identification results for the two-bay, two-storey moment frame (1 in.^{4} = 41.62 cm^4).

Iteration	I_1	I_2	I_3	I_4	I_5	I_6	I_7	I_8	I_9	I_{10}	I_{11}	I_{12}	I_{13}	I_{14}	I_{15}	I_{16}	I_{17}	I_{18}	I_{19}	I_{20}	I_{21}	I_{22}
1	6	6	6	6	6	6	6	6	6	6	6	6	6	6	6	6	6	6	6	6	6	6
2	11.98	11.98	11.98	11.98	11.98	11.98	11.98	11.98	11.98	11.98	11.98	11.98	11.98	11.98	11.98	11.98	11.98	11.98	11.98	11.98	11.98	11.98
3	23.9	23.9	23.89	23.89	23.9	23.9	23.86	23.86	23.86	23.86	23.86	23.86	23.86	23.86	23.86	23.86	23.86	23.86	23.86	23.86	23.86	23.86
4	47.51	47.52	47.49	47.51	47.51	47.52	47.33	47.33	47.34	47.34	47.34	47.34	47.33	47.33	47.33	47.33	47.34	47.34	47.34	47.34	47.33	47.33
5	93.92	93.94	93.83	93.9	93.92	93.94	93.16	93.16	93.19	93.19	93.19	93.19	93.16	93.16	93.17	93.17	93.18	93.18	93.18	93.18	93.17	93.17
6	183.5	183.6	183.2	183.4	183.5	183.6	180.5	180.5	180.6	180.6	180.6	180.6	180.5	180.5	180.5	180.5	180.6	180.6	180.6	180.6	180.5	180.5
7	350.6	350.8	349.3	350.2	350.6	350.8	339.2	339.2	339.6	339.6	339.6	339.6	339.2	339.2	339.3	339.3	339.5	339.5	339.5	339.5	339.3	339.3
8	641.1	641.7	636.6	639.7	641.1	641.7	601.4	601.4	602.6	602.6	602.6	602.6	601.4	601.4	601.6	601.6	602.3	602.3	602.3	602.3	601.6	601.6
9	1,080	1,081	1,068	1,076	1,080	1,081	960.8	960.8	963.9	963.9	963.9	963.9	960.8	960.8	961.4	961.4	963.3	963.3	963.3	963.3	961.4	961.4
10	1,583	1,585	1,561	1,575	1,583	1,585	1,305	1,305	1,310	1,310	1,310	1,310	1,305	1,305	1,306	1,306	1,309	1,309	1,309	1,309	1,306	1,306
11	1,917	1,918	1,901	1,911	1,917	1,918	1,474	1,474	1,477	1,477	1,477	1,477	1,474	1,474	1,474	1,474	1,476	1,476	1,476	1,476	1,474	1,474
12	1,997	1,997	1,995	1,996	1,997	1,997	1,499	1,499	1,500	1,500	1,500	1,500	1,499	1,499	1,500	1,500	1,500	1,500	1,500	1,500	1,500	1,500
13	2,000	2,000	2,000	2,000	2,000	2,000	1,500	1,500	1,500	1,500	1,500	1,500	1,500	1,500	1,500	1,500	1,500	1,500	1,500	1,500	1,500	1,500

10.4.4 Damage detection in a three-bay, five-storey moment frame

Damage will be detected in the three-bay, five-storey steel moment frame shown in Fig. 10.35. Each bay is 40 ft (12.19 m) wide and each storey is 15 ft (4.57 m) tall. The healthy moment of inertia values are shown in Table 10.25.

The moment frame contains a total of 115 degrees of freedom, as shown in Fig. 10.36. Due to the size of the structure, 45 load cases will be used to detect damage. Each load case is also shown in Fig. 10.36. Each load case applies a 10 kip (44.5 kN) gravity load on the structure.

Notice that degrees of freedom exist at the quarter points of each bay (e.g. degrees of freedom 26, 27, 30 and 31, etc.), even though only two beam elements span each bay. To create a model with only 50 elements, but still containing all 115 degrees of freedom, 4 elements must be used to span

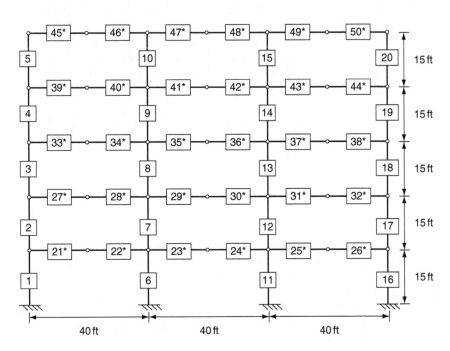

Figure 10.35 Three-bay, five-storey moment frame (1 ft = 0.305 m).

Table 10.25 Moments of inertia for the frame members (1 in.^{4} = 41.62 cm^4).

Member	I_x (in.^{4})
1–20	2,000
21–50	1,500

Load Case	d.o.f. Loaded	d.o.f. Measured
1	26	28
2	28	28
3	30	1, 28
4	32	34
5	34	34
6	36	34
7	38	40
8	40	40
9	42	40
10	44	46
11	46	46
12	48	3, 46
13	50	52
14	52	52
15	54	52
16	56	58
17	58	58
18	60	58
19	62	64
20	64	64
21	66	5, 64
22	68	70
23	70	70
24	72	70
25	74	76
26	76	76
27	78	76
28	80	82
29	82	82
30	84	7, 82
31	86	88
32	88	88
33	90	88
34	92	94
35	94	94
36	96	94
37	98	100
38	100	100
39	102	9, 100
40	104	106
41	106	106
42	108	106
43	110	112
44	112	112
45	114	112

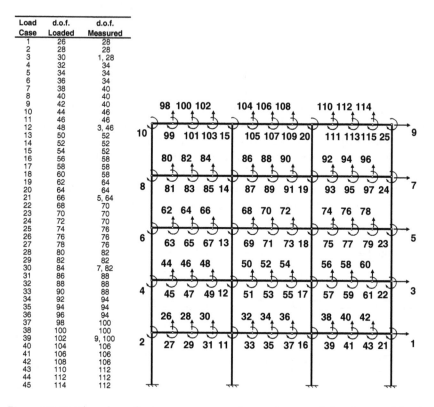

Figure 10.36 Three-bay, five-storey moment frame degrees of freedom and load cases.

each bay. Then, by linking elements, the finite-element model will appear to comprise 50 elements rather than 80 elements, as shown in Fig. 10.37.

The two members on either side of the mid-point of each bay are linked, giving a total of 30 sets of linked members. Linking is employed to provide additional degrees of freedom where displacement measurements can be made and loads can be applied, without adding more design variables that must be identified by the algorithm.

For this example, the axial extension is assumed negligible and therefore structural stiffness is attributed entirely to the bending resistance of each structural element. Therefore, by using the element moments of inertia as the design variables, the algorithm can detect damage by identifying locations of reduced moment of inertia in the structure. The assumed damage state is shown in Table 10.26, which will be detected using generalized optimality-criteria optimization.

The damage state of Table 10.26 contains damage in all columns along the first column line (members 1 through 5) and the first two columns along

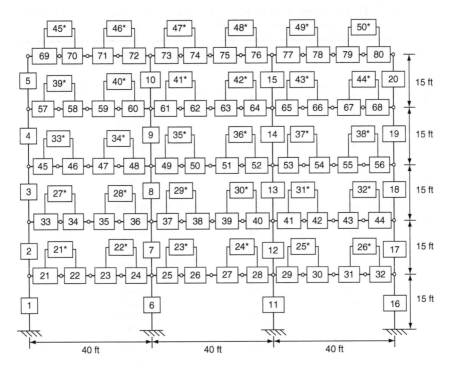

Figure 10.37 Three-bay, five-storey moment frame with linked members (1 ft = 0.305 m).

Table 10.26 Damage state for the three-bay, five-storey damage-detection example (1 in.⁴ = 41.62 cm⁴).

Member	Damage (in.⁴)
1–5	−500
7–8	−300
21–22[a]	−500
27–28[a]	−300
33–34[a]	−100

[a] Linked elements

the second column line (members 7 and 8). All beams on the first three storeys of the left-most bay are damaged as well. This damage state is used to simulate damage to an exterior bay of a structure, which can be common after an extreme loading event such as blast loading.

The measured displacements in the damaged structure are used to form the constraints in the optimization algorithm. These damaged structure displacements were obtained using finite-element analysis; the displacement measurements are shown in Table 10.27.

Table 10.27 Displacement measurements for damage detection of the three-bay, five-storey moment frame (1 in. = 2.54 cm).

Load case	d.o.f. loaded	d.o.f. measured	Measured displacement
			(in.)
1	26	28	0.1255
2	28	28	0.2304
3	30	1	0.0074
3	30	28	0.1208
4	32	34	0.0859
5	34	34	0.1580
6	36	34	0.0848
7	38	40	0.0847
8	40	40	0.1585
9	42	40	0.0869
10	44	46	0.1104
11	46	46	0.1996
12	48	3	0.0105
12	48	46	0.1062
13	50	52	0.0875
14	52	52	0.1599
15	54	52	0.0862
16	56	58	0.0857
17	58	58	0.1599
18	60	58	0.0880
19	62	64	0.0978
20	64	64	0.1748
21	66	5	0.0108
21	66	64	0.0934
22	68	70	0.0864
23	70	70	0.1589
24	72	70	0.0858
25	74	76	0.0858
26	76	76	0.1600
27	78	76	0.0881
28	80	82	0.0934
29	82	82	0.1655
30	84	7	0.0119
30	84	82	0.0886
31	86	88	0.0860
32	88	88	0.1588
33	90	88	0.0860
34	92	94	0.0866
35	94	94	0.1611
36	96	94	0.0889
37	98	100	0.1125
38	100	100	0.1882
39	102	9	0.0129
39	102	100	0.1037
40	104	106	0.0987
41	106	106	0.1757
42	108	106	0.0986
43	110	112	0.1004
44	112	112	0.1816
45	114	112	0.1058

To initialize the optimization algorithm, the design variables are set to the healthy stiffness values given in Table 10.25. The algorithm terminates when optimality is satisfied. Table 10.28 shows the iterative results in reaching the optimization-based damaged members.

The algorithm was able to accurately locate and quantify damage for the given damage state. The damage state was detected in five iterations. The undamaged members, not shown in Table 10.28, were accurately detected and exhibited little variation throughout the optimization process and were therefore excluded from the table. The results show that generalized optimality-criteria optimization is able to locate an optimal solution for systems with many design variables. As long as the algorithm is provided with sufficient information in the form of displacement constraints, the algorithm is able to detect the correct damage state in very few iterations.

10.4.5 *Structural identification for a ten-bar truss*

This example is used to demonstrate the application of the generalized optimality-criteria method to the structural identification of the elemental stiffness parameters of the ten-bar truss shown in Fig. 10.38. By loading the structure with a set of point loads and measuring strategic displacements at key degrees of freedom, the cross-sectional area of each truss member can be identified.

The parameters that will be identified by the algorithm are shown in Table 10.29. The elastic modulus of each member is taken as 29,000 ksi (200,000 MPa). Displacements are created using two load cases. The load cases and measured degrees of freedom are shown in Table 10.30.

Both load cases apply a 10 kip (44.5 kN) load, in the negative y-direction, to the degree of freedom listed. For example, load case 1 will apply a 10 kip (44.5 kN) load to degree of freedom 4 and while the static point load is applied, displacement measurements will be taken at degrees of freedom 2, 3 and 4. Table 10.31 shows the displacement measurements, which were obtained (simulated) using finite-element analysis. The eight displacement constraints will be used to determine the ten stiffness parameters.

Each stiffness parameter is initialized at 1 in.^{2} (6.45 cm^2). As mentioned in the previous example, only the geometry, loads and measured displacements are needed to recreate the elemental stiffnesses (design variables).

The optimization results are shown in Table 10.32. The algorithm converged to an optimal solution in nine iterations. The last iteration of Table 10.32 shows that all calculated parameters are within 1.5 percent of the actual stiffness values given in Table 10.29.

Fig. 10.39 shows the constraint error for all eight displacement constraints reduced to zero, meaning that a feasible solution was reached. The strain energy objective function shows a continual decrease showing the algorithm's minimization of the objective function.

Table 10.28 Iterative moments of inertia results of damaged members (1 in.⁴ = 41.62 cm⁴).

Iteration	I_1	I_2	I_3	I_4	I_5	I_7	I_8	I_{21*}	I_{22*}	I_{27*}	I_{28*}	I_{33*}	I_{34*}
	(in.⁴)	(in.⁴)	(in.⁴)	(in.⁴)	(in.⁴)	(in.⁴)	(in.⁴)	(in.⁴)	(in.⁴)	(in.⁴)	(in.⁴)	(in.⁴)	(in.⁴)
1	2,000	2,000	2,000	2,000	2,000	2,000	2,000	1,500	1,500	1,500	1,500	1,500	1,500
2	1,493.6	1,399.2	1,394	1,362	1,378.9	1,540.1	1,619	720.43	778.955	1,108.5	1,141.2	1,385.9	1,399.7
3	1,533.7	1,502.2	1,496.7	1,489.7	1,492.9	1,671.1	1,692.7	919.89	953.538	1,191.1	1,199.1	1,399.2	1,400.7
4	1,501.9	1,500.9	1,500.2	1,500	1,500	1,698.3	1,699.7	993.4	998.084	1,199.8	1,200.1	1,400	1,400
5	1,500	1,500	1,500	1,500	1,500	1,700	1,700	999.96	999.999	1,200	1,200	1,400	1,400

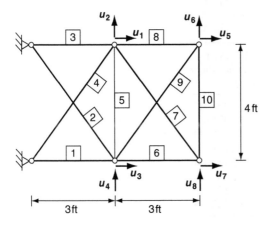

Figure 10.38 Ten-bar truss structure (1 ft = 0.305 m).

Table 10.29 Cross-sectional areas of truss members (1 in.^{2} = 6.45 cm^2).

Member	Area (in.^{2})
1	20
2	10
3	20
4	10
5	10
6	20
7	10
8	20
9	10
10	10

Table 10.30 Load case data for structural identification.

Load case	d.o.f. loaded	d.o.f. measured
1	4	2, 3, 4
2	6	1, 3, 5, 6, 7

10.4.6 *Damage detection in a continuous beam that considers modelling error*

It has been shown in previous examples that the generalized optimality-criteria algorithm is able to detect damage in the absence of data error (simulating the damaged displacements through analysis). Of course, in real life there will always be errors; errors in the computational model, errors

Table 10.31 Displacement measurements for structural identification of the 10-bar truss (1 in. = 2.54 cm).

Load case	d.o.f. measured	Measured displacement (in.)
1	2	−1.499E-03
	3	−2.707E-04
	4	−2.083E-03
2	1	6.912E-04
	3	−7.054E-04
	5	9.618E-04
	6	−5.325E-03
	7	−9.002E-04

in the displacement measurements, errors in the load placement, etc. This example is used to show how the algorithm performs in detecting stiffness reductions in the presence of modelling error. To demonstrate this, an analytical model of a 45 ft (13.72 m) long continuous beam is used as the test structure and is shown in Fig. 10.40. The finite-element model of the beam contains 450 bending elements, resulting in a total of 899 degrees of freedom. The healthy stiffness properties of each element are: $I_x = 1,000$ in.^{4} (41,620 cm^4), $E = 29,000$ ksi (200,000 MPa). Only the in-plane degrees of freedom are considered.

The 899-degrees-of-freedom analytical model is approximated using a 17-degree-of-freedom identification model. The identification model is created using nine bending elements and is shown in Fig. 10.41. Each element of the identification model represents 50 elements of the analytical model.

In order to ensure that modelling error is incorporated, damage is simulated so that it does not span an entire element represented by the identification model. A maximum of five consecutive elements are damaged in the analytical model, representing only 10 per cent of one of the identification model elements. The damage state is shown in Table 10.33 and corresponds to partial damage in elements 2 and 8 in the model with elements 1–9 from left to right, as shown in Fig. 10.41. To detect damage, three load cases are used with nine displacement measurements. The load cases and measured degrees of freedom are shown in Table 10.34. Each load case applies a 10 kip (44.5 kN) gravity load to the degree of freedom listed. The load cases and measured displacements are chosen so that only vertical loads are applied and only vertical displacements are measured. This is done to utilize only measurement data and loadings that are easiest to apply on real structures, as rotational degrees of freedom are difficult to load as well as measure. It is expected that the damage-detection algorithm will calculate damage in members 2 and 8 of the identification model.

Table 10.32 Structural identification results (1 in.^{2} = 6.45 cm^2).

Iter	A$_1$	A$_2$	A$_3$	A$_4$	A$_5$	A$_6$	A$_7$	A$_8$	A$_9$	A$_{10}$	λ$_1$	λ$_2$	λ$_3$	λ$_4$	λ$_5$	λ$_6$	λ$_7$	λ$_8$	T$_1$	T$_2$	T$_3$	T$_4$	T$_5$	T$_6$	T$_7$	T$_8$	T$_9$	T$_{10}$
1	1.00	1.00	1.00	1.00	1.00	1.00	1.00	1.00	1.00	1.00	7E-05	0.043	14	−0.381	0.475	−0.221	14	0.161	2.8995	2.7986	2.9005	2.8019	2.8023	2.9012	2.8024	2.8991	2.7978	2.8024
2	1.96	1.90	1.96	1.90	1.90	1.96	1.90	1.96	1.90	1.90	−8E-04	0.068	13.1	−0.706	0.883	−0.409	13.1	0.297	2.8042	2.6176	2.8058	2.6233	2.6239	2.8074	2.6245	2.8033	2.6161	2.6245
3	3.77	3.44	3.77	3.44	3.44	3.77	3.44	3.70	3.44	3.44	−0.004	0.074	11.6	−1.209	1.528	−0.7	11.56	0.51	2.6278	2.3087	2.6303	2.3171	2.3178	2.6333	2.3196	2.6258	2.3058	2.3196
4	6.73	5.69	6.74	5.72	5.72	6.73	5.72	6.73	5.67	5.72	−0.015	−0.024	9.32	−1.778	2.295	−1.026	9.301	0.753	2.3254	1.8575	2.3283	1.8658	1.8663	2.3341	1.8713	2.3215	1.8522	1.8713
5	11.2	8.12	11.2	8.18	8.20	11.2	8.21	11.2	8.09	8.21	−0.033	−0.35	6.88	−1.946	2.629	−1.116	6.852	0.829	1.8794	1.3701	1.8811	1.3714	1.3709	1.8907	1.3813	1.8727	1.3622	1.3813
6	16.1	9.62	16.1	9.70	9.71	16.3	9.78	16.0	9.56	9.78	−0.04	−0.664	5.37	−1.218	1.809	−0.69	5.345	0.526	1.3875	1.0716	1.3873	1.0646	1.0632	1.3977	1.0752	1.3799	1.0643	1.0752
7	19.2	10.0	19.3	10.0	10.0	19.5	10.1	19.1	9.87	10.1	−0.015	−0.316	5.02	−0.279	0.473	−0.154	5.013	0.123	1.0753	1.004	1.0748	1.0001	0.9995	1.0796	1.0039	1.0718	1.0015	1.0039
8	19.9	10.0	20.0	10.0	10.0	20.3	10.2	19.8	9.87	10.2	−8E-04	−0.018	5	−0.011	0.02	−0.006	5	0.005	1.0028	1.0001	1.0028	0.9999	0.9998	1.0032	1.0001	1.0026	1.0015	1.0001
9	20.0	10.0	20.0	10.0	10.0	20.3	10.2	19.8	9.87	10.2	−2E-06	−3E-05	5	−2E-05	3E-05	−7E-06	5	8E-06	1	1	1	1	1	1	1	1	1	1

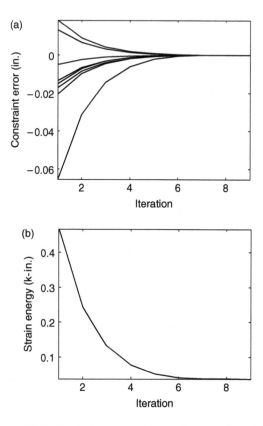

Figure 10.39 Ten-bar truss optimization results: (a) constraint histories and (b) objective function history, for parameter identification (1 in. = 2.54 cm, 1 K = 4.45 kN).

Figure 10.40 Two-span continuous beam to be used for damage detection (1 ft = 0.305 m).

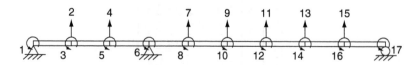

Figure 10.41 Nine-element identification model for the two-span continuous beam.

Table 10.33 Load case data for damage detection.

Member no.	Stiffness reduction (%)
61–65	60
371–375	80

Table 10.34 Load case data for damage detection (1 in. = 2.54 cm).

Load case	d.o.f. loaded	d.o.f. measured	Measured displacement (in.)
1	2	2	−0.031
		4	−0.025
2	4	2	−0.025
		4	−0.027
3	11	7	−0.079
		9	−0.168
		11	−0.221
		13	−0.202
		15	−0.119

The results in Fig. 10.42 reveal that the algorithm required 11 iterations to obtain an optimal solution and noticeable damage was detected in members 1, 2 and 8. It is promising that damage was calculated in member 8 and that the presence of damage was calculated in the vicinity of members 1 and 2. However, the algorithm calculated more damage in member 1 than in member 2, even though member 2 is where the damage was actually concentrated. Even though the stiffness reduction was not pinpointed exactly, the results still provide sufficient information to identify the region of likely damage. The convergence path was not as smooth as those of previous examples. It is observed that the constraint values oscillate before an optimal solution is reached.

Quantifying damage is more difficult since the identification model contains fewer elements than the analytical model. One element of the identification model encompasses 50 elements of the analytical model, therefore the 30 per cent decrease in stiffness of member 8 could represent 30 per cent damage in 50 elements of the analytical model or it could represent highly concentrated damage in only a few elements of the analytical model.

The convergence oscillation is caused by the modelling error incorporated by using a less refined structural model for the damage-detection model. Nonlinearities in the constraints cause the algorithm to oscillate as it converges to an optimal solution. One way to remedy this oscillation is to limit the change in each design variable during each iteration. By reducing the

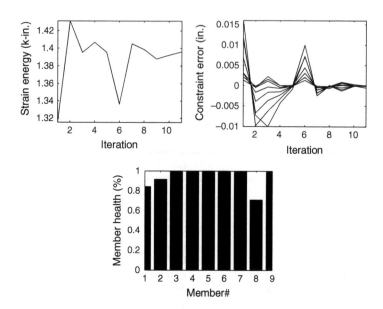

Figure 10.42 Optimality-criteria damage-detection results for the two-span continuous beam (1 in. = 2.54 cm, 1 kip = 4.45 kN).

step length, the algorithm is able to recalculate the constraint and objective gradient at an intermediate step, thereby better compensating for the nonlinearity contained in the problem. For example, during the variable resizing step of the algorithm, the recursion relationship from Chapter 8, Eq. (8.27), can be modified as:

$$\Delta\delta_i = \delta_i^{(k+1)} - \delta_i^{(k)} = \varsigma\frac{1}{r}(1 - T_i)\delta_i^{(k)} \quad \text{for} \quad i = 1, 2, \ldots, n \qquad (10.26)$$

where ς is a scalar value less than 1. Repeating the example with ς set to a value of 0.2 eliminates the oscillation present in Fig. 10.42. The results of this modification are shown in Fig. 10.43. The nonlinearity has much less of an effect on the algorithm as it is now able to calculate gradients at intermediate points. However, this modification comes at the expense of computational effort. As can be seen, the algorithm required 26 iterations, as shown in Fig. 10.43, to reach an optimal solution whereas the results in Figure 10.42 required only 11 iterations. Ultimately, damage was detected in the same members as the previous results, but element 2 now shows more damage compared to the first solution.

Despite the increased number of iterations, the generalized optimality-criteria algorithm efficiently detected areas of stiffness reduction with minimal computational effort. Results show that these stiffness reductions can be detected using a limited number of load cases with a reasonable number of measured displacements. Its efficiency makes the algorithm a

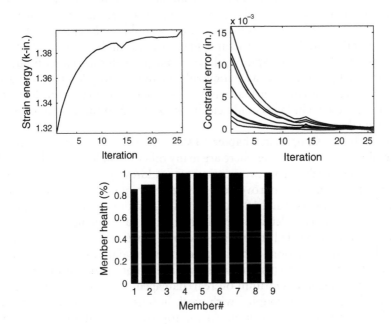

Figure 10.43 Damage-detection results using a modified recursion relationship ($\varsigma = 0.2$).

valuable tool for damage detection. This example shows that there are areas that need to be studied, such as modelling and sensor error, that have an effect on the ability of the numerical, optimality-criteria-based algorithm to consistently produce accurate, meaningful results.

10.5 Concluding remarks

This chapter was used to show the versatility of the optimality-criteria technique. The advantage of using this technique is the efficiency in computational effort. This technique typically requires far less computational effort than traditional mathematical programming and genetic algorithm-based approaches. This efficiency comes at the expense of customization. Each type of problem has to be integrated into the generalized optimality-criteria technique in a manner that will satisfy the constraints while optimizing the objectives as shown in Chapters 8, 9 and 10.

This necessary customization, leading to high computational efficiency, for solving a variety of problems was again evident in the examples shown in Chapter 10. The optimality criterion was customized and used in Chapter 9 to optimize the structural elements in 2-D and 3-D building systems for both static and dynamic loads. In Chapter 10, optimality-criteria techniques were shown to be useful in solving topology problems, pile foundation problems, structural identification problems and damage-detection problems. In the case of Chapter 10 applications, the applied loads were either static or pseudo-dynamic (pile-founded systems) loads, yet there is no reason why the theories for dynamic loads from Chapter 9 could not be extended to the topology, pile foundation, structural identification and damage-detection problems. As shown in Chapter 9, the dynamic problems are either pseudo-static (treated just the same as static loads) or time-history loads which are optimized as a series of static load cases at a specific point in the time history. Therefore, these techniques could easily be extended to the variety of uses shown in Chapter 10.

Although the examples in Chapter 10 show many uses for the generalized optimality-criteria technique, there are many other possible uses. Additional research and development are necessary to address potential new applications as well as the improvement of the techniques for problem types presented in Chapters 9 and 10. For example, overcoming the NIM requirement for the damage-detection problem is key to reducing the amount of measured data needed to guarantee convergence to the damaged state. Also, the optimal locations for load placement and data extraction for the damage-detection problems are areas that need more exploration. Testing the damage-detection algorithm on real 'damaged' structures is still necessary which needs to be coupled with work in sensing and modelling noise and their effect on the algorithm. In the topology area, the optimization 'path' or history using the presented generalized optimality-criteria technique can influence the final topological design as shown in Section 10.2 and needs to be explored further. Several issues regarding discrete variables and practical design for pile foundations can be improved due to the complexity of the designs. Yet, generalized optimality-criteria techniques have been shown to be capable of solving all of these specific types of problem and capable of producing meaningful results for topology, pile-founded structural, damage-detection and structural-identification problems.

References

1. Hendel, M.A., *Weight Based Topology Optimization of 2-D Finite Element Structures using the Optimality Criterion Method*, Department of Mechanical, Aerospace and Structural Engineering Master's Thesis, Washington University in St Louis, 2008.
2. Hendel, M.A. and Truman, K.Z., Topology optimization using the optimality criterion method, in *Proceedings of the Ninth International Conference on Computational Structures Technology*, Civil-Comp Press, paper No. 176, Stirling, UK, 2008.
3. Bendsoe, M.P., Optimal shape design as a material distribution problem, *Structural Optimization*, Vol. 1, pp. 192–202, 1989.
4. Bendsoe, M.P. and Sigmund, O., *Topology Optimization: Theory, Methods and Applications*, Springer, Berlin, Heidelberg, 2003.
5. Rossow, M.P. and Taylor, J.E., A finite element method for the optimal design of variable thickness sheets, *AIAA Journal*, Vol. 11, No. 11, pp. 1566–1569, 1973.

6. Rozvany, G.I.N., Exact analytical solutions for some popular benchmark problems in topology optimization, *Structural Optimization*, Vol. 15, No. 1, pp. 42–48, 1998.

7. Diaz, A. and Sigmund, O., Checkerboard patterns in layout optimization, *Structural Optimization*, Vol. 10, No. 1, pp. 40–45, 1995.

8. Sigmund, O. and Petersson, J., Numerical instabilities in topology optimization: A survey on procedures dealing with checkerboards, mesh-dependencies and local minima, *Structural Optimization*, Vol. 16, No. 1, pp. 68–75, 1998.

9. Zhou, M., Shyy, Y.K. and Thomas, H.L., Checkerboard and minimum member size control in topology optimization, *Structural and Multidisciplinary Optimization*, Vol. 21, pp. 152–158, 2001.

10. Sigmund, O., On the design of compliant mechanisms using topology optimization, *Mechanics of Structures and Machines*, Vol. 25, pp. 493–524, 1997.

11. Hoback, A.S., *Three-Dimensional, Flexible Slab Pile Foundations: Optimal Design*, Department of Civil Engineering Doctoral Dissertation, Washington University in St Louis, 1993.

12. Hoback, A.S. and Truman, K.Z., *Optimization of Steel Pile Foundations using Optimality Criteria*, Department of Army Waterways Experiment Station, USACE, ITL-92-1, Vicksburg, MS, 1992.

13. Hoback, A.S. and Truman, K.Z., A new method for finding the global and discrete optimums for structural systems, *Computers and Structures*, Pergamon Press, Elsevier, Oxford, Vol. 52, No. 1, pp. 127–134, 1994.

14. Hoback, A.S. and Truman, K.Z., *Computerized Layout and Sizing of Pile Foundations Using Structural Optimization*, Department of Army Waterways Experiment Station, USACE, ITL-95, Vicksburg, MS, 1995.

15. Hoback, A.S., Truman, K.Z., Jones, W. and Mosher, R., Optimal design of pile foundations: algorithm, OPTPILE program, in *Proceedings of the U.S. Army Corps of Engineers Structural Engineering Conference*, USACE, 1995.

16. Hurd, A., *Optimal Design of Large Pile Foundations using Optimality Criterion*, Department of Civil Engineering Master's Thesis, Washington University in St Louis, 2006.

17. Hurd, A. and Truman, K.Z., Optimization of pile foundations, in *Proceedings of the International Conference on Advances in Engineering Structures, Mechanics and Construction*, University of Waterloo, Waterloo, CA, pp. 653–661, 2006.

18. Truman, K.Z. and Hoback, A.S., Optimization of steel piles under rigid slab foundations using optimality criteria, *Structural Optimization*, Springer International, Vol. 5, No. 1, pp. 30–36, 1993.

19. Truman, K. Z. and Hoback, A.S., Least weight design of steel pile foundations, *Engineering Structures*, Elsevier Science Ltd, Vol. 15, No.5, pp. 379–386, 1993.

20. U.S. Army Corps of Engineers, Design memorandum No. 21, auxiliary lock & remainder of dam, *Lock and Dam No. 26 (Replacement)Report*, St Louis District, 1987.

21. U.S. Army Corps of Engineers, *User's guide: Pile Group Optimization Program (OPTPILE)*, Department of the Army Waterways Experiment Station, USACE, ITL-93, Vicksburg, MS, 1993.

22. Dawkins, W.P., *Pile Head Stiffness Matrices*, Department of the Army Waterways Experiment Station, USACE, Vicksburg, MS, 1978.

23. U.S. Army Corps of Engineers, *Basic Pile Group Behavior*, Department of the Army Waterways Experiment Station (CASE task Group on Pile Foundations), USACE, K-83-1, 1983.

24. U.S. Army Corps of Engineers, *User's guide: Pile Group Analysis (CPGA) Computer Program*, Department of the Army Waterways Experiment Station, USACE, ITL–88, Vicksburg, MS, 1988.

25. American Institute of Steel Construction Inc., *Manual of Steel Construction*, 8th–13th Editions, Chicago, IL, 1980–2004.

26. Johnson, L., *Damage Detection Using Static Responses and Unconstrained, Nonlinear Optimization*, Department of Civil Engineering Master's Thesis, Washington University in St Louis, 2004.

27. Terlaje, A.S., *Damage Detection in 2-D Steel Moment Frames Using Static Response Data and Optimality Criterion*, Master's Thesis, Department of Mechanical, Aerospace and Structural Engineering, Washington University in St Louis, 2006.

28. Terlaje, A.S. and Truman, K.Z., Earthquake damage detection using static response data and optimality criterion, in *Proceedings of the 8th National Conference on Earthquake Engineering*, EERI, paper No. 435, San Francisco, CA, 2006.

29. Terlaje, A.S. and Truman, K.Z., Damage detection in 2-D frames using static response data and optimality criterion, in *Proceedings of The Tenth East Asia-Pacific Conference on Structural Engineering*, Asian Institute of Technology, Bangkok, Thailand, Vol. 5, pp. 449–454, 2006.

30. Terlaje, A.S. and Truman, K.Z., A static response based parameter identification algorithm using OC optimization, in *Proceedings of the Eleventh International Conference on Civil, Structural, and Environmental Engineering Computing*, 2007. Civil-Comp Press, Stirling, UK, paper No. 88, 2007.

31. Terlaje, A.S. and Truman, K.Z., Parameter identification and damage detection using structural optimization and static response data, *Advances in Structural Engineering – An International Journal*, Vol. 10, No. 6, pp. 607–621, 2007.

32. Terlaje, A.S. and Truman, K.Z., Crack detection in large scale steel moment frames, in *Proceedings for the 14th World Conference in Earthquake Engineering*, Beijing, China, 2008.

33. Sanayei, M. and Onipede, O., Damage assessment of structures using static test data, *AIAA Journal*, Vol. 29, No. 7, pp. 1174–1179, 1991.

11 Nondeterministic structural optimization and parametric assessments

11.1 Introduction and conceptual probability of failure

11.1.1 Introduction

In the structural optimization field as presented in the previous chapters, most of the optimization techniques and the computer programs are generally developed on the premise that the design variables, resistances, responses and loadings are deterministic. The reason for the lack of research advancement with regard to structural optimization of nondeterministic systems may be that (1) the modern structural optimization algorithms rely on the various techniques of sensitivity analysis [1] (rate of change of response quantities with respect to design variables), which can become a pyramidical task in nondeterministic cases, and (2) it is too difficult to establish the objective function involving risk analysis of expected damage costs suitable for an individual structural design at a particular site [2].

In recognition of the random nature of seismic structural problems and the advances that have been made in reliability analyses of structures, this chapter is focused on optimum design of nondeterministic structures with consideration of safety levels of a system with respect to its various failure modes, uncertainties in the dead and live loads as well as seismic forces, and random parameters in responses and resistances. The cost function includes initial construction costs of structural and non-structural elements and expected costs of failure at various safety levels. The structural members are designed on the basis of the global optimum decision, namely, when an increase of the initial cost is balanced by a reduction in the expected cost of failure times the risk.

11.1.2 Reliability and probability of failure

Reliability is a measure of the probability of structural survival during a structure's lifetime. Thus, reliability is a probabilistic measure of the safety of the structure concerned. The probability of failure, which is the opposite of reliability, can also be adopted to represent the safety problem in an alternative way. In a classical formulation, reliability and probability of failure are defined as

$$P_r\,(R \geq S) = \int_{-\infty}^{\infty} P_d\,(S) \left[\int_{S}^{\infty} P_d\,(R)\,\mathrm{d}R \right]\mathrm{d}S \tag{11.1}$$

$$P_f\,(R < S) = 1 - P_r\,(R \geq S) \tag{11.2}$$

where R, S = structural resistance and response; $P_d\,(S), P_d\,(R)$ = probability density function of structural response and resistance, respectively. Hereafter, the reliability and probability of failure will also be represented as P_r and P_f.

Since in practice, a complete knowledge of the exact distributions of resistance and response is impossible to determine, two approximate expressions of the reliability or failure probability have been used in many studies. One is a first-order, second-moment expression that estimates reliability and probability of failure through a safety factor formulation. The other is a first-passage expression that estimates the probability of crossing the specified barrier during the vibration time interval. This chapter focuses on the former.

11.2 Probability distribution of response and resistance

A simple approximation model called a first-order, second-moment expression involves the evaluation of the mean and variance of the structural resistance and response, and represents either reliability or probability of failure [3]. This model along with two probability distributions of structural resistance and response considered as normal and lognormal distributions are used in this chapter.

11.2.1 Normal distribution

Since R and S are independent normally distributed variables, a linear function Q of R and S, where $Q = R - S$, is also normally distributed and thus, $\left(Q - \bar{Q}\right)/\sigma_Q$ is the standard normal variable with zero mean and unit variance. Let P_f be the probability of failure and be given by

$$P_f = P_f\,(R - S \leq 0) = \mathrm{PN}\left(\frac{0 - \bar{Q}}{\sigma_Q}\right) = \mathrm{PN}\left(\frac{-\bar{Q}}{\sigma_Q}\right) = 1 - \mathrm{PN}\left(\frac{\bar{Q}}{\sigma_Q}\right) \tag{11.3}$$

where $\mathrm{PN}\,(\bar{Q}/\sigma_Q)$ is the standard cumulative normal distribution function, $\bar{Q} = \bar{R} - \bar{S}$, $\sigma_Q = \left(\sigma_R^2 + \sigma_S^2\right)^{1/2}$ and $\bar{R}, \bar{S}, \sigma_R^2, \sigma_S^2$ = the mean and variance of the structural resistance, R and response, S, respectively.

If a safety factor (or safety index), β, is defined as the ratio $\bar{Q}/\sigma_Q$, then the probability of failure is

$$P_f = 1 - \text{PN}(\beta) \tag{11.4}$$

and reliability

$$P_r = \text{PN}(\beta) \tag{11.5}$$

where

$$\beta = \frac{\left(\bar{R} - \bar{S}\right)}{\left(\sigma_R^2 + \sigma_S^2\right)^{1/2}} \tag{11.6}$$

11.2.2 *Lognormal distribution*

If R and S are both lognormally distributed, the function $Q = \ln R - \ln S$ is normally distributed. Thus, analogous to the normal distribution case, the probability of failure and the reliability are given as follows: since

$$P_f = P_f\left(\ln R - \ln S \le 0\right) = \text{PN}\left(\frac{0 - \bar{Q}}{\sigma_Q}\right) = \text{PN}\left(\frac{-\bar{Q}}{\sigma_Q}\right) = 1 - \text{PN}\left(\frac{\bar{Q}}{\sigma_Q}\right) \tag{11.7}$$

then

$$P_f = 1 - \text{PN}\left(\frac{\bar{Q}}{\sigma_Q}\right) \tag{11.8}$$

and

$$P_r = \text{PN}\left(\frac{\bar{Q}}{\sigma_Q}\right) = \text{PN}\left(\frac{\ln \bar{R} - \ln \bar{S}}{\left(\sigma_{\ln R}^2 + \sigma_{\ln S}^2\right)^{1/2}}\right) \tag{11.9}$$

where the mean and variance of $\ln R$ and $\ln S$ are

$$\overline{\ln R} = \ln \bar{R} + \left(\frac{-1}{2}\sigma_{\ln R}^2\right), \quad \sigma_{\ln R}^2 = \ln\left(V_R^2 + 1\right) \tag{11.10}$$

$$\overline{\ln S} = \ln \bar{S} + \left(\frac{-1}{2}\sigma_{\ln S}^2\right), \quad \sigma_{\ln S}^2 = \ln\left(V_S^2 + 1\right) \tag{11.11}$$

where

$$V_R^2 = \frac{\sigma_R^2}{\bar{R}^2}, \quad V_S^2 = \frac{\sigma_S^2}{\bar{S}^2} \tag{11.12}$$

The derivation of Eqs. (11.10) and (11.11) are given in Appendix G. The mean and variance of Q are obtained as

$$\bar{Q} = \overline{\ln R} - \overline{\ln S} = \ln \left[\left(\frac{\bar{R}}{\bar{S}} \right) \left(\frac{1 + V_S^2}{1 + V_R^2} \right)^{1/2} \right] \tag{11.13}$$

$$\sigma_Q^2 = \sigma_{\ln R}^2 + \sigma_{\ln S}^2 = \ln \left(V_R^2 + 1 \right) + \ln \left(V_S^2 + 1 \right) = \ln \left[(1 + V_R^2) + (1 + V_S^2) \right] \tag{11.14}$$

Hence, the safety factor, $\beta = \bar{Q}/\sigma_Q$, is of the form

$$\beta = \frac{\ln \left[(\bar{R} - \bar{S}) \sqrt{\frac{1 + V_S^2}{1 + V_R^2}} \right]}{\sqrt{\ln \left[(1 + V_R^2)(1 + V_S^2) \right]}} \tag{11.15}$$

11.3 Uncertainty formulations of response and resistance

According to Eqs. (11.6) and (11.15), finding the safety factor formulation is equivalent to calculating the means and coefficients of variation of the structural resistance and response. Since determining the uncertainties of the structural response and resistance is impossible, a first-order approximation is employed to approximate the uncertainties in terms of their parameter uncertainties. The approximation of the means and variances of the structural response and resistance can be derived as

mean

$$\overline{S(r)} = E \left[S(\bar{r}) + \sum_i \left(\frac{\partial S}{\partial r_i} \right)_{\bar{r}} (r_i - \bar{r}_i) \right]$$

$$= S(\bar{r}) + \sum_i \left(\frac{\partial S}{\partial r_i} \right)_{\bar{r}} E[r_i - \bar{r}_i] \tag{11.16}$$

$$= S(\bar{r}) + \sum_i \left(\frac{\partial S}{\partial r_i} \right)_{\bar{r}} (\bar{r}_i - \bar{r})$$

$$= S(\bar{r})$$

$$\overline{R(r')} = E \left[R(\bar{r}') + \sum_{i'} \left(\frac{\partial R}{\partial r_i'} \right)_{\bar{r}'} (r_i' - r_i') \right] \tag{11.17}$$

$$= R(\bar{r}')$$

variance

$$\sigma_S^2 = E\left[\left(S - \bar{S}\right)\left(S - \bar{S}\right)\right]$$

$$= E\left[\left(S(\bar{r}) + \sum_i \left(\frac{\partial S}{\partial r_i}\right)_{\bar{r}} (r_i - \bar{r}_i) - \bar{S}\right)\left(S(\bar{r}) + \sum_j \left(\frac{\partial S}{\partial r_j}\right)_{\bar{r}} (r_j - \bar{r}_j) - \bar{S}\right)\right]$$

$$= \sum_i \sum_j \left(\frac{\partial S}{\partial r_i}\right)_{\bar{r}} \left(\frac{\partial S}{\partial r_j}\right)_{\bar{r}} \rho_{r_i r_j} V_{r_i} V_{r_j} \bar{r}_i \bar{r}_j \qquad (11.18)$$

$$\sigma_R^2 = E\left[\left(R - \bar{R}\right)\left(R - \bar{R}\right)\right] \qquad (11.19)$$

$$= \sum_{i'} \sum_{j'} \left(\frac{\partial R}{\partial r_{i'}'}\right)_{\bar{r}'} \left(\frac{\partial R}{\partial r_{j'}'}\right)_{\bar{r}'} \rho_{r_{i'}' r_{j'}'} V_{r_{i'}'} V_{r_{j'}'} \bar{r}_{i'}' \bar{r}_{j'}'$$

where r_i, r_j, $r_{i'}'$, $r_{j'}' = $ ith, jth random parameters of response and resistance; $\rho_{r_i r_j}$, $\rho_{r_{i'}' r_{j'}'} = $ the correlation coefficients of r_i and r_j, $r_{i'}'$ and $r_{j'}'$; V_{r_i}, V_{r_j}, $V_{r_{i'}'}$, $V_{r_{j'}'} = $ the coefficients of variation of r_i, r_j, $r_{i'}'$, $r_{j'}'$; $\bar{r}_i$, $\bar{r}_j$, $\bar{r}_{i'}'$, $\bar{r}_{j'}' = $ the means of r_i, r_j, $r_{i'}'$, $r_{j'}'$; $\bar{r}$, $\bar{r}' = $ all mean random parameter values of response and resistance.

A structural response S such as a displacement and an internal force can be determined by

$$S = C_S q' \qquad (11.20)$$

where $C_S = $ an influence coefficient that transforms the load intensity (q') into the desired response; $q' = $ the load intensity which may be dead, live or earthquake load. Two approaches for finding the variance of a response are suggested in the following Sections 11.3.1 and 11.3.2.

11.3.1 1st variance approach

If the random parameters involved in a response are known, the variance of a response is found to be

$$\sigma_S^2 = \sum_i \sum_j \left(\frac{\partial S}{\partial r_i}\right)_{\bar{r}} \left(\frac{\partial S}{\partial r_j}\right)_{\bar{r}} \rho_{r_i r_j} V_{r_i} V_{r_j} \bar{r}_i \bar{r}_j$$

$$= \sum_i \sum_j \left[\left(\frac{\partial C_S}{\partial r_i}\right)_{\bar{r}} \left(\frac{\partial C_S}{\partial r_j}\right)_{\bar{r}} \bar{q}'^2 + \left(\frac{\partial q'}{\partial r_i}\right)_{\bar{r}} \left(\frac{\partial q'}{\partial r_j}\right)_{\bar{r}} \bar{C}_S^2\right] \rho_{r_i r_j} V_{r_i} V_{r_j} \bar{r}_i \bar{r}_j$$

$$(11.21)$$

Once the uncertainties of each random parameter are determined, the variance of a response can be obtained through Eq. (11.21).

11.3.2   2nd variance approach

In many practical situations, finding the means and coefficients of the variation of random parameters in a response is difficult. Therefore, the variance of the structural response can be calculated based on the square of some percentage of the mean value of the influence coefficient $\left(\bar{C}_S\right)$ by

$$\sigma_S^2 = \left(C_P \bar{C}_S\right)^2 \bar{q}'^2 + \left(\bar{C}_S\right)^2 V_{q'}^2 \bar{q}'^2 \tag{11.22}$$

where C_P is a percentage constant of $\bar{C}_S$ and $V_{q'}$ is a coefficient of variation of load intensity, q'. Comparing Eqs. (11.21) and (11.22), the two equations will be the same if the following terms are equal

$$\sum_i \sum_j \left[\left(\frac{\partial C_S}{\partial r_i}\right)_{\bar{r}} \left(\frac{\partial C_S}{\partial r_j}\right)_{\bar{r}} \bar{q}'^2 \right] \rho_{r_i r_j} V_{r_i} V_{r_j} \bar{r}_i \bar{r}_j = \left(C_P \bar{C}_S\right)^2 \bar{q}'^2 \tag{11.23}$$

and

$$\sum_i \sum_j \left[\left(\frac{\partial q'}{\partial r_i}\right)_{\bar{r}} \left(\frac{\partial q'}{\partial r_j}\right)_{\bar{r}} \bar{C}_S^2 \right] \rho_{r_i r_j} V_{r_i} V_{r_j} \bar{r}_i \bar{r}_j = \left(\bar{C}_S\right)^2 V_{q'}^2 \bar{q}'^2 \tag{11.24}$$

11.4   Reliability of structural system

In Section 11.1.2, the derivations of reliability and probability of failure are based on a single failure mode with a single structural response and resistance case. However, a structural system may have many failure modes. Hence, the system reliability or probability of failure may also be the desired quantity for considering the structural safety problem.

   There are two fundamental types of system, namely, a series system and a parallel system. A system is a series system if it is in a state of failure whenever any of its elements fail. Such a system is also called a weakest-link system. For a parallel system, failure in a single mode will not always result in the failure of the total system, because the structural capacity may be able to sustain the loads.

   In this chapter, the structural system is considered to be a series system, which means that the structure fails when any safety criterion is not satisfied. The probability of failure for this system, P_{fT}, may have the form

$$P_{fT}(R < S) = 1 - P_r(R > S) = 1 - P_r(R_1 > S_1 \cap R_2 > S_2 \cap \ldots) \tag{11.25}$$

where $R_1, R_2, S_1, S_2, \ldots$ are the resistances and responses of each failure mode for a structural system. The determination of the exact value of Eq. (11.25) is very difficult since the relationship among failure modes is too complicated. Nevertheless, two extreme bounds can be determined on the basis of the following conditions in Sections 11.4.1 and 11.4.2 [4].

11.4.1 *Perfectly correlated among failure modes*

From Eq. (11.25), the system probability of failure, P_{fT}, may be expressed as

$$P_{fT} = 1 - P_r\left(R_j > S_j\right) = 1 - \left(1 - P_f\left(R_j < S_j\right)\right) = P_f\left(R_j < S_j\right) \qquad (11.26)$$

Therefore, using the maximum value of all failure probability modes yields the system failure probability as

$$P_{fT} = \max\left(P_{f_j}\right) \qquad (11.27)$$

11.4.2 *Perfectly uncorrelated among failure modes*

$$P_{fT} = 1 - P_r\left(R_1 > S_1\right)P_r\left(R_2 > S_2\right)\dots P_r\left(R_j > S_j\right) = 1 - \Pi\left(1 - P_f\left(R_j \le S_j\right)\right)$$
$$= 1 - 1 + \sum_j P_{f_j} + \sum_i \sum_j P_{f_i} P_{f_j} \qquad (11.28)$$

or

$$P_{fT} = \sum_j P_{f_j} \qquad (11.29)$$

where P_{f_j} = the component failure probability of a structural system.

11.5 Approximation formulas for cumulative normal distribution

From previous sections, the reliability and probability of failure are calculated through the cumulative normal distribution. However, finding the exact value of this distribution is very complicated and difficult. Therefore, some approximation formulas have to be used to approximate this function. In this chapter an approximation equation called Formula (1) with two different ranges of safety factor is used [5] as

Formula (1)

$\beta \le 1.6$

$$P_r = \left(\beta - \frac{\beta^3}{6} + \frac{\beta^5}{40}\right)\frac{1}{\sqrt{2\pi}} + 0.5 \qquad (11.30)$$

$$P_f = 1 - P_r \qquad (11.31)$$

$15 > \beta > 1.6$

$$P_r = 0.5\left[1 - \left(1 - \frac{1}{\beta^2} + \frac{3}{\beta^4}\right)e^{-\frac{\beta^2}{2}}\sqrt{\frac{2}{\pi}} \Big/ \beta\right] + 0.5 \qquad (11.32)$$

$$P_f = 1 - P_r \qquad (11.33)$$

Compare this with other published equations identified as Formula (2) and Formula (3) [3] shown below

Formula (2)

$$\beta \geq 1$$
$$P_f = \frac{1+\beta^2}{2+\beta^2} \frac{P_n(\beta)}{\beta} \tag{11.34}$$

where

$$P_n(\beta) = \text{standard normal density function}$$
$$= \frac{1}{\sqrt{2\pi}} e^{-\beta^2/2} \tag{11.35}$$

Formula (3)

$$P_f = \frac{\left(\dfrac{AB}{1+2\underline{y}} + \dfrac{1}{1 + \underline{B}\beta + \underline{C}\underline{y} + \underline{D}\underline{y}\beta + \underline{x} + \underline{E}x\beta}\right)}{2e^{\underline{y}}} \tag{11.36}$$

where $\underline{A} = (2/\pi)^{1/2}$; $\underline{B} = 1.604$; $\underline{C} = 3.91$; $\underline{D} = 4.45$; $\underline{E} = 0.73$; $\underline{y} = \beta^2/2$; $\underline{x} = 2.93\underline{y}^2$ and the comparison is shown in Fig. 11.1 from which one may

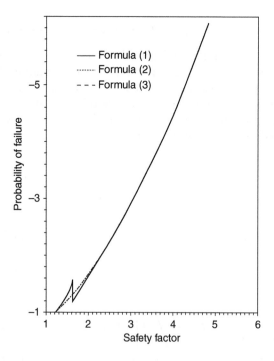

Figure 11.1 Various approximate equations for the standard cumulative normal distribution.

observe that the results are very close except in the range of safety factor between 1.4 and 1.8. Therefore, Formula (1) is used in this work except where the safety factor is between 1.4 and 1.8.

11.6 Uncertainty of loads

Structures may be subjected to dead loads due to the weights of the structural and permanent fixtures; live loads due to the maximum total loads of occupancy and movable furniture; and lateral loads due to earthquakes. In what follows, the models for dead, live and static equivalent lateral (UBC) seismic loads are discussed.

11.6.1 Dead load

The dead load on a structure consists of the weight of the structure and permanent installations. The weight of a structure can be obtained from its geometry and the unit weight of the elements and their dimensions.

The mean value of a dead load is assumed to be the constituent of all its mean component values. The coefficient of variation of a dead load may be caused by the coefficient of variation of the weight of steel members which is 0.05 and the coefficient of variation due to the weight of non-structural elements, which is estimated as 0.1 [6]. By using the above values, the coefficient of variation for the dead load can be obtained as

$$V_D = \sqrt{(0.05)^2 + (0.1)^2} = 0.12 \tag{11.37}$$

11.6.2 Live load

Live loads are loads arising from movable equipment and fixtures, vehicles and stored goods, and other non-permanent loads. Some live loads may be practically permanent, and others may be highly transient. They vary with both time and space, and can be idealized as being composed of two parts, sustained live loads and transient live loads.

The sustained live loads exist on the floor for a relatively long time and include furniture and normal working personnel. The changes in the sustained live loads may occur due to tenancy changes in the use of the floor area. The transient live loads occur infrequently but with a relatively high intensity and short duration. The transient live loads may be due to people gathering in a room in large numbers for a special occasion, due to stocking some goods in a room for short time, or due to the concentration of furniture during remodelling. The probabilistic model for live loads may be assumed to be

$$L(x, y) = L_y + \varepsilon(x, y) \tag{11.38}$$

where $L(x,y) =$ the live load intensity, in pounds per square foot at (x,y); $L_y =$ a random variable modelling the average unit load on the floor; $\varepsilon(x,y) =$ a stochastic process with zero mean describing the deviations from the average.

The equivalent uniform distributed load (EUDL), which is the uniform distributed load and which produces the same load effects as the actual set of loads, is our desired model and has the following meaning:

$$\text{EUDL} = \frac{\int\int L(x,y)I(x,y)\mathrm{d}x\mathrm{d}y}{\int\int I(x,y)\,\mathrm{d}x\mathrm{d}y} \qquad (11.39)$$

where $I(x,y) =$ an influence surface coefficient. The mean and variance of *EUDL* can be determined as follows:

mean

$$E[\text{EUDL}] = E\left[\frac{\int\int (L_y + \varepsilon(x,y))\,I(x,y)\,\mathrm{d}A_I}{\int\int I(x,y)\,\mathrm{d}A_I}\right]$$

$$= \frac{\bar{L}_y A_I \int_0^1 \int_0^1 I(x,y)\,\mathrm{d}x\mathrm{d}y}{A_I \int_0^1 \int_0^1 I(x,y)\,\mathrm{d}x\mathrm{d}y} = \bar{L}_y \qquad (11.40)$$

variance

$$\sigma^2_{\text{EUDL}} = \sigma_r^2 + \frac{\sigma_t^2}{A_I}K_L \qquad (11.41)$$

where $K_L = \int_0^1 \int_0^1 I^2(x,y)\,\mathrm{d}x\mathrm{d}y \Big/ \left(\int_0^1 \int_0^1 I(x,y)\,\mathrm{d}x\mathrm{d}y\right)^2$; $\sigma_r^2, \sigma_t^2 =$ empirical constants to fit the data from a live load survey; $A_I =$ the influence area which covers the influence surface. K_L, σ_r^2, σ_t^2 can be determined from a live load survey. Statistical information on live loads is obtained from live load surveys. These surveys gave the instantaneous values of the live loads on a particular building at the time when the survey was conducted.

The lifetime maximum total live load, which is the maximum value of the live load (sustained plus transient) over the whole lifetime of the structure and which is described in Appendix H, is considered in this study. For the purpose of simplicity, four lifetime maximum total live load models of the office building structure which are independent of time are described and used. The statistics of these four models are:

(A) *ANSI 1980 recommendations (L1)* [7]

The load subcommittee of the American National Standard Committee A58 recommended a 1980 version of a live load model. The statistics of this model were given as

mean

$$\overline{L1} = 50\left(0.25 + 15/\sqrt{A_I}\right) \text{ psf} \tag{11.42}$$

coefficient of variation

$$V_{L1} = 0.14 \tag{11.43}$$

(B) *NBS survey (L2)* [8]

The National Bureau of Standards (NBS) published a load survey as results for office buildings in the United States. The statistics of this live load model were given as

mean

$$\overline{L2} = 18.7 + 520/\sqrt{A_I} \text{ psf} \tag{11.44}$$

coefficient of variation

$$V_{L2} = \left(14.2/18,900\right)/\overline{L2} \tag{11.45}$$

(C) *United Kingdom survey (L3)* [9]

Mitchell and Woodgate proposed a live load model according to survey results for office buildings in the United Kingdom. The analysis resulted in

mean

$$\overline{L3} = 14.9 + 763/\sqrt{A_I} \text{ psf} \tag{11.46}$$

coefficient of variation

$$V_{L3} = \left(\left(11.3 + 15,000\right)/A_I\right)/\overline{L3} \tag{11.47}$$

(D) *UNREDUCED live loads (L4)* [7]

This live load model is a fixed value which does not relate to any space parameters. The statistics of this model are

mean

$$\overline{L4} = 50 \text{ psf} \tag{11.48}$$

coefficient of variation

$$V_{L4} = 0.14 \tag{11.49}$$

In the above models, the influence area parameter A_I is the area that contributes the load determination, and the unit 1 psf (pounds per square foot) equals 992.16 kPa.

11.6.3   UBC seismic load

The seismic load may be represented as the static equivalent lateral load that is constructed in code specifications such as UBC [10]. In the code, the earthquake is first expressed as the ground base shear force and then distributed to the desired location as

$$E = ZI_E K_E C_E S_E W \tag{11.50}$$

where E = the shear force at the base; Z = numerical coefficient depending on the zone where the structure is located; I_E = occupancy importance factor; K_E = numerical coefficient; $C_E = (1/15\sqrt{T})$, the value needed not to exceed 0.12; T = structural fundamental period; S_E = numerical coefficient for site-structure resonance, when the characteristic site period is not properly established, the value of S shall be 1.5; and W = the total dead load. The structural fundamental period, T, can be expressed by the following formula:

$$T = 2\pi \sqrt{\left(\sum_{i=1}^{n} W_i u_i^2\right)} \Big/ \sqrt{\left(g \sum_i F_{Ei} u_i\right)} \tag{11.51}$$

Because the calculation of this period involves structural analysis which is not known in the beginning, the approximate period $T = 0.05 h_n / \sqrt{D_n}$ is needed first; where D_n = the dimension of the structure parallel to the applied force direction; h_n = the height of the highest level; W_i = the weight of the ith level; u_i = the deflection of the ith level; F_{Ei} = the lateral force applied at level i.

After the force E is determined, E shall be distributed over the height of the structure by the following formula:

$$F_{Ex} = \left[(1 - F_{tu}) W_x \frac{h_x}{\sum_i W_i h_i} \right] E = F_{Eux} E \tag{11.52}$$

where F_{Ex} = the lateral force applied to level x; F_{Eux} = the lateral force applied to level x due to a unit seismic load intensity; F_{tu} = the concentrated force at the top due to a unit seismic load intensity = $0.07T$; W_i, W_x = the portion of W which is located at or is assigned to level i or x, respectively; h_i, h_x = the height of the i or xth level. The statistic of the UBC seismic load may be determined as

mean

$$\bar{E} = \bar{Z} \bar{I}_E \bar{K}_E \bar{C}_E \bar{S}_E \bar{W} \qquad (11.53)$$

The coefficient of variation for an earthquake may be assumed to follow a type II extreme value distribution which is described in Appendix H and has a probability distribution of

$$E'(e_1) = e^{\left(\frac{-e_1}{u1}\right)^{-K1}} \qquad (11.54)$$

where E' = earthquake random variable; e_1 = a value of E'; $u1, k1$ = parameters which are determined from a seismological data survey. For this type of distribution, the coefficient of variation of an earthquake, V_E, is

$$V_E = \sqrt{\frac{\Gamma\left(1 - \frac{2}{K1}\right)}{\Gamma^2\left(1 - \frac{1}{K1}\right)} - 1} \qquad (11.55)$$

where $\Gamma(\ )$ is a gamma function.

Since the phenomena of earthquakes are highly complicated and unpredictable, the exact value of $K1$ is impossible to determine. However, many values of $K1$ and coefficient of variation of an earthquake have been proposed in the past. Within them, the value of $K1$, 2.3, which corresponds to a coefficient of variation, 1.38, is reported in the National Bureau of Standards special publication No. 577 [7] and has been used for some seismic designs. The value of $K1$, 2.7, which corresponds to a coefficient of variation, 0.85, is recommended for nuclear power plant design [11].

11.7 Load effects on structural response

The desired load effects for a structure subjected to dead, live or seismic loads may be the structural displacements or members' internal forces. The total load effects may be due to the combination of these loads. In the following sections, the general response formulation for dead, live and UBC seismic loads are described.

11.7.1 Dead load effect

The dead load effect, S_D, can be obtained by

$$S_D = C_D D \qquad (11.56)$$

where C_D is an influence coefficient which transforms the dead load intensity D into the desired load effect (displacement or internal force) through structural analysis. The mean is

mean

$$\bar{S}_D = \bar{C}_D \bar{D} \tag{11.57}$$

and the variance may be expressed in two approaches as

(1) *1st variance approach*

$$\left(\sigma_{S_D}^2\right) = \sum_i \sum_j \left(\frac{\partial S_D}{\partial r_i}\right)_{\bar{r}} \left(\frac{\partial S_D}{\partial r_j}\right)_{\bar{r}} \rho_{r_i r_j} V_{r_i} V_{r_j} \bar{r}_i \bar{r}_j$$

$$= \sum_i \sum_j \left(\frac{\partial C_D}{\partial r_i} D\right)_{\bar{r}} \left(\frac{\partial C_D}{\partial r_j} D\right)_{\bar{r}} \rho_{r_i r_j} V_{r_i} V_{r_j} \bar{r}_i \bar{r}_j \tag{11.58}$$

$$+ \sum_i \sum_j \left(\frac{\partial D}{\partial r_i} C_D\right)_{\bar{r}} \left(\frac{\partial D}{\partial r_j} C_D\right)_{\bar{r}} \rho_{r_i r_j} V_{r_i} V_{r_j} \bar{r}_i \bar{r}_j$$

(2) *2nd variance approach*

$$\left(\sigma_{S_D}^2\right) = \left(\bar{S}_D\right)^2 \left[(0.1)^2 + V_D^2\right] \tag{11.59}$$

where 0.1 is the assumed value to describe the static analysis error [6]; V_D = the coefficient of variation of the dead load intensity which is assumed to be 0.12 in Eq. (11.37).

11.7.2 Live load effect

The live load effect is obtained as

$$S_L = C_L L \tag{11.60}$$

where C_L is an influence coefficient which transforms the live load intensity L into the desired load effect. The mean is

mean

$$\bar{S}_L = \bar{C}_L \bar{L} \tag{11.61}$$

and the variance has the following two approaches:

(1) *1st variance approach*

$$\left(\sigma_{S_L}^2\right) = \sum_i \sum_j \left(\frac{\partial S_L}{\partial r_i}\right)_{\bar{r}} \left(\frac{\partial S_L}{\partial r_j}\right)_{\bar{r}} \rho_{r_i r_j} V_{r_i} V_{r_j} \bar{r}_i \bar{r}_j$$

$$= \sum_i \sum_j \left(\frac{\partial C_L}{\partial r_i} L\right)_{\bar{r}} \left(\frac{\partial C_L}{\partial r_j} L\right)_{\bar{r}} \rho_{r_i r_j} V_{r_i} V_{r_j} \bar{r}_i \bar{r}_j \tag{11.62}$$

$$+ \sum_i \sum_j \left(\frac{\partial L}{\partial r_i} C_L\right)_{\bar{r}} \left(\frac{\partial L}{\partial r_j} C_L\right)_{\bar{r}} \rho_{r_i r_j} V_{r_i} V_{r_j} \bar{r}_i \bar{r}_j$$

(2) *2nd variance approach*

$$\left(\sigma_{S_L}^2\right) = \left(\bar{S}_L\right)^2 \left[(0.1)^2 + V_L^2\right] \tag{11.63}$$

where 0.1 is the assumed value to describe the static analysis error; $V_L =$ the coefficient of variation of the live load intensity which is assumed to be the value in Eqs. (11.43), (11.45), (11.47) or (11.49).

11.7.3 UBC seismic load effect

The UBC seismic load effect is obtained as

$$S_E = C_E E \tag{11.64}$$

where C_E is an influence coefficient which transforms the earthquake load intensity E into the desired load effect. The mean is

mean

$$\bar{S}_E = \overline{C_E}\overline{E} \tag{11.65}$$

and the variance has the following two approaches:

(1) *1st variance approach*

$$\left(\sigma_{S_E}^2\right) = \sum_i \sum_j \left(\frac{\partial S_E}{\partial r_i}\right)_{\bar{r}} \left(\frac{\partial S_E}{\partial r_j}\right)_{\bar{r}} \rho_{r_i r_j} V_{r_i} V_{r_j} \bar{r}_i \bar{r}_j$$

$$= \sum_i \sum_j \left(\frac{\partial C_E}{\partial r_i} E\right)_{\bar{r}} \left(\frac{\partial C_E}{\partial r_j} E\right)_{\bar{r}} \rho_{r_i r_j} V_{r_i} V_{r_j} \bar{r}_i \bar{r}_j \tag{11.66}$$

$$+ \sum_i \sum_j \left(\frac{\partial E}{\partial r_i} C_E\right)_{\bar{r}} \left(\frac{\partial E}{\partial r_j} C_E\right)_{\bar{r}} \rho_{r_i r_j} V_{r_i} V_{r_j} \bar{r}_i \bar{r}_j$$

(2) *2nd variance approach*

$$\left(\sigma_{S_E}^2\right) = \left(\bar{S}_E\right)^2 \left[(0.1)^2 + V_E^2\right] \tag{11.67}$$

where 0.1 is the assumed value to describe the static analysis error; $V_E =$ the coefficient of variation of the earthquake load intensity which is assumed to be the value in Eq. (11.55).

11.7.4 Combined load effect

The total load effect due to the combined action of dead, live and UBC loads is

$$S = S_D + S_L + S_E \tag{11.68}$$

which has the mean

$$\bar{S}_E = \bar{S}_D + \bar{S}_L + \bar{S}_E \tag{11.69}$$

and the variance of the total load effect has two approaches as

(1) *1st variance approach*

$$
\begin{aligned}
\left(\sigma_S^2\right) &= \sum_i \sum_j \left(\frac{\partial S}{\partial r_i}\right)_{\bar{r}} \left(\frac{\partial S}{\partial r_j}\right)_{\bar{r}} \rho_{r_i r_j} V_{r_i} V_{r_j} \bar{r}_i \bar{r}_j \\
&= \sum_{i1} \sum_{j1} \left(\frac{\partial C_D}{\partial r_{i1}} D + \frac{\partial C_L}{\partial r_{i1}} L + \frac{\partial C_E}{\partial r_{i1}} E\right)_{\bar{r}} \\
&\quad \left(\frac{\partial C_D}{\partial r_{j1}} D + \frac{\partial C_L}{\partial r_{j1}} L + \frac{\partial C_E}{\partial r_{j1}} E\right)_{\bar{r}} \rho_{r_{i1} r_{j1}} V_{r_{i1}} V_{r_{j1}} \bar{r}_{i1} \bar{r}_{j1} \\
&\quad + \left(\bar{C}_D \bar{D}\right)^2 V_D^2 + \left(\bar{C}_L \bar{L}\right)^2 V_L^2 + \left(\bar{C}_E \bar{E}\right)^2 V_E^2
\end{aligned} \tag{11.70}
$$

where *i*1 and *j*1 are random parameters related to structural analysis.

(2) *2nd variance approach*

$$
\begin{aligned}
\left(\sigma_S^2\right) &= \sigma_{S_D}^2 + \sigma_{S_L}^2 + \sigma_{S_E}^2 + 2\,(0.1)^2 \left(\bar{S}_D \bar{S}_L + \bar{S}_D \bar{S}_E + \bar{S}_L \bar{S}_E\right) \\
&= \bar{C}_D^2 \bar{D}^2 \left((0.1)^2 + V_D^2\right) + \bar{C}_L^2 \bar{L}^2 \left((0.1)^2 + V_L^2\right) + \bar{C}_E^2 \bar{E}^2 \left((0.1)^2 + V_E^2\right) \\
&\quad + 2\,(0.1)^2 \left(\bar{S}_D \bar{S}_L + \bar{S}_D \bar{S}_E + \bar{S}_L \bar{S}_E\right) \\
&= (0.1)^2 \left[\bar{C}_D^2 \bar{D}^2 + \bar{C}_L^2 \bar{L}^2 + \bar{C}_E^2 \bar{E}^2\right. \\
&\quad \left. + 2\bar{C}_D \bar{D} \bar{C}_L \bar{L} + 2\bar{C}_D \bar{D} \bar{C}_E \bar{E} + 2\bar{C}_L \bar{L} \bar{C}_E \bar{E}\right] \\
&\quad + \left(\bar{C}_D \bar{D}\right)^2 V_D^2 + \left(\bar{C}_L \bar{L}\right)^2 V_L^2 + \left(\bar{C}_E \bar{E}\right)^2 V_E^2
\end{aligned} \tag{11.71}
$$

11.8 Computation of structural displacements with load effects

The computations of the structural response of displacements can be performed as follows:

$$\{u\} = \{C_S\}\, q' = [K]^{-1} \{q\}\, q' \tag{11.72}$$

where $\{u\}$ = displacement matrix; $[K]^{-1}$ = inverse of global stiffness matrix; q = the applied load matrix due to a unit dead, live or UBC seismic load intensity; and q' = the load intensity which may be the dead, live or UBC seismic load intensity. By using the above equation, the uncertainties of load effects for dead load, live load, UBC seismic load, and combination of these loads are given.

11.8.1   With dead load effect

The *mean* of the dead load effect can be expressed as

$$\{\bar{u}_D\} = \left\{\bar{C}_D\right\}\bar{D} = \left(\left[\bar{K}\right]^{-1}\{q_D\}\right)\bar{D} \tag{11.73}$$

and the variance may be expressed as

(1) *1st variance approach*

$$\left(\sigma^2_{\{u_D\}}\right) = \sum_{i1}\sum_{j1}\left(\frac{\partial\{u_D\}}{\partial r_{i1}}\right)_{\bar{r}}\left(\frac{\partial\{u_D\}}{\partial r_{j1}}\right)_{\bar{r}}\rho_{r_{i1}r_{j1}}\,V_{r_{i1}}\,V_{r_{j1}}\bar{r}_{i1}\bar{r}_{j1} + \{\bar{u}_D^2\}\,V_D^2 \tag{11.74}$$

where $i1$ and $j1$ are random parameters related to structural analysis; and Eq. (11.74) is derived from Eq. (11.58) as

$$\left(\sigma^2_{\{u_D\}}\right) = \sum_{i1}\sum_{j1}\left(\frac{\partial\{u_D\}}{\partial r_{i1}}\right)_{\bar{r}}\left(\frac{\partial\{u_D\}}{\partial r_{j1}}\right)_{\bar{r}}\rho_{r_{i1}r_{j1}}\,V_{r_{i1}}\,V_{r_{j1}}\bar{r}_{i1}\bar{r}_{j1}$$

$$+ \left\{\frac{\partial D}{\partial D}C_D\right\}_{\bar{r}2}\bar{D}^2V_D^2$$

$$= \sum_{i1}\sum_{j1}\left(\frac{\partial\{u_D\}}{\partial r_{i1}}\right)_{\bar{r}}\left(\frac{\partial\{u_D\}}{\partial r_{j1}}\right)_{\bar{r}}\rho_{r_{i1}r_{j1}}\bar{r}_{i1}\bar{r}_{j1} + \left\{\bar{C}_D^2\right\}\bar{D}^2V_D^2 \tag{11.75}$$

The derivative of the displacement with respect to the $i1$th or $j1$th random variable of r can be derived as follows: since the derivative of this equation with respect to the random parameter r_{i1} is

$$\frac{\partial(\{q_D\}D)}{\partial r_{i1}} = \frac{\partial[K]}{\partial r_{i1}}\{u_D\} + [K]\frac{\partial\{u_D\}}{\partial r_{i1}} \tag{11.76}$$

Rearranging Eq. (11.76), it can be found that

$$\frac{\partial\{u_D\}}{\partial r_{i1}} = [K]^{-1}\left(\frac{\partial(\{q_D\}D)}{\partial r_{i1}} - \frac{\partial[K]}{\partial r_{i1}}\{u_D\}\right) \tag{11.77}$$

Similarly, we have

$$\frac{\partial \{u_D\}}{\partial r_{j1}} = [K]^{-1} \left(\frac{\partial (\{q_D\} D)}{\partial r_{j1}} - \frac{\partial [K]}{\partial r_{j1}} \{u_D\} \right) \tag{11.78}$$

(2) *2nd variance approach*

$$\sigma^2_{\{u_D\}} = \{\bar{u}^2_D\} \left((0.1)^2 + V^2_D \right) \tag{11.79}$$

which is based on Eq. (11.59).

11.8.2 With live load effect

The *mean* of the live load effect can be expressed as

$$\{\bar{u}_L\} = \{\bar{C}_L\} \bar{L} = \left(\left[\bar{K} \right]^{-1} \{q_L\} \right) \bar{L} \tag{11.80}$$

and the variance may be expressed as

(1) *1st variance approach*

$$\left(\sigma^2_{\{u_L\}} \right) = \sum_{i1} \sum_{j1} \left(\frac{\partial \{u_L\}}{\partial r_{i1}} \right)_{\bar{r}} \left(\frac{\partial \{u_L\}}{\partial r_{j1}} \right)_{\bar{r}} \rho_{r_{i1} r_{j1}} V_{r_{i1}} V_{r_{j1}} \bar{r}_{i1} \bar{r}_{j1} + \{\bar{u}^2_L\} V^2_L \tag{11.81}$$

The derivation of Eq. (11.81) is similar to Eq. (11.74); and the sensitivity analysis of the displacement may be expressed as

$$\frac{\partial \{u_L\}}{\partial r_{i1}} = [K]^{-1} \left(\frac{\partial (\{q_L\} L)}{\partial r_{i1}} - \frac{\partial [K]}{\partial r_{i1}} \{u_L\} \right) \tag{11.82}$$

$$\frac{\partial \{u_L\}}{\partial r_{j1}} = [K]^{-1} \left(\frac{\partial (\{q_L\} L)}{\partial r_{j1}} - \frac{\partial [K]}{\partial r_{j1}} \{u_L\} \right) \tag{11.83}$$

(2) *2nd variance approach*

$$\sigma^2_{\{u_L\}} = \{\bar{u}^2_L\} \left((0.1)^2 + V^2_L \right) \tag{11.84}$$

11.8.3 With UBC seismic load effect

The *mean* of the UBC load effect can be expressed as

$$\{\bar{u}_E\} = \{\bar{C}_E\} \bar{E} = \left(\left[\bar{K} \right]^{-1} \{q_E\} \right) \bar{E} \tag{11.85}$$

Here, q_E has the same meaning as F_{Eux} in Eq. (11.52). Two approaches of variance are

(1) *1st variance approach*

$$\left(\sigma^2_{\{u_E\}}\right) = \sum_{i1}\sum_{j1}\left(\frac{\partial\{u_E\}}{\partial r_{i1}}\right)_{\bar{r}}\left(\frac{\partial\{u_E\}}{\partial r_{j1}}\right)_{\bar{r}}\rho_{r_{i1}r_{j1}}V_{r_{i1}}V_{r_{j1}}\bar{r}_{i1}\bar{r}_{j1} + \{\bar{u}_E^2\}V_E^2$$

$$(11.86)$$

The derivation of Eq. (11.86) is similar to Eq. (11.75); and the sensitivity analysis of the displacement may be expressed as

$$\frac{\partial\{u_E\}}{\partial r_{i1}} = [K]^{-1}\left(\frac{\partial(\{q_E\}E)}{\partial r_{i1}} - \frac{\partial[K]}{\partial r_{i1}}\{u_E\}\right) \qquad (11.87)$$

$$\frac{\partial\{u_E\}}{\partial r_{j1}} = [K]^{-1}\left(\frac{\partial(\{q_E\}E)}{\partial r_{j1}} - \frac{\partial[K]}{\partial r_{j1}}\{u_E\}\right) \qquad (11.88)$$

(2) *2nd variance approach*

$$\sigma^2_{\{u_E\}} = \{\bar{u}_E^2\}\left((0.1)^2 + V_E^2\right)$$

$$= \left([\bar{K}]^{-1}\{q_E\}\bar{E}\right)^2\left((0.1)^2 + V_E^2\right) \qquad (11.89)$$

11.8.4   With combined load effect

mean

$$\{\bar{u}\} = \{\bar{u}_D\} + \{\bar{u}_L\} + \{\bar{u}_E\}$$

$$= [\bar{K}]^{-1}\{q_D\}\bar{D} + [\bar{K}]^{-1}\{q_L\}\bar{L} + [\bar{K}]^{-1}\{q_E\}\bar{E} \qquad (11.90)$$

(1) *1st variance approach*

$$\left(\sigma^2_{\{u\}}\right) = \sum_{i1}\sum_{j1}\left(\frac{\partial\{u_D\}}{\partial r_{i1}} + \frac{\partial\{u_L\}}{\partial r_{i1}} + \frac{\partial\{u_E\}}{\partial r_{i1}}\right)_{\bar{r}}$$

$$\left(\frac{\partial\{u_D\}}{\partial r_{j1}} + \frac{\partial\{u_L\}}{\partial r_{j1}} + \frac{\partial\{u_E\}}{\partial r_{j1}}\right)_{\bar{r}}\rho_{r_{i1}r_{j1}}V_{r_{i1}}V_{r_{j1}}\bar{r}_{i1}\bar{r}_{j1} \qquad (11.91)$$

$$+ \{\bar{u}_D^2\}V_D^2 + \{\bar{u}_L^2\}V_L^2 + \{\bar{u}_E^2\}V_E^2$$

(2) *2nd variance approach*

$$\sigma_{(u)}^2 = \{\bar{u}_D^2\}\left((0.1)^2 + V_D^2\right) + \{\bar{u}_L^2\}\left((0.1)^2 + V_L^2\right) + \{\bar{u}_E^2\}\left((0.1)^2 + V_E^2\right)$$

$$+ 2\,(0.1)^2\left(\{\bar{u}_D\bar{u}_L + \bar{u}_D\bar{u}_E + \bar{u}_L\bar{u}_E\}\right) \tag{11.92}$$

The derivation of Eqs. (11.91) and (11.92) are similar to Eqs. (11.70) and (11.71).

For displacement failure mode, substituting Eqs. (11.90), (11.91) and (11.92) into Eqs. (11.6) or (11.15), then the probability of failure in Eqs. (11.4) or (11.8) can be determined if the statistics of allowable displacements are given.

11.9   Computation of member forces with load effects

The internal forces in a structural member which may be axial forces or bending moments are given by

$$\{F\}_m = [S]_m\,[A]_m^T\,\{u\}_m \tag{11.93}$$

where $\{F\}_m$ = the internal forces in a member; $[S]_m$ = a member stiffness matrix; $[A]_m^T = [B]_m$, $[A]_m$ is the member's equilibrium matrix and $[B]_m$ is a kinematic matrix as introduced in previous chapters including Chapters 5 and 7; $\{u\}_m$ = the corresponding external displacements in a member. The uncertainties of load effects due to dead, live and earthquake loads are:

11.9.1   For dead load

The *mean* and *variance* are

$$\left\{\bar{F}_D\right\}_m = \left[\bar{S}\right]_m [A]_m^T\{\bar{u}_D\}_m \tag{11.94}$$

(1) *1st variance approach*

$$\sigma_{\{F_D\}_m}^2 = \sum_{i1}\sum_{j1}\left(\frac{\partial\{F_D\}_m}{\partial r_{i1}}\right)_{\bar{r}}\left(\frac{\partial\{F_D\}_m}{\partial r_{j1}}\right)_{\bar{r}}\rho_{r_{i1}r_{j1}}V_{r_{i1}}V_{r_{j1}}\bar{r}_{i1}\bar{r}_{j1} + \left(\{\bar{F}_D^2\}\,V_D^2\right) \tag{11.95}$$

where $i1$ and $j1$ are random parameters related to structural analysis; and for which the derivation may be obtained as

$$\sigma_{\{F_D\}_m}^2 = \sum_{i1}\sum_{j1}\left(\frac{\partial\{F_D\}_m}{\partial r_{i1}}\right)_{\bar{r}}\left(\frac{\partial\{F_D\}_m}{\partial r_{j1}}\right)_{\bar{r}}\rho_{r_{i1}r_{j1}}V_{r_{i1}}V_{r_{j1}}\bar{r}_{i1}\bar{r}_{j1} + \left(\frac{\partial\{F_D\}_m}{\partial D}\right)_{\bar{r}2}V_D^2 \tag{11.96}$$

In Eq. (11.95) we have

$$\left(\frac{\partial \{F_D\}_m}{\partial r_{i1}}\right) = \frac{\partial [S]_m}{\partial r_{i1}} [A]_m^T \{u_D\}_m + [S]_m [A]_m^T \frac{\partial \{u_D\}_m}{\partial r_{i1}} \tag{11.97}$$

$$\left(\frac{\partial \{F_D\}_m}{\partial r_{j1}}\right) = \frac{\partial [S]_m}{\partial r_{j1}} [A]_m^T \{u_D\}_m + [S]_m [A]_m^T \frac{\partial \{u_D\}_m}{\partial r_{j1}} \tag{11.98}$$

(2) *2nd variance approach*

$$\sigma_{\{F_D\}_m}^2 = \left\{\bar{F}_D^2\right\}_m ((0.1)^2 + V_D^2) \tag{11.99}$$

11.9.2 For live load

The *mean* and *variance* are

$$\left\{\bar{F}_L\right\} = \left[\bar{S}\right]_m [A]_m^T \{\bar{u}_L\}_m \tag{11.100}$$

(1) *1st variance approach*

$$\sigma_{\{F_L\}_m}^2 = \sum_{i1} \sum_{j1} \left(\frac{\partial \{F_L\}_m}{\partial r_{i1}}\right)_{\bar{r}} \left(\frac{\partial \{F_L\}_m}{\partial r_{j1}}\right)_{\bar{r}} \rho_{r_{i1} r_{j1}} V_{r_{i1}} V_{r_{j1}} \bar{r}_{i1} \bar{r}_{j1} + \left(\left\{\bar{F}_L^2\right\}_m V_L^2\right) \tag{11.101}$$

The derivation of Eq. (11.101) is similar to Eq. (11.95); and the sensitivity analysis of the internal force may be expressed as

$$\left(\frac{\partial \{F_L\}_m}{\partial r_{i1}}\right) = \frac{\partial [S]_m}{\partial r_{i1}} [A]_m^T \{u_L\}_m + [S]_m [A]_m^T \frac{\partial \{u_L\}_m}{\partial r_{i1}} \tag{11.102}$$

$$\left(\frac{\partial \{F_L\}_m}{\partial r_{j1}}\right) = \frac{\partial [S]_m}{\partial r_{j1}} [A]_m^T \{u_L\}_m + [S]_m [A]_m^T \frac{\partial \{u_L\}_m}{\partial r_{j1}} \tag{11.103}$$

(2) *2nd variance approach*

$$\sigma_{\{F_L\}_m}^2 = \left\{\bar{F}_L^2\right\}_m ((0.1)^2 + V_L^2) \tag{11.104}$$

11.9.3 For UBC load

The *mean* and *variance* are

$$\{\bar{F}_E\} = [\bar{S}]_m [A]_m^T \{\bar{u}_E\}_m \tag{11.105}$$

(1) *1st variance approach*

$$\sigma^2_{\{F_E\}_m} = \sum_{i1}\sum_{j1} \left(\frac{\partial \{F_E\}_m}{\partial r_{i1}}\right)_{\bar{r}} \left(\frac{\partial \{F_E\}_m}{\partial r_{j1}}\right)_{\bar{r}} \rho_{r_{i1}r_{j1}} V_{r_{i1}} V_{r_{j1}} \bar{r}_{i1}\bar{r}_{j1} + \left(\left\{\bar{F}_E^2\right\} V_E^2\right)$$

(11.106)

The derivation of Eq. (11.106) is similar to Eq. (11.95); and the sensitivity analysis of the internal force may be expressed as

$$\left(\frac{\partial \{F_E\}_m}{\partial r_{i1}}\right) = \frac{\partial [S]_m}{\partial r_{i1}} [A]_m^T \{u_E\}_m + [S]_m [A]_m^T \frac{\partial \{u_E\}_m}{\partial r_{i1}}$$

(11.107)

$$\left(\frac{\partial \{F_E\}_m}{\partial r_{j1}}\right) = \frac{\partial [S]_m}{\partial r_{j1}} [A]_m^T \{u_E\}_m + [S]_m [A]_m^T \frac{\partial \{u_E\}_m}{\partial r_{j1}}$$

(11.108)

(2) *2nd variance approach*

$$\sigma^2_{\{F_E\}_m} = \left\{\bar{F}_E^2\right\}_m ((0.1)^2 + V_E^2)$$

(11.109)

11.9.4   *For combined load*

The *mean* and *variance* are similar to the previous and are expressed as

$$\left\{\bar{F}\right\}_m = \bar{F}_m + \left\{\bar{F}_L\right\}_m + \left\{\bar{F}_E\right\}_m$$

(11.110)

(1) *1st variance approach*

$$\sigma^2_{\{F\}m} = \sum_{i1}\sum_{j1} \left(\frac{\partial \{F\}_m}{\partial r_{i1}}\right)_{\bar{r}} \left(\frac{\partial \{F\}_m}{\partial r_{j1}}\right)_{\bar{r}} \rho_{r_{i1}r_{j1}} V_{r_{i1}} V_{r_{j1}} \bar{r}_{i1}\bar{r}_{j1} + \left\{\bar{F}_D^2\right\}_m V_D^2$$
$$+ \left\{\bar{F}_L^2\right\}_m V_L^2 + \left\{\bar{F}_E^2\right\}_m V_E^2$$

(11.111)

The derivation of Eq. (11.111) is

$$\sigma^2_{\{F\}m} = \sum_{i1}\sum_{j1} \left(\frac{\partial \{F\}_m}{\partial r_{i1}}\right)_{\bar{r}} \left(\frac{\partial \{F\}_m}{\partial r_{j1}}\right)_{\bar{r}} \rho_{r_{i1}r_{j1}} V_{r_{i1}} V_{r_{j1}} \bar{r}_{i1}\bar{r}_{j1}$$
$$+ \left(\frac{\partial \{F\}_m}{\partial D}\right)_{\bar{r}2} V_D^2 \bar{D}^2 + \left(\frac{\partial \{F\}_m}{\partial L}\right)_{\bar{r}2} V_L^2 \bar{L}^2 + \left(\frac{\partial \{F\}_m}{\partial E}\right)_{\bar{r}2} V_E^2 \bar{E}^2$$

(11.112)

The sensitivity analysis of the internal force may be expressed as

$$\left(\frac{\partial\,\{F\}_m}{\partial r_{i1}}\right)=\left(\frac{\partial\,\{F_D\}_m}{\partial r_{i1}}\right)+\left(\frac{\partial\,\{F_L\}_m}{\partial r_{i1}}\right)+\left(\frac{\partial\,\{F_E\}_m}{\partial r_{i1}}\right) \qquad (11.113)$$

$$\left(\frac{\partial\,\{F\}_m}{\partial r_{j1}}\right)=\left(\frac{\partial\,\{F_D\}_m}{\partial r_{j1}}\right)+\left(\frac{\partial\,\{F_L\}_m}{\partial r_{j1}}\right)+\left(\frac{\partial\,\{F_E\}_m}{\partial r_{j1}}\right) \qquad (11.114)$$

(2) *2nd variance approach*

$$\sigma^2_{(F)m}=\sigma^2_{\{F_D\}_m}+\sigma^2_{\{F_L\}_m}+\sigma^2_{\{F_E\}_m}+2\,(0.1)^2\left(\left\{\bar{F}_D\bar{F}_L+\bar{F}_D\bar{F}_E+\bar{F}_L\bar{F}_E\right\}_m\right)$$

$$(11.115)$$

For column failure modes, substituting Eqs. (11.110), (11.111) and (11.115) which may be the uncertainties of the applied axial load (P) or moments (M) into Eqs. (11.203), (11.205), (11.209), (11.210), (11.214), (11.215), (11.218) and (11.219) yields the mean and variances of the interaction equations. Substituting these statistics, the safety factor in Eqs. (11.6) or (11.15), then the probabilities of failure in Eqs. (11.4) or (11.8) can be determined if the statistics of allowable values of the interaction equations are given.

11.10 Resistance of members

In Eqs. (11.6) and (11.15), it was shown that the uncertainties of structural resistance are required in calculating the safety factor. The steel member resistances considered are outlined in Sections 11.10.1 through 11.10.5.

11.10.1 *Yield moment of beams*

The yield moment, M_y, is $F_y S_c$; where F_y =yield strength of the material; S_c = elastic section modulus. The mean and variance of the yield moment are

mean

$$\bar{M}_y=\bar{F}_y\bar{S}_c \qquad (11.116)$$

variance

$$\sigma^2_{M_y}=\bar{M}_y^2\,\bar{V}^2_{M_y} \qquad (11.117)$$

where $\bar{V}^2_{M_y}$ is the coefficient of variation of the yield moment that equals 0.12 and is assumed to be a sum of the square of the prediction error, elastic section modulus, and yield strength of a steel member (the coefficients of variation of F_y, S_c and the predicted behaviour error are assumed to be 0.1, 0.04 and 0.05, respectively) [12]; i.e.

$$V_{M_y} = \sqrt{(0.1)^2 + (0.04)^2 + (0.05)^2} = 0.12 \tag{11.118}$$

11.10.2   *Euler buckling load of long columns*

For a long column, the capacity of the column is governed by the Euler buckling load and may be expressed as

$$P_E = \frac{\pi^2 E_m I}{(KL)^2} \tag{11.119}$$

where E_m = elastic modulus; I = moment of inertia; KL = effective length. The mean and variance of P_E are

mean

$$\bar{P}_E = \frac{\pi^2 \bar{E}_m \bar{I}}{\left(\overline{KL}\right)^2} \tag{11.120}$$

variance

$$\sigma^2_{P_E} = V^2_{P_E} \bar{P}^2_E \tag{11.121}$$

where V_E is the coefficient of variation of P_E, which equals 0.3 and involves the uncertainties of prediction, elastic modulus, moment of inertia and effective length factor [12].

11.10.3   *Axial buckling load capacity of short columns*

The axial load capacity, P_{cr}, is

$$P_{cr} = \left[1 - \frac{\left(\frac{KL}{r_g}\right)^2}{2C_c^2} \right] F_y A_c, \quad \text{for } \frac{KL}{r_g} < C_c \tag{11.122}$$

or

$$P_{cr} = P_E, \quad \text{for } \frac{KL}{r_g} > C_c \tag{11.123}$$

in which $C_c^2 = 2\pi^2 E_m / F_y$; P_E = Euler buckling load; r_g = the radius of gyration; A_c = cross-sectional area. The mean and variance of P_{cr} are

mean

$$\bar{P}_{cr} = \left[1 - \frac{\lambda^2}{4}\right] \bar{F}_y \bar{A}_c, \quad \text{for } \frac{KL}{r_g} < C_c \tag{11.124}$$

or

$$\bar{P}_{cr} = \bar{P}_E, \quad \text{for } \frac{KL}{r_g} > C_c \tag{11.125}$$

variance

$$\sigma^2_{P_{cr}} = V^2_{P_{cr}} \bar{P}^2_{cr} \tag{11.126}$$

where $V_{P_{cr}}$ may vary from 0.14 to 0.31 which are considered to be the uncertainties of the steel member area and elastic modulus. In Eq. (11.124),

$$\bar{\lambda} = \frac{\overline{KL}}{\bar{r}_g} \frac{1}{\pi} \sqrt{\frac{\bar{F}_y}{\bar{E}_m}} \tag{11.127}$$

11.10.4 *Axial yield load capacity of short columns*

The yield load is given by

$$P_y = F_y A_c \tag{11.128}$$

The mean and variance of P_y are

mean

$$\bar{P}_y = \bar{F}_y \bar{A}_c \tag{11.129}$$

variance

$$\sigma^2_{P_y} = V^2_{P_y} \bar{P}^2_y \tag{11.130}$$

where V_{P_y} is the coefficient of variation of P_y which is assumed to be 0.14 and includes the uncertainties of yield strength and steel member area.

11.10.5 *Critical moment due to lateral-torsional buckling*

The critical moment, i.e. the moment at which lateral-torsional buckling occurs, is

$$M_{cr} = \left\{\left[\frac{C_b \pi E_m I_y}{K_y L}\right]\left[\frac{G_s J}{E_m} + \frac{\pi^2 C_W}{(K_z L)^2}\right]\right\}^2 \tag{11.131}$$

where K_y and K_z are effective length factors which account for effects of end restraints to lateral deflection and twist, respectively; C_b is a coefficient which depends on the variation in moment along the span; the shear modulus, G_s, is assumed to be $0.385E_m$; I_y is the moment of inertia of the weak axis; J = the polar moment of inertia; and C_W is a warping torsional coefficient.

The mean and variance of M_{cr} are

mean

$$\bar{M}_{cr} = \left\{ \left[\frac{\bar{C}_b \pi \bar{E}_m \bar{I}_y}{\bar{K}_y \bar{L}} \right] \left[\frac{\bar{G}_s \bar{J}}{\bar{E}_m} + \frac{\pi^2 \bar{C}_W}{\left(\bar{K}_z \bar{L} \right)^2} \right] \right\}^2 \tag{11.132}$$

variance

$$\sigma^2_{M_{cr}} = V^2_{M_{cr}} \bar{M}^2_{cr} \tag{11.133}$$

where the coefficient of variation $V_{M_{cr}}$ may be from 0.15 to 0.20 and contains the uncertainties of the moment of inertia, elastic modulus and shear modulus.

11.11 Newmark's nondeterministic seismic response spectrum (NNSRS) analysis

The earthquake load is due to the ground acceleration at the support of a structure either in the horizontal and/or vertical directions. As presented in previous chapters, the structural responses due to the ground acceleration may be obtained from the time-history or response-spectrum methods. The time-history response method is based on integration numerical procedures with individual earthquake records. But the response-spectrum method calculates the structural response based on a collection of earthquake records. The response spectra employed in Chapters 5 and 7 are deterministic. Hereafter, the Newmark's Nondeterministic Seismic Response Spectrum (NNSRS) is introduced, which uses a statistical technique to estimate the horizontal or vertical ground spectra from actual earthquake accelerograms. The seismic response spectrum suggested by Newmark was based on 14 earthquakes which had 28 horizontal components and 14 vertical components [13]. By utilizing these earthquake acceleration records, the statistics of spectral displacements, velocities and accelerations for maximum ground acceleration were obtained and used to estimate the structural displacements and internal forces.

11.11.1 Motion equation for structural systems

The general equation of motion for a multi-degree system is similar to the derivation presented in previous chapters as

$$[m]\{\ddot{u}\} + [C]\{\dot{u}\} + [K]\{u\} = \{P(t)\} \tag{11.134}$$

where $[m]$ = mass matrix; $[C]$ = damping matrix; $[K]$ = stiffness matrix; $\{P\}$ = external force matrix; $\{u\}, \{\dot{u}\}, \{\ddot{u}\}$ = structural displacement, velocity and acceleration, respectively. In order to distinguish the deterministic and nondeterministic formulation, some notations are changed even for the similar equations used for both approaches. If the applied external forces are the ground accelerations, Eq. (11.134) becomes

$$[m]\{\ddot{u}_E\} + [C]\{\dot{u}_E\} + [K]\{u_E\} = -[m]\{a\} \tag{11.135}$$

where $\{a\}$ = the ground acceleration matrix which may include horizontal and/or vertical ground accelerations; $\{u_E\}, \{\dot{u}_E\}, \{\ddot{u}_E\}$ = structural displacements, velocities and accelerations due to earthquake excitations, respectively.

The structural displacements may be expressed as a sum of a linear combination of the undamped free vibration mode shapes and the spectral displacements as

$$\{u_E\} = \sum_n \{\Phi\}_n Y_n \tag{11.136}$$

where the nth mode shape matrix $\{\Phi\}_n$ is used to transform from a generalized coordinate Y_n to a geometric coordinate u_E at the nth mode, and the generalized coordinate Y_n, which is the modal magnitude, is called the normal coordinate. Since the mode shapes have orthogonal properties,

$$\{\Phi\}_m^T [m] \{\Phi\}_n = 0, \quad \{\Phi\}_m^T [C] \{\Phi\}_n = 0, \quad \{\Phi\}_m^T [K] \{\Phi\}_n = 0, \quad m \neq n \tag{11.137}$$

Eq. (11.135) can be transformed into a single-degree equation of motion

$$M_n \ddot{Y}_n + 2\zeta_n \omega_n M_n \dot{Y}_n + M_n \omega_n^2 Y_n = -\{\Phi\}_n^T [m] \{I\}_a \tag{11.138}$$

where $M_n = \{\Phi\}_n^T [m] \{\Phi\}_n$; $2\zeta_n \omega_n M_n = \{\Phi\}_n^T [C] \{\Phi\}_n$; ω_n = structural natural frequency for the nth mode = $\left(\{\Phi\}_n^T [K] \{\Phi\}_n / M_n\right)^{1/2}$; and ζ_n = critical damping ratio coefficient for the nth mode.

If the orthonormal mode in which M_n equals 1 is used, Eq. (11.138) becomes

$$\ddot{Y}_n + 2\zeta_n \omega_n \dot{Y}_n + \omega_n^2 Y_n = -\Gamma_n a \tag{11.139}$$

where Γ_n = participation factor for the nth mode = $-\{\Phi\}_n^T [m] \{I\}$.

When the second-order $(P - \Delta)$ effect is considered, the stiffness matrix in Eq. (11.135) becomes

$$[K] = [K]_S - [K]_G \tag{11.140}$$

where $[K]_S$ = elastic stiffness matrix, $[K]_G$ = geometric stiffness matrix, which are extensively explored in previous chapters.

11.11.2   *Structural displacement based on NNSRS*

The desired structural displacements and internal forces for the nondeterministic approach are presented in this section and the next Section
and 11.11.3, respectively. This section consider: (A) NNSRS, (B) the displacement response, and (C) the displacement for the combined load
effect.

(A) *NNSRS* – The normal coordinate at the nth mode, Y_n, can be solved by
the Duhamal integral

$$Y_n = \frac{\Gamma_n}{M_n \omega_{Dn}} \int_0^t ae^{-\zeta_n \omega_n (t-\tau)} \sin \omega_{Dn} (t-\tau) \, d\tau \tag{11.141}$$

where $\omega_{Dn} = \omega_n \sqrt{1-\zeta_n^2}$; $t =$ time; $\tau =$ a time difference. For practical usage, the commonly used spectral displacement, $y_{\omega n}$, which is the
maximum value of Eq. (11.141) has been chosen, that is,

$$y_{\omega n} = \left[\frac{\Gamma_n}{M_n \omega_{Dn}} \int_0^t ae^{-\zeta_n \omega_n (t-\tau)} \sin \omega_{Dn} (t-\tau) \, d\tau \right]_{max} \tag{11.142}$$

In order to find the statistics of displacement, the mean, variance and
coefficient of variation of the spectral displacement for each mode, $y_{\omega n}$,
are needed and can be derived for different frequency ranges listed in (i)
through (v) based on Fig. 11.2.

(i) *In the constant displacement range* $\omega_n / 2\pi < \overline{f}_{\omega 1}$

$$\bar{y}_{\omega n} = \bar{\alpha}_d \bar{d} = \bar{\alpha}_d \left(\frac{v^2}{v^2} \frac{a^2}{a^2} d \right) = \overline{\bar{\alpha}_d \left(\frac{ad}{v^2} \right)} \left(\frac{v}{a} \right)^{-2} \bar{a} \tag{11.143}$$

$$\frac{\partial y_{\omega n}}{\partial r_i} = \frac{\partial \left[\alpha_d \left(\frac{ad}{v^2} \right) \left(\frac{v}{a} \right)^2 a \right]}{\partial r_i} \tag{11.144}$$

$$\sigma_{y_{\omega n}}^2 = \left(\frac{\partial y_{\omega n}}{\partial \alpha_d} \right)_{\bar{r}}^2 V_{\alpha_d}^2 \bar{a}_d^2 + \left(\frac{\partial y_{\omega n}}{\partial \left(\frac{ad}{v^2} \right)} \right)_{\bar{r}}^2 V_{ad/v^2}^2 \overline{\left(\frac{ad}{v^2} \right)}^{-2} + \left(\frac{\partial y_{\omega n}}{\partial a} \right)_{\bar{r}}^2 V_a^2 \bar{a}^2$$

$$+ \left(\frac{\partial y_{\omega n}}{\partial \left(\frac{v}{a} \right)} \right)_{\bar{r}}^2 V_{v/a}^2 \overline{\left(\frac{v}{a} \right)}^{-2}$$

$$= \left[\left(\frac{ad}{v^2} \right) \left(\frac{v}{a} \right)^2 a \right]_{\bar{r}}^2 \bar{a}_d^2 V_{\alpha_d}^2 + \left[\alpha_d \left(\frac{v}{a} \right)^2 a \right]_{\bar{r}}^2 V_{ad/v^2}^2 \overline{\left(\frac{ad}{v^2} \right)}^{-2}$$

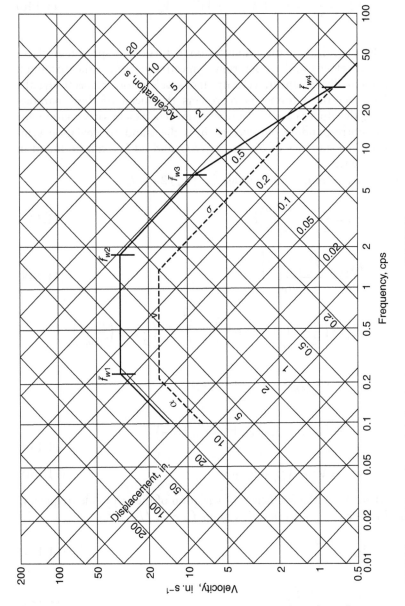

Figure 11.2 Newmark's nondeterministic seismic response spectrum (NNSRS).

$$+\left[2\alpha_d\left(\frac{ad}{v^2}\right)\left(\frac{v}{a}\right)a\right]_{\bar{r}}^2\left(\frac{v}{a}\right)^2 V_{v/a}^2+\left[\alpha_d\left(\frac{ad}{v^2}\right)\left(\frac{v}{a}\right)^2\right]_{\bar{r}}^2\bar{a}^2 V_a^2$$

$$=\bar{y}_{\omega n}^2 V_{\alpha_d}^2+\bar{y}_{\omega n}^2 V_{ad/v^2}^2+4\bar{y}_{\omega n}^2 V_{v/a}^2+\bar{y}_{\omega n}^2 V_a^2 \tag{11.145}$$

$$V_{y_{\omega n}}^2=\frac{\sigma^2}{\bar{y}_{\omega n}^2}=V_{\alpha_d}^2+V_{ad/v^2}^2+4V_{v/a}^2+V_a^2 \tag{11.146}$$

where α_d, α_v, α_a = the amplification factors of the spectral displacement, velocity and acceleration, respectively; d, v, a = the maximum ground displacement, velocity and acceleration, respectively; the bar over the parameters represents the mean of these parameters; the upper-case V represents the coefficient of variation for the corresponding parameters; σ is the standard deviation symbol. The statistics of α_d, α_v, α_a, v/a and ad/v^2 can be found in Reference 15.

(ii) *In the constant velocity range* $\bar{f}_{\omega1}<\bar{\omega}_n/2\pi<\bar{f}_{\omega2}$

$$\bar{y}_{\omega n}=\frac{\bar{\alpha}_v\bar{v}}{\bar{\omega}_n}=\frac{\bar{\alpha}_v}{\bar{\omega}_n}\overline{\left(\frac{v}{a}\right)}\bar{a} \tag{11.147}$$

$$\frac{\partial y_{\omega n}}{\partial r_i}=\frac{\partial\left[\frac{\alpha_v}{\omega_n}\left(\frac{v}{a}\right)a\right]}{\partial r_i} \tag{11.148}$$

$$\sigma_{y_{\omega n}}^2=\left(\frac{\partial y_{\omega n}}{\partial\alpha_v}\right)_{\bar{r}}^2 V_{\alpha_v}^2\bar{\alpha}_v^2+\left(\frac{\partial y_{\omega n}}{\partial\left(\frac{v}{a}\right)}\right)_{\bar{r}}^2 V_{v/a}^2\overline{\left(\frac{v}{a}\right)}^2+\left(\frac{\partial y_{\omega n}}{\partial\omega_n}\right)_{\bar{r}}^2$$

$$V_{\omega n}^2(\bar{\omega}_n)^2+\left(\frac{\partial y_{\omega n}}{\partial a}\right)_{\bar{r}}^2 V_a^2\bar{a}^2$$

$$=\bar{y}_{\omega n}^2 V_{\alpha_v}^2+\bar{y}_{\omega n}^2 V_{v/a}^2+\bar{y}_{\omega n}^2 V_{\omega n}^2+\bar{y}_{\omega n}^2 V_a^2 \tag{11.149}$$

$$V_{y_{\omega n}}^2=V_{\alpha_v}^2+V_{v/a}^2+V_a^2+V_{\omega n}^2 \tag{11.150}$$

where $V_{\omega n}$ = the coefficient of variation of the *n*th mode frequency.

(iii) *In the constant acceleration range* $\bar{f}_{\omega2}<\bar{\omega}_n/2\pi<\bar{f}_{\omega3}$

$$\bar{y}_{\omega n}=\frac{\bar{\alpha}_a\bar{a}}{\bar{\omega}_n^2} \tag{11.151}$$

$$\frac{\partial y_{\omega n}}{\partial r_i}=\frac{\partial\left[\frac{\alpha_a a}{\omega_n^2}\right]}{\partial r_i} \tag{11.152}$$

$$\sigma_{y_{\omega n}}^2 = \left(\frac{\partial y_{\omega n}}{\partial \alpha_a}\right)_{\bar{r}}^2 V_{\alpha_a}^2 \bar{\alpha}_a^2 + \left(\frac{\partial y_{\omega n}}{\partial \omega_n}\right)_{\bar{r}}^2 V_{\omega n}^2 (\bar{\omega}_n)^2 + \left(\frac{\partial y_{\omega n}}{\partial a}\right)_{\bar{r}}^2 V_a^2 \bar{a}^2$$

$$= \bar{y}_{\omega n}^2 V_{\alpha_a}^2 + 4\bar{y}_{\omega n}^2 V_{\omega n}^2 + \bar{y}_{\omega n}^2 V_a^2 \qquad (11.153)$$

$$V_{y_{\omega n}}^2 = V_{\alpha_a}^2 + V_a^2 + 4V_{\omega n}^2 \qquad (11.154)$$

(iv) *In the transition range* $\bar{f}_{\omega 3} < \omega_n/2\pi < \bar{f}_{\omega 4}$

$$\bar{y}_{\omega n} = \frac{\overline{\alpha_a(\omega)a}}{\bar{\omega}_n^2} = \exp\left[\frac{\ln \bar{\alpha}_a \ln\left(\bar{f}_{\omega 1}/\bar{f}_{\omega n}\right)}{\ln\left(\bar{f}_{\omega 4}/\bar{f}_{\omega 3}\right)}\right] \bar{a}/\bar{\omega}_n^2 \qquad (11.155)$$

$$\frac{\partial y_{\omega n}}{\partial r_i} = \partial \left(\exp\left[\frac{\ln \alpha_a \ln\left(f_{\omega 1}/f_{\omega n}\right)}{\ln\left(f_{\omega 4}/f_{\omega 3}\right)}\right] a/\omega_n^2\right) \Big/ \partial r_i \qquad (11.156)$$

$$\sigma_{y_{\omega n}}^2 = \left(\frac{\partial y_{\omega n}}{\partial \alpha_a}\right)_{\bar{r}}^2 V_{\alpha_a}^2 \bar{\alpha}_a^2 + \left(\frac{\partial y_{\omega n}}{\partial (\omega_n)}\right)_{\bar{r}}^2 V_{\omega n}^2 (\bar{\omega}_n)^2 + \left(\frac{\partial y_{\omega n}}{\partial a}\right)_{\bar{r}}^2 V_a^2 \bar{a}^2$$

$$= \bar{y}_{\omega n}^2 \left[\frac{\ln\left(\bar{f}_{\omega 1}/\bar{f}_{\omega n}\right)}{\ln\left(\bar{f}_{\omega 4}/\bar{f}_{\omega 3}\right)}\right]^2 V_{\alpha_a}^2 + \bar{y}_{\omega n}^2 \left[\frac{\ln \bar{\alpha}_a}{\ln\left(\bar{f}_{\omega 4}/\bar{f}_{\omega 3}\right)}\right]^2$$

$$V_{\omega n}^2 + 4\bar{y}_{\omega n}^2 V_{\omega n}^2 + \bar{y}_{\omega n}^2 V_a^2 \qquad (11.157)$$

$$V_{y_{\omega n}}^2 = \left[\frac{\ln\left(\bar{f}_{\omega 1}/\bar{f}_{\omega n}\right)}{\ln\left(\bar{f}_{\omega 4}/\bar{f}_{\omega 3}\right)}\right]^2 V_{\alpha_a}^2 + V_a^2 + \left[\frac{\ln \bar{\alpha}_a}{\ln\left(\bar{f}_{\omega 4}/\bar{f}_{\omega 3}\right)}\right]^2 V_{\omega n}^2 + 4V_{\omega n}^2$$

$$(11.158)$$

(v) *In the constant ground acceleration range* $\bar{f}_{\omega 4} < \bar{\omega}_n/2\pi$

$$\bar{y}_{\omega n} = \frac{\bar{a}}{\bar{\omega}_n^2} \qquad (11.159)$$

$$\frac{\partial y_{\omega n}}{\partial r_i} = \frac{\partial\left[\frac{a}{\omega_n^2}\right]}{\partial r_i} \qquad (11.160)$$

$$\sigma_{y_{\omega n}}^2 = \left(\frac{\partial y_{\omega n}}{\partial \omega_n}\right)_{\bar{r}}^2 V_{\omega n}^2 \bar{\omega}_n^2 + \left(\frac{\partial y_{\omega n}}{\partial (\omega_n)}\right)_{\bar{r}}^2 V_a^2 \bar{a}^2 = \bar{y}_{\omega n}^2 V_a^2 + \bar{y}_{\omega n}^2 V_{\omega n}^2$$

$$(11.161)$$

$$V_{y_{\omega n}}^2 = V_a^2 + 4V_{\omega n}^2 \qquad (11.162)$$

(B) *Displacement Response* – Since displacements due to earthquake excitations can be assumed to be the sum of the squares of displacements
for each mode, the statistics of displacements are determined as follows:

mean

$$\{\bar{u}_E\} = \left[\sum_n \left(\{\bar{\Phi}\}_n \bar{\Gamma}_n \bar{y}_{\omega n}\right)^2\right]^{1/2} \tag{11.163}$$

1st variance approach

$$\sigma^2_{\{u_E\}} = \sum_i \sum_j \left(\frac{\partial\{u_E\}}{\partial r_i}\right)_{\bar{r}} \left(\frac{\partial\{u_E\}}{\partial r_j}\right)_{\bar{r}} \rho_{r_i r_j} V_{r_i} V_{r_j} \bar{r}_i \bar{r}_j \tag{11.164}$$

in which

$$\frac{\partial\{u_E\}}{\partial r_i} = \frac{1}{u_E}\left[\sum_n \{\Phi\}_n \left(\frac{\partial\{\Phi\}_n}{\partial r_i}\right)\Gamma_n^2 y_{\omega n}^2 + \{\Phi\}_n^2 \Gamma_n \left(\frac{\partial\Gamma_n}{\partial r_i}\right)\right.$$

$$\left. y_{\omega n}^2 + \{\Phi\}_n^2 \Gamma_n^2 \left(\frac{\partial y_{\omega n}}{\partial r_i}\right) y_{\omega n}\right]$$

$$\frac{\partial\{u_E\}}{\partial r_j} = \frac{1}{u_E}\left[\sum_n \{\Phi\}_n \left(\frac{\partial\{\Phi\}_n}{\partial r_j}\right)\Gamma_n^2 y_{\omega n}^2 + \{\Phi\}_n^2 \Gamma_n\right.$$

$$\left. \left(\frac{\partial\Gamma_n}{\partial r_j}\right) y_{\omega n}^2 + \{\Phi\}_n^2 \Gamma_n^2 \left(\frac{\partial y_{\omega n}}{\partial r_j}\right) y_{\omega n}\right]$$

where $\partial\{\Phi\}_n/\partial r_i$, $\partial\{\Phi\}_n/\partial r_j$ = the derivatives of mode shape matrices
with respect to the *i*th and *j*th random parameters; $\partial\Gamma_n/\partial r_i$, $\partial\Gamma_n/r_j$ =
the derivatives of participation factors with respect to the *i*th and
*j*th random parameters; $\partial y_{\omega n}/\partial r_i$, $\partial y_{\omega n}/\partial r_j$ = the derivatives of spectral displacements with respect to the *i*th and *j*th random parameters.
$\partial\{\Phi\}_n/\partial r_i$ is assumed to be a linear combination of mode shapes,
i.e. $\partial\{\Phi\}_n/\partial r_i = \sum_k a_{nik}\{\Phi\}_k$, for which the determination of the scalar
multiplier, a_{nik}, can be derived by two conditions of $n \neq k$ and $n = k$ as

(1) *n ≠ k case*
Differentiating the undamped vibration equation with respect to
random parameters, $([K] - \omega_n^2[m])\{\Phi\}_n = \{0\}$, we have

$$\left(\frac{\partial[K]}{\partial r_i} - \frac{\partial\omega_n^2}{\partial r_i}[m] - \omega_n^2\frac{\partial[m]}{\partial r_i}\right)\{\Phi\}_n + ([K] - \omega_n^2[m])\frac{\partial\{\Phi\}_n}{\partial r_i} = \{0\} \tag{11.165}$$

Substituting the linear combination of the mode shapes for $\partial \{\Phi\}_n / \partial r_i$ into the above equation and pre multiplying by $\{\Phi\}_k^T$ for $k \neq n$ yield

$$\{\Phi\}_k^T \left(\frac{\partial [K]}{\partial r_i} - \frac{\partial \omega_n^2}{\partial r_i} [m] - \omega_n^2 \frac{\partial [m]}{\partial r_i} \right) \{\Phi\}_n + \{\Phi\}_k^T \left([K] - \omega_n^2 [m] \right)$$

$$\sum_k a_{nik} \{\Phi\}_k = 0 \qquad (11.166)$$

Because of the mode shape orthogonal properties, Eq. (11.166) becomes

$$\{\Phi\}_k^T \left(\frac{\partial [K]}{\partial r_i} - \frac{\partial \omega_n^2}{\partial r_i} [m] - \omega_n^2 \frac{\partial [m]}{\partial r_i} \right) \{\Phi\}_n + a_{nik} \{\Phi\}_k^T$$

$$\left([K] - \omega_n^2 [m] \right) \{\Phi\}_k = 0 \qquad (11.167)$$

If the mode shapes are orthonormal modes, i.e., $\{\Phi\}_k^T [m] \{\Phi\}_k = 1$, the term $\{\Phi\}_k^T [K] \{\Phi\}_k$ yields ω_k^2. Then the coefficients a_{nik} can be found from Eq. (11.167) as

$$a_{nik} = \frac{\{\Phi\}_k^T \frac{\partial [K]}{\partial r_i} \{\Phi\}_n - \omega_n^2 \{\Phi\}_k^T \frac{\partial [m]}{\partial r_i} \{\Phi\}_n}{\omega_n^2 - \omega_k^2} \qquad (11.168)$$

(2) *$n = k$ case*
In Eq. (11.168), a_{nik} approaches an infinite value as ω_k equals ω_n. As a result, it can not be used in the $n = k$ condition. For the orthonormal modes,

$$\{\Phi\}_n^T [m] \{\Phi\}_n = 1 \qquad (11.169)$$

Differentiating Eq. (11.169) with respect to random parameters yields

$$2 \{\Phi\}_n^T [m] \frac{\partial \{\Phi\}_n}{\partial r_i} + \{\Phi\}_n^T \frac{\partial [m]}{\partial r_i} \{\Phi\}_n = 0 \qquad (11.170)$$

Now, substituting $\partial \{\Phi\}_n / \partial r_i$ into Eq. (11.170) solves the coefficient a_{nin} which is given by

$$a_{nin} = \frac{\{\Phi\}_n^T \frac{\partial [m]}{\partial r_i} \{\Phi\}_n}{2} \qquad (11.171)$$

The derivative of the natural frequency with respect to the design variable, $\partial \omega_n^2 / \partial r_i$, can also be derived as follows:

Differentiating the equation $\{\Phi\}_n^T \left([K] - \omega_n^2 [m]\right) \{\Phi\}_n = 0$, we have

$$\frac{\partial \{\Phi\}_n^T}{\partial r_i} \left([K] - \omega_n^2 [m]\right) \{\Phi\}_n + \{\Phi\}_n^T \left(\frac{\partial [K]}{\partial r_i} - \frac{\partial \omega_n^2}{\partial r_i} [m] - \omega_n^2 \frac{\partial [m]}{\partial r_i}\right) \{\Phi\}_n$$

$$+ \{\Phi\}_n^T \left([K] - \omega_n^2 [m]\right) \frac{\partial \{\Phi\}_n}{\partial r_i} = 0 \qquad (11.172)$$

By observing the above equation, in which the first and third terms are zero, Eq. (11.167) becomes

$$\frac{\partial \omega_n^2}{\partial r_i} = \{\Phi\}_n^T \left(\frac{\partial [K]}{\partial r_i} - \omega_n^2 \frac{\partial [m]}{\partial r_i}\right) \{\Phi\}_n \qquad (11.173)$$

The derivative of the participation factor with respect to the design variable, $\partial \Gamma_n / \partial r_i$, is

$$\frac{\partial \Gamma_n}{\partial r_i} = \frac{\partial \{\Phi\}_n^T}{\partial r_i} [m] \{1\} + \{\Phi\}_n^T \frac{\partial [m]}{\partial r_i} \{1\} \qquad (11.174)$$

2nd variance approach

$$\sigma_{\{u\}_E}^2 = \sum_{n1} \sum_{n2} \left[\left(\frac{\partial \{u_E\}}{\partial y_{\omega n1}}\right)_{\bar{r}} \left(\frac{\partial \{u_E\}}{\partial y_{\omega n2}}\right)_{\bar{r}} \rho_{y_{\omega n1} y_{\omega n2}} V_{y_{\omega n1}} V_{y_{\omega n2}} \bar{y}_{\omega n1} \bar{y}_{\omega n2}\right]$$

$$+ (0.15)^2 \{\bar{u}_E^2\} \qquad (11.175)$$

where

$$\left(\frac{\partial \{u_E\}}{\partial y_{\omega n1}}\right) = \frac{1}{u_E} \left[\{\Phi\}_{n1}^2 \Gamma_{n1}^2 y_{\omega n1}\right] \qquad (11.176)$$

$$\left(\frac{\partial \{u_E\}}{\partial y_{\omega n2}}\right) = \frac{1}{u_E} \left[\{\Phi\}_{n2}^2 \Gamma_{n2}^2 y_{\omega n2}\right] \qquad (11.177)$$

and 0.15 is assumed to be the dynamic analysis error [14].

(C) *Mean and Variance of Displacement for Combined Load Effect* – After the uncertainties of dead, live and earthquake load effects are individually obtained from previous knowledge, the uncertainties of combination of these load effects may be given as

mean

$$\{\bar{u}\} = \{\bar{u}_D\} + \{\bar{u}_L\} + \{\bar{u}_E\} \tag{11.178}$$

1st variance approach

$$\sigma^2_{\{u\}} = \sum_i \sum_j \left(\frac{\partial \{u\}}{\partial r_i}\right)_{\bar{r}} \left(\frac{\partial \{u\}}{\partial r_j}\right)_{\bar{r}} \rho_{r_i r_j} V_{r_i} V_{r_j} \bar{r}_i \bar{r}_j \tag{11.179}$$

where

$$\frac{\partial \{u\}}{\partial r_i} = \frac{\partial \{u_D\}}{\partial r_i} + \frac{\partial \{u_L\}}{\partial r_i} + \frac{\partial \{u_E\}}{\partial r_i} \tag{11.180}$$

2nd variance approach

$$\sigma^2_{\{u\}} = \sigma^2_{\{u_D\}} + \sigma^2_{\{u_L\}} + \sigma^2_{\{u_E\}} + 2(0.1)^2 \{\bar{u}_D\} \{\bar{u}_L\} \tag{11.181}$$

where 0.1 is the static analysis error.

For the displacement failure mode, by substituting Eqs. (11.178), (11.179) and (11.181) into Eqs. (11.6) or (11.15) the probabilities of failure in Eqs. (11.4) or (11.8) can be determined if the statistics of allowable displacement are given.

11.11.3 Response of member forces

(A) *Mean and Variance of Member Forces due to Earthquake Load Effect –* The internal force formulation for the dead and live loads are the same as in the previous chapter. However, for the dynamic seismic load effect considered here, we need the mean and variance as follows:

mean

$$\left\{\bar{F}_E\right\}_m = \left[\sum_n \left([S]_m [A]_m^T \{u_E\}_{mn}\right)^2\right]^{1/2} \tag{11.182}$$

where $[S]_m$ and $[A]_m^T$ are the same notations used in Eq. (11.93); $\{u_E\}_{mn}$ = the seismic displacements for the *n*th mode corresponding to the *m*th member.

1st variance approach

$$\sigma^2_{\{F_E\}_m} = \sum_i \sum_j \left(\frac{\partial \{F_E\}_m}{\partial r_i}\right)_{\bar{r}} \left(\frac{\partial \{F_E\}_m}{\partial r_j}\right)_{\bar{r}} \rho_{r_i r_j} V_{r_i} V_{r_j} \bar{r}_i \bar{r}_j \tag{11.183}$$

where

$$\frac{\partial \{F_E\}_m}{\partial r_i} = \frac{1}{(F_E)_m} \left[\sum_n ([S]_m [A]_m^T \{u_E\}_{mn}) \right.$$

$$\left. \left(\frac{\partial [S]_m}{\partial r_i} [A]_m^T \{u_E\}_{mn} + [S]_m [A]_m^T \frac{\partial \{u_E\}_{mn}}{\partial r_i} \right) \right] \qquad (11.184)$$

$$\frac{\partial \{F_E\}_m}{\partial r_j} = \frac{1}{(F_E)_m} \left[\sum_n ([S]_m [A]_m^T \{u_E\}_{mn}) \right.$$

$$\left. \left(\frac{\partial [S]_m}{\partial r_j} [A]_m^T \{u_E\}_{mn} + [S]_m [A]_m^T \frac{\partial \{u_E\}_{mn}}{\partial r_j} \right) \right] \qquad (11.185)$$

2nd variance approach

$$\sigma^2_{\{F_E\}_m} = \sum_{n1} \sum_{n2} \left[\left(\frac{\partial \{F_E\}_{mn}}{\partial y_{\omega n1}} \right)_{\bar{r}} \left(\frac{\partial \{F_E\}_{mn}}{\partial y_{\omega n2}} \right)_{\bar{r}} \rho_{y_{\omega n1} y_{\omega n2}} V_{y_{\omega n1}} V_{y_{\omega n2}} \bar{y}_{\omega n1} \bar{y}_{\omega n2} \right]$$

$$+ (0.15)^2 \left\{ \bar{F}^2_{Em} \right\} \qquad (11.186)$$

where

$$\left(\frac{\partial \{F_E\}_m}{\partial y_{\omega n1}} \right) = \frac{1}{(F_E)_m} \left[\sum_n ([S]_m [A]_m^T \{u_E\}_{mn}) \left([S]_m [A]_m^T \frac{\partial \{u_E\}_{mn}}{\partial y_{\omega n1}} \right) \right] \qquad (11.187)$$

$$\left(\frac{\partial \{F_E\}_m}{\partial y_{\omega n2}} \right) = \frac{1}{(F_E)_m} \left[\sum_n ([S]_m [A]_m^T \{u_E\}_{mn}) \left([S]_m [A]_m^T \frac{\partial \{u_E\}_{mn}}{\partial y_{\omega n2}} \right) \right] \qquad (11.188)$$

and 0.15 is assumed to be the dynamic analysis error.

(B) *Mean and Variance of Member Forces due to Combined Load Effect –* The internal forces due to dead, live and earthquake loads are given as a combination in the following:

mean

$$\{\bar{F}\}_m = \{\bar{F}_D\}_m + \{\bar{F}_L\}_m + \{\bar{F}_E\}_m \qquad (11.189)$$

1st variance approach

$$\sigma^2_{\{F\}_m} = \sum_i \sum_j \left(\frac{\partial \{F\}_m}{\partial r_i} \right)_{\bar{r}} \left(\frac{\partial \{F\}_m}{\partial r_j} \right)_{\bar{r}} \rho_{r_i r_j} V_{r_i} V_{r_j} \bar{r}_i \bar{r}_j \qquad (11.190)$$

where

$$\frac{\partial \{F\}_m}{\partial r_i} = \frac{\partial \{F_D\}_m}{\partial r_i} + \frac{\partial \{F_L\}_m}{\partial r_i} + \frac{\partial \{F_E\}_m}{\partial r_i} \qquad (11.191)$$

$$\frac{\partial \{F\}_m}{\partial r_j} = \frac{\partial \{F_D\}_m}{\partial r_j} + \frac{\partial \{F_L\}_m}{\partial r_j} + \frac{\partial \{F_E\}_m}{\partial r_j} \qquad (11.192)$$

2nd variance approach

$$\sigma^2_{\{F\}_m} = \sigma^2_{\{F_D\}_m} + \sigma^2_{\{F_L\}_m} + \sigma^2_{\{F_E\}_m} + 2(0.1)^2 \left\{\bar{F}_D\right\}_m \left\{\bar{F}_L\right\}_m \qquad (11.193)$$

where 0.1 is assumed to be the static analysis error.

For column failure modes, substituting Eqs. (11.189), (11.190) and (11.193) which may be the uncertainties of the applied axial load (*P*) or moments (*M*) into Eqs. (11.204), (11.205), (11.209), (11.210), (11.214), (11.215), (11.218) and (11.219) yields the mean and variances of the interaction equations. The safety factor in Eqs. (11.6) or (11.15), and the probabilities of failure in Eqs. (11.4) or (11.8) can be determined if the statistics of allowable values of the interaction equations are given.

11.11.4 Member resistance

The resistances of members are (A) yield moment, (B) Euler buckling load, (C) axial load capacity of columns, (D) yield load, and (E) critical moment. The mean and variances of these resistances, using the same notations as in Section 11.10, are summarized as follows.

(A) *Yield Moment*

   mean

$$\bar{M}_y = \bar{F}_y \bar{S}_c$$

   variance

$$\sigma^2_{M_y} = (0.12)^2 \bar{M}_y^2$$

(B) *Euler Buckling Load*

   mean

$$\bar{P}_E = \frac{\pi^2 \bar{E}_m \bar{I}}{\left(\overline{KL}\right)^2}$$

variance

$$\sigma_{P_E}^2 = (0.3)^2 \, \bar{P}_E^2$$

(C) *Axial Load Capacity*

mean

$$\bar{P}_{cr} = \left[1 - \frac{\lambda^2}{4} \right] \bar{F}_y \bar{A}_c, \quad \text{for} \frac{KL}{r_g} < C_c$$

or

$$\bar{P}_{cr} = \bar{P}_E, \quad \text{for} \frac{KL}{r_g} > C_c$$

Variance

$$\sigma_{P_{cr}}^2 = V_{P_{cr}}^2 \, \bar{P}_{cr}^2$$

where $V_{P_{cr}}$ varies from 0.14 to 0.31.

(D) *Yield Load*

mean

$$\bar{P}_y = \bar{F}_y \bar{A}_c$$

variance

$$\sigma_{P_y}^2 = (0.14)^2 \, \bar{P}_y^2$$

(E) *Critical Moment*

mean

$$\bar{M}_{cr} = \left\{ \left[\frac{\bar{C}_b \pi \bar{E}_m \bar{I}_y}{\bar{K}_y \bar{L}} \right] \left[\frac{\bar{G}_s \bar{J}}{\bar{E}_m} + \frac{\pi^2 \bar{C}_W}{\left(\bar{K}_z \bar{L} \right)^2} \right] \right\}^2$$

variance

$$\sigma_{M_{cr}}^2 = V_{M_{cr}}^2 \, \bar{M}_{cr}^2$$

where $V_{M_{cr}}$ varies from 0.15 to 0.2.

11.12 Optimization formulations

In engineering for the optimum design, the goal is to produce a best solution which provides not only a safety margin but also a best-objective value. An optimum structural problem can be formulated in a similar form to that presented previously as

Minimize   Objective Function
Subject to   Constraints

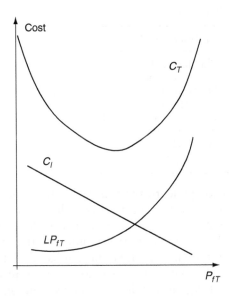

Figure 11.3 Cost versus system probability of failure.

11.12.1 Objective function of structural weight and cost

The objective function of a structural design problem may be the weight or cost functions which are

(A) *Weight*

Weight (W) is the summation of the constituent of structural members' weights of a system and can be expressed as

$$W = \sum_i r_{di} \ell_i A_i \tag{11.194}$$

where r_{di}, ℓ_i, A_i = the mass density, length and area of a member, respectively.

(B) *Cost*

Total estimated structural cost (C_T) which is shown in Fig. 11.3 consists of two parts: the initial construction cost (C_I) and the expected future failure loss$(L_f P_{ft})$; i.e.,

$$C_T = C_I + L_f P_{ft} \tag{11.195}$$

where L_f = expected failure cost; P_{ft} = system failure probability. The initial construction cost C_I comprises the structural material cost and the miscellaneous cost. The structural material cost is the cost of structural members. The miscellaneous cost may be the product of the total unit floor price as shown in Table 11.1 [4] and the total area. The

Table 11.1 Unit floor cost for office building.

Items average $ ft^{-2}	Items average $ ft^{-2}
Foundations 2.27	Plumbing 2.13
Exterior walls 5.96	Fire protection 0.27
Partitions 3.07	HVAC 6.20
Interior wall finishes 1.36	Electrical 3.99
Floor finishes 1.22	General conditions 2.20
Ceilings 0.99	Equipment 0.69
Specialities 0.38	Site work 2.00
Conveying systems 0.56	Total 33.29

expected future loss has two components: the expected failure cost L_f and the system probability of failure P_{ft}. The external failure cost (L_f) is the total loss incurred in a structural failure state. This includes an additional replacement cost, damage to property, liabilities due to death and injury, and business interruption. They are estimated based on the following assumptions: the additional replacement cost is taken to be twice the initial structural cost [15]; the loss due to property damage, in which no loss of expensive equipment is assumed, is 50 per cent of the initial construction cost [2]; liabilities due to death and injury include only the people present in the building failure.

Loss due to death is calculated based on an average death age of 30 and is the sum of a person's salary until he reaches the retirement age of 65 years. Thus, this loss is 35 times the average annual net income [2]. The loss due to serious injuries is assumed to be $350,000 per person [15]. The loss due to minor injury is $5,000 per person. Business interruption is estimated as the income of employees during a 4-year reconstruction period. The loss due to legal services may be assumed to be 15 per cent of the failure loss [16]. Note that the dollar amounts cited above are presented only for discussion purposes; the realistic loss should be constantly updated as time changes. The estimation of a system failure probability is approximated by Eq. (11.29) which is expressed as

$$P_{fT} = \sum_{i=1}^{nf} P_{F_j} \tag{11.196}$$

where nf = the number of failure modes.

11.12.2 *Failure constraints of structural displacements*

In a structural design, one requires the probabilities of a desired response failure are less than the allowable probabilities of failures or the safety factors for those failures greater than the allowable values. These requirements become the constraints in optimum structural design problems. The desired response failures may be structural displacement failures, and members' failures including column failures and beam failures which are described in this section and Section 11.12.3. In addition to the above individual failure mode constraints, the system failure probability, which is less than the allowable probability value, can also be added to the constraints. Here, using the normal distribution for the safety factor, Eq. (11.6), and the probability of failure, the Eq. (11.4), we express a structural failure due to an excessive displacement as

$$\beta_u = \frac{\bar{u} - \bar{u}_a}{\sqrt{\sigma_u^2 + \sigma_{ua}^2}} \tag{11.197}$$

where $\bar{u}$, $\sigma_u^2 =$ the mean and variance of a displacement; $\bar{u}_a$, $\sigma_{ua}^2 =$ the mean and variance of the allowable displacement. And the probability of failure

$$P_f(\beta_u) = 1 - \text{PN}(\beta_u) \tag{11.198}$$

11.12.3   Failure constraints of members

The safety factor and probability failure are also based on a normal distribution model as illustrated here. The failure constraints of beams and columns are presented as follows:

(A) *Beam Failure* – A beam may fail in a number of different modes. However, it is assumed that the beam fails when the applied moment exceeds the flexural yield capacity of the beam, i.e., β_M or $P_f(\beta_M)$. The safety factor is

$$\beta_M = \frac{\bar{M} - \bar{M}_y}{\sqrt{\sigma_{Mx}^2 + \sigma_{My}^2}}, \quad P_f(\beta_m) = 1 - \text{PN}(\beta_m) \tag{11.199, 11.200}$$

where $\bar{M}$, $\sigma_{Mx}^2 =$ the computed mean and variance of a moment; $\bar{M}_y$, $\sigma_{My}^2 =$ the mean and variance of the allowable yielding moment.

(B) *Column Failure* – Four cases may occur in column failure. The representations of them are:

(1) *Failure by yielding at the ends f_1*
The safety factor and probability of failure in this case may be expressed as β_{f_1} and $P_f(\beta_{f_1})$. Thus

$$\beta_{f_1} = \frac{1.0 - \bar{f}_1}{\sqrt{\sigma_{f_1}^2}} \tag{11.201}$$

$$P_{f_1} = 1 - \text{PN}(\beta_{f_1}) \tag{11.202}$$

where

$$f_1 = \frac{P}{P_y} + \frac{M}{M_y} \tag{11.203}$$

and P = applied axial load; M = applied moment at one end of the column; P_y = yield axial load; M_y = yield moment. To determine the safety factor and probability of failure, the mean and variance of f_1 are needed. These uncertainties can be determined from Eqs. (11.16) and (11.18) as

mean

$$\bar{f}_1 = \frac{\overline{P}}{\overline{P_y}} + \frac{\overline{M}}{\overline{M_y}} \tag{11.204}$$

variance

$$\sigma_{f_1}^2 = \left(\frac{\partial f_1}{\partial P}\right)_{\bar{r}}^2 V_P^2 \overline{P}^2 + \left(\frac{\partial f_1}{\partial P_y}\right)_{\bar{r}}^2 V_{P_y}^2 \overline{P}_y^2 + \left(\frac{\partial f_1}{\partial M}\right)_{\bar{r}}^2 V_M^2 \overline{M}^2$$

$$+ \left(\frac{\partial f_1}{\partial M_y}\right)_{\bar{r}}^2 V_{M_y}^2 \overline{M}_y^2 + 2\left(\frac{\partial f_1}{\partial P_y}\right)_{\bar{r}} \left(\frac{\partial f_1}{\partial M_y}\right)_{\bar{r}} \rho_{P_y M_y} V_{M_y} V_{P_y} \overline{P}_y \overline{M}_y$$

$$= \left(\frac{1}{P_y}\right)_{\bar{r}}^2 V_P^2 \overline{P}^2 + \left(\frac{-P}{P_y}\right)_{\bar{r}}^2 V_{P_y}^2 P_y^2 + \left(\frac{1}{M_y}\right)_{\bar{r}}^2 V_M^2 \overline{M}^2 + \left(\frac{-M}{M_y}\right)_{\bar{r}}^2 V_{M_y}^2 M_y^2$$

$$+ 2\left(\frac{\overline{P}}{\overline{P_y}}\right) \left(\frac{\overline{M}}{\overline{M_y}}\right) \rho_{P_y M_y} V_{P_y} V_{M_y}$$

$$= \left(\frac{\overline{P}}{\overline{P_y}}\right)^2 \left(V_P^2 + V_{P_y}^2\right) + \left(\frac{\overline{M}}{\overline{M_y}}\right)^2 \left(V_M^2 + V_{M_y}^2\right) + 2\left(\frac{\overline{P}}{\overline{P_y}}\right) \left(\frac{\overline{M}}{\overline{M_y}}\right) \rho_{P_y M_y} V_{P_y} V_{M_y} \tag{11.205}$$

where $\overline{P}$, $\overline{P_y}$, V_P, V_{P_y} = the means and coefficients of the variation of P and P_y; $\rho_{P_y M_y}$ = the correlation coefficient of P_y and M_y; the value of $\rho_{P_y M_y}$ is assumed to be 0.70 in the study.

(2) *Instability in the plane of bending f_2*

The safety factor and probability of failure may be expressed as β_{f_2} and $P_f(\beta_{f_2})$. Thus

$$\beta_{f_2} = \frac{1.0 - \bar{f}_2}{\sqrt{\sigma_{f_2}^2}} \tag{11.206}$$

$$P_{f_2}(\beta_{f_2}) = 1 - PN(\beta_{f_2}) \tag{11.207}$$

where

$$f_2 = \frac{P}{P_{cr}} + \frac{M}{M_y\left(1 - P/P_E\right)} \tag{11.208}$$

where P_{cr} = critical axial load; P_E = Euler buckling load. The mean and variance of f_2 which can be determined from Eqs. (11.16) and (11.18) are

mean

$$\bar{f}_2 = \frac{\overline{P}}{\overline{P}_{cr}} + \frac{\overline{M}}{\left(1 - \bar{P}/\bar{P}_E\right)\overline{M}_y} \tag{11.209}$$

variance

$$\sigma_{f_2}^2 = \left(\frac{\overline{P}}{\overline{P}_{cr}} + \frac{\overline{M}}{\left(1 - \bar{P}/\bar{P}_E\right)^2 \overline{M}_y} \frac{\overline{P}}{\overline{P}_E} \right)^2 V_P^2 + \left(\frac{\overline{P}}{\overline{P}_{cr}} \right)^2 V_{P_{cr}}^2$$

$$+ \left(\frac{\overline{M}}{\left(1 - \bar{P}/\bar{P}_E\right)^2 \overline{M}_y} \frac{\overline{P}}{\overline{P}_E} \right)^2 V_{P_E}^2$$

$$+ \left(\frac{\overline{M}}{\left(1 - \bar{P}/\bar{P}_E\right)\overline{M}_y} \right)^2 \left(V_{M_y} + V_M^2 \right) + 2 \left(\frac{\overline{P}}{\overline{P}_{cr}} \right)$$

$$\left(\frac{\overline{M}}{\left(1 - \bar{P}/\bar{P}_E\right)^2 \overline{M}_y} \frac{\overline{P}}{\overline{P}_E} \right) \rho_{P_{cr}P_E} V_{P_{cr}} V_{P_E}$$

$$+ 2 \left(\frac{\overline{P}}{\overline{P}_{cr}} \right) \left(\frac{\overline{M}}{\left(1 - \bar{P}/\bar{P}_E\right)\overline{M}_y} \right) \rho_{P_{cr}M_y} V_{P_{cr}} V_{M_y}$$

$$+ 2 \left(\frac{\overline{M}}{\left(1 - \bar{P}/\bar{P}_E\right)\overline{M}_y} \right) \left(\frac{\overline{M}}{\left(1 - \bar{P}/\bar{P}_E\right)^2 \overline{M}_y} \frac{\overline{P}}{\overline{P}_E} \right) \rho_{P_{M_y}P_E} V_{M_y} V_{P_E} \tag{11.210}$$

where $\overline{P}_{cr}, \overline{P}_E, V_{P_{cr}}, V_{P_E}$ = the means and coefficients of the variation of P_{cr} and P_E; $\rho_{P_{cr}P_E}, \rho_{P_{cr}M_y}, \rho_{P_E M_y}$ = the correlation coefficients of P_{cr} and P_E, P_{cr} and M_y, M_y and P_E.

(3) *Lateral-torsional buckling f_3*

The safety factor and probability of failure may be expressed as β_{f_3} and $P_f\left(\beta_{f_3}\right)$. Thus

$$\beta_{f_3} = \frac{1.0 - \bar{f}_3}{\sqrt{\sigma_{f_3}^2}} \tag{11.211}$$

$$P_{f_3} = 1 - \text{PN}\left(\beta_{f_3}\right) \tag{11.212}$$

and

$$f_3 = \frac{P}{P_{cr}} + \frac{M}{M_{cr}\left(1 - P/P_E\right)} \tag{11.213}$$

where M_{cr} = critical moment. The mean and variance of f_3 which can be determined from Eqs. (11.16) and (11.18) are

mean

$$\bar{f}_3 = \frac{\bar{P}}{\bar{P}_{cr}} + \frac{\bar{M}}{\bar{M}_{cr}\left(1 - \bar{P}/\bar{P}_E\right)} \tag{11.214}$$

variance

$$\sigma_{f_3}^2 = \left(\frac{\bar{P}}{\bar{P}_{cr}} + \frac{\bar{M}}{\left(1 - \bar{P}/\bar{P}_E\right)^2 \bar{M}_{cr}} \frac{\bar{P}}{\bar{P}_E}\right)^2 V_P^2 + \left(\frac{\bar{P}}{\bar{P}_{cr}}\right)^2 V_{P_{cr}}^2$$

$$+ \left(\frac{\bar{M}}{\left(1 - \bar{P}/\bar{P}_E\right)^2 \bar{M}_{cr}} \frac{\bar{P}}{\bar{P}_E}\right)^2 V_{P_E}^2$$

$$+ \left(\frac{\bar{M}}{\left(1 - \bar{P}/\bar{P}_E\right) \bar{M}_{cr}}\right)^2 \left(V_{M_{cr}} + V_M^2\right) + 2\left(\frac{\bar{P}}{\bar{P}_{cr}}\right)$$

$$\left(\frac{\bar{M}}{\left(1 - \bar{P}/\bar{P}_E\right)^2 \bar{M}_{cr}} \frac{\bar{P}}{\bar{P}_E}\right) \rho_{P_{cr}P_E} V_{P_{cr}} V_{P_E} \tag{11.215}$$

$$+ 2\left(\frac{\bar{P}}{\bar{P}_{cr}}\right) \left(\frac{\bar{M}}{\left(1 - \bar{P}/\bar{P}_E\right) \bar{M}_{cr}}\right) \rho_{P_{cr}M_{cr}} V_{P_{cr}} V_{M_{cr}}$$

$$+ 2\left(\frac{\bar{M}}{\left(1 - \bar{P}/\bar{P}_E\right) \bar{M}_{cr}}\right) \left(\frac{\bar{M}}{\left(1 - \bar{P}/\bar{P}_E\right)^2 \bar{M}_{cr}} \frac{\bar{P}}{\bar{P}_E}\right)$$

$$\rho_{M_{cr}P_E} V_{M_{cr}} V_{P_E}$$

where $\bar{M}_{cr}, V_{M_{cr}}$ = the mean and coefficient of the variation of M_{cr}; $\rho_{P_{cr}P_E}, \rho_{P_{cr}M_{cr}}, \rho_{M_{cr}P_E}$ = the correlation of P_{cr} and P_E, P_{cr} and M_{cr}, M_{cr} and P_E.

(4) *Buckling about the weak axis f_4*

The safety factor and probability of failure may be expressed as β_{f_4} and $P_f\left(\beta_{f_4}\right)$. Thus

$$\beta_{f_4} = \frac{1.0 - \bar{f}_4}{\sqrt{\sigma_{f_4}^2}}, \quad P_{f_4} = 1 - \text{PN}\left(\beta_{f_4}\right) \qquad (11.216a, 11.216b)$$

where

$$f_4 = \frac{P}{P_{cry}} \qquad (11.217)$$

The mean and variance of f_4 which can be determined from Eqs. (11.16) and (11.18) are

mean

$$\bar{f}_4 = \frac{\bar{P}}{\bar{P}_{cry}} \qquad (11.218)$$

variance

$$\sigma_{f_4}^2 = \left(\frac{\bar{P}}{\bar{P}_{cry}}\right)^2 \left(V_P^2 + V_{P_{cry}}^2\right) \qquad (11.219)$$

where $\bar{P}_{cry}, V_{P_{cry}}^2$ = the mean and coefficient of the variation of P_{cry}.

11.13   Optimization algorithms

11.13.1   *Interior-penalty-function method*

As introduced in Chapter 4, the penalty-function methods are those mathematical programming techniques which are used to search the best moving route of design variables and to reduce the objective functional value until no objective functional value can further be minimized; the procedure is then terminated where an optimum solution is obtained. In this chapter, the interior penalty function is used. The formulation of the method can be defined as:

$$P_P\left(X^{(k)}, r_P^{(k)}\right) = OX^{(k)} - r_P^{(k)} \sum_i \frac{1}{g_j^{(k)}}, \quad g_j^{(k)} < 0 \qquad (11.220)$$

where $P_P\left(X^{(k)}, r_P^{(k)}\right)$ = penalty function; $OX^{(k)}$ = objective function; $X^{(k)}$ = design variables; $r_P^{(k)}$ = a penalty value; $g_j = j$th constraint; k = the number of completed stages for the subsequent change of penalty values $r_P^{(k)}$.

At an interior point, the sum of the inversing constraints in a penalty term is negative, and a positive $r_P^{(k)}$ will result in a positive penalty term to be added to the objective function. As a boundary of the feasible region is reached, some constraints will approach zero and the inverse constraints will approach infinity. By successively reducing the parameter $r_P^{(k)}$, the optimal solution for the constrained problem will be found. The process for finding a solution is as follows:

(1) The initial design variables $X^{(0)}$ are assumed to be in a feasible region and the penalty value $r_P^{(0)}$ is chosen; then the unconstrained optimization algorithm is used to find the minimum solution of the penalty function $P_P\left(X^{(0)}, r_P^{(0)}\right)$.

(2) Reduce $r_P^{(k)}$ by using the rule $r_P^{(k+1)} = C_r r_P^{(k)}$, where $C_r < 1$, and find the minimum of $P_P\left(X^{(k+1)}, r_P^{(k+1)}\right)$.

(3) Check whether the convergence criterion is satisfied, if not, repeat step (2); if yes, an optimal solution is obtained.

In order to solve the penalty function, the unconstrained optimization algorithm is needed. An algorithm used in this study is based on Powell's method with Goggin's one-dimensional search technique which is described in Appendix J.

11.13.2   Generalized optimality-criteria method

The generalized optimality-criteria method has been extensively discussed in Chapters 7, 8, 9 and 10 for deterministic structures and it is now further developed for nondeterministic structural design. The method is briefly reviewed as follows:

> Minimize $O(X)$
> Subject to $g_j(X) \le 0,\quad j=1,2,3,\ldots,\text{nc}$

where nc = the number of constraints, the other notations are defined as in Eq. (11.220). The criterion for optimality of the method may be expressed as:

$$\left(\frac{\partial O(X)}{\partial x_i}\right)_{x^*} + \sum_{j=1}^{\text{nc}} \lambda_j \left(\frac{\partial g_j(X)}{\partial x_i}\right)_{x^*} = 0, \text{ and } \lambda_j \ge 0 \qquad (11.221)$$

where $x_i = i$th design variable; $\lambda_j = j$th Lagrangian multiplier; $x^* =$ an optimal solution. Based on the above criterion, the design variable $x_i^{(k+1)}$ at the $(k+1)$th iteration can be expressed in terms of $x_1^{(k)}$ at the kth iteration as

$$x_i^{(k+1)} = [\alpha + (1-\alpha) T_i] x_i^{(k)} \qquad (11.222)$$

in which T_i is called the *recurrence equation* and can be written as

$$T_i = \left(-\sum_{j=1}^{act} \lambda_j \left(\frac{\partial g_j(X)}{\partial x_i} \right) \right) \Big/ \left(\frac{\partial O(X)}{\partial x_i} \right) \qquad (11.223)$$

$\alpha =$ a relaxation constant; and act = the number of active constraints. The partial derivation of constraints with respect to the design variables, $\partial g_j/\partial x_i$, can be expressed in terms of the safety factor, β_j, as

$$\frac{\partial (\beta_a - \beta)}{\partial x_i} = \frac{-\partial \beta_j}{\partial x_i} \qquad (11.224)$$

or in terms of the probability of failure as

$$\frac{\partial g_j}{\partial x_i} = \frac{\partial (P_f - P_{fa})}{\partial x_i} = \frac{\partial P_f(\beta)}{\partial x_i} = \frac{-1}{\sqrt{2\pi}} \exp\left(\frac{-\beta_j^2(X)}{2} \right) \frac{\partial \beta_j(X)}{\partial x_i} \qquad (11.225)$$

where β_a and P_{fa} are the allowable safety factor and probability of failure, respectively.

At each iteration, one has to scale the initial design and to decide the active constraints for accelerating convergence before the value of T_i (Eq. (11.223)) is found. For each constraint, the scaling factor is computed by a linearized approximation to the allowable safety factor or probability of failure as follows:

$$\Delta \beta_j = \sum_{i=1}^{act} \left(\frac{\partial \beta_j}{\partial x_i} \right) \Delta x_i + \sum_{i'=1}^{passive} \left(\frac{\partial \beta_j}{\partial x_{i'}} \right) (x_{i' \, min} - x_{i'}) \qquad (11.226)$$

in which $x_{i' \, min}$ represents passive elements. Substituting $\Delta x_i = x_1^{(k+1)} - x_1^{(k)} = \Lambda x_1^{(k)} - x_1^{(k)}$ into the above equation yields

$$\beta_a - \beta_j = \sum_{i=1}^{act} \left(\frac{\partial \beta_j}{\partial x_i} \right) (\Lambda - 1) x_i + \sum_{i'=1}^{passive} \left(\frac{\partial \beta_j}{\partial x_{i'}} \right) (x_{i' \, min} - x_{i'}) \qquad (11.227)$$

Thus, the scaling factor for the safety factor can be calculated as follows:

$$\Lambda_j = \frac{(\beta_a - \beta_j) + \sum_{i=1}^{act}\left(\frac{\partial \beta_j}{\partial x_i}\right)x_i - \sum_{i'=1}^{passive}\left(\frac{\partial \beta_j}{\partial x_{i'}}\right)(x_{i'\,min} - x_{i'})}{\sum_{i=1}^{act}\left(\frac{\partial \beta_j}{\partial x_i}\right)x_i} \tag{11.228}$$

For the probability of failure, the scaling factors become

$$\Lambda_j = \frac{\left(P_{f_a} - P_{f_j}\right) + \sum_{i=1}^{act}\left(\frac{\partial P_{f_j}}{\partial x_i}\right)x_i - \sum_{i'=1}^{passive}\left(\frac{\partial P_{f_j}}{\partial x_{i'}}\right)(x_{i'\,min} - x_{i'})}{\sum_{i=1}^{act}\left(\frac{\partial P_{f_j}}{\partial x_i}\right)x_i} \tag{11.229}$$

In order to determine T_i, the active Lagrangian multipliers λ_p must be determined from the following simultaneous equations which are derived from linearized approximations to the zero constraint values:

$$\Delta g_j = g_j^{(k+1)} - g_j^{(k)} = \sum_{i=1}^{act}\left(\frac{\partial g_j}{\partial x_i}\right)\Delta x_i + \sum_{i'=1}^{passive}\left(\frac{\partial g_j}{\partial x_{i'}}\right)\Delta x_{i'}\,(P) \tag{11.230}$$

in which $\Delta x_{i'}\,(P)$ represents passive elements. Let $g_j^{(k+1)} = 0$,

$$\Delta x_i = [\alpha + (1 - \alpha)\,T_i]\,x_i^{(k)} - x_{i'}^{(k)} \tag{11.231}$$

and

$$T_i = \left(-\sum_{j=1}^{act}\lambda_j\left(\frac{\partial g_j\,(x)}{\partial x_i}\right)\right)\Big/ (\partial O\,(x)/\partial x_i) \tag{11.232}$$

then Eq. (11.230) becomes

$$-g_j = \sum_{i=1}^{act}\left(\frac{\partial g_j}{\partial x_i}\right)\alpha x_i + (1 - \alpha)\sum_{i=1}^{act}\frac{\left(-\sum_{P}^{ncl}\lambda_P\frac{\partial g_j'}{\partial x_i}\frac{\partial g_P}{\partial x_i}\right)}{\left(\frac{\partial O}{\partial x_i}\right)}x_i$$

$$-\sum_{i=1}^{act}\frac{\partial g_j}{\partial x_i}x_i + \sum_{i'=1}^{passive}\frac{\partial g_j}{\partial x_{x'}}\Delta x_{i'}\,(P) \tag{11.233}$$

In which nc1 = the number of active constraints, g_P = the active constraints, λ_P = the Lagrangian multipliers of the active constraints and can be found from the following equation:

$$(1-\alpha)\sum_{p=1}^{nc}\left(\sum_{i=1}^{act}\left(\frac{\partial g_i}{\partial x_i}\right)\left(\frac{\partial g_P}{\partial x_i}\right)\middle/\left(\frac{\partial O}{\partial x_i}\right)\lambda_P x_i\right)$$

$$= g_j+(\alpha-1)\sum_{i=1}^{act}\left(\frac{\partial g_i}{\partial x_i}\right)x_i+\sum_{i'=1}^{passive}\frac{\partial g_i}{\partial x_{x'}}\Delta x_{i'}\ (P)$$

(11.234)

The recurrence equation, T_i, derived from $\partial C_T/\partial x_i = 0$ for the cost function in the unconstrained algorithm, has the following form:

$$T_i = \frac{-L_f P_{fT}}{C_I}$$

(11.235)

11.14 Numerical illustrations and design parameters assessment for dead and live loads

11.14.1 Optimum design based on generalized optimality-criteria method and interior-penalty-function method

Example 11.14.1 – The unsymmetric three-bar truss in Fig. 11.4 is designed for minimum weight subject to displacement and stress constraints. The given conditions are:

Allowable displacements: u_1 (x-axis) $=u_2$ (y-axis) $=0.125$ in. (0.3175 cm);
Allowable stress for members 1,2,3: $\tau_1 = \tau_2 = \tau_3 = 20$ ksi $(1.37 \times 10^5$ kPa);
Variance of resistance for displacements: $S_{u_1}^2 = S_{u_2}^2 = 0$;

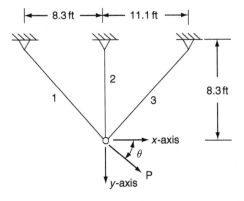

Figure 11.4 Unsymmetric three-bar truss of Example 11.14.1 (1 ft = 30.48 cm).

Variance of resistance for stresses: $S_{\tau_1}^2 = S_{\tau_2}^2 = S_{\tau_3}^2 = 2.25$; $P_{r_r r_r} = 1$;

Random parameters: $\{A_1, A_2, A_3, E_1, E_2, E_3, P, \theta\}$;

Variance: $\{V_{A1} = V_{A2} = V_{A3} = 0.05; V_{E1} = V_{E2} = V_{E3} = 0.015; V_P = 0.05;$
$V_\theta = 0.05\}$,

Loading: $\bar{P} = 20\,\text{kips}\,(89\,\text{kN})$; $\bar{\theta} = \frac{\pi}{3}$;

Modulus of elasticity: $\bar{E}_m = 30,000\,\text{ksi}\,(20.67 \times 10^7\,\text{kPa})$;

Allowable safety factors: $\beta_{u_{10}} = \beta_{u_{20}} = \beta_{\tau_{10}} = \beta_{\tau_{20}} = \beta_{\tau_{30}} = 3.09$;

Displacement constraints: $g_1(A) = \beta_{u_{10}} - \beta_{u1} \leq 0$; $g_2(A) = \beta_{u2} - \beta_{u2} \leq 0$;

Stress constraints: $g_3(A) = \beta_{\tau 1} - \beta_{\tau 1} \leq 0$; $g_4(A) = \beta_{\tau 2} - \beta_{\tau 2} \leq 0$; $g_5(A) = \beta_{\tau 3} - \beta_{\tau 3} \leq 0$.

Solve the problem by using (A) the generalized optimality-criteria method and (B) the interior-penalty-function method.

Solution:

(A) *Generalized Optimality-Criteria Method* – The mean displacement, $\bar{u}$, and its variance of displacement S_u^2, are

$$\bar{u} = \sum_{i=1}^{3} \frac{\bar{T}_i' \bar{t}' \bar{\ell}_i}{\bar{E}_m \bar{A}_i} \tag{a}$$

and

$$
\begin{aligned}
S_u^2 &= \sum \left(\frac{\partial u}{\partial r_r} \right)_{\bar{r}}^2 V_{r_r}^2 \bar{r}_r^2 \\
&= \sum_{i=1}^{3} \left(\frac{\partial u}{\partial A_i} \right)_{\bar{r}}^2 V_{A_i}^2 \bar{A}_i^2 + \left(\frac{\partial u}{\partial E_m} \right)_{\bar{r}}^2 V_{E_m}^2 \bar{E}_m^2 + \left(\frac{\partial u}{\partial P} \right)_{\bar{r}}^2 V_P^2 \bar{P}^2 \\
&\quad + \left(\frac{\partial u}{\partial \theta} \right)_{\bar{r}}^2 V_\theta^2 \bar{\theta}^2
\end{aligned}
\tag{b}
$$

in which $T_i' = $ the internal force of the ith member; $t_i' = $ the virtual internal force of the ith member; $\ell_i = $ the length of the ith member; and the other terms are expressed in Eqs. (c) through (f) as

$$\frac{\partial u}{\partial A_i} = \frac{T_i' t_i' \ell_i}{E_m A_i^2} \tag{c}$$

$$\frac{\partial u}{\partial E_m} = \sum_{i=1}^{3} \frac{-T_i' t_i' \ell_i}{E_m^2 A_i} \tag{d}$$

$$\frac{\partial u}{\partial P} = \frac{\partial \left(\{P\}^T \{S\} \right)}{\partial P} = S_x \cos \theta + S_y \sin \theta \tag{e}$$

$$\frac{\partial u}{\partial \theta} = \frac{\partial \left(\{P\}^T \{S\} \right)}{\partial \theta} = -P \sin \theta S_x + P \cos \theta S_y \tag{f}$$

In Eqs. (e) and (f), $\{P\}^T$ = transpose of the column matrix of external forces $= [P\cos\theta, P\sin\theta]$; $\{S\}$ = the column matrix of the external virtual displacement due to a virtual load in the x or y direction $= [S_x, S_y]^T$. The variance of displacement is

$$S_{u_j}^2 = \sum_{r=1}^{8} \left(\frac{\partial u_j}{\partial r_r}\bigg|_{R=\bar{R}}\right)^2 \rho_{r_r r_r} V_{r_r} V_{r_r} \bar{r}_r \bar{r}_r$$

$$= \left(\frac{\partial u_j}{\partial A_1}\bigg|_{R=\bar{R}}\right)^2 \rho_{A_1 A_1} V_{A_1} V_{A_1} \bar{A}_1 \bar{A}_1$$

$$+ \left(\frac{\partial u_j}{\partial A_2}\bigg|_{R=\bar{R}}\right)^2 \rho_{A_2 A_2} V_{A_2} V_{A_2} \bar{A}_2 \bar{A}_2$$

$$+ \left(\frac{\partial u_j}{\partial A_3}\bigg|_{R=\bar{R}}\right)^2 \rho_{A_3 A_3} V_{A_3} V_{A_3} \bar{A}_3 \bar{A}_3 \qquad\text{(g)}$$

$$+ \left(\frac{\partial u_j}{\partial E_1}\bigg|_{R=\bar{R}}\right)^2 \rho_{E_1 E_1} V_{E_1} V_{E_1} \bar{E}_1 \bar{E}_1$$

$$+ \left(\frac{\partial u_j}{\partial E_2}\bigg|_{R=\bar{R}}\right)^2 \rho_{E_2 E_2} V_{E_2} V_{E_2} \bar{E}_2 \bar{E}_2$$

$$+ \left(\frac{\partial u_j}{\partial E_3}\bigg|_{R=\bar{R}}\right)^2 \rho_{E_3 E_3} V_{E_3} V_{E_3} \bar{E}_3 \bar{E}_3$$

$$+ \left(\frac{\partial u_j}{\partial P}\bigg|_{R=\bar{R}}\right)^2 \rho_{PP} V_P V_P \bar{P}\bar{P} + \left(\frac{\partial u_j}{\partial \theta}\bigg|_{R=\bar{R}}\right)^2 \rho_{\theta\theta} V_\theta V_\theta \overline{\theta\theta}$$

If $\bar{A}_i' = \Lambda A_i$, Λ = scaling factor, then we have

$$\frac{\partial u_j'}{\partial r_r'} = \frac{\partial \{u'\}^T}{\partial r_r} [k'] \{\delta u_j'\} + \{u'\}^T \frac{\partial [k']}{\partial r_r} \{\delta u_j'\} + \{u'\}^T [k'] \frac{\partial \{\delta u_j'\}}{\partial r_r}$$

which is for various random parameters considered for this problem; for $r_r = A_i$, one may write

$$\frac{\partial u_j}{\partial r_r'} = \frac{1}{\Lambda^2} \frac{\partial \{u\}^T}{\partial r_r} \Lambda [k] \frac{1}{\Lambda} \{\delta u_j\} + \frac{1}{\Lambda} \{u\}^T \frac{\partial [k]}{\partial r_r} \frac{1}{\Lambda} \{\delta u_j\}$$

$$+ \frac{1}{\Lambda} \{u\}^T \Lambda [k] \frac{1}{\Lambda^2} \frac{\partial \{\delta u_j\}}{\partial r_r}$$

$$= \frac{1}{\Lambda^2} \left[\frac{\partial \{u\}^T}{\partial r_r} [k] \{\delta u_j\} + \{u\}^T \frac{\partial [k]}{\partial r_r} \{\delta u_j\} + \{u\}^T [k] \frac{\partial \{\delta u_j\}}{\partial r_r} \right]$$

and

$$\left(\frac{\partial u_j}{\partial A_i'}\Big|_{R=R'}\right)^2 \rho_{A_i'A_i'} V_{A_i'} V_{A_i} \overline{A_i'A_i'}$$

$$= \frac{1}{\Lambda^2} \left(\frac{\partial u_j}{\partial A_i}\Big|_{R=\bar{R}}\right)^2 \rho_{A_iA_i} V_{A_i} V_{A_i} \Lambda \bar{A}_i \bar{A}_i$$

Therefore, for all the random parameters, we have the following new displacement variance

$$S_{u'j}^2 = \sum_{r=1}^{8} \left(\frac{\partial u_j}{\partial r_r'}\Big|_{R=R'}\right)^2 \rho_{r_r'r_r'} V_{r_r'} V_{r_r'} \bar{r}'_r \bar{r}'_r$$

$$= \frac{1}{\Lambda^2} \left(\sum_{r=1}^{8} \left(\frac{\partial u_j}{\partial r_r}\Big|_{R=\bar{R}}\right)^2 \rho_{r_rr_r} V_{r_r} V_{r_r} \bar{r}_r \bar{r}_r\right)$$

in which $\rho_{r_rr_r} = 1$. Similarly, the new stress variance can be derived from the following variance equation:

$$S_{\tau_1}^2 = \left(\frac{\partial \tau_1}{\partial A_1}\right)^2 V_{A_1}^2 \bar{A}_1^2 + \left(\frac{\partial \tau_1}{\partial A_2}\right)^2 V_{A_2}^2 \bar{A}_2^2 + \ldots + \left(\frac{\partial \tau_1}{\partial \theta}\right)^2 V_\theta^2 \bar{\theta}^2 \qquad \text{(h)}$$

Using the above equations and the derivatives of uncertainties, one can determine the scaling factors (Eq. (11.228)), Lagrangian multipliers (Eq. (11.234)), and recurrence equations (Eq. (11.232)). The detailed expressions of Lagrangian multiplier and recurrence are given in items (a) and (b) as follows:

(1) *Lagrangian multiplier*

$$\sum_{P=1}^{m} - \left(\sum_{l=1}^{n_1} \frac{\partial g_j}{\partial A_l} \frac{\partial g_P}{\partial A_l} A_l^{(k)} / \rho_l l_l\right) \lambda_P = r\left(\beta_j - \beta_{j0}\right)$$

$$+ \sum_{l=1}^{n_1} \frac{\partial g_j}{\partial A_l} A_l^{(k)} - r \sum_{l=n_1+1}^{n} \frac{\partial g_j}{\partial A_l} \Delta A_l (P) \qquad \text{(i)}$$

If two constraints (λ_3, λ_4) and one element (element 3) are active, then

$$-\left(\sum_{l=1}^{2} \frac{\partial g_3}{\partial A_l} \frac{\partial g_3}{\partial A_l} A_l^{(k)} / \rho l_l\right) \lambda_3 - \left(\sum_{l=1}^{2} \frac{\partial g_3}{\partial A_l} \frac{\partial g_4}{\partial A_l} A_l^{(k)} / \rho l_l\right)$$

$$\lambda_4 = r\left(\beta_3 - \beta_{30}\right) + \sum_{l=1}^{2} \frac{\partial g_3}{\partial A_l} A_l^{(k)} - r \frac{\partial g_3}{\partial A_3} \Delta A_3 (P) \qquad \text{(ia)}$$

$$-\left(\sum_{l=1}^{2}\frac{\partial g_4}{\partial A_l}\frac{\partial g_3}{\partial A_l}A_l^{(k)}\bigg/\rho l_l\right)\lambda_3-\left(\sum_{l=1}^{2}\frac{\partial g_4}{\partial A_l}\frac{\partial g_4}{\partial A_l}A_l^{(k)}\bigg/\rho l_l\right)$$

(ib)

$$\lambda_4=r\left(\beta_4-\beta_{40}\right)+\sum_{l=1}^{2}\frac{\partial g_4}{\partial A_l}A_l^{(k)}-r\frac{\partial g_4}{\partial A_3}\Delta A_3\,(P)$$

Let

$$C_{11}=-\sum_{l=1}^{2}\left(\frac{\partial g_3}{\partial A_l}\right)^2 A_l^{(k)}\bigg/\rho l_l,\quad C_{12}=-\sum_{l=1}^{2}\frac{\partial g_3}{\partial A_l}\frac{\partial g_4}{\partial A_l}A_l^{(k)}\bigg/\rho l_l$$

$$C_{21}=-\sum_{l=1}^{2}\frac{\partial g_4}{\partial A_l}\frac{\partial g_3}{\partial A_l}A_l^{(k)}\bigg/\rho l_l,\quad C_{22}=-\sum_{l=1}^{2}\left(\frac{\partial g_4}{\partial A_l}\right)^2 A_l^{(k)}\bigg/\rho l_l$$

$$R_1=r\left(\beta_3-\beta_{30}\right)+\sum_{l=1}^{2}\frac{\partial g_3}{\partial A_l}A_l^{(k)}-r\frac{\partial g_3}{\partial A_3}\Delta A_3\,(P)$$

$$R_2=r\left(\beta_4-\beta_{40}\right)+\sum_{l=1}^{2}\frac{\partial g_4}{\partial A_l}A_l^{(k)}-r\frac{\partial g_4}{\partial A_3}\Delta A_3\,(P)$$

we can solve Eqs. (ia) and (ib) to find

$$\lambda_3=\frac{R_1 C_{22}-R_2 C_{12}}{C_{11}C_{22}-C_{12}C_{21}}$$

(j)

$$\lambda_4=\frac{C_{11}R_2-C_{12}R_1}{C_{11}C_{22}-C_{12}C_{21}}$$

(k)

(2) *Recurrence equation* $(\lambda_3,\lambda_4$ active) The general expression is

$$A_l^{(k+1)}=A_l^{(k)}\left(1+\frac{1}{r}\left(\frac{\sum\limits_{j=1}^{m}-\lambda_j\frac{\partial g_3}{\partial A_l}}{\rho l_l}-1\right)\right)$$

(l)

from which

$$A_1^{(k+1)}=A_1^{(k)}\left(1+\frac{1}{r}\left(\frac{-\lambda_3\frac{\partial g_3}{\partial A_1}-\lambda_4\frac{\partial g_4}{\partial A_1}}{\rho l_1}-1\right)\right)$$

(m)

$$A_2^{(k+1)}=A_2^{(k)}\left(1+\frac{1}{r}\left(\frac{-\lambda_3\frac{\partial g_3}{\partial A_2}-\lambda_4\frac{\partial g_4}{\partial A_2}}{\rho l_2}-1\right)\right)$$

(n)

$$A_3^{(k+1)}=A_3^{(k)}\left(1+\frac{1}{r}\left(\frac{-\lambda_3\frac{\partial g_3}{\partial A_3}-\lambda_4\frac{\partial g_4}{\partial A_3}}{\rho l_3}-1\right)\right)$$

(o)

The solution with detailed calculations is tabulated in Table 11.2.

Table 11.2 Detailed calculations for generalized optimality-criteria method.

Details		Cycles				Note
		1	...	13	14	
1	A_1	1.0		1.005	1.009	I
2	A_2	1.0		0.619	0.620	
3	A_3	1.0		0.134	0.1	
4	u_1	0.04206		0.0492	0.05	
5	u_2	0.0350		0.042	0.041	
6	τ_1	11.5595		13.677	13.70	
7	τ_2	10.5023		12.596	12.41	
8	τ_3	−2.2752		−2.549	−2.73	
9	S_{u1}^2	0.258×10^{-4}		0.158×10^{-3}	0.000	II
10	S_{u2}^2	0.68×10^{-5}		0.372×10^{-4}	0.000	
11	$S_{\tau1}^2$	0.7054		1.72	1.86	III
12	$S_{\tau2}^2$	0.6051		3.33	3.76	
13	$S_{\tau3}^2$	0.6263		5.53	6.39	
14	β_{u1}	16.33		6.03	5.581	IV
15	β_{u2}	34.51		13.61	12.89	
16	$\beta_{\tau1}$	4.91		3.17	3.10	V
17	$\beta_{\tau2}$	5.62		3.14	3.09	
18	$\beta_{\tau3}$	10.45		6.26	5.87	VI
19	Λ_{u1}	0.211		0.0829	0.083	
20	Λ_{u2}	0.216		0.185	0.186	
21	$\Lambda_{\tau1}$	0.416		0.455	0.456	VII
22	$\Lambda_{\tau2}$	0.377		0.455	0.455	
23	$\Lambda_{\tau3}$	−0.0085		−0.24	−0.241	
24	Λ_{max}	0.8053		0.994	0.9989	VIII
25	$A_1^* = \Lambda_{max}A_1$	0.8053		0.998	1.006	
26	$A_2^* = \Lambda_{max}A_2$	0.8053		0.615	0.620	
27	$A_3^* = \Lambda_{max}A_3$	0.8053		0.133	0.1	
28	u_1^*	0.0522		0.0495	0.05	IX
29	u_2^*	0.0435		0.042	0.041	
30	τ_1^*	14.355		13.76	13.72	
31	τ_2^*	13.042		12.675	12.42	
32	τ_3^*	−2.825		−2.565	−2.73	
33	$S_{u_1^*}^2$	0.398×10^{-4}		0.16×10^{-3}	0.18×10^{-3}	X
34	$S_{u_2^*}^2$	0.105×10^{-4}		0.38×10^{-4}	0.42×10^{-4}	
35	$S_{\tau_1^*}^2$	1.088		1.742	1.866	
36	$S_{\tau_2^*}^2$	0.9332		3.369	3.768	
37	$S_{\tau_3^*}^2$	0.9658		5.601	6.403	
38	$\beta_{u_1^*}$	11.5379		5.97	5.57	XI
39	$\beta_{u_2^*}$	25.1754		13.48	12.88	
40	$\beta_{\tau_1^*}$	3.09		3.12	3.09	
41	$\beta_{\tau_2^*}$	3.9		3.09	3.09	
42	$\beta_{\tau_3^*}$	9.577		6.22	5.87	
43	$\partial g_3^* / \partial A_1$	−7.281		−7.373	−7.401	
44	$\partial g_3^* / \partial A_2$	−2.775		−0.858	−0.639	

#		Col1	Col2	Col3	
45	$\partial g_3^* / \partial A_3$	$-0.9469\ldots$	-2.660	-3.052	
46	$\left(\partial g_3^* / \partial A_1\right)^2 A_1^* / \rho l_1$	3.0183	3.838	3.90	
47	$\left(\partial g_3^* / \partial A_2\right)^2 A_2^* / \rho l_2$	0.6200	0.045	0.025	
48	$\left(\partial g_3^* / \partial A_3\right)^2 A_3^* / \rho l_3$	0.0433	0.0564	0	
49	$C_{11} = -\sum_{i=1}^{n_1} \left(\partial g_3^* / \partial A_i\right)^2 A_i^* / \rho l_i$	-3.6816	-3.940	-3.9253	XII
50	r	2	2	2	
51	$r(\beta_3 - \beta_{30})$	0	0.06	0	
52	$\left(\partial g_3^* / \partial A_1\right) A_1^*$	-5.8633	-7.358	-7.452	
53	$\left(\partial g_3^* / \partial A_2\right) A_2^*$	-2.2347	-0.528	-0.396	
54	$\left(\partial g_3^* / \partial A_3\right) A_3^*$	-0.7624	-0.354	0.0	
55	$\sum_{i=1}^{n_1} \left(\partial g_3^* / \partial A_i\right) A_i^*$	-8.8604	-8.24	-7.85	
56	$TA1$	-8.8604	-8.18	-7.85	
57	$\Delta A_3' = A_{min} - A_3$	0	0	0	
58	$\left(\partial g_3^* / \partial A_3\right) \Delta A_3'$	0	0	0	
59	$R_3' = r \sum_{i=n_1}^{n} \left(\partial g_3^* / \partial A_i\right) \Delta A_1'$	0	0	0	
60	$R_3 - R_3'$	-8.8604	-8.18	-7.85	
61	$\partial g_4^* / \partial A_1$	14.228	-1.130	-0.848	
62	$\partial g_4^* / \partial A_2$	-6.650	-9.530	-9.59	
63	$\partial g_4^* / \partial A_3$	0.628	-1.555	-1.88	
64	$\left(\partial g_4^* / \partial A_1\right)^2 A_1^* / \rho l_1$	1.219	0.090	0.05	
65	$\left(\partial g_4^* / \partial A_2\right)^2 A_2^* / \rho l_2$	3.8883	5.583	5.709	
66	$\left(\partial g_4^* / \partial A_3\right)^2 A_3^* / \rho l_3$	0.0173	0.019	0	
67	$C_{22} = -\sum_{i=1}^{n_1} \left(\partial g_4^* / \partial A_i\right)^2 A_i^* / \rho l_i$	-5.152	-5.695	-5.763	XIII
68	$r(\beta_4 - \beta_{40})$	1.095	0	0	
69	$\left(\partial g_4^* / \partial A_1\right) A_1^*$	-3.726	-1.128	-0.855	
70	$\left(\partial g_4^* / \partial A_2\right) A_2^*$	-5.596	-5.858	-5.94	
71	$\left(\partial g_4^* / \partial A_3\right) A_3^*$	0.483	-0.207	-0.187	
72	$\sum_{i=1}^{n_1} \left(\partial g_4^* / \partial A_i\right) A_i^*$	-8.839	-7.193	-6.802	
73	$TA2$	-7.743	-7.196	-6.802	XIV
74	$\Delta A_3' = A_{min} - A_3$	0	0	0	
75	$\left(\partial g_4^* / \partial A_3\right) \Delta A_3'$	0	0	0	
76	$R_4' = r \sum_{i=n+1}^{n} \left(\partial g_4 / \partial A_i\right) \Delta A_i$	0	0	0	
77	$R_2 - R_2'$	-7.743	-7.193	-6.802	XV
78	$\left(\partial g_3^* / \partial A_1\right)\left(\partial g_4^* / \partial A_1\right) A_1^* / \rho l_1$	1.956	0.588	0.447	
79	$\left(\partial g_3^* / \partial A_2\right)\left(\partial g_4^* / \partial A_2\right) A_2^* / \rho l_2$	1.410	0.503	0.38	
80	$\left(\partial g_3^* / \partial A_3\right)\left(\partial g_4^* / \partial A_3\right) A_3^* / \rho l_3$	-0.026	0.033	0	
81	$C_{34} = -\sum_{i=1}^{n_1} \left(\partial g_3^* / \partial A_i\right)\left(\partial g_4^* / \partial A_i\right) A_i^* / \rho l_i$	-3.341	-1.124	-0.83	
82	$C_{43} = C_{34}$	-3.341	-1.124	-0.83	
83	If λ_3 active, $\lambda_3 = (60)/(49)$	2.4065	0	0	
84	If λ_4 active, $\lambda_4 = (73)/(67)$	-1.503	0	0	
85	If λ_3 and λ_4 active	-45.646	46.569	45.23	
86	$(81)(77)$	25.869	8.085	5.65	

Table 11.2 (Continued)

Details		Cycles				Note
		1	...	13	14	
87	$(85)-(86)$	-71.515		38.484	39.58	
88	$(49)(67)-(81)^2$	-30.131		21.107	21.94	
89	$\lambda_3 = (87)/(88)$	2.373		1.819	1.805	XVI
90	$(49)(77)$	28.506		28.34	26.98	
91	$(82)(60)$	29.601		9.19	6.76	
92	$(90)-(91)$	-1.905		19.15	20.21	
93	$\lambda_4 = (92)/(88)$	0.036		0.905	0.922	XVI
94	$\lambda_3 \left(\partial g_3 / \partial A_1 \right)$	-17.5217		-13.419	-13.32	
95	$\lambda_4 \left(\partial g_4 / \partial A_1 \right)$	0		-1.023	-0.782	
96	$((94)+(95))/\rho l_1$	1.239		1.021	1.0	
97	$1 + \frac{1}{r}((96)-1)$	1.1195		1.011	1.0	
98	$A_1^{(k+1)} = A_1^*(97)$	0.9015		1.010	1.006	XVII
99	$\lambda_3 \left(\partial g_3 / \partial A_2 \right)$	-6.677		-1.5616	-1.15	
100	$\lambda_4 \left(\partial g_4 / \partial A_2 \right)$	0		-8.625	-8.83	
101	$-((99)+(100))/\rho l_2$	0.6677		1.019	1.0	
102	$1 + 1/r((101)-1)$	0.8339		1.009	1.0	
103	$A_2^{(k+1)} = A_2^* (102)$	0.6715		0.621	0.62	XVIII
104	$\lambda_3 \left(\partial g_3 / \partial A_3 \right)$	-2.2728		-4.839	-5.49	
105	$\lambda_4 \left(\partial g_4 / \partial A_3 \right)$	0		-1.407	-1.73	
106	$-((104)+(105))/\rho l_3$	0.1367		0.375	0.43	
107	$1 + 1/r((106)-1)$	0.5684		0.687	0.72	
108	$A_3^{(k+1)} = A_3^*(107)$	0.4577		0.1	0.1	XIX
109	Weight $= \sum \rho_i l_i A_i$	27.905		22.14	22.11	

Note: Some notations in Table 11.2 are $TA1 = R_3 = \sum\limits_{i=1}^{n_1} \left(\partial g_3^* / \partial A_i \right)$ $A_i^* + r(\beta_3 - \beta_{30})$, $TA2 = R_2 = \sum\limits_{i=1}^{n_1} \left(\partial g_4^* / \partial A_i \right) A_i^* + r(\beta_4 - \beta_{40})$. Notes I through XIX in Table 11.2 are for the detailed calculations of the items signified by the individual note: three major cycles of the 1st, 13th and 14th are illustrated because they reveal the details at the initial numerical procedure. These notes are explained in Appendix K.

(B) *Interior-Penalty-Function Method* – The interior penalty function is given in Eq. (p):

$$P(A) = W(A) - r \sum \frac{1}{g_i(A)} \tag{p}$$

The detailed calculations are shown in Table 11.3 with intervals of five cycles.

The comparison of the results is shown in Fig 11.5 from which one may observe that both methods yield the same results; however, the generalized

Table 11.3 Detailed calculations for penalty-function method.

Cycle No	r	A_1	A_2	A_3	u_1	βu_1	u_2	βu_2	τ_1	$\beta \tau_1$	τ_2	$\beta \tau_2$	τ_3	$\beta \tau_3$	$P(A)$	$W(A)$
0	1.0	1.0	1.0	1.0	0.0421	16.329	0.035	34.50	11.559	4.910	10.505	5.619	−2.275	10.451	160.75	40.70
5	2.40	1.223	0.867	0.282	0.0409	9.929	0.0318	23.931	10.911	5.073	9.560	5.498	−2.450	8.192	61.368	30.596
10	0.404	1.087	0.722	0.169	0.0466	7.166	0.0368	17.259	12.514	3.917	11.042	4.181	−2.745	6.868	30.460	25.356
15	0.0678	1.038	0.662	0.127	0.0488	6.185	0.0393	14.651	13.22	3.430	11.803	3.556	−2.785	6.278	24.795	23.374
20	0.0114	1.019	0.638	0.111	0.0496	5.812	0.0405	13.611	13.519	3.229	12.158	3.284	−2.774	6.034	23.053	22.590
25	0.0021	1.012	0.627	0.105	0.0499	5.665	0.0410	13.191	13.642	3.148	12.311	3.170	−2.764	5.935	22.442	22.274
30	0.00032	1.008	0.623	0.102	0.050	5.604	0.041	13.018	13.693	3.113	12.375	3.123	−2.760	5.893	22.237	22.144

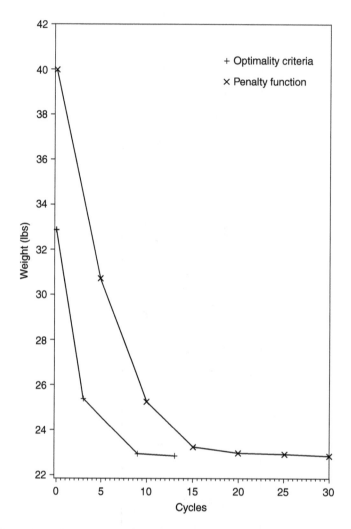

Figure 11.5 Comparison of optimum weight based on optimality-criteria method and penalty-function method (1 lb = 4.45 N).

optimality-criteria method needs much more formulation than the interior-penalty-function method which seems to be more favourable for nondeterministic structural design.

11.14.2   *Optimal parameter studies of ANSI, NBS, UK, and UNREDUCED models*

Example 11.14.2 – The two-storey steel frame of Fig. 11.6 is used to demonstrate the design with consideration of four live load models for optimal weight, moment of inertia, and probability of failure. Let the notations

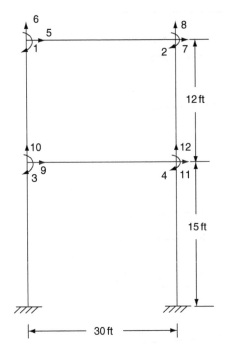

Figure 11.6 Two-storey building structure (1 ft = 30.48 cm).

be: *L1*, *L2*, *L3* and *L4* signifying the live load models of ANSI, NBS, UK and UNREDUCED, respectively; 1st and 2nd representing the first and second variance approach, respectively; and *N* and *LN* corresponding to normal and lognormal, respectively.

The parameters used in the example are $A_1 = 900\,\text{ft}^2\,(264.32\,\text{m}^2)$, $\overline{D} = 80\,\text{psf}\,(3.82\,\text{kPa})$, $V_D = 0.12$, allowable individual failure probabilities = 0.0001, allowable joint rotations = 0.05 rad, allowable joint displacements = 0.5 in. (1.27 cm), allowable variances of joint rotations and displacements are assumed to be zero, $\overline{F}_y = 36\,\text{ksi}\,(2.448 \times 10^5\,\text{kPa})$, $\overline{E}_m = 30{,}000\,\text{ksi}$ $(2.067 \times 10^8\,\text{kPa})$, $V_{My} = 0.12$, $V_{PE} = 0.3$, $V_{Pcr} = 0.31$, $V_{Py} = 0.14$, $V_{Mcr} = 0.20$, $V_{Pcry} = 0.3$, $\rho_{P_yM_y} = 0.8$, $\rho_{P_{cr}M_y} = 0.8$, $\rho_{P_{cr}P_E} = 0.8\,(=1\text{ if }P_{cr} = P_E)$, $\rho_{P_EM_y} = 0$, $\rho_{P_{cr}M_y} = 0.8$, $\rho_{P_yM_{cr}} = 0.15$, $\rho_{P_EM_{cr}} = 0.15$. The structural members are assumed to be rectangular sections and have the relationship area $= 2\sqrt{2}\,(\text{moment of inertia})^{1/2}$. The minimum moment of inertia of all the members is $50\,\text{in.}^4\,(2{,}081.16\,\text{cm}^4)$.

Solution:

Using the generalized optimality-criteria method as presented in detail earlier, we obtain the optimum solution given in Fig. 11.7 and corresponding normal distribution which reveals the magnitude of optimum weight and

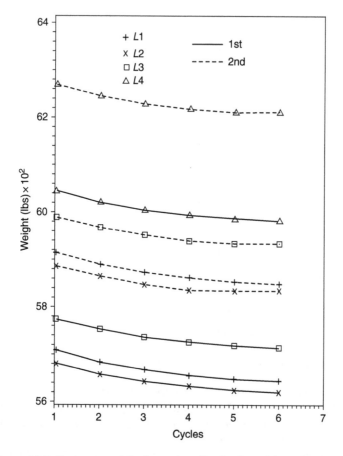

Figure 11.7 Optimum weight for various live load models with N and $D + L$ cases (1 lb = 4.45 N).

shows that the British model ($L3$) demands a heavier structural design than the U.S. models ($L1$ and $L2$); ANSI ($L1$) demands a heavier design than NBS ($L2$); UNREDUCED model ($L4$) demands the heaviest design simply because it does not reduce the live load in terms of influence area A_I. The 2nd variance approach yields more weight than the first. The moment of inertia of a typical beam (I_1) and a column (I_4) are sketched in Figs. 11.8 and 11.9, respectively. The effects of the four live load models and the two variance formulations on the design are similar to the optimum weight in Fig. 11.7. The probability of failure of the constituent members is given in Fig. 11.10 in which b_1 and b_2 correspond to the failures due to the yielding moment given in Eq. (11.199) of beams 1 and 2, respectively; and $(f_1)_4$ and $(f_2)_4$ represent the column failure (element 4) due to yielding at the member end (Eq. (11.203)) and instability in the plane of bending (Eq. (11.208)), respectively. The supporting column, member 3, reaches the minimum

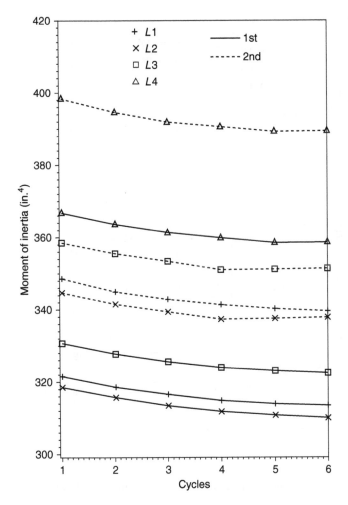

Figure 11.8 I_1 for various live load models with N and $D + L$ cases (1 in.^{4} = 41.62 cm^4).

moment of inertia before any failure modes. The failure modes at the optimum solution are close to the allowable 10×10^{-5}. The design was determined at the 6th cycle, because the next cycle could not further improve the optimum solution of the structural weight or the moment of inertia.

The optimum design parameters of weight, moment of inertia and failure of modes are also studied for lognormal distribution and are sketched in Figs. 11.11 and 11.12. The effects of the four live load models and that of the two variance approaches on the optimum design results are similar to those observed on the bases of normal distribution.

Comparison of the optimum solutions resulting from normal and lognormal distributions reveals that the normal distribution requires a heavier

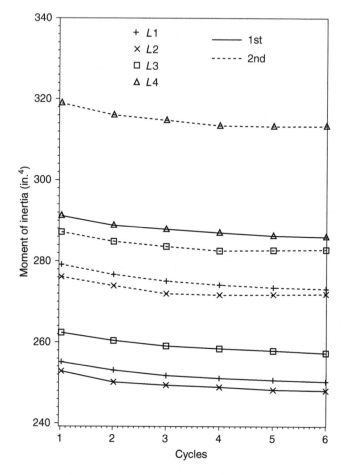

Figure 11.9 I_4 for various live load models with N and $D + L$ cases (1 in.^{4} = 41.62 cm^4).

structural design than the lognormal distribution for the first approach, but a lighter structural design for the second approach. The comparison may be observed from Figs. 11.7 and 11.11. Similar observations may be concluded for the moments of inertia. The optimum solution may be summarized as follows:

1. For the $D + L$ case, the magnitude order of optimum weight from large to small is UNREDUCED, UK, ANSI, NBS for both normal and lognormal distributions.
2. For the $D + L$ case, the lognormal distribution requires a heavier structural design than normal distribution for the 2nd variance approach, but a lighter structural design for the 1st variance approach.

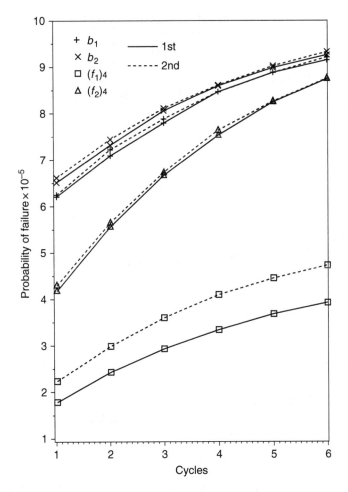

Figure 11.10 Probability of failure for ANSI live load models with N and $D + L$ cases.

3. For the $D + L$ case, the 2nd variance approach yields a heavier structural design than the 1st variance approach for both normal and lognormal distributions.
4. For the $D + L$ case, the structural failures are mainly due to the failures of beams and the top column.

11.14.3   *Optimization with various cost parameters*

Example 11.14.3 – The two-storey building shown in Fig. 11.6 subjected to the NBS live load model (*L2*) is designed with consideration of the cost function of $C_T = C_I + L_f P_{fT}$ as expressed in Eq. (11.195). The initial

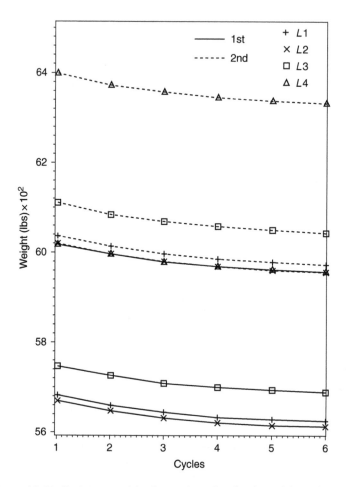

Figure 11.11 Optimum weight for various live load models with LN and $D+L$ cases (1 lb = 4.45 N).

cost is assumed to be $C_I = 1.55$ (dollars/volume) times the structural volume and the expected loss is $L_f = 1.15 \, (2.5 C_I + 3, 194, 100)$ dollars, where 3,194,100 is the combination of the liability due to death ($700,000 for 1 person of annual income $20,000), serious injuries ($1,400,000 for 4 persons), minor injuries ($75,000 for 15 persons), and the business inter-ruption ($1,019,100). Note that the expected loss is dependent on the structure's function and its type; for instance, a five-storey building of the same type of construction may also have the initial construction cost $C_I = 1.55$(dollars/volume) times the structural volume, but the expected loss could be $L_f = 1.15 \, (2.5 C_I + 14, 362, 500)$ dollars, where 14,362,500 is the combination of the liability due to death ($3,500,000 for 5 persons of annual income $20,000), serious injuries ($3,500,000 for 10 persons),

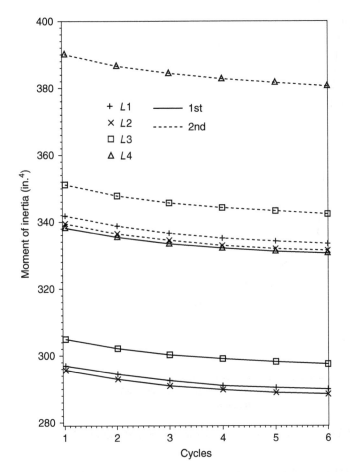

Figure 11.12 I_1 for various live load models with LN and $D + L$ cases (1 in.^{4} = 41.62 cm^4).

minor injuries ($375,000 for 75 persons), and the business interruption ($6,987,500). The calculations of liability cost due to death, serious injuries, and minor injuries for each person are described in Section 11.12.1. The business interruption has been calculated in the remaining amount of 3,194,100 or 14,362,500 excluding the liability cost. The optimum solution is shown in Fig. 11.13 where cost is 1.56×10^4 dollars at the allowable profitability failure of $P_{f0} = 10^7$. As pointed out in Section 11.12.1, the cost function is undoubtedly dependent on when and where the accident occurs; the presentation herewith is only to reveal what possible expected loss could be included. Since it is difficult to reasonably estimate the cost parameters, the next example emphasizes the sensitivity study of the cost function.

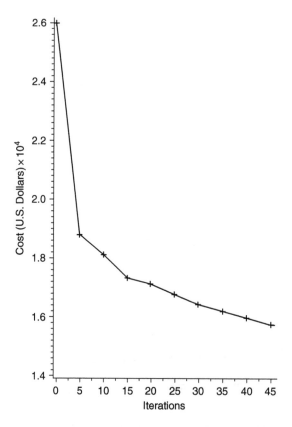

Figure 11.13 Optimum cost of 2-storey structure with N and $D+L$ cases.

Example 11.14.4 – This example is to address the sensitivity of the cost function used in Example 11.14.3. The function has three components: initial construction cost (C_I), future failure cost (L_f), and system probability of failure (P_{fT}); and can be expressed in more detail as:

$$C_T = C_I + L_f P_{ft} \tag{a}$$

$$C_I = C_u \sum_i L_i A_i + C_n \text{ and } L_f = C_V C_I + C_L \tag{b, c}$$

where C_u = unit steel volume cost, C_n = non-structural members' cost, C_V = a coefficient to describe the ratio of repair cost to initial cost, and C_L = the business and human losses. Although the initial construction cost and future failure cost can be classified into many items, these quantities are difficult to estimate. Therefore, two coefficients of the ratio of initial cost to members' cost (C_{in}) and the ratio of future failure cost to initial cost (C_{VL}) are used to represent the various magnitudes of initial construction cost and future failure cost which may now be expressed as:

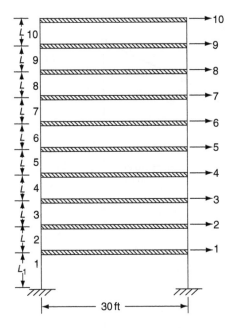

Figure 11.14 Ten-storey shear building structure ($L_1 = 15$ ft, $L = 12$ ft, 1 ft = 30.48 cm).

$$C_I = C_{in} C_u \sum_i L_i A_i \qquad \text{(d)}$$

$$L_f = C_{VL} C_I \qquad \text{(e)}$$

Through these two coefficients, the influences of non-structural cost and future failure on the optimum cost design may be observed as sensitivity studies. The ten-storey shear building shown in Fig. 11.14 is used for this study including two probability models of normal (N) and lognormal (LN) distributions. The failure modes are displacements' failures of the structure, and yielding and torsional buckling failures of columns. Two cases are illustrated as follows:

Case (A) – As indicated in Eq. (b), the initial cost has two components: structural member cost and non-structural member cost. Since the terms involved in the non-structural member cost are not easy to define, it is reasonable to assume that the structural member cost may be a percentage of initial cost. That means that the initial cost is assumed as the product of C_{in} and member cost. Three values of C_{in}, 2, 5 and 20, are used to investigate the relationship between the structural and non-structural costs for which the value of C_{VL} is assumed to be constant as 1.0. The designs are based on the 2nd variance approach for both N and LN distribution models.

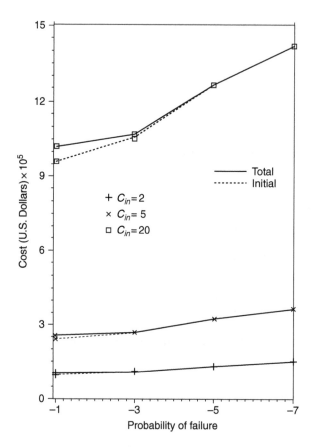

Figure 11.15 Optimum cost for various C_{in} and N of ten-storey building.

Fig. 11.15 shows optimum cost versus probability of failure (N) for various C_{in} from which the design results of the moments of inertia of columns 1 through 5 are shown in Fig. 11.16. Similarly, the optimum solutions based on lognormal (LN) are given in Figs. 11.17 and 11.18. From the figures, one may observe that there are some differences between total cost and initial cost at low reliability $(P_{f0} < 10^{-3})$; the differences are practically zero as the reliability levels increase. The total costs increase as C_{in} increases. But the moments of inertia are not much different or different C_{in} values. This is because the change in the non-structural member cost does not affect the change in the structural responses under the same safety criterion. Therefore, the increase of the non-structural member cost does not influence the design of the structural members. The difference between the total costs and initial costs decreases for higher reliability levels because the increase of reliability results from the decrease of future failure losses.

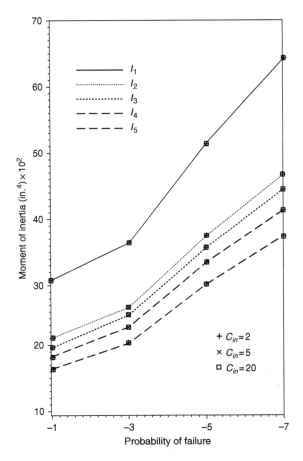

Figure 11.16 $I_1 - I_5$ for various C_{in} with N of ten-storey building (1 in.^{4} = 41.62 cm^4).

Case (B) – The expected failure cost has two components which are structural losses (repair costs) and non-structural losses (business or human losses). However, this cost is also not clearly defined. So herein the expected failure costs are assumed to have a relationship with the initial cost. Three values of C_{VL}, 0.5, 1.0 and 10, are used to investigate the influence of different expected failure costs on the optimum solutions. The value of C_{in} is assumed to be constant as 5.

Figs. 11.19 and 11.20 reveal the design results of the building. Again, all the designs are based on the 2nd variance approach with normal and lognormal distributions. The unit cost is 0.15 dollars in.$^{-4}$. From all the figures, one may observe that there are some differences between total costs at low reliability $(P_{f0} < 10^{-5})$; the differences practically become zero as the reliability level increases. The total cost increases as C_{VL}

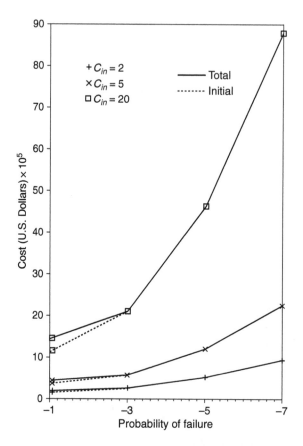

Figure 11.17 Optimum cost for various C_{in} with *LN* of 10-storey building.

increases at low reliability $(P_{f0} < 10^{-5})$. But the differences are not much different for different C_{VL} values as reliability increases. This is because the values of future failure losses are small at high reliability; therefore, the increase of future failure cost does not change the design of structural members significantly. The difference between the total costs and initial costs decreases for higher reliability levels because the increase of reliability results from the decrease of future failure losses.

11.15   Numerical illustrations and design parameters assessment for seismic excitations

11.15.1   *Optimum design of truss based on Newmark's inelastic design spectrum with a ductility of 3*

Example 11.15.1 – The truss shown in Fig. 11.21 is designed using Newmark's inelastic design spectrum with a ductility of 3 as shown in Fig. E.1 of

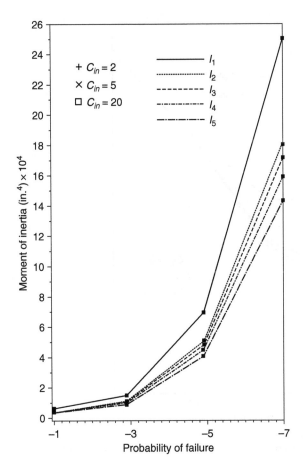

Figure 11.18 $I_1 - I_5$ for various C_{in} with *LN* of 10-storey building (1 in.^{4} = 41.62 cm^4).

Appendix E. The structural member properties are $\gamma = 0.1$ lb in.^{3}, $E_1 = E_2 = E_3 = E = 1,000$ ksi, $L = 100$ in., and $A_1 = A_3$.

Use the interior-penalty-function method to find the minimum weight based on the design requirement given as follows:

Safety index of displacement: $\beta_D = 3.09$;
Safety index of stress: $\beta_S = 3.09$;
Allowable displacement: $\bar{u}_R = 0.02$ in.;
Allowable stress: $\overline{T}_R = 1$ ksi;
Allowable upper frequency: $\omega_{upper} = 100$ rad s^{-1}; and
Allowable lower frequency: $\omega_{lower} = 10$ rad s^{-1}.

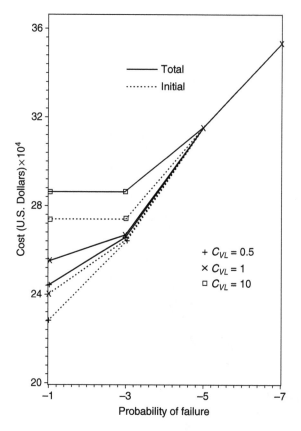

Figure 11.19 Optimum cost for various C_{VL} with N of ten-storey building.

Solution:

Objective function

$$W(A) = \gamma (L_1 A_1 + L_2 A_2 + L_3 A_3) = \gamma \left(2\sqrt{2} L A_1 + L A_2 \right)$$
$$= 10 \left(2\sqrt{2} A_1 + A_2 \right) \tag{a}$$

Constraints

$$\left.\begin{array}{ll}
\text{Displacement}: & \overline{U}_{1S} + \beta \sqrt{S_{u_1}^2} - \overline{U}_{1R} \leq 0 \\[4pt]
\text{Stress}: & \overline{T}_{1S} + \beta \sqrt{S_{T_1}^2} - \overline{T}_{1R} \leq 0 \\[4pt]
\text{Frequency}: & \omega_{\text{lower}} \leq \omega \leq \omega_{\text{upper}} \\[4pt]
\text{Member size}: & A_1 \geq 0, \ A_2 \geq 0
\end{array}\right\} \tag{b}$$

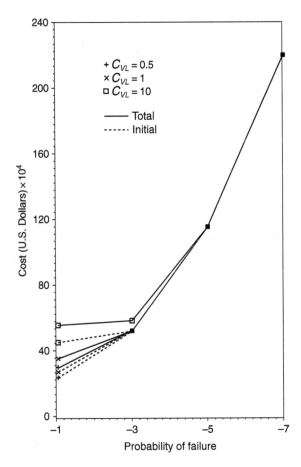

Figure 11.20 Optimum cost for various C_{VL} with LN of ten-storey building.

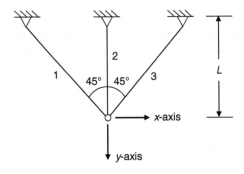

Figure 11.21 Symmetric truss of Example 11.15.1.

Penalty function

$$F(A) = W(A) - \gamma \left(\frac{1}{G_1} + \frac{1}{G_2} + \frac{1}{G_3} - \frac{1}{G_4} - \frac{1}{G_5} - \frac{1}{G_6} \right) \qquad (c)$$

in which

$$
\left.
\begin{aligned}
G_1 &= \overline{U}_{1S} + \beta_D \sqrt{S_{u_1}^2} - \overline{U}_{1R} \\
G_2 &= \overline{T}_{1S} + \beta_S \sqrt{S_{T_1}^2} - \overline{T}_{1R} \\
G_3 &= \omega - \omega_{\text{upper}} \\
G_4 &= \omega - \omega_{\text{lower}} \\
G_5 &= A_1 - 0 \\
G_6 &= A_2 - 0
\end{aligned}
\right\} \qquad (d)
$$

In order to use Eq. (d), one needs to find the mean and variance which are calculated in the following four major parts: A, B, C and D. Part A includes three detailed steps of A_1, A_2 and A_3; similarly, C has four steps of C_1 through C_4.

Part (A) Mean Displacement $\overline{U}$
(A_1) Find expressions for the mean, $\overline{U}$, frequencies ω, and modes $\{\Phi\}$

$$\overline{U}_i = \left[\sum_{j=1}^{2} \left(\bar{u}_{ij}^* \right)^2 \right]^2 \qquad (e)$$

where i = the ith component of displacement $\overline{U}$; j = the number of modes, and

$$\bar{u}_{ij}^* = \{\Phi\}^{ij} (Y_i)_{\max} \qquad (f)$$

in which $\{\Phi\}^{ij}$ = the jth mode of the ith component of the displacement; $(Y_i)_{\max}$ = the maximum value of the normal coordinate of $u_i = \Gamma_j y_{w_j}$; Γ_j = the participation factor = $\{\Phi\}^{ij} \{P_0\} / \{\Phi\}^{ij} [m] \{\Phi\}^j$; and y_{w_j} = the maximum displacement which can be obtained from the response spectrum. The natural frequency ω and orthonormal mode shape $\{\Phi\}$ are determined from the following procedures which were similarly presented in previous chapters as:

$$[K - \omega^2 m] \{u\} = \{0\} \qquad (g)$$

$$
[K] = \begin{bmatrix} \dfrac{EA_1}{\sqrt{2}L} & 0 \\[2ex] 0 & \dfrac{E}{\sqrt{2}L} \left(A_1 + \sqrt{2}A_2 \right) \end{bmatrix}
$$

$$
= \begin{bmatrix} 70.7107A_1 & 0 \\[1ex] 0 & 70.7107 \left(A_1 + \sqrt{2}A_2 \right) \end{bmatrix} \qquad (h)
$$

$$[m] = \begin{bmatrix} \dfrac{\gamma}{2g}(L_1A_1 + L_2A_2 + L_3A_3) & 0 \\ 0 & \dfrac{\gamma}{2g}(L_1A_1 + L_2A_2 + L_3A_3) \end{bmatrix} \tag{i}$$

$$= \begin{bmatrix} 0.0366A_1 + 0.1294A_2 & 0 \\ 0 & 0.0366A_1 + 0.01294A_2 \end{bmatrix}$$

$$|K - \omega^2 m| = 0 \tag{j}$$

from which the frequencies and modes are

$$\omega_1^2 = \frac{70.7107A_1}{0.0366A_1 + 0.01294A_2} \tag{k}$$

$$\omega_2^2 = \frac{70.7107\left(A_1 + \sqrt{2}A_2\right)}{0.0366A_1 + 0.01294A_2} \tag{l}$$

$$\{\hat{\Phi}\}^1 = \begin{Bmatrix} \hat{\Phi}_{11} \\ \hat{\Phi}_{21} \end{Bmatrix} \begin{Bmatrix} 1 \\ 0 \end{Bmatrix}; \qquad \{\hat{\Phi}\}^2 = \begin{Bmatrix} \hat{\Phi}_{12} \\ \hat{\Phi}_{22} \end{Bmatrix} \begin{Bmatrix} 0 \\ 1 \end{Bmatrix} \tag{m, n}$$

The orthogonality yields

$$\{\hat{\Phi}\}^{jT} [m] \{\hat{\Phi}\}^j = \begin{bmatrix} \hat{\Phi}_{1j} \\ \hat{\Phi}_{2j} \end{bmatrix}^T$$

$$\begin{bmatrix} 0.0366A_1 + 0.01294A_2 & 0 \\ 0 & 0.0366A_1 + 0.01294A_2 \end{bmatrix}$$

$$\begin{bmatrix} \hat{\Phi}_{1j} \\ \hat{\Phi}_{2j} \end{bmatrix}$$

$$= (0.0366A_1 + 0.01294A_2)\hat{\Phi}_{1j}^2$$

$$+ (0.0366A_1 + 0.01294A_2)\hat{\Phi}_{2j}^2 = M_j \tag{o}$$

If $[\Phi]$ is an orthonormal set, $M_j = 1$, thus

$$\{\Phi\}^j = \frac{1}{(M_j)^{1/2}} \{\Phi\}^j$$

$$\{\Phi\}^1 = \frac{1}{\left[(0.0366A_1 + 0.01294A_2)\,\hat{\Phi}_{11}^2 + (0.0366A_1 + 0.01294A_2)\,\hat{\Phi}_{21}^2\right]^{1/2}}$$

$$\begin{Bmatrix} \hat{\Phi}_{11} \\ \hat{\Phi}_{21} \end{Bmatrix}$$

$$= \begin{Bmatrix} (0.0366A_1 + 0.01294A_2)^{-1/2} \\ 0 \end{Bmatrix} \tag{p}$$

$$\{\Phi\}^2 = \begin{Bmatrix} 0 \\ (0.0366A_1 + 0.01294A_2)^{-1/2} \end{Bmatrix} \tag{q}$$

$$[\Phi] = \begin{bmatrix} (0.0366A_1 + 0.01294A_2)^{-1/2} & 0 \\ 0 & (0.0366A_1 + 0.01294A_2)^{-1/2} \end{bmatrix} \tag{r}$$

(A_2) Find the normal coordinates Y_j based on the response spectrum

$$\{\Phi\}^{j^T} [m] \{\Phi\}^j \{\ddot{Y}\} + \{\Phi\}^{j^T} [c] \{\Phi\}^j \{\dot{Y}\} + \{\Phi\}^{j^T} [K] \{\Phi\}^j = \{\Phi\}^j \{P_0\} \hat{p}(t) \tag{s}$$

$$\{\ddot{Y}\} + 2\zeta_i \omega_i \{\dot{Y}\} + \omega_i^2 \{Y\} = \{\Phi\}^j \{P_0\} \hat{p}(t) \tag{t}$$

Thus,

$$\left(Y_j\right)_{\max} = \Gamma_j y_{\omega_j} \tag{u}$$

where $\Gamma_j = \{\Phi\}^{j^T} \{P_0\}$. Assume $\{P_0\} = \begin{Bmatrix} 1 \\ 0 \end{Bmatrix}$, then

$$\Gamma_1 = \{\Phi\}^{1^T} \{P_0\} = \left\{ (0.0366A_1 + 0.01294A_2)^{-1/2} \quad 0 \right\} \begin{Bmatrix} 1 \\ 0 \end{Bmatrix}$$

$$= (0.0366A_1 + 0.01294A_2)^{-1/2} \tag{v}$$

$$\Gamma_2 = \{\Phi\}^{2^T} \{P_0\} = \left\{ 0 \quad (0.0366A_1 + 0.01294A_2)^{-1/2} \right\} \begin{Bmatrix} 0 \\ 1 \end{Bmatrix}$$

$$= 0 \tag{w}$$

where y_{ω_j} = the maximum displacement of which the values are obtained from the equation derived from Newmark's inelastic design spectrum with ductility = 3 in Fig. E.1 in Appendix E as

$$T = \frac{2\pi}{\omega} \leq 0.03$$

$$\omega^2 y_\omega = 0.16; \quad \omega^2 \frac{dy_\omega}{dT} = 0 \tag{x, y}$$

$$T \le 0.1$$

$$\omega^2 y_\omega = 0.142857 + 0.5714286T; \quad \omega^2 \frac{dy_\omega}{dT} = 0.5714286 \qquad \text{(z,a1)}$$

$$T \le 0.5$$

$$\omega^2 y_\omega = 0.2; \quad \omega^2 \frac{dy_\omega}{dT} = 0 \qquad \text{(a2)}$$

$$T \le 0.4$$

$$\omega^2 y_\omega = 0.47352195 - 0.78964406T + 0.53285116T^2$$
$$- 0.15230572T^3 + 0.015344568T^4 \qquad \text{(a3)}$$

$$\omega^2 \frac{dy_\omega}{dT} = -0.78964406 + 1.06570232T - 0.45691716T^2$$
$$+ 0.061378272T^3 \qquad \text{(a4)}$$

$$T > 0.4$$

$$\omega^2 y_\omega = 0.0090749264 + 0.021564577T - 0.0075195804T^2$$
$$+ 0.00085075316T^3 - 0.000031932839T^4 \qquad \text{(a5)}$$

$$\omega^2 \frac{dy_\omega}{dT} = 0.021564577 - 0.015023916T + 0.0025522594T^2$$
$$- 0.00012773135T^3 \qquad \text{(a6)}$$

Therefore, $(Y_j)_{\max}$ can be expressed as

$$(Y_1)_{\max} = \Gamma_1 y_{\omega_1} = (0.0366A_1 + 0.01294A_2)^{-1/2} y_{\omega_1} \quad \text{for } T_1 = \frac{2\pi}{\omega_1} \qquad \text{(a7)}$$

$$(Y_2)_{\max} = \Gamma_1 y_{\omega_2} = 0 \quad \text{for } T_2 = \frac{2\pi}{\omega_2} \qquad \text{(a8)}$$

(A$_3$) Find u_{ij}^* and U_i

$$u_{i1}^* = \begin{Bmatrix} u_{11}^* \\ u_{21}^* \end{Bmatrix} = \{\Phi\}^1 (Y_1)_{\max}$$

$$= \begin{Bmatrix} (0.0366A_1 + 0.01294A_2)^{-1/2} \\ 0 \end{Bmatrix} (0.0366A_1 + 0.01294A_2)^{-1/2} y_{\omega_1}$$

$$= \begin{Bmatrix} (0.0366A_1 + 0.01294A_2)^{-1/2} y_{\omega_1} \\ 0 \end{Bmatrix} \qquad \text{(a9)}$$

$$u_{i2}^* = \begin{Bmatrix} u_{12}^* \\ u_{22}^* \end{Bmatrix} = \{\Phi\}^2 (Y_2)_{max} = \begin{Bmatrix} 0 \\ (0.0366A_1 + 0.01294A_2)^{-1/2} \end{Bmatrix} \times 0$$

$$= \begin{Bmatrix} 0 \\ 0 \end{Bmatrix} \tag{a10}$$

$$U_i = \begin{Bmatrix} U_1 \\ U_2 \end{Bmatrix} = \left\{ \begin{array}{c} \left[\displaystyle\sum_{j=1}^{2} \left(u_{1j}^*\right)^2 \right]^{1/2} \\ \left[\displaystyle\sum_{j=1}^{2} \left(u_{2j}^*\right)^2 \right]^{1/2} \end{array} \right\} = \begin{Bmatrix} \left(u_{11}^{*2} + u_{12}^{*2}\right)^{1/2} \\ \left(u_{21}^{*2} + u_{22}^{*2}\right)^{1/2} \end{Bmatrix}$$

$$= \left\{ \begin{array}{c} \left\{ \left[(0.0366A_1 + 0.01294A_2)^{-1} y_{\omega_1} \right]^2 + 0 \right\}^{1/2} \\ 0 \end{array} \right\} \tag{a11}$$

$$= \begin{Bmatrix} (0.0366A_1 + 0.01294A_2)^{-1} y_{\omega_1} \\ 0 \end{Bmatrix}$$

Part (B) Mean Stress

$$\tau_j^* = Q \times u_j^* = \frac{1}{2L} \begin{bmatrix} E & E \\ 0 & 2E \\ -E & E \end{bmatrix} \{u_j^*\} = \begin{bmatrix} 50 & 50 \\ 0 & 100 \\ -50 & 50 \end{bmatrix} \begin{Bmatrix} u_{1j}^* \\ u_{2j}^* \end{Bmatrix} \tag{a12}$$

$$\tau_{i1}^* = \begin{Bmatrix} \tau_{11}^* \\ \tau_{21}^* \\ \tau_{31}^* \end{Bmatrix} = \begin{bmatrix} 50 & 50 \\ 0 & 100 \\ -50 & 50 \end{bmatrix} \begin{Bmatrix} (0.0366A_1 + 0.01294A_2)^{-1} y_{\omega_1} \\ 0 \end{Bmatrix}$$

$$= \begin{Bmatrix} 50 (0.0366A_1 + 0.01294A_2)^{-1} y_{\omega_1} \\ 0 \\ 50 (0.0366A_1 + 0.01294A_2)^{-1} y_{\omega_1} \end{Bmatrix} \tag{a13}$$

$$\tau_{i2}^* = \begin{Bmatrix} \tau_{12}^* \\ \tau_{22}^* \\ \tau_{32}^* \end{Bmatrix} = \begin{bmatrix} 50 & 50 \\ 0 & 100 \\ -50 & 50 \end{bmatrix} \begin{Bmatrix} 0 \\ 0 \end{Bmatrix} = \begin{Bmatrix} 0 \\ 0 \\ 0 \end{Bmatrix} \tag{a14}$$

$$T_{i1} = \begin{Bmatrix} T_{1j} \\ T_{2j} \\ T_{3j} \end{Bmatrix} = \left\{ \begin{array}{c} \displaystyle\sum_{j=1}^{2} \tau_{1j}^{*2} \\ \displaystyle\sum_{j=1}^{2} \tau_{2j}^{*2} \\ \displaystyle\sum_{j=1}^{2} \tau_{3j}^{*2} \end{array} \right\} = \begin{Bmatrix} \left(\tau_{11}^{*2} + \tau_{12}^{*2}\right)^{1/2} \\ \left(\tau_{21}^{*2} + \tau_{22}^{*2}\right)^{1/2} \\ \left(\tau_{31}^{*2} + \tau_{32}^{*2}\right)^{1/2} \end{Bmatrix}$$

$$= \left\{ \begin{array}{c} \left\{ \left[50\left(0.0366A_1 + 0.01294A_2\right)^{-1} y_{\omega_1} \right]^2 + 0 \right\}^{1/2} \\ 0+0 \\ \left\{ \left[-50\left(0.0366A_1 + 0.01294A_2\right)^{-1} y_{\omega_1} \right]^2 \right\}^{1/2} \end{array} \right. \tag{a15}$$

$$= \left\{ \begin{array}{c} 50\left(0.0366A_1 + 0.01294A_2\right)^{-1} y_{\omega_1} \\ 0 \\ 50\left(0.0366A_1 + 0.01294A_2\right)^{-1} y_{\omega_1} \end{array} \right.$$

Part (C) Variance of Displacement

$$S_{U_i}^2 = \sum_{k=1}^{n_p} \sum_{s=1}^{n_p} \frac{\partial U_i}{\partial r_k} \frac{\partial U_i}{\partial r_s} \rho_{ks} V_k V_s \bar{r}_k \bar{r}_s$$

$$(n_p = \text{the number of random parameters})$$

$$= \sum_{k=1}^{n_p} \sum_{s=1}^{n_p} \frac{\partial \left[\sum_{j=1}^{n_m} \left(u_{ij}^*\right)^2 \right]^{1/2}}{\partial r_k} \frac{\partial \left[\sum_{j=1}^{n_m} \left(u_{ij}^*\right)^2 \right]^{1/2}}{\partial r_s} \rho_{ks} V_k V_s \bar{r}_k \bar{r}_s$$

$$= \sum_{K=1}^{n_p} \sum_{S=1}^{n_p} \left\{ \frac{1}{2} \left[\sum_{j=1}^{n_m} \left(u_{ij}^*\right)^2 \right]^{-1/2} 2 \sum_{j=1}^{n_m} \left(u_{ij}^* \frac{\partial u_{ij}^*}{\partial r_k} \right) \right\} \tag{a16}$$

$$\left\{ \frac{1}{2} \left[\sum_{j=1}^{n_m} \left(u_{ij}^*\right)^2 \right]^{-1/2} 2 \sum_{j=1}^{n_m} \left(u_{ij}^* \frac{\partial u_{ij}^*}{\partial r_s} \right) \right\} \rho_{ks} V_k V_s \bar{r}_k \bar{r}_s$$

$$= \sum_{k=1}^{n_p} \sum_{s=1}^{n_p} \left\{ U_i^{-1} \sum_{j=1}^{n_m} \left(u_{ij}^* \frac{\partial u_{ij}^*}{\partial r_k} \right) U_i^{-1} \sum_{j=1}^{n_m} \left(u_{ij}^* \frac{\partial u_{ij}^*}{\partial r_s} \right) \right\} \rho_{ks} V_k V_s \bar{r}_k \bar{r}_s$$

where

$$\frac{\partial u_j^*}{\partial r_k} = \frac{\partial \{\Phi\}^j}{\partial r_k} \Gamma_j y_{\omega_i} + \{\Phi\}^j \frac{\partial \Gamma_j}{\partial r_k} y_{\omega_i} + \{\Phi\}^j \Gamma_j \frac{\partial y_{\omega_i}}{\partial r_k} \tag{a17}$$

(C1) Find $\dfrac{\partial \{\Phi\}^j}{\partial r_k}$

$$\frac{\partial \{\Phi\}^j}{\partial r_k} = \sum_{r=1}^{n_m} a_{jkr} \{\Phi\}^r$$

For $j = r$

$$\{\Phi\}^{jT} [m] \{\Phi\}^j = 1, \quad \frac{\partial}{\partial r_k} \left\{ \{\Phi\}^{jT} [m] \{\Phi\}^j \right\} = 0 \tag{a18, a19}$$

$$2 \{\Phi\}^{jT} [m] \frac{\partial \{\Phi\}^j}{\partial r_k} + \{\Phi\}^{jT} \frac{\partial [m]}{\partial r_k} \{\Phi\}^j = 0 \tag{a20}$$

$$2 \{\Phi\}^{jT} [m] \sum_{r=1}^{n_m} a_{jkr} \{\Phi\}^r + \{\Phi\}^{jT} \frac{\partial [m]}{\partial r_k} \{\Phi\}^j = 0 \tag{a21}$$

$$a_{jkr} = \frac{- \{\Phi\}^{jT} \dfrac{\partial [m]}{\partial r_k} \{\Phi\}^j}{2 \{\Phi\}^{jT} [m] \{\Phi\}^j}$$

$$= \frac{- \{\Phi\}^{jT} \dfrac{\partial [m]}{\partial r_k} \{\Phi\}^j}{2} \tag{a22}$$

$$\frac{\partial [m]}{\partial r_1} = \frac{\partial [m]}{\partial A_1} = \begin{bmatrix} \dfrac{r\sqrt{2}L}{g} & 0 \\ 0 & \dfrac{r\sqrt{2}L}{g} \end{bmatrix} = \begin{bmatrix} 0.0366 & 0 \\ 0 & 0.0366 \end{bmatrix} \tag{a23}$$

$$\frac{\partial [m]}{\partial r_2} = \frac{\partial [m]}{\partial A_2} = \begin{bmatrix} \dfrac{rL}{2g} & 0 \\ 0 & \dfrac{rL}{2g} \end{bmatrix} = \begin{bmatrix} 0.01294 & 0 \\ 0 & 0.01294 \end{bmatrix} \tag{a24}$$

$$a_{111} = \frac{- \{\Phi\}^{1T} \dfrac{\partial [m]}{\partial A_1} \{\Phi\}^1}{2}$$

$$= \frac{-1}{2} \left\{ \begin{matrix} (0.0366A_1 + 0.01294A_2)^{-1/2} \\ 0 \end{matrix} \right\}^T \begin{bmatrix} 0.0366 & 0 \\ 0 & 0.0366 \end{bmatrix} \tag{a25}$$

$$\left\{ \begin{matrix} (0.0366A_1 + 0.01294A_2)^{-1/2} \\ 0 \end{matrix} \right\}$$

$$= -0.0183 (0.0366A_1 + 0.01294A_2)^{-1}$$

$$a_{121} = \frac{- \{\Phi\}^{1T} \dfrac{\partial [m]}{\partial A_2} \{\Phi\}^1}{2}$$

$$= \frac{-1}{2} \left\{ \begin{matrix} (0.0366A_1 + 0.01294A_2)^{-1/2} \\ 0 \end{matrix} \right\}^T \begin{bmatrix} 0.01294 & 0 \\ 0 & 0.01294 \end{bmatrix}$$

$$\left\{ \begin{matrix} (0.0366A_1 + 0.01294A_2)^{-1/2} \\ 0 \end{matrix} \right\}$$

$$= -0.00647 (0.0366A_1 + 0.01294A_2)^{-1} \tag{a26}$$

$$a_{212} = \frac{-\{\Phi\}^{2^T} \frac{\partial [m]}{\partial A_1} \{\Phi\}^2}{2}$$

$$= \frac{-1}{2} \left\{ \begin{array}{c} 0 \\ (0.0366A_1 + 0.01294A_2)^{-1/2} \end{array} \right\}^T \begin{bmatrix} 0.0366 & 0 \\ 0 & 0.0366 \end{bmatrix} \quad \text{(a27)}$$

$$\left\{ \begin{array}{c} 0 \\ (0.0366A_1 + 0.01294A_2)^{-1/2} \end{array} \right\}$$

$$= -0.0183 \, (0.0366A_1 + 0.01294A_2)^{-1}$$

$$a_{222} = \frac{-\{\Phi\}^{2^T} \frac{\partial [m]}{\partial A_2} \{\Phi\}^2}{2} \quad \text{(a28)}$$

$$= -0.00647 \, (0.0366A_1 + 0.01294A_2)^{-1}$$

for $j \neq r$

$$\left[K - \omega_j^2 m \right] \{\Phi\}^j = 0 \quad \text{(a29)}$$

$$\frac{\partial \left[K - \omega_j^2 m \right]}{\partial r_k} \{\Phi\}^j + \left[K - \omega_j^2 m \right] \sum_{r=1}^{n_m} a_{jkr} \{\Phi\}^r = 0 \quad \text{(a30)}$$

$$\{\Phi\}^r \frac{\partial \left[K - \omega_j^2 m \right]}{\partial r_k} \{\Phi\}^j + \{\Phi\}^r \left[K - \omega_j^2 m \right] \sum_{r=1}^{n_m} a_{jkr} \{\Phi\}^r = 0 \quad \text{(a31)}$$

$$a_{jkr} = \frac{-\{\Phi\}^{r^T} \frac{\partial \left[K - \omega_j^2 m \right]}{\partial r_k} \{\Phi\}^j}{\{\Phi\}^{r^T} [K] \{\Phi\}^r - \omega_j^2 \{\Phi\}^{r^T} [m] \{\Phi\}^r}$$

$$\quad \text{(a32)}$$

$$= \frac{\{\Phi\}^{r^T} \frac{\partial \left[K - \omega_j^2 m \right]}{\partial r_k} \{\Phi\}^r}{\omega_j^2 - \omega_r^2}$$

$$\frac{\partial [K]}{\partial r_1} = \frac{\partial [K]}{\partial A_1} = \begin{bmatrix} \frac{E}{\sqrt{2}L} & 0 \\ 0 & \frac{E}{\sqrt{2}L} \end{bmatrix} = \begin{bmatrix} 70.7107 & 0 \\ 0 & 70.7107 \end{bmatrix} \quad \text{(a33)}$$

$$\frac{\partial [K]}{\partial r_2} = \frac{\partial [K]}{\partial A_2} = \begin{bmatrix} 0 & 0 \\ 0 & \frac{E}{L} \end{bmatrix} = \begin{bmatrix} 0 & 0 \\ 0 & 100 \end{bmatrix} \quad \text{(a34)}$$

for $j=1$, $k=1$, $r=2$

$$a_{112} = \frac{\{\Phi\}^{2^T} \left\{ \dfrac{\partial [K]}{\partial A_1} - \omega_1^2 \dfrac{\partial [m]}{\partial A_1} \right\} \{\Phi\}^1}{\omega_1^2 - \omega_2^2} = 0 \tag{a35}$$

for $j=2$, $k=1$, $r=1$

$$a_{211} = \frac{\{\Phi\}^{1^T} \left\{ \dfrac{\partial [K]}{\partial A_1} - \omega_2^2 \dfrac{\partial [m]}{\partial A_1} \right\} \{\Phi\}^2}{\omega_2^2 - \omega_1^2} = 0$$

for $j=1$, $k=2$, $r=2$

$$a_{122} = \frac{\{\Phi\}^{2^T} \left\{ \dfrac{\partial [K]}{\partial A_2} - \omega_1^2 \dfrac{\partial [m]}{\partial A_2} \right\} \{\Phi\}^1}{\omega_1^2 - \omega_2^2} = 0 \tag{a36}$$

for $j=2$, $k=2$, $r=1$

$$a_{221} = \frac{\{\Phi\}^{1^T} \left\{ \dfrac{\partial [K]}{\partial A_2} - \omega_2^2 \dfrac{\partial [m]}{\partial A_2} \right\} \{\Phi\}^2}{\omega_1^2 - \omega_2^2} = 0 \tag{a37}$$

$$\frac{\partial \{\Phi\}^1}{\partial r_1} = \frac{\partial \{\Phi\}^1}{\partial A_1} = \sum_{r=1}^{2} a_{11r} \{\Phi\}^r = a_{111} \{\Phi\}^1 + a_{112} \{\Phi\}^2$$

$$= \left\{ \begin{array}{c} -0.0183\,(0.0366A_1 + 0.01294A_2)^{-3/2} \\ 0 \end{array} \right\} \tag{a38}$$

$$\frac{\partial \{\Phi\}^1}{\partial r_2} = \frac{\partial \{\Phi\}^1}{\partial A_2} = \sum_{r=1}^{2} a_{12r} \{\Phi\}^r = a_{121} \{\Phi\}^1 + a_{122} \{\Phi\}^2$$

$$= \left\{ \begin{array}{c} -0.00647\,(0.0366A_1 + 0.01294A_2)^{-3/2} \\ 0 \end{array} \right\} \tag{a39}$$

$$\frac{\partial \{\Phi\}^2}{\partial r_1} = \frac{\partial \{\Phi\}^2}{\partial A_1} = \sum_{r=1}^{2} a_{21r} \{\Phi\}^r = a_{211} \{\Phi\}^1 + a_{212} \{\Phi\}^2$$

$$= \left\{ \begin{array}{c} 0 \\ -0.0183\,(0.0366A_1 + 0.01294A_2)^{-3/2} \end{array} \right\} \tag{a40}$$

$$\frac{\partial \{\Phi\}^2}{\partial r_2} = \frac{\partial \{\Phi\}^2}{\partial A_2} = \sum_{r=1}^{2} a_{22r} \{\Phi\}^r = a_{221} \{\Phi\}^1 + a_{222} \{\Phi\}^2$$

$$= \left\{ \begin{array}{c} 0 \\ (0.0366A_1 + 0.01294A_2)^{-3/2} \end{array} \right\} \tag{a41}$$

(C2) Find $\dfrac{\partial \Gamma_0}{\partial r_k} = \dfrac{\partial \left[\{\Phi\}^{j^T} \{P_0\} \right]}{\partial r_k} = \dfrac{\partial \{\Phi\}^j}{\partial r_k} \{P_0\}$

$$\dfrac{\partial \Gamma_1}{\partial r_1} = \dfrac{\partial \Gamma_1}{\partial A_1} = \dfrac{\partial \{\Phi\}^{1T}}{\partial A_1} \{P_0\}$$

$$= \left\{ \begin{matrix} -0.0183 \, (0.0366 A_1 + 0.01294 A_2)^{-3/2} \\ 0 \end{matrix} \right\}^T \left\{ \begin{matrix} 1 \\ 0 \end{matrix} \right\} \qquad (a42)$$

$$= -0.0183 \, (0.0366 A_1 + 0.01294 A_2)^{-3/2}$$

$$\dfrac{\partial \Gamma_1}{\partial r_2} = \dfrac{\partial \Gamma_1}{\partial A_2} = \dfrac{\partial \{\Phi\}^{1T}}{\partial A_2} \{P_0\}$$

$$= -0.00647 \, (0.0366 A_1 + 0.01294 A_2)^{-3/2} \qquad (a43)$$

$$\dfrac{\partial \Gamma_2}{\partial r_1} = \dfrac{\partial \Gamma_2}{\partial A_1} = \dfrac{\partial \{\Phi\}^{2T}}{\partial A_1} \{P_0\} = 0 \qquad (a44)$$

$$\dfrac{\partial \Gamma_2}{\partial r_2} = \dfrac{\partial \Gamma_2}{\partial A_2} = \dfrac{\partial \{\Phi\}^{2T}}{\partial A_2} \{P_0\} = 0 \qquad (a45)$$

(C3) Find $\dfrac{\partial y_{\omega_j}}{\partial r_k} = \dfrac{dy_{\omega_j}}{dT_j} \dfrac{dT_j}{d\omega_j} \dfrac{\partial \omega_j}{\partial r_k}$

$$\dfrac{dy_{\omega_j}}{dT_j} = \dfrac{\omega_j^2 y_{\omega_j}}{\omega_j^2}$$

$$\dfrac{dT_j}{d\omega_j} = \dfrac{d \, (2\pi / \omega)}{d\omega} = -2\pi \omega^{-2} \qquad (a46, a47)$$

$$\dfrac{\partial \omega_j}{\partial r_k} = \dfrac{\partial \omega_j}{\partial \omega_j^2} \dfrac{\partial \omega_j^2}{\partial r_k} = \dfrac{1}{2\omega_j} \dfrac{\partial \omega_j^2}{\partial r_k} \qquad (a48)$$

Since $\{\Phi\}^{j^T} \left[K - \omega_j^2 m \right] \{\Phi\}^j = 0$

$$\{\Phi\}^{j^T} \dfrac{\partial \left[K - \omega_j^2 m \right]}{\partial r_k} \{\Phi\}^j + 2 \{\Phi\}^{j^T} \left[K - \omega_j^2 m \right] \dfrac{\partial \{\Phi\}^j}{\partial r_k} = 0 \qquad (a49)$$

$$\{\Phi\}^{j^T} \dfrac{\partial [K]}{\partial r_k} \{\Phi\}^j - \omega_j^2 \{\Phi\}^{j^T} \dfrac{\partial [m]}{\partial r_k} \{\Phi\}^j - \{\Phi\}^{j^T} [m] \{\Phi\}^j \dfrac{\partial \omega_j^2}{\partial r_k} = 0 \qquad (a50)$$

$$\dfrac{\partial \omega_j^2}{\partial r_k} = \{\Phi\}^{j^T} \dfrac{\partial [K]}{\partial r_k} \{\Phi\}^j - \omega_j^2 \{\Phi\}^{j^T} \dfrac{\partial [m]}{\partial r_k} \{\Phi\}^j \qquad (a51)$$

Therefore

$$\frac{\partial \omega_j}{\partial r_k} = \frac{1}{2\omega_i} \left(\{\Phi\}^{jT} \frac{\partial [K]}{\partial r_k} \{\Phi\}^j - \omega_j^2 \{\Phi\}^{jT} \frac{\partial [m]}{\partial r_k} \{\Phi\}^j \right) \tag{a52}$$

$$\frac{\partial y_{\omega_1}}{\partial r_1} = \frac{\partial y_{\omega_1}}{\partial A_1} = \frac{dy_{\omega_1}}{dT_1} \frac{dT_1}{d\omega_1} \frac{\partial \omega_1}{\partial r_1}$$

$$= \frac{dy_{\omega_1}}{dT_1} \left(-2\pi \frac{(0.0366A_1 + 0.01294A_2)}{70.7107A_1} \right) \frac{\partial \omega_1}{\partial r_1}$$

$$\frac{\partial \omega_1}{\partial r_1} = \frac{1}{2\omega_1} \left\{ \left\{ \begin{matrix} (0.0366A_1 + 0.01294A_2)^{-1/2} \\ 0 \end{matrix} \right\}^T \begin{bmatrix} 70.7107 & 0 \\ 0 & 70.7107 \end{bmatrix} \right.$$

$$\left\{ \begin{matrix} (0.0366A_1 + 0.01294A_2)^{-1/2} \\ 0 \end{matrix} \right\}$$

$$-\omega_1^2 \left\{ \begin{matrix} (0.0366A_1 + 0.01294A_2)^{-1/2} \\ 0 \end{matrix} \right\}^T \begin{bmatrix} 0.0366 & 0 \\ 0 & 0.0366 \end{bmatrix}$$

$$\left. \left\{ \begin{matrix} (0.0366A_1 + 0.01294A_2)^{-1/2} \\ 0 \end{matrix} \right\} \right\}$$

$$= \frac{1}{2\omega_1} (70.7107) (0.0366A_1 + 0.01294A_2)^{-1}$$

$$- \frac{\omega_1}{2} (0.0366) (0.0366A_1 + 0.01294A_2)^{-1}$$

$$= 4.20448273A_1^{-1/2} (0.0366A_1 + 0.01294A_2)^{-1/2}$$

$$- 0.15388407A_1^{1/2} (0.0366A_1 + 0.01294A_2)^{-3/2}$$

$$\tag{a53}$$

$$\frac{\partial y_{\omega_1}}{\partial A_1} = \frac{\partial y_{\omega_1}}{\partial T_1} (-2\pi) \frac{(0.0366A_1 + 0.01294A_2)}{70.7107A_1} \left[4.20448273A_1^{-1/2} \right.$$

$$(0.0366A_1 + 0.01294A_2)^{-1/2}$$

$$\left. -0.15388407A_1^{1/2} (0.0366A_1 + 0.01294A_2)^{-3/2} \right] \tag{a54}$$

$$= \frac{\partial y_{\omega_1}}{\partial T_1} (-0.08885763) \left[4.20448273A_1^{-1/2} \right.$$

$$(0.0366A_1 + 0.01294A_2)^{-1/2}$$

$$\left. -0.15388407A_1^{1/2} (0.0366A_1 + 0.01294A_2)^{-3/2} \right]$$

$$\frac{\partial y_{\omega_1}}{\partial r_2} = \frac{\partial y_{\omega_1}}{\partial A_2} = \frac{dy_{\omega_1}}{dT_1} \frac{dT_1}{d\omega_1} \frac{\partial \omega_1}{\partial A_2}$$

$$= \frac{dy_{\omega_1}}{dT_1} \left(-2\pi \frac{(0.0366A_1 + 0.01294A_2)}{70.7107A_1} \right) \frac{\partial \omega_1}{\partial A_2} \tag{a55}$$

$$\frac{\partial \omega_1}{\partial A_2} = \frac{1}{2\omega_1} \left\{ \left\{ \begin{matrix} (0.0366A_1 + 0.01294A_2)^{-1/2} \\ 0 \end{matrix} \right\}^T \begin{bmatrix} 0 & 0 \\ 0 & 100 \end{bmatrix} \right.$$

$$\left\{ \begin{matrix} (0.0366A_1 + 0.01294A_2)^{-1/2} \\ 0 \end{matrix} \right\}$$

$$- \omega_1^2 \left\{ \begin{matrix} (0.0366A_1 + 0.01294A_2)^{-1/2} \\ 0 \end{matrix} \right\}^T \begin{bmatrix} 0.01294 & 0 \\ 0 & 0.01294 \end{bmatrix}$$

$$\left. \left\{ \begin{matrix} (0.0366A_1 + 0.01294A_2)^{-1/2} \\ 0 \end{matrix} \right\} \right\}$$

$$= -70.7107A_1 \left(0.0366A_1 + 0.01294A_2\right)^{-2} \tag{a56}$$

So

$$\frac{\partial y_{\omega_1}}{\partial A_2} = \frac{dy_{\omega_1}}{dT_1} \left(-2\pi \frac{(0.0366A_1 + 0.01294A_2)}{70.7107A_1} \right) - 70.7107A_1$$

$$\left(0.0366A_1 + 0.01294A_2\right)^{-2} \tag{a57}$$

$$= \frac{dy_{\omega_1}}{dT_1} (-2\pi) (0.0366A_1 + 0.01294A_2)^{-1}$$

$$\frac{\partial y_{\omega_2}}{\partial r_1} = \frac{\partial y_{\omega_2}}{\partial A_1} = \frac{dy_{\omega_2}}{dT_2} \frac{dT_2}{d\omega_2} \frac{\partial \omega_2}{\partial A_1}$$

$$= \frac{dy_{\omega_2}}{dT_1} \left(-2\pi \frac{(0.0366A_1 + 0.01294A_2)}{70.7107\left(A_1 + \sqrt{2}A_2\right)} \right) \frac{\partial \omega_2}{\partial A_1}$$

$$\frac{\partial \omega_2}{\partial A_1} = \frac{1}{2\omega_2} \left\{ \left\{ \begin{matrix} 0 \\ (0.0366A_1 + 0.01294A_2)^{-1/2} \end{matrix} \right\}^T \begin{bmatrix} 71.7107 & 0 \\ 0 & 71.7107 \end{bmatrix} \right.$$

$$\left\{ \begin{matrix} 0 \\ (0.0366A_1 + 0.01294A_2)^{-1/2} \end{matrix} \right\}$$

$$- \omega_2^2 \left\{ \begin{matrix} 0 \\ (0.0366A_1 + 0.01294A_2)^{-1/2} \end{matrix} \right\}^T \begin{bmatrix} 0.0366 & 0 \\ 0 & 0.0366 \end{bmatrix}$$

$$\left. \left\{ \begin{matrix} 0 \\ (0.0366A_1 + 0.01294A_2)^{-1/2} \end{matrix} \right\} \right\}$$

$$= 4.20448273 \left(A_1 + \sqrt{2}A_2\right)^{-1/2} (0.0366A_1 + 0.01294A_2)^{-1/2}$$

$$- 0.15388407 \left(A_1 + \sqrt{2}A_2\right)^{1/2} (0.0366A_1 + 0.01294A_2)^{-3/2} \tag{a58}$$

$$\frac{\partial y_{\omega_2}}{\partial A_2} = \frac{dy_{\omega_2}}{dT_2} \left(-2\pi \frac{(0.0366A_1 + 0.01294A_2)}{70.7107\left(A_1 + \sqrt{2}A_2\right)} \right)$$

$$\left[4.20448273 \left(A_1 + \sqrt{2}A_2\right)^{-1/2} (0.0366A_1 + 0.01294A_2)^{-1/2} \right.$$

$$\left. -0.15388407 \left(A_1 + \sqrt{2}A_2\right)^{1/2} (0.0366A_1 + 0.01294A_2)^{-3/2} \right]$$

$$= \frac{dw_2}{dT_2} (-0.088857631) \left[4.20448273 \left(A_1 + \sqrt{2}A_2\right)^{-3/2} \right.$$

$$(0.0366A_1 + 0.01294A_2)^{1/2}$$

$$\left. -0.15388407 \left(A_1 + \sqrt{2}A_2\right)^{-1/2} (0.0366A_1 + 0.01294A_2)^{-1/2} \right]$$

$$\tag{a59}$$

$$\frac{\partial y_{\omega_2}}{\partial r_2} = \frac{\partial y_{\omega_2}}{\partial A_2} = \frac{dy_{\omega_2}}{dT_2} \frac{dT_2}{d\omega_2} \frac{\partial \omega_2}{\partial A_2}$$

$$\tag{a60}$$

$$= \frac{dy_{\omega_2}}{dT_2} (-2\pi) \left(\frac{(0.0366A_1 + 0.1294A_2)}{70.7107\left(A_1 + \sqrt{2}A_2\right)} \right) \frac{\partial \omega_2}{\partial A_2}$$

$$\frac{\partial \omega_2}{\partial A_2} = \frac{1}{2\omega_2} \left\{ \left\{ \begin{matrix} 0 \\ (0.0366A_1 + 0.01294A_2)^{-1/2} \end{matrix} \right\}^T \begin{bmatrix} 0 & 0 \\ 0 & 100 \end{bmatrix} \right.$$

$$\left\{ \begin{matrix} 0 \\ (0.0366A_1 + 0.01294A_2)^{-1/2} \end{matrix} \right\}$$

$$-\omega_2^2 \left\{ \begin{matrix} 0 \\ (0.0366A_1 + 0.01294A_2)^{-1/2} \end{matrix} \right\}^T \begin{bmatrix} 0.0366 & 0 \\ 0 & 0.0366 \end{bmatrix}$$

$$\left. \left\{ \begin{matrix} 0 \\ (0.0366A_1 + 0.01294A_2)^{-1/2} \end{matrix} \right\} \right\}$$

$$= \frac{1}{2\omega_2} \left\{ 100 \left(0.0366A_1 + 0.01294A_2\right)^{-1} \right.$$

$$\left. -\omega_2^2 0.0366 \left(0.0366A_1 + 0.01294A_2\right)^{-1/2} \right\}$$

$$\tag{a61}$$

$$\frac{\partial y_{\omega_2}}{\partial A_2} = \frac{dy_{\omega_2}}{dT_2} \left(-2\pi \frac{(0.0366A_1 + 0.01294A_2)}{70.7107\left(A_1 + \sqrt{2}A_2\right)} \right)$$

$$\left[\frac{(0.0366A_1 + 0.01294A_2)^{1/2}}{2\left(70.7107\left(A_1 + \sqrt{2}A_2\right)\right)^{1/2}} \times 100 \left(0.0366A_1 + 0.01294A_2\right)^{-1} \right.$$

$$\frac{\left(70.7107\left(A_1+\sqrt{2}A_2\right)\right)^{1/2}}{2\left(0.0366A_1+0.01294A_2\right)^{1/2}}(0.0366)$$

$$(0.0366A_1+0.01294A_2)^{-1/2}\Bigg]$$

(a62)

$$=\frac{d\omega_2}{dT_2}(-0.088857631)$$

$$\left\{\begin{array}{l}2.95221531\left(A_1+\sqrt{2}A_2\right)^{-3/2}(0.0366A_1+0.01294A_2)^{1/2}\\-0.15388407\left(A_1+\sqrt{2}A_2\right)^{-1/2}\end{array}\right\}$$

(C4) Find $\dfrac{\partial u_j^*}{\partial r_k}$

$$\frac{\partial u_1^*}{\partial r_1}=\frac{\partial u_1^*}{\partial A_1}=\frac{\partial\{\Phi\}^1}{\partial A_1}\Gamma_1 y_{\omega_1}+\{\Phi\}^1\frac{\partial\Gamma_1}{\partial A_1}y_{\omega_1}+\{\Phi\}^1\Gamma_1\frac{\partial y_{\omega_1}}{\partial A_1}$$

$$=\left\{\begin{array}{c}-0.0183\,(0.0366A_1+0.01294A_2)^{-3/2}\\0\end{array}\right\}$$

$$(0.0366A_1+0.01294A_2)^{-1/2}\,y_{\omega_1}$$

$$+\left\{\begin{array}{c}(0.0366A_1+0.01294A_2)^{-1/2}\\0\end{array}\right\}$$

(a63)

$$(-0.0183)\,(0.0366A_1+0.01294A_2)^{-3/2}\,y_{\omega_1}$$

$$+\left\{\begin{array}{c}(0.0366A_1+0.01294A_2)^{-1/2}\\0\end{array}\right\}$$

$$(0.0366A_1+0.01294A_2)^{-1/2}\frac{\partial y_{\omega_1}}{\partial A_1}$$

$$\frac{\partial u_1^*}{\partial r_2}=\frac{\partial u_1^*}{\partial A_2}=\frac{\partial\{\Phi\}^1}{\partial A_2}\Gamma_1 y_{\omega_1}+\{\Phi\}^1\frac{\partial\Gamma_1}{\partial A_2}y_{\omega_1}+\{\Phi\}^1\Gamma_1\frac{\partial y_{\omega_1}}{\partial A_2}$$

$$=\left\{\begin{array}{c}-0.00647\,(0.0366A_1+0.01294A_2)^{-3/2}\\0\end{array}\right\}$$

$$(0.0366A_1+0.01294A_2)^{-1/2}\,y_{\omega_1}$$

$$+\left\{\begin{array}{c}(0.0366A_1+0.01294A_2)^{-1/2}\\0\end{array}\right\}$$

(a64)

$$(-0.00647)\,(0.0366A_1+0.01294A_2)^{-3/2}\,y_{\omega_1}$$

$$+ \left\{ \begin{array}{c} (0.0366A_1 + 0.01294A_2)^{-1/2} \\ 0 \end{array} \right\}$$

$$(0.0366A_1 + 0.01294A_2)^{-1/2} \frac{\partial y_{\omega_1}}{\partial A_2}$$

$$\frac{\partial u_2^*}{\partial r_1} = \frac{\partial u_2^*}{\partial A_1} = \frac{\partial \{\Phi\}^2}{\partial A_1} \Gamma_2 y_{\omega_2} + \{\Phi\}^2 \frac{\partial \Gamma_2}{\partial A_1} y_{\omega_2} + \{\Phi\}^2 \Gamma_2 \frac{\partial y_{\omega_2}}{\partial A_1}$$

$$= \left\{ \begin{array}{c} -0.0183 (0.0366A_1 + 0.01294A_2)^{-3/2} \\ 0 \end{array} \right\} (0) y_{\omega_2} \qquad \text{(a65)}$$

$$+ \left\{ \begin{array}{c} 0 \\ (0.0366A_1 + 0.01294A_2)^{-1/2} \end{array} \right\} (0) y_{\omega_2}$$

$$+ \left\{ \begin{array}{c} 0 \\ (0.0366A_1 + 0.01294A_2)^{-1/2} \end{array} \right\} (0) \frac{\partial y_{\omega_1}}{\partial A_2} = 0$$

$$\frac{\partial u_2^*}{\partial r_2} = \frac{\partial u_2^*}{\partial A_2} = \frac{\partial \{\Phi\}^2}{\partial A_2} \Gamma_2 y_{\omega_2} + \{\Phi\}^2 \frac{\partial \Gamma_2}{\partial A_2} y_{\omega_2} + \{\Phi\}^2 \Gamma_2 \frac{\partial y_{\omega_2}}{\partial A_2}$$

$$= \left\{ \begin{array}{c} -0.0183 (0.0366A_1 + 0.01294A_2)^{-3/2} \\ 0 \end{array} \right\} (0) y_{\omega_2}$$

$$+ \left\{ \begin{array}{c} 0 \\ (0.0366A_1 + 0.01294A_2)^{-1/2} \end{array} \right\} (0) y_{\omega_2} \qquad \text{(a66)}$$

$$+ \left\{ \begin{array}{c} 0 \\ (0.0366A_1 + 0.01294A_2)^{-1/2} \end{array} \right\} (0) \frac{\partial y_{\omega_1}}{\partial A_2} = 0$$

$$\frac{\partial u_1^*}{\partial r_1} = \frac{\partial u_1^*}{\partial A_1} = \left\{ \begin{array}{c} \frac{\partial u_{11}^*}{\partial A_1} \\ \frac{\partial u_{21}^*}{\partial A_1} \end{array} \right\} \left\{ \begin{array}{c} -0.0366 (0.0366A_1 + 0.01294A_2)^{-2} \\ 0 \end{array} \right\} y_{\omega_1}$$

$$+ \left\{ \begin{array}{c} (0.0366A_1 + 0.01294A_2)^{-1} \\ 0 \end{array} \right\} \frac{\partial y_{\omega_1}}{\partial A_1} \qquad \text{(a67)}$$

$$\frac{\partial u_1^*}{\partial r_2} = \frac{\partial u_1^*}{\partial A_2} = \left\{ \begin{array}{c} \frac{\partial u_{11}^*}{\partial A_2} \\ \frac{\partial u_{21}^*}{\partial A_2} \end{array} \right\} \left\{ \begin{array}{c} -0.01294 (0.0366A_1 + 0.01294A_2)^{-2} \\ 0 \end{array} \right\} y_{\omega_1}$$

$$+ \left\{ \begin{array}{c} (0.0366A_1 + 0.01294A_2)^{-1} \\ 0 \end{array} \right\} \frac{\partial y_{\omega_1}}{\partial A_2} \qquad \text{(a68)}$$

$$\frac{\partial u_2^*}{\partial r_1} = \frac{\partial u_2^*}{\partial A_1} = \left\{ \begin{array}{c} 0 \\ 0 \end{array} \right\} \qquad \text{(a69)}$$

$$\frac{\partial u_2^*}{\partial r_2} = \frac{\partial u_2^*}{\partial A_2} = \left\{ \begin{array}{c} 0 \\ 0 \end{array} \right\} \qquad \text{(a70)}$$

Assume

$$\rho_{ks} = 0, V_k = 0.05, V_s = 0.05, \bar{r}_1 = \bar{A}_1, \bar{r}_2 = \bar{A}_2 \tag{a71}$$

$$S^2_{U_i} = \sum_{j=1}^{2} \frac{1}{U_i^2} \left[\sum_{k=1}^{2} \sum_{s=1}^{2} (u^*_{ij})^2 \frac{\partial u^*_{ij}}{\partial r_k} \frac{\partial u^*_{ij}}{\partial r_s} \rho_{ks} V_k V_s \bar{r}_k \bar{r}_s \right] \tag{a72}$$

$$S^2_{U_1} = \sum_{j=1}^{2} \frac{1}{U_1^2} \left\{ (u^*_{1j})^2 \left[\frac{\partial u^*_{1j}}{\partial r_1} \frac{\partial u^*_{1j}}{\partial r_1} \rho_{11} V_1 V_1 \bar{r}_1 \bar{r}_1 \right] \right.$$

$$\left. + 2 \left(\frac{\partial u^*_{1j}}{\partial r_2} \frac{\partial u^*_{1j}}{\partial r_1} \right) \rho_{12} V_1 V_2 \bar{r}_1 \bar{r}_2 + \left(\frac{\partial u^*_{1j}}{\partial r_2} \frac{\partial u^*_{1j}}{\partial r_2} \right) \rho_{22} V_2 V_2 \bar{r}_1 \bar{r}_2 \right\}$$

$$= \frac{1}{U_1^2} \left\{ (u^*_{11})^2 \left[\frac{\partial u^*_{11}}{\partial r_1} \frac{\partial u^*_{11}}{\partial r_1} \rho_{11} V_1 V_1 \bar{A}_1 \bar{A}_1 + \frac{\partial u^*_{11}}{\partial r_2} \frac{\partial u^*_{11}}{\partial r_2} \rho_{22} V_2 V_2 \bar{A}_2 \bar{A}_2 \right] \right.$$

$$\left. + (u^*_{12})^2 \left[\frac{\partial u^*_{12}}{\partial r_1} \frac{\partial u^*_{12}}{\partial r_1} \rho_{11} V_1 V_1 \bar{A}_1 \bar{A}_1 + \frac{\partial u^*_{12}}{\partial r_2} \frac{\partial u^*_{12}}{\partial r_2} \rho_{22} V_2 V_2 \bar{A}_2 \bar{A}_2 \right] \right\}$$

$$= \frac{1}{\left\{ (0.0366A_1 + 0.01294A_2)^{-1} y_{\omega_1} \right\}^2}$$

$$\left[\left\{ (0.0366A_1 + 0.01294A_2)^{-1} y_{\omega_1} \right\}^2 \right]$$

$$\left\{ \left[-0.0366 (0.0366A_1 + 0.01294A_2)^{-2} y_{\omega_1} \right. \right.$$

$$\left. + (0.0366A_1 + 0.01294A_2)^{-1} \frac{\partial y_{\omega_1}}{\partial A_1} \right]^2 V^2_{A_1} A_1^2$$

$$+ \left[-0.01294 (0.0366A_1 + 0.01294A_2)^{-2} y_{\omega_1} \right.$$

$$\left. \left. + (0.0366A_1 + 0.01294A_2)^{-1} \frac{\partial y_{\omega_1}}{\partial A_2} \right]^2 V^2_{A_2} A_2^2 \right\} + 0$$

$$= \left[-0.0366 (0.0366A_1 + 0.01294A_2)^{-2} y_{\omega_1} + \right.$$

$$\left. + (0.0366A_1 + 0.01294A_2)^{-1} \frac{\partial y_{\omega_1}}{\partial A_1} \right]^2 V^2_{A_1} A_1^2$$

$$+ \left[-0.01294 (0.0366A_1 + 0.01294A_2)^{-2} y_{\omega_1} \right.$$

$$\left. + (0.0366A_1 + 0.01294A_2)^{-1} \frac{\partial y_{\omega_1}}{\partial A_2} \right]^2 V^2_{A_2} A_2^2 \tag{a73}$$

$$S^2_{U_2} = 0 \tag{a74}$$

Part (D) Variance of Stress

$$S_{T_i}^2 = \sum_{j=1}^{2} \frac{\tau_{ij}^* \tau_{ij}^*}{T_i^2} \left[\sum_{k=1}^{2} \sum_{s=1}^{2} (\tau_{ij}^*)^2 \frac{\partial \tau_{ij}^*}{\partial r_k} \frac{\partial \tau_{ij}^*}{\partial r_s} \rho_{ks} V_k V_s \bar{r}_k \bar{r}_s \right] \tag{a75}$$

$$S_{T_1}^2 = \sum_{j=1}^{2} \frac{\tau_{1j}^* \tau_{1j}^*}{T_1^2} \left\{ \left(\frac{\partial \tau_{1j}^*}{\partial r_1} \right)^2 \rho_{11} V_1 V_1 \bar{r}_1 \bar{r}_1 \right.$$

$$\left. + 2 \left(\frac{\partial \tau_{1j}^*}{\partial r_1} \right) \left(\frac{\partial \tau_{1j}^*}{\partial r_2} \right) \rho_{12} V_1 V_2 \bar{r}_1 \bar{r}_2 \left(\frac{\partial \tau_{1j}^*}{\partial r_2} \right)^2 \rho_{22} V_2 V_2 \bar{r}_2 \bar{r}_2 \right\}$$

$$= \frac{1}{T_1^2} \left\{ (\tau_{11}^*)^2 \left[\left(\frac{\partial \tau_{11}^*}{\partial A_1} \right)^2 \rho_{11} V_1^2 A_1^2 + \left(\frac{\partial \tau_{11}^*}{\partial A_2} \right)^2 \rho_{22} V_2^2 A_2^2 \right] \right. \tag{a76}$$

$$\left. + (\tau_{12}^*)^2 \left[\left(\frac{\partial \tau_{12}^*}{\partial A_1} \right)^2 \rho_{11} V_1^2 A_1^2 + \left(\frac{\partial \tau_{12}^*}{\partial A_2} \right)^2 \rho_{22} V_2^2 A_2^2 \right] \right\}$$

$$= \left(\frac{\partial \tau_{11}^*}{\partial A_1} \right)^2 \rho_{11} V_1^2 A_1^2 + \left(\frac{\partial \tau_{11}^*}{\partial A_2} \right)^2 \rho_{22} V_2^2 A_2^2$$

$$S_{T_2}^2 = 0 \tag{a77}$$

$$S_{T_3}^2 = \sum_{j=1}^{2} \frac{\tau_{3j}^* \tau_{3j}^*}{T_3} \left\{ \left(\frac{\partial \tau_{3j}^*}{\partial r_1} \right)^2 \rho_{11} V_1 V_1 \bar{r}_1 \bar{r}_1 + 2 \left(\frac{\partial \tau_{3j}^*}{\partial r_1} \right) \left(\frac{\partial \tau_{3j}^*}{\partial r_2} \right) \rho_{12} V_1 V_2 \bar{r}_1 \bar{r}_2 \right.$$

$$\left. + \left(\frac{\partial \tau_{3j}^*}{\partial r_2} \right)^2 \rho_{22} V_2 V_2 \bar{r}_2 \bar{r}_2 \right\}$$

$$= \sum_{j=1}^{2} \frac{\tau_{3j}^* \tau_{3j}^*}{T_3} \left[\left(\frac{\partial \tau_{3j}^*}{\partial A_1} \right)^2 \rho_{11} V_1^2 \overline{A}_1^2 + \left(\frac{\partial \tau_{3j}^*}{\partial r_2} \right)^2 \rho_{22} V_2^2 \overline{A}_2^2 \right]$$

$$= \frac{1}{T_3^2} \left\{ (\tau_{31}^*)^2 \left[\left(\frac{\partial \tau_{31}^*}{\partial A_1} \right)^2 \rho_{11} V_1^2 A_1^2 + \left(\frac{\partial \tau_{31}^*}{\partial A_2} \right)^2 \rho_{22} V_2^2 A_2^2 \right] \right.$$

$$\left. + \tau_{32} \left[\left(\frac{\partial \tau_{32}^*}{\partial A_1} \right)^2 \rho_{11} V_1^2 A_1^2 + \left(\frac{\partial \tau_{32}^*}{\partial A_2} \right)^2 \rho_{22} V_2^2 A_2^2 \right] \right\}$$

$$= \left(\frac{\partial \tau_{31}^*}{\partial A_1} \right)^2 \rho_{11} V_1^2 A_1^2 + \left(\frac{\partial \tau_{31}^*}{\partial A_2} \right)^2 \rho_{22} V_2^2 A_2^2 \tag{a78}$$

$$\frac{\partial \tau_{ij}^*}{\partial r_k} = \frac{\partial (Q U_{ij}^*)}{\partial r_k} = \frac{\partial Q}{\partial r_k} u_{ij}^* + Q \frac{\partial u_{ij}^*}{\partial r_k}$$

$$\frac{\partial \tau_{i1}^*}{\partial r_1} = \frac{\partial \tau_{i1}^*}{\partial A_1} = \begin{Bmatrix} \dfrac{\partial \tau_{11}}{\partial A_1} \\[2mm] \dfrac{\partial \tau_{21}}{\partial A_1} \\[2mm] \dfrac{\partial \tau_{31}}{\partial A_1} \end{Bmatrix} = 0 + \begin{bmatrix} 50 & 50 \\ 0 & 100 \\ -50 & 50 \end{bmatrix} \begin{Bmatrix} \dfrac{\partial u_{11}^*}{\partial A_1} \\[2mm] \dfrac{\partial u_{21}^*}{\partial A_1} \end{Bmatrix} = \begin{Bmatrix} 50\dfrac{\partial u_{11}^*}{\partial A_1} + 50\dfrac{\partial u_{21}^*}{\partial A_1} \\[2mm] 100\dfrac{\partial u_{21}^*}{\partial A_1} \\[2mm] -50\dfrac{\partial u_{11}^*}{\partial A_1} + 50\dfrac{\partial u_{21}^*}{\partial A_1} \end{Bmatrix}$$

$$= \begin{Bmatrix} 50\left(-0.0366\,(0.0366A_1 + 0.01294A_2)^{-2}\,y_{\omega_1} + (0.0366A_1 + 0.01294A_2)^{-1}\dfrac{\partial y_{\omega_1}}{\partial A_1}\right) + 0 \\[3mm] 0 \\[3mm] -50\left(-0.0366\,(0.0366A_1 + 0.01294A_2)^{-2}\,y_{\omega_1} + (0.0366A_1 + 0.01294A_2)^{-1}\dfrac{\partial y_{\omega_1}}{\partial A_1}\right) + 0 \end{Bmatrix}$$

$$\text{(a79)}$$

$$\frac{\partial \tau_{i1}^*}{\partial r_2} = \frac{\partial \tau_{i1}^*}{\partial A_2} = \begin{Bmatrix} \dfrac{\partial \tau_{11}}{\partial A_2} \\[2mm] \dfrac{\partial \tau_{21}}{\partial A_2} \\[2mm] \dfrac{\partial \tau_{31}}{\partial A_2} \end{Bmatrix} = 0 + \begin{bmatrix} 50 & 50 \\ 0 & 100 \\ -50 & 50 \end{bmatrix} \begin{Bmatrix} \dfrac{\partial u_{11}^*}{\partial A_2} \\[2mm] \dfrac{\partial u_{21}^*}{\partial A_2} \end{Bmatrix} = \begin{Bmatrix} 50\dfrac{\partial u_{11}^*}{\partial A_2} + 50\dfrac{\partial u_{21}^*}{\partial A_2} \\[2mm] 100\dfrac{\partial u_{21}^*}{\partial A_2} \\[2mm] -50\dfrac{\partial u_{11}^*}{\partial A_2} + 50\dfrac{\partial u_{21}^*}{\partial A_2} \end{Bmatrix}$$

$$= \begin{Bmatrix} 50\left(-0.01294\,(0.0366A_1 + 0.01294A_2)^{-2}\,y_{\omega_1} + (0.0366A_1 + 0.01294A_2)^{-1}\dfrac{\partial y_{\omega_1}}{\partial A_2}\right) \\[3mm] 0 \\[3mm] -50\left(-0.01294\,(0.0366A_1 + 0.01294A_2)^{-2}\,y_{\omega_1} + (0.0366A_1 + 0.01294A_2)^{-1}\dfrac{\partial y_{\omega_1}}{\partial A_2}\right) \end{Bmatrix}$$

$$\text{(a80)}$$

$$S_{T_1}^2 = \left(\frac{\partial \tau_{11}^*}{\partial A_1}\right)^2 \rho_{11} V_1^2 A_1^2 + \left(\frac{\partial \tau_{11}^*}{\partial A_2}\right)^2 \rho_{22} V_2^2 A_2^2$$

$$= \left\{50\left(-0.0366\,(0.0366A_1 + 0.01294A_2)^{-2}\,y_{\omega_1}\right.\right.$$

$$\left.\left. + (0.0366A_1 + 0.01294A_2)^{-1}\frac{\partial y_{\omega_1}}{\partial A_1}\right)\right\}^2 V_1^2 A_1^2 \qquad \text{(a81)}$$

$$+ \left\{50\left(-0.01294\,(0.0366A_1 + 0.01294A_2)^{-2}\,y_{\omega_1}\right.\right.$$

$$\left.\left. + (0.0366A_1 + 0.01294A_2)^{-1}\frac{\partial y_{\omega_1}}{\partial A_2}\right)\right\}^2 V_2^2 A_2^2$$

$$S_{T_2}^2 = 0 \qquad \text{(a82)}$$

$$S_{T_3}^2 = \left(\frac{\partial \tau_{31}^*}{\partial A_1}\right)^2 \rho_{11} V_1^2 A_1^2 + \left(\frac{\partial \tau_{31}^*}{\partial A_2}\right)^2 \rho_{22} V_2^2 A_2^2 \qquad \text{(a83)}$$

$$= S_{T_1}^2$$

Substituting the given parameters of the problem and the derived results into Eqs. (c) and (d) yields the optimization results of 120 cycles shown in Table 11.4.

Table 11.4 Optimization results of example 11.15.1.

Cycle No	r	A_1	A_2	$F(A)$	$W(A)$	G_1	β_D	G_2	β_s	G_3	G_4
0	10.0	1.0	1.0	664.8958	38.2843	−0.01683	154.9513	−0.8415	154.9514	−62.2198	27.7802
1	10.0	1.9355	1.0371	636.4165	65.1301	−0.01835	296.5093	−0.9173	296.5093	−59.6978	30.3022
5	2.40	1.0220	0.5074	186.4095	33.9814	−0.01686	144.1849	−0.8432	144.1850	−59.4601	30.5399
10	0.40	0.5105	0.2099	49.3057	16.5286	−0.01370	59.1618	−0.6852	59.1618	−58.9291	31.0709
15	0.23	0.4237	0.1587	34.4920	13.5706	−0.01241	44.8140	−0.6203	44.8141	−58.6950	31.3050
20	0.1780	0.3917	0.1398	29.6279	12.4790	−0.01178	39.5472	−0.5890	39.5472	−58.5808	31.4192
25	0.1378	0.3640	0.1234	25.6329	11.5233	−0.01115	35.0017	−0.5575	35.0017	−58.4645	31.5355
30	0.1066	0.3392	0.1085	22.3388	10.6786	−0.01049	30.9338	−0.5248	30.9339	−58.3385	31.6615
35	0.08248	0.3176	0.0955	19.6119	9.9393	−0.0098	27.4145	−0.4922	27.4146	−58.2113	31.7887
40	0.06382	0.2987	0.0842	17.3457	9.2918	−0.0092	24.3456	−0.4598	24.3456	−58.0849	31.9151
45	0.04938	0.2822	0.07411	15.4551	8.7226	−0.0085	21.6621	−0.4277	21.6621	−57.9544	32.0456
50	0.03821	0.2677	0.06530	13.8719	8.2241	−0.0079	19.3238	−0.3963	19.3238	−57.8267	32.1733
55	0.02957	0.2549	0.05753	12.5414	7.7859	−0.0073	17.2790	−0.3656	17.2790	−57.7007	32.2993
60	0.02288	0.2438	0.05067	11.4193	7.4022	−0.0067	15.4998	−0.3362	15.4998	−57.5767	32.4233
65	0.01770	0.2340	0.04466	10.4697	7.0645	−0.0062	13.9411	−0.3079	13.9411	−57.4577	32.5423
70	0.01370	0.2253	0.03919	9.6636	6.7656	−0.0056	12.5743	−0.2809	12.5743	−57.3377	32.6623
75	0.01060	0.2178	0.03459	8.9771	6.5091	−0.0051	11.3979	−0.2557	11.3979	−57.2298	32.7702
80	0.00820	0.2112	0.03044	8.3910	6.2792	−0.0046	10.3638	−0.2320	10.3638	−57.1242	32.8758
85	0.00635	0.2055	0.02681	7.8891	6.0792	−0.0042	9.4629	−0.2099	9.4629	−57.0259	32.9741
90	0.00491	0.2003	0.02360	7.4583	5.9021	−0.0038	8.6711	−0.1893	8.6711	−56.9335	33.0665
95	0.00380	0.1959	0.02078	7.0876	5.7474	−0.0034	7.9813	−0.1703	7.9813	−56.8475	33.1525
100	0.00294	0.1919	0.01831	6.7680	5.6118	−0.0031	7.3789	−0.1529	7.3789	−56.7688	33.2312
105	0.00228	0.1885	0.01616	6.4918	5.4920	−0.0027	6.8503	−0.1370	6.8503	−56.6970	33.3030
110	0.00176	0.1854	0.01418	6.2528	5.3866	−0.0024	6.3307	−0.1207	6.3307	−56.6233	33.3767
115	0.00136	0.1828	0.01250	6.0456	5.2940	−0.0022	5.9857	−0.1094	5.9857	−56.6555	33.4449
120	0.00105	0.1804	0.01100	5.8656	5.2116	−0.0019	5.6254	−0.0973	5.6254	−56.5121	33.4879

11.15.2 *Optimum design of shear bending based on Newmark's nondeterministic seismic response spectrum*

Example 11.15.2 – The two-storey shear building shown in Fig. 11.22 is optimized by using Newmark's nondeterministic seismic response spectrum (NNSRS) of Fig. 11.2 to investigate: (A) the sensitivities of the coefficients of structural resistance parameters, and (B) the comparison of the 1st and 2nd variance approaches. The statistics of the amplification factors for the response spectrum in the 50 percentile are adopted here. These statistics for horizontal ground acceleration are $\bar{f}_{\omega 1} = 0.4$, $\bar{f}_{\omega 2} = 2.0$, $\bar{f}_{\omega 3} = 6.0$, $\bar{f}_{\omega 4} = 20.0$, $\bar{\alpha}_d = 1.4$, $\bar{\alpha}_v = 1.66$, $\bar{\alpha}_a = 2.11$, $\bar{\alpha}_{(ad/v^2)} = 6.04$, $\bar{\alpha}_{(v/a)} = 48/g$, $\sigma_{\alpha_d} = 0.64$, $\sigma_{\alpha_v} = 0.66$, $\sigma_{\alpha_a} = 0.56$, $V_{(ad/v^2)} = 0.65$, $V_{(v/a)} = 0.45$, $V_{\alpha_d} = \sigma_{\alpha_d}/\bar{\alpha}_d$, $V_{\alpha_v} = \sigma_{\alpha_v}/\bar{\alpha}_v$, $V_{\alpha_a} = \sigma_{\alpha_a}/\bar{\alpha}_a$.

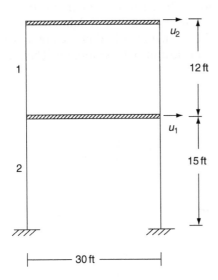

Figure 11.22 Two-storey shear building structure (1 ft = 30.48 cm).

The peak ground accelerations are assumed to be 0.2g. The given data for the building are:

Lumped mass for each storey : 0.27 k-s² in⁻¹ (0.47 kN-s² cm⁻¹);
Allowable displacements : 0.005 times the corresponding height relative to structural base;
Allowable variance of displacements: 0;
Mean yielding strength: $\bar{F}_y = 36$ ksi $(2.448 \times 10^5 \text{ kPa})$;
Mean elastic modulus: $\bar{E}_m = 30,000$ ksi $(2.067 \times 10^8 \text{ kPa})$;
Coefficient of variation of elastic modulus: $V_{E_m} = 0.06$;
Coefficient of variation of moment of inertia: $V_I = 0.05$;
Coefficient of variation of yield moment and critical moment: 0.12 and 0.2, respectively.

Solution:

(A) *Sensitivities of variation of column resistance parameters*
In order to see the sensitivities of the coefficient of variation of the yield moment and critical moment, no variation of peak ground acceleration will be assumed. The failure modes considered are the yielding or torsional buckling failures for each column member.

1.   *Sensitivity of variation of yield moment* Let the coefficient of variation of the yield moment be varied from 0.05 to 0.2, and the coefficient of variation of the critical moment be constant at 0.2, then the optimum weights obtained are shown in Figs. 11.23 and 11.25 for the normal and lognormal distributions, respectively; similarly, the moments of inertia corresponding to the optimum weight are revealed in Figs. 11.24 and 11.26 when the coefficient of variation of yield moment varies. These changes in weight and moment of inertia associated with the normal distribution are not as sensitive as those associated with the lognormal distribution. The weight

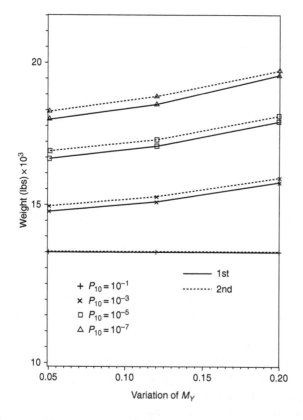

Figure 11.23 Optimum weight for various V_{M_y} with N of two-storey building (1 lb = 4.45 N).

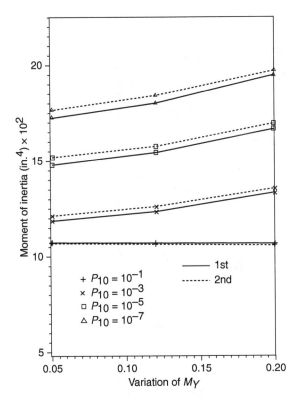

Figure 11.24 I_1 for various V_{M_y} with N of two-storey building $(1\,\text{in.} = 2.54\,\text{cm})$.

percentages for the 2nd variance approach between $V_{M_y} = 0.05$ and $V_{M_y} = 0.2$ are 6.8 per cent at $P_{f0} = 10^{-5}$ and 7.3 per cent at $P_{f0} = 10^{-7}$ with the normal distribution; 14.5 per cent at $P_{f0} = 10^{-5}$ and 18.3 per cent at $P_{f0} = 10^{-7}$ with the lognormal distribution.

The discrepancies of weights and moments of inertia between the 1st and 2nd variance approaches are not sensitive to the change of V_{M_y}. The weight differences between the 1st and 2nd variance approaches at $P_{f0} = 10^{-5}$ for V_{M_y} varying from 0.05 to 0.2 change from 1,708 lbs (7,600.6 N) to 1,852 lbs (7,685.8 N) with the normal distribution and from 22,310 lbs (99,219.5 N) to 25,071 lbs (104,044 N) with the lognormal distribution.

2. *Sensitivity of variation of critical moment* The coefficient of variation of the critical moment is varied from 0.1 to 0.3. The coefficient of variation of the yield moment used is constant at 0.12. The optimum weights given in Figs. 11.27 and 11.28, with the normal and lognormal distributions for the two variance approaches, show that they are not sensitive to the change of coefficient of variation of the critical moment. The moments of inertia are consequently not given here.

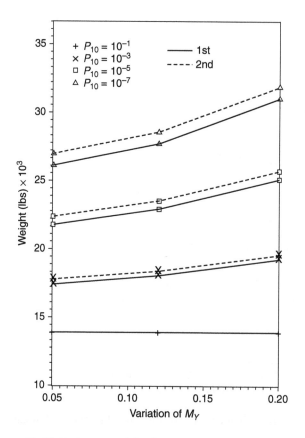

Figure 11.25 Optimum weight for various V_{M_y} with *LN* of two-storey building (1 lb = 4.45 N).

(B) *Comparison of 1st and 2nd variance approach*

In the 2nd variance approach, the recommended value of the coefficient of variation of the natural frequency, V_ω, is 0.16 [14]. The determination of this value is based on the following equation. Since $\omega = (M/K)^{1/2}$, then

$$V_\omega = \sqrt{(0.1)^2 + \frac{1}{\bar{\omega}^2}\left[\left(\frac{\partial \omega}{\partial K}\right)_{\bar{r}}^2 V_K^2 + \left(\frac{\partial \omega}{\partial M}\right)_{\bar{r}}^2 V_M^2\right]}$$

$$= \sqrt{(0.1)^2 + \frac{1}{\bar{\omega}^2}\left\{\left[\frac{1}{2}\left(\frac{K}{M}\right)^{-1/2}\frac{1}{M}\right]_{\bar{r}}^2 \bar{K}^2 V_K^2 + \left[\frac{1}{2}\left(\frac{K}{M}\right)^{-1/2}\frac{K}{M^2}\right]_{\bar{r}}^2 \bar{M}^2 V_M^2\right\}}$$

$$= \sqrt{(0.1)^2 + \frac{1}{4}\left(V_K^2 + V_M^2\right)} = 0.16 \tag{a}$$

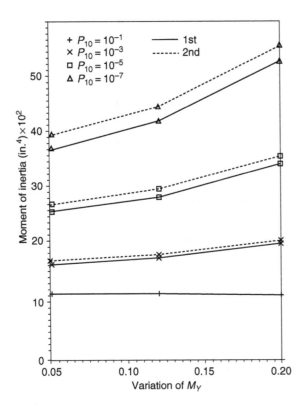

Figure 11.26 I_1 for various V_{M_y} with *LN* of two-storey building (1 in. = 2.54 cm).

where 0.1 is the estimation of the natural frequency which may reflect the influence of non-structural elements, soil structure interactions; $\bar{\omega}$, V_ω = the mean and coefficient of the variation of the natural frequency; V_M = the coefficient of variation of mass = 0.12; V_K = the coefficient of variation of the stiffness value and is assumed to be

$$V_K = \sqrt{(0.2)^2 + V_{E_m}^2 + V_I^2} = 0.21 \tag{b}$$

in which 0.2 is the stiffness formulation error, V_{E_m} = the coefficient of variation of elastic modulus = 0.06, V_I = the coefficient of variation of moment of inertia = 0.05 [12].

Therefore, there are three assumed values for formulation errors which are not clearly defined for the 2nd variance approach. The first value, 0.2, is the stiffness formulation error. The second value, 0.1, is the natural frequency formulation error. The third is the value of 0.15, C_E, is a coefficient of variation of dynamic analysis

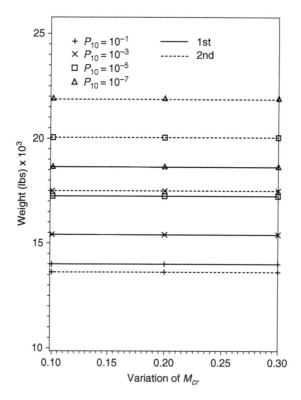

Figure 11.27 Optimum weight for various $V_{M_{cr}}$ with N of two-storey building (1 lb = 4.45 N).

error. If the first value is excluded, the coefficient of variation of stiffness becomes

$$V_K = \sqrt{V_{E_m}^2 + V_I^2} = 0.08 \qquad (c)$$

If the second one is also excluded, the coefficient of variation of natural frequency is

$$V_\omega = \left\{ \frac{1}{4} \left(V_K^2 + V_M^2 \right) \right\}^{1/2} = 0.072 \qquad (d)$$

In addition to the recommended values of 0.16 for the coefficient of variation of natural frequency and 0.15 for the coefficient of variation of dynamic analysis error, C_E, two other assumed values for the 2nd variance approach can be used to see the comparison of the 1st and 2nd variance approaches. One assumes the value of 0.072 for the coefficient of variation of natural frequency and zero value for the dynamic analysis error. The other assumes the zero value for the coefficient of variation

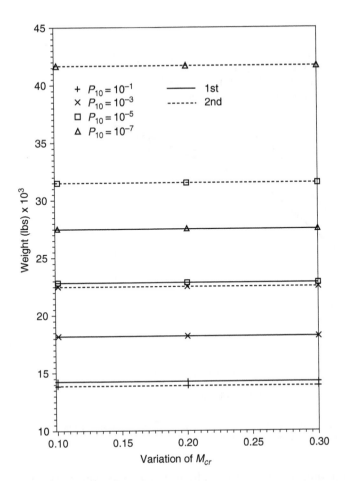

Figure 11.28 Optimum weight for various $V_{M_{cr}}$ with LN of two-storey building (1 lb = 4.45 N).

of natural frequency and 0.15 for the coefficient of variation of dynamic analysis error.

The optimum weights in Figs. 11.29 and 11.30 reveal that the 1st variance approach is close to the 2nd variance approach when $V_\omega = 0.072$ and $C_E = 0$ as well as $V_\omega = 0$ and $C_E = 0.15$ for the normal and lognormal distributions. When reliability increases or probability failure decreases, the differences among them increase. The 2nd variance approach with $V_\omega = 0.16$ and $C_E = 0.15$ has the largest optimum design results.

11.16 Concluding remarks

Reliability-based optimization techniques are developed for steel structures subjected to static load and various live load models as well as seismic

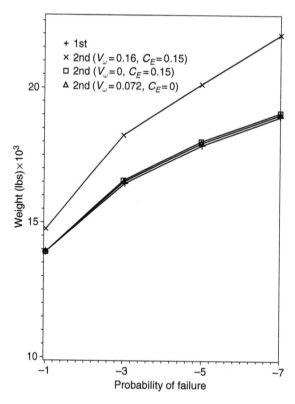

Figure 11.29 Optimum weight for various variance expressions with N of two-storey building (1 lb = 4.45 N).

excitation including the Uniform Building Code, and Newmark's seismic inelastic design spectrum as well as a nondeterministic seismic response spectrum. The optimization techniques are derived from the generalized optimality-criteria method and the interior-penalty-function method and the reliability is based on two mathematical models of normal and lognormal distributions with two different variance approaches. The objective function can be either minimum weight or minimum total cost consisting of construction and expected failure costs at various safety levels. Observations of numerical studies illustrate some interesting results: for example, (1) for the dead and live load case, the model of Great Britain requires heavier design than the models of U.S.; (2) at a high-reliability criterion, the lognormal distribution requires a heavier design than the normal distribution. Because of the complexity of nondeterministic problems, the penalty function is more convenient than the generalized optimality-criteria method and is thus recommended [17–20].

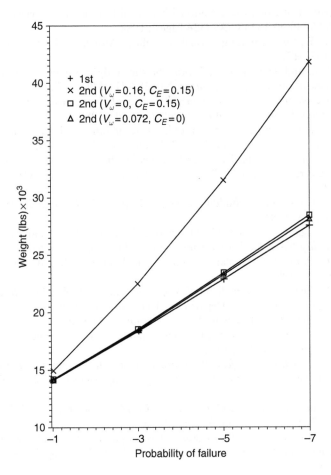

Figure 11.30 Optimum weight for various variance expressions with *LN* of two-storey building (1 lb = 4.45 N).

References

1. Chang, C.C., Ger, J.R. and Cheng, F.Y., Reliability-based optimum design for UBC and non-deterministic seismic spectra, *Journal of Structural Engineering*, ASCE, Vol. 120, No. 1, pp. 139–160, 1994.
2. Grandori, G. and Benedetti, D., On the choice of the acceptable seismic risk, *International Journal of Earthquake Engineering and Structural Dynamics*, Vol. 2, No. 1, pp. 3–10, 1973.
3. Ang, A.H.-S. and Cornell, C.A., Reliability bases of structural safety and design, *Journal of the Structural Division*, ASCE, Vol. 100, No. ST9, pp. 1755–1769, 1974.
4. Cheng, F.Y. and Chang, C.C., *Safety-based Optimum Design of Nondeterministic Structures Subjected to Various Types of Seismic Loading*, NSF Report,

U.S. Department of Commerce, National Technical Information Service, Virginia, NTIS No. PB90-133489/AS (326 Pages), 1988.

5. Rosenblueth, E., On computing normal reliabilities, *Structural Safety*, Vol. 2, No. 3, pp. 165–167, 1985.

6. Ellingwood, B. and Ang, A.H.-S., A probabilistic study in safety criteria for design, *Structural Engineers Registration*. No. 387, U. of Ill., 1972.

7. NBS 577, *Development of a Probability Based Load Criterion for American National Standard A58*, National Bureau of Standards, 1980.

8. Ellingwood, B. and Culver, C., Analysis of live loads in office buildings, *Journal of the Structural Division*, ASCE, Vol. 103, No. ST8, pp. 1551–1560, 1977.

9. Mitchell, G. and Woodgate, R., Floor loadings in office buildings the result of a survey, Current Paper 3/71, *Building Research Station*, Gartson, Watford, England, 1971.

10. International Conference of Building Officials, *Uniform Building Code*, 1984 edn, Whittier, CA, 1984.

11. Hwang, H., Kagami, S., Reich, M., Ellingwood, B., Shinozuka, M., and Kao, C.S., *Probability based load combination criteria for design of concrete containment structures*, U.S. regulatory Commission Report NUREG/CR-3876, Washington, DC, 1985.

12. Rojiani, K.B., *Evaluation of the Reliability of Steel Buildings to Wind Loadings*, Ph.D. Thesis, Civil Engineering Department, University of Illinois, 1978.

13. Mohraz, B., Hall, W. and Newmark, N., *A Study of Vertical and Horizontal Earthquake Spectra*, Division of Reactor Standards, U.S. Atomic Energy Commission Washington, DC, 1972.

14. Gallo, M.P., *Evaluation of safety of Reinforced Concrete Buildings to Earthquakes*, Ph.D. Thesis, Department of Civil Engineering, University of Illinois, 1971.

15. Roos, N.R. and Gerber, J. S., *Government Risk Management Manual*, Risk Management Publishing, Tuscon, Arizona, 1976.

16. Surahman, A. and Rojiani, K.B., Reliability based optimum design of concrete frames, *Journal of the Structural Division*, ASCE, Vol. 109, No. 3, pp. 741–757, 1983.

17. Cheng, F.Y., Introduction of optimum structural design and its recent developments, *Frontier Technologies for Infrastructure Engineering*, Ang, A.H.-S. (ed.), CRC Press, Taylor and Francis Group, Boca Raton, FL, Chapter 4, 2009.

18. Davidson, J.W., Felton, L.P. and Hart, G.C., On reliability-based optimization for earthquake, *Computers and Structures*, Vol. 12, pp. 99–105, 1980.

19. Lev, O.E. (ed.), *Structural Optimization: Recent Development and Applications*, ASCE Publications, 1981.

20. Liu, S.C., Dougherty, M.R. and Neghabat, F., Optical aseismic design of building and equipment, *Journal of the Engineering Mechanics Division, American Society of Civil Engineers*, Vol. 102, No. EM3, Proceeding Paper 12208, pp. 395–414, 1976.

12 Multi-objective optimization with genetic algorithm, fuzzy logic and game theory

12.1 Introduction

This chapter develops a constrained multi-objective optimization method in the form of a robust, practical, problem-independent algorithm, and investigates the effect of multi-objective optimization on structural design. A single-objective optimization formulation can produce an optimum design with respect to a specified optimal objective, but the design may not always be 'good' [1]. Consider a hypothetical example. If a structure is optimized for minimum cost subject to constraints on stress, displacement, buckling and vibration period, then an economical design is obtained. However, the structure may exhibit a poor dynamic response under seismic loading. If minimum earthquake input energy is also included in the optimum criteria, then a more rational, compromise design may be achieved. Through the analytical procedure, it is clear that multi-objective formulations can better serve the engineering design process when several conflicting objectives must be satisfied.

A multi-objective optimum procedure comprises constructing an analysis model, deciding objective functions and constraints, and selecting mathematical programming techniques. The solution technique presented hereafter is based on the Pareto genetic algorithm. Genetic algorithm search procedures display the intelligent characteristics of utilizing and/or learning from information generated from previous stages. New designs are produced based on features of existing designs. The Pareto genetic algorithm requires no gradient information and locates multiple optima rather than a single, local optimum. It is able to cope with ill-behaved functions and to discontinue optimization problems. The Pareto genetic algorithm is robust, stable and computationally feasible even for extremely difficult problems.

When a structural optimal design includes more than one objective, it is in the realm of game theory. Game theory is applied to provide a method for understanding and guiding the optimization process, and finding a solution that can be accepted by all players. Fuzzy theory can effectively use the vague and imprecise information presented in objective functions and constraints to formulate a fuzzy optimization model. This presentation treats optimization problems with 'precise' (determined) specifications for objectives and constraints, and searches for the 'exact' optimum solution. Instead of using fuzzy set theory as a tool to solve a design problem, the theory is used to construct a *fuzzy constrained environment*.

12.2   Fundamentals of multi-objective formulation

12.2.1   *Multi-objective optimization*

As discussed in the introduction, multi-objective optimization involves more than one objective (criterion) to consider in a design option, and these objectives usually conflict with each other. Multi-objective optimization is defined here as determining a vector of design variables within a feasible region and minimizing (or maximizing) a vector of objective functions. It can be expressed as follows:

$$\left.\begin{array}{l} \text{Minimize } \{f_1(x),\dots,f_n(x)\} \\ \text{Subject to } g_i(x) \leq 0 \quad i=1,2,\dots,m \\ \quad\quad x_i^{(l)} \leq x_i \leq x_i^{(u)} \quad i=1,2,\dots,L \end{array}\right\} \qquad (12.1)$$

where $f_i(x)$ is the objective function which defines the merit of a set of design variables; x is the design vector; $g_i(x)$ is the behavioural constraint (it is assumed that constraints are normalized to place them in the same general order of magnitude); and $x_i^{(l)}$ and $x_i^{(u)}$ are the side constraints. When $n=1$, it is a single-objective (scalar) optimization problem. Otherwise, it is a multi-objective (vector) optimization problem. Any point (set of design variables) in the design space defines a possible design for the structure. If the design satisfies the given constraints, it is called a *feasible* or *admissible design*. Constraints divide the design space into feasible and infeasible regions.

The main feature of a multi-objective optimization problem is the appearance of an objective conflict; that is, none of the feasible solutions allows simultaneously minimizing all objectives. Therefore, solutions to a multi-objective optimization problem can be defined as follows. If vector x^* is a solution to the problem in Eq. (12.1), there exists no feasible vector x which would decrease some objective functions without causing a simultaneous increase in at least one objective function. This is the same as the definition of *Pareto optimum (nondominated solution)*.

A feasible vector x^* is a Pareto optimum for the problem defined in Eq. (12.1) if and only if there exists no feasible vector x such that [2]

$$\left.\begin{array}{l} f_i(x) \leq f_i(x^*) \quad \text{for all } i \in \{1,2,\dots,m\} \\ f_i(x) < f_i(x^*) \quad \text{for all least one } i \in \{1,2,\dots,m\} \end{array}\right\} \qquad (12.2)$$

Fig. 12.1 shows a two-objective function, two-design-variable maximization problem, where the Pareto optimal set lies on the solid line *AB*. To

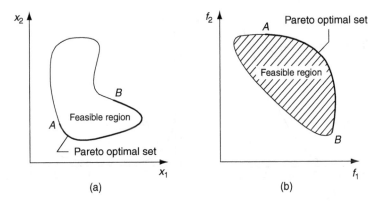

Figure 12.1 Two-objective maximization problem: (a) Pareto optimal set in decision space, (b) Pareto optimal set in objective space.

demonstrate the basic concepts of multi-objective optimization, consider the following example:

$$\left.\begin{array}{l} \text{Minimize } F(x) = \{f_1(x) = (x-1)^2,\ f_2(x) = (x-3)^2\} \\ \text{Subject to } x \geq -1 \end{array}\right\} \quad (12.3)$$

Obviously, the ideal situation would be a single design variable x which minimizes both functions simultaneously. But no such solution exists. The optimum procedure from Eq. (12.3) is illustrated in Fig. 12.2, where P stands for the Pareto optimal set. In Fig. 12.2a, $(F\text{-}x)$ space, f_1 has its minimum $f_1 = 0$ at $\tilde{x}_1 = 1$ while $f_2 = 4$, and f_2 has its minimum $f_2 = 0$ at $\tilde{x}_2 = 3$ while $f_1 = 4$. A compromise solution, $\tilde{x} \in P$ ($x \in [1,3]$), should be chosen, where P is the Pareto optimal set. Note that choosing as a solution $\hat{x} = 4 \notin P$ with $f_1 = 9$ and $f_2 = 1$ is not a proper choice. This is because both criteria can be improved by making an alternative choice $\hat{x} = 3.5$ with $f_1 = 6.25$ and $f_2 = 0.25$. If a solution in the Pareto optimal set is chosen, one objective cannot be improved without detracting from another. For example, $\tilde{x} = 2$ is a Pareto optimum with $f_1 = 1$ and $f_2 = 1$. If an alternative solution $\tilde{x} = 2.1$ is chosen with $f_1 = 1.21$ and $f_2 = 0.81$, the objective f_2 is improved from 1 to 0.81; however, the objective f_1 deteriorates from 1 to 1.21. In Fig. 12.2b, the solution Q is in the Pareto optimal set P, and is a compromise solution. Any point in the set can be a solution candidate for the problem. It depends on the choice of the decision-maker. In this example, if f_1 is taken as the main factor and f_2 is ignored, point A is the ideal result since it represents the lowest cost for f_1. Yet point B represents the lowest cost for f_2. Since all the criteria should be considered, the optimum result falls between A and B of the Pareto optimal set, such as point Q. As noted, multi-objective optimization is a trade-off procedure.

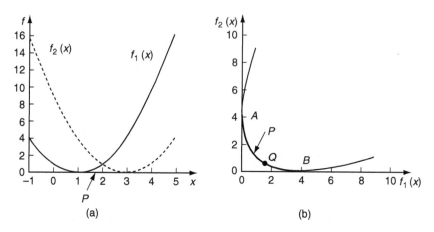

Figure 12.2 Two-objective minimization problem: (a) objectives in *F-x* space, (b) Pareto optimal set.

12.2.2 *Objective functions*

As presented in previous chapters, objective (cost, criterion) function is a function of design variables. It is the merit measurement (criterion) of a design. When an engineer designs a structure, numerous acceptable (feasible) designs can be chosen. Among them, some need less monetary investment, some offer greater safety factors, some have less dynamic response than others, and so on. To make the best choice, appropriate criteria to compare various designs should be constructed. Selection of objective functions in an optimum design is highly important. This selection decides the optimum (search) direction and optimum result of a design. For example, minimizing structural weight usually decreases structural stiffness while minimizing top-storey displacement increases structural stiffness. Objective functions are constructed according to various optimization purposes. Several objective functions are discussed as follows.

1 *Structural Cost* – This function consists of material, fabrication and installation costs (see Chapter 7). Minimizing structural cost makes a design more competitive, and is generally the aim of an investor. There are four ways to estimate the cost of a structure (or construct the cost function) [3]: unit price estimate, system estimate, square foot and cubic foot estimate, and order of magnitude estimate. The first method of unit price estimate requires working drawings and specifications, and has the best rate of accuracy (±5 per cent). However, this method can only be executed after completion of the drawings, which makes it impractical for the initial design stage and requires extremely detailed work. Calculations of the unit price estimate take time, and the process

is so tedious that an optimum design with the unit price estimate as the cost objective may lose its attraction in the initial design stage. The second method of system estimate involves grouping the whole structure into component parts. In terms of structural optimization, the super-structure can be divided into floors, roof, frames and walls (concrete, steel, wood or a combination of the three). The system estimate has an accuracy rate of ± 10 per cent and usually takes several days to complete. This method is well-suited for optimum design with cost as one of the objective functions. The third method, square foot and cubic foot estimate, is another candidate to construct the structural cost function for an optimum design. This type of estimate uses square foot and cubic foot tables to calculate the cost. It has a ± 15 per cent rate of accuracy, and can be finished within a day. The fourth method, the order of magnitude estimate, is based on building type, area and location, applies only to advance planning of a project, and has no direct connection with structural members. This method is not suitable for the construction of an optimization objective. With any approach, required data for the structural cost function can be found in various publications. Fig. 12.3 [3, 4] shows the relative time and accuracy for estimating a $2,000,000 building with the four different methods.

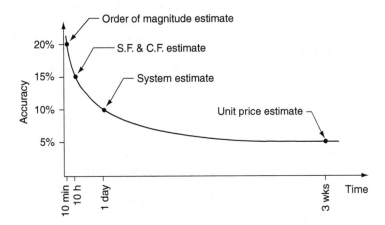

Figure 12.3 Time and accuracy of estimates for a $2,000,000 building.

The structural cost function can be expressed as

$$f_1 = \sum C_{i1} V_i + \sum C_{i2} U_i \tag{12.4}$$

where C_{i1} is volume cost; V_i is volume of the structural member; C_{i2} is installation cost; and U_i is an installed structural member. Cost parameters vary with estimate method, type of material (steel, concrete and wood) and structural system.

2   *Structural Weight* – To better utilize materials and reduce costs of a structure, structural weight can be chosen as the objective function and written as

$$f_2 = W = \sum \rho_i A_i l_i \qquad (12.5)$$

where ρ_i, A_i and l_i are weight density, cross-sectional area and length of member *i*, respectively. If the weight density ρ_i is the same for all members, then structural volume can be taken as the objective function. In general for optimum design, the structural weight function acts in a similar manner to the cost function (Eq. (12.4)) when only the structural member cost is considered and when structural members use the same material for the entire structural system. This function has been extensively used in previous chapters.

3   *Strain Energy* – In a given structure, when the *internal work (strain energy)* done by the stresses and the strains has a minimum value, the structure may have an optimal result. Strain energy is given by

$$f_3 = E_s = \sum \sigma_i \varepsilon_i v_i \qquad (12.6)$$

where σ_i, ε_i and v_i are respectively stress, strain and volume of member i.

4   *Potential Energy* – Minimizing the structural *potential energy* can reduce the effects of external forces and increase the safety level of a structure. If the given loads are $\{\bar{P}\} = \{p_1, p_2, \ldots, p_n\}$ and the corresponding displacements are $\{\Delta\} = \{\delta_1, \delta_2, \ldots, \delta_n\}$, the potential energy of the structure is

$$f_4 = E_p = \{\bar{P}\}^T \{\Delta\} = \sum p_i \delta_i \qquad (12.7)$$

5   *Displacements* – To minimize changes in the shape of a structure under the action of different loading conditions, displacements at selected points or regions of the structure are taken as the objectives. For example, when required to minimize the *i*th node displacement, the objective function will be

$$f_5 = \delta_i \qquad (12.8)$$

At the same time, δ_i can still be taken as a constraint; $\delta_i \leq \delta_{i,a}$ is the allowable displacement at the *i*th node.

6   *Seismic Structure (Input Energy)* – When a structure is subjected to earthquake excitations, reducing the dynamic response and damage caused by seismic loads is the main design consideration. Decreasing dynamic response can be done by optimizing such factors as acceleration of mass, fundamental period and earthquake input energy. The

earthquake input energy of a structure represents the work done by the base shear on the structure as that shear moves through ground displacement, and can be expressed as

$$f_6 = E_i = \sum D_i^2 \left(\frac{1}{2} MS_{vi}^2 \right) \tag{12.9}$$

where $D_i = \{\Phi_i\}^T \{m\}/\sqrt{M}$ is the ith mode energy parameter. $\{\Phi_i\}$ is the normalized ith mode. $\{m\} = \{m_1, m_2, \ldots, m_n\}$, $M = \Sigma m_i$ is the total mass of the structure, and S_{vi} is the response spectrum of velocity relative to the ith mode.

7   *Control Structure* – For a *control system* optimization problem, design variables include feedback gain and parameters of a control system. Constraints are put on the closed-loop damping factor, frequencies and design requirements. Usually, optimization of a control system minimizes a specified performance index for the purpose of reducing control energy or effective damping response time. In order to design a controller using a linear quadratic regulator, a *performance index* (PI) can be defined as [5]

$$PI = \int_0^{t_f} (\{z\}^T [Q] \{z\} + \{u\}^T [R] \{u\}) \, dt \tag{12.10}$$

where $[Q]$ is the positive semi-definite state weighting matrix, $[R]$ is the positive definite control weighting matrix, $\{z\}$ is the state displacement vector, and $\{u\}$ is the control input vector. Minimizing the *quadratic performance index* and satisfying the structural system state equation gives the state feedback control law

$$\{u\} = -[R]^{-1} [B]^T [P] \{z\} = -[G] \{z\} \tag{12.11}$$

where $[G]$ is the closed-loop gain matrix, and $[P]$ is the *Riccati matrix*. The objective function to minimize the performance index can be taken as

$$f_7 = \{z_0\}^T [P] \{z_0\} \tag{12.12}$$

Here, z_0 is the initial state vector (disturbance vector). The effective damping response time is given by

$$f_8 = \frac{\{z_0\}^T [P] \{z_0\}}{\{z_0\}^T [Q] \{z_0\}} \tag{12.13}$$

where $[Q]$ is the weighting matrix. Note the special requirements of earthquake structural engineering control. Due to the nature of seismic input data, the *generalized optimal active control* algorithm (GOAC) should be used [5].

In an optimum design problem, a maximization problem can be treated as a minimization one. This entails no loss of generality since the minimum of $-f(x)$ occurs where the maximum of $f(x)$ takes place, i.e., $\max f(x) = -\min[-f(x)]$. Similarly, a minimization problem can be optimized as a maximization one.

12.2.3 Multi-objective optimum algorithms

As noted, multi-objective algorithms can be classified into two fundamental categories. First, the original multi-objective optimization problem is transformed into a scalar substitute one, which contains contributions from all the objectives. Second, vector-optimization techniques are used. Optimum results can also be divided into two types. The first is to search for the 'best' compromise solution by optimizing a scalar substitute function. This solution varies with the search methods in most multi-objective optimization problems. The solution can be determined when a scalar interpretation algorithm is selected. The second is to locate the Pareto optimal set (nondominated solutions). The result is a collection of solutions instead of only one. A final design is chosen from that collection through a trade-off procedure. Among several approaches, two that are better known for handling multi-objective design problems are briefly described here. For this purpose, the problem in Eq. (12.1) is taken as the optimum design formulation. It is assumed that the separate objectives $f_i(x)$ are formulated to be consistent with the direct optimization for each of them, and X defines the feasible region.

$$\left. \begin{array}{l} \text{Minimize } \{f_1(x), ..., f_n(x)\} \\ \text{Subject to } x \in X \end{array} \right\} \tag{12.14}$$

1 *Weighting Method* – The weighting method [6, 7] belongs to the scalar interpretation. It assumes that a set of non-negative, not all zero numbers, $w_1, ..., w_n$, which reflect the relative importance of the objectives, is given. Using the given set, the problems (Eq. (12.14)) can be reduced to a single-objective problem:

$$\left. \begin{array}{l} \text{Minimize } \sum_{i=1}^{n} w_i f_i(x) \\ \text{Subject to } x \in X \\ w_i \geq 0 \quad i = 1, ..., n \end{array} \right\} \tag{12.15}$$

Eq. (12.15) is a single-objective optimization problem. The solution obtained by solving this problem depends only on the choice of weights w_i, and the solution is at least a weak, nondominated one. A Pareto optimal set can be determined by varying the weights w_i of the objective functions. The main disadvantages of the weighting method are the difficulty of choosing the weight (especially in cases where the optimal solution is very sensitive to variation of weights) and the method's inability to produce the whole Pareto optimal set for non-convex problems. Fig. 12.4 illustrates the weighting method for a two-objective optimization case, where $F(X)$ is the image of the feasible region in the objective space. Points A, B and C are nondominated solutions. They are candidates for the optimum solution. Point A is obtained by letting $w_2 = 0$. Points B and C are obtained by taking different combinations of w_1 and w_2.

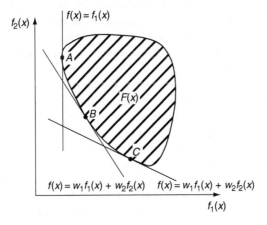

Figure 12.4 Geometrical interpretation of weighting method in two-objective case.

2 *Genetic Algorithm* (GA) [8] – The genetic algorithm is a parallel and evolutionary search technique. It is typical of large-scale optimization techniques. The genetic algorithm is based on biological principles and operates analogous to evolution. Unlike many other search techniques which maintain a single 'current best' solution and try to improve it, a GA maintains a set of possible solutions called *population*. This population is improved by a cyclic evolutionary search process. Each cycle is called a *generation*. Ultimately, the population reaches the optimum goals. GAs have successfully been applied to optimization of nonlinear programming [9]. They require no gradient information and produce multiple optima rather than a single optimum, and are capable of searching for optima in function spaces which cause difficulty for gradient techniques. The Pareto GA [10, 11] is an extension of traditional GAs which apply to single-objective optimization problems. The Pareto GA allows the search of parameter spaces where multiple objectives are optimized simultaneously. By maintaining a population of solutions, the Pareto GA can conduct a parallel search for a group of nondominated

solutions and find a representative set of such solutions. More on this algorithm will be presented later along with detailed developments.

12.2.4 General procedures and design decisions on multi-objective optimization of structures

The procedure for the multi-objective optimization of building structures is shown in Fig. 12.5. Step 1 (a) is the initial design which includes defining the structural system type, i.e., frame, frame-tube, shear wall or frame-shear wall system; position of structural members; material used for structural members; cross-sectional area of structural members; structural dimensions, i.e., storey heights and bay lengths. Step 2 covers (b), (c) and (d): determining the optimum objectives or goals, such as minimizing materials, structural costs, and dynamic responses; choosing design variables which can be changed to achieve the optimum objectives, such as cross-sectional areas of structural members, or type of material; and setting the constraint conditions so that the optimum design of a structure satisfies code specifications, materials' limitations, owner's requirements, and so on. Step 3 (e) is the structural-optimization procedure which comprises optimization analysis and structural static and/or dynamic analysis. Structural static and/or dynamic analytical procedures use finite-element methods to find structural displacements, internal forces, stresses and dynamic characteristics which generate the values of objective and constraint functions. The optimum analytical procedure guides any change in the direction of the different variables. Some variables may be increased, some may be decreased. These analytical procedures

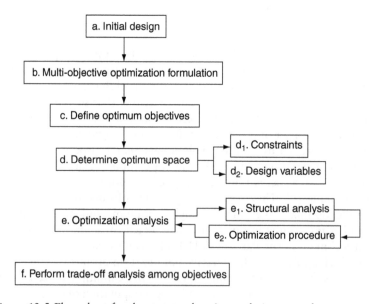

Figure 12.5 Flow chart for the structural optimum design procedure.

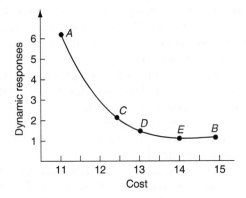

Figure 12.6 Trade-off procedure between two objectives.

operate as cycles. A proper optimum algorithm can generate a complete set of nondominated solutions (Pareto optimal set). Step 4 (f) performs trade-off analysis, based on the nondominated solution set obtained, to choose the final optimal design. A trade-off procedure is illustrated in Fig. 12.6.

A two-objective optimization problem is shown. Optimum objectives are to minimize structural cost and dynamic response. The solid line in Fig. 12.6 locates the nondominated (Pareto optimal) solutions for which improving one objective worsens another objective. Any point in the nondominated solutions represents a possible optimum design for the multi-objective optimization problem. Solution choice is a decision-making procedure involving a trade-off. Choosing point *A* allows minimum structural cost, but dynamic responses would increase which might cause serious structural damage under strong earthquake action. Choosing point *B* gives greater safety but higher structural cost. Point *C* offers a compromise solution between the two objectives. Compared to point *E*, point *D* allows significant cost savings with a small increase in dynamic response from 1.15 to 1.5. The final selection depends on the perspective of the decision-maker. To make a trade-off procedure feasible, the complete nondominated set of solutions of a given problem should be generated. This set gives greater insight into the array of competing objectives and facilitates the trade-off analysis so that decision-makers can see the benefit gained from one objective by making concessions in another.

12.3   Game theory

12.3.1   *Cooperative concepts in game theory*

Game theory was developed for both cooperative and non-cooperative games. *Cooperative games* are those in which participants have the opportunity to communicate with one another and to form binding and enforceable agreements. In *non-cooperative games*, each player acts independently in an effort to maximize his/her own pay-off, which produces an outcome that may be favourable for one player but unfavourable for another. The concept of player cooperation therefore becomes important when considering compromise game outcomes. The measurement of success of cooperative play is embodied in the concept of the Pareto optimum; a Pareto optimum has the property that if any other solution is used, at least one player does worse or they all do the same. A cooperative game theory consists of ways to analyze conflicts existing in objectives or interest groups (players), to provide a neutral forum for discussion and negotiations among players, and then to suggest a compromise solution acceptable to all players.

Assume that there are n decision-makers, and the decision variables controlled by the ith decision-maker are denoted by x_i. Then x_i is called the strategy of the decision-maker or player i. Note that set X of all possible decision vectors, $x_i = \{x_1, x_2, \ldots, x_n\}_i$, is called the set of simultaneous strategies. Let the objective function of the ith individual player be f_i; then f_i is called the *pay-off function* of player i. In game theory, players are assumed to be rational in the sense that they attempt to maximize their pay-off, or objective. A pay-off vector is $f = \{f_1, f_2, \ldots, f_n\}$ for n players. The problem facing each player in an n-person bargaining game is the selection of one pay-off vector from the set of pay-off vectors. This choice must be unanimously agreed by all n players, and is an equilibrium point. A vector $(x_1^*, x_2^*, \ldots, x_n^*) \in x$ is called the equilibrium point of game A with vector-valued pay-off, if, for all and fixed $x_1^*, \ldots, x_{i-1}^*, x_i, x_{i+1}^*, \ldots, x_n^*$, the strategy $x_i = x_i^*$ is a nondominated solution of the following problem:

$$\left. \begin{array}{l} \text{Maximize} f_i \left(x_1^*, \ldots, x_{i-1}^*, x_i, x_{i+1}^*, \ldots, x_n^* \right) \\ \text{Subject to} \left(x_1^*, \ldots, x_{i-1}^*, x_i, x_{i+1}^*, \ldots, x_n^* \right) \in x \end{array} \right\} \tag{12.16}$$

For cooperative game theory, Nash [12] developed a solution concept for two-person bargaining games. Harsanyi [13] extended the Nash procedure to the n-person game. In two-person bargaining games, a unique bargaining solution exists. The Nash–Harsanyi model takes the following form:

$$\left. \begin{array}{l} \text{Maximize} \prod_{i=1}^{n} (f_i - f_{di}) \\ \text{Subject to} f_i \geq f_{di}; \ f \in F \end{array} \right\} \tag{12.17}$$

where $f_d = \{f_{d1}, f_{d2}, \ldots, f_{dn}\}$ is a disagreement pay-off vector and F is the set of pay-off vectors. This vector essentially represents the pay-off that players would receive if they could not agree on a final pay-off. Their agreement must be unanimous. Disagreement values are regarded as given quantities. There is no feasible solution which yields every player a higher pay-off than the solution $f = \{f_1, f_2, \ldots, f_n\}$. It is a symmetrical game; every player receives the same pay-off.

12.3.2 *Multi-objective optimization with game theory*

A multi-objective optimization problem can be cast as a cooperative game problem which assumes that each player is associated with an objective. The objective function f_i can be regarded as the pay-off of the ith player. Fig. 12.7 shows a two-objective maximization problem with two scalar design variables x_1 and x_2. Fig. 12.7a shows the objective function space, and Fig. 12.7b shows the design variable space. In Fig. 12.7a, the feasible zone is indicated. Solid line A-Q_0-Q-B represents the nondominated optimal set. In this set, improving any objective worsens at least one other objective. Moving point Q_0 to Q increases the pay-off of f_1; however, it decreases the pay-off of f_2 at the same time. Point P is not a nondominated point; therefore, P moves to Q_0 which improves the pay-offs of f_1 and f_2 simultaneously.

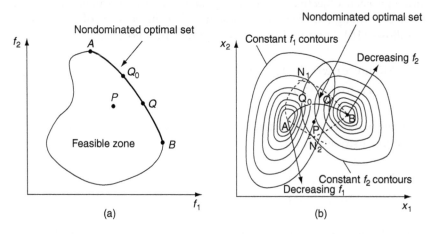

Figure 12.7 Two-objective maximization problem: (a) objective function shape, (b) cooperative and non-cooperative game solutions (design variable space).

In Fig. 12.7b, the two-objective optimum problem can be illustrated by a two-person bargaining game. The dotted lines passing through A and B stand for the loci of maximizing f_1 and f_2 for fixed values x_1 and x_2, respectively. Intersection points, N_1 and N_2, of the lines are candidates for the two-objective maximization problem assuming that the players, whose pay-offs are f_1 and f_2, respectively, are in a non-cooperative game. These points are the equilibrium points (Eq. (12.16)) from which no player can deviate unilaterally for further improvement of his/her own criterion. If both players are in a cooperative game, the nondominated set (line A-Q_0-Q-B) will provide a better solution than their respective equilibrium solutions. Every point on this line has the property of not being dominated by any other point in its neighbourhood. For this reason, the following equation always holds (see Fig. 12.7b).

$$\left.\begin{array}{l} f_1(Q_0) \geq f_1(P) \\ f_2(Q_0) \geq f_2(P) \end{array}\right\} \tag{12.18}$$

Thus, only the points in the nondominated set can be candidates for a multi-objective optimum problem. Since a cooperative game can produce a *nondominated set* (also known as a *Pareto optimal set*), the main task in a multi-objective optimum problem is to locate the set. After locating the set, the optimum solution for a multi-objective optimization problem can be picked up from the set based on the preference of a decision-maker. In the following procedure, the two steps are converted into one step: searching for the best compromise solution in the nondominated set.

With cooperative multi-objective optimization, the compromise solution should ascertain that each objective obtains its maximum possible value although each objective cannot arrive at its own best value. An optimal trade-off among the objectives is sought using the concept of game theory as follows [10]. For a multi-objective optimization problem,

$$\left.\begin{array}{l} \text{Minimize } f_i(x) \quad i = 1,\ldots,m \\ \text{Subject to } g(x) \leq 0 \end{array}\right\} \tag{12.19}$$

First, the m individual objective functions are minimized respectively subject to given constraints.

$$\left.\begin{array}{l} \text{Minimize } f_i(x) \\ \text{Subject to } g(x) \leq 0 \end{array}\right\} \tag{12.20}$$

For each objective function f_i, an optimal solution x_i^* is obtained; then a pay-off matrix is constructed as

$$[P_0] = \begin{pmatrix} f_1(x_1^*), & f_2(x_1^*), & \ldots, & f_m(x_1^*) \\ f_1(x_2^*), & f_2(x_2^*), & \ldots, & f_m(x_2^*) \\ \ldots & & \ldots & \\ f_1(x_m^*), & f_2(x_m^*), & \ldots, & f_m(x_m^*) \end{pmatrix} \tag{12.21}$$

For each objective function, its best and worst values in the Pareto set can be obtained from the above matrix

$$\left.\begin{array}{l} f_{i,\min} = f_i(x_i^*), \quad i = a,\ldots,m \\ f_{i,\max} = \max\left[f_i\left(x_j^*\right)\right], \quad j = 1,\ldots,m, \quad i = a,\ldots,m \end{array}\right\} \tag{12.22}$$

From the pay-off matrix, it can be observed that during cooperative optimization, the ith objective function should not expect a value better than

$f_{i,\min}$ but this value should not be worse than $f_{i,\max}$. Based on the above consideration, a substitute objective function can be constructed as

$$S = \prod_{i=1}^{m} \frac{[f_{i,\max} - f_i(x)]}{[f_{i,\max} - f_{i,\min}]} = \prod_{i=1}^{m} \overline{f}_i(x) \tag{12.23}$$

In Eq. (12.23), when $f_i(x)$ is equal to $f_{i,\max}$, S becomes zero and f_i achieves its worst value. When objective function $f_i(x)$ is equal to $f_{i,\min}$ ($i=1,\ldots,m$), S becomes 1 and all objectives achieve their best values. For a rationally defined multi-objective optimization problem, the second case in which all objectives achieve their best values cannot occur. Therefore, the value of S is $0 \le S \le 1$.

Maximizing *surrogate function* S produces a solution that results in optimal compliance with multiple objectives subject to given constraints. The solution is a Pareto optimum [10] and stands for a rational compromise among conflicting objectives. The multi-objective optimization of Eq. (12.19) becomes

$$\left. \begin{array}{l} \text{Maximize} \quad S = \displaystyle\prod_{i=1}^{m} \overline{f}_i(x) \\[2mm] \text{Subject to} \quad g(x) \le 0 \end{array} \right\} \tag{12.24}$$

The solution of Eq. (12.24) is a Pareto optimal one and can be proved as follows. Assuming x^* is the solution of Eq. (12.24), but is not a Pareto optimal one, then there exists an $\overline{x}$, which belongs to the feasible zone, such that $f_j(\overline{x}) < f_j(x^*)$, and for $i \ne j$, $f_i(\overline{x}) \le f_i(x^*)$, implying

$$\begin{aligned} S(\overline{x}) &= \prod_{i=1}^{m} \frac{[f_{i,\max} - f_i(\overline{x})]}{[f_{i,\max} - f_{i,\min}]} = \frac{[f_{j,\max} - f_j(\overline{x})]}{[f_{j,\max} - f_{j,\min}]} \prod_{\substack{i=1 \\ i \ne j}}^{m} \frac{[f_{i,\max} - f_i(\overline{x})]}{[f_{i,\max} - f_{i,\min}]} \\[2mm] &> \frac{[f_{j,\max} - f_j(x^*)]}{[f_{j,\max} - f_{j,\min}]} \prod_{\substack{i=1 \\ i \ne j}}^{m} \frac{[f_{i,\max} - f_i(x^*)]}{[f_{i,\max} - f_{i,\min}]} \\[2mm] &= \prod_{i=1}^{m} \frac{[f_{i,\max} - f_i(x^*)]}{[f_{i,\max} - f_{i,\min}]} = S(x^*) \end{aligned} \tag{12.25}$$

This contradicts the assumption that x^* is the optimal solution of Eq. (12.24), so x^* must be a Pareto optimal solution.

12.3.3 Comparison of a structural design with nine objective functions

Consider a three-storey steel shear frame, as shown in Fig. 12.8; the floor diaphragms are rigid and axial deformations are neglected. Thus, the system has only one degree of freedom (in the lateral direction) at each floor.

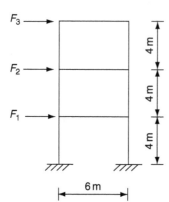

Figure 12.8 Three-storey shear frame.

The live and dead load at each storey is 56 kN/m, which does not include the weight of the columns. Lateral forces ($F_1 = 34$ kN, $F_2 = 52$ kN and $F_3 = 36$ kN) are shown in Fig. 12.8. The weight density and elastic modulus of steel are $\rho = 7,800$ kg m^{-3} and $E = (200)10^3$ MPa, respectively. The design variables are the column moment of inertia I_i. The mass m_i of the ith story can be calculated as

$$m_i = \frac{(56)6}{g} + (2)(4)A_i\rho \tag{12.26}$$

where A_i is the column cross-sectional area of the ith storey and g is the acceleration of gravity. For AISC standard wide-flange sections, sectional properties can be expressed in terms of the moment of inertia as [11]

$$A = 0.80\, I^{1/2}; \quad S = 0.78\, I^{3/4} \tag{12.27, 28}$$

where S is the section modulus. Optimum constraint conditions are

$$\left.\begin{aligned}
&|\sigma_i| \leq 165\,\text{MPa} \\
&|\delta_i| \leq \frac{1}{400}h_i \\
&T_1 \geq 0.3 \quad (s) \\
&0.00001 \leq I_i \leq 0.002\,\text{m}^4 \\
&0.5 \leq \frac{k_{i+1}}{k_i} \leq 1
\end{aligned}\right\} \tag{12.29}$$

where σ_i, δ_i and k_i are column stress, relative storey displacement and storey stiffness of the ith storey, respectively, and $i = 1, 2, 3$. T_1 is the fundamental natural period of the structure.

To investigate the impact of different objectives as well as scalar optimization and multi-objective optimization on structural design, the following nine cases are analyzed.

Case 1 Minimize W (weight);   *Case* 2 Minimize E_s (strain energy)

$$(12.30, 31)$$

Case 3 Minimize E_p (potential energy);

Case 4 Minimize E_i (input energy) $(12.32, 33)$

Case 5 Minimize $\{W, E_s\}$;   *Case* 6 Minimize $\{W, E_p\}$ $(12.34, 35)$

Case 7 Minimize $\{W, E_i\}$;   *Case* 8 Minimize $\{W, E_s, E_i\}$ $(12.36, 37)$

Case 9 Minimize $\{W, E_s, E_p, E_i\}$ (12.38)

Solutions for these cases are presented in Table 12.1 and in Figs. 12.9–12.11. The earthquake wave records used in this example are those of El Centro, N-S, 18 May 1940. The following observations can be made from the results.

Table 12.1 Active constraints.

Case	1	2	3	4	5	6	7	8	9
$|\sigma_i| \leq \sigma_{ia}$									
$|\delta_1| \leq \delta_{1a}$	*								
$|\delta_2| \leq \delta_{2a}$	*								
$|\delta_3| \leq \delta_{3a}$									
$T_1 \geq T_a$		*	*	*					
$I_{i1} \leq I_i \leq I_{iu}$									
$k_{i+1} \leq k_i \leq 1$							*		
$k_{i+1} \leq k_i \leq 0.5$	*	*	*	*	*	*		*	*

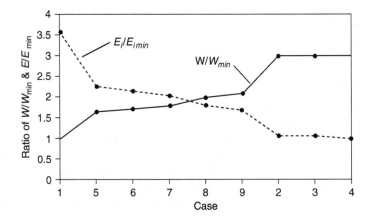

Figure 12.9 Single- and multi-objective optimization.

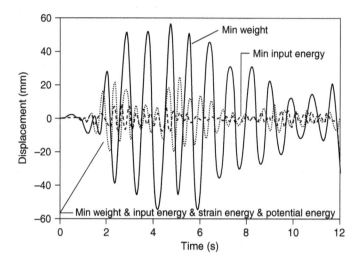

Figure 12.10 Comparison of top-storey displacement.

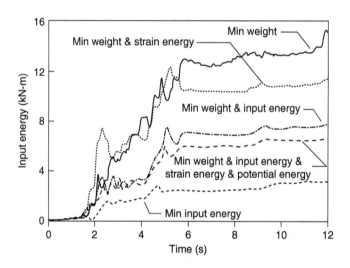

Figure 12.11 Comparison of seismic input energy.

(1) It may not be possible to produce a good compromise design within the context of single optimization (see Figs. 12.9–12.11). For example, when a structure is optimized for minimum weight subject to imposed constraints, the dynamic response of the structure is strong. When this structure is optimized for earthquake input energy, a small dynamic response is accompanied by a large structural weight. Multi-objective optimization can achieve a complementary solution from conflicting objectives.

(2) In Table 12.1, observe that single-objective optimization often produces a design in which two constraint types are active, which means that the final design is bound by several constraint modes. Therefore, the solution may be sensitive to changes in constraints.

(3) For single-objective structural optimization, if the physical characteristics (such as elastic modulus) of the members are fixed and the cross-sectional areas of the members serve as the design variables, there are only two groups of objectives. One increases the cross-sectional areas while the other decreases them. In multi-objective optimization, taking more objectives from one group means exerting greater influence on that group.

12.4 Fuzzy set theory

12.4.1 *Basic concept of fuzzy set*

Fuzzy set theory, introduced by Zadeh [14], is aimed at providing a model which is approximate rather than exact. It derives from the fact that classes and concepts for natural phenomena tend to be fuzzy instead of crisp. In a fuzzy set, a point is identified with its degree of membership in that set; there is no strict definition of membership and non-membership. The transition between full membership and non-membership is gradual rather than abrupt. In a fuzzy environment, the membership function can be defined as follows: it has the value of zero for the worst possible case, the value of 1 for the best possible case, and intermediate values for those cases in-between. The membership function of a fuzzy set characterizes the membership status of an element in a given universe. The function can be either a discrete or continuous formulation. Equation (12.39) and Fig. 12.12 illustrate a linear member function $\mu_L(x)$.

$$\mu_L(x) = \begin{cases} 0 & x \le 0 \\ \dfrac{x}{d} & x \le d \\ 1 & x > d \end{cases} \tag{12.39}$$

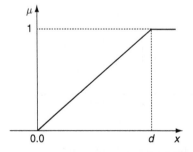

Figure 12.12 Linear membership function.

Let X denote a collection of objects; then fuzzy set A in X is a set of ordered pairs

$$A = \{[x, \mu_A(x)], \quad x \in X\} \tag{12.40}$$

where μ_A is a mapping from X to the unit interval $[0,1]$; μ_A is the membership function; and $\mu_A(x)$ is the membership grade of x in A. The membership function measures the extent to which an element is in a fuzzy set. The closer the value of $\mu_A(x)$ is to 1, the more x belongs to A. Obviously, A is a subset of X that has no sharp boundary. When $\mu_A(x)$ only has a value of 0 or 1, fuzzy set theory reverts to classical crisp set theory, since $\mu_A(x) = 1$ stands for $x \in A$ and $\mu_A(x) = 0$ stands for $x \notin A$. The membership function of a crisp set A of X can be viewed as a characteristic function μ_A from X to $\{0, 1\}$ such that

$$\mu_A(x) = \begin{cases} 0 & \text{if } x \in A \\ 1 & \text{if } x \notin A \end{cases} \tag{12.41}$$

Consider a discrete set X, which includes ten elements $x_1, x_2, \ldots, x_{10}$. Degrees of membership of the ten elements are defined as $\mu_1, \mu_2, \ldots, \mu_{10}$, respectively. Among them, only μ_3 has a degree of membership 1. Crisp and fuzzy sets of X are shown in Figs. 12.13a and 12.13b. The crisp set only has one element while the fuzzy set has ten. The intersection of two fuzzy sets A and B, with corresponding membership functions $\mu_A(x)$ and $\mu_B(x)$, is denoted by the symbol $A \cap B$ and defined as

$$\mu_A(x) \cap \mu_B(x) = \min\{\mu_A(x); \mu_B(x)\} \tag{12.42}$$

Figure 12.14a illustrates Eq. (12.42). In Eq. (12.42) the symbol $\cap$ can be considered as the greatest lower bound (the overlapping area of A and B in Fig. 12.14a). The union of fuzzy sets A and B is denoted by the symbol $A \cup B$ and defined as

$$\mu_A(x) \cup \mu_B(x) = \max\{\mu_A(x); \mu_B(x)\} \tag{12.43}$$

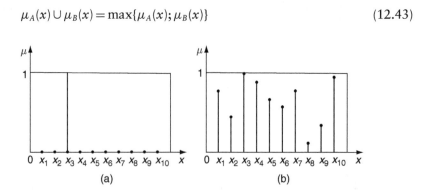

Figure 12.13 Discrete crisp and fuzzy sets: (a) crisp set, (b) fuzzy set.

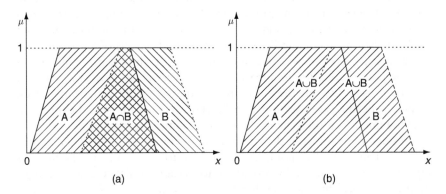

Figure 12.14 Fuzzy intersection and union operation: (a) intersection, (b) union.

Figure 12.14b shows the union of fuzzy sets A and B. The symbol $\cup$ can be interpreted as the least upper bound (the area of A plus B in Fig. 12.14b). These two operations will be used to construct a fuzzy environment in the fuzzy optimum design procedure.

12.4.2 *Fuzzy penalty-function scheme*

A fuzzy-logic-based penalty-function scheme is to solve a constrained optimization problem with genetic algorithms (see Section 12.5) in a fuzzy environment; the search space is redefined in terms of fuzzy set theory. There is no crisp boundary between feasible and infeasible regions. Each point is associated with the degree of membership corresponding to the fuzzy feasible set. The closer a point is to the feasible region, the more it belongs to the set. When a point is in the feasible region, it has the highest degree of membership in the fuzzy feasible set. In a fuzzy environment, a constraint is defined by fuzzy set G in X, and μ_G shows to what degree a point satisfies the constraint G. In classical optimization programming, however, an infeasible point can be accepted as an incomplete feasible solution. According to fuzzy set theory, classical constraints can be adapted to a fuzzy model. When a point is in the feasible zone, the membership function $\mu_G(x)$ equals 1. Otherwise, this function is $0 \le \mu_G(x) < 1$. The maximum constrained violation of a point, defined as $\mu_c(x)$ (Eq. (12.44)), is taken as the constraint measurement of that point in the fuzzy environment. Thus

$$\mu_c(x) = \mu_{G_1}(x) \cap \mu_{G_2}(x) \cap \ldots \cap \mu_{G_L}(x) \qquad (12.44)$$

where L is the number of constrained conditions. To define the fitness function (see Section 12.5.4) of each point for a genetic algorithm optimum procedure in terms of the fuzzy environment, consider the following: (A) The genetic algorithm optimum procedure, as distinct from traditional optimization methods, is random and probabilistic rather than numerically precise and calculable. (B) A genetic algorithm requires no input on the type of problem, gradient or other internal aspects of a problem. (C) When treating a point's violated amount for constraints, a fuzzy quantity – e.g., the point in relation to a feasible zone is very close, close, far, very far, and so on – can provide the information required for the genetic algorithm ranking procedure (see Section 12.5.3). (D) An exact numerical expression has little impact on the genetic algorithm evolutionary progress.

Consider an example of the relationship definition of points U, V and T. A fuzzy definition that point U is superior to point V, and point V is superior to point T, actually provides information similar to a genetic algorithm's numerical fitness definition that $U_{\text{fitness}} = 3$, $V_{\text{fitness}} = 2.3$ and $T_{\text{fitness}} = 1.2$. In both cases, point U has the highest probability of proceeding to the next generation. Introducing the fuzzy-logic concept into the definition of fitness may affect the convergence velocity of a GA. However, the GA's convergence direction and goal do not change. To use fuzzy set theory to redefine the constraints, a proper membership function must be determined. Fig. 12.15 and Eq. (12.45) define a linear membership function, where ε_j is the minimum allowable error that determines the feasible/infeasible status of a point. $\bar{d}_j$ is the distance for which the boundary of the jth constraint is moved.

$$
\mu_{G_j}(x) = \begin{cases} 0, & \text{if } g_j(x) \geq \bar{d}_j \\ 1 - \dfrac{g_j(x) - \varepsilon_j}{\bar{d}_j}, & \text{if } \varepsilon_j < g_j(x) < \bar{d}_j \\ 1, & \text{if } g_j(x) \leq \varepsilon_j \end{cases} \tag{12.45}
$$

Although various types of membership function can be employed, this chapter presents only a discrete membership function in practical applications. For a minimization optimization problem, the jth objective function $\bar{f}_j(x)$ of

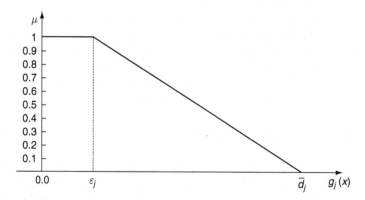

Figure 12.15 Linear membership function.

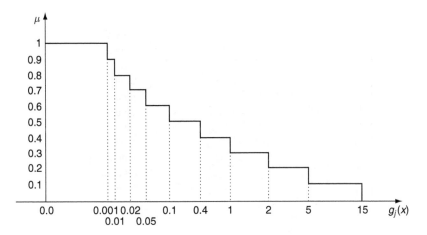

Figure 12.16 Discrete membership function.

a point in a fuzzy environment with a discrete membership function can be defined as

$$\bar{f}_j(x) = \frac{f_j(x)}{1+f_j(x)} + R_k \tag{12.46}$$

where R_k is the *fuzzy penalty function* determined by a discrete member function (Fig. 12.16). In Eq. (12.46), objective function $f_j(x)$ is normalized so that a feasible point can be decided when its membership function is less than or equal to 1. R_k is expressed as

$$R_k = \begin{cases} \text{int}\{(1 - \mu_C(x))^* \text{KD}\} & \mu_C(x) > 0 \\ 100 & \mu_C(x) = 0 \end{cases} \tag{12.47}$$

where $KD = 10$ for the discrete membership function used here. In Fig. 12.16, it is assumed that $\in = 0.001$ and $\bar{d}_j = 15$. Based on the fuzzy function (Eq. (12.46)), the fitness of each point can be decided. When the value of the function is larger than or equal to KD, it is assumed that the corresponding point has no fitness or minimal fitness. In Eq. (12.47), the item $R_k = \text{int}\{(\mu_C(x)) \times \text{KD}\}$ always takes an integer value. For the discrete membership function shown in Fig. 12.16, if $0.1 < g(x) = \max\{g_1(x),\dots,g_m(x)\} \le 0.4$, then $R_k = \text{int}\{(1-0.5) \times 10\} = 5$. If $g(x) \le \in = 0.001$, then $R_k = \text{int}\{(1-1) \times 10\} = 0$.

Based on this fuzzy definition, the entire search space is divided into 11 zones. Zone 1 $[\max(d_{k1},\dots,d_{kL}) \le D_1]$ is the feasible space, where penalty term R_k equals zero. Zones 2 through 10 [zone i: $D_{i-1} < \max(d_{k1},\dots,d_{kL}) \le D_i$; $i=2,\dots,10)$] are the penalty space, where penalty term R_k equals i. Zone 11 $[D_{10} < \max(d_{k1},\dots,d_{kL})]$ is the decline space, where penalty term

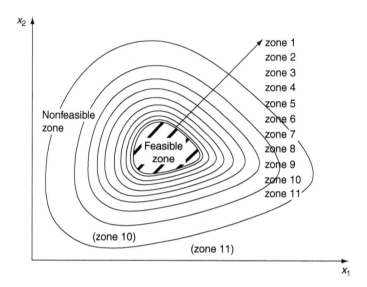

Figure 12.17 Definition of revised penalty function.

R_k equals 100. If a point is in zone 11, its reproduction ratio is zero. In the above statements, d_{ki} is the violated amount of point k to the ith constraint (Eq. (12.48)), L is the number of constraint conditions, and D_i is the tolerance interval of the ith infeasible zone. Here D_i is set as equal to 0.001, 0.01, 0.02, 0.05, 0.1, 0.4, 1.0, 2.0, 5.0 and 15.0 for $i=1$ to 10. Points in the same tolerance interval zone have the same penalty terms.

$$d_{ki} = \begin{cases} 0, \ g_i(x_k) \leq 0 \\ g_i(x_k), \ \text{otherwise} \end{cases} \tag{12.48}$$

In Eq. (12.48) constraints are normalized to place them in the same general order of magnitude. Eq. (12.47) is illustrated in Fig. 12.17. In this fuzzy environment, fuzzy formulation is applied using region definition instead of value definition, which gives a violated constraint a fuzzy quantity instead of an exact value. Assuming there are two infeasible points Q and P, point Q makes the ith constraint $g_i(x_Q) = 1.2$ and point P makes the ith constraint $g_i(x_P) = 1.4$. For this case, penalty terms $R_Q = R_P = 7$ (see Fig. 12.16 and Eq. (12.47), $1 < g_i(x_Q)$, $g_i(x_P) < 2$, $\mu_c(x) = 0.3$; $R = \text{int}(1-0.3) \times 10 = 7$) because both of them are in the infeasible zone 8; here, the exact values of the violations have not been used. Moreover, this method greatly reduces rank levels and satisfies the requirements of a nondominated rank process in a Pareto genetic algorithm [11]. In terms of penalty concept, the transformed function value of a point in an infeasible zone is a comprehensive representation of its objective function plus its penalty function. Observing the fuzzy objective function, $\bar{f}_i(x)$ stands for the degree of membership (violated measure) of a point while the decimal represents the position (objective value) of a point in the search space. For multi-objective optimization problems, this function can satisfy the requirements of the rank-based Pareto genetic algorithm: it indicates a point's status ($\bar{f}_i(x) < 1$, feasible, $\bar{f}_i(x) \geq 1$, infeasible); it represents a point's violated degree (integer) in the infeasible zone; and it defines a point's distance (decimal) to the Pareto optimal set. This is the essential information required for a genetic algorithm's evolutionary mechanism.

Using the fuzzy penalty-function scheme, a multi-objective constrained optimization problem can be transformed into a non-constrained one (Eq. (12.49)).

$$\left.\begin{array}{l} \text{Minimize } \{\bar{f}_1, \ldots, \bar{f}_m\} \\ \text{Subject to } x_i^{(l)} \leq x_i \leq x_i^{(u)}, \quad i = 1, \ldots, n \end{array}\right\} \tag{12.49}$$

This problem (Eq. (12.49)), now non-constrained, can be solved by using the Pareto genetic algorithm discussed in the next section, Section 12.5. Note that fuzzy value $\bar{d}_j$, which determines the attribution of a point in a set and the search environment of an adaptive search strategy, is introduced. Analyzing $\bar{d}_j$, two points should be noted: (1) If the value of $\bar{d}_j$ is too small, most members of a population may be rejected from the set and the population thus loses its variety; most members now have similar genotype characteristics. (2) If the value of $\bar{d}_j$ is too large, the search pressure on evolutionary populations will decrease and convergence of the evolutionary processes may be slow.

12.5 Genetic algorithm (GA) for multi-objective optimization

12.5.1 Basic concept of GA

In nature, evolution is a natural selection or self-optimizing process within a specific environment. More successful species survive and propagate while less successful ones decline. In a natural environment, the best-suited organisms tend to live long enough to reproduce whereas less-suited organisms often die young. After a certain length of time, most of the less-suited organisms die and most of the best-suited organisms survive. The survivors combine and produce new organisms. For only the best-suited organisms to survive, their offspring should be better adapted to the environment. Clearly, this mechanism is an optimal one. Fig. 12.18 illustrates a simple biological evolutionary process. After natural selection over many generations, the new population is better suited to the environment as well as being healthier and stronger than its predecessors.

GA, a computer algorithm simulating the evolutionary process, is a search procedure with randomized yet structured information exchange in a finite space. GA uses random processes to produce an initial population, and simple operators are then applied to produce a new population. The new

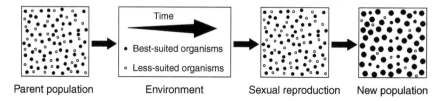

Figure 12.18 Simple biological evolutionary process.

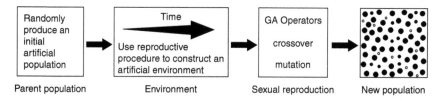

Figure 12.19 Computer-simulated evolutionary process.

population is referred to as offspring, and the original population as parents. Fig. 12.19 shows a computer-simulated process. For a computer simulation, the environment can be defined in terms of special requirements. Any change in environment causes a change in evolutionary direction. A specified artificial environment is tailored to promote the survival of individuals who most closely approximate the solutions being sought. Therefore, different search goals can be achieved within varying environments through an evolutionary process.

GA represents complex models by simple encoding. It works with a coding of the parameter set, not the parameters themselves. One encoding method is the representation of a model by *binary bit strings*. Consider an optimal design problem in which design variables are $\{x\} = \{x_1, x_2, x_3\}$ and each variable has a specified range so that $x_{i,\min} \leq x_i \leq x_{i,\max}$. A variable can be encoded by mapping in a range $x_{\min}$ to $x_{\max}$ using an n bit binary unsigned integer. If a 4-bit code is chosen to map a variable, then $x_{\min} \rightarrow (0000)$ and $x_{\max} \rightarrow (1111)$. There are 2^n values in this range (for the present example, $n=4$). The following equation can be used to decode a string

$$x_j = x_{j,\min} + \frac{B_j}{2^n - 1}(x_{j,\max} - x_{j,\min}) \tag{12.50}$$

where B_j is the decimal integer value of the binary string for variable x_j (such as binary string 1111, its $B_j = (1)(2)^0 + (1)(2)^1 + (1)(2)^2 + (1)(2)^3 = 15$). Eq. (12.50) is established according to the linear interpolation principle. A string usually stands for a point in a design variable space. In the above example, let $x_{1,\min} = x_{2,\min} = x_{3,\min} = 0$ and $x_{1,\max} = x_{2,\max} = x_{3,\max} = 4$. A design point $\{x\} = \{1.8667, 2.4, 3.2\}$ could then be represented by the

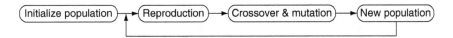

Figure 12.20 Simple genetic algorithm.

string determined by a random process as (011110011100) where the first 4 bits $(0111, B_j = 2^2 + 2^1 + 2^0 = 7)$ equal $x_1 = 0 + 7(4-0)/(2^4-1) = 1.8667$, the middle 4 bits $(1001, B_j = 9)$ equal $x_2 = 0 + 9(4-0)/(2^4-1) = 2.4$, and the last 4 bits $(1100, B_j = 2^3 + 2^2 = 12)$ equal $x_3 = 0 + 12(4-0)/(2^4-1) = 3.2$.

In contrast to traditional optimization which usually starts its search from a point $\{x\}^0 = \{x_1, x_2, x_3\}^0$, a GA search works from a population of points, $\{x\}_1^0, \{x\}_2^0, \ldots, \{x\}_m^0$. In the traditional method, a search moves gingerly from one point in the decision space to the next, using certain rules to proceed to the next point. In GA, the process works generation by generation (iteration) using probabilistic, not deterministic, transition rules and successively generating a new population of strings. In a mathematical model, a simple genetic algorithm can be expressed by the following four steps: initialize population, reproduction, crossover and mutation, and new population (Fig. 12.20). Since GA is a simulation of biological evolution, all GA operators are random probabilistic. An example is illustrated as follows

$$\left.\begin{array}{l} \text{Maximize } f(x) = x_1^2 + x_1 x_2 \\ \text{Subject to } 0 \le x_1 \le 40 \\ \qquad\quad 0 \le x_2 \le 30 \end{array}\right\} \tag{12.51}$$

Initializing a population randomly produces m points in a finite space, where m is the population size. For the example in Eq. (12.51), $0 \le x_1 \le 40$ and $0 \le x_2 \le 30$ define a finite space. The m points are the initial population (see Fig. 12.21).

For our above-mentioned example, the traditional optimization methods usually start their search from a point A^0 as shown in Fig. 12.22a.

$$A^0 = \{A_1, A_2, A_3\}^0 \tag{12.52}$$

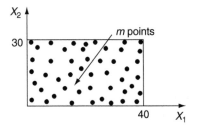

Figure 12.21 Initial population.

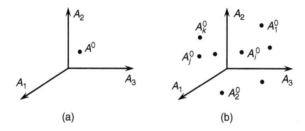

Figure 12.22 Optimization process: (a) traditional approach, (b) genetic approach.

The GA's search works from a population of points, as sketched in Fig. 12.22b, usually the number is $50 \sim 200$, that is

$$
\left.\begin{array}{l}
A_1^0 = \{A_1, A_2, A_3\}_1^0 \\
\vdots \\
A_i^0 = \{A_1, A_2, A_3\}_i^0 \\
\vdots \\
A_n^0 = \{A_1, A_2, A_3\}_n^0
\end{array}\right\} \tag{12.53}
$$

If we still use a 4-bit binary string, the samples can be

$$
\left.\begin{array}{c c|c c c c c}
 & & A_1^0 & A_2^0 & A_3^0 & \cdots & \cdots \\
\hline
 & & 0 & 0 & 1 & & \\
 & & 0 & 0 & 1 & & \\
A_1 & & 1 & 0 & 1 & \cdots & \cdots \\
 & & 0 & 1 & 1 & & \\
\hline
 & & 0 & 0 & 0 & & \\
 & & 1 & 1 & 0 & & \\
A_2 & & 1 & 0 & 0 & \cdots & \cdots \\
 & & 0 & 1 & 0 & & \\
\hline
 & & 1 & 1 & 0 & & \\
 & & 1 & 1 & 0 & & \\
A_3 & & 1 & 1 & 1 & \cdots & \cdots \\
 & & 0 & 0 & 0 & &
\end{array}\right\} \tag{12.54}
$$

Example 12.5.1

$$
\left.\begin{array}{l}
\text{Maximize } f(x) = (1 - x_1)^2 + (2 - x_2)^2 \\
0 \leq x_i \leq 4
\end{array}\right\} \tag{a}
$$

Assuming we only choose $m = 4$ (population size), and using an 8-bit binary string to map the design variables, we have

Table 12.2 Determination of copy number.

String number	x_1	x_2	$Point(x_1, x_2)$	$f(x)$	$f_i/\sum f_i/m$	Number of copy
A_1^0	00000000	00000000	(0, 0)	5	0.86	1
A_2^0	00000000	11110000	(0, 3.765)	4.115	0.71	1
A_3^0	11111111	00000000	(4, 0)	13	2.25	2
A_4^0	10000000	10000000	(2.008, 2.008)	1.016	0.18	0
$\sum$				23.131		4
$\sum/m$				5.783		

From the above, Table 12.2, string 3 has the maximum fitness, it is copied into the next population by two, and string 4 has the minimum fitness, it is not transformed to the new population. After reproduction, a mating pool is formed as shown in Table 12.3.

Table 12.3 Mating pool.

	String number	String	
1	A_1^0	00000000	00000000
2	A_2^0	00000000	11110000
3	A_3^0	11111111	00000000
4	A_4^0	11111111	00000000

The *crossover* operator follows the reproduction procedure. It may proceed in two steps. First, newly reproduced strings in the mating pool are mated at random. Second, each pair of strings undergoes crossing-over based on some rules of number of parents and number of points for the crossover operator. For example, we may generate a new population from Table 12.3, as shown in Table 12.4.

Table 12.4 Crossover results.

	Mating pool (Table 12.3) by random	New population	Point (x_1, x_2)
1	00000:000 00000000	11111000 00000000	(3.8902, 0)
2	00000000 11:110000	00000111 00000000	(0.1098, 0)
3	11111:111 00000000	11111111 00110000	(4, 0.7529)
4	11111111 00:000000	00000000 11000000	(0, 3.0118)

The above is based on a 2-parent and 2-point crossover operator. We can also construct the 3-parent, 2-point crossover operator. The principle is that at the first part, parent 1 changes information with parent 2; at the second part, parent 2 changes with parent 3; and at the last part, parent 3 changes with parent 1. The n-parent crossover $(n > 2)$ operators lay the foundation for a new family of GAs. In theory, the offspring which are produced by n-parents have more changing characters than the offspring reproduced only by 2 parents. On other hand, the n-parent crossover $(n > 2)$ may slow down the velocity of convergence for it increases the possibility of diversity distribution of the design variables.

Mutation plays a decidedly secondary role in the operation of genetic algorithms. After the procedures of reproduction and crossover, or the defects of the initial population strings, occasionally the population strings may become over-zealous and lose some potentially useful genetic material (1's or 0's at particular locations).

For example, we have population strings

$$
\left.
\begin{array}{l}
1\ 1000001 \\
2\ 0100100 \\
3\ 0011000 \\
4\ 1001001 \\
5\ 0000100
\end{array}
\right\}
\tag{12.55}
$$

The 6th position (from left) of the population strings is always zero. To make up the defect, we make a mutation at string 2 (random), the new population strings are (after mutation):

$$
\left.
\begin{array}{l}
1\ 1000001 \\
\boxed{2}\ 0100111 \\
3\ 0011000 \\
4\ 1001001 \\
5\ 0000101
\end{array}
\right\}
\tag{12.56}
$$

Example 12.5.2 – Use the GA to find the optimal results of the given function and then compare with the exact solution.

$$
\text{Minimize } f(x) = 2x^2 + \frac{16}{x}; \quad 1 \le x \le 5
\tag{a}
$$

Solution:

Let the number of the population be 8; the size of population be 10; and the rate of crossover and mutation be set as 0.8 and 0.005, respectively. The calculations are given in Table 12.5.

Table 12.5 GA optimization.

	Population string	$x = 1 + \dfrac{B}{2^{10}} \times 4$	Fitness	Mating pool
Initial	1111100000	4.8750	50.8133	0000011111
	1111101000	4.9063	51.4047	0000011111
	1111100100	4.8905	51.1075	0000011111
	1111100010	4.8828	50.9602	1000011111
	1111100001	4.8789	50.8862	1000011111
	0000011111	1.1211	16.7854	0000000001
	1000011111	3.1211	24.6089	0000000001
	0000000001	1.0039	17.9535	0000000001
Offspring	0000011111	1.1211	16.7854	0000011111
	0000000001	1.0039	17.9535	0000000001
	1000011111	3.1211	24.6089	0010011111
	1000000111	3.0273	23.6143	0010011111
	0000011001	1.0977	16.9858	0100011001
	0000000111	1.0273	17.6854	0100011001
	0100011001	2.0977	16.4281	0000011001
	0010011111	1.6211	15.1258	0000000111
Offspring	0100011001	2.0977	16.4281	0010011111
	0010011111	1.6211	15.1258	0010011111
	0000111001	1.2227	16.0758	0000111001
	0000000111	1.0273	17.6854	0010000111
	0010000111	1.5273	15.1413	0010000111
	0100000001	2.0039	16.0157	0100000001
	0001011001	1.3477	15.5047	0001011001
	0010011111	1.6211	15.1258	0010011111
Offspring	0010111111	1.7461	15.2610	0010111111
	0000011001	1.0977	16.9858	0010000111
	0010000111	1.5273	15.1413	0010011111
	0010011111	1.6211	15.1258	0010000111
	0010000111	1.5273	15.1413	0100000001
	0100000001	2.0039	16.0157	0011011111
	0000011001	1.0977	16.9858	0010111111
	0011011111	1.8711	15.5532	0010000111
Offspring	0010000111	1.5273	15.1413	0010000111
	0010111111	1.7461	15.2610	0010111111
	0010011111	1.6211	15.1258	0010011111
	0010000111	1.5273	15.1413	0010000111
	0100111001	2.2227	17.0792	0010100111
	0010100111	1.6523	15.1436	0010010111
	0011011111	1.8711	15.5532	0010011111
	0010010111	1.5898	15.1191	0010010111

From the above, the optimal solution is $f(x) = 15.1191$ at $x = 1.5898$. The exact solution is

$$x^* = 1.5874; \; f^* = 15.1191$$

Note that the result is close to the exact solution, usually it is not equal to it. The algorithm is only applied to non-constraint optimization.

12.5.2   *Pareto GA for multi-objective optimization*

Using GAs to optimize a multi-objective optimization (MOP) problem is totally different from a single-objective optimization one. This difference is due to the characteristics of the former. A multi-objective optimization searches a group of points which simultaneously optimizes multi-objectives, and a single objective searches a point which optimizes the single objective. To solve this type of problem and locate its Pareto optimal set, a genetic multi-objective optimization algorithm is constructed, called the *Pareto genetic algorithm* (Pareto GA) based on the Pareto solutions. It proceeds as follows: introduction of the basic theory and properties of a multi-objective optimization problem; presentation of the main frame of the Pareto GA; and description of the details of the technique.

A multi-objective optimization problem may be stated as

$$\left. \begin{array}{l} \text{Minimize } \{f_1(x), f_2(x), \ldots, f_n(x)\} \\ \text{Subject to } g(x) \leq 0 \end{array} \right\} \tag{12.57}$$

Its solution, as noted, is always situated in the Pareto optimal set. A Pareto optimization gives a set of nondominated solutions, i.e., solutions for which no criterion (objective) can be improved without worsening at least one other criterion. Fig. 12.23 shows two optimization problems: a single-objective optimization and a multi-objective optimization with two design variables. In Fig. 12.23a, point A is the global optimal solution for the single-objective optimization problem; it is a unique solution. In Fig. 12.23b, the dashed line indicates the Pareto optimal set in the decision space, which is a group of solutions.

A typical characteristic of an MOP is the absence of a unique point which would optimize all criteria simultaneously. Any point in the Pareto optimal set can become an optimum solution which varies with different

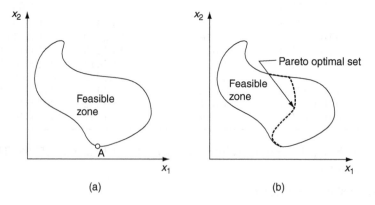

Figure 12.23 Feasible region and Pareto optimal set: (a) single-objective optimization, (b) multi-objective optimization.

decision-makers. A Pareto optimal set conveys information on the trade-off. Thus, a multi-objective optimization formulation can be ideally handled by obtaining an evenly distributed subset of the Pareto optimal set, and searching this subset to find the decision-maker's preference. Most traditional multi-objective optimization algorithms seek the best compromise solution. This solution varies according to the search method used for multi-objective optimization problems. Furthermore, a solution from a traditional method tends to be trapped by the first local minimum encountered, which may not belong in the Pareto optimal set. Clearly, these methods limit designers of multi-objective optimization; they cannot verify a solution as a truly optimal one with a conventional algorithm.

Genetic algorithm procedures are designed to locate global optima. As noted, a GA is able to maintain a population of solutions and to conduct a parallel search for many nondominated solutions. This ability and search process meet the requirements for seeking a Pareto optimal set to solve a multi-objective optimization problem. The basic concept of multi-objective optimization via GAs is that in each generation, the fitness function of each individual is decided according to its nondominated property. *Nondominated individuals* always have a higher probability of proceeding to the next generation because they have the highest fitness values. As evolution continues, the convergence of a population goes to its Pareto optimal set zone. Utilizing the parallel search and group output properties of GAs, a Pareto GA for multi-objective optimization is powerful. Unlike a simple GA, the goal of a Pareto GA's search is a region (Pareto optimal zone) rather than an optimal point. Furthermore, this Pareto GA has two operators – *niche* and *Pareto-set filter* – besides the three basic operators: reproduction, crossover and mutation. The analysis procedure for this Pareto GA consists of finding an evenly distributed Pareto optimal set and choosing a solution from this set.

$$F_i = (N_r - i + 1)/SS \qquad (12.58a)$$

$$SS = \frac{\sum_{i=1}^{N_r}(N_r - i + 1)P_{si}}{M} \qquad (12.58b)$$

where M is population size; N_r is the highest rank in the population; P_{si} is population size of rank i; and F_i expresses the fitness of a string ranked i. In the fitness definition given above, the reproduction ratio of each rank relies on both its rank level and population size. F_i also defines the reproduction ratio of each point for $\sum F_i = M$. Eq. (12.58) leads to lower rank with greater fitness. SS is the average value of the entire population. $(N_r - i + 1)$ transforms a lower rank to a larger value. When $i = 1$, $N_r - i + 1 = N_r$. When $i = N_r$, $N_r - i + 1 = 1$.

Consider a model population. Its size is $M = 40$ and the maximum rank at generation K is $N_r = 4$. Two cases with different population sizes assumed at each rank are investigated.

Rank (i)	Population size (P_{si}) (Case I)	Population size (P_{si}) (Case II)
1	20	4
2	10	6
3	6	10
4	4	20

Using Eq. 12.58, we have

$$SS_1 = \frac{[(4)20 + (4-2+1)10 + (4-3+1)6 + (4-4+1)4]}{40} = 3.15$$

$$SS_2 = \frac{[(4)4 + (4-2+1)6 + (4-3+1)10 + (4-4+1)20]}{40} = 1.85$$

and the fitness of a point at each rank (reproduction ratio) is

Rank (i)	Fitness of point (F_i) (Case I)	Fitness of point (F_i) (Case II)
1	$(4-1+1)/3.15 = 1.27$	$(4-1+1)/1.85 = 2.16$
2	$(4-2+1)/3.15 = 0.95$	$(4-2+1)/1.85 = 1.62$
3	$(4-3+1)/3.15 = 0.63$	$(4-3+1)/1.85 = 1.08$
4	$(4-4+1)/3.15 = 0.32$	$(4-4+1)/1.85 = 0.54$

12.5.3   Nondominated solutions and rank

In this Pareto GA, a point's fitness depends on its rank and a point's rank depends on its nondominated nature. An algorithm selecting nondominated solutions from a given solutions' set $\{F\}$ can be defined as follows. A vector f^* is a nondominated solution for a multi-objective optimization problem in a population if and only if (maximization problem)

$$\{f^*|f^* \in F\} \cap \left\{\overline{f_i};\, i = 1, 2, \ldots,\, P_n | \overline{f_i} \in F\right\} < f^* \left(\overline{f_i} \neq f^*\right) \tag{12.59}$$

or (minimization problem)

$$\{f^*|f^* \in F\} \cap \{\overline{f_i};\, i = 1, 2, \ldots,\, P_n | \overline{f_i} \in F\} \neq f^* \,(\overline{f_i} \neq f^*) \tag{12.60}$$

where P_n is population size. An illustration of the definition (Eq. (12.59)) for a two-criterion maximization problem is shown in Fig. 12.24. Figs. 12.24a and 12.24b illustrate Eq. (12.59) when $P_n = 2$ and $P_n = n$, respectively.

The nondominated procedure operates in the criterion domain. In practical programming, the procedure (for a minimization problem) can be

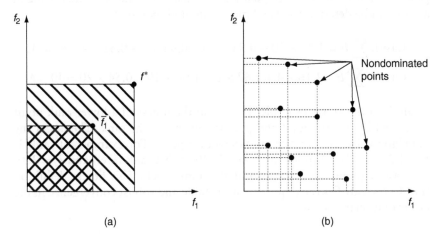

Figure 12.24 Nondominated points for maximization problem: (a) illustration, (b) nondominated points in a population.

implemented as follows. All items are arranged in ascending order with respect to the first criterion f_1. Where the value of f_1 is the same, relevant items are arranged in ascending order with respect to the second criterion f_2, the third criterion f_3, and so on. After organizing the entire population and eliminating those items dominated by others, the remaining items are nondominated points belonging to rank 1 in the present population.

Ranking a population is a continuous labelling process. At each generation, nondominated points are identified and assigned rank 1. From the remaining population, nondominated points are identified and assigned rank 2. This process continues until the entire population is ranked. In reproduction, strings of rank 1 have more copies while strings of higher rank have fewer. Fig. 12.25 illustrates ranking for a two-objective maximization

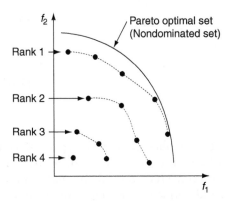

Figure 12.25 Population rank (two objectives).

problem. When the whole population is ranked, the fitness of each string in a rank can be determined. The fitness sum of all points is

Case I: $\sum F_i = 1.27 \times 20 + 0.95 \times 10 + 0.63 \times 6 + 0.32 \times 4 = 40 = M$

Case II: $\sum F_i = 2.16 \times 4 + 1.62 \times 6 + 1.08 \times 10 + 0.54 \times 20 = 40 = M$

Analysis of the above example shows that the fitness of a lower-rank point is decided by its rank level and population size of the rank. The larger the population size, the smaller the fitness of a point. For example, at rank 1, $F_i = 1.27$ for Case I (population size is 20 at this rank level) and $F_i = 2.16$ for Case II (population size is 4 at this rank level). This fitness definition (Eq. 12.58) satisfies the requirement of stable and robust convergence in the GA evolutionary process.

12.5.4 Objective and fitness functions in GA

In a GA for single-objective optimization, the fitness function connects with the objective function in a linear or nonlinear formulation. In a Pareto GA dealing with a vector function space, fitness expresses the relative position of a point in the rank space rather than the values of objective functions. The nondominated property of a point determines its fitness value. Nondominated procedures for points operate in the objective function domain. For a constrained optimization problem, constraint conditions of a point must be included in the point's fitness function. Using a fuzzy penalty-function scheme (see Section 12.4.2) with the stipulation that a feasible point is always superior to an infeasible one, a process of refining objective functions can be constructed as follows. First, the n objective values $(\min\{f_1,\dots,f_m\}_j, \max\{f_{m+1},\dots,f_n\}_j)$ of point j are normalized; then, integrated with the point's fuzzy penalty function, they are transformed into a minimization problem $(\min\{f_1'',\dots,f_n''\}_j)$ according to Eq. (12.61).

$$f_i'' = \begin{cases} f_i' & (f_i' \in F) \\ f_i' + R_j & (f_i' \notin F) \end{cases} \quad i = 1,\dots,n \qquad (12.61a)$$

$$f_i' = \begin{cases} \dfrac{f_i}{(1+f_i)} & (\min f_i) \\ \dfrac{1}{(1+f_i)} & (\max f_i) \end{cases} \qquad (12.61b)$$

where F stands for the feasible zone in an objective function space; R_j is the penalty value of point j; and $R_j > 1$ (see Section 12.5.5). In Eq. (12.61b), it is assumed that $f_i \geq 0$. Equation (12.61) converts a general optimization problem into a minimization problem, and normalizes the objective functions so that $0 \leq f'_i < 1$. Eq. (12.61b) normalizes the objective functions. When $f_i = 0$, $f'_i = 0$ (min) and $f'_i = 1$ (max). When $f_i \rightarrow \infty$, $f'_i \rightarrow 1$ (min) and $f'_i \rightarrow 0$ (max). In Eq. (12.61a), f''_i is the penalty function which comprises the normalized function f'_i and the penalty value R_j.

Now, the ranking procedure is applied to the new function value set $(\{f''_1, \ldots, f''_n\}_j, j = 1, \ldots, P_n$, where P_n is the population size) of the population. From Eq. (12.61), it can be seen that $0 \leq f''_i \leq 1$ when the point is in the feasible zone; otherwise $f''_i > 1$. Enforcing this principle, it is easy to separate feasible points from infeasible points. Note that infeasible points always hold higher rank than feasible points. This is because any feasible point ($f''_i \leq 1$) is a nondominated one with respect to infeasible points ($f''_i > 1$). The fitness strategy is demonstrated in Fig. 12.26.

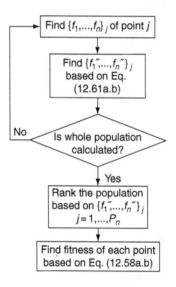

Figure 12.26 Relationship between fitness and objective functions.

12.5.5 *Pareto optimal set and Pareto-set filter*

The Pareto GA tries to locate a Pareto optimal set in an objective space. Each point in the set should be equally important and treated as an optimum goal. However, reproducing a new generation cannot guarantee that the best traits of the parents will be inherited by their offspring. It is possible that some of these traits may never appear in future phases of evolution due to a limited population size. In the evolutionary process, many points appear once or twice and then disappear forever. Some of them may be the sought-after optimum goals: the Pareto optimal points. To avoid missing the Pareto optimal points, a new concept called the *Pareto-set filter* is introduced. A Pareto-set filter pools nondominated points ranked 1 at each generation, and drops dominated points.

At each generation, the points designated rank 1 are put into a filter. When new points are added to the Pareto-set filter, all points in the filter are subjected to a nondominated check (filtering process) and dominated points are discharged. Thus, only nondominated points are stored in the filter. The filter size can be set as equal to the population size or any reasonable value. When the number of points in a filter surpasses a given size, the points at a minimum distance from other points are removed in order to maintain an even distribution of points. Figs. 12.27 and 12.28 explain filter operations. In Fig. 12.28b, the points in a Pareto-set filter are composed of the points of rank 1 in generations 1 and 2. Dominated points are dropped from the filter.

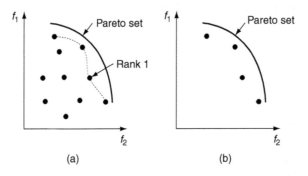

Figure 12.27 Population distribution at generation 1: (a) generation 1, (b) Pareto-set filter.

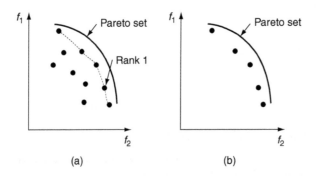

Figure 12.28 Population distribution at generation 2: (a) generation 2, (b) Pareto-set filter.

While the solution of a GA without a Pareto-set filter comes only from the last generation p_n (Eq. (12.62)), the solution of a Pareto GA is picked from the entire evolution strategy P (Eq. (12.63)).

$$p_n = \left\{ \overline{f}_{i,n}; i = 1, \ldots, L \, | \overline{f}_{i,n} \in F \right\} \tag{12.62}$$

$$P = \left\{ \overline{f}_{i,j}; i = 1, \ldots, m; j = 1, \ldots, n \, | \overline{f}_{i,j} \in F \right\} \tag{12.63}$$

where L is the number of points assigned rank 1 in generation n; m is the filter size; n is the generation number; $\bar{f}_{i,j}$ is a nondominated solution in generation j; and F is the feasible zone of an optimization problem. Collection P explicitly includes collection p_n and dominates the latter. See Section 12.5.7 for sample illustrations, Figs. 12.32 and 12.33, which show the results of points from rank 1 at generation 500. In Fig. 12.33, the points are evenly distributed along the Pareto optimum set and cover more of the region than the points in Fig. 12.32 do. This new technique avoids missing any Pareto set points; all Pareto set points appearing in the evolutionary process are recorded. The final filter gives a set of points significantly closer to or actually in the Pareto set zone.

12.5.6 *Niche technique for Pareto GA*

The development of GA operators is based on the principles of biological evolution. In nature, evolution maintains a variety of species. Due to the finite size of an artificial population and the stochastic errors associated with GA operators, a phenomenon known as genetic drift occurs in GA evolutionary processes. *Genetic drift* makes a converging population become nearly identical and cluster at certain optimum regions. Niche techniques force individuals to share available resources and maintain appropriate diversity [15]. As noted, the goal of a Pareto GA is to locate the Pareto optimal set of an MOP. Thus, an effective niche technique is the key to the success of GAs for MOPs.

A *niche technique* presented here is derived from the concept that an offspring replaces its parent if the offspring's fitness exceeds that of the inferior parent. After a pair of new offspring is produced, the parents are replaced only when the rank of one offspring in the parent population is no worse than the best rank of its parents. Otherwise, the parents go to the next generation. Applying this limitation, new offspring (children) always hold gene characteristics superior or equivalent to their parents. Children are only produced around their parent positions or at new positions not dominated by the old ones. Note two other characteristics of this niche method. It prevents the formation of a 'lethal' (note the special meaning of 'lethal' here: it means preventing two lower-rank parents from producing two children with higher rank), and its reproduction procedure is a steady-state one. Eq. (12.64) outlines the niche technique.

Parent 1+Parent 2 = >Child 1 and Child 2   (12.64a)

Parent.Rank = min(Parent1.Rank, Parent2.Rank)   (12.64b)

Child.Rank = min(Child1.Rank, Child2.Rank)   (12.64c)

Test = Child.Rank $\leq$ Parent.Rank   (12.64d)

New.Child 1 = if (test) Child 1 else Parent 1   (12.64e)

New.Child 2 = if (test) Child 2 else Parent 2   (12.64f) where two children's ranks (Child1.Rank and Child2.Rank) are determined in the domain of the parent population in which they are treated as members of their paternal population. Alternatively, if the min in Eq. (12.64b) is changed to max, then the control condition is relaxed. However, numerical experiments show the original choice (min) is better than the alternative one (max). Eq. (12.64a) expresses that the combination of two parents produces two children. Eq. (12.64b) finds the lower rank of the two parents. Eq. (12.64d) tests whether the lower rank of the two children is less than the lower rank of their parents. Eqs. (12.64e, f) explain that the two children will go to the next generation if the test is true; otherwise, the two parents will go to the next generation. Eq. (12.64) is illustrated in Figs. 12.29 and 12.30 for a two-objective minimization problem. In Fig. 12.29, child 1 belongs to rank 1 of its paternal population. The best rank of its parents is rank 1 of parent 2. 'Test' is equal to 'True'. Therefore, the two children go to the next generation. In Fig. 12.30, the best rank of the two children is 3 which is larger than rank 2 of parent 1. 'Test' is equal to 'False'. Therefore, the two parents go to the next generation. The proposed method has been tested in numerical experiments of multi-objective optimization problems. See Figs. 12.34 and 12.35 of Section 12.5.7 for a 25-bar space truss (with/without niche technique). The niche technique helps prevent genetic drift and maintains a uniformly distributed population along the Pareto optimal set.

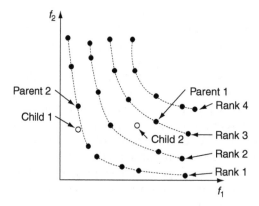

Figure 12.29 Children go to next generation.

12.5.7 *Comparative studies of benchmarked structures*

The Pareto GA techniques outlined above were applied to two examples. The first is a 25-bar space truss. Some points of its Pareto optimal set can be determined by a traditional optimization method. The second is an integrated optimum design of structural and control systems for a four-bar

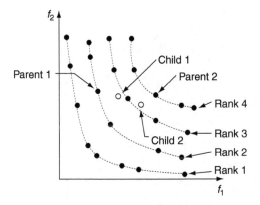

Figure 12.30 Parents go to next generation.

pyramid truss. Here, a traditional optimization algorithm is usually trapped in a local solution. For both examples, GA parameters are chosen as follows. The *uniform crossover* [16] and stochastic reminder selection are used for the crossover and selection operating procedures. The crossover probability is 0.60, and the mutation probability is 0.01. These probabilities are two basic parameters for the genetic algorithm. Usually, the crossover probability is between 0.5 and 1.0 while the mutation probability is taken as less than 0.1 because it is a small probabilistic case.

Example 12.5.3 – A 25-bar space truss shown in Fig. 12.31 is doubly symmetric. This condition divides all truss members into eight groups. They are A_1, $A_2 = A_3 = A_4 = A_5$, $A_6 = A_7 = A_8 = A_9$, $A_{10} = A_{11}$, $A_{12} = A_{13}$, $A_{14} = A_{15} = A_{16} = A_{17}$, $A_{18} = A_{19} = A_{20} = A_{21}$ and $A_{22} = A_{23} = A_{24} = A_{25}$.

Acting loads, $P_1 = 20\,\text{kip}$, $P_2 = -5\,\text{kip}$, $P_3 = 20\,\text{kip}$ and $P_4 = -5\,\text{kip}$, are also shown in Fig. 12.31. The weight density and elastic modulus of the structure are $\rho = 0.01\,\text{lbs in.}^{-3}$ and $E = 10,000\,\text{ksi}$, respectively. This case is optimized by some researchers for single-objective optimization [17]. The optimum objective is to minimize the structural weight W and the displacement Δ of point A at the direction of loading P_1 as

$$\left. \begin{aligned} f_1 &= W = \sum A_i L_i \rho_i \\ f_2 &= \Delta \end{aligned} \right\} \tag{a}$$

where A_i, L_i and ρ_i are the cross-sectional area, length and weight density of the *i*th bar, respectively. Constraint conditions are

$$\left. \begin{aligned} |\sigma_i| &\leq 25\,\text{ksi}, \quad i = 1, \ldots, 25 \\ 0.1\,\text{in.}^2 &\leq A_i \leq 0.5\,\text{in.}^2, \quad i = 1, \ldots, 25 \end{aligned} \right\} \tag{b}$$

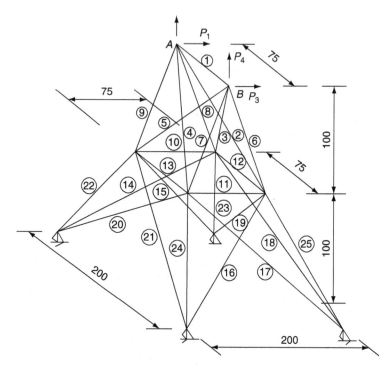

Figure 12.31 Twenty-five-bar truss.

Solution:

In the GA optimization procedure, the population size is 400 and the *chromosome* (string) length is 120. Optimal results are displayed in Figs. 12.32–12.35.

Figs. 12.32 and 12.33 show the results of points in a filter and points from rank 1 at generation 500. In Fig. 12.32, the points are evenly distributed along the Pareto optimum set and cover more of the region than the points in Fig. 12.31 do. Apparently, the filter brings much improvement in the optimum result set. Figs. 12.34 and 12.35 reveal the results with/without the niche technique; the niche technique prevents genetic drift and maintains a uniform distribution along the Pareto optimal set. Fig. 12.36 has the optimum solutions (Min W, Min Δ, and Min W, Δ) which are in the Pareto-set filter. Two optimum solutions (Min W, $W = 11.570$; Min Δ, $\Delta = 0.152$) are found by using a traditional optimal method [18]; the multi-objective optimum solution (Min W and Δ, $W = 40.611$, $\Delta = 0.339$) is solved by using a game theory algorithm [10, 11]. Results indicate that the points in the Pareto-set filter in generation 800 are a subset of the Pareto optimal set.

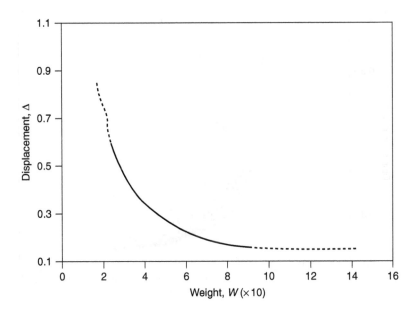

Figure 12.32 Points of rank 1 (generation 500).

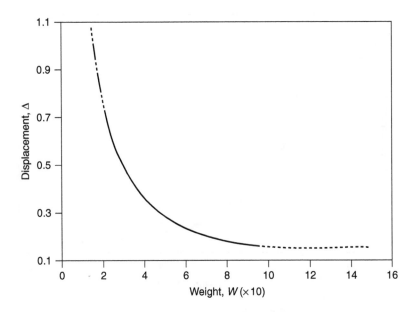

Figure 12.33 Points of rank 1 in Pareto-set filter (generation 500).

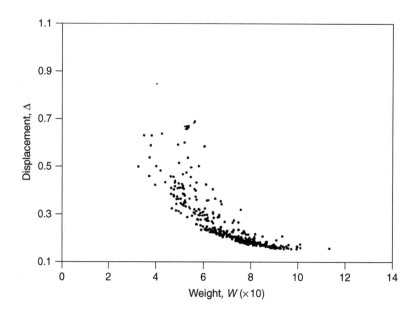

Figure 12.34 Generation 800 without niche.

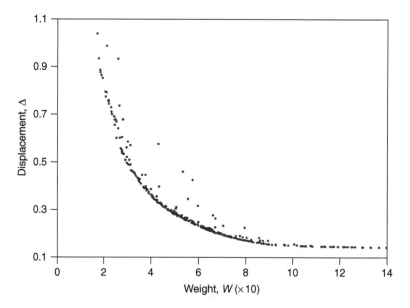

Figure 12.35 Generation 800 with niche.

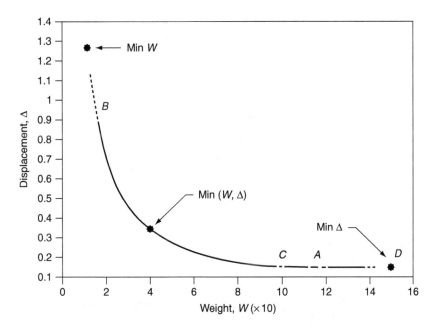

Figure 12.36 Pareto-set filter in generation 800.

Optimal results of a Pareto GA allow designers to select the best compromise from the solution set and to be able to determine whether the compromise solution is robust. In the selection process, a decision-maker should not pick a solution from segments whose slopes are close or equal to 0 (such as point A) or $\pi/2$ (such as point B) because a small change in one objective can cause a large improvement in another objective. Note the right-end segment of the Pareto optimal set in Fig. 12.36; when the displacement decreases from 0.155 to 0.152 (−2%) (point C to point D), the structural weight increases from 100.0 to 150.0 (+50%). Clearly, this segment is not a good trade-off region. In this example, a rational compromise design is situated in the middle segment of the set.

Example 12.5.4 – A four-bar pyramid truss [19] is shown in Fig. 12.37. The weight density and elastic modulus are $\rho = 0.0001$ and $E = 1$, respectively. An active control system is located on bar 1. The passive damping ratio ξ of the structure is assumed to be zero. A non-structural mass of two units is attached at the top node. Minimize structural weight W and control energy E_p as

$$\left. \begin{array}{l} f_1 = W = \sum A_i L_i \rho_i \\ E_p = \text{trace}[P] \end{array} \right\} \tag{a}$$

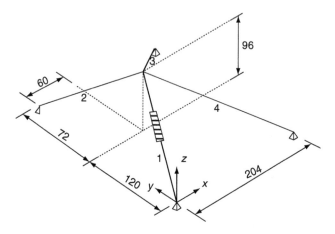

Figure 12.37 Four-bar pyramid truss.

where trace $[P] = \Sigma p_{ii}$; p_{ii} is the main dialogue element of the Riccati matrix $[P]$ (Eq. (12.11)). Constraint conditions are

$$
\left.\begin{array}{l}
0.7 \leq \bar{\omega}_1 \leq 0.8 \\
0.9 \leq \bar{\omega}_2 \leq 1.0 \\
1.1 \leq \bar{\omega}_3 \\
10.0 \leq A_i \leq 1{,}000 \quad i = 1, \ldots, 4
\end{array}\right\} \tag{b}
$$

Solution:

In this example, the population size is 700 and chromosome length is 60. Optimal results are shown in Figs. 12.38 and 12.39. Points W, E_p and WE_p in Fig. 12.38 are optimum solutions of traditional optimization algorithms [18], which minimize the weight ($W = 145.83$), input energy ($E_p = 48.218$), and weight and input energy (W and E_p, $W = 249.83$, $E_p = 52.388$) subject to the given constraints. Obviously, the single-objective optimization results (W,E_p) are local ones and the multi-objective optimum solution (WE_p) is not in the Pareto optimal set. Point WE_p is not an optimum solution for this MOP because weight W and input energy E_p can be simultaneously improved without damage to either one. Fig. 12.39 displays the point change in the filter from generation 100 to generation 700. Numerical results show that this Pareto GA is efficient and robust, and displays global convergence.

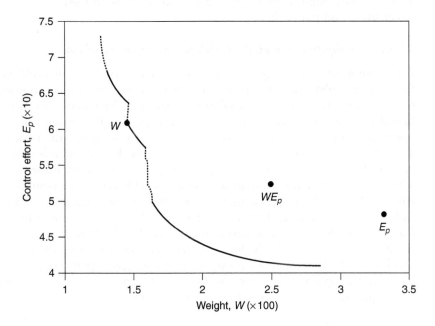

Figure 12.38 Pareto-set filter in generation 700.

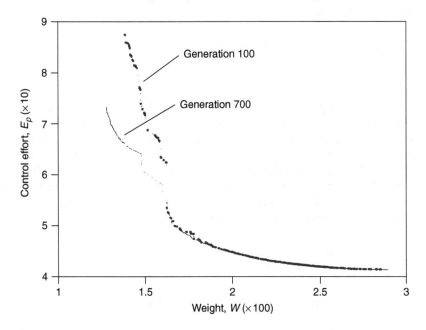

Figure 12.39 Filters in generations 100 and 700.

12.6 Constrained multi-objective optimization with GA, fuzzy logic and game theory

12.6.1 *Penalty-function method with fuzzy constraint scheme*

GAs cannot be directly applied to a constrained optimization problem. One way to solve such problems via GAs is to transform a constrained optimization into a nonconstrained one by using penalty-function methods. In order to do the transformation, we need to check a few features developed previously including rank. Remember that to determine the rank of a point, it is necessary to know the point's status (feasible or infeasible), distance from the Pareto optimal set, and the position in an infeasible zone. Three rules should guide establishment of the input function of a point for a Pareto GA to implement ranking. (A) The point's status as feasible or infeasible should be indicated by the function. (B) The closer a point is to the feasible zone, the higher its fitness evaluation is. (C) The closer a point is to the Pareto optimal set, the higher its fitness evaluation is. Rule (B) should dominate Rule (C) because an infeasible point is an unacceptable solution in a constrained optimization problem. The parabolic penalty-function is used for the penalty-function method in an MOP as follows:

$$f''_{Qi} = f_{Qi} + \lambda R_Q \tag{12.65}$$

where f_{Qi} is the ith objective function of point Q, λ is the penalty parameter, and R_Q is the penalty term given by

$$R_Q = \sum \{g_i^*\}^2; \ \{g_i^*\} = \begin{cases} 0, & g_i \leq 0 \\ g_i, & g_i > 0 \end{cases} \tag{12.66}$$

where g_i is the ith constraint function. In GA optimization, the only input desired from a problem being solved is the evaluation of each point in a population. For point Q, the input is $\{f''_{Q1}, \dots, f''_{Qm}\}$. Observation of function values f''_{Qi} (Eq. (12.65)) shows that they do not carry the knowledge required to find the point's rank or even its status. To find a point's feasible or infeasible status, the above penalty-function method is revised as follows. First, the objective function value f_{Qi} is normalized as f'_{Qi} (Eq. (12.61b)) so that $0 \leq f'_{Qi} \leq 1$; then the penalty term R_Q is adjusted to be larger than 2 so that an infeasible point is always larger than 1.

$$R_Q = \begin{cases} 0, & R_Q = 0 \\ R_Q + 2, & R_Q > 0 \end{cases} \tag{12.67}$$

The ith input function of point Q is expressed as

$$f''_{Qi} = f'_{Qi} + R'_Q \tag{12.68}$$

where f'_{Qi} is the ith normalized objective function of point Q.

New input functions $\{f''_{q1}, \ldots, f''_{qm}\}$ can indicate whether a point is feasible $(f''_{Qi} \leq 1)$ or infeasible $(f''_{Qi} > 1)$. They can also guarantee that a feasible point gets a lower rank than an infeasible one. However, they cannot distinguish between objective function values and violated amounts for constraints. A penalty term R is not in the same domain as an objective function. Their combination has no explicit meaning. Therefore, it is not possible to rank a point correctly by input functions defined in Eq. (12.68). This is because the rank of a point is decided by its position in the objective function domain and by the violated amounts for constraints. Such information cannot be obtained from input functions (Eq. (12.68)).

Fig. 12.40 shows a two-variable and two-objective optimization problem with three arbitrarily chosen points, P, Q and N. Infeasible points P and Q are equidistant from the feasible region $(g_{2P} = g_{2Q})$, and the normalized objective values and violated amounts of the three points are

$$\left. \begin{array}{l} f_{1N} = 0.6, \ f_{2N} = 0.74, \ g_{3N} = 0.4 \\ f_{1P} = 0.4, \ f_{2P} = 0.55, \ g_{1P} = 0.3 \, g_{2P} = 0.5 \\ f_{1Q} = 0.44, \ f_{2Q} = 0.6, \ g_{2Q} = 0.5 \end{array} \right\} \tag{12.69}$$

Using the revised penalty-function method (Eqs. (12.67) and (12.68)), the input functions for the three points are $\{2.76, 2.90\}_N$ $(f''_{N1} = f'_{N1} + R'_N = 0.6 + 2 + 0.4^2 = 2.76;$ $f''_{N2} = f'_{N2} + R'_N = 0.74 + 2 + 0.4^2 = 2.90)$, $\{2.74, 2.89\}_p$ and $\{2.69, 2.85\}_Q$. Note that this result disobeys the above three rules, (A), (B) and (C), for the ranking procedure. Point N, being nearest to the feasible zone, should dominate the other two points. Point P, being closer to the

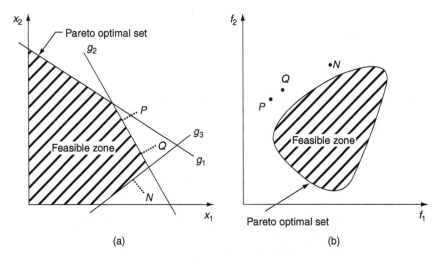

Figure 12.40 Feasible zone and Pareto-set filter in variable space: (a) design variable space, (b) objective function space.

Pareto optimal set, should be superior to point Q. Thus, the revised penalty-function method fails to work properly in a Pareto GA for a multi-objective constrained optimization problem. Further development is given hereafter. In Section 12.4.2, a fuzzy-logic-based penalty-function scheme is proposed. It derives from two facts: (A) classes and concepts for natural phenomena tend to be fuzzy instead of crisp; and (B) the GA procedure is random and probabilistic rather than numerically precise and calculable. Setting the tolerance intervals D_i to be equal to 0.001, 0.01, 0.02, 0.05, 0.1, 0.4, 1.0, 2.0, 5.0, and 15.0 for $i = 1$ to 10, the discrete fuzzy penalty-function formulation of point k is given as follows:

$$R_K = \begin{cases} 0 & \max(d_{K1}, \ldots, d_{KM}) \le 0.001 \text{ (zone 1)} \\ 2 & 0.001 < \max(d_{K1}, \ldots, d_{KM}) \le 0.01 \text{ (zone 2)} \\ 3 & 0.01 < \max(d_{K1}, \ldots, d_{KM}) \le 0.02 \text{ (zone 3)} \\ 4 & 0.02 < \max(d_{K1}, \ldots, d_{KM}) \le 0.05 \text{ (zone 4)} \\ 5 & 0.05 < \max(d_{K1}, \ldots, d_{KM}) \le 0.1 \text{ (zone 5)} \\ 6 & 0.1 < \max(d_{K1}, \ldots, d_{KM}) \le 0.4 \text{ (zone 6)} \\ 7 & 0.4 < \max(d_{K1}, \ldots, d_{KM}) \le 1.0 \text{ (zone 7)} \\ 8 & 1.0 < \max(d_{K1}, \ldots, d_{KM}) \le 2.0 \text{ (zone 8)} \\ 9 & 2.0 < \max(d_{K1}, \ldots, d_{KM}) \le 5.0 \text{ (zone 9)} \\ 10 & 5.0 < \max(d_{K1}, \ldots, d_{KM}) \le 15.0 \text{ (zone 10)} \\ 100 & 15.0 < \max(d_{K1}, \ldots, d_{KM}) \text{ (zone 11)} \end{cases} \tag{12.70}$$

where d_{Ki} is the violation amount of point K to the ith constraint, and M is the number of constraint conditions. d_{Ki} may be expressed as

$$d_{Ki} = \begin{cases} 0, & g_i(x_k) \le 0 \\ g_i(x_k), & \text{otherwise} \end{cases} \tag{12.71}$$

where the constraints are normalized (for example, $g_i(x_k) = \sigma(x_k) - 25 \le 0$ becomes $g_i(x_k) = \sigma(x_k)/25 \le 0$, to give them the same general order of magnitude.

Applying Eq. (12.70) to Fig. 12.40, the new values of the three points are N (6.6, 6.74) $(f''_{N1} = f'_{N1} + R_N = 0.6 + 6 = 6.6; f''_{N2} f'_{N2} + R_N = 0.74 + 6 = 6.74)$ where $R = 6$, $P(7.4, 7.55)$ and $Q(7.44, 7.6)$ where $R = 7$. Point N dominates point P, and point P dominates point Q, which satisfies the three rules defined for a ranking procedure. Thus, the *fuzzy-logic-based penalty-function* scheme can express the rank order of points in a Pareto GA and transform a multi-objective constrained optimization into a nonconstrained one.

12.6.2 Trade-off with game theory

An important property of a multi-objective optimization problem is that it allows the designer to participate in the design selection process even after formulation of the mathematical optimization model. It does this by providing a set of solutions (Pareto set) instead of a single solution. The optimal solution for a multi-objective optimization problem must be nondominated. For a given problem, there will be many nondominated solutions which could be regarded as acceptable, valid, optimum designs. A decision on the choice of the 'best' design is usually based on preferential ranking. In this case, the decision-maker has already set priorities based on objective functions or system reliability of the structure.

One method that can help a decision-maker find the 'best' solution is to discard the most redundant points and retain a subset of the most dissimilar points from the nondominated set. This generates a representative subset of points from the nondominated solution space. Then this subset is filtered down to a certain number of points (usually fewer than 10). From these points, the decision-maker will identify the preferred solution from the filtered points, and make the final choice. The problem of choosing the preferred nondominated solution from the nondominated solution set may also be viewed as a search for the point which optimizes the decision-maker's value function. Game theory programming discussed in Section 12.3 can be used to specify a priority function representation of the decision-maker's preference structure. This functional representation can then be optimized to obtain a single 'best' nondominated solution. The game theory decision-maker method which is designed for the Pareto GA can be stated as follows. At each generation, the maximum and minimum values of each objective are obtained from the filter. Then the *substitute (surrogate) function S* is constructed as

$$S = \prod_{i=1}^{n} \left[\frac{f_{i\max} - f_i(j)}{f_{i\max} - f_{i\min}} \right] \tag{12.72}$$

where n is the dimension of the objective function, m is the Pareto-set filter size, $f_{i\max}$, $f_{i\min}$ are the maximum and the minimum values of function f_i in the filter. If S does not change, or change less than a given limitation in K generations, then the convergence can be accepted.

$$S = \max(\beta_1, \beta_2, \ldots, \beta_m) \tag{12.73}$$

Generally, if maximum S remains constant for 10 or more generations, it can be concluded that the optimal zone is reached, and the analysis ends. At that time, the points in the Pareto-set filter should be on or close to the Pareto optimal set. This decision-making method produces a compromise solution among conflicting objective functions. This solution stands for the 'best' outcome of a cooperative game. It provides the players (objectives) with a better solution than their respective equilibrium solutions. In the Pareto GA optimization process, the game theory decision-making method gives designers a reference design by which they can judge their preference for the final choice. Also, game theory decision-making defines a convergence criterion for the Pareto GA.

12.6.3 Summary of constrained multi-objective optimization algorithm

The focus of this section is to outline the optimizer and decision-maker with the Pareto GA search algorithm, fuzzy penalty function, and game theory decision-making [20]. For a multi-objective optimization problem, the optimization procedure for the Pareto GA comprises two steps: (A) obtain an evenly distributed subset of the Pareto optimal set; and (B) choose a reasonable solution from the set. Fig. 12.41 shows the flow chart of this Pareto GA. In the figure, P_n is the population size and two-parent crossover is assumed.

The niche procedure of the GA optimizer, which prevents genetic drift and keeps a uniformly distributed representation of a population along a Pareto optimal set, is shown in Fig. 12.42. In the figure, P-Rank = Min(Parent1.Rank, Parent2.Rank), and C-Rank = Min(Child1.Rank, Child2.Rank). The ranks of two children (Child1.Rank and Child2.Rank) are decided in the domain of the parent population where they are treated as members of their paternal population.

A flow chart in Fig. 12.43 illustrates operation of the Pareto-set filter which pools nondominated points ranked 1 at each generation and drops dominated points. In the figure, NNP is the number of nondominated points and PFS is the Pareto-set filter size.

12.6.4 Illustration of optimization results

Two examples are studied by means of the proposed Pareto GA programming techniques. Uniform crossover [16] and stochastic remainder selection [9] are used in crossover and selection procedures. The crossover probability is 0.60 and mutation probability is 0.01.

Example 12.6.1 – Optimize the three-storey frame in Fig. 12.8 using Pareto GA with game theory guiding the decision procedure. Objective functions are structural weight and input energy. Chromosome length is 60 and population size is 400.

Solution:

Figs. 12.44 and 12.45 show the points in the Pareto-set filter which give the trade-offs between two conflicting objectives. From generation 60 through generation 100, the value of S (Eq. (12.34)) changes very little. Therefore, the evolutionary procedure is stopped and point *A* is taken as the

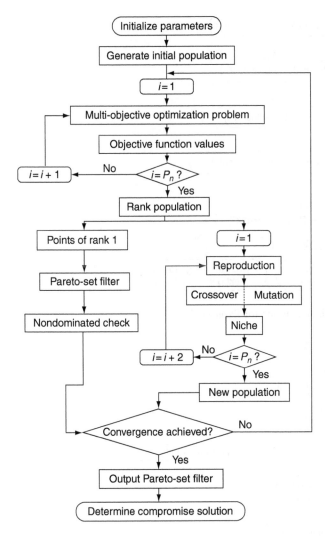

Figure 12.41 Pareto GA for multi-objective optimization.

compromise solution. Two points (Min W and Min E_i) in Fig. 12.44 are determined by using a traditional optimal method. In Fig. 12.45, they are drawn as reference points because they are end points of the Pareto optimal set. As shown in Fig. 12.45, the compromise solution point ($W = 2.8110$, $E_i = 5.1635$) yields the decision-making result. Note that this solution is in accord with evolutionary progress. At each generation, the magnitude of S (Eq. (12.72)) relates to the minimum and maximum values of each objective function and distribution of points in the filter. When the population uniformly converges to a Pareto optimal set, S will become steady.

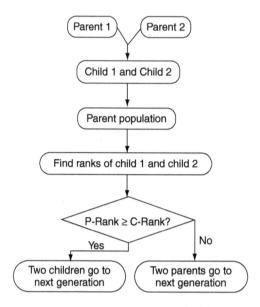

Figure 12.42 Niche technique for Pareto GA.

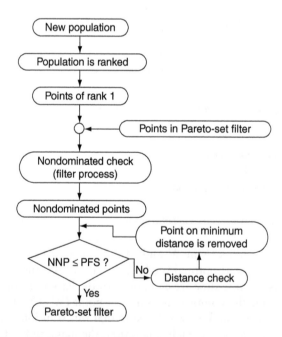

Figure 12.43 Pareto-set filter operation.

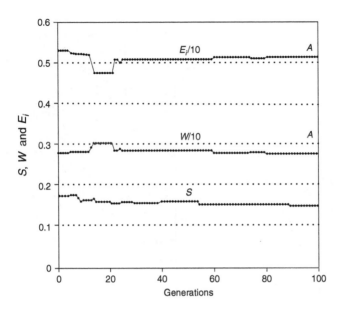

Figure 12.44 Compromise solution.

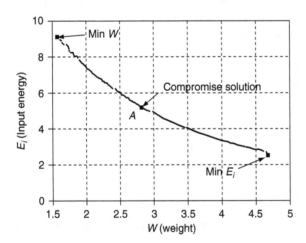

Figure 12.45 Points in Pareto-set filter (generation 100).

Example 12.6.2 – A seventy-two-bar space truss shown in Fig. 12.46 needs to be doubly symmetric. This condition divides all truss members into 16 groups. At each storey, four vertical bars (1, 2, 3, 4) become a group. The eight slanting support bars (5, 6, 7, 8, 9, 10, 11, 12), four horizontal bars (13, 14, 15, 16), and two horizontal slanting supporting bars (17, 18) form the other three groups. Acting loads $P_x = 10$, $P_y = 10$ and $P_z = -15$ are also

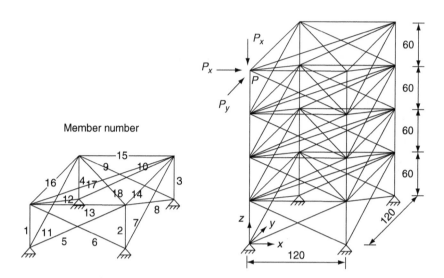

Figure 12.46 Seventy-two-bar space truss.

shown in Fig. 12.46. The weight density and elastic modulus are assumed to be $\rho = 0.1$ and $E = 10,000$, respectively. Some researchers [21] have optimized this structure by taking the structural weight as the only objective. Optimum constraint conditions are

$$\left. \begin{array}{l} |\sigma_i| \leq 25 \\ |\delta_i| \leq .25 \\ 0.1 \leq A_i \leq 5.0 \end{array} \right\} \tag{a}$$

The optimum objective functions are structural weight W and strain energy E_w as

$$\left. \begin{array}{l} W = \sum A_i L_i \rho_i \\ E_w = \sum \sigma_i \delta_i v_i \end{array} \right\} \tag{b}$$

For Pareto GA optimization, the population size is 400 and chromosome length is 240.

Solution:

Fig. 12.47 gives the design results for generation 500 where the optimum solutions (Min W, Min E_w and Min W, E_w) are in the Pareto-set filter. Both single-objective optimum solutions (Min W, $W = 6,149$ and Min E_w, $E_w = 1.4235$), found by using a traditional method, are the end points of the Pareto optimal set for this problem. The multi-objective optimum solution (Min W, E_w, $W = 13,666$ and $E_w = 3.0380$) is achieved by using a game

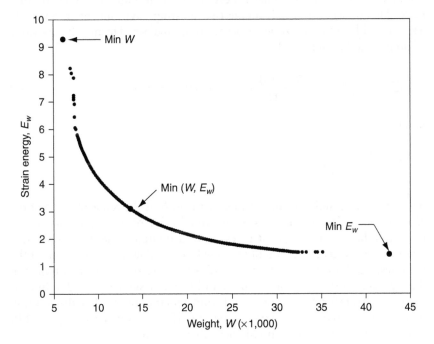

Figure 12.47 Pareto-set filter (generation 500).

theory algorithm. Points in the Pareto-set filter in generation 500 comprise a subset of the Pareto optimal set.

Note that points in Fig. 12.47 do not extend to the region at either end of the Pareto optimal set. This is because the population size is not large enough, and the feasible zones at each end are much narrower than the zone in the centre. If some zones in a feasible space are narrow or not easily accessible, or if the whole feasible zone is small and quite sparse, then finding a feasible point in these zones is difficult. It is also difficult for points to enter these zones in the evolutionary process if the model population size is not large enough. To correct this phenomenon, the niche technique is revised by adding uniform distribution pressure on a model population so that the population evenly converges along the entire Pareto optimal set. Derivation of this algorithm is based on a bi-objective optimization problem. For this kind of problem, the points in a rank form a line composed of segments. Continuous points make up a line segment. Two points next to each other are considered non-continuous when the distance between them is three times larger than the average distance between points in the present rank. By increasing fitness values of points in the end zones and decreasing fitness values in the centre of each segment, uniform convergence can be achieved. At the same time, points close to the end zones of the Pareto optimal set always have higher fitness values than points in the centre of a rank. This offsets the disadvantage that existing environments are usually less accessible in the end zones of a Pareto optimal set. The new fitness value of points can be calculated as

$$f_{\text{fitness},i}^{(\text{new})} = f_{\text{fitness},i}^{(\text{old})} + \frac{4(h_1 + h_0)}{(m-1)^2}[i^2 - (m+1)i + m] + h_1 \qquad (c)$$

in which

$$h_1 = \frac{2(m-2)}{(m+1)}h_0 \qquad (d)$$

where m is the number of points in a segment, h_0 is the reduced value of the central point's fitness, and h_1 is the enlarged value of the end points. When point number m in a segment is less than a limited value, the adjustment process is neglected. The sum of the new fitness values equals the sum of the old ones. Eq. (c) is illustrated in Fig. 12.48, where the number of points in the segment is 40; h_0 is equal to 0.04 and $f^{(\text{old})}$ is assumed to be 0.8. In fact, Eq. (c) is a transformation formula. It transfers the straight line distribution of points into a parabolic distribution with offsets h_0 and h_i (see Fig. 12.48). The derivation of Eq. (c) is briefly described hereafter.

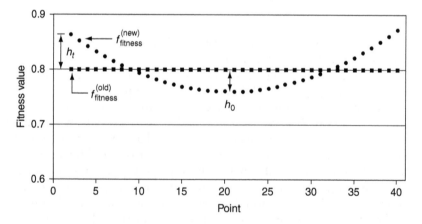

Figure 12.48 New fitness values under revised niche method.

Let the new fitness function be expressed as

$$f_{\text{fitness},i}^{(\text{new})} = f_{\text{fitness},i}^{(\text{old})} + h_1 + ai^2 + bi + c \qquad (e)$$

where i stands for the ith point. Assuming that the number of a population is m and $m > 1$, we have

$$i=1; \quad f_{\text{fitness}}^{\text{new}} = f_{\text{fitness}}^{\text{old}} + h_1 \tag{f}$$

$$i=m; \quad f_{\text{fitness}}^{\text{new}} = f_{\text{fitness}}^{\text{old}} + h_1 \tag{g}$$

$$i=\frac{m+1}{2}; \quad f_{\text{fitness}}^{\text{new}} = f_{\text{fitness}}^{\text{old}} - h_0 \tag{h}$$

Using the known condition to Eqs. (f), (g) and (h) yields Eqs. (i), (j) and (k), respectively

$$a+b+c=0, \quad am^2 + bm + c = 0 \tag{i, j}$$

$$a\frac{(m+1)^2}{4} + b\frac{m+1}{2} + c + h_1 = -h_0 \tag{k}$$

Solving Eqs. (i, j and k), we have

$$a = \frac{4(h_1 + h_0)}{(m-1)^2}; \quad b = -a(m+1); \quad c = am \tag{l, m, n}$$

Substituting Eqs. (l, m and n) into Eq. (e) yields Eq. (c). Using the following condition

$$\sum f_{\text{fitness},i}^{(\text{new})} = \sum f_{\text{fitness},i}^{(\text{old})} \tag{o}$$

we may rewrite Eq. (c) as

$$m f_{\text{fitness},i}^{(\text{old})} + m h_1 + \frac{4(h_1 + h_0)}{(m-1)^2} \sum [i^2 - (m+1)i + m] = m f_{\text{fitness},i}^{(\text{old})} \tag{p}$$

or

$$m h_1 + \frac{4(h_1 + h_0)}{(m-1)^2} \sum [i^2 - (m+1)i + m] = 0 \tag{q}$$

Since

$$\sum_{i=1}^{m} i^2 = \frac{1}{6} m(m+1)(2m+1); \quad \sum_{i=1}^{m} i = \frac{1}{2} m(m+1) \tag{r, s}$$

then Eq. (q) can be condensed as Eq. (d).

The modified results of Fig. 12.47 are illustrated in Fig. 12.49 which are obtained by using the revised niche method ($h_0 = 0.08$), and evenly cover the entire Pareto optimal set. Optimal solutions from a Pareto GA give decision-makers a full view of possible optimization designs. Designers can see if a chosen solution is robust, and what trade-off can be made between conflicting design objectives. Decision-makers should avoid choosing a sensitivity solution. For this example, they should not pick a solution

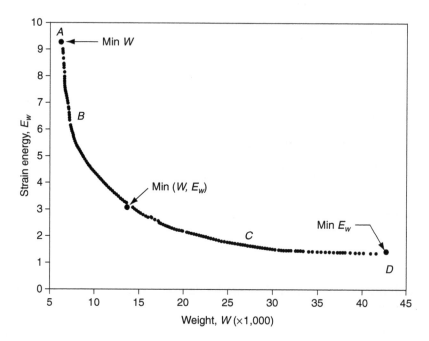

Figure 12.49 Pareto-set filter (generation 500, new niche method).

from either end segment of the Pareto optimal set (see Fig. 12.49). In the left end segment *AB*, a small change in weight causes a large change in the strain energy. In the right end segment *CD*, a small decrease in the strain energy causes a large increase in weight. Obviously, the best compromise solution comes from central segment *BC*.

12.7   Optimum design of various structural systems subjected to earthquake loads

12.7.1   *Construction of building structural models including cost functions and constraints*

Problems studied here belong to the class of planar frame design associated with a building structure of the frame system. In this case, the main challenge is to produce optimal designs which combine lower structural cost and higher earthquake resistance for steel, RC and steel–RC building systems. Design variables are the cross-sectional areas of structural members, namely, columns and beams. For a reinforced concrete (RC) member, it is assumed that the ratios of cross-sectional width b and height h of column and beam are 1/1 and 3/5, respectively (Fig. 12.50). The cross-sectional properties of a reinforced concrete column and beam can be represented as

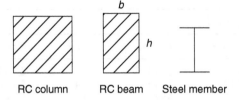

RC column   RC beam   Steel member

Figure 12.50 Cross-sections of structural members.

$$\left.\begin{array}{l} A = b^2; \quad I = \dfrac{1}{12}A^2; \quad S = \dfrac{1}{6}A^{3/2} \text{ for a column} \\[3mm] A = \dfrac{3}{5}b^2; \quad I = \dfrac{5}{36}A^2; \quad S = 0.125A^{3/2} \text{ for a beam} \end{array}\right\} \tag{12.74}$$

where A is the cross-sectional area; I is the sectional moment of inertia; and S is the sectional modulus. For a steel member, the following relationship (Eq. 12.75) is used (see Chapter 5):

$$A = 0.80I^{\frac{1}{2}}; \quad S = 0.78I^{\frac{3}{4}} \tag{12.75}$$

Objectives are minimum structural potential energy and minimum structural cost. In structural dynamic analysis, the assumed earthquake forces are based on UBC and the earthquake records of El Centro, N-S, 18 May 1940. For a given structure, the total structural cost is obtained by summing the cost of each member. For steel structural members, the cost of members is related to the location of a building, and covers fabrication and delivery, transportation, and installation as well as the subcontractor's overheads and profit [4]. Table 12.6 gives sample data for the total cost per tonne of a steel-frame structure.

Cost parameters of the reinforced concrete members include material, mix, form, location, reinforcement, transportation and the subcontractor's overheads and profit. These costs are shown in Tables 12.7 and 12.8. It is assumed that a column has an average reinforcement ratio, the formwork is used four times, and the concrete strength is 4,000 psi.

Table 12.6 Cost data for structure metal framing.

Structural types	Total cost per tonne	
	High-strength bolting connection	*Field-welded connection*
2 to 6-storey frame	$1,676	$1,736
7 to 15-storey frame	$1,704	$1,764
Over 15-storey frame	$1,732	$1,792

Table 12.7 Cost data for RC columns.

Size (in.)	Total cost per cubic yard				
	Material	Labour	Equip.	O & P	Total
12 × 12	$288	$430	$410	$317	$1,175
16 × 16	$272	$350	$114	$254	$990
24 × 24	$224	$247	$81	$183	$735
36 × 36	$213	$187	$61.5	$143.5	$605

Note: O & P = Overheads and Profit.

Table 12.8 Cost data for RC beams.

Span (ft)	Total cost per cubic yard			
	Live load in kips per linear foot			
	Under 1 kip	2–3 kips	4–5 kips	6–7 kips
10	$639	$475	$460	$431
16	$604	$457	$423	$409
20	$534	$421	$383	$338

Tables 12.7 and 12.8 give discrete values. If a value does not appear in the tables, it can be found by using linear interpolation. For example, the cost per cubic yard for a column with cross-sectional sizes 20 × 20 in. can be calculated by

$$C_{\text{Col}20} = \frac{C_{\text{Col}16} - C_{\text{Col}24}}{24 - 16}(20 - 16) + C_{\text{Col}24} \tag{12.76}$$

where $C_{\text{Col}k}$ is the cost parameter of a column with cross-sectional sizes, $k \times k$ (in.), given in Table 12.7.

The cost function f_{si} of steel member i can be expressed as

$$f_{si} = C_s \rho_s v_i \tag{12.77}$$

where C_s is the cost parameter given in Table 12.6; $\rho_s = 490$ pcf is the weight density of steel; and v_i is the volume of member i.

The cost function f_{ci} of RC column i can be expressed as

$$f_{ci} = C_{\text{Col}h_i} b_i h_i L_i = C_{\text{Col}h_i} A_i L_i \tag{12.78}$$

where $C_{\text{Col}h_i}$ is the cost parameter given in Table 12.8; b_i, h_i and A_i are the cross-sectional width, height and area of the RC column i, respectively; and

L_i is the length of the RC column i. The cost function f_{bi} of RC beam i can be expressed as

$$f_{bi} = C_{b\,sipi}h_i b_i L_i = C_{b\,sipi}A_i L_i \qquad (12.79)$$

where $C_{b\,sipi}$ is the cost parameter given in Table 12.8 reflecting the span and loading of the beam; h_i and b_i are the cross-sectional height and width of RC beam i, respectively; and L_i is the length of RC beam i.

Constraints include storey drift and member stresses. The former applies to each storey. These storey-drift (displacements) constraints have two important roles. First, they enable the attachments and auxiliary structures of a building to work well under the action of design loading. Second, they eliminate the occupants' uncomfortable feeling caused by large vibration under the action of dynamic loadings. This feeling occurs even though such displacements do not damage the building. Under the action of earthquake loading, storey-drift limitations are defined in Eq. (12.80) according to UBC provisions [22].

$$\Delta_a = \begin{cases} \min\left(\dfrac{0.04h_{sx}}{R_w},\ 0.005h_{sx}\right) & T < 0.7\,s \\[4mm] \min\left(\dfrac{0.03h_{sx}}{R_w},\ 0.004h_{sx}\right) & T \geq 0.7\,s \end{cases} \qquad (12.80)$$

where T is the fundamental period of the structure; h_{sx} is the storey height at level x; and R_w is the numerical coefficient given in the UBC based on structural type. Stress constraints apply to structural members and are determined by the material properties of these members. If the stress on a structural member is larger than allowable, this member may not perform as efficiently as expected at the design stage, and structural damage could occur.

Stress constraints apply only to steel members, not to reinforced concrete members, because of the difference in their design. For an RC member, the design consists of two steps: (A) finding the internal forces of a member, and (B) designing the reinforcement according to these forces. Internal forces are borne by concrete and reinforcement together, and the force distribution to both parts can be adjusted by changing the reinforcement ratio. On the basis of design experience, reinforcement ratios of the columns and beams of a frame tend to be lower than the maximum code-permitted values if displacements and storey drifts of a frame are less than the allowable values. Consequently, reinforced members can bear the internal forces, and no structural damage occurs under normal loading conditions. Furthermore, the design of a reinforced concrete member needs only to meet the basic limitation on reinforcement ratio and cross-sectional area. It is therefore reasonable to put drift constraints on reinforced concrete members at the initial design stage. This makes the optimum analysis procedure simpler and optimum design more practical.

To simplify the analysis procedure, the design of steel beam members only considers bending stress checks. For compression steel columns, it is assumed that the axial compression ratio of a column is $f_a/F_a \leq 0.15$ where f_a is axial compression stress and F_a is allowable axial compression stress. Also, there is no shear stress check for steel columns. Since this is an initial design due to a lateral seismic force, member-size determination is simplified by not conforming to all code requirements.

Optimum procedures for each structural system consist of two parts: the first is single-objective optimization with structural cost as the objective; the second is multi-objective optimization with structural cost and potential energy as the objectives.

12.7.2 *Optimum design data for steel, RC and RC–steel frames*

The optimum procedure is also a simulation procedure for the initial design stage of a structure. After an architect determines the shape of a building based on its future functions and the owner's requirements, and a structural engineer decides the loads according to design codes, the next step is to select a proper structural system. This system must support the acting loads and the weight of the structure. The engineer then determines the cross-sectional dimensions of the structural members. For the ten-storey building in Fig. 12.51, a frame structural system with the lower storeys being wider is chosen. In the direction of earthquake excitation shown, the force-resistant system consists of five identical planar frames. One of them

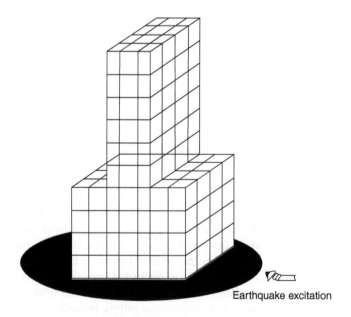

Earthquake excitation

Figure 12.51 Ten-storey building.

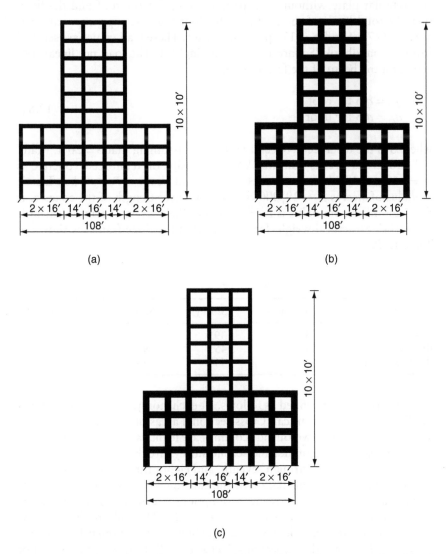

Figure 12.52 Three planar frames: (a) steel frame, (b) RC frame, (c) RC and steel frame.

will be used for analysis. To achieve an economical design and good structural performance, the multi-objective optimization procedure is applied to evaluate three choices: steel frame (see Fig. 12.52a), reinforced concrete frame (see Fig. 12.52b), and composite steel and reinforced concrete frame (see Fig. 12.52c).

All three have the same structural topology. Assume that the total superimposed load is 40 psf for the roof and 75 psf for the floor. If the reinforced concrete flat plate without drop panels is used for the roof and the floor, then the total loads of the roof and the floor plus the superimposed loads are $q_r = 127$ psf and $q_f = 175$ psf, respectively. These load values are selected from the handbook 'Square Foot Estimating' [3]. The load per linear foot acting on the beam of the frames is equal to

$$\left. \begin{array}{l} q_{rb} = q_r S \text{ for roof} \\ q_{fb} = q_f S \text{ for floor} \end{array} \right\} \tag{12.81}$$

where S is the distance between two frames, assuming that $S = 20$ ft. Material properties are as follows. Weight density of reinforced concrete is $\rho_c = 150$ pcf. Yield stress of steel column and beam is $F_y = 36$ ksi. Elastic modulus of steel is $E_s = 29,000$ ksi. Compressive strength of the concrete is $f'_c = 4$ ksi. Elastic modulus is $E_c = 3,605$ ksi. Yield stress of reinforcement is $f_y = 60$ ksi. Structural weight for each floor of the three frames is shown in Table 12.9.

Table 12.9 Structural weight (kips).

Structural type	Storey 1–3	Storey 4	Storey 5–9	Storey 10
Steel Frame	395	393	158	114
RC Frame	467	455	195	139
RC & Steel Frame	467	447	158	114

Design variables are defined as follows. At each storey, cross-sectional areas of all the beams are identical. Cross-sectional areas of the internal columns are the same, and the two outside columns are identical. Structural members and member locations of first and second storeys are the same as those of the third and fourth, fifth through seventh, and eighth through tenth storeys.

For seismic design, calculation of the earthquake force is based on UBC requirements, and the following assumptions are made. The building is located in seismic zone 2B with factor $Z = 0.20$. The soil profile type is S_2 which leads to site coefficient $S = 1.2$. The building is a standard occupancy structure; therefore, the importance factor $I = 1.0$. Bearing force system of the building is a moment-resisting frame type with factor $R_w = 6$ for the steel frame, and $R_w = 5$ for the RC and the composite frame. Factor C_t is 0.035 for the steel frame, and $C_t = 0.030$ for the RC frame and the composite frame. Since the building is in the occupancy category of seismic zone 2, the static lateral force procedure in UBC can be used. The vertical distribution of the design base shear is derived from

$$F_x = \frac{(V - F_t)w_x h_x}{\displaystyle\sum_{i=1}^{n} w_i h_i} \tag{12.82}$$

where F_t is the concentration force, and is determined by

$$F_t = \begin{cases} 0.07TV, & T \leq 0.7\,\text{s} \\ 0, & T > 0.7\,\text{s} \end{cases} \tag{12.83}$$

12.7.3 Optimum design results from single-objective and multi-objective functions

In this section, the procedures consist of optimizing the three frames with single-objective optimization of structural cost, analyzing the optimal results, and comparing the dynamic responses of the three optimal frames. Then the frames with multi-objective functions are optimized, and results are compared with the single-objective ones. Finally, an optimum design which has rational structural cost and rational dynamic response is recommended.

1. *Single-Objective Optimization*
   Here, the objective is to minimize structural cost. For the steel frame, the single-objective optimization formulation is

$$\left. \begin{aligned} & \text{Minimize } F_1 = \sum_{i=1}^{m} f_{si} \\ & \text{Subject to } f_{bi} \leq F_b, \quad i = 1, \ldots, m_2 \\ & \frac{f_a}{F_a} + \frac{f_b}{F_b} \leq 1, \quad i = 1, \ldots, m_1 \\ & \Delta_j \leq \Delta_a, \quad j = 1, \ldots, n \\ & A_i^L \leq A_i \leq A_i^u, \quad i = 1, \ldots, m \end{aligned} \right\} \tag{12.84}$$

where m is the number of members, $m = 102$; m_1 is the number of columns, $m_1 = 56$; m_2 is the number of beams, $m_2 = 46$, $m_1 + m_2 = m$; n is the number of structural storeys, $n = 10$; F_a and F_b [AISC] are the allowable axial compression and bending stress, respectively; Δ_a is the allowable storey drift defined in Eq. (12.80) (for this case $\Delta_a = 0.48$ in.); F_1 is the structural cost and f_{si} is defined in Eq. (12.77) (high-strength bolting connection); the lower bound and upper bound of the cross-sectional area of the steel members are $A_i^L = 8$ in.^{2} and $A_i^u = 125$ in.^{2}, respectively. When the optimum variables reach the lower bound A_i^L, it means that the corresponding member can be eliminated from the structure.

For the RC frame, the single-objective optimization formulation is

$$\left. \begin{aligned} & \text{Minimize } F_1 = \sum_{i=1}^{m_1} f_{ci} + \sum_{i=1}^{m_2} f_{bi} \\ & \text{Subject to } \Delta_j \leq \Delta_a, \quad j = 1, \ldots, n \\ & A_i^L \leq A_i \leq A_i^u, \quad i = 1, \ldots, m \end{aligned} \right\} \tag{12.85}$$

where f_{ci} and f_{bi} are defined in Eqs. (12.78) and (12.79); the lower bound and upper bound of the cross-sectional area of the RC column and beam are $A_i^l = 100$ in.^{2} and $A_i^u = 1,500$ in.^{2}, respectively.

For the composite frame, the single-objective optimization formulation is

$$
\left.
\begin{aligned}
&\text{Minimize } F_1 = \sum_{i=1}^{n_1} f_{si} + \sum_{i=1}^{n_2} f_{ci} + \sum_{i=1}^{n_3} f_{bi} \\
&\text{Subject to } f_{bi} \leq F_b, \quad i = 1, \ldots, m_4 \\
&\frac{f_a}{F_a} + \frac{f_b}{F_b} \leq 1, \quad i = 1, \ldots, m_3 \\
&\Delta_j \leq \Delta_a, \quad j = 1, \ldots, n \\
&A_i^L \leq A_i \leq A_i^u, \quad i = 1, \ldots, m
\end{aligned}
\right\}
\qquad (12.86)
$$

where m_3 is the number of steel columns, $m_3 = 24$; m_4 is the number of steel beams, $m_4 = 18$; n_1 is the number of steel members, $n_1 = 42$; n_2 is the number of concrete columns, $n_2 = 32$; and n_3 is the number of concrete beams, $n_3 = 28$.

Optimal design results under the action of earthquake forces for the three cases are given in Table 12.10. It shows optimum structural costs and cross-sectional areas of the structural members. Note that the reinforced concrete frame has the lowest structural cost. The structural cost of the composite frame is close to that of the steel frame. An RC frame is clearly the first choice if structural cost is the prime consideration. However, the structural volume of the RC frame is the largest. Thus, it occupies the maximum space out of the three structural systems. A steel frame would be preferable if spatial dimensions were more important.

Table 12.10 Optimal results of single-objective optimization.

Cost		Steel frame	RC frame	Composite frame
		$58,946	$48,302	$58,890
1–2 Storey (Area in.²)	Internal Col.	16.01	221.81	305.45
	Side Column	8.86	100.00	188.93
	Beam	16.69	234.78	213.33
3–4 Storey (Area in.²)	Internal Col.	15.97	219.94	293.16
	Side Column	9.52	100.00	179.14
	Beam	14.96	206.91	205.63
5–7 Storey (Area in.²)	Internal Col.	18.71	250.49	19.47
	Side Column	18.81	262.01	19.79
	Beam	20.64	274.40	23.73
8–10 Storey (Area in.²)	Internal Col.	14.65	191.75	16.84
	Side Column	15.24	199.32	17.56
	Beam	11.09	162.63	13.89
Volume ft²		136.40	1,863.31	1,294.70

Table 12.11 shows the active constraints. Drift constraints play a major role in the optimum procedure. For stress constraints, only the two side columns at the top storey of the steel frame reach the limitation. For the composite frame, sixth-storey drift also reaches the limitation.

Table 12.11 Active constraints.

Structural system	Drift constraints										Stress constraints
	1	2	3	4	5	6	7	8	9	10	Side Columns
Steel Frame		#	#		#			#			# – top storey
RC Frame		#	#		#			#			
RC & Steel		#	#		#	#		#			

Next, structural performances under seismic loading are compared. This comparison includes displacement responses at the top storey and the fifth storey, and earthquake input energies of all three structural systems. Analytical results are shown in Figs. 12.53–12.55. For dynamic analysis, the maximum acceleration of El Centro, N-S, 18 May 1940 earthquake is adjusted to 28 in. s^{-2} to match the dynamic displacement responses at the same magnitude obtained from the pseudo-dynamic analysis.

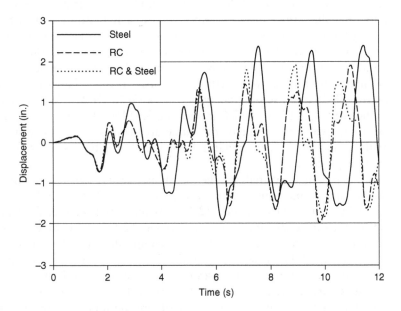

Figure 12.53 Displacements of fifth storey.

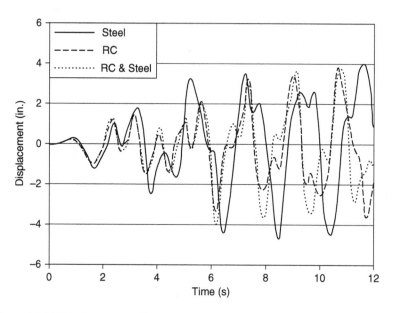

Figure 12.54 Displacements of top storey.

The displacement responses (see Figs. 12.53 and 12.54) of the three frames do not differ greatly. This is because the frames are designed such that some of their storey drifts are at the drift limitation which has the same value for all three frames (0.004 h_{sx} controls). The steel frame has the largest displacement response. The reason is that it has a longer period and more flexible stiffness than the other two frames. The maximum displacement responses of the RC frame and the composite frame are approximately the same. Maximum displacements calculated by time-history analysis are close to the values calculated by the UBC lateral seismic analysis method: 4.714/4.276 for the steel frame, 3.854/4.227 for the RC frame, and 3.924/4.300 for the composite frame, where the value of the numerator comes from time-history analysis and of the denominator from the UBC lateral seismic force analysis procedure.

In the time-history analysis, the maximum sway distance (from minimum value to maximum value) at the top storey of the steel frame is 8.532 in. around 11 s (see Fig. 12.54). At the initial stage of dynamic action, the earthquake input energies (see Fig. 12.55) of the three frames are the same. The input energy of the composite frame becomes the largest around 10 s. However, when the time is close to 12 s, The input energy of the steel frame increases rapidly. Alterations of input energy for the RC frame are consistent with those of the composite frame.

Observing the dynamic responses of all three optimum design frames, it can be concluded that the dynamic characteristics of the composite

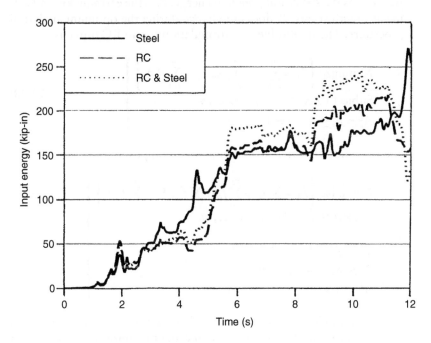

Figure 12.55 Input energies of structures.

frame are close to those of the RC frame. If a designer wants a lower-cost structure as well as a structure with small displacements and structural members of small cross-sectional sizes at the upper storeys, then a composite structural system is a good candidate.

As noted, single-objective optimization results in an optimal design with respect to a specific problem formulation, say, structural cost. This type of optimization leads to a flexible structure which may not be desirable in terms of structural performance factors, such as protection of equipment and comfort of occupants.

2. *Multi-Objective Optimization* Multi-objective formulation for a steel frame is

$$
\left.
\begin{aligned}
& \text{Minimize } \left\{ F_1 = \sum_{i=1}^{m} f_{si}, \ F_2 = \sum_{i=1}^{n} P_i \Delta_i \right\} \\
& \text{Subject to } f_{bi} \le F_b \quad i = 1, \ldots, m_2 \\
& \frac{f_a}{F_a} + \frac{f_b}{F_b} \le 1 \quad i = 1, \ldots, m_1 \\
& \Delta_j \le \Delta_a \quad j = 1, \ldots, n \\
& A_i^L \le A_i \le A_i^u \quad i = 1, \ldots, m
\end{aligned}
\right\}
\tag{12.87}
$$

where F_2 is the structural potential energy; P_i is the earthquake load acting at the ith storey. P_i does not change during the optimum analytical procedure. The multi-objective formulation for an RC frame is

$$
\left.
\begin{aligned}
&\text{Minimize } \left\{ F_1 = \sum_{i=1}^{m_1} f_{ci} + \sum_{i=1}^{m_2} f_{bi},\ F_2 = \sum_{i=1}^{n} P_i \Delta_i \right\} \\
&\text{Subject to } \Delta_j \leq \Delta_a \quad j = 1, \ldots, n \\
&A_i^L \leq A_i \leq A_i^u \quad i = 1, \ldots, m
\end{aligned}
\right\}
\tag{12.88}
$$

and the multi-objective formulation for a composite frame is

$$
\left.
\begin{aligned}
&\text{Minimize } \left\{ F_1 = \sum_{i=1}^{n_1} f_{si} + \sum_{i=1}^{n_2} f_{ci} + \sum_{i=1}^{n_3} f_{bi},\ F_2 = \sum_{i=1}^{n} P_i \Delta_i \right\} \\
&\text{Subject to } f_{bi} \leq F_b \quad i = 1, \ldots, m_4 \\
&\frac{f_a}{F_a} + \frac{f_b}{F_b} \leq 1 \quad i = 1, \ldots, m_3 \\
&\Delta_j \leq \Delta_a \quad j = 1, \ldots, n \\
&A_i^L \leq A_i \leq A_i^u \quad i = 1, \ldots, m
\end{aligned}
\right\}
\tag{12.89}
$$

These multi-objective formulations seek a compromise solution between the minimization of structural cost and the minimization of structural response under loading action. Results from these formulations are presented in Figs. 12.56 (a, b and c). Any point in the Pareto optimal set can be selected as an optimum design which represents a designer preference. A designer has many choices instead of only one. Figs. 12.5 (a, b and c) provide a trade-off between structural cost and structural potential energy. If a designer wants to compromise between lower cost and higher safety, the middle part of a Pareto optimal set provides the choices. For the steel frame, the lowest structural cost is $58,946, while structural potential energy is 303.98 kip-in. If the cost is approximately doubled to $118,000, the potential energy will be 78.3 kip-in (point A, Fig. 12.56a), which is decreased approximately fourfold. If the building is an essential structure, such as a hospital, and safety is the main consideration, a steel-frame design with a cost of $150,000 and a potential energy of 49 kip-in may be the optimum selection. A further improvement in potential energy leads to a larger increase in structural cost but with a small decrease in structural potential energy (see Fig. 12.56a). Without the Pareto optimal set, it is impossible for a designer to have an overview of the trade-offs.

Next, the dynamic responses of single-objective optimum designs and multi-objective optimum designs are compared. A composite frame design, which about doubles the structural cost of the single-objective optimum design, is selected from the Pareto optimal set (see Fig. 12.56c) to compare structural performance under the action of El Centro, N-S,

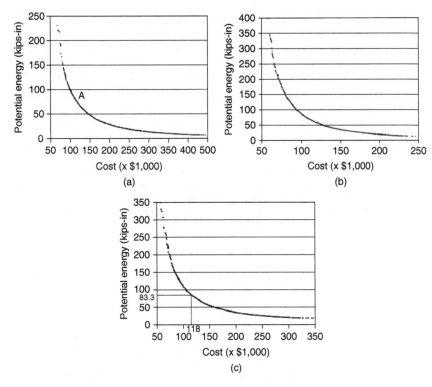

Figure 12.56 (a) Pareto optimal set for steel, (b) Pareto optimal set for RC frame, (c) Pareto optimal set for composite frame.

18 May 1940 earthquake. Table 12.12 gives cross-sectional sizes of the optimum design. Figs. 12.57–12.59 show the results. Fig. 12.57 gives a comparison of the base shear. Base shear of the multi-objective design increases due to increased structural stiffness. Although base shear of the multi-objective optimum design is larger than that of the single-objective design, displacements are much smaller (see Fig. 12.58): the maximum value at the top storey decreases from 3.924 to 2.237 in., and the maximum sway distance at the top storey decreases from 7.384 to 4.412 in. (see Fig. 12.58). Observing the curve of input energy versus time for the new design, the peak value is lowered to about 20 per cent (see Fig. 12.59). The multi-objective optimum design thus has a much better earthquake-resistant capacity since it has greater structural stiffness and lower earthquake input energy. Note that maximum values of the dynamic response (see Figs. 12.57–12.59) for the multi-objective optimum design occur earlier than for the single-objective optimum design.

Table 12.12 Optimal results of multi-objective optimization of composite frame.

Cost	Cross-sectional area (in.²)					
	1–2 Storey			3–4 Storey		
	Int. Col.	Side Col.	Beam	Int. Col.	Side Col.	Beam
$118,000	642.92	100.00	579.77	492.65	242.62	556.95
Potential Energy	Cross-sectional Area (in.²)					
	5–7 Storey			8–10 Storey		
	Int. Col.	Side Col.	Beam	Int. Col.	Side Col.	Beam
83.3 k-in	48.30	63.10	51.90	18.64	25.21	22.91

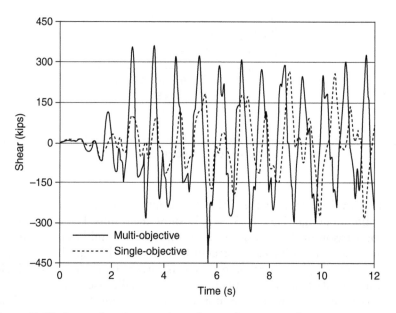

Figure 12.57 Comparison between base shears of composite frame.

As a final step, this analysis compares the structural displacement response at the top storey of all three multi-objective optimum designs. Note that the selected optimum designs have the same structural cost of $118,000. Fig. 12.60 shows the top-storey displacement results. The RC frame has the smallest displacement response, and the steel frame has the largest.

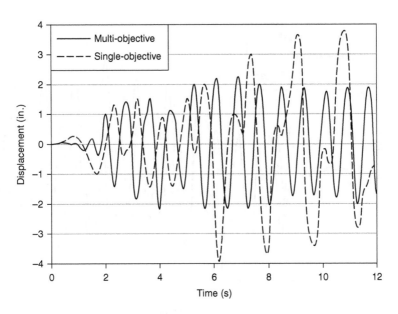

Figure 12.58 Comparison between top-storey displacements of composite frame.

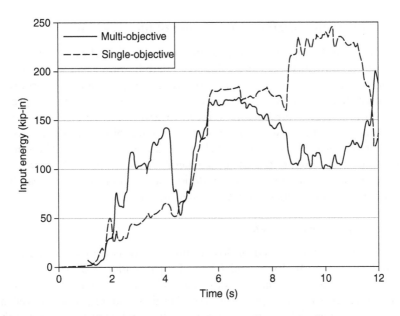

Figure 12.59 Comparison between input energies of composite frame.

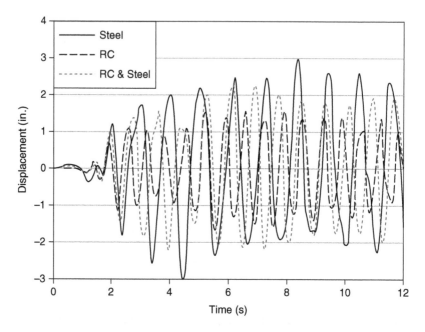

Figure 12.60 Comparison between top-storey displacements.

12.8   GA multi-objective optimization of life-cycle cost

Loss of life and property from possible future earthquakes as well as the expense and difficulty of post-earthquake rehabilitation and reconstruction strongly suggest the need for proper structural design with damage control. Design criteria should balance the initial cost of the structure with expected losses from potential earthquake-induced structural damage. *Life-cycle cost* design addresses these issues. Such a design methodology can be developed using multi-objective and multi-level optimization techniques. These techniques are well-suited to the use of a genetic algorithm (GA). GAs have the characteristics of maintaining a population of solutions, and can search in a parallel manner for many nondominated solutions.

12.8.1   *Life-cycle cost model*

Recent studies have focused on loss estimation for earthquakes and other natural hazards [23]. Jones and Chang [24] provide an overview of current work on the economic impact of natural disasters. Nigg [25] presents community-level disaster preparedness planning and response as well as mitigation actions. These studies are generally conducted to determine the after-effects of an event, or to estimate gross regional losses expected in a future event, and were intended primarily for post-earthquake recovery and response planning [26, 6].

The platform for the systematic integration of seismic reliability engineering and socioeconomics is the expected life-cycle cost function which can be summarized as follows:

$$C_T = C_I + C_D \tag{12.90}$$

where C_I is the *initial cost* of a structure, C_D is the *damage cost* over the life of the structure composed of the following (for buildings):

$$C_D = C_r + C_c + C_e + C_s + C_f \tag{12.91}$$

in which C_r is the *repair* or *replacement cost* of the structure; $C_c = $ loss of contents; $C_e = $ economic impact of structural damage; C_s is the cost of injuries caused by structural damage; $C_f = $ cost of fatalities from structural damage or collapse. Items in Eq. (12.91) comprising the total damage cost are functions of the structural damage level, x; for example, item C_r has an expected repair cost of

$$C_r = \int_0^\infty C_r(x) fx(x) dx \tag{12.92}$$

where $x = $ damage level, and $fx(x) = $ probability density function (PDF) of X. Structural damage, x, and its PDF, $fx(x)$ are calculated for a given earthquake, from which the particular damage costs, $C_r(x)$, can be determined as a function of damage level, x.

Direct losses caused by earthquakes include human fatality and injury, as well as physical damage to structures and their contents. Concerning human fatality, Wiggins [7] correlated the number of fatalities to building damage; Shiono and Krimgold [27] reported fatality data for a hospital that collapsed during the 1985 Mexico City earthquake; and Coburn et al. [28] suggested a simulation model to estimate the number of fatalities in a building of a given structural type. Besides immediate or direct losses from an earthquake, there is also secondary or indirect impact: e.g., economic loss associated with the disruption of business due to structural damage [25]. Ang and De Leon [29] examined the minimum expected life-cycle cost vis-à-vis structural damage, and the reliability analysis of an RC-frame building based on actual cost data from Mexico City buildings damaged in the 1985 earthquake.

12.8.2 Illustration of sample case

The general approach described above is illustrated for a class of five-storey RC office buildings located in a soft-soil site in the Los Angeles area [30]. The structure of each model building consists of parallel frames similar to that shown in Fig. 12.61. Only the building response in the short direction

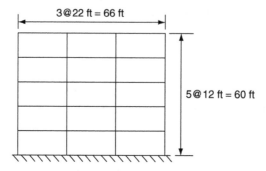

Figure 12.61 Model building.

is considered for this study. Several model buildings (SMRF frames) were designed according to the UBC code for service level seismic base shear coefficients, C_b, ranging from 0.04 to 0.13. Currently, C_b is equal to 0.08 in the UBC. A finite-element model is constructed for each model building and used for the nonlinear time-history analyses required for structural damage and reliability assessments.

A regression relationship between the computed median global damage index and the actual or estimated damage repair costs was obtained on the basis of reported damage repair costs for RC buildings damaged under previous earthquakes, together with damage-assessment analyses. The regression relationship for the normalized repair cost is formulated and expressed in terms of the median global damage index, leading to the following equations for the damage repair cost function:

$$\left.\begin{array}{l} C_R = \alpha_1 \, (d_m)^{\alpha_2}, \; 0 \le d_m \le d_0 \\ C_R = C_I, \; d_m \ge d_0 \end{array}\right\} \tag{12.93}$$

where d_m = median global damage index of the structure and $\alpha_1 = 0.57C$, $\alpha_2 = 1.0$ and $d_0 = 0.5$. The value of contents of the building, which is an office building, is assumed to be 40 per cent of the initial building cost, and the loss of contents is assumed to reach its maximum for a median global damage index $d_m = 1.0$. Making use of available data, the loss-of-contents function for all values of the median structural damage is as follows:

$$\left.\begin{array}{l} C_C = 0.4C_R, \; 0 \le d_m \le 0.5 \\ = 0.114C_I + 0.572(d_m - 0.5)C_I, \; 0.5 \le d_m \le 1.0 \\ = 0.4C_I, \; d_m \ge 1.0 \end{array}\right\} \tag{12.94}$$

Expected life-cycle costs and structural reliabilities are obtained for a seismic hazard at a site in downtown Los Angeles. Initial costs and total expected life-cycle costs for the five designs of the model building are shown in Figs. 12.62 and 12.63 as a function of the probability that a specified damage level is exceeded over the 50-year lifespan of the structure at a given site for an average annual discount rate $q = 4\%$. The work based on a genetic algorithm for multi-objective optimization with target reliability and minimum expected life-cycle cost as multi-level optimization is summarized in Fig. 12.64.

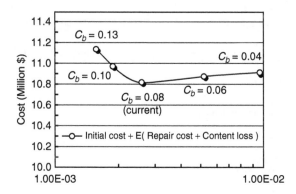

Figure 12.62 Initial damage cost–life-cycle probability of damage, $P(D > 0.5)$.

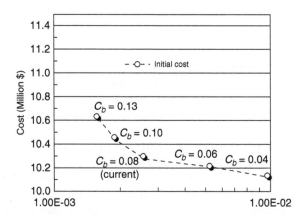

Figure 12.63 Expected life-cycle cost–life-cycle probability of damage, $P(D > 0.5)$.

12.9 Concluding remarks

This chapter presents the properties and characteristics of multi-objective optimization and solutions for various design objectives and constraints. Formulations include: a multi-objective optimization algorithm based on game theory, fuzzy constraints based on fuzzy set theory, and a Pareto genetic algorithm using a simple genetic algorithm. Two new genetic

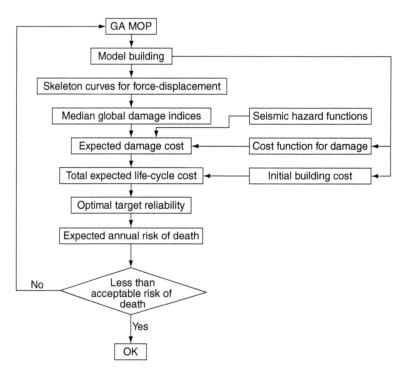

Figure 12.64 Multi-level optimization scheme.

algorithm operators – the multi-objective fitness function and the niche method – are proposed. A new genetic algorithm process – the Pareto-set filter – is added to the genetic algorithms. The multi-objective optimization algorithm presented is robust and performs uniformly well on a broad range of problems.

Six frame structures and five truss structures have been optimally designed with static load, seismic excitation and active control. Materials for the structures are steel, reinforced concrete and a combination of both. Optimization goals cover weight, structural cost, displacement, strain energy, seismic input energy, potential energy and the performance index. Constraints include stress, displacement, frequency and ratio of storey stiffness. The numerical results show that the newly developed algorithms can locate a global solution with the Pareto optimum set which are usually difficult to find using traditional optimization algorithms. The life-cycle cost optimization is introduced with the GA multi-objective and multi-level optimization.

References

1. Frangpool, D. and Cheng, F.Y. (eds.), *Advances in Structural Optimization*, ASCE, Reston, Virginia, 1996.

2. Eschenauer, H., Koski, J. and Osyczka, A. (eds.), *Multicriteria Design Optimization: Procedures and Applications,* Springer-Verlag, 1990.
3. Cox, B.J. and Horsley, F.W., *Square Foot Estimating,* Robert Snow Means Co., Inc., Kingston, MA, 1983.
4. R.S. Means Co., Inc., *Means Building Construction Cost Data,* Robert Snow Means Co., Inc., Kingston, MA, 1995.
5. Cheng, F.Y., Jiang, H.P. and Lou, K.Y., *Smart Structures – Innovative Systems for Seismic Response,* CRC Press, Boca Raton, FL, 2008.
6. Geoffrion, A.M., Proper efficiency and the theory of vector maximization, *Mathematical Analysis and Applications,* Vol. 22, No. 3, pp. 618–630, 1968.
7. Wiggins, J.H., Estimated building losses from U.S. earthquakes, *Proceedings of 2nd National Conference on Earthquake Engineering,* pp. 253–262, 1979.
8. Holland, J.H., *Adaptation in Natural and Artificial Systems,* University of Michigan Press, Ann Arbor, MI, 1975.
9. Goldberg, D.E., *Genetic Algorithms in Search, Optimization, and Machine Learning,* Addison-Wesley, Reading, MA, 1989.
10. Cheng, F.Y. and Li, D., Multiobjective optimization of structures with/without control, *Journal of Guidance, Control, and Dynamics,* Vol. 19, No. 2, pp. 392–397, 1996.
11. Cheng, F.Y. and Li, D., PGAMOP – Pareto genetic algorithm for multiobjective optimization problems, *UMR Structural Engineering 97-12,* NSF Final Report, 1997.
12. Nash, J., Two-person cooperative games, *Econometrica,* Vol. 21, pp. 128–140, 1953.
13. Harsanyi, J.C., Games with randomly disturbed payoffs: A new rationale for mixed strategy equilibrium points, *International Journal Game Theory,* Vol. 2, No. 1–23, 1973.
14. Zadeh, L.A., Fuzzy set, *Information and Control,* Vol. 8, pp. 338–353, 1965.
15. Cavicchio, D.J., Reproductive adaptive plans, *Proceedings of ACM 1972 Annual Confernce,* Boston, pp. 1–11, 1972.
16. Davis, L. (ed.), *Handbook of Genetic Algorithms,* Van Nostrand Reinhold, New York, 1991.
17. Venkayya, V.B., Khot, N.S. and Reddy, V.S., *Energy Distribution in an Optimal Structural Design, AFFDLTR-68-165,* Flight Dynamics Laboratory, Wright-Patterson AFB, OH 1969.
18. Vanderplaats, G.N., *DOT: Design Optimization Tools User Manual,* 1994.
19. Morris, A.J., The optimization of statically indeterminate structures by means of approximate geometric programming, *2nd Symposium on Structural Optimization,* preprint-123, AGARD, Milan, Italy, 1973.
20. Cheng, F.Y. and Li, D., Multiobjective optimization design with pareto genetic algorithm, *Journal of Structural Engineering,* ASCE, Vol. 123, No. 9, pp. 1252–1261, 1997.
21. Adeli, H. and Kamal, O., Efficient optimization of space trusses, *Computers and Structures,* Vol. 24, No. 3, pp. 502–511, 1986.
22. UBC, Uniform Building Code, International Conference of Building Officials, Whittier, Calif, 1997.
23. FEMA-2149, *Assessment of the State-of-the-art Earthquake Loss Estimation Methodologies,* Washington, DC, 1994.

24. Jones, B.G. and Chang, S.E., Economic aspects of urban vulnerability and disaster mitigation, *Urban Disaster Mitigation: The Role of Engineering and Technology*, Cheng, F.Y. and Sheu, M.S. (eds.), Elsevier Science Ltd., pp. 311–320, 1995.

25. Nigg, J.M., Social science approaches in disaster research: selected research issues and findings on mitigating natural hazards in the urban environment, *Urban Disaster Mitigation: The Role of Engineering and Technology*, Cheng, F.Y. and Sheu, M.S. (eds.), Elsevier Science Ltd., London, pp. 303–310, 1995.

26. Cheng, F.Y. and Wang, Y.Y. (eds.), *Post-Earthquake Rehabilitation and Reconstruction*, Elsevier Science Ltd, London, 1996.

27. Shiono, K. and Krimgold, F., A computer model for the recovery of trapped people in a collapsed building: development of a theoretical framework and direction for future data collection, *International Workshop on Earthquake Injury Epidemiology for Mitigation and Response*, 1989.

28. Coburn, A.W., Spence, R.J.S. and Promonis, A., Factors determining human casualty levels in earthquakes: mortality prediction in building collapse, *Proceedings of 10th World Conference on Earthquake Engineering*, Madrid, Spain, pp. 5894–5994, 1992.

29. Ang, A.H.-S. and De Leon, D., Basis for cost-effective decision on upgrading existing structures for earthquake protection, *Post-Earthquake Rehabilitation and Reconstruction*, Cheng, F.Y. and Wang, Y.Y. (eds.), Elsevier Science Ltd, London, pp. 69–84, 1996.

30. Cheng, F.Y., Ang, A.H.-S., Lee, J.H. and Li, D., Genetic algorithm for multiobjective optimization and life-cycle cost, *Structural Engineering in the 21st Century*, ASCE, Reston, Virginia, pp. 484–489, 1999.

Appendix A: Illustration of linear programming for feasible direction vector of Example 5.6.1

Substituting Eqs. (t), (u), (w) and (x) into Eqs. (5.27a) and (5.27b), we have

$$\text{Maximize } \beta \tag{A.1}$$

$$\text{Subject to } \{s_1' \quad s_2'\} \left\{ \begin{array}{c} -1.100 \times 10^{-1} \\ -3.999 \times 10^{-1} \end{array} \right\} + \beta \leq -0.4299 \tag{A.2}$$

$$\{s_1' \quad s_2'\} \left\{ \begin{array}{c} 1.198 \times 10^{-2} \\ 7.990 \times 10^{-3} \end{array} \right\} + \beta \leq 1.997 \times 10^{-2} \tag{A.3}$$

$$s_1' \leq 2, \quad s_2' \leq 2 \tag{A.4}$$

Due to the negative on the R.H.S. of the constraint Eq. (A.3), an artificial variable must be added (with a large coefficient in the maximization equation to force it to zero).

The linear programming problem:

$$\text{Maximize } \beta + 0x_1 + 0x_2 + 0x_3 + 0x_4 - 10{,}000A_1 \tag{A.5}$$

$$\text{Subject to } \{s_1' \quad s_2'\} \left\{ \begin{array}{c} 1.100 \times 10^{-1} \\ 3.999 \times 10^{-1} \end{array} \right\} - \beta - x_1 + A_1 \leq 0.4299 \tag{A.6}$$

$$\{s_1' \quad s_2'\} \left\{ \begin{array}{c} 1.198 \times 10^{-2} \\ 7.990 \times 10^{-3} \end{array} \right\} + \beta + x_2 \leq 1.997 \times 10^{-2} \tag{A.7}$$

$$s_1' \leq 2 \Rightarrow s_1' + x_3 = 2; \quad s_2' \leq 2 \Rightarrow s_2' + x_4 = 2 \tag{A.8}$$

For the normalized form for the optimization procedure, we express the above as

$$\text{Maximize } \beta + 0x_1 + 0x_2 + 0x_3 + 0x_4 - 10{,}000A_1 \tag{A.9}$$

$$\{s_1' \quad s_2'\} \left\{ \begin{array}{c} 1.100 \times 10^{-1} \\ 3.999 \times 10^{-1} \end{array} \right\} - \beta - x_1 + A_1 \leq 1.2710 \tag{A.10}$$

$$\{s_1' \;\; s_2'\}\left\{\begin{array}{c} 1.198 \times 10^{-2} \\ 7.990 \times 10^{-3} \end{array}\right\} + \beta + x_2 \le 1.3870 \qquad\qquad \text{(A.11)}$$

$$s_1' \le 2; \quad s_2' \le 2 \qquad\qquad \text{(A.12)}$$

Following the optimization procedures of (A) through (F) illustrated in Section 2.4 of Chapter 2, the results are summarized as follows:

	s_1'	s_2'	β	x_1	x_2	x_3	x_4	A_1	
var\c	0	0	1.0	0	0	0	0	$-10,000$	x_B
A_1	0.3254	0.9456	-1	-1	0	0	0	1	1.2710
x_2	0.8320	0.5547	1	0	1	0	0	0	1.3870
x_3	1	0	0	0	0	1	0	0	2
x_4	0	1	0	0	0	0	1	0	2
$z-c$	$-3,254.0$	$-9,456.0$	$9,999.0$	$10,000$	0	0	0	0	$-12,710$

$$\text{(A.13)}$$

Eq. (A.13) indicates that s_2' should enter and A_1 should leave. The results of the next cycle are:

	s_1'	s_2'	β	x_1	x_2	x_3	x_4	A_1		
var\c	0	0	1.0	0	0	0	0	$-10,000$	x_B	x_B/y
s_2'	0.3441	1	-1.0575	-1.0575	0	0	0	1.0575	1.3441	
x_2	0.64112	0	1.58661	0.58661	1	0	0	-0.58661	0.6414	0.404
x_3	1	0	0	0	0	1	0	0	2	
x_4	-1	0	1.0575	1.0575	0	0	1	-1.0575	0.65588	0.62
$z-c$	0	0	-1	0	0	0	0	$10,000$	0	

$$\text{(A.14)}$$

Eq. (A.14) reveals that x_2 leaves and β enters

	s_1'	s_2'	β	x_1	x_2	x_3	x_4	A_1	
var\c	0	0	1.0	0	0	0	0	−10,000	x_B
s_2'	0.77142	1	0	−0.66652	0.66652	0	0	0.66652	1.77160
β	0.40408	0	1	0.36973	0.63027	0	0	−0.36973	0.40426
x_3	1	0	0	0	0	1	0	0	2
x_4	−1.4273	0	0	0.66652	−0.66652	0	1	−0.66652	0.22838
$z-c$	0.40408	0	0	0.36973	0.63027	0	0	9,999.63	0.40426

$$\text{(A.15)}$$

which gives the solution as

$$s_1' = 0; \quad s_2' = 1.77; \quad \beta = 0.404 \tag{A.16}$$

The feasible direction is

$$s_1 = s_1' - 1 = -1.0; \quad s_2 = s_2' - 1 = 0.77 \tag{A.17}$$

Eq. (A.17) is identical to Eq. (y)

Appendix B: Newton's backward difference for calculating the acceleration and velocity at time t_n

Let $\nabla y(t_n) = y(t_n) - y(t_{n-1})$ (B.1)

where

$$y(t_{n-1}) = y(t_n - \Delta t) \tag{B.2}$$

then

$$\nabla^2 y(t_n) = \nabla(\nabla y(t_n)) = y(t_n) - y(t_{n-1}) - y(t_{n-1}) + y(t_{n-2}) \tag{B.3}$$

$$\nabla^3 y(t_n) = y(t_n) - 3y(t_{n-1}) + 3y(t_{n-2}) - y(t_{n-3}) \tag{B.4}$$

The displacement, velocity and acceleration at time t are obtained as shown in Eqs. (B.5) through (B.7), respectively.

$$y(t) = y(t_n) + (t - t_n) \frac{\nabla y(t_n)}{1!\Delta t} + (t - t_n)(t - t_{n-1}) \frac{\nabla^2 y(t_n)}{2!\Delta t^2}$$

$$+ (t - t_n)(t - t_{n-1})(t - t_{n-2}) \frac{\nabla^3 y(t_n)}{3!\Delta t^3}$$

$$= y(t_n) + (t - t_n) \frac{\nabla y(t_n)}{\Delta t} + \left(t^2 - tt_n - tt_{n-1} + t_n t_{n-1}\right) \frac{\nabla^2 y(t_n)}{2\Delta t^2}$$

$$+ \left(t^3 - t^2 t_n - t^2 t_{n-1} - t^2 t_{n-2} + tt_n t_{n-1} + tt_n t_{n-2} + tt_{n-1} t_{n-2} - t_n t_{n-1} t_{n-2}\right)$$

$$\frac{\nabla^3 y(t_n)}{6\Delta t^3} \tag{B.5}$$

$$\dot{y}(t) = \frac{\nabla y(t_n)}{\Delta t} + (2t - t_n - t_{n-1}) \frac{\nabla^2 y(t_n)}{2\Delta t^2}$$

$$+ \left(3t^2 - 2tt_n - 2tt_{n-1} - 2tt_{n-2} + t_n t_{n-1} + t_n t_{n-2} + t_{n-1} t_{n-2}\right) \frac{\nabla^3 y(t_n)}{6\Delta t^3} \tag{B.6}$$

$$\ddot{y}(t) = \frac{\nabla^2 y(t_n)}{\Delta t^2} + (6t - 2t_n - 2t_{n-1} - 2t_{n-2}) \frac{\nabla^3 y(t_n)}{6\Delta t^3} \tag{B.7}$$

The above equations can be formulated to be expressed in terms of t_{n-1}, t_{n-2} and t_{n-3} as

$$\dot{y}(t) = \frac{\nabla y(t_n)}{\Delta t} + (2t_n - t_n - t_{n-1})\frac{\nabla^2 y(t_n)}{2\Delta t^2}$$

$$+ \left(3t_n^2 - 2t_n^2 - 2t_n t_{n-1} - 2t_n t_{n-2} + t_n t_{n-1} + t_n t_{n-2} + t_{n-1}t_{n-2}\right)\frac{\nabla^3 y(t_n)}{6\Delta t^3}$$

$$= \frac{\nabla y(t_n)}{\Delta t} + \Delta t\frac{\nabla^2 y(t_n)}{2\Delta t^2}$$

$$+ \left(t_n^2 - 2t_n t_{n-1} - 2t_n t_{n-2} + t_n t_{n-1} + t_n t_{n-2} + t_{n-1}t_{n-2}\right)\frac{\nabla^3 y(t_n)}{6\Delta t^3}$$

$$= \frac{\nabla y(t_n)}{\Delta t} + \frac{\nabla^2 y(t_n)}{2\Delta t} + (t_n - t_{n-1})(t_n - t_{n-2})\frac{\nabla^3 y(t_n)}{6\Delta t^3}$$

$$= \frac{\nabla y(t_n)}{\Delta t} + \frac{\nabla^2 y(t_n)}{2\Delta t} + (\Delta t)(2\Delta t)\frac{\nabla^3 y(t_n)}{6\Delta t^3}$$

$$= \frac{\nabla y(t_n)}{\Delta t} + \frac{\nabla^2 y(t_n)}{2\Delta t} + \frac{\nabla^3 y(t_n)}{3\Delta t}$$

$$= \frac{1}{6\Delta t}\left(6\nabla y(t_n) + 3\nabla^2 y(t_n) + 2\nabla^3 y(t_n)\right)$$

$$= \frac{1}{6\Delta t}\left(6y(t_n) - 6y(t_{n-1}) + 3y(t_n) - 6y(t_{n-1}) + 3y(t_{n-2}) + 2y(t_n)\right.$$

$$\left. -6y(t_{n-1}) + 6y(t_{n-2}) - 2y(t_{n-3})\right)$$

$$= \frac{1}{6\Delta t}\left(11y(t_n) - 18y(t_{n-1}) + 9y(t_{n-2}) - 2y(t_{n-3})\right)$$

$$= \frac{1}{6\Delta t}\left(11y(t_n) - B\right) \tag{B.8}$$

where

$$B = 18y(t_{n-1}) + 9y(t_{n-2}) - 2y(t_{n-3}) \tag{B.9}$$

$$\ddot{y}(t) = \frac{\nabla^2 y(t_n)}{\Delta t^2} + (6t_n - 2t_n - 2t_{n-1} - 2t_{n-2})\frac{\nabla^3 y(t_n)}{6\Delta t^3}$$

$$= \frac{\nabla^2 y(t_n)}{\Delta t^2} + (4t_n - 2t_{n-1} - 2t_{n-2})\frac{\nabla^3 y(t_n)}{6\Delta t^3}$$

$$= \frac{\nabla^2 y\,(t_n)}{\Delta t^2} + 2\,(2t_n - t_{n-1} - t_{n-2})\,\frac{\nabla^3 y\,(t_n)}{6\Delta t^3}$$

$$= \frac{\nabla^2 y\,(t_n)}{\Delta t^2} + 2\,(\Delta t + 2\Delta t)\,\frac{\nabla^3 y\,(t_n)}{6\Delta t^3}$$

$$= \frac{\nabla^2 y\,(t_n)}{\Delta t^2} + \frac{\nabla^3 y\,(t_n)}{\Delta t^2}$$

$$= \frac{1}{\Delta t^2}\,(y\,(t_n) - 2y\,(t_{n-1}) + y\,(t_{n-2}) + y\,(t_n) - 3y\,(t_{n-1})$$

$$+ 3y\,(t_{n-2}) - y\,(t_{n-3}))$$

$$= \frac{1}{\Delta t^2}\,(2y\,(t_n) - 5y\,(t_{n-1}) + 4y\,(t_{n-2}) - y\,(t_{n-3}))$$

$$= \frac{1}{\Delta t^2}\,(2y\,(t_n) - A) \tag{B.10}$$

where

$$A = 5y\,(t_{n-1}) + 4y\,(t_{n-2}) - y\,(t_{n-3}) \tag{B.11}$$

Appendix C: Interpolation formulas

For any numerical curve to be useful at points not expressly stored as data, some method of interpolation must be utilized. For this investigation, a five-point central difference interpolator called Sterling's formula was used and was found to approximate the points very accurately. The formula can be expressed as:

$$f_x = f_0 + u\frac{\Delta f_{-1} + \Delta f_0}{2} + \frac{u^2}{2}\Delta^2 f_{-1} + \frac{u^2(u-1)}{12}\left(\Delta^3 f_{-1} + \Delta^3 f_{-2}\right)$$
$$+ \frac{u^2(u^2-1)}{24}\Delta^4 f_{-2} \tag{C.1}$$

where

$$u = \frac{x - x_0}{h} \tag{C.2}$$

$$\Delta f_0 = f_1 - f_0$$
$$\Delta f_{-1} = f_0 - f_1$$
$$\Delta^2 f_{-1} = f_1 - 2f_0 + f_{-1}$$
$$\Delta^3 f_{-1} = f_2 - 3f_1 + 3f_0 - f_{-1}$$
$$\Delta^3 f_{-2} = f_1 - 3f_0 + 3f_{-1} - f_{-2}$$
$$\Delta^4 f_{-2} = f_2 - 4f_1 + 6f_0 - 4f_{-1} + f_{-2}$$

The $\Delta^r f_k$ terms are called *the rth difference at point k*. f_k are the ordinates at the respective points x_k. h is the uniform spacing for all points. All unknowns are shown in Fig. C.1.

It will be shown later that the slope of the shock spectrum at any point will be needed and this can be approximated by differentiating Eq. (C.1) with respect to x.

$$\frac{df_x}{dx} = \frac{du}{dx} \times \frac{df(u)}{du} = \frac{1}{h} \times \frac{df(u)}{du} \tag{C.3}$$

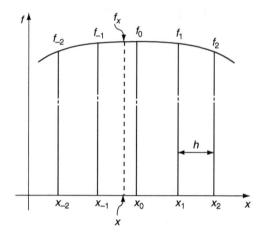

Figure C.1 Graphical representation of the five-point 'Stirling's' method of interpolation.

in which du/dx can be obtained from Eq. (C.2). Carrying out the derivative yields

$$\frac{df_x}{dx} = \frac{1}{h}\left[\frac{\Delta f_{-1} + \Delta f_0}{2} + u\Delta^2 f_{-1} + \frac{3u^2 - 2u}{12}\left(\Delta^3 f_{-1} + \Delta f_{-2}\right)\right.$$
$$\left. + \frac{4u^3 - 2u}{24}\Delta^4 f_{-2}\right] \tag{C.4}$$

Appendix D: Illustration of linear programming for feasible direction vector of Example 5.8.1

Using Eqs.(c), (oo) and (pp), we have

$$\{\nabla W\} = \begin{bmatrix} 0.00786 & 0.00943 \end{bmatrix}^T \tag{D.1}$$

$$\{\nabla G\} = \begin{bmatrix} -0.7479 \times 10^{-3} & -0.7820 \times 10^{-3} \end{bmatrix}^T \tag{D.2}$$

of which the normalized results are

$$\{\nabla W\} = 1.227 \times 10^{-3} \begin{bmatrix} 0.640 & 0.768 \end{bmatrix}^T \tag{D.3}$$

$$\{\nabla G\} = 1.082 \times 10^{-3} \begin{bmatrix} -0.6912 & -0.7226 \end{bmatrix}^T \tag{D.4}$$

Similar to the illustration in Appendix A,

Maximize β (D.5)

Subject to $\{s_1' \quad s_2'\} \begin{Bmatrix} -0.6912 \\ -0.7226 \end{Bmatrix}$

$$+ \beta \leq (-0.6912 \quad -0.7226) = -1.4138 \tag{D.6}$$

$$\{s_1' \quad s_2'\} \begin{Bmatrix} 0.640 \\ 0.768 \end{Bmatrix} + \beta \leq (0.640 \quad 0.768) = 1.408 \tag{D.7}$$

$$s_1' \leq 2; \quad s_2' \leq 2; \quad s_1' \geq 0; \quad s_2' \geq 0 \tag{D.8}$$

The linear optimization problem is written as

Maximize β (D.9)

Subject to $\{s_1' \quad s_2'\} \begin{Bmatrix} -0.6912 \\ -0.7226 \end{Bmatrix} + \beta + x_1 = -1.4138 \tag{D.10}$

$$\{s_1' \quad s_2'\} \begin{Bmatrix} 0.640 \\ 0.768 \end{Bmatrix} + \beta + x_2 = 1.408 \tag{D.11}$$

$$s_1' + x_3 = 2; \quad s_2' + x_4 = 2 \tag{D.12}$$

Due to the negative value on the R.H.S. of Eq. (D.10), we then rewrite the problem as

$$\text{Maximize } \beta + 0x_1 + 0x_2 + 0x_3 + 0x_4 - 10,000A_1 \tag{D.13}$$

$$\text{Subject to } \{s_1' \quad s_2'\} \begin{Bmatrix} 0.6912 \\ 0.7226 \end{Bmatrix} + \beta - x_1 + A_1 = 1.4138 \tag{D.14}$$

$$\{s_1' \quad s_2'\} \begin{Bmatrix} 0.640 \\ 0.768 \end{Bmatrix} + \beta + x_2 = 1.408 \tag{D.15}$$

$$s_1' + x_3 = 2; \quad s_2' + x_4 = 2 \tag{D.16}$$

The first tableau is

	s_1'	s_2'	β	x_1	x_2	x_3	x_4	A_1	
var\c	0	0	1.0	0	0	0	0	−10,000	x_B
A_1	0.6912	0.7226	−1	−1	0	0	0	1	1.4138
x_2	0.640	0.768	1	0	1	0	0	0	1.408
x_3	1	0	0	0	0	1	0	0	2
x_4	0	1	0	0	0	0	1	0	2
$z-c$	−6,912.0	−7,226.0	9,999.0	10,000	0	0	0	0	−14,138

$$\tag{D.17}$$

which indicates that s_2' enters the basis and x_2 leaves the basis.
Then, the second tableau is

	s_1'	s_2'	β	x_1	x_2	x_3	x_4	A_1		
var\c	0	0	1.0	0	0	0	0	−10,000	x_B	x_B/y
A_1	0.089033	0	−1.940885	−1.0	−0.940885	0	0	1.0	0.089033	1
s_2'	5/6	1.0	1.302083	0	1.302083	0	0	0	15/6	2.2
x_3	1	0	0	0	0	1	0	0	2.0	2
x_4	−5/6	0	−1.302083	0	1.302083	0	1	0	1/6	—
$z-c$	−890.333	0	19,407.85	10,000	9,408.85	0	0	0	−890.333	—

$$\tag{D.18}$$

From which we observe that s'_1 enters the basis and A_1 leaves the basis. Thus, the third tableau becomes

	s'_1	s'_2	β	x_1	x_2	x_3	x_4	A_1		
var\c	0	0	1.0	0	0	0	0	$-10,000$	x_B	x_B/y
s'_1	1.0	0	-21.79953	-11.23175	-10.56778	0	0	11.23175	1.0	
s'_2	0	1.0	19.46836	9.35979	10.10857	0	0	-9.35979	1.0	Leaves
x_3	0	0	21.79953	11.23175	10.56778	1.0	0	-11.23175	1.0	By Inspection
x_4	0	0	-19.46836	-9.35979	-10.10857	0	1	9.35979	1.0	
$z-c$	0	0	-1.0	0	0	0	0	$10,000$	0	—

$$(D.19)$$

From Eq. (D.19), β enters the basis and x_3 leaves the basis. The fourth tableau is

	s'_1	s'_2	β	x_1	x_2	x_3	x_4	A_1	
var\c	0	0	1.0	0	0	0	0	$-10,000$	x_B
s'_1	1.0	0	0	0	0	1.0	0	0	2.0
s'_2	0	1.0	0	-0.67087	0.67087	-0.89306	0	0.67087	0.10694
β	0	0	1.0	0.51523	0.48771	0.04587	0	-0.51523	0.04587
x_4	0	0	0	0.67087	-0.67087	0.89306	1.0	-0.067087	1.89306
$z-c$	0	0	0	0.51523	0.48771	0.04587	0	$9,999.48477$	0.04587

$$(D.20)$$

Since $z-c$ is positive, the feasible optimal solution is found as

$$s'_1 = 2.0; \quad s'_2 = 0.10694; \quad \beta_{mx} = 0.04587 \qquad (D.21)$$

and

$$s_1 = s'_1 - 1 = 1.0; \quad s_2 = s'_2 - 1 = -0.893; \qquad (D.22)$$

Eq. (D.22) is given in Eq. (qq).

Appendix E: Newmark's spectra

Let Eq. (7.77) be rewritten as

$$S_d = \left| \frac{1}{\omega} \int_0^t a_g(\tau) e^{-\beta\omega(t-\tau)} \sin\omega(t-\tau) d\tau \right|_{max} \tag{E.1}$$

where $a_g(\tau)$ is the same as $x_g(\tau)$ in the original equation. The spectral pseudo-velocity, S_v, and spectral acceleration, S_a, are given as

$$S_v = \omega S_d \tag{E.2}$$

and

$$S_a = \omega^2 S_d \tag{E.3}$$

On a logarithmic tripartite chart, the relationships between the coordinates of displacement, velocity and acceleration are given in Eqs. (E.2) and (E.3). As shown in Fig. E.1, when the frequency $f \le f_1$, or the period $T \ge T_1$, the spectral curve is perpendicular to the displacement axis, which means that it corresponds to a constant displacement value, d_1. Therefore, the spectral acceleration can be obtained by using Eq. (E.3), that is, when $T \ge T_1$

$$S_a = \omega^2 d_1 = \left(\frac{2\pi}{T}\right)^2 d_1 \tag{E.4}$$

When $f_1 < f \le f_2$, that is, $T_1 > T \ge T_2$, the spectral curve corresponds to a constant velocity, v_1, therefore, S_a can be obtained by using

$$S_a = \omega v_1 = \left(\frac{2\pi}{T}\right) v_1 \tag{E.5}$$

When $f_2 < f \leq f_3$ or $f > f_4$, that is, $T_2 > T \geq T_3$ or $T < T_4$, the spectral curve corresponds to a constant acceleration, a_1 or a_2, therefore

$$S_a = a_1, \quad \text{for } T_2 > T \geq T_3 \tag{E.6}$$

or,

$$S_a = a_2, \quad \text{for } T < T_4 \tag{E.7}$$

When $T_3 > T > T_4$, that is, $f_3 < f \leq f_4$, one may generate a curve-fitted polynomial function or use the linearized relationship of a natural logarithm to interpolate the spectral acceleration. For the latter method, any spectral acceleration, S_a, corresponds to a natural frequency, f, in the range given as

$$\frac{\ln(S_a) - \ln(a_2)}{\ln(f) - \ln(f_4)} = \frac{\ln(a_2) - \ln(a_1)}{\ln(f_4) - \ln(f_3)} \tag{E.8}$$

from which one obtains

$$S_a = \exp\left(\frac{\ln\left(\frac{a_1}{a_2}\right)\ln\left(\frac{f_4}{f}\right)}{\ln\left(\frac{f_4}{f_3}\right)}\right) a_2 \tag{E.9}$$

Inelastic design spectrum with 5% critical damping and a ductility factor of 3 – The elastic response spectrum is reasonable to be used to design structures subjected to earthquake excitations with moderate intensity. However, for severe earthquakes, it is not economical to design the structures for elastic behaviour. In order to design the structures for elasto-plastic behaviour, the elastic response spectrum has been extended by Newmark and Hall to include the inelastic range. In general, the inelastic acceleration spectra have a similar appearance to the elastic response spectra, but the curves are moved downward by an amount related to the ductility factor, μ, which is defined as the ratio of the maximum permissible displacement, r_m, to the yield displacement, r_y.

The elastic response spectrum shown in Fig. E.1 is constructed on the basis of a maximum ground acceleration of 0.16 g, a damping ratio (β) of 0.05, and the 84.1 percentile. A percentile of 84.1 means that the spectrum amplification is obtained on the basis of a cumulative probability of 84.1%. That is, 84.1 per cent of the actual spectral values corresponding to the specified damping ratio can be expected to fall at or below the smoothed maximum ground motion values multiplied by these particular amplification factors. A ductility factor of 3 is introduced into the elastic design spectrum to construct the inelastic acceleration spectrum. According to Newmark's procedure, the maximum spectral displacement approaches

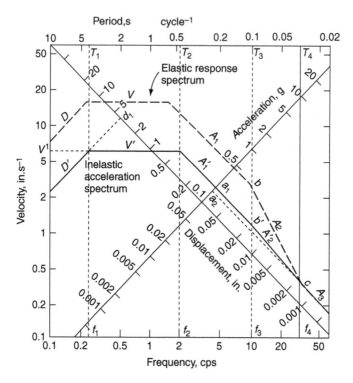

Figure E.1 Inelastic design spectrum with 5% critical damping and a ductility factor of 3 (1 in. = 2.54 cm).

the maximum ground displacement in the low-frequency region. Because the force in the inelastic range does not increase, the spectral acceleration is reduced by dividing ordinate values of the regions D and V by μ. In the very high-frequency region A_3, the maximum spectral acceleration approaches the maximum ground acceleration, therefore A_3 was not reduced. For the region between these two extremes, the energy is preserved. The research indicated that a reduction factor of $1/\sqrt{2\mu - 1}$, which was derived on the basis of an equivalence of energy between the elasto-plastic system and an elastic system having the same frequency, provides better agreement with the actual earthquake response spectra. By dividing ordinate values of region A_1 by $\sqrt{2\mu - 1}$, the inelastic spectral acceleration is located. For region A_2, the inelastic acceleration spectrum is simply obtained by connecting points b' and c together. Following this procedure, the inelastic acceleration spectrum, $D'V'A_1'A_2'A_3'$, is constructed.

Although the inelastic acceleration spectrum is different from the elastic response spectrum, the mathematical relationships between the coordinates of displacement, velocity and acceleration are not changed, which means that Eqs. (E.4) through (E.9) can be used to generate the spectral acceleration for both elastic and inelastic spectra in a computer program. The inelastic spectral accelerations in Fig. E.1 correspond to the natural period of mode *i*. T_i can be found by using the following equations:

$$(S_a)_i = 0.16g, \quad \text{for } T_i \leq 0.03 \text{ s} \tag{E.10}$$

$$(S_a)_i = (0.142857 + 0.5714286 \, T_i)g, \quad \text{for } 0.03 \text{ s} < T_i \leq 0.1 \text{ s} \tag{E.11}$$

$$(S_a)_i = 0.2g, \quad \text{for } 0.1 \text{ s} < T_i \leq 0.5 \text{ s} \tag{E.12}$$

$$(S_a)_i = 6.15 \left(\frac{2\pi}{gT_i} \right) g, \quad \text{for } 0.5 \text{ s} < T_i \leq 4.0 \text{ s} \tag{E.13}$$

$$(S_a)_i = 3.92 \frac{1}{g} \left(\frac{2\pi}{T_i} \right)^2 g, \quad \text{for } 4.0 \text{ s} < T_i \leq 10.0 \text{ s} \tag{E.14}$$

and when $T_i > 10.0$ s, Eq. (E.14) is used by assuming T_i is equal to 10.0 s.

50 percentile design spectra with 5% critical damping for alluvium – Figs. E.2 and E.3 show the design spectra for the horizontal and vertical directions of the 50 percentile, respectively. All spectra are normalized to 1.0 g. These spectra are developed with regard to the soil effect. The soil type considered in these spectra is alluvium. To find the normalized

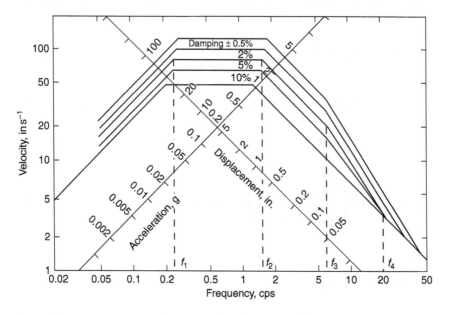

Figure E.2 Design spectra for horizontal motion, alluvium and 50 percentile (1 in. = 2.54 cm).

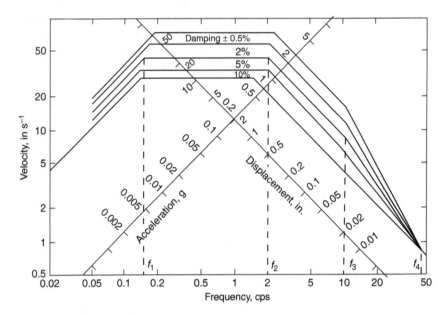

Figure E.3 Design spectra for vertical motion, alluvium and 50 percentile (1 in. = 2.54 cm).

spectra accelerations, one can use the following equations that correspond to different frequency regions:

$$(S_a)_i = \alpha_d \frac{ad}{v^2} \left(\frac{v}{a}\right)^2 \omega_i^2 a, \quad \text{for } \frac{\omega_i}{2\pi} \leq f_1 \tag{E.15}$$

$$(S_a)_i = \alpha_v \left(\frac{v}{a}\right) \omega_i a, \quad \text{for } f_1 < \frac{\omega_i}{2\pi} \leq f_2 \tag{E.16}$$

$$(S_a)_i = \alpha_d \frac{ad}{v^2} \left(\frac{v}{a}\right)^2 \omega_i^2 a, \quad \text{for } f_2 < \frac{\omega_i}{2\pi} \leq f_3 \tag{E.17}$$

$$(S_a)_i = \exp\left(\frac{\ln(\alpha_a)\ln\left(\frac{2\pi f_4}{\omega_i}\right)}{\ln\left(\frac{f_4}{f_3}\right)}\right) a, \quad \text{for } f_3 < \frac{\omega_i}{2\pi} \leq f_4 \tag{E.18}$$

and

$$(S_a)_i = a, \quad \text{for } \frac{\omega_i}{2\pi} \geq f_4 \tag{E.19}$$

in which ω_i is the circular frequency, and α_a, α_v and α_d are the amplification factors of the spectral acceleration, velocity and displacement, respectively.

Table E.1 The amplification factors of Newmark's design spectra for alluvium.

Direction	Percentile	Amplification factor		
		α_a	α_v	α_d
Horizontal	50	2.11	1.66	1.40
Vertical	50	2.05	1.51	1.40

The values α_a, α_v and α_d are listed in Table E.1. For horizontal ground motion, the acceleration, a, velocity, v, and displacement, d, are 1.0 g, 48 in. s^{-1}, and 36 in., respectively. For vertical ground motion, a, v and d are respectively given as 2/3 g, 29 in. s^{-1} and 33 in.

Appendix F: Equivalent lateral force procedure

The equivalent lateral force procedures are commonly available in various building codes in different countries; presented hereafter is the original work that leads to the current building code provisions in the U.S. The original specified requirements are somewhat more complicated than the current ones in practice after a few revisions. In order to evaluate the optimum design results and to assess the response parameters, the original requirements are employed here. In general, the procedures are mainly used for the most common constructions of regular buildings with certain limits of irregular structural configurations.

With this analysis procedure, one can view the effect of earthquake excitations as static lateral forces. The design base shear, V, is defined as

$$V = C_s W \tag{F.1}$$

in which C_s is the seismic design coefficient, W the total gravity load of the building including the structural weight and the weight of partitions, permanent equipment, and the effective snow load. For storage and warehouse structures, at least 25 per cent of the live load on the floor should be included as well. The seismic coefficient, C_s, is the lower value computed from the following formulas:

$$C_s = \frac{1.2A_v S}{RT^{2/3}} \tag{F.2}$$

and

$$C_s = \frac{2.5A_a}{R} \tag{F.3}$$

or

$$C_s = \frac{2.0A_a}{R} \tag{F.4}$$

Table F.1 Coefficients A_a and A_v.

U.S. area number in map	Coeff. A_a	Coeff. A_v
1	0.05	0.05
2	0.05	0.05
3	0.10	0.10
4	0.15	0.15
5	0.20	0.20
6	0.30	0.30
7	0.40	0.40

Table F.2 Site coefficients, *S*.

Soil profile type	Site Coeff, S
1	1.0
2	1.2
3	1.5

for soil profile type S_3 and $A_a \geq 3$, in which A_a and A_v are the effective peak acceleration coefficient and effective peak velocity coefficient, respectively (both values are determined from Table F.1), S is the site coefficient, which is a value that represents the effect of site conditions on the building response and is given in Table F.2, R is the response modification factor with various values for different structural systems to account for the ductility of the system, and T is the fundamental period of the building [4].

The fundamental period, T, may be taken as the approximate fundamental period of the building, T_a, and is determined by either of the following formulas:

(a) For moment-resisting frame structures without rigid components to resist seismic forces,

$$T_a = C_T(h_n)^{3/4} \tag{F.5}$$

in which C_T is equal to 0.025 for concrete frames, C_T is equal to 0.035 for steel frames, and h_n is the height of the highest level above the base, in feet.

(b) For all other buildings,

$$T_a = \frac{0.05\, h_n}{\sqrt{L}} \tag{F.6}$$

in which L is the overall length of the building at the base in the direction being considered, in feet. Alternatively, the fundamental period of the building, T, may be determined by using the established methods of mechanics and assuming that the base of the building is fixed with $1.2T_a$ as the upper bound.

The distribution of lateral forces on each floor level, F_x, is determined in accordance with the formula:

$$F_x = C_{vx} V \tag{F.7}$$

in which

$$C_{vx} = \frac{w_x h_x^k}{\sum\limits_{i=1}^{n_s} w_i h_i^k} \tag{F.8}$$

and w_i, w_x are weights at level i and x, h_i, h_x are the heights of levels i and x above the base, k is equal to 1 for $T \leq 0.5$ s, k is equal to 2 for $T \geq 2.5$ s, and k is equal to $0.75 + T/2$ for 0.5 s $< T < 0.5$ s.

At any level, the storey shear, V_x, is related to the lateral forces by the equation of statics

$$V_x = \sum\limits_{i=1}^{n_s} F_i \tag{F.9}$$

The overturning moment, M_x, at level x is determined by the following formula:

$$M_x = \kappa \sum\limits_{i=1}^{n_s} F_i(h_i - h_x) \tag{F.10}$$

in which κ is equal to 1.0 for the top ten storeys, to 0.8 for the 20th storey from the top and below, and is a value determined by linear interpolation between 1.0 and 0.8 for storeys between the 20th and 10th storeys below the top. Except for inverted pendulum structures, the foundation overturning design moment, M_f, at the foundation–soil interface may be determined by using Eq. (F.10) with κ equal to 0.75 for all building heights.

The deflection at any level, δ_x, should be determined according to the formula:

$$\delta_x = C_d \delta_{xe} \tag{F.11}$$

in which C_d is the deflection amplification factor, and δ_{xe} is the building's elastic displacement, which is occasioned by the seismic design forces, F_x. The base of the building should be considered as fixed. The storey drift, Δ, is the difference between deflection, δ_x, at the top and the bottom of the

Table F.3 Allowable storey drift.

Seismic Hazard exposure group		
I[a]	II	III
$0.015h_{sx}$	$0.015h_{sx}$	$0.010h_{sx}$

[a] if there are no brittle-type finishes in buildings three storeys or fewer in height, these limits may be increased by one-third.

storey under consideration and the maximum storey drift of the building is limited to the allowable drift given in Table F.3.

The $P - \Delta$ effects define the stability coefficient, θ, as follows:

$$\theta = \frac{P_x \Delta}{V_x h_{sx} C_d} \tag{F.12}$$

in which Δ is the design storey drift, V_x the seismic shear force at level x, which is determined from Eq. (F.9), h_{sx} the storey height below level x, and P_x the total unfactored vertical design load at and above level x. The stability coefficient at any level cannot exceed 0.10. If it is greater than 0.10, then the design storey drift must be adjusted by a multiplying factor of $0.9/(1 - \theta)$, which must be equal to or greater than 1.0 to account for the $P - \Delta$ effect. The effect of $P - \Delta$ forces on storey shears should be adjusted on the same rational analysis procedure.

Appendix G: Derivation of mean and variance of LNR or LNS

If x and y are random variables, they have a relationship

$$x = \ln y \tag{G.1}$$

and

$$\frac{dx}{dy} = \frac{1}{y} \tag{G.2}$$

For a one-to-one transformation, $f_y(y)\, dy = f_x(x)\, dx$, as shown in Fig. G.1, it induces

$$f_y(y) = f_x(x) \frac{1}{y} \tag{G.3}$$

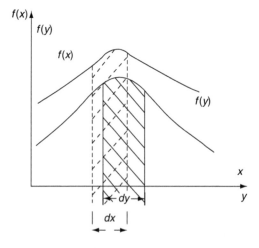

Figure G.1 The equal area of $f(x)\, dx$ and $f(y)\, dy$.

where $f_x(x)$ and $f_y(y)$ are the probability density function of x and y. Since x is normally distributed, $f_x(x)$ is

$$f_x(x) = \frac{1}{\sigma_x\sqrt{2\pi}} \exp\left[\frac{-1}{2}\left(\frac{x-\bar{x}}{\sigma_x}\right)^2\right] \tag{G.4}$$

where $\bar{x}$ and σ_x are mean and standard deviation of x.

Substituting Eq. (G.4) into Eq. (G.3), it yields

$$f_y(y) = \frac{1}{y\sigma_x\sqrt{2\pi}} \exp\left[\frac{-1}{2}\left(\frac{\ln y-\bar{x}}{\sigma_x}\right)^2\right] \tag{G.5}$$

considering

$$0.5 = P[y \le \hat{y}] = P[\ln y \le \ln \hat{y}] = P[x \le \ln \hat{y}] = F_x(\ln \hat{y}) = F_x(\hat{x}) \tag{G.6}$$

where the symbol () of $F()$ signifies cumulative distribution function, $\hat{x}, \hat{y} =$ medians of x and y; and normal distribution is a symmetric distribution, we have

$$0.5 = F_x(\hat{x}) = F_x(\bar{x}) \tag{G.7}$$

$$\ln \hat{y} = \bar{x} \tag{G.8}$$

and

$$E[y^r] = \int_0^\infty y^r f_r(y)dy = (\hat{y})^r \exp\left(\frac{1}{2}r^2\sigma_{\ln y}^2\right) \tag{G.9}$$

The 1st and 2nd moment can be determined to be

$$\bar{y} = E[y] = \hat{y}\exp\left(\frac{1}{2}\sigma_{\ln y}^2\right) \tag{G.10}$$

$$\text{and}\, \sigma_y^2 = E[y^2] - \bar{y}^2 = \hat{y}^2\exp\left(2\sigma_{\ln y}^2\right) - \hat{y}^2\exp\left(\sigma_{\ln y}^2\right)$$
$$= \hat{y}^2\exp\left(\sigma_{\ln y}^2\right)\left(\exp\left(\sigma_{\ln y}^2\right) - 1\right) = \bar{y}^2\left(\exp\left(\sigma_{\ln y^2}\right) - 1\right) \tag{G.11}$$

Thus,

$$V_y^2 = \frac{\sigma_y^2}{\bar{y}^2} = \exp\left(\sigma_{\ln y}^2\right) - 1 \tag{G.12}$$

Therefore,

$$\sigma_{\ln y}^2 = \ln\left(1 + V_y^2\right) \tag{G.13}$$

If $y = R$ or S, Eqs. (G.13) and (G.14) become Eqs. (11.10) and (11.11).

Appendix H: Equivalent uniform distributed load

To derive an equivalent uniform distributed load (EUDL), an influence surface coefficient is needed. An influence surface coefficient is a two-dimensional extension of the principle of influence line. The ordinate $I(x,y)$ of an influence surface at any point (x,y) is the influence on some desired load effect due to a unit load at (x,y). An influence surface may be constructed by multiplying appropriate influence lines.

For example, a desired load effect is an axial load on a typical interior column (one storey). From the principle of influence line, this axial load will be determined by assuming a unit displacement at one corner of a panel. In Fig. H.1 a panel subjected to a unit displacement at one corner will be the product of two influence lines in the x- and y-axes. The determination of this influence surface is as follows.

Assuming a deflection curve in the x-axis is $U_z = a + bx + cx^2 + dx^3$, the deflection in terms of non-dimensional variable, $x' = x/\ell$, is solved by substituting boundary conditions $U_z = 0$, $\dot{U}_z = 0$ at $x' = 0$, and $U_z = 1$, $\dot{U}_z = 0$ at $x' = 1$. Therefore, the influence line along the x-axis has the form:

$$U_z = 3x'^2 - 2x'^3, \quad 0 \le x' \le 1$$

In a similar manner, the influence line along the y-axis, which has the same boundary conditions, is

$$U_z = 3y'^2 - 2y'^3, \quad 0 \le y' \le 1$$

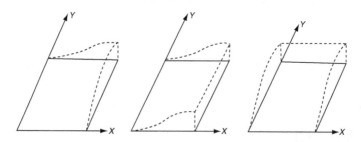

Figure H.1 Panel subjected to a unit displacement at one corner.

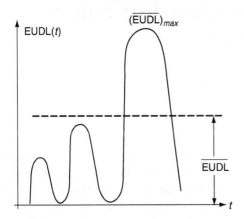

Figure H.2 The time history of EUDL.

where $y' = y/\ell$. Thus, the influence surface $I(x,y)$ is approximately the product of the influence lines in the x- and y-axes.

$$I\left(x',y'\right) = \left(3x'^2 - 2x'^3\right)\left(3y'^2 - 2y'^3\right), \quad 0 \le x' \le 1 \text{ and } 0 \le y' \le 1$$

In the design, the maximum EUDL as shown in Fig. H.2 is our concern. The statistics of the maximum EUDL can be assumed to be

$$\left(\overline{EUDL}\right)_{max} = \overline{EUDL} + r_1 \left(\sigma_{EUDL}^2\right)^{1/2} \tag{H.1}$$

and

$$\sigma_{(EUDL)_{max}}^2 = r_2 \sigma_{EUDL}^2 \tag{H.2}$$

where $\overline{EUDL}$, σ_{EUDL} = mean and variance of EUDL; r_1, r_2 are constants.
  Substituting Eqs. (11.40) and (11.41) into Eqs. (H.1) and (H.2), they become

$$\left(\overline{EUDL}\right)_{max} = \overline{L}_y + r_1 \left(\sigma_r^2 + \frac{\sigma_t^2}{A_I}K_L\right)^{1/2} = C_1 + \frac{C_2}{A_I^{1/2}} \tag{H.3}$$

and

$$\sigma_{EUDL}^2 = r_2 \left(\sigma_r^2 + \frac{\sigma_t^2}{A_I}K_L\right) = C_3 + \frac{C_4}{A_I} \tag{H.4}$$

where C_1, C_2, C_3, C_4 are constants which can be determined from a live load survey. Eq. (H.4) corresponds to Eq. (11.41).

Appendix I: Probability distribution of peak acceleration

To determine the mean and variance of the maximum peak ground acceleration, the probability distribution of this acceleration has to be known. Since the energy, E, released by an earthquake has the relationship $\log_{10} E = a_0 + b_0 M$ which was proposed by Ritcher, some researchers, such as Kanai, Esteva and Rosenblueth, suggested the following relationship among peak ground acceleration (A), magnitude (M) and focal distance (R) which is determined to be

$$A = b1 e^{b2M} R^{-b3} \tag{I.1}$$

where $b1$, $b2$, $b3$ are constants which are obtained from field data. Esteva and Rosenblueth used $b1 = 2,000$, $b2 = 0.8$ and $b3 = 2$ with units of A (cm s^{-2}) and R (kilometres) to represent the southern California condition as shown in Fig. I.1.

The formulation of Eq. (I.1) may be explained as follows: since E = mass (m) times acceleration (A) times distance (R) $= a_0 + b_0 M$, then $mAR = e^{a_0 \ln 10} + e^{Mb_0 \ln 10}$. Consequently, A is exponentially proportional to M but inversely proportional to R. From Eq. (I.1), the alternative relationship is

$$M = \frac{-\ln b1}{b2} + \frac{b3}{b2} \ln R + \frac{\ln A}{b2} \tag{I.2}$$

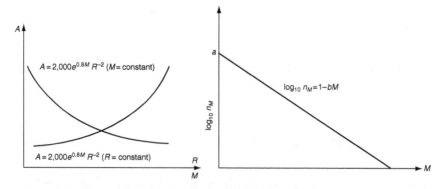

Figure I.1 The relationship of A and M, A and R, n and M.

The conditional probability of $\ln A$ of actual A which is greater than any number $\ln a$ of the allowable peak ground acceleration at focal distance $R = r$ is

$$P[\ln A \geq \ln a | R = r] = P\left[M \geq \frac{\ln a + b3 \ln r - \ln b1}{b2}\right]$$

$$= 1 - F\left(\frac{\ln a + b3 \ln r - \ln b1}{b2}\right) \tag{I.3}$$

where $F(M) =$ the cumulative probability distribution function of M. Since Ritcher also suggested the following relationship which is shown in Fig. I.1

$$\log_{10} n_M = a - bM \tag{I.4}$$

where $n_M =$ frequency of earthquake, a and b are constants obtained from field data, the frequency of earthquakes can be expressed as the form

$$n_M = \exp\left[a \ln 10 - b \ln 10M\right] = \exp\left[-b \ln 10 \left(M - \frac{a}{b}\right)\right] \tag{I.5}$$

Thus, $1 - F(M)$ is computed as

$$1 - F(M) = e^{-\beta(M - M_0)} \tag{I.6}$$

where $\beta = b \ln 10$ and M_0 is the smallest magnitude that will be considered for the design.

Therefore, from Eq. (I.3)

$$P[\ln A \geq \ln a | R = r] = \exp\left[-\beta\left(\frac{\ln a + b3 \ln r - \ln b1}{b2} - M_0\right)\right] \tag{I.7}$$

Based on Eq. (I.7) and Fig. I.2, the cumulative distribution of $\ln A$ for the distance from the site to the central point of the line fault source (d) and to the furthest point of the line fault source (r_0) yields

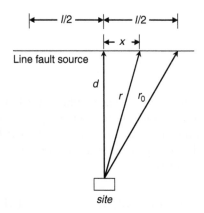

Figure I.2 The configuration of site and line earthquake source.

$$1 - F_{\ln A}(\ln a) = P[\ln A \geq \ln a] = \int_{d}^{r_0} P[\ln A \geq \ln a | R = r] f_R(r)\, dr \qquad (I.8)$$

where $f_R(r)$ is the probability density function of R. The cumulative distribution of local distance R is

$$f_R(r) = P[R \leq r] = P[R^2 \leq r^2]$$
$$= P[X^2 + d^2 \leq r^2] = P\left[|X| \leq \sqrt{r^2 - d^2}\right]$$

Since the cumulative distribution is the probability of focal length from 0 to $\sqrt{r^2 - d^2}$ in a half length of the line source $(\ell/2)$, then

$$F_R(r) = \frac{2\sqrt{r^2 - d^2}}{\ell}, \quad d \leq r \leq r_0 \qquad (I.9)$$

the probability density function of R is

$$f_R(r) = \frac{dF_R(r)}{dr} = \frac{2r}{\ell\sqrt{r^2 - d^2}} \quad d \leq r \leq r_0 \qquad (I.10)$$

After substituting Eq. (I.10) into Eq. (I.8) its integration is very complicated. However, in the region of the greatest interest, namely, larger values of $\ln a$, the result is

$$F_{\ln A}(\ln a) = \frac{1}{\ell} CG \exp\left[-\frac{\beta}{b2} \ln a\right] \qquad (I.11)$$

where $C = a$ constant $= \exp[\beta(b1/b2) + m_0]$,

$$G = a \text{ constant} = \frac{2\pi}{(2d)^{r1}} \frac{\Gamma(r1)}{\left[\Gamma\left(\frac{r1+1}{2}\right)\right]^2},$$

$$r1 = \beta\frac{b3}{b2} - 1,$$

$\Gamma(\ )$ signifies Gamma function.

Eq. (I.11) is also the probability of an earthquake event in which A is larger than a. This probability can be substituted into the following equation to determine a probability distribution of the random number, N, of this earthquake event which is assumed to be a Poisson process in a time interval of t years with average occurrence rate of v per year:

$$P_N(n) = P(N = n) = \frac{e^{-P_i v t}(P_i v t)^n}{n!}, \quad n = 0, 1, 2, 3, \ldots \qquad (I.12)$$

where $P_i = 1 - F_A(a)$.

If the zero number of an earthquake event is less than an allowable value, the probability becomes

$$F_A(a) = P[N = 0] = \exp\left[-\tilde{\nu}CGta^{-\beta/b2}\right] \tag{I.13}$$

where $\tilde{\nu} = \nu/\ell$.

Comparing Eq. (I.13) with a type II distribution

$$F_A(a) = e^{-\left(\frac{a}{u1}\right)^{-K1}} \tag{I.14}$$

The coefficients $u1$ and $K1$ will be determined to be

$$u1 = (\tilde{\nu}CGt)^{b2/\beta} \tag{I.15}$$

$$K1 = \frac{\beta}{b2} \tag{I.16}$$

which are used in Eq. (11.54).

Appendix J: Flow chart
of interior-penalty-function algorithm

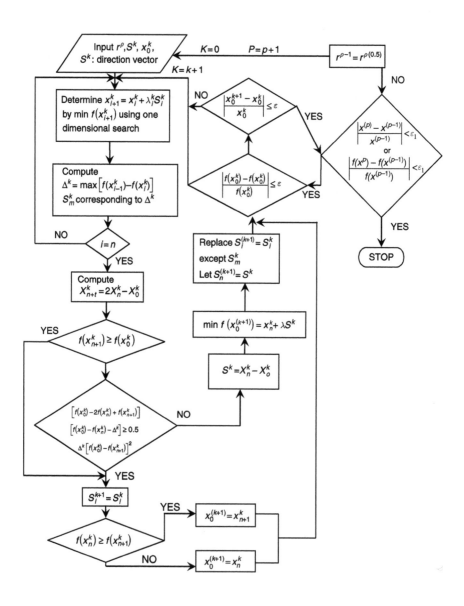

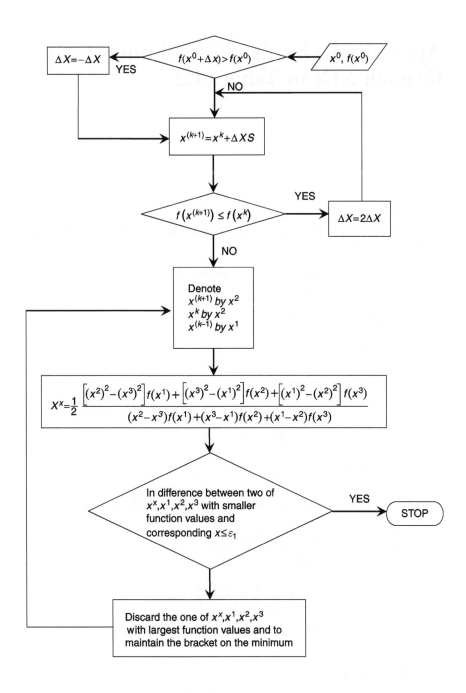

Appendix K: Explanation of notes I through XIX in Table 11.2

(A) The following explanation is for Notes I through XIX associated with Cycle No. 1.

    (I) Calculation of displacements and stresses;

    (II) Based on Eq. (g);

    (III) Based on Eq. (h);

    (IV) $\beta_{u1} = \dfrac{\mu_{Ru1} - \mu_1}{\left(S_{Ru1}^2 + S_{u1}^2\right)^{1/2}};$

    (V) $\beta_{\tau1} = \dfrac{\mu_{R\tau1} - \tau_1}{\left(S_{R\tau1}^2 + S_{\tau1}^2\right)^{1/2}};$

    (VI) $\Lambda_{u1} = \dfrac{-2\mu_{Ru1}\mu_1 \pm \left(SQT_{u1}\right)^{1/2}}{2\left(AA_{u1}S_{Ru1}^2 - \mu_{Ru1}^2\right)}$

        where $AA_{u1} = \left(\dfrac{\beta_{u10}}{\beta_{u1}}\right)^2 \dfrac{\left(\mu_{Ru1} - \mu_1\right)^2}{S_{Ru1}^2 + S_{u1}^2}$, and

        $SQT_{u1} = 4\mu_{Ru1}^2 u_1^2 - 4\left(AA_{u1}S_{Ru1}^2 - \mu_{Ru1}^2\right)\left(AA_{u1}S_{u1}^2 - u_1^2\right);$

    (VII) $\Lambda_{\tau1} = \dfrac{-2\mu_{R\tau1}\mu_1 \pm \left(SQT_{\tau1}\right)^{1/2}}{2\left(AA_{\tau1}S_{R\tau1}^2 - \mu_{R\tau1}^2\right)}$

        where $AA_{\tau1} = \left(\dfrac{\beta_{\tau10}}{\beta_{\tau1}}\right)^2 \dfrac{\left(\mu_{R\tau1} - \tau_1\right)^2}{S_{R\tau1}^2 + S_{\tau1}^2}$ and

        $SQT_{\tau1} = 4\mu_{R\tau1}\tau_1 - 4\left(AA_{\tau1}S_{R\tau1} - \mu_{R\tau1}^2\right)\left(AA_{\tau1}S_{\tau1}^2 - \tau_1^2\right);$

(VIII) From Line 21;

    (IX) Similar to (I) based on revised areas;

    (X) Similar to (II) based on revised area as $S_{u_1^*}^2 = \displaystyle\sum_{i=1}^{8} \dfrac{\partial u_1^*}{\partial r_r^*} V_{r_r^*}^2 \bar{V}_{r_r}^{*2};$

(XI) Similar to (IV) and (V) based on revised areas as $\beta_{u1}^* = \dfrac{u_{Ru_1}^* - u_1^*}{\left(S_{Ru1}^2 + S_{u1}^2\right)^{1/2}}$;

(XII), (XIII), ..., (XIX) are for Cycles No. 13 and 14. The equations are detailed as follows:

(B) *Cycle 13* (λ_3, λ_4 active and no passive elements)

$$C_{11} = -\sum_{l=1}^{3}\left(\frac{\partial g_3}{\partial A_l}\right)^2 \frac{A_l^{(k)}}{\rho l_l}$$

$$= -\left[\left(\frac{\partial g_3}{\partial A_1}\right)^2 \frac{A_1}{\rho l_1} + \left(\frac{\partial g_3}{\partial A_2}\right)^2 \frac{A_2}{\rho l_2} + \left(\frac{\partial g_3}{\partial A_3}\right)^2 \frac{A_3}{\rho l_3}\right]$$

$$\left(\frac{\partial g_3}{\partial A_1}\right)^2 \frac{A_1}{\rho l_1} = \frac{(-7.373)^2\,(0.998)}{(0.1 \times 141.421)} = 3.836;$$

$$\left(\frac{\partial g_3}{\partial A_2}\right)^2 \frac{A_2}{\rho l_2} = \frac{(-0.858)^2\,(0.615)}{(0.1 \times 100)} = 0.0453;$$

$$\left(\frac{\partial g_3}{\partial A_3}\right)^2 \frac{A_3}{\rho l_3} = \frac{(-2.660)^2\,(0.133)}{(0.1 \times 166.667)} = 0.05646$$

(XII) Thus, $C_{11} = -3.940$

$$C_{12} = -\sum_{l=1}^{3}\left(\frac{\partial g_3}{\partial A_l}\right)\left(\frac{\partial g_4}{\partial A_l}\right)\frac{A_l}{\rho l_l}$$

$$= -\left[\left(\frac{\partial g_3}{\partial A_1}\right)\left(\frac{\partial g_4}{\partial A_1}\right)\frac{A_1}{\rho l_1} + \left(\frac{\partial g_3}{\partial A_2}\right)\left(\frac{\partial g_4}{\partial A_2}\right)\frac{A_2}{\rho l_2}\right.$$

$$\left. + \left(\frac{\partial g_3}{\partial A_3}\right)\left(\frac{\partial g_4}{\partial A_2}\right)\frac{A_3}{\rho l_3}\right]$$

$$\left(\frac{\partial g_3}{\partial A_1}\right)\left(\frac{\partial g_4}{\partial A_1}\right)\frac{A_1}{\rho l_1} = \frac{(-7.373)\,(-1.130)\,(0.998)}{(0.1 \times 141.421)} = 0.5879;$$

$$\left(\frac{\partial g_3}{\partial A_2}\right)\left(\frac{\partial g_4}{\partial A_2}\right)\frac{A_2}{\rho l_2} = \frac{(-0.858)\,(-9.530)\,(0.615)}{(0.1 \times 100)} = 0.5029;$$

$$\left(\frac{\partial g_3}{\partial A_3}\right)\left(\frac{\partial g_4}{\partial A_3}\right)\frac{A_3}{\rho l_3} = \frac{(-2.660)\,(-1.555)\,(0.133)}{(0.1 \times 166.667)} = 0.0330;$$

Thus, $C_{12} = -1.1238$; $C_{21} = C_{12} = -1.1238$

$$C_{22} = -\sum_{l=1}^{3} \left(\frac{\partial g_4}{\partial A_l}\right)^2 \frac{A_l}{\rho l_l}$$

$$= -\left[\left(\frac{\partial g_4}{\partial A_1}\right)^2 \frac{A_1}{\rho l_1} + \left(\frac{\partial g_4}{\partial A_2}\right)^2 \frac{A_2}{\rho l_2} + \left(\frac{\partial g_4}{\partial A_3}\right)^2 \frac{A_3}{\rho l_3}\right]$$

$$\left(\frac{\partial g_4}{\partial A_1}\right)^2 \frac{A_1}{\rho l_1} = \frac{(-1.130)^2 (0.998)}{(0.1 \times 141.421)} = 0.0901;$$

$$\left(\frac{\partial g_4}{\partial A_3}\right)^2 \frac{A_3}{\rho l_3} = \frac{(-1.555)^2 (0.133)}{(0.1 \times 166.667)} = 0.0193;$$

(XIII) $C_{22} = 5.695$

$$R_1 = r(\beta_3 - \beta_{30}) + \sum_{l=1}^{3} \frac{\partial g_3}{\partial A_l} A_l^{(k)} - r\frac{\partial g_3}{\partial A_3} \Delta A_3 (P)$$

$$r(\beta_3 - \beta_{30}) = 2(3.12 - 3.09) = 0.06;$$

$$\frac{\partial g_3}{\partial A_1} A_1 = (-7.373)(0.998) = -7.3583;$$

$$\frac{\partial g_3}{\partial A_2} A_2 = (-0.858)(0.615) = -0.5277;$$

$$\frac{\partial g_3}{\partial A_3} A_3 = (-2.660)(0.133) = -0.3538;$$

Thus, $R_1 = -8.1798$

$$R_2 = r(\beta_4 - \beta_{40}) + \sum_{l=1}^{3} \frac{\partial g_4}{\partial A_l} A_l^{(k)} - r\frac{\partial g_4}{\partial A_3} \Delta A_3 (P)$$

$$r(\beta_4 - \beta_{40}) = 0; \quad \frac{\partial g_4}{\partial A_1} A_1 = (-1.130)(0.998) = -1.1277$$

$$\frac{\partial g_4}{\partial A_2} A_2 = (-9.530)(0.615) = -5.8610;$$

$$\frac{\partial g_4}{\partial A_3} A_3 = (-1.555)(0.133) = -0.2068$$

(XIV) Thus, $R_2 = -7.196$

Lagrangian multiplier – λ_3, λ_4

$$\lambda_3 = \frac{(R_1 C_{22} - R_2 C_{12})}{(C_{11} C_{22} - C_{12} C_{21})}$$

$R_1 C_{22} = 46.5831;\quad R_2 C_{12} = 8.0863;$

$R_1 C_{22} - R_2 C_{12} = 46.5831 - 8.0863 = 38.4968$

$C_{11} C_{22} = 22.4254;\quad C_{12} C_{12} = 1.2629$

$C_{11} C_{22} - C_{12} C_{21} = 22.4254 - 1.2629 = 21.1625$

(XV) $\lambda_3 = \dfrac{(38.4968)}{(21.1625)} = 1.819$

$$\lambda_4 = \frac{(C_{11} R_2 - C_{12} R_1)}{(C_{11} C_{22} - C_{12} C_{21})}$$

$C_{11} R_2 = 28.3344;\quad C_{12} R_1 = 9.1925$

$C_{11} R_2 - C_{12} R_1 = 28.3344 - 9.1925 = 19.1419$

(XVI) $\lambda_4 = \dfrac{(19.1419)}{(21.1625)} = 0.905$

Recurrence equation

$$A_l^{(k+1)} = A_l^{(k)} \left(1 + \frac{1}{r} \left(\frac{-\lambda_3 \dfrac{\partial g_3}{\partial A_l} - \lambda_4 \dfrac{\partial g_4}{\partial A_l}}{\rho l_l} - 1 \right) \right)$$

$- \lambda_3 \dfrac{\partial g_3}{\partial A_1} = -(1.8191)(-7.373) = 13.4122;$

$- \lambda_4 \dfrac{\partial g_4}{\partial A_1} = -(0.9045)(-1.130) = 1.0221$

$\dfrac{-\lambda_3 \dfrac{\partial g_3}{\partial A_1} - \lambda_4 \dfrac{\partial g_4}{\partial A_1}}{\rho l_1} = \dfrac{(13.4122) + (1.0221)}{14.1421} = 1.0207;$

$1 + \dfrac{1}{r}(1.0207 - 1) = 1.0104$

(XVII) $A_1^{(k+1)} = A_1^{(k)}(1.0104) = (0.998)(1.0104) = 1.0084$

$$-\lambda_3 \frac{\partial g_3}{\partial A_2} = -(1.8191)(-0.858) = 1.5608;$$

$$-\lambda_4 \frac{\partial g_4}{\partial A_2} = -(0.9045)(-9.530) = 8.6199;$$

$$\frac{-\lambda_3 \dfrac{\partial g_3}{\partial A_2} - \lambda_4 \dfrac{\partial g_4}{\partial A_2}}{\rho l_2} = 1.0181; \quad 1 + \frac{1}{r}(1.0181 - 1) = 1.0090;$$

(XVIII) $A_2^{(k+1)} = A_2^{(k)}(1.0090) = (0.615)(1.0090) = 0.621$

$$-\lambda_3 \frac{\partial g_3}{\partial A_3} = -(1.8191)(-2.660) = 4.8388;$$

$$-\lambda_4 \frac{\partial g_4}{\partial A_3} = -(0.9045)(-1.555) = 1.4065;$$

$$\frac{-\lambda_3 \dfrac{\partial g_3}{\partial A_3} - \lambda_4 \dfrac{\partial g_4}{\partial A_3}}{\rho l_3} = 0.3747; \quad 1 + \frac{1}{r}(0.3747 - 1) = 0.6874$$

$$A_3^{(k+1)} = A_3^{(k)}(0.6874) = (0.133)(0.6874) = 0.091 < A_{\min} = 0.1$$

(XIX) $A_3^{(k+1)} = 0.1$ (passive element)

(C) *Cycle 14* (λ_3, λ_4 active, element 3 passive)

$$C_{11} = -\sum_{l=1}^{2} \left(\frac{\partial g_3}{\partial A_l}\right)^2 \frac{A_l^{(k)}}{\rho l_l} = -\left[\left(\frac{\partial g_3}{\partial A_1}\right)^2 \frac{A_1}{\rho l_1} + \left(\frac{\partial g_3}{\partial A_2}\right)^2 \frac{A_2}{\rho l_2}\right]$$

$$\left(\frac{\partial g_3}{\partial A_1}\right)^2 \frac{A_1}{\rho l_1} = \frac{(-7.401)^2 (1.006)}{(14.1421)} = 3.900;$$

$$\left(\frac{\partial g_3}{\partial A_2}\right)^2 \frac{A_2}{\rho l_2} = \frac{(-0.639)^2 (0.620)}{(10)} = 0.025;$$

(XII) $C_{11} = -3.9253$

$$C_{12} = -\sum_{l=1}^{2} \left(\frac{\partial g_3}{\partial A_l}\right) \left(\frac{\partial g_4}{\partial A_l}\right) \frac{A_l}{\rho l_l}$$

$$= -\left[\left(\frac{\partial g_3}{\partial A_1}\right) \left(\frac{\partial g_4}{\partial A_1}\right) \frac{A_1}{\rho l_1} + \left(\frac{\partial g_3}{\partial A_2}\right) \left(\frac{\partial g_4}{\partial A_2}\right) \frac{A_2}{\rho l_2}\right]$$

$$\left(\frac{\partial g_3}{\partial A_1}\right) \left(\frac{\partial g_4}{\partial A_1}\right) \frac{A_1}{\rho l_1} = \frac{(-7.401)(-0.848)(1.006)}{(14.1421)} = 0.4469;$$

$$\left(\frac{\partial g_3}{\partial A_2}\right) \left(\frac{\partial g_4}{\partial A_2}\right) \frac{A_2}{\rho l_2} = \frac{(-0.639)(-9.598)(0.620)}{(10)} = 0.3803;$$

$C_{12} = C_{21} = -0.8272$

$$C_{22} = -\sum_{l=1}^{2} \left(\frac{\partial g_4}{\partial A_l}\right)^2 \frac{A_l}{\rho l_l} = -\left[\left(\frac{\partial g_4}{\partial A_1}\right)^2 \frac{A_1}{\rho l_1} + \left(\frac{\partial g_4}{\partial A_2}\right)^2 \frac{A_2}{\rho l_2}\right]$$

$$\left(\frac{\partial g_4}{\partial A_1}\right)^2 \frac{A_1}{\rho l_1} = \frac{(-0.848)^2 (1.006)}{(14.1421)} = 0.0512;$$

$$\left(\frac{\partial g_4}{\partial A_2}\right)^2 \frac{A_2}{\rho l_2} = \frac{(-9.598)^2 (0.620)}{(10)} = 5.7115;$$

(XIII) $C_{22} = -5.763$

$$R_1 = r(\beta_3 - \beta_{30}) + \sum_{l=1}^{2} \frac{\partial g_3}{\partial A_l} A_l^k - r\frac{\partial g_3}{\partial A_3} \Delta A_3 \, (P)$$

$$r(\beta_3 - \beta_{30}) = 0; \quad \frac{\partial g_3}{\partial A_1} A_1 = (-7.401)(1.006) = -7.4528$$

$$\Delta A_3 \, (p) = (0.1 - 0.1) = 0$$

$$\frac{\partial g_3}{\partial A_2} A_2 = (-0.639)(0.620) = -0.3962$$

Therefore, $R_1 = -7.8490$

$$R_2 = r(\beta_4 - \beta_{40}) + \sum_{l=1}^{2} \frac{\partial g_4}{\partial A_l} A_l - r\frac{\partial g_4}{\partial A_3} \Delta A_3 \, (P)$$

$$r(\beta_4 - \beta_{40}) = 0; \quad \frac{\partial g_4}{\partial A_1} A_1 = (-0.848)(1.006) = -0.8539$$

$$\frac{\partial g_4}{\partial A_2} A_2 = (-9.598)(0.620) = -5.9508;$$

$$\Delta A_3\,(P) = (0.1 - 0.1) = 0; \quad r\left(\frac{\partial g_4}{\partial A_2}\Delta A_3\,(P)\right) = 0$$

(XIV) $R_2 = -6.8047$

Lagrangian multiplier – $\lambda_3,\ \lambda_4$

$$\lambda_3 = \frac{(R_1 C_{22} - R_2 C_{12})}{(C_{11} C_{22} - C_{12} C_{21})}$$

$R_1 C_{22} = 45.2314; \quad R_2 C_{12} = 5.6288;$

$R_1 C_{22} - R_2 C_{12} = 39.6026$

$C_{11} C_{22} = 22.6203; \quad C_{12} C_{12} = 0.6843;$

$C_{11} C_{22} - C_{12} C_{21} = 21.9360$

(XV) $\lambda_3 = 1.805$

$$\lambda_4 = \frac{(C_{11} R_2 - C_{12} R_1)}{(C_{11} C_{22} - C_{12} C_{21})}$$

$C_{11} R_2 = 26.7105; \quad C_{12} R_1 = 6.4927;$

$C_{11} R_2 - C_{12} R_1 = 20.2178$

(XVI) $\lambda_4 = 0.922$

Recurrence equation

$$A_1^{(k+1)} = A_1^{(k)}\left(1 + \frac{1}{r}\left(\frac{-\lambda_3\dfrac{\partial g_3}{\partial A_1} - \lambda_4\dfrac{\partial g_4}{\partial A_1}}{\rho l_1} - 1\right)\right)$$

$$-\lambda_3\frac{\partial g_3}{\partial A_1} = -\,(1.8054)\,(-7.401) = 13.3618;$$

$$-\lambda_4\frac{\partial g_4}{\partial A_1} = -\,(0.9217)\,(-0.848) = 0.7816$$

$$\frac{-\lambda_3\dfrac{\partial g_3}{\partial A_1} - \lambda_4\dfrac{\partial g_4}{\partial A_1}}{\rho l_1} == 1.0001; \quad 1 + \frac{1}{r}(1.0001 - 1) = 1.00005$$

(XVII) $A_1^{(k+1)} = A_1^{(k)} (1.0005) = 1.006$

$$-\lambda_3 \frac{\partial g_3}{\partial A_2} = -(1.8054)(-0.639) = 1.1537;$$

$$-\lambda_4 \frac{\partial g_4}{\partial A_2} = -(0.9217)(-9.958) = 8.8465$$

$$\frac{-\lambda_3 \dfrac{\partial g_3}{\partial A_2} - \lambda_4 \dfrac{\partial g_4}{\partial A_2}}{\rho l_2} = 1.000018; \quad 1 + \frac{1}{r}(1.000018 - 1) = 1$$

(XVIII) $A_2^{(k+1)} = A_2^{(k)} (1) = 0.620$

$$-\lambda_3 \frac{\partial g_3}{\partial A_3} = -(1.8054)(-3.052) = 5.510;$$

$$-\lambda_4 \frac{\partial g_4}{\partial A_3} = -(0.9217)(-1.880) = 1.7328$$

$$\frac{-\lambda_3 \dfrac{\partial g_3}{\partial A_3} - \lambda_4 \dfrac{\partial g_4}{\partial A_3}}{\rho l_3} = 0.4346; \quad 1 + \frac{1}{r}(0.4346 - 1) = 0.7173$$

$$A_3^{(k+1)} = 0.131 (0.7173) = 0.0939 < A_{min}$$

(XIX) $A_3^{(k+1)} = 0.1$

Index

Milton Keynes UK
Ingram Content Group UK Ltd.
UKHW021935071024
449327UK00022B/1822

9 780367 865139